T0189316

Lecture Notes in Computer Science 13943

Founding Editors

Gerhard Goos
Juris Hartmanis

The series Lecture Notes in Computer Science (LNCS), including its subseries Lecture Notes in Artificial Intelligence (LNAI) and Lecture Notes in Bioinformatics (LNBI), has established itself as a medium for the publication of new developments in computer science and information technology research, teaching, and education.

LNCS enjoys close cooperation with the computer science R & D community, the series counts many renowned academics among its volume editors and paper authors, and collaborates with prestigious societies. Its mission is to serve this international community by providing an invaluable service, mainly focused on the publication of conference and workshop proceedings and postproceedings. LNCS commenced publication in 1973.

Xin Wang · Maria Luisa Sapino ·
Wook-Shin Han · Amr El Abbadi · Gill Dobbie ·
Zhiyong Feng · Yingxiao Shao · Hongzhi Yin
Editors

Database Systems for Advanced Applications

28th International Conference, DASFAA 2023
Tianjin, China, April 17–20, 2023
Proceedings, Part I

Springer

Editors
Xin Wang 🆔
Tianjin University
Tianjin, China

Wook-Shin Han
POSTECH
Pohang, Korea (Republic of)

Gill Dobbie 🆔
University of Auckland
Auckland, New Zealand

Yingxiao Shao 🆔
Beijing University of Posts
and Telecommunications
Beijing, China

Maria Luisa Sapino 🆔
University of Torino
Turin, Italy

Amr El Abbadi
University of California Santa Barbara
Santa Barbara, CA, USA

Zhiyong Feng
Tianjin University
Tianjin, China

Hongzhi Yin 🆔
The University of Queensland
Brisbane, QLD, Australia

ISSN 0302-9743   ISSN 1611-3349 (electronic)
Lecture Notes in Computer Science
ISBN 978-3-031-30636-5   ISBN 978-3-031-30637-2 (eBook)
https://doi.org/10.1007/978-3-031-30637-2

This Springer imprint is published by the registered company Springer Nature Switzerland AG
The registered company address is: Gewerbestrasse 11, 6330 Cham, Switzerland

Preface

It is our great pleasure to present the proceedings of the 28th International Conference on Database Systems for Advanced Applications (DASFAA 2023), organized by Tianjin University and held during April 17–20, 2023, in Tianjin, China. DASFAA is an annual international database conference which showcases state-of-the-art R&D activities in database systems and advanced applications. It provides a premier international forum for technical presentations and discussions among database researchers, developers, and users from both academia and industry.

This year we received a record high number of 652 research paper submissions. We conducted a double-blind review following the tradition of DASFAA, and constructed a large committee consisting of 31 Senior Program Committee (SPC) members and 254 Program Committee (PC) members. Each valid submission was reviewed by at least three PC members and meta-reviewed by one SPC member, who also led the discussion with the PC members. We, the PC co-chairs, considered the recommendations from the SPC members and investigated each submission as well as its reviews to make the final decisions. As a result, 125 full papers (acceptance ratio of 19.2%) and 66 short papers (acceptance ratio of 29.3%) were accepted. The review process was supported by the Microsoft CMT system. During the three main conference days, these 191 papers were presented in 17 research sessions. The dominant keywords for the accepted papers included model, graph, learning, performance, knowledge, time, recommendation, representation, attention, prediction, and network. In addition, we included 15 industry papers, 15 demo papers, 5 PhD consortium papers, and 7 tutorials in the program. Finally, to shed light on the direction in which the database field is headed, the conference program included four invited keynote presentations by Sihem Amer-Yahia (CNRS, France), Kyuseok Shim (Seoul National University, South Korea), Angela Bonifati (Lyon 1 University, France), and Jianliang Xu (Hong Kong Baptist University, China).

Four workshops were selected by the workshop co-chairs to be held in conjunction with DASFAA 2023, which were the 9th International Workshop on Big Data Management and Service (BDMS 2023), the 8th International Workshop on Big Data Quality Management (BDQM 2023), the 7th International Workshop on Graph Data Management and Analysis (GDMA 2023), and the 1st International Workshop on Bundle-based Recommendation Systems (BundleRS 2023). The workshop papers are included in a separate volume of the proceedings also published by Springer in its Lecture Notes in Computer Science series.

We are grateful to the general chairs, Amr El Abbadi, UCSB, USA, Gill Dobbie, University of Auckland, New Zealand, and Zhiyong Feng, Tianjin University, China, all SPC members, PC members, and external reviewers who contributed their time and expertise to the DASFAA 2023 paper-reviewing process. We would like to thank all the members of the Organizing Committee, and the many volunteers, for their great support in the conference organization. Lastly, many thanks to the authors who submitted their papers to the conference.

March 2023

Xin Wang
Maria Luisa Sapino
Wook-Shin Han

Organization

Steering Committee Members

Chair

Lei Chen — Hong Kong University of Science and Technology (Guangzhou), China

Vice Chair

Stéphane Bressan — National University of Singapore, Singapore

Treasurer

Yasushi Sakurai — Osaka University, Japan

Secretary

Kyuseok Shim — Seoul National University, South Korea

Members

Zhiyong Peng	Wuhan University of China, China
Zhanhuai Li	Northwestern Polytechnical University, China
Krishna Reddy	IIIT Hyderabad, India
Yunmook Nah	DKU, South Korea
Wenjia Zhang	University of New South Wales, Australia
Zi Huang	University of Queensland, Australia
Guoliang Li	Tsinghua University, China
Sourav Bhowmick	Nanyang Technological University, Singapore
Atsuyuki Morishima	University of Tsukuba, Japan
Sang-Won Lee	SKKU, South Korea
Yang-Sae Moon	Kangwon National University, South Korea

Organizing Committee

Honorary Chairs

Christian S. Jensen	Aalborg University, Denmark
Keqiu Li	Tianjin University, China

General Chairs

Amr El Abbadi	UCSB, USA
Gill Dobbie	University of Auckland, New Zealand
Zhiyong Feng	Tianjin University, China

Program Committee Chairs

Xin Wang	Tianjin University, China
Maria Luisa Sapino	University of Torino, Italy
Wook-Shin Han	POSTECH, South Korea

Industry Program Chairs

Jiannan Wang	Simon Fraser University, Canada
Jinwei Zhu	Huawei, China

Tutorial Chairs

Jianxin Li	Deakin University, Australia
Herodotos Herodotou	Cyprus University of Technology, Cyprus

Demo Chairs

Zhifeng Bao	RMIT, Australia
Yasushi Sakurai	Osaka University, Japan
Xiaoli Wang	Xiamen University, China

Workshop Chairs

Lu Chen	Zhejiang University, China
Xiaohui Tao	University of Southern Queensland, Australia

Panel Chairs

Lei Chen — Hong Kong University of Science and Technology (Guangzhou), China

Xiaochun Yang — Northeastern University, China

PhD Consortium Chairs

Leong Hou U. — University of Macau, China

Panagiotis Karras — Aarhus University, Denmark

Publicity Chairs

Yueguo Chen — Renmin University of China, China

Kyuseok Shim — Seoul National University, South Korea

Yoshiharu Ishikawa — Nagoya University, Japan

Arnab Bhattacharya — IIT Kanpur, India

Publication Chairs

Yingxiao Shao — Beijing University of Posts and Telecommunications, China

Hongzhi Yin — University of Queensland, Australia

DASFAA Steering Committee Liaison

Lei Chen — Hong Kong University of Science and Technology (Guangzhou), China

Local Arrangement Committee

Xiaowang Zhang — Tianjin University, China

Guozheng Rao — Tianjin University, China

Yajun Yang — Tianjin University, China

Shizhan Chen — Tianjin University, China

Xueli Liu — Tianjin University, China

Xiaofei Wang — Tianjin University, China

Chao Qiu — Tianjin University, China

Dong Han — Tianjin Academy of Fine Arts, China

Ying Guo — Tianjin University, China

| Hui Jiang | Tianjin Ren'ai College, China |
| Kun Liang | Tianjin University of Science and Technology, China |

Web Master

| Zirui Chen | Tianjin University, China |

Program Committee Chairs

Xin Wang	Tianjin University, China
Maria Luisa Sapino	University of Torino, Italy
Wook-Shin Han	POSTECH, South Korea

Senior Program Committee (SPC) Members

Baihua Zheng	Singapore Management University, Singapore
Bin Cui	Peking University, China
Bingsheng He	National University of Singapore, Singapore
Chee-Yong Chan	National University of Singapore, Singapore
Chengfei Liu	Swinburne University of Technology, Australia
Haofen Wang	Tongji University, China
Hong Gao	Harbin Institute of Technology, China
Hongzhi Yin	University of Queensland, Australia
Jiaheng Lu	University of Helsinki, Finland
Jianliang Xu	Hong Kong Baptist University, China
Jianyong Wang	Tsinghua University, China
K. Selçuk Candan	Arizona State University, USA
Kyuseok Shim	Seoul National University, South Korea
Lei Li	Hong Kong University of Science and Technology (Guangzhou), China
Lina Yao	University of New South Wales, Australia
Ling Liu	Georgia Institute of Technology, USA
Nikos Bikakis	Athena Research Center, Greece
Qiang Zhu	University of Michigan-Dearborn, USA
Reynold Cheng	University of Hong Kong, China
Ronghua Li	Beijing Institute of Technology, China
Vana Kalogeraki	Athens University of Economics and Business, Greece
Vincent Tseng	National Yang Ming Chiao Tung University, Taiwan
Wang-Chien Lee	Pennsylvania State University, USA

Xiang Zhao	National University of Defense Technology, China
Xiaoyong Du	Renmin University of China, China
Ye Yuan	Beijing Institute of Technology, China
Yongxin Tong	Beihang University, China
Yoshiharu Ishikawa	Nagoya University, Japan
Yufei Tao	Chinese University of Hong Kong, China
Yunjun Gao	Zhejiang University, China
Zhiyong Peng	Wuhan University, China

Program Committee (PC) Members

Alexander Zhou	Hong Kong University of Science and Technology, China
Alkis Simitsis	Athena Research Center, Greece
Amr Ebaid	Google, USA
An Liu	Soochow University, China
Anne Laurent	University of Montpellier, France
Antonio Corral	University of Almería, Spain
Baoning Niu	Taiyuan University of Technology, China
Barbara Catania	University of Genoa, Italy
Bin Cui	Peking University, China
Bin Wang	Northeastern University, China
Bing Li	Institute of High Performance Computing, Singapore
Bohan Li	Nanjing University of Aeronautics and Astronautics, China
Changdong Wang	SYSU, China
Chao Huang	University of Notre Dame, USA
Chao Zhang	Tsinghua University, China
Chaokun Wang	Tsinghua University, China
Chenyang Wang	Aalto University, Finland
Cheqing Jin	East China Normal University, China
Chih-Ya Shen	National Tsing Hua University, Taiwan
Christos Doulkeridis	University of Pireaus, Greece
Chuan Ma	Zhejiang Lab, China
Chuan Xiao	Osaka University and Nagoya University, Japan
Chuanyu Zong	Shenyang Aerospace University, China
Chunbin Lin	Amazon AWS, USA
Cindy Chen	UMass Lowell, USA
Claudio Schifanella	University of Torino, Italy
Cuiping Li	Renmin University of China, China

Damiani Ernesto	University of Milan, Italy
Dan He	University of Queensland, Australia
De-Nian Yang	Academia Sinica, Taiwan
Derong Shen	Northeastern University, China
Dhaval Patel	IBM Research, USA
Dian Ouyang	Guangzhou University, China
Dieter Pfoser	George Mason University, USA
Dimitris Kotzinos	ETIS, France
Dong Wen	University of New South Wales, Australia
Dongxiang Zhang	Zhejiang University, China
Dongxiao He	Tianjin University, China
Faming Li	Northeastern University, USA
Ge Yu	Northeastern University, China
Goce Trajcevski	Iowa State University, USA
Gong Cheng	Nanjing University, China
Guandong Xu	University of Technology Sydney, Australia
Guanhua Ye	University of Queensland, Australia
Guoliang Li	Tsinghua University, China
Haida Zhang	WorldQuant, USA
Hailong Liu	Northwestern Polytechnical University, China
Haiwei Zhang	Nankai University, China
Hantao Zhao	ETH, Switzerland
Hao Peng	Beihang University, China
Hiroaki Shiokawa	University of Tsukuba, Japan
Hongbin Pei	Xi'an Jiaotong University, China
Hongxu Chen	Commonwealth Bank of Australia, Australia
Hongzhi Wang	Harbin Institute of Technology, China
Hongzhi Yin	University of Queensland, Australia
Huaijie Zhu	Sun Yat-sen University, China
Hui Li	Xidian University, China
Huiqi Hu	East China Normal University, China
Hye-Young Paik	University of New South Wales, Australia
Ioannis Konstantinou	University of Thessaly, Greece
Ismail Hakki Toroslu	METU, Turkey
Jagat Sesh Challa	BITS Pilani, India
Ji Zhang	University of Southern Queensland, Australia
Jia Xu	Guangxi University, China
Jiali Mao	East China Normal University, China
Jianbin Qin	Shenzhen Institute of Computing Sciences, China
Jianmin Wang	Tsinghua University, China
Jianqiu Xu	Nanjing University of Aeronautics and Astronautics, China

Jianxin Li	Deakin University, Australia
Jianye Yang	Guangzhou University, China
Jiawei Jiang	Wuhan University, China
Jie Shao	University of Electronic Science and Technology of China, China
Jilian Zhang	Jinan University, China
Jilin Hu	Aalborg University, Denmark
Jin Wang	Megagon Labs, USA
Jing Tang	Hong Kong University of Science and Technology, China
Jithin Vachery	NUS, Singapore
Jongik Kim	Chungnam National University, South Korea
Ju Fan	Renmin University of China, China
Jun Gao	Peking University, China
Jun Miyazaki	Tokyo Institute of Technology, Japan
Junhu Wang	Griffith University, Australia
Junhua Zhang	University of New South Wales, Australia
Junliang Yu	The University of Queensland, Australia
Kai Wang	Shanghai Jiao Tong University, China
Kai Zheng	University of Electronic Science and Technology of China, China
Kangfei Zhao	The Chinese University of Hong Kong, China
Kesheng Wu	LBNL, USA
Kristian Torp	Aalborg University, Denmark
Kun Yue	School of Information and Engineering, China
Kyoung-Sook Kim	National Institute of Advanced Industrial Science and Technology, Japan
Ladjel Bellatreche	ISAE-ENSMA, France
Latifur Khan	University of Texas at Dallas, USA
Lei Cao	MIT, USA
Lei Duan	Sichuan University, China
Lei Guo	Shandong Normal University, China
Leong Hou U.	University of Macau, China
Liang Hong	Wuhan University, China
Libin Zheng	Sun Yat-sen University, China
Lidan Shou	Zhejiang University, China
Lijun Chang	University of Sydney, Australia
Lin Li	Wuhan University of Technology, China
Lizhen Cui	Shandong University, China
Long Yuan	Nanjing University of Science and Technology, China
Lu Chen	Swinburne University of Technology, Australia

Lu Chen   Zhejiang University, China
Makoto Onizuka   Osaka University, Japan
Manish Kesarwani   IBM Research, India
Manolis Koubarakis   University of Athens, Greece
Markus Schneider   University of Florida, USA
Meihui Zhang   Beijing Institute of Technology, China
Meng Wang   Southeast University, China
Meng-Fen Chiang   University of Auckland, New Zealand
Ming Zhong   Wuhan University, China
Minghe Yu   Northeastern University, China
Mizuho Iwaihara   Waseda University, Japan
Mo Li   Liaoning University, China
Ning Wang   Beijing Jiaotong University, China
Ningning Cui   Anhui University, China
Norio Katayama   National Institute of Informatics, Japan
Noseong Park   George Mason University, USA
Panagiotis Bouros   Johannes Gutenberg University Mainz, Germany
Peiquan Jin   University of Science and Technology of China,
                              China
Peng Cheng   East China Normal University, China
Peng Peng   Hunan University, China
Pengpeng Zhao   Soochow University, China
Ping Lu   Beihang University, China
Pinghui Wang   Xi'an Jiaotong University, China
Qiang Yin   Shanghai Jiao Tong University, China
Qianzhen Zhang   National University of Defense Technology,
                              China
Qing Liao   Harbin Institute of Technology (Shenzhen), China
Qing Liu   CSIRO, Australia
Qingpeng Zhang   City University of Hong Kong, China
Qingqing Ye   Hong Kong Polytechnic University, China
Quanqing Xu   A*STAR, Singapore
Rong Zhu   Alibaba Group, China
Rui Zhou   Swinburne University of Technology, Australia
Rui Zhu   Shenyang Aerospace University, China
Ruihong Qiu   University of Queensland, Australia
Ruixuan Li   Huazhong University of Science and Technology,
                              China
Ruiyuan Li   Chongqing University, China
Sai Wu   Zhejiang University, China
Sanghyun Park   Yonsei University, South Korea

Sanjay Kumar Madria	Missouri University of Science & Technology, USA
Sebastian Link	University of Auckland, New Zealand
Sen Wang	University of Queensland, Australia
Shaoxu Song	Tsinghua University, China
Sheng Wang	Wuhan University, China
Shijie Zhang	Tencent, China
Shiyu Yang	Guangzhou University, China
Shuhao Zhang	Singapore University of Technology and Design, Singapore
Shuiqiao Yang	UNSW, Australia
Shuyuan Li	Beihang University, China
Sibo Wang	Chinese University of Hong Kong, China
Silvestro Roberto Poccia	University of Turin, Italy
Tao Qiu	Shenyang Aerospace University, China
Tao Zhao	National University of Defense Technology, China
Taotao Cai	Macquarie University, Australia
Thanh Tam Nguyen	Griffith University, Australia
Theodoros Chondrogiannis	University of Konstanz, Germany
Tieke He	State Key Laboratory for Novel Software Technology, China
Tieyun Qian	Wuhan University, China
Tiezheng Nie	Northeastern University, China
Tsz Nam (Edison) Chan	Hong Kong Baptist University, China
Uday Kiran Rage	University of Aizu, Japan
Verena Kantere	National Technical University of Athens, Greece
Wei Hu	Nanjing University, China
Wei Li	Harbin Engineering University, China
Wei Lu	RUC, China
Wei Shen	Nankai University, China
Wei Song	Wuhan University, China
Wei Wang	Hong Kong University of Science and Technology (Guangzhou), China
Wei Zhang	ECNU, China
Wei Emma Zhang	The University of Adelaide, Australia
Weiguo Zheng	Fudan University, China
Weijun Wang	University of Göttingen, Germany
Weiren Yu	University of Warwick, UK
Weitong Chen	Adelaide University, Australia
Weiwei Sun	Fudan University, China
Weixiong Rao	Tongji University, China

Wen Hua	Hong Kong Polytechnic University, China
Wenchao Zhou	Georgetown University, USA
Wentao Li	University of Technology Sydney, Australia
Wentao Zhang	Mila, Canada
Werner Nutt	Free University of Bozen-Bolzano, Italy
Wolf-Tilo Balke	TU Braunschweig, Germany
Wookey Lee	Inha University, South Korea
Xi Guo	University of Science and Technology Beijing, China
Xiang Ao	Institute of Computing Technology, CAS, China
Xiang Lian	Kent State University, USA
Xiang Zhao	National University of Defense Technology, China
Xiangguo Sun	Chinese University of Hong Kong, China
Xiangmin Zhou	RMIT University, Australia
Xiangyu Song	Swinburne University of Technology, Australia
Xiao Pan	Shijiazhuang Tiedao University, China
Xiao Fan Liu	City University of Hong Kong, China
Xiaochun Yang	Northeastern University, China
Xiaofeng Gao	Shanghai Jiaotong University, China
Xiaoling Wang	East China Normal University, China
Xiaowang Zhang	Tianjin University, China
Xiaoyang Wang	University of New South Wales, Australia
Ximing Li	Jilin University, China
Xin Cao	University of New South Wales, Australia
Xin Huang	Hong Kong Baptist University, China
Xin Wang	Southwest Petroleum University, China
Xinqiang Xie	Neusoft, China
Xiuhua Li	Chongqing University, China
Xiulong Liu	Tianjin University, China
Xu Zhou	Hunan University, China
Xuequn Shang	Northwestern Polytechnical University, China
Xupeng Miao	Carnegie Mellon University, USA
Xuyun Zhang	Macquarie University, Australia
Yajun Yang	Tianjin University, China
Yan Zhang	Peking University, China
Yanfeng Zhang	Northeastern University, China, and Macquarie University, Australia
Yang Cao	Hokkaido University, Japan
Yang Chen	Fudan University, China
Yang-Sae Moon	Kangwon National University, South Korea
Yanjie Fu	University of Central Florida, USA

Yanlong Wen	Nankai University, China
Ye Yuan	Beijing Institute of Technology, China
Yexuan Shi	Beihang University, China
Yi Cai	South China University of Technology, China
Ying Zhang	Nankai University, China
Yingxia Shao	BUPT, China
Yiru Chen	Columbia University, USA
Yixiang Fang	Chinese University of Hong Kong, Shenzhen, China
Yong Tang	South China Normal University, China
Yong Zhang	Tsinghua University, China
Yongchao Liu	Ant Group, China
Yongpan Sheng	Southwest University, China
Yongxin Tong	Beihang University, China
You Peng	University of New South Wales, Australia
Yu Gu	Northeastern University, China
Yu Yang	Hong Kong Polytechnic University, China
Yu Yang	City University of Hong Kong, China
Yuanyuan Zhu	Wuhan University, China
Yue Kou	Northeastern University, China
Yunpeng Chai	Renmin University of China, China
Yunyan Guo	Tsinghua University, China
Yunzhang Huo	Hong Kong Polytechnic University, China
Yurong Cheng	Beijing Institute of Technology, China
Yuxiang Zeng	Hong Kong University of Science and Technology, China
Zeke Wang	Zhejiang University, China
Zhaojing Luo	National University of Singapore, Singapore
Zhaonian Zou	Harbin Institute of Technology, China
Zheng Liu	Nanjing University of Posts and Telecommunications, China
Zhengyi Yang	University of New South Wales, Australia
Zhenya Huang	University of Science and Technology of China, China
Zhenying He	Fudan University, China
Zhipeng Zhang	Alibaba, China
Zhiwei Zhang	Beijing Institute of Technology, China
Zhixu Li	Fudan University, China
Zhongnan Zhang	Xiamen University, China

Industry Program Chairs

Jiannan Wang   Simon Fraser University, Canada
Jinwei Zhu   Huawei, China

Industry Program Committee Members

Bohan Li   Nanjing University of Aeronautics and
                                    Astronautics, China
Changbo Qu   Simon Fraser University, Canada
Chengliang Chai   Tsinghua University, China
Denis Ponomaryov   The Institute of Informatics Systems of the
                                    Siberian Division of Russian Academy of
                                    Sciences, Russia
Hongzhi Wang   Harbin Institute of Technology, China
Jianhua Yin   Shandong University, China
Jiannan Wang   Simon Fraser University, Canada
Jinglin Peng   Simon Fraser University, Canada
Jinwei Zhu   Huawei Technologies Co. Ltd., China
Ju Fan   Renmin University of China, China
Minghe Yu   Northeastern University, China
Nikos Ntarmos   Huawei Technologies R&D (UK) Ltd., UK
Sheng Wang   Alibaba Group, China
Wei Zhang   East China Normal University, China
Weiyuan Wu   Simon Fraser University, Canada
Xiang Li   East China Normal University, China
Xiaofeng Gao   Shanghai Jiaotong University, China
Xiaoou Ding   Harbin Institute of Technology, China
Yang Ren   Huawei, China
Yinan Mei   Tsinghua University, China
Yongxin Tong   Beihang University, China

Demo Track Program Chairs

Zhifeng Bao   RMIT, Australia
Yasushi Sakurai   Osaka University, Japan
Xiaoli Wang   Xiamen University, China

Demo Track Program Committee Members

Benyou Wang	Chinese University of Hong Kong, Shenzhen, China
Changchang Sun	Illinois Institute of Technology, USA
Chen Lin	Xiamen University, China
Chengliang Chai	Tsinghua University, China
Chenhao Ma	Chinese University of Hong Kong, Shenzhen, China
Dario Garigliotti	Aalborg University, Denmark
Ergute Bao	National University of Singapore, Singapore
Jianzhong Qi	The University of Melbourne, Australia
Jiayuan He	RMIT University, Australia
Kaiping Zheng	National University of Singapore, Singapore
Kajal Kansal	NUS, Singapore
Lei Cao	MIT, USA
Liang Zhang	WPI, USA
Lu Chen	Swinburne University of Technology, Australia
Meihui Zhang	Beijing Institute of Technology, China
Mengfan Tang	University of California, Irvine, USA
Na Zheng	National University of Singapore, Singapore
Pinghui Wang	Xi'an Jiaotong University, China
Qing Xie	Wuhan University of Technology, China
Ruihong Qiu	University of Queensland, Australia
Tong Chen	University of Queensland, Australia
Yile Chen	Nanyang Technological University, Singapore
Yuya Sasaki	Osaka University, Japan
Yuyu Luo	Tsinghua University, China
Zhanhao Zhao	Renmin University of China, China
Zheng Wang	Huawei Singapore Research Center, Singapore
Zhuo Zhang	University of Melbourne, Australia

PhD Consortium Track Program Chairs

Leong Hou U.	University of Macau, China
Panagiotis Karras	Aarhus University, Denmark

PhD Consortium Track Program Committee Members

Anton Tsitsulin	Google, USA
Bo Tang	Southern University of Science and Technology, China

Hao Wang   Wuhan University, China
Jieming Shi   Hong Kong Polytechnic University, China
Tsz Nam (Edison) Chan   Hong Kong Baptist University, China
Xiaowei Wu   State Key Lab of IoT for Smart City, University of
                                     Macau, China

Contents – Part I

Query Processing

Semantic-Driven Instance Generation for Table Question Answering   3
  Shuai Ma, Wenbin Jiang, Xiang Ao, Meng Tian, Xinwei Feng,
  Yajuan Lyu, Qiaoqiao She, and Qing He

Fine-Grained Tuple Transfer for Pipelined Query Execution on CPU-GPU
Coprocessor 19
  Zhenhua Yang, Qingfeng Pan, and Chen Xu

Efficient Index-Based Regular Expression Matching with Optimal Query
Plan Tree 35
  Tao Qiu, Xiaochun Yang, Bin Wang, Chuanyu Zong, Rui Zhu,
  and Xiufeng Xia

PFtree: Optimizing Persistent Adaptive Radix Tree for PM Systems
on eADR Platform 46
  Rui Zhang, Yao Wu, Shangyi Sun, Lulu Chen, Yibo Huang, Ming Yan,
  and Jie Wu

Learned Bloom Filter for Multi-key Membership Testing 62
  Yunchuan Li, Ziwei Wang, Ruixin Yang, Yan Zhao, Rui Zhou,
  and Kai Zheng

ACR-Tree: Constructing R-Trees Using Deep Reinforcement Learning 80
  Shuai Huang, Yong Wang, and Guoliang Li

A Scalable Query Pricing Framework for Incomplete Graph Data 97
  Huiwen Hou, Lianpeng Qiao, Ye Yuan, Chen Chen, and Guoren Wang

Answering Label-Constrained Reachability Queries via Reduction
Techniques 114
  Yuzheng Cai and Weiguo Zheng

JG2Time: A Learned Time Estimator for Join Operators Based
on Heterogeneous Join-Graphs 132
  Hao Miao, Jiazun Chen, Yang Lin, Mo Xu, Yinjun Han, and Jun Gao

Time Series Data

Long-Tailed Time Series Classification via Feature Space Rebalancing 151
Pengkun Wang, Xu Wang, Binwu Wang, Yudong Zhang, Lei Bai,
and Yang Wang

Flow-Based End-to-End Model for Hierarchical Time Series Forecasting
via Trainable Attentive-Reconciliation 167
Shiyu Wang, Yinbo Sun, Yan Wang, Fan Zhou, Lin-Tao Ma,
James Zhang, and YangFei Zheng

Orthrus: A Dual-Branch Model for Time Series Forecasting with Multiple
Exogenous Series 177
Ziang Yang, Biyu Zhou, Xuehai Tang, Ruixuan Li, and Songlin Hu

A Predictive Coding Approach to Multivariate Time Series Anomaly
Detection 188
Zhi Qi, Hong Xie, and Mingsheng Shang

RpDelta: Supporting UCR-Suite on Multi-versioning Time Series Data 205
Xiaoyu Han, Fei Ye, Zhenying He, X. Sean Wang, Yingze Song,
and Clement Liu

GP-HLS: Gaussian Process-Based Unsupervised High-Level Semantics
Representation Learning of Multivariate Time Series 221
Chengyang Ye and Qiang Ma

Towards Time-Series Key Points Detection Through Self-supervised
Learning and Probability Compensation 237
Mingxu Yuan, Xin Bi, Xuechun Huang, Wei Zhang, Lei Hu,
George Y. Yuan, Xiangguo Zhao, and Yongjiao Sun

SNN-AAD: Active Anomaly Detection Method for Multivariate Time
Series with Sparse Neural Network 253
Xiaoou Ding, Yida Liu, Hongzhi Wang, Donghua Yang, and Yichen Song

Spatial Data

Continuous k-Similarity Trajectories Search over Data Stream 273
Rui Zhu, Meichun Xiao, Bin Wang, Xiaochun Yang, Xiufeng Xia,
Chuanyu Zong, and Tao Qiu

Towards Effective Trajectory Similarity Measure in Linear Time 283
Yuanjun Liu, An Liu, Guanfeng Liu, Zhixu Li, and Lei Zhao

Accurate and Efficient Trajectory-Based Contact Tracing with Secure
Computation and Geo-Indistinguishability 300
  Maocheng Li, Yuxiang Zeng, Libin Zheng, Lei Chen, and Qing Li

Efficient and Accurate Range Counting on Privacy-Preserving Spatial
Data Federation 317
  Maocheng Li, Yuxiang Zeng, and Lei Chen

Predicting Where You Visit in a Surrounding City: A Mobility Knowledge
Transfer Framework Based on Cross-City Travelers 334
  Shuai Xu, Jianqiu Xu, Bohan Li, and Xiaoming Fu

Approximate *k*-Nearest Neighbor Query over Spatial Data Federation 351
  Kaining Zhang, Yongxin Tong, Yexuan Shi, Yuxiang Zeng, Yi Xu,
  Lei Chen, Zimu Zhou, Ke Xu, Weifeng Lv, and Zhiming Zheng

TAGnn: Time Adjoint Graph Neural Network for Traffic Forecasting 369
  Qi Zheng and Yaying Zhang

Adversarial Spatial-Temporal Graph Network for Traffic Speed Prediction
with Missing Values 380
  Pengfei Li, Junhua Fang, Wei Chen, An Liu, and Pingfu Chao

Trajectory Representation Learning Based on Road Network Partition
for Similarity Computation 396
  Jiajia Li, Mingshen Wang, Lei Li, Kexuan Xin, Wen Hua,
  and Xiaofang Zhou

ISTNet: Inception Spatial Temporal Transformer for Traffic Prediction 414
  Chu Wang, Jia Hu, Ran Tian, Xin Gao, and Zhongyu Ma

SimiDTR: Deep Trajectory Recovery with Enhanced Trajectory Similarity 431
  Yupu Zhang, Liwei Deng, Yan Zhao, Jin Chen, Jiandong Xie,
  and Kai Zheng

Fusing Local and Global Mobility Patterns for Trajectory Recovery 448
  Liwei Deng, Yan Zhao, Hao Sun, Changjie Yang, Jiandong Xie,
  and Kai Zheng

Explicit Assignment and Dynamic Pricing of Macro Online Tasks
in Spatial Crowdsourcing 464
  Lin Sun, Yeqiao Hou, and Zongpeng Li

Systems and Optimization

InstantChain: Enhancing Order-Execute Blockchain Systems
for Latency-Sensitive Applications 483
  Jianfeng Shi, Heng Wu, Diaohan Luo, Heran Gao, and Wenbo Zhang

DALedger: Towards High-Performance Transaction Processing
for Collaborative Decentralized Applications 499
  *Junkai Wang, Zhiwei Zhang, Shuai Zhao, Jiang Xiao, Ye Yuan,
  and Guoren Wang*

Efficient Execution of Blockchain Transactions Through Deterministic
Concurrency Control 509
  Huahui Xia, Jinchuan Chen, Nabo Ma, Jia Huang, and Xiaoyong Du

Anole: A Lightweight and Verifiable Learned-Based Index for Time
Range Query on Blockchain Systems 519
  Jian Chang, Binhong Li, Jiang Xiao, Licheng Lin, and Hai Jin

Xenos : Dataflow-Centric Optimization to Accelerate Model Inference
on Edge Devices 535
  *Runhua Zhang, Hongxu Jiang, Fangzheng Tian, Jinkun Geng,
  Xiaobin Li, Yuhang Ma, Chenhui Zhu, Dong Dong, Xin Li,
  and Haojie Wang*

Accelerating Recommendation Inference via GPU Streams 546
  Yuean Niu, Zhizhen Xu, Chen Xu, and Jiaqiang Wang

Mitigating Data Stalls in Deep Learning with Multi-times Data Loading
Rule 562
  *Derong Chen, Shuang Liang, Gang Hu, Han Xu, Xianqiang Luo,
  Hao Li, and Jie Shao*

Internet Public Safety Event Grading and Hybrid Storage Based
on Multi-feature Fusion for Social Media Texts 578
  Die Hu, Yulai Xie, Dan Feng, Shixun Zhao, and Pengyu Fu

A Self-decoupled Interpretable Prediction Framework for Highly-Variable
Cloud Workloads 588
  Bingchao Wang, Xiaoyu Shi, and Mingsheng Shang

TieComm: Learning a Hierarchical Communication Topology Based
on Tie Theory 604
  *Ming Yang, Renzhi Dong, Yiming Wang, Furui Liu, Yali Du,
  Mingliang Zhou, and Leong Hou U*

Privacy Computing

Rewriting-Stego: Generating Natural and Controllable Steganographic
Text with Pre-trained Language Model 617
  Fanxiao Li, Sixing Wu, Jiong Yu, Shuoxin Wang, BingBing Song,
  Renyang Liu, Haoseng Lai, and Wei Zhou

Towards Defending Against Byzantine LDP Amplified Gain Attacks 627
  Yukun Yan, Qingqing Ye, Haibo Hu, Rui Chen, Qilong Han,
  and Leixia Wang

Authenticated Ranked Keyword Search over Encrypted Data with Strong
Privacy Guarantee 644
  Ningning Cui, Zheli Deng, Man Li, Yuliang Ma, Jie Cui, and Hong Zhong

Combining Autoencoder with Adaptive Differential Privacy for Federated
Collaborative Filtering 661
  Xuanang Ding, Guohui Li, Ling Yuan, Lu Zhang, and Qian Rong

Robust Clustered Federated Learning 677
  Tiandi Ye, Senhui Wei, Jamie Cui, Cen Chen, Yingnan Fu, and Ming Gao

Privacy Preserving Federated Learning Framework Based on Multi-chain
Aggregation 693
  Yingchun Cui and Jinghua Zhu

FedGR: Federated Learning with Gravitation Regulation for Double
Imbalance Distribution 703
  Songyue Guo, Xu Yang, Jiyuan Feng, Ye Ding, Wei Wang, Yunqing Feng,
  and Qing Liao

Federated Learning with Emerging New Class: A Solution Using
Isolation-Based Specification 719
  Xin Mu, Gongxian Zeng, and Zhengan Huang

A Static Bi-dimensional Sample Selection for Federated Learning
with Label Noise 735
  Qian Rong, Ling Yuan, Guohui Li, Jianjun Li, Lu Zhang,
  and Xuanang Ding

An Information Theoretic Perspective for Heterogeneous Subgraph
Federated Learning 745
  Jiayan Guo, Shangyang Li, and Yan Zhang

Author Index 761

Query Processing

Semantic-Driven Instance Generation for Table Question Answering

Shuai Ma[1,2], Wenbin Jiang[4], Xiang Ao[2,3(✉)], Meng Tian[4], Xinwei Feng[4], Yajuan Lyu[4], Qiaoqiao She[4], and Qing He[1,2(✉)]

[1] Henan Institutes of Advanced Technology, Zhengzhou University, Zhengzhou 450052, People's Republic of China
[2] Key Lab of Intelligent Information Processing of Chinese Academy of Sciences (CAS), Institute of Computing Technology, CAS, Beijing 100190, China
{aoxiang,heqing}@ict.ac.cn
[3] Institute of Intelligent Computing Technology, CAS, Suzhou, China
[4] Baidu inc., Beijing, China
{jiangwenbin,tianmeng03,fengxinwei,lvyajuan,sheqiaoqiao}@baidu.com

Abstract. Recent studies exhibit that generating sufficient samples can improve the performance of Table QA, especially in complex cross-domain applications. However, most existing data augmentation approaches for Table QA adopt a top-down paradigm and rely on pre-defined rules designed by human experts. In this paper, we aim to generate training instances for Table QA while mitigating the dependence on the human experience. We propose an approach coined SIG-TQA in which an off-shelf parsing tool is utilized to extract semantic patterns of SQL queries from text-to-SQL corpus. Then, these semantic patterns are underpinned to generate question/SQL pair with a SQL generator and a natural language question generator, respectively. Both the semantic pattern extraction and question/SQL pair generation are performed based on the original text-to-SQL corpus with few manual efforts injected, our proposed SIG-TQA is making a different bottom-up paradigm. Extensive experiments on a widely-used benchmark and online experiments on a practical industry system demonstrate the superiority of SIG-TQA. Currently, our SIG-TQA has been applied to a real-world Table QA system, and its code is available on https://github.com/DHms2020/SIG-TQA.

Keywords: Table QA · Text2SQL · Data Augmentation

1 Introduction

Table question-answering (a.k.a Table QA) [13,18,19,37] aims at getting result answering user questions for a given table. It can be converted into a text-to-SQL parsing task [2,4,7,11,22,24,27,30,32]. Recent studies exhibit that generating sufficient samples can improve the performance of Table QA [9,34], especially in

---

S. Ma—Work done during an internship at Baidu Inc.

X. Wang et al. (Eds.): DASFAA 2023, LNCS 13943, pp. 3–18, 2023.
https://doi.org/10.1007/978-3-031-30637-2_1

complex cross-domain applications [33]. The scarcity of labeled instances may render the text-to-SQL parsing ineffective.

Existing data augmentation approaches for Table QA can be categorized into two groups, namely template-based and grammar-based methods. The former generates SQL queries through manually constructed templates and then translates the SQL queries into natural language questions [10]. The latter group usually generates SQL queries by context-free grammar and back-translates them to natural language questions [25], or simultaneously generates SQL queries and questions by synchronous context-free grammar [31,32].

Though existing methods achieve remarkable success in improving Table QA's performance, both of them generate instances according to predefined abstract rules by human experts, e.g., manually created templates or grammars, which can be regarded as heuristic top-down approaches. These top-down approaches may have their limitations considering the inadequate coverage and inefficient management of human experience.

In this paper, we aim to generate training instances for Table QA while mitigating the dependence on the human experience. To this end, we propose a simple-yet-effective approach named SIG-TQA (Semantic-driven Instance Generation for Table Question Answering). First, an off-shelf tool is adopted to extract semantic patterns of SQL queries from cross-domain text-to-SQL datasets, considering SQL queries toward Table QA might have intrinsic regularity[1] [10]. These semantic patterns can be viewed as a kind of abstraction extracted from raw data but without human efforts in designing rules or grammar. Then we utilize the semantic patterns to underpin new question/SQL pair generation, i.e., implementing the data augmentation. A SQL generator is trained in which the semantic patterns are combined with the table schema to generate SQL queries. Meanwhile, a natural language question generator is fine-tuned on pre-trained language model (PLM) via consuming the semantic patterns and the SQL queries to predict corresponding questions.

Since both the semantic pattern extraction and question/SQL pair generation are performed based on the original text-to-SQL corpus with few manual efforts injected, our proposed SIG-TQA is following a bottom-up paradigm for data augmentation on Table QA, which is orthogonal to the existing top-down approaches[2]. We verify our method on a widely-used cross-domain Table QA benchmark Spider [33]. The results demonstrate that SIG-TQA can generate high-quality labeled instances, resulting in clear improvements for two base models compared with other data augmentation methods. In addition, we have applied our method to a practical industry online Table QA system and gained significant improvement.

---

[1] Such regularity is also the basis of template-based data augmentation methods for Table QA.

[2] We call the existing template/grammar-based methods top-down because most of them design rule first then to generalize following specific regularity. In contrast, our approach, which extracts patterns from the data and finds the regularity corresponding with the raw data distribution by means of training, is thus called bottom-up.

2   Related Work

Data augmentation strategies aim at enlarging the training dataset from existing data using various translations, such as flipping, cutout [6], and erasing [36] in the computer vision field; Or word-translation [8], text-generation [1] in the natural language processing field. Moreover, Li et al. (2022) [16] designed a learning policy scheduling for data augmentation to search the best augmentation schedule based on specific datasets. Liu et al. (2020) [17] proposed a novel adversarial data augmentation method to solve the class imbalance problem for classifier in financial domain.

Besides, data augmentation strategies have been successfully exploited in Table QA in recent years. To name some, Guo et al. (2018) [10] sampled SQL queries based on SQL templates [35] and converted SQL queries into questions by the Seq2Seq model. Wang et al. (2021) [25] used abstract syntax trees (ASTs) [29] to model the underlying grammar of SQL, generated referring SQL queries by probabilistic context-free grammars of abstract syntax trees, and translated the generated SQL queries into questions by a pre-trained language model. Yu et al. (2018b, 2020) [31,32] followed the synchronous context-free grammar (SCFG) [12], designed rules to assemble corresponding fragments of SQL queries and questions simultaneously, and then generated both SQL queries and questions. Unlike the above, the semantic patterns in our work are more like generalized templates containing grammar information, which denote the abstraction of the question semantics.

In addition to generating parallel training instances, i.e., question/SQL pairs, there are also works only generating natural language questions toward tables. For example, with the help of pre-defined control codes, Shi et al. (2021) [21] controlled a neural network model to generate corresponding questions given the tabular data. In contrast to this work, the control symbols of our approach are question patterns summarized from existing samples rather than defined by human experts.

3   SIG-TQA Methodology

3.1   Preliminaries

Database-oriented structured query language (SQL) can be considered a natural semantic abstraction of natural language questions since different natural language questions can be represented as the same SQL query, as illustrated in Fig. 4. We propose the semantic question pattern based on SQL query, which could be regarded as a further abstraction of semantics to generalize among databases. Intuitively, for scenarios where human could acquire information in need properly by Table QA, we think there are limited semantic patterns of questions that humans can ask, as verified by statistical analysis on Spider as illustrated in Fig. 1, which show that as the proportion of the data set increases, the number of semantic patterns will gradually converge to a certain value.

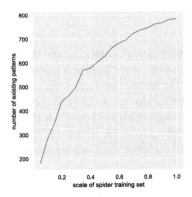

Fig. 1. Quantitative relationship curves between the number of patterns and different proportions of corresponding Spider training sets.

We represent the semantic question pattern by a sequence consisting of the non-terminal node symbols set $\mathcal{U}$ of the parse tree, keywords set $\mathcal{S}$, column type set $\mathcal{C}$, special token *TABLE*, and bracket. In the following sections, we will describe how to obtain semantic patterns.

Problem Definition. Given the semantic question patterns, coupled with new table schemas, even in an unseen domain, the SIG-TQA model should generate new question/SQL pairs that can be used for training in downstream tasks.

3.2   Model Framework

The Pipeline of the SIG-TQA is depicted in Fig. 2. The Basic idea is that, with the modified pattern constructor, we extract semantic question patterns from widely-used cross-domain Text-to-SQL benchmark, coupled with schema items/SQL queries to train SQL/Question generator respectively. Besides, with the extracted and summarized patterns set, existing table schemas, and pre-processing of rematching[3], the model could generate new question/SQL pairs. Because of the uncovered question patterns on each database or schema, we can expand the existing dataset, may benefit few-shot learning scenarios. Given the unseen table schemas, especially in a new domain, the trained SIG-TQA model could generate new question/SQL pairs of that domain, benefiting zero-shot learning scenarios.

In detail, to capture the representatives of kinds of questions, we propose a tree-structure semantic question pattern that could be extracted from existing text-to-SQL datasets and then be fed to the generator as input of training flow as a guiding signal. The SQL/question generator will take schemas/SQL from original datasets as input in training flow, then use SQL/question from original datasets as ground truth for loss computation. In the generation flow, the trained

---

[3] Performing a Cartesian product between the two sets, then eliminating the existing combinations to get new input for SIG-TQA.

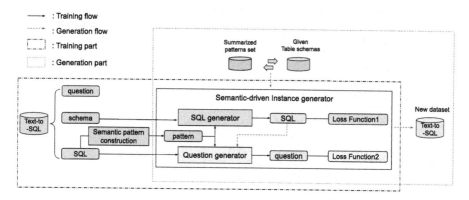

Fig. 2. Training and generation flow of SIG-TQA approach. The green question-SQL pairs consist new text-to-SQL dataset. (Color figure online)

SIG-TQA model will take the rematched semantic pattern/schema item pairs as input, then output the new question/SQL pairs as the new text-to-SQL dataset. Meanwhile, the SQL generator's output will be the question generator's input inside SIG-TQA.

3.3 Semantic Pattern Extraction

First we introduce how to extract semantic patterns from a text-to-SQL corpus. Recall that Guo et al. (2019) [11] designed a semi-structured query language (SemQL), together with a tool parsing SQL clauses and matching corresponding production rules to build a tree-like structural SemQL. Specifically, we know that a single SQL query can be split into multiple clauses, like `FROM`, `SELECT`, `WHERE`, `GROUPBY` clauses, etc [33]. Besides, a compound SQL query consists of several single queries, and the components `UNION`, `EXCEPT`, `INTERSECT`. Therefore, the parsing tool can traverse each clause and component to match the corresponding production rule. In detail: (1) If there exist one of the components `UNION`, `EXCEPT`, `INTERSECT` in SQL query, the tool attach the corresponding keywords and two R node under root node Z, otherwise a single R node. (2) If the `ORDERBY` clause contains the keyword `LIMIT`, there will be a *Superlative* node under the R node, otherwise an *Order* node. (3) The number of A node is determined by the number of search columns in `SELECT` clause. For each column, the tool attaches its aggregate function node, a C node, and a T node under A. (4) If it has a nested query in `WHERE` clause or `HAVING` clause, the tool recursively attaches the R node under *Filter* as the sub-tree, instead of A node. The production rules are depicted in Fig. 3. More detail can refer to [11].

In our SIG-TQA, we adopt the original SemQL to adapt more complex queries with abstraction. First, we abstract the table names as a defined token *TABLE*, also generalize specific column names to corresponding data type tokens, such as *numeric*, *date*, *text*, etc. Since most of the column and table nodes appear as the lowermost leaf nodes of the subtree in SemQL, we can

$Z ::= intersect\ R\ R\ |\ union\ R\ R\ |\ except\ R\ R\ |\ R$
$R ::= Select\ |\ Select\ Filter\ |\ Select\ Order\ |\ Select\ Superlative\ |\ Select\ Order\ Filter$
$\qquad |Select\ Superlative\ Filter$
$Select ::= A\ |\ A\ A\ |\ A\ A\ A\ |\ A\ A\ A\ A\ |\ A\ A\ ...\ A\ A$
$Order ::= asc\ A\ |\ desc\ A$
$Superlative ::= most\ A\ |\ least\ A$
$Filter ::= and\ Filter\ Filter\ |\ or\ Filter\ Filter$
$\qquad |> A\ |> A\ R\ |< A\ |< A\ R$
$\qquad |>= A\ |>= A\ R\ |= A\ |= A\ R$
$\qquad |\neq A\ |\neq A\ R\ |\ between\ A$
$\qquad |\ like\ A\ |\ not\ like\ A\ |\ in\ A\ R\ |\ not\ in\ A\ R$
$A ::= max\ C\ T\ |\ min\ C\ T\ |\ count\ C\ T$
$\qquad |\ sum\ C\ T\ |\ avg\ C\ T\ |\ none\ C\ T$

Fig. 3. The SemQL construction node production rules.

directly make the abstraction within a parsed tree. Second, we integrate the operators whose objects are similar. Five additional tokens are defined, including (1) AGG_OP aggregates keywords like *sum*, *min*, *max*, *avg*; (2) CMP_OP compares keywords like $=, !=, <, >, <=, >=$; (3) $ORDER\_OP$ combines keywords *des*, *asc*; (4) $CONDI1\_OP$ subordinates keywords *in* and *not in*; (5) $CONDI2\_OP$ contains keywords *like* and *not like*.

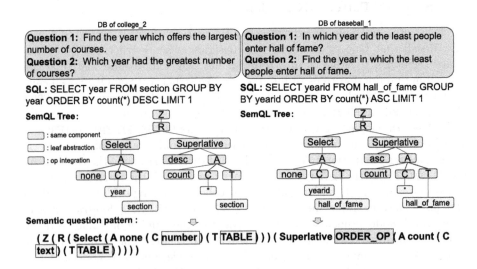

Fig. 4. Example of Semantic question pattern construction process.

With these operations, we can aggregate a semantic pattern given a SemQL tree. Figure 4 shows an example of an extracted semantic pattern from different text-to-SQL instances. From this figure, we can see that the leaf nodes are converted to TABLE and column type, respectively; *desc* and *asc* are aggregated to

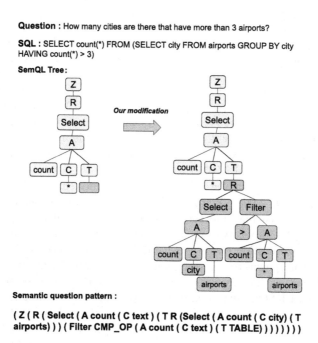

Question : How many cities are there that have more than 3 airports?

SQL : SELECT count(*) FROM (SELECT city FROM airports GROUP BY city HAVING count(*) > 3)

SemQL Tree:

Semantic question pattern :

(Z (R (Select (A count (C text) (T R (Select (A count (C city) (T airports))) (Filter CMP_OP (A count (C text) (T TABLE))))))))

Fig. 5. Example of additional production rules to ensure the correct nested structure.

ORDER_OP. After these operations, different questions toward different tables can be aggregated as the same semantic pattern.

On top of that, we adjust the production rule of SemQL on T node to ensure that if there exists a nested query in FROM clause, the tree structure would have a corresponding sub-tree starting with R node. Figure 5 depicts the effect that, after supplementing the production rules, we have the correct nested structure of the semantic question pattern given the specific nested SQL query.

3.4 Semantic-Driven Instance Generation

So far, the remaining is how to generate new instances for Table QA prompted by the extracted semantic patterns. In our model, a SQL generator and a question generator are devised to resolve.

The SQL generator is modified from a text-to-SQL parser RATSQL [24]. We adopt the encoder-decoder framework, keep the tree-based decoder, but use the BART [14] encoder instead of BERT [5] inspired by Shi et al. (2021) [21]. Firstly, its inputs are the semantic patterns and the table schemas. Each semantic question pattern is represented as a sequence of symbols, and the table schema information is denoted as a sequence consisting of table names and column names/types. Both the pattern and schema will then be concatenated following the format below after tokenizing:

$$INPUT = \left(p_1 \cdots p_{|\mathcal{P}|} \mid c_1 \cdots c_{|\mathcal{C}|} \mid t_1 \cdots t_{|\mathcal{T}|}\right) \tag{1}$$

where $\mathcal{P}$ is the tokens set of semantic question pattern, $\mathcal{C}$ is the tokens set of all column names/types from the specified database, and $\mathcal{T}$ is the tokens set of all table names from the specified database. Secondly, the outputs are SQL queries, which were generated as an abstract syntax tree in depth-first traversal order. That generating is implemented by using an LSTM to output a sequence of decoder actions that either (1) expand the last generated node into a grammar rule, called *APPLYRULE*, or (2) choose a column/table from the schema, called *SELECTCOLUMN* and *SELECTTABLE*. We refer the reader to Yin et al. (2017) [28] and Wang et al. (2019) [24] for detail.

The training sample of the SQL generator is derived from Table QA samples (*question, SQL, table*), being a triple consisting of the semantic question pattern, the table schema, and the corresponding SQL query according to the original Table QA sample. We can obtain various SQL queries by rematching all kinds of semantic question patterns with different table schemas.

The question generator is implemented in Transformer-based encoder-decoder architecture. The inputs of the encoder are the extracted semantic patterns together with SQL queries, and the outputs are natural language questions. The generator is fine-tuned based on parameters of the large-scale pre-trained language model BART-large [14] to take advantage of the existing pre-trained text generation model. Its training samples are also derived from the Table QA samples, i.e. (*question, SQL, table*), being another kind of triple consisting of the semantic question pattern, the SQL query, and the question. Here the *question* is taken as the output, the *SQL* and the semantic pattern are the inputs.

Both the two generators can be trained with cross-entropy loss [3] as follows.

$$CE = -\sum_{i=1} p_i \log\left(\hat{p}_i\right) \tag{2}$$

For the question generator, p corresponds with the ground truth question tokens from the original text-to-SQL dataset. While in SQL generator, it represents the ground truth action decomposed from SQL query on the text-to-SQL dataset, which could refer to [24].

4   Experiments

In this section, we aim to answer the following research questions:

- **(RQ1.)** How much improvement does SIG-TQA bring to the base models on data augmentation?
- **(RQ2.)** How does SIG-TQA perform compared to other traditional data augmentation methods?
- **(RQ3.)** How about the quality of generated data compared to other data augmentation methods?
- **(RQ4.)** Is the SIG-TQA helpful in few-shot learning or zero-shot learning scenarios?

4.1 Datasets

We conduct experiments on a cross-domain Table QA benchmark Spider [33]. Its training set is used to extract the semantic patterns set and train our SIG-TQA. We evaluate the effectiveness of the method on the dev set of Spider following [9]. Specifically, we use the dev set to assess the text-to-SQL parser augmented by generated samples compared with the original parser. We follow the common practice on text-to-SQL to report the exact set match accuracy [33], which decomposes SQL query into clauses and then conducts the evaluation.

The training set of Spider consists of $8,659$ cross-domain question/SQL pair records. After standard SemQL transforming from raw SQL queries, we obtain $4,011$ SemQL syntactic trees. We then get 783 semantic question patterns following the method in Sect. 3.3.

Considering that the Spider dataset does not have a publicly released test set, we re-split the Spider dataset to facilitate the comparison of different base models with/without different data augmentation methods. Most importantly, that can simulate the few-shot and zero-shot scenarios with known table schemas but few or no relevant in-domain question/SQL pairs as training data. We use 90% of the original training set in base models, the remaining 10% is used as the validation set, and the original dev set is used as the test set. The division is set according to the out-of-domain principle, i.e., to ensure that the table schemas of the training set, validation set, and test set do not overlap.

4.2 Base Model and Baseline

We use two popular text-to-SQL parsers as our base models, namely IRnet [11] and RATSQL [24]. IRnet adopts a grammar-based neural model to synthesize a SemQL query, an intermediate representation to bridge NL and SQL, then infers SQL query. RATSQL uses relation-aware self-attention to jointly learn schema and question representations based on their alignment with each other and schema relations. GAP [21] as a pre-training framework can better model the connections between schema and question, thus further enhancing RATSQL parser.

We replicate back-translation and PCFG as two baseline methods to compare with our approach. Back-translation [20] is a general data augmentation method mainly used to handle monolingual data. We adjusted the application to generate new questions given the SQL queries. PCFG [25] approach reassembles production rules based on probability, which is decomposed and calculated from text-to-SQL datasets. Obtained new SQL queries are then translated to natural language questions by BART. It is also a two-phase and grammar-based instance generation method.

4.3 Experimental Settings

SQL Generator Setting. We implement the SQL generator with the encoder of pre-trained language model, i.e., the 12-layer transformer-based BART encoder, and tree-based decoder from RATSQL parser. We follow most of the hyperparameters for training RATSQL-BART [21] except that we use batch size of 8, learning rate of 1e-5, and 4 GPU for Distributed Data Parallel [15], with the max step as 13K.

Question Generator Setting. We fine-tune the pre-trained text generation model BART-large to implement the question generator. We use batch size of 8, learning rate of 5e-5, and the Adam optimizer to fine-tune. The ultimate training epoch is 20.

Base Model and Baseline Setting. We use the default hyperparameters in the released code for each base parser model and data augmentation method baseline.

4.4   Experiment Results

Main Result (RQ1&RQ2). We evaluate the performance of two base models without/with different data augmentation methods containing ours on Spider. According to the main results of our experiments shown in Table 1, we have the following observations: (1) Our approach can effectively expand the existing dataset, and the data augmentation with generated instances could significantly improve the performance of the base models on our replication. That demonstrates our proposed SIG-TQA could effectively exploit the potential information under the table schemas to expand existing datasets and augment base models. (2) Besides, our SIG-TQA outperforms the back-translation and PCFG approaches. That indicates the advantage of our bottom-up paradigm for data augmentation. Meanwhile, the PCFG approach could reduce the performance of RATSQL+GAP when we directly merge the generated data and the original training set as data augmentation. We think it could stem from the uncertainty of the PCFG method that relies on grammatical splicing and reorganization

Table 1. Main results on Spider.

Model	EM
IRnet (Guo et al., 2019) [11]	53.20%
RATSQL+GAP (Shi et al., 2021) [21]	71.80%
IRnet (our replicate)	53.30%
+Aug (BackTranslation)	51.80%
+Aug (PCFG)	54.40%
+Aug (Ours)	**54.70%**
RATSQL+GAP (our replicate)	69.24%
+Aug (BackTranslation)	69.73%
+Aug (PCFG)	67.89%
+Aug (Ours)	**72.34%**

to form a new corpus. In other words, the generated question/SQL pairs would over-expand the semantic patterns, which does not match the distribution of the actual application, thus introducing too much noise. To verify that, we extract semantic question patterns specifically from the generated data of PCFG. After eliminating the duplication, we find the amount of semantic question patterns from PCFG generation is 2073, compared with the amount of 783 from Spider. That confirms what we thought.

Data Quality Comparison Experiment (RQ3). In this subsection, we aim to demonstrate the higher usability and quality of the data generated by SIG-TQA compared to other methods on TableQA tasks, especially text-to-SQL. In order to have a quantifiable and comparable metric, we trained different text-to-SQL parsers on the synthetic question/SQL pairs of our SIG-TQA as well as the synthetic data of other data augmentation method, and evaluate the exact match of different trained parsers on the test set. From Table 2, we can observe that the model trained on data generated from SIG-TQA performs better than that from other methods. That indicates the synthetic data of SIG-TQA can help the parser learn the distribution of target text-to-SQL dataset (i.e., Spider) better, proving the higher usability and quality of our generated data and the effectiveness of SIG-TQA.

Table 2. Performance on different base models using only data generated from different methods as training sets.

Method	IRnet	RATSQL+GAP
BackTranslation	23.8%	24.5%
PCFG	15.7%	16.8%
SIG-TQA(ours)	**25.1%**	**27.0%**

Auxiliary Experiments (RQ4). Next, we conduct a simulation experiment to approximate the effect on new scenarios, i.e., with few table schemas and labeled samples, or even without labeled samples. We construct the semantic patterns, build the instance generation system with 50% of the Spider training set, and automatically generate large-scale samples on the unseen table schemas corresponding to the remaining training set. Table 3 shows the experimental results under different split ratios on the remaining training set. The division is also set according to the out-of-domain principle we mentioned above in Sect. 4.1. We can conclude as follows: (1) Because after training on the completely manually labeled Spider original dataset, the model results (i.e. EM_{ori} on Table 2) reflect an upper bound on the quality of the data, therefore the performance in zero-shot learning (i.e. EM_{gen} on Table 2) seems inadequate, which means there is still room for improvement in synthetic data, and that will be our future work. (2) The generated samples can significantly improve the final results when overlaid with the original samples (i.e. EM_{merge} on Table 2). That indicates that our approach is also suitable for few-shot learning scenarios. Experimentally, even if getting unseen table schema items as input, the trained SIG-TQA can still synthesize question/SQL pairs with summarized semantic question patterns, which means our approach can generalize semantic question patterns to the different databases, thus effectively expanding existing labeled text-to-SQL data.

Table 3. Simulation experiment on different split ratios. The 'ori' means training data split from Spider directly, the 'gen' means training data generated by the schema of former split data, and the 'merge' means combining both of them. We use IRnet as the base parser.

Split	EM_{ori}	EM_{gen}	EM_{merge}
20%	31.00%	15.40%	**32.80%**
40%	39.80%	17.50%	**40.20%**
60%	43.90%	18.60%	**45.30%**

Online Experiments (RQ4). Finally, to further confirm the advantage of SIG-TQA in few-shot learning, we have applied the SIG-TQA approach to a Chinese Table QA system, especially for bank and electricity domains. We train the SIG-TQA on a Chinese cross-domain Table QA dataset DuSQL [26], then use the industry schema to generate related question/SQL pairs.

Table 4. Online experiment results on Chinese Table QA system based on industry (i.e. bank/electricity) data.

Version	precision	recall	F1
Online Table-QA system	91.85%	78.16%	84.44%
+SIG-TQA	91.83%	**83.27%**	**87.34%**

The generated pairs are utilized in further pretraining on a Chinese PLM, i.e., Ernie [23] with text-to-SQL specific task GenSQL [21]. We then plug the pre-trained model into the Online Chinese Table QA system to replace the existing Ernie encoder. Table 4 shows the results. We can see that with the help of our SIG-TQA, our online Chines Table QA system have better performance on industry-specific tasks. For example, the F1 increases 2.9% compared with the online running system without data augmentation techniques. That indicates that our SIG-TQA can effectively generate training data on the specific industry, which facilitates the implementation of customized domain data augmentation, and alleviate the data scarcity problem in industry applications. Besides, it also proves that the framework we proposed is applicable to different language scenarios, like Chinese scenarios in our online experiment.

4.5   Case Study

In this subsection, we aim to conduct case studies to compare the data quality from different data augmentation methods and the original Spider. It is worth noting that: (1) The SQL queries of back-translation are the source corpus, so they keep the same with original spider data; (2) And the data generated by this method is much smaller in magnitude than other methods. Therefore, we did not use the back-translation case as a comparison specifically.

Table 5. Original data vs. SIG-TQA generated data.

Spider data	SIG-TQA generated data
Example 1:	**Example 1:**
Question: How many problems are there for product voluptatem?	**Question**: How many products are in the sigir?
SQL: SELECT Count(*) FROM Product AS T1 JOIN Problems AS T2 ON T1.product_id = T2.product_id WHERE T1.product_name = "voluptatem"	**SQL**: SELECT Count(*) FROM Product WHERE Product.product_name = "sigir"
Pattern: (Z (R (Select (A count (C text) (T TABLE))) (Filter text CMP_OP (A none (C text) (T TABLE)))))	**Pattern**: (Z (R (Select (A count (C text) (T TABLE))) (Filter text CMP_OP (A none (C text) (T TABLE)))))
Example 2:	**Example 2:**
Question: What is the id of the problem log that is created most recently?	**Question**: Which problem log has the most corresponding problem? give me the log id and number of times the problem log
SQL: SELECT problem_log_id FROM Problem_log ORDER BY log_entry_date DESC LIMIT 1	**SQL**: SELECT Problem_Log.problem_id, Count(*) FROM Problem_Log GROUP BY Problem_Log.problem_id ORDER BY Count(*) Desc LIMIT 1
Pattern: (Z (R (Select (A none (C number) (T TABLE))) (Superlative ORDER_OP (A none (C time) (T TABLE)))))	**Pattern**: (Z (R (Select (A none (C number) (T TABLE))) (A count (C text) (T TABLE)))) (Superlative ORDER_OP (A count (C text) (T TABLE)))))

First, we compare the original spider data and generated data from SIG-TQA to explain the effectiveness of data augmentation. The case we selected comes from the same domain (i.e. the 'tracking_software_problems' database of Spider). As shown in Table 5, Example 1 has the same semantic question pattern. From the perspective of semantic alignment between the question and SQL query, the generated data is comparable with the original Spider data which was manually labeled. Besides, from Example 2 we can observe that, under the same domain, our approach offers more diversity of questioning/SQL. That diversity helps downstream tasks to learn the correlation between questions and SQL [24].

Table 6. PCFG vs. SIG-TQA on the same semantic question pattern : (Z (R (Select (A count (C text) (T TABLE))))).

PCFG generated data	SIG-TQA generated data
Example 3:	**Example 3:**
Question: how many staff last names do we have?	**Question**: how many products are there?
SQL: SELECT Count(Count(staff_last_name)) FROM Staff	**SQL**: SELECT Count(*) FROM Product

Meanwhile, the diversity is limited to the range of semantic question patterns summarized from the original dataset so as to avoid bringing too much noise.

Next, we compare the generated data from PCFG and SIG-TQA, respectively. They also come from the 'tracking_software_problems' domain. We deliberately selected the generated data from the same domain, then chose the question/SQL pairs with the same pattern by semantic question pattern extraction, in order to clearly compare their differences in syntax. As shown in Table 6, despite having the same semantics of question structure, the case from PCFG seems improper and redundant in syntax compared with that from SIG-TQA, which can also explain the performance difference on data quality comparison experiment described in Table 2.

5 Conclusion

In this paper, we proposed SIG-TQA that can drive a neural network generator to synthesize high-quality training instances for a given table schema, with the help of extracting semantic patterns from raw text-to-SQL datasets. No human experience is injected into the generation process, making the proposed model take a different bottom-up paradigm. Experiments on a challenge cross-domain Table QA benchmark show the effectiveness of the method, outperforming previous instance generation methods on two base models. We also applied this approach to our Chinese Table QA system based on the banking/electricity industry and achieved significant improvement.

Acknowledgments. The research work supported by National Key R&D Plan No. 2022YFC3303302, the National Natural Science Foundation of China under Grant No. 61976204. This work was also supported by the National Natural Science Foundation of China (No. 31900979, U1811461) and Zhengzhou Collaborative Innovation Major Project (No. 20XTZX11020). Xiang Ao is also supported by the Project of Youth Innovation Promotion Association CAS and Beijing Nova Program Z201100006820062.

References

1. Bayer, M., Kaufhold, M.A., Buchhold, B., Keller, M., Dallmeyer, J., Reuter, C.: Data augmentation in natural language processing: a novel text generation approach for long and short text classifiers. Int. J. Mach. Learn. Cybern. **14**, 135–150 (2022). https://doi.org/10.1007/s13042-022-01553-3

2. Cao, R., Chen, L., Chen, Z., Zhao, Y., Zhu, S., Yu, K.: LGESQL: line graph enhanced text-to-SQL model with mixed local and non-local relations. arXiv preprint arXiv:2106.01093 (2021)

3. Cybenko, G., O'Leary, D.P., Rissanen, J.: The Mathematics of Information Coding, Extraction and Distribution (2012)

4. Date, C.J., Darwen, H.: A Guide to the SQL Standard: A User's Guide to the Standard Database Language SQL. Addison-Wesley (1997)

5. Devlin, J., Chang, M.W., Lee, K., Toutanova, K.: BERT: pre-training of deep bidirectional transformers for language understanding. arXiv preprint arXiv:1810.04805 (2018)

6. DeVries, T., Taylor, G.W.: Improved regularization of convolutional neural networks with cutout. arXiv preprint arXiv:1708.04552 (2017)

7. Dong, L., Lapata, M.: Language to logical form with neural attention. arXiv preprint arXiv:1601.01280 (2016)

8. Fadaee, M., Bisazza, A., Monz, C.: Data augmentation for low-resource neural machine translation. arXiv preprint arXiv:1705.00440 (2017)

9. Gan, Y., et al.: Towards robustness of text-to-SQL models against synonym substitution. In: Proceedings of the 59th Annual Meeting of the Association for Computational Linguistics and the 11th International Joint Conference on Natural Language Processing (Volume 1: Long Papers), pp. 2505–2515 (2021)

10. Guo, D., et al.: Question generation from SQL queries improves neural semantic parsing. arXiv preprint arXiv:1808.06304 (2018)

11. Guo, J., et al.: Towards complex text-to-SQL in cross-domain database with intermediate representation. arXiv preprint arXiv:1905.08205 (2019)

12. Jia, R., Liang, P.: Data recombination for neural semantic parsing. arXiv preprint arXiv:1606.03622 (2016)

13. Jin, N., Siebert, J., Li, D., Chen, Q.: A survey on table question answering: recent advances. arXiv preprint arXiv:2207.05270 (2022)

14. Lewis, M., et al.: BART: denoising sequence-to-sequence pre-training for natural language generation, translation, and comprehension. arXiv preprint arXiv:1910.13461 (2019)

15. Li, S., et al.: PyTorch distributed: experiences on accelerating data parallel training. arXiv preprint arXiv:2006.15704 (2020)

16. Li, S., Ao, X., Pan, F., He, Q.: Learning policy scheduling for text augmentation. Neural Netw. **145**, 121–127 (2022)

17. Liu, Y., Ao, X., Zhong, Q., Feng, J., Tang, J., He, Q.: Alike and unlike: resolving class imbalance problem in financial credit risk assessment. In: Proceedings of the 29th ACM International Conference on Information & Knowledge Management, pp. 2125–2128 (2020)

18. Müller, T., Piccinno, F., Nicosia, M., Shaw, P., Altun, Y.: Answering conversational questions on structured data without logical forms. arXiv preprint arXiv:1908.11787 (2019)

19. Pasupat, P., Liang, P.: Compositional semantic parsing on semi-structured tables. arXiv preprint arXiv:1508.00305 (2015)

20. Sennrich, R., Haddow, B., Birch, A.: Improving neural machine translation models with monolingual data. arXiv preprint arXiv:1511.06709 (2015)
21. Shi, P., et al.: Learning contextual representations for semantic parsing with generation-augmented pre-training. In: AAAI (2021)
22. Sun, Y., et al.: Semantic parsing with syntax-and table-aware SQL generation. arXiv preprint arXiv:1804.08338 (2018)
23. Sun, Y., et al.: ERNIE: enhanced representation through knowledge integration. arXiv preprint arXiv:1904.09223 (2019)
24. Wang, B., Shin, R., Liu, X., Polozov, O., Richardson, M.: RAT-SQL: relation-aware schema encoding and linking for text-to-SQL parsers. arXiv preprint arXiv:1911.04942 (2019)
25. Wang, B., Yin, W., Lin, X.V., Xiong, C.: Learning to synthesize data for semantic parsing. arXiv preprint arXiv:2104.05827 (2021)
26. Wang, L., et al.: DuSQL: a large-scale and pragmatic Chinese text-to-SQL dataset. In: Proceedings of the 2020 Conference on Empirical Methods in Natural Language Processing (EMNLP), pp. 6923–6935 (2020)
27. Xu, X., Liu, C., Song, D.: SQLNet: generating structured queries from natural language without reinforcement learning. arXiv preprint arXiv:1711.04436 (2017)
28. Yin, P., Neubig, G.: A syntactic neural model for general-purpose code generation. In: Proceedings of the 55th Annual Meeting of the Association for Computational Linguistics (Volume 1: Long Papers), pp. 440–450 (2017)
29. Yin, P., Neubig, G.: TRANX: a transition-based neural abstract syntax parser for semantic parsing and code generation. arXiv preprint arXiv:1810.02720 (2018)
30. Yu, T., Li, Z., Zhang, Z., Zhang, R., Radev, D.: TypeSQL: knowledge-based type-aware neural text-to-SQL generation. arXiv preprint arXiv:1804.09769 (2018)
31. Yu, T., et al.: GraPPa: grammar-augmented pre-training for table semantic parsing. arXiv preprint arXiv:2009.13845 (2020)
32. Yu, T., et al.: SyntaxSQLNet: syntax tree networks for complex and cross-domaintext-to-SQL task. arXiv preprint arXiv:1810.05237 (2018)
33. Yu, T., et al.: Spider: a large-scale human-labeled dataset for complex and cross-domain semantic parsing and text-to-sql task. arXiv preprint arXiv:1809.08887 (2018)
34. Zhang, A., et al.: Data augmentation with hierarchical SQL-to-question generation for cross-domain text-to-SQL parsing. arXiv preprint arXiv:2103.02227 (2021)
35. Zhong, V., Xiong, C., Socher, R.: Seq2SQL: generating structured queries from natural language using reinforcement learning. arXiv preprint arXiv:1709.00103 (2017)
36. Zhong, Z., Zheng, L., Kang, G., Li, S., Yang, Y.: Random erasing data augmentation. In: Proceedings of the AAAI Conference on Artificial Intelligence, vol. 34, pp. 13001–13008 (2020)
37. Zhu, F., et al.: TAT-QA: a question answering benchmark on a hybrid of tabular and textual content in finance. arXiv preprint arXiv:2105.07624 (2021)

Fine-Grained Tuple Transfer for Pipelined Query Execution on CPU-GPU Coprocessor

Zhenhua Yang[1], Qingfeng Pan[1], and Chen Xu[1,2(✉)]

[1] Shanghai Engineering Research Center of Big Data Management,
East China Normal University, Shanghai 200062, China
{zhyang,qfpan}@stu.ecnu.edu.cn, cxu@dase.ecnu.edu.cn
[2] Guangxi Key Laboratory of Trusted Software, Guilin University of Electronic
Technology, Guilin 541004, China

Abstract. To leverage the massively parallel capability of GPU for query execution, GPU databases have been studied for over a decade. Recently, researchers proposed to execute queries with both CPU and GPU in a pipelined approach. In the pipelined query execution, the cross-processor tuple transfer plays a crucial role for the overall query execution performance. The state-of-the-art solution achieves cross-processor tuple transfer using a queue-like data structure. However, it is coarse-grained due to the use of a single spin lock to achieve thread-safety. This design causes performance issues as it prevents the threads from accessing the queue simultaneously. In this paper, we propose a fine-grained tuple transfer mechanism. It employs decoupled enqueue/dequeue to enable two threads on different processors to access the queue at the same time. Moreover, this mechanism explores subqueue-based locking to enable the threads on the same processor to access the queue at the same time. In particular, we implement a prototype system, namely πQC, which adopts fine-grained tuple transfer. Our experiments show that πQC achieves an order of magnitude better performance than existing GPU databases such as HeavyDB.

Keywords: GPU Database · Tuple Transfer · Pipelined Execution

1 Introduction

GPU provides massive parallelism, which is suitable for data-intensive workloads. With the development of the GPU-oriented general purpose programming frameworks (e.g., OpenCL, CUDA and DPC++), GPU is adopted as a general computing accelerator in many fields including database systems.

GPU databases like RateupDB [10] focus on fully utilizing the GPU on a system with heterogeneous processors. They do not co-utilize CPU and GPU to execute a single query. Actually, the CPU in modern high-end servers also provides considerable computing power which can be used to further improve the query execution efficiency. To co-utilize CPU and GPU to execute queries, an efficient approach is to migrate the Volcano [9] query execution model designed for

© The Author(s), under exclusive license to Springer Nature Switzerland AG 2023
X. Wang et al. (Eds.): DASFAA 2023, LNCS 13943, pp. 19–34, 2023.
https://doi.org/10.1007/978-3-031-30637-2_2

CPU to the heterogeneous CPU-GPU environment [4]. This approach achieves cross-processor pipelining, overlapping CPU computation and GPU computation to utilize the heterogeneous processors. Moreover, it avoids expensive material- ization of tuples between the processors.

To implement cross-processor pipelined query execution, a proper tuple trans- fer mechanism is indispensable. One approach employed by the systems including HetExchange [4], PTask [15] and XKaapi [8] is to implement the tuple trans- fer mechanism by `cudaMemcpy()`, a built-in function from CUDA to copy data from CPU memory to GPU memory. After a call to `cudaMemcpy()`, a GPU ker- nel is launched to process the copied data. Finally, the result may be copied back to CPU memory. This "copy-kernel-copy" workflow is argued to have a high overhead due to the repeated and costly cross-processor invocations [11]. Another approach to implement the tuple transfer mechanism is to build a queue- like data structure for cross-processor one-way tuple transfer. A long-running kernel consumes blocks continuously from the queue without the overhead of repeated cross-processor invocations. To ensure thread-safety, the queue uses cross-processor spin locks [11] backed by the cross-processor atomic instructions, since mainstream operating systems do not provide support for a cross-processor mutex. The single spin lock couples enqueue and dequeue, which means two threads running on different processors are not able to perform enqueue and dequeue operations simultaneously. Moreover, the single spin lock prevents the threads running on the same processor from performing the same operations simultaneously. Hence, this *coarse-grained* tuple transfer mechanism causes a performance bottleneck as massive threads are used in query execution for high performance.

Our goal is to enable the massive threads to simultaneously operate the cross-processor concurrent queue to achieve *fine-grained* tuple transfer. In this work, we implement πQC, a cross-processor pipelined query execution engine that employs the fine-grained tuple transfer mechanism. In specific, πQC first integrates the decoupled enqueue/dequeue optimization. Instead of the single spin lock, this optimization explores two spin locks, so threads on different pro- cessors can acquire different spin locks and perform enqueue/dequeue operations at the same time. It reduces the overhead of the coarse-grained tuple transfer mechanism by up to 94.2%. Then, πQC integrates the subqueue-based locking optimization. This optimization splits a single queue into many subqueues. Each subqueue has its own spin locks. Hence, threads running on the same processor are able to perform operations on different subqueues simultaneously. It further reduces the overhead of tuple transfer by up to 96.3%. Our experiments show that πQC outperforms existing GPU databases such as HeavyDB [1] by an order of magnitude in performance.

In the rest of this paper, we introduce essential background about GPU and its adoption in cross-processor pipelined query (in Sect. 2.1) execution followed by a motivation (in Sect. 2.2). Then, we make the following contributions:

- We propose the *decoupled enqueue/dequeue* optimization in Sect. 3.1, which enables two threads on different processors to perform operations on the cross- processor concurrent queue at the same time.

- We propose the *subqueue-based locking* optimization in Sect. 3.2, which enables the threads on the same processor to perform operations on the cross-processor concurrent queue simultaneously.
- We describe the implementation of πQC in Sect. 4 and the evaluation in Sect. 5 demonstrates the efficiency of our proposed techniques.

Finally, we review related work in Sect. 6 and make a conclusion about this paper in Sect. 7.

2   Background and Motivation

This section introduces the background about GPU and its adoption in cross-processor pipelined query execution, and highlights our research motivation.

2.1   Background

GPU Hardware. A discrete GPU is comprised of an array of Streaming Multiprocessors (SMs) and an off-chip memory module called global memory. Global memory provides higher bandwidth and lower latency than CPU memory, but the size is usually limited (e.g., up to 32GB for Nvidia Tesla V100). Each SM is capable of running hundreds of threads and is equipped with KBs of on-chip shared memory, which provides better performance than global memory. The threads are organized into thread blocks, which are scheduling units to be placed on SMs. Thread blocks are further divided into warps to run in an SIMT (Single Instruction, Multiple Thread) fashion. A GPU program is organized as kernels, which are basically functions executed on GPU. To launch a kernel, the code needs to specify execution configuration defining the number of thread blocks and the number of threads within each thread block. Notably, the launch overhead of a kernel is orders of magnitude more expensive than an ordinary CPU function call [5]. To facilitate the programming of kernels, GPU provides atomic instructions to perform operations like atomic compare-and-swap or atomic add. GPU is typically connected with CPU by a PCIe bus, of which the bandwidth is a performance bottleneck in GPU databases [18].

Cross-Processor Pipelined Query Execution. Pipelined execution is a query execution model pioneered by Volcano [9], and is widely used in both commercial (e.g., Oracle and Microsoft SQL Server) and open source (e.g., PostgreSQL, SQLite [7] and MonetDB [2]) relational database systems today. The core idea of pipelined execution is to let the tuples passing throught the relational operators in a pipelining approach, so it is also called the tuple-at-a-time model. Pipelined execution avoids the materialization overhead between operators, and is considered high-performance for complex OLAP queries.

GPU databases also leverage the advantages provided by pipelined execution. HetExchange [4] migrates the exchange operator in Volcano into the heterogeneous CPU-GPU environment to achieve cross-processor pipelined execution. Figure 1 provides an example of cross-processor pipelined execution. Here, we

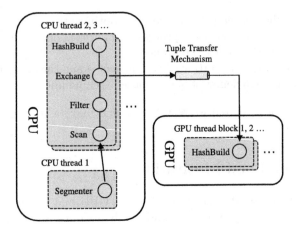

Fig. 1. An example of cross-processor pipelined execution

use a simple pipeline including three operators (i.e., Scan, Filter and Hash-Build), which is a recurring part of OLAP queries, for illustration. First, due to the capacity limit of GPU memory, GPU database systems typically store the table data in CPU memory [5,10,16]. To support multi-threaded execution, an auxiliary Segmenter operator is running in a standalone thread to split the table data into blocks. Then, the blocks are passed to the pipeline instances running on CPU. Each instance runs the three operators which can be fused by query compilation [12]. Among the three operators, HashBuild is compute-intensive. Hence, it also runs on GPU, which requires to route some of the output tuples from the Filter operator to GPU. To fulfill this requirement, an auxiliary Exchange operator is inserted between Filter and HashBuild for the pipeline instance running on CPU. This operator routes the tuples according to a hash function. Moreover, as CPU and GPU are separate processors, the tuples routed to GPU must be carried by a tuple transfer mechanism to the destination, which is typically implemented as a cross-processor concurrent queue.

2.2   Motivation

The concurrent queue is a widely adopted thread-safe CPU data structure used for cross-thread message passing in a FIFO (First In First Out) fashion. It is even provided by mainstream programming languages (e.g., Java and Go) as a built-in concurrent programming utility. A concurrent queue is backed by a thread-unsafe queue which can be implemented as a linked list or an array. To support thread safety, concurrent queues employ synchronization mechanisms (such as a mutex or a condition variable) provided by the operating systems. However, no operating system provides cross-processor synchronization mechanisms for the heterogeneous CPU-GPU environment. Hence, cross-processor concurrent queues instead rely on spin locks to achieve thread-safety, which are implemented based on system-wide atomic instructions.

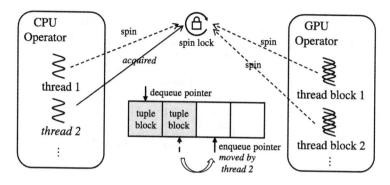

Fig. 2. The structure of the coarse-grained cross-processor concurrent queue

Figure 2 illustrates the structure of the coarse-grained cross-processor concurrent queue. It is "coarse-grained" since it relies on a single spin lock to achieve cross-processor thread-safety. The single spin lock becomes a performance bottleneck due to the lock contention. This is especially a concern for the heterogeneous CPU-GPU environment as the massive parallel threads running on GPU exacerbate the lock contention. In Fig. 2, the two threads running on CPU produce the tuples, which are then consumed by the two GPU thread blocks. The tuples are transferred via a queue composed of three components: the backing-array, the enqueue pointer and the dequeue pointer. To make tuple transfer more efficient, several tuples are grouped into tuple blocks. The minimum transfer unit is a tuple block. To achieve thread-safety, before a thread on CPU or a thread block on GPU accesses the queue, it must acquire the spin lock. In Fig. 2, thread 2 on CPU has acquired the spin lock. Hence, it can write a tuple block into the array and move the enqueue pointer. Before thread 2 releases the spin lock, the other three threads (or thread blocks) just spin on the lock and waste their processor cycles. We can further divide the lock contention into two types. *First, two threads running on different processors cannot perform dequeue and enqueue operations simultaneously.* Due to the single spin lock, thread 2 and thread block 1 (or thread block 2) cannot acquire it and perform enqueue and dequeue operations simultaneously. *Second, the threads running on the same processor cannot perform operations simultaneously on the queue.* As depicted in Fig. 2, thread 1 and thread 2 cannot acquire the spin lock at the same time to write their tuple blocks despite the fact that they may want to put their tuples into different array locations.

3   Fine-Grained Tuple Transfer Mechanism

This section proposes the fine-grained tuple transfer mechanism. In specific, Sect. 3.1 describes the decoupled enqueue/dequeue optimization, and Sect. 3.2 illustrates the subqueue-based locking optimization.

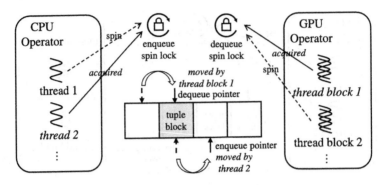

Fig. 3. The decoupled enqueue/dequeue optimization

3.1 Decoupled Enqueue/Dequeue

The coarse-grained cross-processor concurrent queue leads to a situaltion that the threads on CPU and the thread blocks on GPU cannot perform enqueue and dequeue operations at the same time. Moreover, as the spin lock needs to be accessed by both CPU and GPU, its underlying variable is tranferred back and forth frequently through the bottlenecking PCIe bus. The additional PCIe overhead makes the performance even worse.

However, the single spin lock is not necessary to ensure cross-processor thread-safety in the pipelined query execution scenario. Considering the example in Fig. 2, if we enable thread block 1 on GPU to dequeue a tuple block while thread 2 is enqueuing its tuple block, then there is no thread-safety issue. It is because thread 2 and thread block 1 do not need to modify the same memory location in the data structure. When thread 2 performs an enqueuing operation, it will first write the tuple block into the backing array of the queue, and then move the enqueue pointer to the next array index. When thread block 1 performs a dequeuing operation, it will first read the tuple block, and then move the dequeue pointer to the next array index. In summary, they both need to read/write a tuple block and move a pointer. However, in a practical queue, the backing-array contains many elements as a buffer to bridge the enqueue and the dequeue rates which may not be compatible strictly. When the array is not empty, thread 2 and thread block 1 will access different array elements. Besides, thread 2 and thread block 1 will move a different pointer.

Follow these observations, we propose to decouple the enqueue/dequeue operations by splitting the single spin lock into two locks, i.e., an enqueue spin lock and a dequeue spin lock (as depicted in Fig. 3). Before enqueue or dequeue a tuple block, threads (or thread blocks) acquire the corresponding spin lock. Here, the enqueue operation and the dequeue operation can be performed at the same time, which improves the tuple transfer efficiency and the efficiency of pipelined query execution. For example, in Fig. 3, thread 2 on CPU and thread block 1 on GPU are able to acquire different spin locks at the same time and perform the

---

Algorithm 1 Implementation of the decoupled enqueue/dequeue optimization

---

1: $arrayLength \leftarrow N$
2: $backingArray[arrayLength]$
3: $enqueuePointer \leftarrow 0, \ dequeuePointer \leftarrow 0$ ▷ *The pointers are actually indices*
4: $enqueueLock \leftarrow 0, \ dequeueLock \leftarrow 0$
5: **function** ENQUEUE($tupleBlock$)
6:     $breakFlag \leftarrow false$
7:     **while** $breakFlag \neq true$ **do**
8:         $oldValue \leftarrow compareAndSwap(\&enqueueLock, 0, 1)$
9:         **if** $oldValue = 0$ **then**
10:             **if** $(enqueuePointer + 1)\%arrayLength \neq dequeuePointer$ **then**
11:                 $backingArray[enqueuePointer] = tupleBlock$
12:                 $enqueuePointer \leftarrow (enqueuePointer + 1)\%arrayLength$
13:                 $breakFlag \leftarrow true$
14:             $enqueueLock \leftarrow 0$
15: **function** DEQUEUE()
16:     $breakFlag \leftarrow false$
17:     $result \leftarrow null$
18:     **while** $breakFlag \neq true$ **do**
19:         $oldValue \leftarrow compareAndSwap(\&dequeueLock, 0, 1)$
20:         **if** $oldValue = 0$ **then**
21:             **if** $enqueuePointer \neq dequeuePointer$ **then**
22:                 $result \leftarrow backingArray[dequeuePointer]$
23:                 $dequeuePointer \leftarrow (dequeuePointer + 1)\%arrayLength$
24:                 $breakFlag \leftarrow true$
25:             $dequeueLock \leftarrow 0$
26:     **return** $result$

---

enqueue and the dequeue operations in parallel. Further, since the tuple transfer between two operators in cross-processor pipelined query execution is one-way transfer, each spin lock will only be acquired by either CPU or GPU. That is, the underlying variable of the spin locks will only be accessed from one processor. This alleviates the repeated PCIe transfer overhead of the underlying variable and improves the tuple transfer efficiency.

Algorithm 1 shows the implementation of the decoupled enqueue/dequeue optimization. Notably, the algorithm is written in CPU code. Its GPU variant has minor differences due to the SIMT execution approach. The algorithm contains two functions, `enqueue()` and `dequeue()`. Take the `enqueue()` function as an example. Before actually enqueuing the tuple block, the function must call the wrapper function `compareAndSwap()` of the corresponding atomic instruction to acquire the enqueue lock (Line 8). If the result of `compareAndSwap()` (i.e., the `oldValue`) is zero, the thread has acquired the lock (Line 9). Then, it need to check if the queue is full by comparing the enqueue pointer and the dequeue pointer (Line 10). If either of these two checks fails, the loop (Line 7) will repeatedly retry them. Otherwise, the thread can enqueue the tuple block into the backing array (Line 11~12). Notably, whether the thread passes the full-queue check, it needs to release the enqueue lock (Line 14) before continuing to the

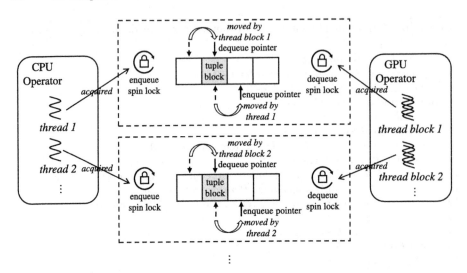

Fig. 4. The subqueue-based locking optimization

next loop. The `dequeue()` function has a similar construction, except for the use of the dequeue lock (Line 19) and the empty-queue check (Line 21).

3.2   Subqueue-Based Locking

The decoupled enqueue/dequeue optimization has successfully enabled the two threads (or thread blocks) on difference processors to perform operations on the cross-processor concurrent queue at the same time. However, the threads (or thread blocks) on the same processor still cannot mutate the cross-processor concurrent queue in parallel due to the queue-based locking. Considering the example in Fig. 3, after thread 2 on CPU has acquired the enqueue lock, thread 1 can only spin on this lock. Typically, to improve the query execution efficiency, both CPU and GPU will launch many threads to execute the operators, which make this effect even worse.

Similar to the reasons in Sect. 3.1, the single enqueue spin lock here is also unnecessary for thread-safe tuple transfer. In Fig. 3, if we allow both thread 1 and thread 2 to enqueue a tuple block in parallel, we actually expect that they write tuple blocks into different locations in the backing array. There is no thread-safety risk on the backing array. However, to ensure the correct semantic of the enqueue operation, writing the backing array and moving the enqueue pointer together must be atomic. In other words, if one thread writes its tuple block to the location pointered by the enqueue pointer in the backing array, then, before this thread moves the enqueue pointer to the next position, other threads are not allowed to write their tuple blocks. Clearly, this cannot be satisfied if there is only one enqueue pointer for parallel enqueuing. Hence, we need to design a solution that equips a queue with multiple enqueue pointers (and also multiple dequeue pointers) correctly.

Algorithm 2 Implementation of the subqueue locking optimization

---

1: *subqueueNumber ← M*
2: *subqueues[subqueueNumber]*
3: *enqueueSubqueueId ← 0, dequeueSubqueueId ← 0*
4: **function** ENQUEUE_SUBQUEUE(*tupleBlock*)
5:     *id ← atomicFetchAndInc(&enqueueSubqueueId)%subqueueNumber*
6:     *subqueues[id].enqueue(tupleBlock)*
7: **function** DEQUEUE_SUBQUEUE()
8:     *id ← atomicFetchAndInc(&dequeueSubqueueId)%subqueueNumber*
9:     **return** *subqueues[id].dequeue()*

From the above observations, we propose subqueue-based locking to enable the threads (or thread blocks) on the same processor to mutate the queue in at the same time (illustrated in Fig. 4). The rationale of this optimization is to break the queue into many subqueues. Each subqueue has separate pointers and spin locks. Hence, different threads on the same processor mutates different subqueues with thread-safety ensurance. In Fig. 4, while thread 1 on CPU has acquired the enqueue spin lock in the first subqueue and is enqueuing the tuple block, thread 2 can acquire the enqueue spin lock in the second subqueue and perform enqueue operations in parallel.

Algorithm 2 shows the implementation of the subqueue-based locking optimization. Each subqueue actually has the same construction with the cross-processor concurrent queue in Algorithm 1. These subqueues form an array (Line 2), of which the length is predefined (Line 1). In this version of the enqueue and the dequeue function, a subqueue must be selected from the subqueue array to perform operations. Here, we use a wrapper function `atomicFetchAndInc` of the corresponding atomic instruction to achieve this. For example, when it is called in the enqueue function (Line 5), it increases `enqueueSubqueueId` by one and returns the old value. Initially, the value of `enqueueSubqueueId` is zero (Line 3). Then, we get the index of the subqueue which we will access this time by a modulo operation of the old value to the length of the subqueue array.

4   System Implementation

We have implemented a prototype query engine, namely πQC, based on the compiler framework provided by DogQC [6]. Figure 5 shows the architecture of πQC, which employs the fine-grained tuple transfer mechanism. The system mainly consists of three components, the just-in-time (JIT) compiler, the executor and the table manager.

The JIT compiler takes a fully optimized query plan [14] as input and translates the plan into optimized executable code. In detail, the pipeline divider divides the query plan into separated pipelines. The code generator fuses the operators, translates each pipeline into C++/CUDA code, and compiles the code to an executable file by calling the NVCC compiler. The executor launches the

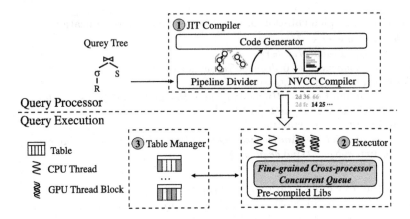

Fig. 5. The architecture of πQC

threads on both CPU and GPU according to the executable file. The table manager maintains the tables in memory for query execution. During query execution, the CPU threads communicate with the GPU threads using the fine-grained cross-processor concurrent queue. Notably, the queue is compiled in advance in the pre-compiled libraries. It avoids additional compilation overheads at query compilation time.

5   Experimental Evaluation

This section describes the experimental evaluation. We conduct both micro-benchmarks and macro-benchmarks to evaluate our techniques in detail. In addition, we evaluate other state-of-the-art solutions as references.

5.1   Experimental Settings

We conduct the experiments on a server with two Intel Xeon Gold 6126 CPUs (with 48 physical threads in total) and a Nvidia Tesla V100 GPU, where we deploy the systems including πQC, HeavyDB [1], and MonetDB [2]. The server also has 256GB main memory to store the relational tables in memory. The server has CentOS 7 and CUDA 11.4 installed as the software environment.

The Micro-benchmark. The micro-benchmark focuses on measuring the improvements of our techniques to the efficiency of cross-processor tuple transfer. Here, we use a hand-written loop that performs cross-processor tuple transfer as the workload. We use simulated tuples and tuple blocks of different sizes for the micro-benchmark. The tuple blocks are transfer units which contain the same number of tuples. Then, we measure the execution time of tuple transfer from CPU to GPU and compare the performance of different techniques.

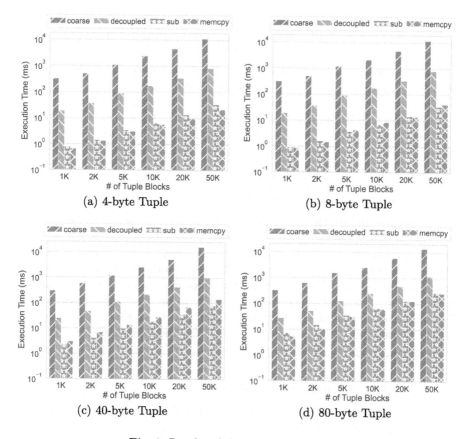

Fig. 6. Results of the micro-benchmark

The Macro-benchmark. The macro-benchmark focuses on measuring the improvements in the overall query execution performance. We use the TPC-H test suite as the workload. In particular, we select Q1, Q5, Q10 and Q14 to cover single-table queries, multi-table queries and subqueries. Notably, the aggregation operators in the queries are not implemented yet due to the current restrictions of πQC. Besides, we generate datasets with different sizes by varing the SF (Scale Factor) parameter of the TPC-H database generator.

5.2   Efficiency of Decoupled Enqueue/Dequeue

In this section, we analyze the results from benchmarks to show the efficiency of the decoupled enqueue/dequeue optimization.

The Micro-benchmark. Figure 6 illustrates the results of the micro-benchmark. The execution time here is the tuple transfer time. Here, we focus on the *coarse* and the *decoupled* bars in Fig. 6, which means the coarse-grained cross-processor concurrent queue and the concurrent queue with our decoupled enqueue/dequeue optimization, respectively. Clearly, *decoupled* always outperforms *coarse* by about a magnitude. In specific, *decoupled* reduces the overhead of *coarse* by up to 94.2% in Fig. 6(a) when 1K blocks are transferred. This means the decoupled enqueue/dequeue optimization improves the tuple transfer efficiency significantly.

Also, we notice that the number of transfer units (i.e., the tuple blocks) plays a more important role in the performance of tuple transfer than the number of bytes transferred. For example, transferring 10K tuple blocks of 4-byte tuples (in Fig. 6(a)) and transferring 1K tuple blocks of 40-byte tuples (in Fig. 6(c)) have the same transferred data volume. But the performance for these two cases are different due to the different number of transfer units. As an example, The latter case reduces 86.1% overheads of the former case for *sub*. This is similar for *coarse*. It means the additional overhead of cross-processor tuple transfer (including lock contention) plays a important role for both *coarse* and *decoupled*.

The Macro-benchmark. Fig. 7 depicts the results of the macro-benchmark. Here, we focus on the *coarse* and *decoupled* bars. Clearly, *decoupled* achieves lower execution time than *coarse* for all the cases. For example, *decoupled* achieves a speedup of two magnitudes than *coarse* in Fig. 7(b) when SF is 20. This result shows that *decoupled* has better performance than *coarse* in practical cross-processor pipelined query execution. In addition, *decoupled* scales linearly when SF increases, but the performance of *coarse* deteriorates rapidly when SF reaches a threshold. As an example, in Fig. 7(a), the execution time of *coarse* increases 147x when SF increases from 5 to 10.

5.3   Efficiency of Subqueue-based Locking

In this section, we analyze the results from the benchmarks to show the efficiency of the decoupled enqueue/dequeue optimization.

The Micro-benchmark. Figure 6 provides the micro-benchmark results of the subqueue-based locking optimization. Here, we focus on the *decoupled*, the *sub* and the *memcpy* bars. In particular, *sub* represents the subqueue-based locking optimization (based on the decoupled enqueue/dequeue optimization), and *memcpy* represents the native tuple transfer mechanism with `cudaMemcpy()`. Figure 6 shows that *sub* always outperforms *decoupled*. As an example, *sub* reduces the overhead of *decoupled* in Fig. 6(a) by up to 96.3% when transferring 10K tuple blocks. Also, *sub* achieves near-*memcpy* performance and even outperforms *memcpy* in some cases (e.g., all cases in Fig. 6(c)). This demonstrates the effectiveness of the subqueue-based locking optimization. Despite the performance of *sub* and *memcpy* are similar for the micro-benchmark, *sub* is superior than *memcpy* in practical query execution. If `cudaMemcpy()` is used for cross-processor tuple transfer, the overlapping benefit of tuple transfer and GPU kernel execution cannot be achieved since they are executed in order. And the GPU kernels are repeatedly launched for each tuple block.

Similarly, it is also a fact for *sub* that the number of transfer units is more important for the tuple transfer performance than the number of bytes transferred. Comparing Fig. 6(a) and Fig. 6(c), transferring 1K tuple blocks of 40-byte tuples reduces the overhead of transferring 10K tuple blocks of 4-byte tuples by 64.7% for *sub*. The difference is smaller than what we have discussed in the

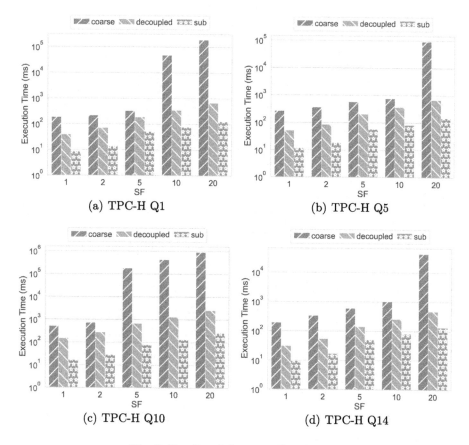

Fig. 7. Results of the macro-benchmark

micro-benchmark in Sect. 5.2. Considering that *sub* and *memcpy* are similar in performance, the remaining overhead for transferring small tuple blocks comes from the PCIe hardware which cannot be avoided.

The Macro-benchmark. We continue the discussion on Fig. 7 for the macro-benchmark. Here, we focus on the *decoupled* and the *sub* bars. In all the four plots, *sub* outperforms *decoupled*. For example, *sub* reduces the overhead of *decoupled* by up to 89.8% in Fig. 7(c). This result shows that the subqueue-based locking optimization is effective for TPC-H query execution. Moreover, *sub* also scales linearly for all the tested queries.

5.4 Comparison with Other Systems

In this section, we compare πQC integrated with both the decoupled enqueue/dequeue and the subqueue-based locking optimization with other systems under macro-benchmark. In particular, we use a in-memory CPU database MonetDB [2] and a mainstream GPU database HeavyDB [1] (previously known as

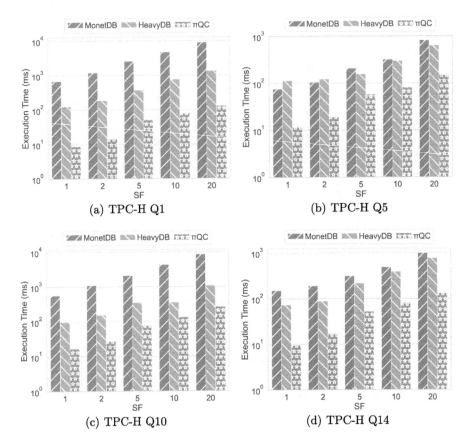

Fig. 8. Comparison with other systems

OmniSciDB and MapD) as baselines. Figure 8 depicts the results of the experiment. Moreover, our πQC always have a lower execution time than both MonetDB and HeavyDB. For example, πQC achieves 78x and 14x better performance than MonetDB and HeavyDB respectively for TPC-H Q1 when SF is 1. Thanks to the GPU, πQC achieves orders of magnitude speedup over MonetDB. Due to the use of the cross-process pipelined query execution with the fine-grained tuple transfer mechanism, πQC significantly outperforms HeavyDB. This result shows the overall efficiency of the cross-processor pipelined query execution and our fine-grained tuple transfer mechanism.

6   Related Work

GPU-Accelerated Query Execution. GPU has been studied in the query execution for a long time. Researches [3,10,17] focus on implementing and optimizing relational operators on GPU, especially the time-consuming ones. For example, HAPE [17] utilize GPU features like shared memory and warp-level instructions to optimize the hash join operator on GPU. CoGaDB [3] and Rate-upDB [10] discusses the design and the trade-offs of the relational operators, including join and aggregation. They are complementary to our work since we focus on optimizing the tuple transfer between operators. Some work considers to optimize the data-path between operators. DogQC [6] and Pyper [13] fuses multiple GPU operators to generate a single kernel, which prevents the tuple transfer between kernels. Besides, they optimize the code generation phase to generate GPU-friendly code with lower warp divergence. Hence, these works optimize the data-path between the GPU operators, whereas we focus on the cross-processor tuple transfer.

Cross-Processor Tuple Transfer. The cross-processor tuple transfer mechanism is essential for GPU databases to migrate the tuples from CPU memory to GPU. Many GPU databases using the operator-at-a-time execution model (e.g., RateupDB [10] and CoGaDB [3]). In these systems, before a GPU operator is executed, all its input tuples needs to be transferred to GPU memory. Our work achieves cross-processor pipelined query execution instead, enabling the over-lapping of the tuple transfer and the operator computation. HetExchange [4] uses the pipelined query execution model but employs repeated cross-processor invocations for tuple transfer. However, our work uses a cross-processor concurrent queue which prevents the overhead of repeated cross-processor invocations. XeFlow [11] employs a queue-like data structure to achieve cross-processor data transfer. But it is coarse-grained due to the single spin lock. On the contrary, our work employs fine-grained tuple transfer by breaking the single spin lock.

7 Conclusion

In this paper, we propose a fine-grained tuple transfer mechanism for pipelined query execution in the heterogeneous CPU-GPU environment. It is designed as a cross-processor concurrent queue. In specific, we first propose the decoupled enqueue/dequeue optimization to enable two threads running on different processors to perform enqueue and dequeue operations simultaneously on the queue. Then, we propose the subqueue-based locking optimization to enable the threads running on the same processor to perform operations simultaneously on the queue. Based on the fine-grained tuple transfer mechanism, we implement a query compiler πQC for cross-processor pipelined query execution. Our experiments show that the fine-grained tuple transfer mechanism significantly outperforms the state-of-the-art solutions.

Acknowledgments. This work was supported by the National Natural Science Foundation of China (No. 62272168), and Guangxi Key Laboratory of Trusted Software.

References

1. HeavyDB. https://www.heavy.ai/product/heavydb
2. Boncz, P.A., Zukowski, M., Nes, N.: Monetdb/x100: hyper-pipelining query execution. In: CIDR, pp. 225–237 (2005)
3. Breß, S.: The design and implementation of CoGADB: a column-oriented GPU-accelerated DBMS. Datenbank-Spektrum **14**(3), 199–209 (2014)
4. Chrysogelos, P., Karpathiotakis, M., Appuswamy, R., Ailamaki, A.: Hetexchange: encapsulating heterogeneous CPU-GPU parallelism in JIT compiled engines. PVLDB **12**(5), 544–556 (2019)
5. Funke, H., Breß, S., Noll, S., Markl, V., Teubner, J.: Pipelined query processing in coprocessor environments. In: SIGMOD, pp. 1603–1618 (2018)
6. Funke, H., Teubner, J.: Data-parallel query processing on non-uniform data. PVLDB **13**(6), 884–897 (2020)
7. Gaffney, K.P., Prammer, M., Brasfield, L.C., Hipp, D.R., Kennedy, D.R., Patel, J.M.: Sqlite: past, present, and future. PVLDB **15**(12), 3535–3547 (2022)
8. Gautier, T., Lima, J.V.F., Maillard, N., Raffin, B.: Xkaapi: a runtime system for data-flow task programming on heterogeneous architectures. In: IPDPS, pp. 1299–1308 (2013)
9. Graefe, G.: Volcano - an extensible and parallel query evaluation system. IEEE Trans. Knowl. Data Eng. **6**(1), 120–135 (1994)
10. Lee, R., et al.: The art of balance: a rateupdb experience of building a CPU/GPU hybrid database product. PVLDB **14**(12), 2999–3013 (2021)
11. Li, Z., Peng, B., Weng, C.: Xeflow: streamlining inter-processor pipeline execution for the discrete CPU-GPU platform. IEEE Trans. Comput. **69**(6), 819–831 (2020)
12. Neumann, T.: Efficiently compiling efficient query plans for modern hardware. PVLDB **4**(9), 539–550 (2011)
13. Paul, J., He, B., Lu, S., Lau, C.T.: Improving execution efficiency of just-in-time compilation based query processing on gpus. PVLDB **14**(2), 202–214 (2020)
14. Pedreira, P., et al.: Velox: meta's unified execution engine. PVLDB **15**(12), 3372–3384 (2022)
15. Rossbach, C.J., Currey, J., Silberstein, M., Ray, B., Witchel, E.: PTask: operating system abstractions to manage GPUs as compute devices. In: SOSP, pp. 233–248 (2011)
16. Shanbhag, A., Madden, S., Yu, X.: A study of the fundamental performance characteristics of GPUs and CPUs for database analytics. In: SIGMOD, pp. 1617–1632 (2020)
17. Sioulas, P., Chrysogelos, P., Karpathiotakis, M., Appuswamy, R., Ailamaki, A.: Hardware-conscious hash-joins on gpus. In: ICDE, pp. 698–709 (2019)
18. Yuan, Y., Lee, R., Zhang, X.: The yin and yang of processing data warehousing queries on GPU devices. PVLDB **6**(10), 817–828 (2013)

Efficient Index-Based Regular Expression Matching with Optimal Query Plan Tree

Tao Qiu[1](✉), Xiaochun Yang[2], Bin Wang[2], Chuanyu Zong[1], Rui Zhu[1], and Xiufeng Xia[1]

[1] School of Computer Science, Shenyang Aerospace University, Shenyang, China
{qiutao,zongcy,zhurui,xiaxf}@sau.edu.cn
[2] School of Computer Science and Engineering, Northeastern University, Shenyang, China
{yangxc,binwang}@mail.neu.edu.cn

Abstract. The problem of matching a regular expression (regex) on a text exists in many applications such as entity matching, protein sequences matching, and shell commands. Classical methods to support regex matching usually adopt the finite automaton which has a high matching cost. Recent methods solve the regex matching problem by utilizing the positional q-gram inverted index – one of the most widely used index schemes, and all matching results can be matched directly based on this index. The efficiency of these methods depends critically on the query plan tree, which is built from the query with some heuristic rules. However, these methods could become inefficient when an improper rule is used for building the query plan tree. To remedy this issue, this paper aims to build a good query plan tree with an efficiency guarantee. We propose a novel method to build an optimal query plan tree with the *minimal expected matching cost* for the index-based regex matching method. While computing an optimal query plan tree is an NP-hard problem even with strong assumptions, we propose a pseudo-polynomial time algorithm to build an *optimal* query plan tree. Finally, extensive experiments have been conducted on real-world data sets and the results show that our method outperforms state-of-the-art methods.

Keywords: Regular expression · Positional inverted index · Pseudo-polynomial algorithm · Optimal query plan

1 Introduction

The regular expression (regex) is a pattern with certain syntax rules usually used to match targeting substrings from the text data sets. Regular expression matching has been widely used in important areas such as information extraction [2], entity matching [4], protein sequences matching [7] and intrusion detection systems (IDSs) [18].

The traditional algorithms convert the regex queries to finite automata, then use them to locate the matching occurrences of the query from the text data set [11–13, 16]. However, the finite automaton has to scan all characters (positions) of the text data set to

This work is partly supported by the National Natural Science Foundation of China (Nos. 62002245, U22A2025, 62072088, 62232007, 61802268), Ten Thousand Talent Program (No. ZX20200035), Liaoning Distinguished Professor (No. XLYC1902057), and the Natural Science Foundation of Liaoning Province (Nos. 2022-BS-218, 2022-MS-303, 2022-MS-302).

X. Wang et al. (Eds.): DASFAA 2023, LNCS 13943, pp. 35–45, 2023.
https://doi.org/10.1007/978-3-031-30637-2_3

match all results of the query, which is obviously costly. To this end, some research utilizes the prebuilt index to quickly jump to *candidate locations* where a query result may appear. FREE [2] and CodeSearch [3] locate the candidate documents from the document collection by building the document-level gram inverted index, then all contents of the candidate documents needed to be verified by the automaton-based algorithm. N-Factor [17] further reduces the scope of automaton-based verification by computing the candidate positions on the document with the negative factors (i.e., filters of the regex query), and a bit-vector index is used to accelerate the bit-parallel-based filtering algorithm. The limitation of these methods is that the prebuilt index is not fully utilized and they still need to perform the costly automaton-based verification on the document. TreeMatch [15] solves this issue by building the positional q-gram inverted index and designing a query plan tree to match regex results based on the positional index. The efficiency of TreeMatch depends critically on the plan tree built by a heuristic rule, which results in the high query efficiency is not guaranteed, especially when the heuristic rule is improper for the application of regex matching.

In this paper, we follow the technical route of TreeMatch that matches the regex query using the positional q-gram inverted index, so that the automaton-based verification on the document can be avoided, while we aim to build a good query plan tree with an efficiency guarantee. We show the *expected matching cost* of regex matching can be quantified by the plan tree and an *optimal plan tree* with the minimal expected matching cost can achieve efficient regex query processing with the efficiency guarantee. While computing an optimal plan tree for a regex query is proved to be NP-hard, we propose a pseudo-polynomial time algorithm to build an optimal plan tree.

The contributions of this paper can be summarized as follows. i) For the positional inverted index-based regex matching methods, a new metric is proposed to quantify the matching cost, named expected matching cost. ii) Building an optimal plan tree with the minimal expected matching cost is proved to be an NP-hard problem. We design a pseudo-polynomial time algorithm to build an optimal plan tree, which is the first effort to solve the NP-hard problem in this field. iii) We have implemented the index-based regex matching method by building an optimal plan tree. Experimental results on real data sets show that the proposed method outperforms previous state-of-the-art methods.

2   Preliminaries

(1) Problem Formulation.

In this paper, we consider a widely used definition of the regular expression (regex) [14, 15, 17]. Let Σ be a finite alphabet, a *regular expression* is defined recursively. i) Each string $s \in \Sigma^*$ is a regular expression, which denotes the string set $\{s\}$; ii) $(e_1)^+$ is a regular expression that denotes a set of strings $x = x_1 \cdots x_k$ $(k \geq 1)$, and e_1 matches each string x_i $(1 \leq i \leq k)$; iii) $(e_1|e_2)$ is a regular expression that denotes a set of strings x such that x matches e_1 or e_2; iv) (e_1e_2) is a regular expression that denotes a set of strings x that can be written as $x = x_1 \cdot x_2$, where e_1 and e_2 match x_1 and x_2 respectively, and $\cdot$ denotes string concatenation. Many other syntactical sugars of regex can be represented by the above definition [14], e.g., the Kleene closure $e^* = (\epsilon|e^+)$.

For a regex query Q, the *regex language* of Q is the set of strings matched by Q, denoted as $L(Q)$. The below regex is used as the running example in this paper.

Example 1. *Consider a regex query* Q=td(␣|␣␣)al(ign|t), *in which the space character is explicitly represented by* ␣ *It is used to match the html tag* td *with properties* align *or* alt, *and* $L(Q) = \{$td␣alt, td␣align, td␣␣alt, td␣␣align$\}$.

Given the text T on the alphabet set Σ, the matching occurrence on T of any string in $L(Q)$ is the matching result of the regex query Q. The *problem of regular expression matching* on the text T is to locate all matching occurrences of strings in $L(Q)$ on T.

In general, a regex query Q is converted to the equivalent nondeterministic finite automata (NFA) when processing a query Q, so that the regex matching algorithm can be designed based on the NFA. Thompson NFA [13] (character-driven NFA) and GNFA [15] (gram-driven NFA) are typical instances of NFA, which accept all and only strings in $L(Q)$. Figure 1 shows the example of GNFA for the query Q in Example 1.

The transitions of GNFA are all q-grams of strings in $L(Q)$. A *path* of NFA is a sequence of transitions (as well as the relevant nodes) that starts from the initial node to the final node. For any string s in language $L(Q)$ of Q, there exists a path P in GNFA that corresponds to s. For example, the path V_0-V_1-V_3-V_4-V_7 in the GNFA (Fig. 1) correspond to the string td␣alt in $L(Q)$, and the path contains all 3-grams td␣, d␣a, ␣al and alt for the string td␣alt.

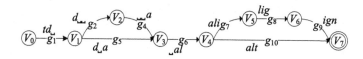

Fig. 1. Gram-driven NFA A_g for regex Q.

(2) Matching Regular Expression Over Positional Inverted Index.

Classical regex matching methods generally utilize the finite automaton(e.g., NFA and DFA) to match the regex query, while GNFA introduces a new way to directly match the regex over the positional inverted index of the text [15], which avoids the costly automaton-based matching on the text and significantly improves the query efficiency.

Given the text T, it can be decomposed to a set of q-grams with the number $|T| - q + 1$. For a q-gram g, the *positional gram list* $l(g)$ for g records all positions where g appears on T, e.g., the 3-gram ign has a list with 4 positions in Table 1 since ign appears at positions 5, 62, 93, and 185 on T. *Positional q-gram inverted index* is the collection of positional gram lists for all distinct q-grams of T, as shown in Table 1.

With the q-gram positions, we can find the occurrences of a string s directly, since the matching positions of q-grams from the matching occurrence must satisfy a set of **positional constraints**, i.e., *for any two q-gram of s, their position offset on s must equal to the offset of positions on the matching occurrence.* For example, td␣, d␣a, ␣al, and alt are the 3-grams for the string $s = $ td␣alt and they have positions 35, 36, 37, and 38 on T (as shown in Table 1), since the offset for these positions between any two grams equals the offset on s (e.g., td␣ and alt have position offset 3 on s which equals to the offset for their positions 35 and 38), then there has a matching occurrence starting at position 35 on T for s.

Based on the above property, all results of the regex query can be matched by checking the q-gram positions with the positional constraints on GNFA. At first, we need to select the *base grams* from the GNFA, such that any string in $L(Q)$ contain one of the base grams. Therefore, the positions of the base grams can be the candidate positions for the matching results. The candidate positions are validated by finding a path from the GNFA such that its grams contain the positions satisfying the positional constraints. For example, if g_6 is used as base gram and 37 will be a candidate position, 37 is turned out to be a matching position when we can find positions 35, 36, and 38 from the gram lists of g_1, g_5 and g_{10}, respectively.

Table 1. An example of positional q-gram lists for a text T (only the grams w.r.t. Q are shown).

td␣	d␣␣	␣␣␣	␣␣a	d␣a	␣al	ali	lig	ign	alt
$l(g_1)$	$l(g_2)$	$l(g_3)$	$l(g_4)$	$l(g_5)$	$l(g_6)$	$l(g_7)$	$l(g_8)$	$l(g_9)$	$l(g_{10})$
11	27	1	16	12	17	41	4	5	18
35	65	28	51	36	37	60	61	62	38
84	104	82	66	58	59	78	79	93	74
103	199	83	84	72	73	145	184	185	94
198	277	125	105	85		239	252		174
		161	121	205		251			223
		217	235						275

(3) Related Work.

The typical regex matching problem is to match the regex query on the text by the online algorithms [1, 10, 11]. They usually utilize an equivalent finite automaton (NFA or DFA) computed from the regex query to match the results from the text. To find all matching results, all positions of the text are checked by the automaton which leads to a heavy time cost. To this end, some methods employ a two-phase framework for this problem – filtering and verification [5, 12, 16]. The candidate positions of the text are located by the online multi-string matching algorithms, then use the automaton verifies these positions to pinpoint the true results, such as Gnu Grep [5], MultiStringRE [16], and NRgrep [12]. Besides, the multi-regex matching problem is studied which aims to reduce the memory requirements of DFAs using rewriting techniques [8, 9, 18].

TreeMatch [15] is the first work using positional q-gram inverted index to match regex without the automaton-based verification on the raw data. The N-Factor technique [17] utilizes the substrings that cannot exist in the matching results as filters for the regex query, then designs the bit-parallel-based algorithms to filter the irrelative positions on the text. CodeSearch [3] is the first work utilizing document-level q-gram inverted index for the regex matching problem.

3   A Pseudo-Polynomial Method for Computing Optimal Plan Tree

3.1   The Optimal Plan Tree with Minimal Expected Matching Cost

(1) Plan Tree for Regular Expression Matching.

To efficiently validate the candidate position, the query plan tree (abbr. plan tree) is introduced in [15], which represents the checking order for the positional constraints of the transition grams, as follows. For the regex with the repeating operators (* or +), an auxiliary node is created for the repeating operator in the plan tree [15]. To ease the presentation, we discuss the regex without repeating operators directly in this paper.

Definition 1 *(Plan tree). A plan tree T is a rooted binary tree, the leaf node represents the validation result for a candidate position and the internal node represents the required position checking of a transition gram for the candidate position.*

The internal node indicates the checking for the required position of a transition gram. The checking result is true when the required position is found from the corresponding gram list; otherwise, the result is false. For each internal node n of T, the left child of n will be checked if the result of n is true; otherwise, the right child of n is checked. Upon a candidate position π_b is given, π_b is validated starting from the root of the plan tree T until a leaf node is reached. There are two types of leaf nodes: *T-leaf* and *F-leaf*, which respectively represent the successful and failed results for validating a candidate position. The binary plan tree T can be used to check if a base position π_b is the matching position of Q by checking the tests starting from the root of T until a leaf node is reached. If a T-leaf is reached, thus π_b is the position of a matching result.

To facilitate the illustration of the optimal plan tree, we define the position checking for a transition gram by *positional test*, as follows.

Definition 2 *(Positional test). Given a transition gram g_i and the base gram g_b in a path P of A_g, π_b is a candidate position, the positional offset between g_i and g_b is δ, a positional test t_i w.r.t. g_i is to check if a position π_i exists in $l(g_i)$ s.t. $\pi_i = \pi_b - \delta$.*

For simplicity, we call it **test** directly in this paper. For the grams g_i and g_b in a path P, their positional offset δ on P can be easily obtained, e.g., g_6 and g_9 have an offset 3, while g_9 and g_6 have an offset -3 in the GNFA of Fig. 2. If the transition grams g_i, g_b and the candidate position π_b are given, then the checked position π_i in the test t_i is $\pi_b - \delta$, called *expected position* for g_i. For the test t_i, the result of t_i is true if and only if the expected position $\pi_i \in l(g_i)$ for the given candidate position π_b. A test t_i can be checked by searching the expected position of t_i on the corresponding gram list $l(g_i)$.

(2) Optimal Plan Tree.

To minimize the number of checked tests, we define a new metric of matching cost by considering the expected value of the number of checked tests for a candidate position, named *expected matching cost*. Given a plan tree T, let S_P be the collection of the root-to-leaf paths of T, and the cost of checking a test is c_t. For a path $P_t \in S_P$, $|P_t|$ indicates the number of internal nodes (i.e., tests) in P_t, if $|P_t|$ is used to validating a candidate position, the validation cost can be calculated by $|P_t| \cdot c_t$.

Definition 3 *(Expected matching cost). Assuming that the probability of P_t used for candidate position validation is $\mathbb{P}(P_t)$, the expected matching cost $E(T)$ of the plan tree T is defined by the expected value on the cost of validating a candidate position.*

$$E(T) = \sum_{P_t \in S_P} (|P_t| \cdot c_t \cdot \mathbb{P}(P_t))$$

For a test t_i, let p_{t_i} be the probability of t_i gets the result true, called *success proba-bility*. In this paper, we calculate p_{t_i} by $\frac{|l(g_i)|}{|T|}$, where $|T|$ is the number of characters in T and $|l(g_i)|$ is the size of the gram list $l(g_i)$ (i.e., the number of occurrences of g_i on T). To calculate $E(\mathcal{T})$, a key issue is to calculate the probability $\mathbb{P}(P_t)$ for a path P_t.

We assume that, *for two tests t_i and t_j existing in a root-to-leaf path of $\mathcal{T}$, t_i and t_j are independent*, so that $\mathbb{P}(P_t)$ can be calculated by the probabilities of related tests. That is, $\mathbb{P}(P_t) = \mathbb{P}(t_1 \cdot t_2 \cdots t_m) = \mathbb{P}(t_1) \cdot \mathbb{P}(t_2) \cdots \mathbb{P}(t_m)$, in which t_i is the internal node on P_t. For a test $t_i \in P_t$, if t_i is true, then $\mathbb{P}(t_i) = p_{t_i}$; otherwise, $\mathbb{P}(t_i) = 1 - p_{t_i}$. Based on this assumption, $E(\mathcal{T})$ can be expanded to Eq. 1, where $\mathcal{T}_l$ and $\mathcal{T}_r$ are the left and right subtrees of $\mathcal{T}$, and p_{t_r} is the success probability of the root test t_r.

$$E(\mathcal{T}) = \begin{cases} c_t + p_{t_r} E(\mathcal{T}_l) + (1 - p_{t_r})E(\mathcal{T}_r) & \text{if } \mathcal{T} \neq \text{ leaf node,} \\ 0 & \text{otherwise.} \end{cases} \tag{1}$$

We omit the proof for the conversion of Eq. 1 since the same conversion is proved for the optimal and-or tree construction problem in [6]. We call the plan tree with the minimal expected matching cost as the *optimal plan tree* and use $\mathcal{T}_{opt}$ to denote it.

Suppose any test is not exclusive to others in the test collection, then the candidate position validation can be presented by a boolean formula. Computing an optimal binary tree with the minimal cost for the boolean formula has been proved to be an NP-hard problem in [6], so computing $\mathcal{T}_{opt}$ is also NP-hard for our problem.

3.2 Computing Optimal Plan Tree in Pseudo-Polynomial Time

(1) Computing Plan Tree Recursively.

We utilize the recursive framework of the plan tree construction initially introduced in [15]. The idea is to compute all positional tests from GNFA first, then iterative pick the internal node (test) from the collection of tests. The initial positional test collection can be computed by bidirectionally traversing A_g starting from the base gram g_b. We use C_t to denote the *test collection* for the transition grams in A_g. For example, the initial test collection C_t for the running example is shown in Fig. 2, there are 9 tests in C_t.

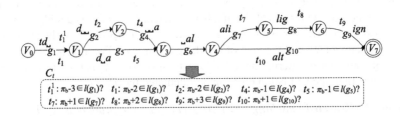

Fig. 2. The initial test collection computed from GNFA

Then, we can pick a test t_i from C_t as the internal node of the plan tree. The next task is to compute the test collections for the left and right subtrees of t_i so that the plan tree is recursively constructed in the same way. For a test t_i, the tests needed to be checked under t_i holds are called *compatible positional tests* of t_i (i.e., the tests for the left subtree of t_i), denoted by $C_{t_i}^+$. On the contrary, the tests needed to be checked under t_i fails are called *incompatible positional tests* of t_i, denoted by $C_{t_i}^-$.

We borrow the idea from [15] to compute $C_{t_i}^+$ and $C_{t_i}^-$, as shown in Eq. 2, where $P^{(all)}$ and $P^{(t_i)}$ are the collection of all paths in GNFA and the paths containing t_i, $P^{(ex(t_i))}$ represents the path collection containing the grams of the exclusive tests of t_i.

$$C_{t_i}^- = C_t \cap C(P^{(all)} - P^{(t_i)}),$$
$$C_{t_i}^+ = C_t \cap (C(P^{(all)} - P^{(ex(t_i))}) - \{t_i\}) \tag{2}$$

(2) Reducing Test Collection to Group Collection.

Next, we present a grouping technique that enables us to avoid enumerating all tests when picking a test as an internal node from the test collection C_t.

Definition 4 *(Commutable tests). Two tests t_i and t_j in C_t are called commutable tests, if and only if $C_{t_i}^+ - t_j = C_{t_j}^+ - t_i$ and $C_{t_i}^- = C_{t_j}^-$.*

For commutable tests t_i and t_j, their incompatible test collections are the same, and the compatible tests excluding t_i and t_j are also the same. With the commutable tests, we define the *test group* by the set of tests in which any tests are commutable, denoted by G_i. For example, $\{t_8, t_9\}$ is a test group since t_8 and t_9 are commutable tests.

Theorem 1. *Given a test collection C_t and a test $t_i \in C_t$, G_i is a test group in C_t s.t. $t_i \notin G_i$. The collection $C_{t_i}^+$ (resp. $C_{t_i}^-$) of t_i includes either all or none of the tests of G_i.*

Due to the space limitation, we omit the formal proof for the theorems in this section. Given the tests for A_g, we can group the tests by computing the $C_{t_i}^+$ and $C_{t_i}^-$ for each test t_i. Let S_G be the group collection of A_g. For tests t_i and t_j in group G_i, we use $S_G^+(t_i)$ and $S_G^-(t_i)$ to represent the groups (except G_i) contained by $C_{t_i}^+$ and $C_{t_i}^-$. According to Definition 4, we have $S_G^+(t_i) = S_G^+(t_j)$ and $S_G^-(t_i) = S_G^-(t_j)$ for the commutable tests t_i and t_j. Therefore, for the tests in same group G_i, their compatible test collections $C_{t_i}^+$ (resp. $C_{t_i}^-$) contains the same groups $S_G^+(t_i)$ (resp.,$S_G^-(t_i)$). We call such group collections contained by $C_{t_i}^+$ and $C_{t_i}^-$ as *compatible and incompatible group collections* of G_i, and denoted by $S_{G_i}^+$ and $S_{G_i}^-$.

For the commutable tests t_i and t_j, we have $C_{t_i}^+ - t_j = C_{t_j}^+ - t_i$. Replacing $C_{t_i}^+$ and $C_{t_j}^+$ by Eq. 2, we get $C(P^{(ex(t_i))}) = C(P^{(ex(t_j))})$. Moreover, since t_j exists in the paths $\{P^{(all)} - P^{(ex(t_j))}\}$, thus we can get $t_j \in C(P^{(all)} - P^{(ex(t_i))})$, which means the commutable tests of t_i also belongs to $C_{t_i}^+$. In summary, $C_{t_i}^+$ consists of the tests in $S_{G_i}^+$ and the commutable tests of t_i, $C_{t_i}^-$ is equal to $S_{G_i}^-$, as follows. By computing $S_{G_i}^+$ and $S_{G_i}^-$ for each group G_i, we can get $C_{t_i}^+$ and $C_{t_i}^-$ for each test t_i, i.e., $S_{G_i}^+$ and $S_{G_i}^-$ are the sub-group collections used for building subtrees.

$$C_{t_i}^+ = S_{G_i}^+ \cup \{G_i - t_i\}, C_{t_i}^- = S_{G_i}^- \tag{3}$$

(3) Computing Optimal Plan Tree based on Test Groups.

So far, the test collection can be reduced to test group collection, then the group collection can be used as the input when recursively computing the optimal plan tree.

Theorem 2. *Given a group collection S_G, let $\mathcal{T}_{opt}$ be the optimal plan tree on S_G, and t_i, t_j are the commutable tests within same group G_i in S_G, if $p_{t_i} < p_{t_j}$ and $S_{G_i}^+ = S_G - G_i$, then t_j is not performed before t_i on any root-to-leaf path of $\mathcal{T}_{opt}$.*

Theorem 2 shows that, for a group G_i such that $S_{G_i}^+ = S_G - G_i$, only the test with minimal probability is considered as the root, which avoids enumerating all tests in G_i. Based on Theorem 2, we design the algorithm to compute $\mathcal{T}_{opt}$, named OptimalTree. The algorithm uses S_G as the input, rather than test collection C_t. Initially, we compute S_G from A_g and sort the tests in each group by the success probability of the test.

Consider the recursion in OptimalTree, for each group G_i, $S_{G_i}^+$ and $S_{G_i}^-$ are computed firstly by $S_{G_i}^+ = S_G \cap S_{G_i}^{i+}$ and $S_{G_i}^- = S_G \cap S_{G_i}^{i-}$. Then, the test t_i in G_i is processed, note that t_i is removed from G_i when computing the left subtree, so the group collection used for the left subtree is $S_{G_i}^+ \cup \{G_i - t_i\}$, as described in Eq. 3. Next, we compute the expected cost of the plan tree in which t_i is used as the root for current S_G through Eq. 1. According to Theorem 2, for the group G_i s.t. $S_{G_i}^+ = S_G - G_i$, we only consider the test with minimal probability. Since the tests in G_i are ordered by the probability, the first accessed test has the minimal probability, if condition $S_{G_i}^+ = S_G - G_i$ holds, the computation terminates after the first test in G_i is processed.

An additional set C_t' is used to record the tests with true results, and C_t' is initialized by $\emptyset$. The algorithm recursively computes until $S_G = \emptyset$. At that time, if the tests in C_t' compose a path of A_g, then a T-leaf is created; otherwise, an F-leaf is created.

At the end of this section, we give the complexity analysis. In OptimalTree, for the group G_i such that $S_{G_i}^+ \neq S_G - G_i$ for current S_G, each test t_i in G_i needs to be processed. However, for the recursive computation where $S'_G = S_{G_i}^+ \cup G'_i$, we have $S'^+_{G_i} = S'_G - G'_i$. It means that for the G_i s.t. $S_{G_i}^+ \neq S_G - G_i$, only the first recursion considers all tests in G_i among all recursions of OptimalTree. Therefore, for any group G_i in OptimalTree, we consider at most $(\frac{(r+1)r}{2} + 1)$ subsets of G_i.

Now, let us consider the number of computed group collection A_g in OptimalTree, let θ be the group number of A_g, so OptimalTree computes at most $(\frac{(r+1)r}{2} + 1)^\theta$ different group collections. For each group collection, the algorithm considers every group, and it processes all tests in a group in the worst case, so the time complexity is $\mathcal{O}(\theta r (\frac{(r+1)r}{2} + 1)^\theta)$. Hence, for the GNFA A_g with the bounded number of disjunctive units (which means θ is bounded by a constant), it runs in time polynomial in r.

4   Experiments

(1) Experimental Setting.

In this section, we present the results of our proposed algorithm by comparing it with state-of-the-art algorithms, including NRGrep [12], RE2, TreeMatch [15], N-Factor [17], CodeSearch [3], and FREE [2]. Our proposed algorithm, named OptTree, shares a similar technical route with TreeMatch, but OptTree builds the optimal plan tree.

We used two real data sets in the experiments. 1) *Web pages* include javascript, PHP, asp.net, and HTML files with a total file number 425,104 and the total size is 2.09 GB.

2) *Source codes* were obtained from Github, including python, ruby, java, C, and C++ files. This data set contains 697,485 files, and the total size is 3.69 GB. The positional 3-gram inverted index is built for the experiments. We collected the real regex queries from the regex library and forums[1], and obtained a query workload with 111 regex queries, where 41 queries contain the repeating operators(+ or *). The regex queries are classified according to repeating operators and the size of language set $L(Q)$. Finally, we get 6 categories (C_1–C_6) containing 9, 9, 25, 16, 31, and 21 queries, respectively.

(2) Effect of Optimal Plan Tree.
We compare the optimal plan tree to another two strategies for building a query plan tree. The first strategy builds the optimal plan tree by enumerating all possible plan trees (OptTree-Enum). Another strategy utilizes a heuristic rule employed by TreeMatch to build the plan tree, that is to preferentially check the test shared by more GNFA paths (HeuTree). We record the average running time of matching regexes on a query set, the results are plotted in Fig. 3(a)–(b). Also, we test the average time of building the query plan tree for the three methods, denoted by TreeBuild. It is obvious that OptTree improves the efficiency of query processing, e.g., for the queries in C_2 on the data set of web pages, OptTree spends a running time 87 ms, against the time 153 ms and 145 ms used by OptTree-Enum and HeuTree. OptTree-Enum spends more time to match regexes than HeuTree for most of query sets such as C_1–C_6 in the web pages.

Besides, we study the effect of the optimal plan tree by testing the number of checked tests for validating candidate positions. Here, the recorded number is the average number for the queries in a query set. The results are shown in Fig. 3(c)–(d), OptTree and OptTree-Enum check the same number of tests for any query set since they actually build the same optimal plan tree. Also, HeuTree checks more tests for the candidate position validation, e.g., for C_2 in source codes, HeuTree checks 157653 tests per regex query, against 86964 tests checked by the optimal plan tree.

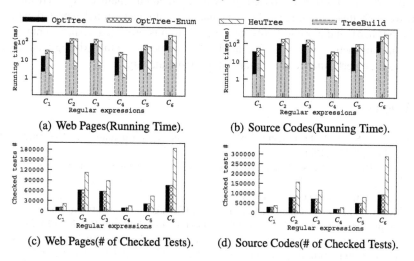

(a) Web Pages(Running Time).   (b) Source Codes(Running Time).

(c) Web Pages(# of Checked Tests).   (d) Source Codes(# of Checked Tests).

Fig. 3. Evaluating plan tree-based matching algorithms on different datasets.

---

[1] http://www.regexlib.com/.

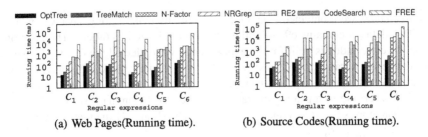

Fig. 4. Performance comparison of alternative algorithms on different datasets.

(3) Performance Comparison with Alternative Algorithms.
Next, we conduct an experiment comparing the running time of our proposed algorithm with the alternative algorithms. We plot the *average running time* of different algorithms in Fig. 4(a)–(b). In both data sets, we can see that the positional index-based algorithms (OptTree and TreeMatch) outperform other algorithms with a prominent superiority, e.g., for the queries in C_1 (simplest regexes), OptTree and TreeMatch only spend 15 ms and 28 ms. This can attribute to that the positional index-based algorithms only consider the occurrence positions of q-grams of the queries. OptTree has a better performance than TreeMatch since OptTree utilizes the optimal query plan.

5   Conclusion

In this paper, we study the problem of regular expression matching by utilizing the pre-built index of the data set. We solve this problem by designing a new metric of matching cost for the positional inverted index-based methods, named expected matching cost. We show that computing an optimal plan tree with the minimal expected matching cost is NP-hard, and a pseudo-polynomial algorithm is proposed. Experimental results show that our method with the optimal plan tree outperforms the existing methods and achieves high performance for regular expression matching.

References

1. Berry, G., Sethi, R.: From regular expressions to deterministic automata. Theoret. Comput. Sci. **48**, 117–126 (1986)
2. Cho, J., Rajagopalan, S.: A fast regular expression indexing engine. In: ICDE, vol. 2, p. 0419 (2002)
3. Cox, R.: https://swtch.com/rsc/regexp/regexp4.html
4. DeRose, P., Shen, W., Chen, F., Lee, Y., et al.: DBLife: a community information management platform for the database research community. In: CIDR, pp. 169–172 (2007)
5. GNUgrep. http://reality.sgiweb.org/freeware/relnotes/fw-5.3/fw_gnugrep/gnugrep.html
6. Greiner, R., Hayward, R., et al.: Finding optimal satisficing strategies for and-or trees. Artif. Intell. **170**(1), 19–58 (2006)
7. Hofmann, K., Bucher, P., Falquet, L., Bairoch, A.: The PROSITE database. Nucleic Acids Res. **27**(1), 215–219 (1999)

8. Kandhan, R., Teletia, N., Patel, J.M.: SigMatch: fast and scalable multi-pattern matching. VLDB **3**(1–2), 1173–1184 (2010)
9. Majumder, A., Rastogi, R., Vanama, S.: Scalable regular expression matching on data streams. In: SIGMOD, pp. 161–172. ACM (2008)
10. McNaughton, R., Yamada, H.: Regular expressions and state graphs for automata. IEEE Trans. Electron. Comput. **1**(EC-9), 39–47 (1960)
11. Mohri, M.: String-matching with automata. Nord. J. Comput. **4**(2), 217–231 (1997)
12. Navarro, C.: NR-grep: a fast and flexible pattern matching tool. Softw. Pract. Experience (SPE) **31**, 1265–1312 (2001)
13. Navarro, G., Raffinot, M.: Flexible Pattern Matching in Strings: Practical Online Search Algorithms for Texts and Biological Sequences. Cambridge University Press, Cambridge (2002)
14. Navarro, G., Raffinot, M.: New techniques for regular expression searching. Algorithmica **41**(2), 89–116 (2005). https://doi.org/10.1007/s00453-004-1120-3
15. Qiu, T., Yang, X., Wang, B., Wang, W.: Efficient regular expression matching based on positional inverted index. IEEE Trans. Knowl. Data Eng. **34**, 1133–1148 (2020)
16. Watson, B.W.: A new regula grammar pattern matching algorithm. In: Diaz, J., Serna, M. (eds.) ESA 1996. LNCS, vol. 1136, pp. 364–377. Springer, Heidelberg (1996). https://doi.org/10.1007/3-540-61680-2_68
17. Yang, X., Qiu, T., Wang, B., Zheng, B., Wang, Y., Li, C.: Negative factor: improving regular-expression matching in strings. ACM Trans. Database Syst. **40**(4), 25 (2016)
18. Yu, F., Chen, Z., Diao, Y., Lakshman, T., Katz, R.H.: Fast and memory-efficient regular expression matching for deep packet inspection. In: ANCS, 2006, pp. 93–102. IEEE (2006)

PFtree: Optimizing Persistent Adaptive Radix Tree for PM Systems on eADR Platform

Rui Zhang, Yao Wu, Shangyi Sun, Lulu Chen, Yibo Huang, Ming Yan, and Jie Wu$^{(\boxtimes)}$

School of Computer Science, Fudan University, Shanghai, China
{zhangrui21,shangyisun21}@m.fudan.edu.cn,
{yao_wu20,llchen18,huangbb16,myan,jwu}@fudan.edu.cn

Abstract. Persistent memory (PM) provides byte-addressability, low latency as well as data persistence. Recently, a new feature called eADR is available on the 3rd generation Intel Xeon Scalable Processors with the 2nd generation Intel Optane PM. eADR ensures that data stored within the CPU caches will be flushed to PM upon the power failure.

In the eADR platform, previous PM-based work suffered more read/write amplification and random access problems, and memory allocations on PM are still expensive. The persistence ways on the eADR platform are still unclear. Therefore, we propose PFtree (PM Line Accesses Friendly Adaptive Radix Tree), a persistent index optimized for the eADR platform. PFtree reduces PM line access with two optimizations: stores key-value pair in leaf array directly to reduce pointer chasing and stores necessary metadata with key-value pair closely and auxiliary metadata in DRAM. PFtree reduces memory allocations in critical paths by allocating bulk memory when creating a leaf array. Then, we design an adaptive persistence way based on data block size for PFtree to fully use PM bandwidth. Experimental results show that our proposed PFtree outperforms the radix tree by up to 1.2× and B+-Trees by 1.1−7× throughput, respectively, with multi-threads.

Keywords: Persistent Memory · Adaptive Radix Tree · eADR

1 Introduction

Emerging byte-addressable persistent memory (PM), such as the Intel Optane DC Persistent Memory Module (DCPMM) [8], is now commercially available. PM is attractive because it offers DRAM-comparable performance and disk-like endurance. In the first generation, PM-equipped platforms support the asynchronous DRAM refresh (ADR) feature [9]. The write content in the CPU cache is still unstable. Therefore, we need to use explicit cache line flush instructions and memory barriers to ensure the persistence of PM writes.

With the arrival of 3rd generation Intel Xeon Scalable processors and 2nd generation Intel Optane DCPMM, extended ADR (eADR) becomes available

X. Wang et al. (Eds.): DASFAA 2023, LNCS 13943, pp. 46–61, 2023.
https://doi.org/10.1007/978-3-031-30637-2_4

[10]. Compared to ADR, eADR further ensures that data in the CPU cache is flushed back to the PM after a crash. It ensures the persistence of globally visible data in the CPU cache and eliminates the need to issue expensive synchronous flushes. The advent of eADR not only omits the flush instruction but also allows us to revisit the design of PM-based data structures.

Building efficient index structures in PM promises high performance and data durability for in-memory databases [4,15,21]. Most existing persistent indexes [1,3–5,12] are designed only for ADR-based PM systems. They work on achieving crash consistency, and the optimization for performance generally lies in reducing the number of flushes, because data flushing is expensive in ADR-based PM systems. However, in the second generation PM, these optimizations are no longer evident as the flush instruction is no longer needed, therefore, these optimizations are not evident when the previous work is applied directly to the second generation PM.

ART(Adaptive Radix Tree) was widely used in database systems because it supported range query and variable-sized keys. Although there have been some works of ART on NVM [11,13,18], there have been some problems. First, the previous work had more PM read/write amplification problems. The previous PM-based data structure usually stored the metadata in the header, separately from the data. Therefore, reading and writing the metadata in the header and the data will cause more PM write amplification problems. Second, the throughput of PM is still slower than DRAM, especially in random access. Separating metadata from data leads to more random accesses [16,17] because the metadata needs to be accessed before the data, and leading to more PM line accesses. Third, the persistence ways on the eADR platform are still unclear. There is no need for persistence instructions on eADR, so there are various ways to achieve data persistence. It is not clear what scenarios these persistence ways are applicable to.

This paper presents PFtree, a PM Line Accesses Friendly Adaptive Radix Tree, to deliver high scalability and low PM overhead. PFtree proposed a leaf array to compress leaf nodes in ART(Adaptive Radix Tree) to solve the problems above. To our knowledge, PFtree is the first PM-based ART optimized for eADR-enabled PM systems.

To reduce PM line accesses, PFtree stores the payload and metadata of each key-value data closely in the leaf array so that when reading or writing a key-value data, it can be done in the same XPline and reduce the PM read/write amplification. In addition, PFtree does not store a pointer to key-value data like others [16,18]. Instead, key-values are stored directly in the data area of the leaf array. For variable key and value, PFtree uses metadata to help store and recovery from the leaf array. The directly stored key-values not only reduce one pointer chasing but also reduce PM line accesses. Meanwhile, PFtree stores auxiliary metadata for accelerated lookup and insertion into DRAM, further reducing PM read and write accesses. These auxiliary metadata are violate and can be reconstructed from other data.

To reduce the overhead of PM memory allocation, PFtree uses bulk memory allocation. The leaf array in PFtree allocates a large chunk of memory at creation time. It then stores the key-value data directly into the memory of the leaf array at insertion time, avoiding the overhead of allocating new memory every time the key value is inserted.

To choose the appropriate persistence way, the performance of the new generation CPU and the new generation PM has been tested in detail. In the new scenario, there are three persistence methods for data persistence to PM. We analyze the latency and bandwidth of the three persistence methods in detail under different data sizes and different threads. Finally, a suitable persistence method was selected for PFtree in different scenarios.

The main contributions are as follows.

- We propose PFtree, the first persistent range index based on eADR platform, to take full advantage of eADR. PFtree not only focuses on data crash consistency but also focuses on reducing PM line accesses to reduce read and write amplification.
- We provide an in-depth analysis of the persistent ways in eADR platforms. Then we propose an adaptive persistent way to fully utilize the PM bandwidth.
- We perform experiments to compare PFtree with state-of-the-art tree-based indexes, including ROART [18], P-ART [13], FAST&FAIR [7], and BzTree [1]. PFtree outperforms the existing solutions by $1.1-7\times$ throughput under YCSB workloads.

2   Background

2.1   PM and eADR

Compared to DRAM, The most significant difference of PM is the non-volatile property. Data written to PM will persist and still exist after power failure. Figure 1 shows the architecture of PM systems and the internal architecture of Optane PM. We assume that the system consists of one or more NUMA-enabled multicore cpus, each with local registers, storage buffers, and caches and that the last level of cache (LLC) is shared between all cores of the CPU. Each CPU has its own memory (DRAM and PM), which is connected to other CPUs through mesh interconnect. PM and Write Pending Queues (WPQ) connect to mesh connect through iMC (integrated memory controller).

Extended ADR (eADR), which is supported in the 3rd generation Intel^{TM} Xeon^{TM} Scalable Processors, solves data consistency problems by making sure that CPU caches are also included in the so called "power fail protected Domain" [10]. In an eADR environment, the CPU cache is also a part of the persistent domain, so there is no need to flush data from the cache to the ADR domain. The data in the CPU cache will be automatically persisted to the PM after a power failure or software crashes. However, the data in the CPU register is still volatile. Because of the nature of eADR, the operations and designs that ensure crash consistency in applications do not need to be performed.

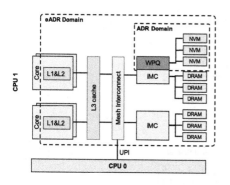

Fig. 1. Architecture of PM.

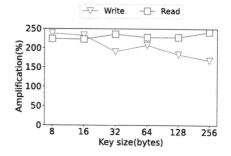

Fig. 2. Read and write amplification in ROART.

2.2 ART and Its Persistent Variants

Adaptive Radix Tree (ART) is a space-efficient radix tree that can dynamically change the node size according to node utilization. To reduce the tree height, ART sets the number of child Pointers in the node to 2^8 and uses a one-byte partial key for each node. At the same time, ART reduces memory consumption by using one of four different node types. Figure 3 illustrates the node structures of ART. Node4 and Node16 store the keys in order together with the corresponding pointers. Sequential search is enough for Node4 because of its small size. SIMD instructions can be used to accelerate the search in Node16. Node48 has a child index with 256 slots to quickly locate the corresponding pointer. Node256 is the normal node in a radix tree whose radix is 256. Starting with Node4, ART adaptively converts nodes to larger or smaller types when the number of entries exceeds or falls behind the capacity of the node type. This requires additional metadata and more memory manipulation than a traditional radix tree, but still shows better performance than other cache conscious in-memory index constructs.

P-ART is a persistent version of the concurrent ART that uses instructions in RECIPE [13] for transformation. For Crash Consistency, P-ART re-used a helper mechanism to detect and fix crash inconsistencies during restarts. However, p-ART does not guarantee dural linearizability-that is, it is possible to read volatile data even if the corresponding operation has returned. ROART [18] is an improved version of P-ART for supporting efficient range queries, lower memory allocation overhead and correctness. However, because ROART inherits the rebalancing algorithm and index structure from ART [14], it still incurs a high allocation overhead during SMOs (structural modification operations), as it needs to allocate more than two leaf arrays for each split operation.

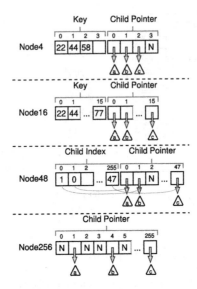

Fig. 3. Node structures in ART

2.3 Motivation

Reduce PM Reading and Writing. In the ADR platform, since the CPU cache is a volatile domain, persistence instructions are needed to ensure data persistence. Much previous work [1,2,7,17,19,20,22] has been devoted to achieving crash consistency and recoverability of data while neglecting optimization of PM line accesses. By analyzing the PM-based indexes, as shown in Fig. 2, ROART has significant read and write amplification to PM under different workloads. One reason is that the metadata in these structures are stored in the head and the key-value data are stored in the tail, and known research work shows that there is 256B buffer inside the PM, so this operation across multiple buffers causes significant read and write amplification.

Reduce PM Memory Allocations on Critical Paths. To improve the efficiency of range queries in ART, ROART [18] uses a leaf array to store pointers to leaf nodes in a compressed manner, in order to eliminate the need to traverse different levels of the trees and pointer chasing. ROART stores each complete key and value in the leaf node, therefore, a memory allocation on PM is made when a key-value pair is inserted. However, the memory allocation in PM incurs high overhead, increasing the query's latency and decreasing the throughput.

Leverage the Features of eADR. The new eADR has multiple methods for persisting data. We are motivated to reveal the effects of these methods on different data sizes, to make full advantage of the eADR features in the real environment.

3   Design

PFtree aims to further optimize query and insertion of ART on PM. Specifically,

- (1) PFtree stores key-value pairs directly in leaf arrays instead of pointers to key-value pairs. It can not only reduce pointer chasing but also reduce PM line accesses.
- (2) PFtree stores metadata and key-value data close together, PM line accesses can be reduced.
- (3) By taking advantage of the latest eADR hardware platform, PFtree's persistence strategy cleverly supports an adaptive persistence way that automatically selects the best persistence strategy according to the block size of the persistent data.

3.1   Block-Based Leaf Array

Motivated by ROART [18], PFtree uses leaf array to delay the leaf split and reduces pointer chasing to improve the range queries in ART. Unlike ROART's leaf array, PFtree stores key-value pairs directly in the leaf array instead of pointers to key-value pairs to further reduce pointer chasing and PM line accesses. In addition to leaf nodes, PFtree does not change the types of other nodes in ROART, that is, the internal nodes will still use Node4, Node16, Node48 and Node256, as mentioned before.

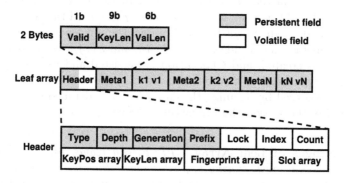

Fig. 4. The leaf array structure of PFtree

Overview of Leaf Array. Figure 4 shows the basic structure of the leaf array in PFtree. The header of each leaf array holds the metadata necessary to identify the characteristics of the leaf array, which can be used for quick recovery and concurrency control. The following is a large data area for storing key-value pairs. In addition to the raw data of key-value pair, there are two bytes of metadata before them, which are used to reconstruct and update key-value from the data area. In the two-byte metadata, the first bit represents the validity of key-value pair, 0 means deleted and 1 means normal. The next nine bits represent the length of the key. The last six bits record the value length of this key-value pair. After the metadata, we store the raw key-value pair. Although we can traverse and lookup in leaf arrays using only two bytes of metadata, this is slow and inefficient. It is tedious and inefficient to parse the metadata and calculate the position of the next key-value pair each time, so we store the *KeyPosarray* and *KeyLenarray* in the header to speed up the traversal and update. Except for volatile metadata, all other data of the leaf array is stored in PM, and the data of the internal nodes are also stored in PM.

The Header of Leaf Array. The header stores the necessary metadata and some other data to speed up queries and updates. *Type* indicates that the type of the node is Node4, Node16, Node48, Node256, or leaf array. *Depth* indicates the depth of the node in the entire PFtree, which facilitates traversal of the entire ART. *Generation* records the generation of each node, for quick recovery after startup. *Lock* is used to mark whether the leaf array is locked, which is used for concurrent access control. *Index* records the index of the inserted key-value pairs in the subsequent 4 arrays. *Prefix* records the common prefixes in the leaf array. *Count* records how many key-value pairs have been inserted into the current leaf array. *KeyPosarray* and *KeyLenarray* record the offset of each key-value pair in the data area and the length of the key, respectively. Similarly, *Fingerprintarray* records the fingerprint of each key-value pair. None of these three arrays needs to be persistent because we can rebuild and update key-value pairs by using the metadata of the first two bytes of each key-value pair in the data area. In addition, we also use *slotarray* to record the size relationship between leaf nodes, which can further improve the efficiency of range queries. Among these metadata of header, *Type*, *Depth*, and *Prefix* are initialized when the leaf array is created, and *Generation* is persisted and restored after each restart. We do not need to modify them after creation. The remaining violate data can be recovered by scanning the data area. The detailed recovery process can be found in the section Data Structure Recovery.

The Advantages of Leaf Array. *PFtree can reduce pointer chasing and PM line accesses through storing key-value pairs directly.* Key-value pairs are stored directly in PM, we do not need to get the pointer and then get the value according to the address as before. Directly stored key-value pairs can reduce a PM-based pointer chasing and also reduce random access on PM because the data and pointer to it are stored in different places.

Adjacent Storage of Metadata and Key-Value Pair Can Reduce PM Line Accesses. The previous data structure tended to keep the metadata in the header and the payload of key-value pair in the back part of leaf nodes. This can make the metadata storage more compact, but in each lookup or modification, we need to update the metadata in the header and then turn to the tail to manipulate the key-value pairs. When there are many key-value pairs, more PM lines will be accessed during persistence. In the leaf array of PFtree, as shown in Fig. 4, we store the necessary metadata in the first two bytes of each key-value pair, followed immediately by the key-value pair. The advantage of such adjacent storage is that we do not need to read or write data across multiple PM lines, and the number of reads and modifications in the same PM line can reduce the PM read/write amplification. Also in ADR environment, storing metadata and key-value pair close together can reduce flush times.

Volatile Metadata Stored in DRAM can Reduce PM Line Accesses. In the leaf array of PFtree, metadata such as *KeyPosarray*, *KeyLenarray*, can be recovered by the 2-byte metadata in front of each key-value pair, so no persistence is required. Considering that the read/write latency of PM is still higher than that of DRAM, we store this recoverable metadata in DRAM, thus reducing the read/write access to read PM.

Bulk Memory in Leaf Array Can Reduce the PM Memory Allocation on Critical Path. The leaf array has already allocated a large chunk of memory when it is created, so there is no need to allocate memory again when inserting key-value pair into the leaf array, and only need to copy key-value pair to the free area, thus reducing the PM memory allocation on the critical path.

3.2   PM-Aware Flush Mode

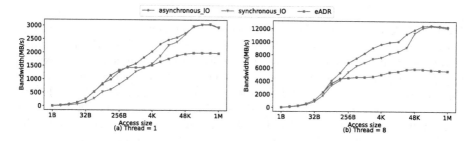

Fig. 5. Random write bandwidths varying with access size for the three persistence ways

With the application of new processors and the new generation of PM, there are three ways to persist data: ❶ **synchronous IO**: Use the persistence instructions the same as before. We can use *clwb* and *fence* to flush data from the L1 cache to PM or use *NTStore* and *fence* to store data directly to PM without going through the CPU cache. ❷ **eADR approach**: Use the features provided by eADR to persist data without using any instructions. In the eADR-enabled system, the CPU cache is located in the persistent domain. As a result, we can persist the data in the L1 cache to PM without instructions like *clwb* and *fence*. ❸ **asynchronous IO**: Use the *flush* to force the data from the cache to PM without calling the *fence* for synchronization. Traditional PM can not support this way, but we can avoid using the fence while keeping the program correct in the eADR system.

To explore which persistence way can make the most of the PM bandwidth and achieve the lowest persistence latency, we first test the basic performance of PM, explored in different threads, different access sizes, form of persistence time latency and the change of bandwidth. Since most programs are based on random access to PM (such as ART insertion and update operations), we will focus on the latency and bandwidth of random read and write to PM under the three different persistence ways in the following experiments. The hardware configuration of the experiment will be introduced in Sect. 4.

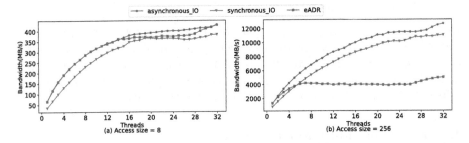

Fig. 6. Random write bandwidths varying with thread for the three persistence ways

Different Access Sizes. We first explore the latency and bandwidth of different persistence ways under different access sizes. We explore the bandwidth of random write with single-thread and multi-thread (8 threads in our experiments) respectively. Figure 5a shows the latency and bandwidth changes of the three persistence ways under different access sizes in the case of a single thread. When the access size is smaller than 128B, the bandwidth of asynchronous IO and eADR persistence is almost the same. In the range of 128B to 256B, the persistent bandwidth in eADR mode is slightly higher than that in asynchronous IO mode. After 256B, the persistent data bandwidth in asynchronous IO mode is higher than that in eADR mode. Especially, when the access size is large, the persistent bandwidth in asynchronous IO mode is significantly higher than that in eADR mode. As for the traditional synchronous IO mode, it has been proven that the fence instruction incurs a significant persistence overhead, so the persistence bandwidth is always lower than the asynchronous IO mode.

In the multi-thread case, the results show similar conclusions to the single-thread case. For example, in the case of 8 threads as shown in Fig. 5b, when the access size is small, the bandwidth of persistent data in asynchronous IO and eADR modes are similar. When the access size exceeds 64B, asynchronous IO bandwidth is higher than eADR bandwidth. As access size continues to grow, the bandwidth gap between the two becomes more and more apparent.

Therefore, it can be seen that when the access size is small, eADR can achieve or even exceed the bandwidth of synchronous and asynchronous IO with less overhead, and as the access size becomes larger, asynchronous IO can get more bandwidth. It still keeps the same pattern in multi-thread. We speculate that the reason for this phenomenon is as follows: when using eADR way for persistence, without the help of any persistence instructions, the persistent data in PM needs to be passively evicted from the cache, from L1 cache to L2 cache to L3 cache, and finally to PM. However, the flush instruction can flush data directly from the L1 cache to PM, which is expected to reduce the IO path of data and reduce the latency of data persistence. For this reason, we propose an asynchronous IO persistence approach that uses flush instructions to omit this part of the data persistence path to reduce data persistence latency.

Different Threads. As for the influence of different persistence modes on PM bandwidth under multi-thread, we also conducted some experiments to analyze. For example, when the access size is 8 bytes, asynchronous IO throughput is similar to eADR throughput when the number of threads is less than 12. Only when it exceeds 14, asynchronous IO performance is slightly better than eADR.

Figure 6b shows the variation of PM bandwidth with different persistence ways when the access size is 256B. When the number of threads is 2, the PM bandwidth in asynchronous IO persistence is higher than that in eADR. As the number of threads increases, The asynchronous IO mode is significantly higher than the bandwidth using eADR mode. This is similar to the previous conclusion: when the access is large, using asynchronous IO persistence can obtain higher PM bandwidth.

Based on Fig. 6, we find that eADR can still obtain high persistence bandwidth without additional persistence instructions when the number of threads and access size both are small. When access size exceeds 256B, asynchronous IO can obtain higher persistence bandwidth. As the number of threads increases, it is more efficient to use asynchronous IO persistence when the access size is large. Based on these findings, we design an adaptive data persistence approach in the eADR environment: the data persistence way is determined by the size of the data.

3.3   Adaptive Flush in Insert and Update

Since eADR platform is gradually on the market, we can take full advantage of eADR features to reduce the overhead of implementing persistent instructions. Based on our previous observations, not all persistence cases that use the eADR approach yield the best performance. Therefore, we design an adaptive persistence approach to take advantage of the low latency of eADR-style while allowing larger chunks of data to achieve higher bandwidth. As shown in Algorithm 1, when persisting data, we first judge the size of the data. If it is a large data block (larger than 256B), the persistence instruction CLWB is invoked to forcibly refresh the data to PM so as to make better use of the bandwidth of

Algorithm 1. flush_data(void *addr, size_t len)

```
1: if support_eADR() then
2:    if len > 256 then
3:       for each cacheline in data do
4:          clwb(cacheline)
5:       end for
6:    else
7:       eadr_auto_flush(data)
8:    end if
9: else
10:    for each cacheline in data do
11:       clwb(cacheline)
12:    end for
13:    fence();
14: end if
```

PM. If it is a small data block (smaller than 256 bytes), eADR mode is used to refresh the data automatically. That is, no persistence command is invoked and the data is passively swapped out of the cache to achieve persistence.

4 Evaluation

Our evaluations consist of four parts to reflect the performance improvements of each proposed design. We evaluate and compare PFtree with some available and usable state-of-the-art representative indexes. For ART, we choose two other state-of-the-art PM ART, ROART [18] and P-ART [13]. We also evaluated the other state-of-the-art B+tree indexes and tire indexes such as BzTree [1] (lock-free B+tree), FAST&FAIR [7](logless crash consistency).

4.1 Experimental Setup

We run experiments on a server with an Intel(R) Xeon(R) Silver 4314 CPU clocked at 2.40 GHz, 512 GB of Optane DCPMM per socket (4 × 128 GB DIMMs on four channels per socket) in AppDirect mode, and 256 GB of DRAM (8 × 32 GB DIMMs). It is important to note that we are using a second generation DCPMM. The CPU has 16 cores (32 hyperthreads) and 48 MB of L3 cache. The server runs Ubuntu 20.04.3 LTS with kernel 5.13.0. For the workload, we use micro-benchmarks and YCSB [6] workload. Each test firstly warms up using 30 million key-value pairs [7,13,17]. Each test lasts 60 s. The total size exceeds the size of the L3-cache and can truly reflect the performance of PM. The micro-benchmarks contain the operations of lookup, insert, update, remove and scan. They are performed using 4 threads. For a fair comparison, we modified the persistence policy in other persistent indexes for the adaptation of eADR features.

4.2   Microbenchmark and Scalability Evaluation

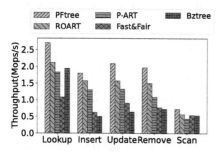

Fig. 7. Microbench (4 threads)

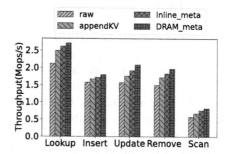

Fig. 8. Performance improvement of each optimization.

The results of micro-benchmark are in Fig. 7. The lookup performance of PFtree is 2.708 Mop/s, faster than ROART (2.118 Mop/s) and P-ART (1.832 Mop/s). This is because PFtree stores key-value pairs directly in the leaf array, whereas ROART stores pointers to leaf nodes in the leaf node array, so we need more pointer chasing in ROART. The reason why PFtree and ROART are faster than P-ART is that N16 in P-ART cannot use SIMD instructions for accelerated lookup. BzTree is fast because it used slotted-page node layout which can have good cache locality. In addition, FastFair does not use binary lookup in internal nodes, so query performance is low. PFtree also has better insertion performance (1.8 Mops/s) than the other trees. There are many reasons for this rise. (1) PFtree stores the inserted key-value pairs in the form of a leaf node array and allocates memory of multiple leaf nodes at a time, which reduces the overhead of memory allocation. (2) PFtree stores the necessary metadata and key-value together, reducing PM line accesses. And the auxiliary metadata is stored in DRAM, further reducing PM line accesses. (3) Compared with PART, the array of leaf nodes can store more key-value pairs, reducing the number of segments.

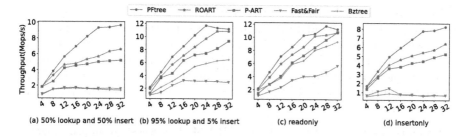

Fig. 9. The performance varying thread under YCSB workload.

For update operations, PFtree shows better for similar reasons. Key-value pairs directly stored reduces a pointer chasing and speeds up the search. More metadata stored in DRAM can help updates and they can update themselves quickly. For the remove operations, ROART needs to free the memory of the node after each deletion, while PFtree free all the memory after all key-value pairs in the leaf array are deleted.

Figure 9 shows the performance and scalability of PFtree and the state-of-the-art ART and B+tree indexes. For the write-intensive workloads A (50% lookup and 50% insert) and D (insert only), PFtree performs up to 2X better than all the other indexes; this can be attributed to PFtree storing key-value pair directly, reducing pointer jumps. And storing metadata and key-value pair closely reduces the access to PM line every time when we access the data. In addition, PFtree apply a bulk request memory policy avoids memory allocation overhead on critical paths. Other B+Tree indexes experience high latency on the critical path due to SMOs. For the read-intensive workloads B (95% lookup and 5% insert) and C (read only), PFtree outperforms all the other indexes by 1.2–7x using 32 threads. The primary reason is metadata stored in efficient DRAM speeds up the traversal and search of leaf nodes, and key-value pair stored directly reduces pointer chasing.

4.3   Factor Analysis of Each Design

Figure 8 presents the factor analysis on PFtree. We start with ROART and add proposed design features. The experiment setup is the same as in microbenchmark, using 4 threads for 60 s of test.

+ Store Key-Value Directly. We saw that this optimization worked in all of our tests. It improves lookup efficiency by 15% and delete efficiency by 11%, as well as in processes such as inserts and updates. In the lookup and scan, efficiency is improved by reducing the number of pointer chasing. In the insert and update, key-value pairs directly stored in the leaf array can not only reduce a pointer chasing, but also avoid memory allocation overhead when constructing a leaf node each time, and reduce flush times. In the remove, the original method requires the removed leaf node be reclaimed every time it is deleted, whereas in PFtree only metadata needs to be changed without memory reclamation.

+ Metadata Stores Closely with Key-Value Pair. Unlike previous data structures that store metadata in the header, PFtree stores the metadata associated with each key-value pair directly with the actual data. In this way, when accessing key-value pair each time, there is no need to read both the PM line where the metadata is located and the PM line where the key-value pair data is located as before, and fewer PM line accesses can reduce the PM read/write amplification.

+ Allocate Violate Metadata from DRAM. This is a slight improvement in all operations because each operation involves reading or writing volatile metadata, and DRAM is more efficient than PM. The improvement is even more pronounced in insert and update operations because more metadata is written and DRAM has a higher write bandwidth than PM.

4.4   Recovery

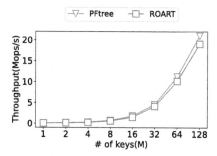

Fig. 10. Data structure recovery.

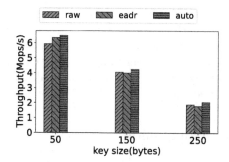

Fig. 11. Insert throughput varies with key length.

In Fig. 10, we test recovery time with different key numbers in PFtree; With 128M keys, data structure recovery takes about 21 s to reconstruct the whole metadata, which is a little slower than ROART (19 s). The main reason is that PFtree stores more volatile metadata in the leaf array to assist in search and update. During recovery, it needs to reconstruct the metadata in Fig. 4 according to the 2-byte metadata in front of each key-value pair, so it takes more time to recover.

4.5   Auto-flush Benchmark

Auto-flush benchmark focus on the persistent ways with PFtree. We have three versions of PFtree: raw-PFtree, eADR-PFtree, and auto-PFtree. Raw-PFtree persists data using the traditional flush and fence instructions. It executes the flush and fence instructions after each data writen to the PM. Based on raw-PFtree, eADR-PFtree removes flush and fence instructions directly, and takes advantage of the nature of cache persistence to ensure data consistency. Auto-PFtree is based on the discovery of the new generation of PM hardware in Sect. 3.2 and uses the adaptive persistence algorithm proposed in Sect. 3.3.

Since the length of the key and value will affect the persistent data size, and the adaptive persistence algorithm is related to the persistent data size, we explored the effects of three persistence modes under different key lengths, and the results are shown in Fig. 11. We tested insert throughput with 8 threads at key lengths of 50, 150, 250 respectively. When the key length is small, eADR mode can improve a little, and auto-PFtree mode improves less than eADR mode. With the increase of key length, auto-PFtree insert throughput is gradually better than eADR-PFtree. This is consistent with the previous observation in Sect. 3.2. Because the persistence method used by auto-PFtree varies according to the amount of data being persisted. As the key length increases, the amount of data required for each persistent operation increases. According to the results in Sect. 3.2, when the amount of data exceeds 256B, eADR persistence cannot fully use the PM bandwidth. Flush instruction assisted persistence can obtain more bandwidth. Therefore, auto-PFtree can have large throughput when the key length is large.

Unfortunately, auto flush does not provide much performance improvement because tree traversal and lookup consume a lot of time during ART insertion, and persistence only takes a small fraction of that time, so auto flush has a limited effect.

5 Conclusion

This paper presents PFtree, a PM Line Accesses Friendly Adaptive Radix Tree optimized for eADR platform. PFtree is proposed with several optimizations. (1) PFtree store key-value pair directly in leaf arrays, reducing pointer chasing, and metadata is stored closely with key-value pair, reducing PM line accesses. (2) PFtree propose an adaptive refresh algorithm to take full advantage of the new PM hardware, based on the detailed analysis of the new generation of PM hardware.

Acknowledgements. This research is supported in part by the National Key Research and Development Program of China (2021YFC3300600).

References

1. Arulraj, J., Levandoski, J., Minhas, U.F., Larson, P.A.: BzTree: a high-performance latch-free range index for non-volatile memory. Proc. VLDB Endow. **11**(5), 553–565 (2018)
2. Chen, S., Jin, Q.: Persistent B+-trees in non-volatile main memory. Proc. VLDB Endow. **8**(7), 786–797 (2015)
3. Chen, Y., Lu, Y., Fang, K., Wang, Q., Shu, J.: uTree: a persistent B+-tree with low tail latency. Proc. VLDB Endow. **13**(12), 2634–2648 (2020)
4. Chen, Y., Lu, Y., Yang, F., Wang, Q., Wang, Y., Shu, J.: FlatStore: an efficient log-structured key-value storage engine for persistent memory. In: Proceedings of the Twenty-Fifth International Conference on Architectural Support for Programming Languages and Operating Systems, pp. 1077–1091 (2020)
5. Chen, Z., Hua, Y., Ding, B., Zuo, P.: Lock-free concurrent level hashing for persistent memory. In: 2020 USENIX Annual Technical Conference (USENIX ATC 2020), pp. 799–812 (2020)
6. Cooper, B.F., Silberstein, A., Tam, E., Ramakrishnan, R., Sears, R.: Benchmarking cloud serving systems with YCSB. In: Proceedings of the 1st ACM Symposium on Cloud Computing, pp. 143–154 (2010)

7. Hwang, D., Kim, W.H., Won, Y., Nam, B.: Endurable transient inconsistency in byte-addressable persistent B+-tree. In: 16th {USENIX} Conference on File and Storage Technologies ({FAST} 2018), pp. 187–200 (2018)

8. Intel: Intel optane dc persistent memory module. https://www.intel.com/content/www/us/en/architecture-and-technology/optane-dc-persistent-memory.html

9. Intel: Deprecating the pcommit instruction (2016). https://software.intel.com/content/www/us/en/develop/blogs/deprecat-epcommit-instruction.html

10. Intel: eADR: new opportunities for persistent memory applications (2021). https://www.intel.com/content/www/us/en/developer/articles/technical/eadr-new-opportunities-for-persistent-memory-applications.html

11. Kim, W.H., Krishnan, R.M., Fu, X., Kashyap, S., Min, C.: PacTree: a high performance persistent range index using PAC guidelines. In: Proceedings of the ACM SIGOPS 28th Symposium on Operating Systems Principles, pp. 424–439 (2021)

12. Krishnan, R.M., et al.: {TIPS}: making volatile index structures persistent with {DRAM-NVMM} tiering. In: 2021 USENIX Annual Technical Conference (USENIX ATC 2021), pp. 773–787 (2021)

13. Lee, S.K., Mohan, J., Kashyap, S., Kim, T., Chidambaram, V.: RECIPE: converting concurrent dram indexes to persistent-memory indexes. In: Proceedings of the 27th ACM Symposium on Operating Systems Principles, pp. 462–477 (2019)

14. Leis, V., Kemper, A., Neumann, T.: The adaptive radix tree: artful indexing for main-memory databases. In: 2013 IEEE 29th International Conference on Data Engineering (ICDE), pp. 38–49. IEEE (2013)

15. Lepers, B., Balmau, O., Gupta, K., Zwaenepoel, W.: KVell: the design and implementation of a fast persistent key-value store. In: Proceedings of the 27th ACM Symposium on Operating Systems Principles, pp. 447–461 (2019)

16. Liu, J., Chen, S., Wang, L.: LB+ trees: optimizing persistent index performance on 3DXPoint memory. Proc. VLDB Endow. **13**(7), 1078–1090 (2020)

17. Liu, M., Xing, J., Chen, K., Wu, Y.: Building scalable NVM-based B+ tree with HTM. In: Proceedings of the 48th International Conference on Parallel Processing, pp. 1–10 (2019)

18. Ma, S., et al.: {ROART}: range-query optimized persistent {ART}. In: 19th {USENIX} Conference on File and Storage Technologies ({FAST} 21), pp. 1–16 (2021)

19. Oukid, I., Lasperas, J., Nica, A., Willhalm, T., Lehner, W.: FPTree: a hybrid SCM-dram persistent and concurrent B-tree for storage class memory. In: Proceedings of the 2016 International Conference on Management of Data, pp. 371–386 (2016)

20. Venkataraman, S., Tolia, N., Ranganathan, P., Campbell, R.H.: Consistent and durable data structures for {Non-Volatile}{Byte-Addressable} memory. In: 9th USENIX Conference on File and Storage Technologies (FAST 2011) (2011)

21. Xia, F., Jiang, D., Xiong, J., Sun, N.: {HiKV}: a hybrid index {Key-Value} store for {DRAM-NVM} memory systems. In: 2017 USENIX Annual Technical Conference (USENIX ATC 2017), pp. 349–362 (2017)

22. Yang, J., Wei, Q., Chen, C., Wang, C., Yong, K.L., He, B.: {NV-Tree}: reducing consistency cost for {NVM-based} single level systems. In: 13th USENIX Conference on File and Storage Technologies (FAST 2015), pp. 167–181 (2015)

Learned Bloom Filter for Multi-key Membership Testing

Yunchuan Li[2], Ziwei Wang[2], Ruixin Yang[2], Yan Zhao[4], Rui Zhou[5], and Kai Zheng[1,2,3(✉)]

[1] Yangtze Delta Region Institute (Quzhou), University of Electronic Science and Technology of China, Chengdu, China
zhengkai@uestc.edu.cn
[2] School of Computer Science and Engineering, University of Electronic Science and Technology of China, Chengdu, China
{liyunchuan,ziwei,youngrx}@std.uestc.edu.cn
[3] Shenzhen Institute for Advanced Study, University of Electronic Science and Technology of China, Chengdu, China
[4] Aalborg University, Aalborg, Denmark
yanz@cs.aau.dk
[5] Cloud Database Innovation Lab of Cloud BU, Huawei Technologies Co., Ltd., Chengdu, China
zhourui24@huawei.com

Abstract. Multi-key membership testing refers to checking whether a queried element exists in a given set of multi-key elements, which is a fundamental operation for computing systems and networking applications such as web search, mail systems, distributed databases, firewalls, and network routing. Most existing studies for membership testing are built on Bloom filter, a space-efficient and high-security probabilistic data structure. However, traditional Bloom filter always performs poorly in multi-key scenarios. Recently, a new variant of Bloom filter that has combined machine learning methods and Bloom filter, also known as Learned Bloom Filter (LBF), has drawn increasing attention for its significant improvements in reducing space occupation and False Positive Rate (FPR). More importantly, due to the introduction of the learned model, LBF can well address some problems of Bloom filter in multi-key scenarios. Because of this, we propose a Multi-key LBF (MLBF) data structure, which contains a value-interaction-based multi-key classifier and a multi-key Bloom filter. To reduce FPR, we further propose an Interval-based MLBF, which divides keys into specific intervals according to the data distribution. Extensive experiments based on two real datasets confirm the superiority of the proposed data structures in terms of FPR and query efficiency.

1 Introduction

Membership testing is a problem of judging whether a queried element e_q exists in a certain element set S, which is common in computer-related fields, e.g.,

---

Y. Li, Z. Wang and R. Yang—Both authors contribute equally to this paper.

database systems, web search, firewalls, network routing, etc. The proliferation of massive raw data brings with it a series of storage and processing challenges.

To enable fast membership testing among massive raw data, effective data structures are needed that can store raw data using small space and answer queries efficiently. As a common data structure, hash table works well in membership testing, where data is mapped on a hash table using random hash functions. Although it helps to ensure fast query to some extent, it has long been blamed for its low space efficiency. To achieve high space efficiency and high-speed query processing, Bloom filter (BF) [1], a probabilistic data structure, is proposed to perform approximate membership testing.

However, it is always necessary to make a trade-off between FPR and memory usage when using Bloom filter. So, in recent years, extensive effort has been devoted to exploring low FPR and low memory occupancy of Bloom filter, but most of existing studies [2,6,7,17,21] have not achieved significant improvement.

In order to further improve the space efficiency, Bloom filter empowered by machine learning techniques are proposed [5,13,16]. Kraska et al. [13] regard membership testing as a binary classification problem, and use a learned classification model combined with traditional Bloom filter. Such a data structure is called Learned Bloom filter (LBF). Based on LBF, Dai et al. [5] propose an Ada-Bloom filter, which divides the scores learned in the machine learning stage into different intervals and uses different discrimination strategies for elements in different score intervals. Considering that when the amount of data in the data stream is close to infinity, the FPR of BF/LBF may be close to 1. Adapting LBF to the extremely large amount of data, Liu et al. [16] propose a Stable LBF, which combines the classifier with an updatable stable Bloom filter to make the filter's performance decay effect more controllable to achieve a satisfying FPR.

Previous studies on the learned Bloom filter mainly focus on single-key membership testing, that is, their methods are mainly for scenarios where each element has only one key-value pair. In a multi-key scenario, for example, data in a distributed database is often stored in multiple physical nodes, and when a query with filtering conditions arrives, it needs to access the disks of all physical nodes and increase the latency, if we set up an efficient and reliable multi-key Bloom filter for each physical node, we can largely reduce the fan-out of nodes. At the same time, the traditional Bloom filter can only deal with the query for which all keys are specified, and existing methods concatenate all keys into one key to convert the multi-key scenarios into the single-key scenarios. However, in most cases, when processing a query in which not all keys are specified, the traditional Bloom filter will have high FPR because each individual key-value pair exists but their combinations do not.

In order to achieve lower FPR and lower space consumption in multi-key membership testing, in this paper, we propose a method, called Multi-key Learned Bloom Filter (MLBF), to tackle the problem. Similar to other LBF methods, MLBF first classifies the queried elements through a value-interaction-based multi-key classifier, and the elements determined to be absent by the classifier will enter a multi-key Bloom filter for further determination. We propose an optimization strategies for multi-key Bloom filters, to improve the performance of MLBF.

Our contributions can be summarized as follows:

1) To the best of our knowledge, this is the first work that adopts LBF to systematically solve the multi-key membership testing problem.
2) For multi-key membership testing, we propose a Value-Interaction-based Multi-key Classifier (VIMC) model, which does not rely on feature engineering, to learn value interactions between keys to classify multi-key elements. In addition, we propose an adaptive negative weighted cross-entropy loss function to limit the FPR of the LBF's learning model, thereby reducing the FPR of the entire LBF.
3) We also propose an optimization strategies for multi-key Bloom Filter, i.e., interval-based optimization, making our proposed method more applicable.
4) We report on experiments using real data, offering evidence of the good performance and applicability of the paper's proposals.

The rest of the paper is structured as follows. We give the problem definition and background of LBF in Sect. 2. Section 3 details the proposed data structures, where their feasibility and performance are analyzed. We report the results from an empirical study in Sect. 4, and Sect. 6 concludes this paper.

2   Preliminaries

In this section, we first formalize the problem of multi-key membership testing and then introduce multi-key Bloom filter that we exploit in the proposed data structures.

2.1   Problem Statement

Multi-key Element Set. A multi-key element set (a.k.a. a multi-key member set), S, is a set of elements $S = \{e_1, e_2, ..., e_n\}$, where n is the number of elements, and $e_i \in S$ denotes a multi-key element containing c key-value pairs, i.e., $e_i = \{key_1 : v_1, key_2 : v_2, ..., key_c : v_c\}$.

Multi-key Membership Testing. Give a multi-key element set $S = \{e_1, e_2, ..., e_n\}$, and a queried element $e_q = \{key_{q1} : v_{q1}, key_{q2} : v_{q2}, ..., key_{qc} : v_{qc}\}$, where $1 \leq q1 < q2 < ... < qc \leq c$. If there is a multi-key element e_i in the multi-key set S, and $v_{qi} = v_i$ for all $key_{qi} = key_i$, element e_q exists. That is, query e_q exists in S if its values of keys are the same with those of an element in S.

2.2   Multi-key Bloom Filter

A Multi-key Bloom filter consists of c hash function families and a bitmap of size m. When inserting a multi-key element e, e is input into these hash function families to calculate a bit vector, and the mapped positions in the bitmap are set to 1. When querying whether an multi-key element exists, the element is also input into the hash function family to obtain a bit vector, and each bit of the vector is mapped to the bitmap. If all the mapped bits in the bitmap are 1, it means that the queried element exists, otherwise it does not exist.

Given a multi-key element set containing four elements, e_1, e_2, e_3, e_4. Initially, all bits of the bitmap are set to 0. Taking the inserting multi-key $e_1 = \{key_1:$ "aa", $key_2:$ "ab", $key_3:$ "ac"$\}$ as an example, e_1 will be hashed to multiple bit positions first. When a query e_q is proposed and the same hashing strategy is used for inserting, if all the positions after the hash are 1, the query e_q is determined to exist; otherwise it does not exist. It can be seen that the basic operations of the multi-key bloom filter and the single-key bloom filter are very similar. However, we need to consider some additional issues, such as how to hash multi-keys into the bitmap, how to reduce FPR, etc., which we will introduce in detail in Sect. 3.

3   Methodology

3.1   MLBF Framework Overview

The framework has two components: a value-interaction-based multi-key classifier and a multi-key Bloom filter. In the first component, a queried multi-key element is input into a value-interaction-based multi-key classifier, which is trained by a given multi-key element set. The classifier can learn the score of the element, where the queried element exists in the given multi-key element set if the score is greater that a given threshold. When the score is less than the threshold, we will put the queried element into a multi-key Bloom filter for further judgment.

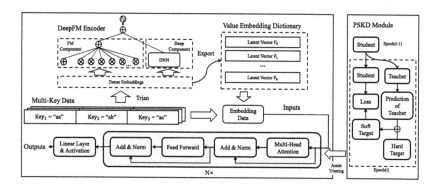

Fig. 1. VIMC Model Overview

3.2   Value-Interaction-Based Multi-key Classifier

To solve the multi-key membership testing problem effectively, it is critical to learn sophisticated key interactions, which can capture the latent correlation between keys. Therefore, we propose a Value-Interaction-based Multi-key Classifier (VIMC) model to learn the value interactions directly from values of keys. The framework of Value-Interaction-based Multi-key Classifier (VIMC) is shown in Fig. 1. Specifically, VIMC encodes the values of the inputs with a DeepFM encoder and then learns the value interactions from the encoded data through a Multi-layer Multi-head Self-Attention (MMSA) module, the outputs of which are fed into a linear layer.

DeepFM Encoder. DeepFM [10] is an efficient end-to-end learning model that can learn both low-order and high-order feature interactions.

Given a multi-key element set, each element has one or more values in each key. Firstly, the continuous values in the set will be discretized, i.e., for key_i, its values are divided into u_i different intervals according to a given segmentation standard st_i. Secondly, we create dictionaries for each key key_j with discrete value types, where the length of its dictionary is dl_j. Then, an element will be embedded into a one-hot vector $X = \{x_1, x_2, ..., x_L\}$, where $L = \sum_i u_i + \sum_j dl_j$. When an element contains $value_i$, the corresponding x_i will be set to 1, otherwise it will be set to 0.

In order to emphasize the low-order and high-order value interactions, DeepFM adopts two components, an FM component and a deep component The FM component is a factorization machine, which is used to learn order-1 and order-2 value interactions. Specifically, a linear regression model is introduced to learn the order-1 value interaction, as shown in Eq. 1.

$$y_{order1} = \langle w, x \rangle + b \tag{1}$$

where w is the parameter vector of the linear regression model and b is the bias.

To learn the order-2 value interaction, for $value_i$, a latent vector V_i is introduced to measure its impact of interactions with other values. Specifically, for each value pair $(value_i, value_j)$ in an element, the value interaction between $value_i$ and $value_j$ is calculated by the inner product of the latent vectors V_i and V_j. The total order-2 value interaction is calculated as follows:

$$y_{order2} = \sum_{i=1}^{L} \sum_{j=i+1}^{L} \langle V_i, V_j \rangle x_i \cdot x_j \tag{2}$$

The output of FM is the summation of the order-1 and order-2 value interaction, which is shown in the following equation:

$$y_{FM} = y_{order1} + y_{order2} \tag{3}$$

The deep component is a feed-forward neural network, which is used to learn high-order feature interactions. The latent vector matrix containing interactive information is used to encode the value in the deep component. Put differently, each value in an element is mapped to the corresponding latent vector, and the concatenation of latent vectors will be used as the input of the feed-forward network. We take FN as the feed-forward network, and the output of the deep component for a given query is shown as follows:

$$y_{Deep} = FN(v) \qquad (4)$$

where $v = [V_1, V_2, ..., V_m]$, and m is the number of values in the query.

After getting the outputs of the two components, the output of DeepFM is shown follows:

$$y_{DeepFM} = sigmoid(y_{FM} + y_{Deep}) \qquad (5)$$

where $y_{DeepFM} \epsilon (0, 1)$ is the predicted membership testing result. All parameters are trained jointly for the combined prediction.

After training, the latent vectors in the FM component of DeepFM contain order-1, order-2 and high-order value interaction information. So, we use the latent vectors as the embeddings of the corresponding values.

Multi-layer Multi-head Self-attention. Self-Attention [24] uses attention mechanism to calculate the value interaction between each value in the query and other values, and uses attention scores to show the degree of interaction between values.

As shown in Fig. 1, each value in a query is encoded by the DeepFM encoder as a vector containing value interaction information. For simple self-attention, the representation matrix of a query will be mapped into $Q(Query)$, $K(Key)$ and $V(Value)$ matrices by three different mapping matrices W_Q, W_K and W_V. Similar to DeepFM, self-attention measures the correlation between values by calculating the inner product of Q and K, which can well retain the value interaction learned by the DeepFM encoder. We compute the matrix of outputs as:

$$Attention(Q, K, V) = softmax(\frac{Q \odot K^T}{\sqrt{d_K}})V \qquad (6)$$

where d_K is the dimension of each row vector in K, in order to prevent the result from being too large.

Multi-head attention is a combination of multiple self-attention structures, each head learning features in different representation spaces. A multi-head self-attention layer consists of a multi-head self-attention model and a feedforward network, the output of multi-head self-attention will pass through the feedforward network to extract features.

Progressive Self-knowledge Distillation. To increase the generalization performance of VIMC, we use a simple yet effective regularization method, namely Progressive Self-Knowledge Distillation (PSKD) [12]. PSKD enables the student model to distill its own knowledge, which means that the student becomes teacher itself. Specifically, PSKD utilizes the past predictions of the student model as a teacher to obtain more information during training. The details of this process can be seen in the right part of Fig. 1. Suppose that $P_t^{stu}(x)$ is the prediction of the student model about the input x at the t-th epoch. Then, the objective at t-th epoch can be written as:

$$\mathcal{L}_{KD,t}(x,y) = H\left((1 - \alpha_t)\, y + \alpha_t P_{t-1}^{stu}(x), P_t^{stu}(x)\right)$$

$$H(y,p) = \frac{1}{N} \sum_i -\left[y_i \cdot \log(p_i) + (1 - y_i) \cdot \log(1 - p_i)\right] \tag{7}$$

where H is the binary cross-entropy loss function, α_t is a hyperparameter that determines how much knowledge from the teacher (i.e., $P_{t-1}^{stu}(x)$) to be accepted, y is the hard target, and $(1 - \alpha_t)y + \alpha_t P_{t-1}^{stu}(x)$ is the softening target for the input x in the t-th epoch of training.

However, there is a problem with α_t that the model does not have enough knowledge in the early stages of training. In such a situation, α_t should be tuned down. For this reason, PSKD increases the value of α_t as t grows. Therefore, α_t is computed as follows:

$$\alpha_t = \alpha_T \times \frac{t}{T}, \tag{8}$$

where T is the total number of epochs for model training, and α_T is the value of α_t at the last epoch that is determined via validation process.

Adaptive Negative Weighted Cross-Entropy Loss Function. As with traditional binary classification tasks, we initially considered the balanced cross-entropy loss function. Suppose there are N input samples to be predicted. Balanced loss is shown as follows:

$$Loss_{balanced} = \frac{1}{N} \sum_i -[\alpha \cdot y_i \cdot log(p_i) + (1 - \alpha) \cdot (1 - y_i) \cdot log(1 - p_i)] \tag{9}$$

where y_i is the label of the input, p_i is the probability of being predicted to be positive, α is a weight parameter. Suppose C_p and C_n are the number of positive and negative examples, respectively. So, the value of α can be calculated from $\frac{\alpha}{1-\alpha} = \frac{C_n}{C_p}$.

Focal loss [15] allows the binary classification model to focus on learning samples that are difficult to learn and solves the problem of class imbalance to a certain extent. Focal loss is shown as follows:

$$Loss_{focal} = \frac{1}{N} \sum_i -[(1 - p_i)^\gamma \cdot y_i \cdot log(p_i) + (p_i)^\gamma \cdot (1 - y_i) \cdot log(1 - p_i)] \tag{10}$$

where γ is a hyperparameter, p_i reflects how close the prediction is to the ground truth, i.e., y_i. Focal loss increases the weight of the hard samples in the loss function through the modulation of γ and p_i, so that the loss function tends to solve the hard samples, which helps to improve the accuracy of the hard samples.

However, neither the balanced loss nor the focal loss is suitable for the MLBF data structure. As discussed above, for a standard LBF, a query judged by the learned model as a positive example will not be checked again, which leads to false positive examples. Therefore, for the learned model in LBF, false positives are often more unacceptable than false negatives. At the same time, in many scenarios, negative examples that can be used for training are more difficult to obtain than positive examples, which leads to more inaccurate predictions for negative examples.

In order to solve the above problem, we propose an adaptive negative weighted cross-entropy loss function. Specifically, we give each negative example an adaptive weight, which is shown as follows:

$$Loss_{ada} = \frac{1}{N} \sum_i -[y_i \cdot log(p_i) + exp(\gamma \cdot p_i) \cdot (1 - y_i) \cdot log(1 - p_i)], \quad (11)$$

where y_i is the label of the input element e_i, p_i is the probability that e_i is predicted to be positive, γ is a hyperparameter, and γ and p_i determine the weight of the negative element e_i. Suppose that the label y_i of the input element e_i is 0. When the predicted value p_i is larger, the greater weight (i.e., $exp(\gamma \cdot p_i)$) is added to the loss of this negative example, which can make the prediction more accurate and can greatly reduce the FPR.

3.3   Multi-key Bloom Filter

In this section, we present the details of the multi-key bloom filter part of the model. The Multi-key Bloom Filter (MBF) contains a bitmap of m-bit size. When the processed multi-key contains c keys, MBF will create c hash function families, and all hash functions are independent of each other. The most basic operations of MBF are $Insert(S, e)$ and $Query(S, e_q)$, where S is the element member set.

$Insert(S, e)$: For each element $e = \{k_1, k_2, ..., k_c\}$ that needs to be inserted into S, each key is associated with a family of hash functions. For the i-th key value in e, we use the i-th hash function family to hash it to the specific positions $MBF[h_{i1}(e[i])], MBF[h_{i2}(e[i])], ..., MBF[h_{ik}(e[i])]$ in the bitmap, and set corresponding positions to 1.

$Query(S, e_q)$: The Query operation is similar to the Insert operation. It also calculates the positions $h_{i1}(e_i), h_{i2}(e_i), ..., h_{ik}(e_i)$ of e_q after being hashed by multiple hash function families. Then MBF will check whether the corresponding bits $MBF[h_{i1}(e_i)], MBF[h_{i2}(e_i)], ..., MBF[h_{ik}(e_i)]$ are all 1 in the bitmap. If they are, e_q exists, otherwise it will return it does not exist.

Interval-Based Optimization. MBF uses the same number of hash functions for all keys, without considering the data distribution of each key. In fact, the data distribution of each key is not consistent, e.g., some keys are uneven, and others are uniform. Considering this situation, we propose an Interval-based Multi-key Bloom Filter (IMBF). IMBF divides the key into specific intervals according to the data distribution, where different intervals use different numbers of hash functions.

More specifically, suppose that each element has c keys. For the i-th key, the probability that the bit is still not set to 1 after being hashed k times is:

$$(1 - \frac{1}{m})^k$$

After inserting n elements, the probability that a bit is still 1 is:

$$1 - (1 - \frac{1}{m})^{\sum_{j=1}^{c} n_j k_j}$$

where n_j is the number of distinct elements in the i-th key. After inserting n multi-key elements, the total number of hashes is $\sum_{j=1}^{c} n_j k_j$. Therefore, for the i-th key, the expected value of its FPR is expressed as follows:

$$E(FRP_i) = (1 - (1 - \frac{1}{m})^{\sum_{j=1}^{c} n_j k_j})^{k_i} \tag{12}$$

We assume that the data distributions of all keys are independent with each other, which is reasonable in practice. We can get the FPR of all keys as follows, where $P_i = \frac{n_j}{n}$, and n is the total number of multi-keys. For simplicity, we assume that the total number of hash functions for all keys is a constant K_{sum}, i.e., $K_{sum} = \sum_{i=1}^{c} k_i$.

$$E(FRP) = \prod_{i=1}^{p} E(FPR_i) = (1 - (1 - \frac{1}{m})^{\sum_{j=1}^{c} n_j k_j})^{\sum_{i=1}^{c} k_i}$$

$$\approx (1 - e^{-\frac{1}{m} \sum_{j=1}^{c} n_j k_j})^{K} \approx (1 - e^{-\frac{n}{m} \sum_{j=1}^{c} P_j k_j})^{K_{sum}} \tag{13}$$

At this point, we can transform the problem that aims to minimize $\{E(FPR)\}$ into the problem that is to minimize $\{\sum_{j=1}^{c} n_j k_j\}$. According to the AM-GM inequality, when $n_j k_j \geq 0$, there is an inequality $\sum_{j=1}^{c} n_j k_j \geq \sqrt[c]{\prod_{i=1}^{c} n_j k_j}$, and the two items are equal if and only if each item on the left side is equal. Therefore, in order to achieve the optimization goal $min\{\sum_{j=1}^{c} n_j k_j\}$, we need to set a smaller k_j for a larger n_j, and a larger k_j for a smaller n_j.

Empirically, for a small amount of data, we can directly calculate the size of n_j. If the amount of data is huge, we can use sampling or some cardinality estimation methods such as HyperLogLog Counting [8] and Adaptive Counting [3]. In the experiments of this paper, we use the method of sampling estimation.

Based on this idea, we can further simplify our model. The ratio of the number of unique elements to the total number of elements is P. We can divide P into intervals, and use different numbers of hash functions for different intervals. For simplicity, we give an interval division parameter I. When $I = 4$, we divide the interval of P into four parts, namely $I_1 = [0, 25\%)$, $I_2 = [25\%, 50\%)$, $I_3 = [50\%, 75\%)$ and $I_4 = [75\%, 100\%]$. Based on the previous derivation, I_4 uses fewer hash functions, and I_1 uses more hash functions. We set the number of hash functions between the two intervals to differ by 1. Combining with the learned classifier (i.e., VIMC in Sect. 3.2), we have an improved MLBF called Interval-based Multi-key Learned Bloom Filter (IMLBF).

3.4   Analysis

We proceed to analyze the false negative rate (FNR) and false positive rate (FPR) of the proposed data structures.

FNR Analysis. MLBF and IMLBF have no false negatives. The MLBF and IMLBF feed data determined by the classifier to be non-existent into an alternate multi-key Bloom filter for further validation. At the same time, their filter parts is based on the standard Bloom filter. Besides, their insertion and query operations are essentially the same as that of the standard Bloom filter. Since the standard Bloom filter has zero false negatives, both MLBF and IMLBF have zero false negatives.

FPR Analysis. IMBF has a lower false positive rate than MBF. Assume the proportion of the i-th key distinct element in IMBF is p'_i, and the number of corresponding hash functions is k'_i, we can get the FPR of IMBF, i.e., $FPR' = (1 - e^{-\frac{n}{m} \sum_{j=1}^{c} P'_j k'_j})^K$. Consider that we use relatively few hash functions for keys with more unique elements, and relatively more hash functions for keys with fewer unique elements in IMBF. According to the AM-GM inequality, this setting makes it possible to make the FPR of MBF closer to its lower limit. Therefore, the FPR of IMBF is lower than that of MBF.

MLBF and IMLBF has a lower false positive rate than MBF. Assume that the expected FPR of MLBF and IMLBF are p^* ($0 \leq p^* \leq 1$). For the classifiers in those three filters, we set its threshold to τ, then $FPR_P = \frac{\sum_{x \in \tilde{N}} I(f(x) > \tau)}{|\tilde{N}|}$, where FPR_P denotes their probability of false positives in the predictors, and $\tilde{N}$ is a held-out set of non-keys. Denoting the FPR of the filter part is FPR_B, the overall FPR of MLBF and IMLBF are $FRP_P + (1 - FPR_P)FPR_B$. We only need to simply set FPR_P or FPR_B, e.g., $FPR_P = FPR_B = \frac{p^*}{2}$. Then we get FPR $= \frac{p^* + p^{*2}}{2}$, which is no greater than p^*.

4   Experimental Evaluation

We evaluate the performance of the multi-key classifier, and the multi-key Bloom filter on real data, respectively. All experiments are implemented on Intel(R) Core(TM) i7-10700 CPU @ 2.90 GHz with 32 GB RAM.

4.1   Datasets

We use the *IMDBmovies* and *Criteo CTR* datasets to simulate multi-key insertion and query for experiments. Since there is a large amount of missing data in both datasets, we use the two datasets to simulate the partial key prediction scenario.

IMDBmovies. *IMDBmovies* comes from IMDB[1], an online database of movies, television programs, etc. *IMDBmovies* has 80,000 movie reviews, each with information such as movie length, average rating, number of directors/actors, and so on. Each data element contains string, number, and other types of fields, which meets the multi-key membership testing scenarios. We remove redundant elements and the columns with massive missing values. We treat the original data as positive samples. Since the dataset does not contain explicit negative data, we randomly select an element from each column to form a new element as a negative sample.

Criteo CTR. *Criteo CTR* is an online advertising dataset released by Criteo Labs[2]. It contains feature values and click feedback of display ads, and this dataset can be used as a benchmark for click-through rate (CTR) prediction. Each advertisement has the function of describing data. To simulate multi-key membership testing, the advertisements with a label of 0 are regarded as negative samples, and others are regarded as positive samples.

4.2   Experiment Results

Performance of VIMC. In this set of experiments, we evaluate the performance of the multi-key classifier. Firstly, we compare our method with the baseline methods on the real-world datasets. Secondly, we verify the effectiveness of each component of the proposed model through ablation experiments. Then we show the FPR of VIMC at different thresholds to demonstrate the significant effect of the proposed model on reducing FPR. Finally, we analyze the hyperparameter settings of γ for different loss functions to prove that our loss function can further reduce FPR.

Accuracy of VIMC. In order to verify the effectiveness of our proposed Value-Interaction-based Multi-key Classifier (VIMC) model, we introduce five mainstream models as baselines, including DeepFM [10], DNN [14], LSTM [11], Linear [19] and Random Forest (RF) [23].

We use *accuracy* to evaluate the performance of the six methods, which is a commonly-used metric to measure the performance of deep learning models and can be calculated as follows:

$$Accuracy = (TP + TN)/(P + N) \tag{14}$$

where TP is the number of true positives, TN is the number of true negatives, P is the number of positives, and N is the number of negatives.

We report the accuracy of the methods in Fig. 2. It is obvious that, on both datasets, our proposed VIMC has the best performance. Respectively, VIMC outperforms the best (i.e., DeepFM) among the baseline methods by 2.00% and

---

[1] https://www.imdb.com/.
[2] https://labs.criteo.com/.

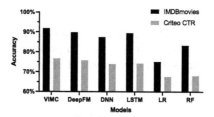

Fig. 2. Model Accuracy Comparison

0.89% on *IMDBmovies* and *Criteo CTR*. Since both datasets contain a large portion of missing attributes, it is a good evidence for the effectiveness of VIMC in reducing FPR of MLBF and IMLBF for partial key queries.

Table 1. Ablation Study Results Based VIMC

Dataset	IMDBmovies		Criteo CTR	
Models	ACC(%)	BCE	ACC(%)	BCE
VIMC-DeepFM	91.11	0.2657	76.28	0.4818
VIMC-pskd	90.93	0.2529	76.38	0.4892
VIMC-DeepFM-pskd	90.49	0.2577	75.84	0.4977
VIMC	**91.86**	**0.2276**	**76.75**	**0.4790**

Ablation Study. In this experiment, we conduct detailed ablation study by comparing the accuracy(ACC) and binary cross-entropy loss (BCE) of different the proposed model variants on both datasets.

We first introduce the three variants each with an optimization component removed. VIMC-DeepFM is a variant based on VIMC, removing DeepFM Encoder. The variant named VIMC-pskd removes the progressive self-knowledge distillation from VIMC. And VIMC-DeepFM-pskd means both components are removed.

As expected, all three variants perform worse than VIMC. As shown in Table 1, no matter which part is removed, the accuracy decreases and the loss increases. Specifically, after deleting both DeepFM and the progressive self-knowledge distillation from VIMC, the accuracy drops by about 1.4% and 0.9% on both datasets, while the binary cross-entropy loss increases by about 0.03 and 0.02. Therefore, we verify that both components of our proposed model are able to learn value interactions from the data and enhance the performance of multi-key classifier.

FPR of VIMC. We now evaluate accuracy (ACC) and false positive rate (FPR) of VIMC at different thresholds. Results are reported in Table 2. Through this experiment, we can effectively evaluate the ability of VIMC to reduce false positives. The threshold varies in the range of [0.5, 0.9]. It can be seen from the table that as the threshold increases, the accuracy of VIMC decreases, but the FPR of VIMC also decreases. A higher threshold can effectively reduce the false positives produced by VIMC, which is very important for LBF. At the same time, we can also see that even if the threshold is set to 0.9, the attenuation of model capability is still acceptable.

Table 2. FPR of VIMC at Different Thresholds

Dataset	IMDBmovies		Criteo CTR	
Threshold τ	ACC(%)	FPR(%)	ACC(%)	FPR(%)
0.50	91.86	10.02	76.75	22.38
0.55	91.75	9.19	76.46	17.95
0.60	91.46	8.46	75.93	13.97
0.65	91.13	7.48	74.38	10.47
0.70	90.50	6.45	72.58	7.56
0.75	89.29	5.67	70.07	5.05
0.80	87.69	4.59	67.07	3.14
0.85	84.99	3.61	62.60	1.56
0.90	80.41	2.15	57.71	0.74

Effect of γ. γ is the hyperparameter of loss function in the weighted methods. We compare the level of FPR reduction in focal loss and adaptive loss at different γ. In this experiment, we train the model at different γ (0, 0.2, 0.5, 1.0, 1.5, 2.0, 5.0) while keeping other parameters unchanged. In addition, VIMC means that no weighted loss function is used. So there is no change in VIMC when γ increases. As illustrated in Fig. 3, both focal loss and adaptive loss methods show decreasing FPR with increasing γ. But the FPR of our proposed method on both datasets drops significantly faster than focal loss. In particular, when $\gamma = 5$, our method is 4.95% lower than focal loss on *IMDBmovies* and 11.71% lower on *Criteo CTR*. Because our method can adaptively weight negative examples for different prediction probability intervals, it can effectively reduce FPR. In another word, VIMC based on adaptive negative weighted method can be used as a multi-key classifier with extremely low FPR level, which is well adapted to bloom filter.

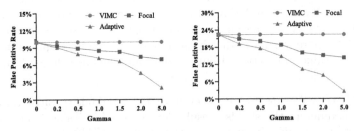

Fig. 3. Effect of γ

Performance of Multi-key Learned Bloom Filter. In this set of experiments, we evaluate the performance of multi-key Bloom filter. Two main metrics are compared for the above methods, i.e., False Positive Rate (FPR), and CPU time for a query. We perform all negative queries and report the average CPU time for a query. We study the following data structures.

1) SMBF: The Standard Multi-Key Bloom Filter data structure that uses only a single Bloom filter, which has the same structure and settings with the Bloom filter in our MLBF.
2) IMBF: Our Interval-based Multi-key Bloom Filter. This data structure uses the interval-based optimization method on the basis of SMBF
3) MLBF: Our Multi-Key Learned Bloom Filter. This variant of SMBF includes the predictor part, i.e., VIMC.
4) IMLBF: Our Interval-based Multi-Key Learned Bloom Filter, which uses the interval-based optimization method on the basis of MLBF.

In the following, we first study the effect of I, a parameter in the interval-based optimization method denoting the number of intervals, to validate the sensitivity of both metrics to I. Then we evaluate the performance of our proposed data structures after using two optimization methods. All the above experiments are performed on both datasets.

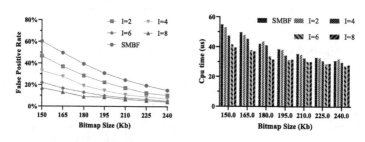

Fig. 4. Effect of I on IMDBmovies

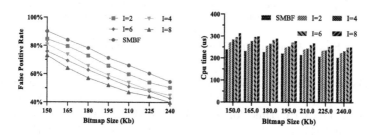

Fig. 5. Effect of I on Criteo CTR

Effect of I. In IMBF, we obtain lower FPR by using different number of hash functions for different keys, and we use the hyperparameter I to adjust the number of hash functions. To facilitate the experiment, we compare only MBF and IMBF in this section, which are the Bloom filter parts of MLBF and IMLBF, respectively. As shown in Fig. 4 and Fig. 5, we compare the FPR and CPU time of MBF and IMBF with different I by changing the bitmap size from 150 Kb to 240 Kb. As we increase the bitmap size, all methods show a decreasing trend in FPR, because a larger bitmap implies a smaller hash function collision probability. At the same time, the CPU time of all methods shows a decreasing trend when the bitmap size is larger since it means that the probability of a bit being 0 is also greater. When querying whether the element exists, if the bit of a mapping is 0, the methods directly return the result that the data do not exist, i.e., a larger bitmap leads to a higher probability of taking less time to return the result. Also, when fixing the size of the bitmap, it can be seen that a larger hyperparameter I corresponds to a smaller FPR, which is consistent with the results derived from Eq. 13.

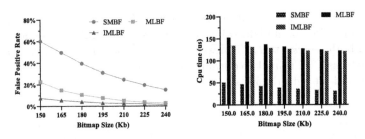

Fig. 6. Comparison of Different Methods on IMDBmovies

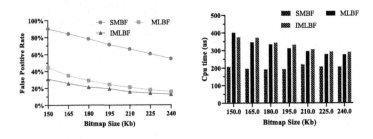

Fig. 7. Comparison of Different Methods on Criteo CTR

Comparison of Different Methods. We proceed to compare the FPR and CPU time for three methods at different bitmap sizes, namely SMBF, MLBF and IMLBF. The CPU time for all LBF-related data structures includes the time spent on VIMC predictions. This experiment also strictly adhered to the parameter settings stated at the beginning of Sect. 4.2. As shown in Fig. 6 and Fig. 7, with our proposed multi-key classifier (i.e., VIMC), the FPR of all learned BF reduced remarkably, at the price of minor extra CPU cost. We also observe that IMLBF always performs better than others in terms of FPR, regardless of the bitmap size, which shows the superiority of our optimization strategies.

Moreover, with the growing bitmap size, the CPU time decreases for all methods, which is due to the fact that larger bitmaps are more likely to return results earlier in a query. And the FPR of all methods shows a decreasing trend as the bitmap size increases. For the same reason as in the previous experiment, because a larger bitmap means that the hash function is more likely to be collision-free, there is a smaller probability of a false positive occurring. This experiment also shows that the running time of the proposed method is in the order of 0.1 ms when querying an element, which demonstrates their feasibility for on-line membership testing scenarios.

5 Related Work

5.1 Bloom Filter

Bloom Filter (BF) [1] was designed by Bloom in 1970 and is widely used for membership testing that is deployed in various domains. For example, some website servers use BF to lock malicious IP addresses [9]. The distributed databases, such as Google Bigtable [4] and Apache Cassandra, use BF to avoid unnecessary disk access to optimize their space efficiency. Even Bitcoin [20] uses BF to determine whether the wallet is synchronized successfully. To meet different requirements (e.g., high lookup performance and low memory consumption), various BF variants have been proposed. Compressed BF [17] uses arithmetic encoding to further compress the space of BF.

5.2 Learned Bloom Filter

Kraska et al. [13] improve the traditional BF by adding a learning classifier before the BF, where the classifier first learns the data distribution when a new queried element arrives and then determines whether the element exists in a given element set. For non-existent elements, an additional BF is used to determine whether they exist or not. This improved BF is called Learned Bloom Filter (LBF). A large number of studies [5,18,22] have proved that LBF can optimize the traditional BF, especially in reducing FPR and memory consumption.

6 Conclusion

We propose and offer solutions to a novel multi-key membership testing problem. In order to achieve low False Positive Rate (FPR) and low memory consumption, we give a Multi-key Learned Bloom Filter (MLBF) data structure that combines a value-interaction-based multi-key classifier and a tailor-made multi-key Bloom filter. Further, a improved MLBF data structure, i.e., Interval MLBF, is proposed to improve the multi-key membership testing performance. To the best of our knowledge, this is the first study that considers multi-key membership testing and multi-key learning Bloom filter. An extensive empirical study with real data offers evidence that the paper's proposals can significantly reduce the FPR during membership query while offering acceptable query efficiency.

Acknowledgements. This work is partially supported by NSFC (No. 61972069, 61836007 and 61832017), Shenzhen Municipal Science and Technology R&D Funding Basic Research Program (JCYJ20210324133607021), and Municipal Government of Quzhou under Grant No. 2022D037.

References

1. Bloom, B.H.: Space/time trade-offs in hash coding with allowable errors. Commun. ACM **13**(7), 422–426 (1970)
2. Bonomi, F., Mitzenmacher, M., Panigrahy, R., Singh, S., Varghese, G.: An improved construction for counting bloom filters. In: Azar, Y., Erlebach, T. (eds.) ESA 2006. LNCS, vol. 4168, pp. 684–695. Springer, Heidelberg (2006). https://doi.org/10.1007/11841036_61
3. Cai, M., Pan, J., Kwok, Y.K., Hwang, K.: Fast and accurate traffic matrix measurement using adaptive cardinality counting. In: SIGCOMM Workshop, pp. 205–206 (2005)
4. Chang, F., et al.: Bigtable: A distributed storage system for structured data. TOCS **26**(2), 1–26 (2008)
5. Dai, Z., Shrivastava, A.: Adaptive learned bloom filter (ADA-BF): efficient utilization of the classifier with application to real-time information filtering on the web. NIPS **33**, 11700–11710 (2020)
6. Fan, B., Andersen, D.G., Kaminsky, M., Mitzenmacher, M.D.: Cuckoo filter: practically better than bloom. In: CoNEXT, pp. 75–88 (2014)
7. Fan, L., Cao, P., Almeida, J., Broder, A.Z.: Summary cache: a scalable wide-area web cache sharing protocol. SIGCOMM **28**(4), 254–265 (1998)
8. Flajolet, P., Fusy, É., Gandouet, O., Meunier, F.: HyperLogLog: the analysis of a near-optimal cardinality estimation algorithm. In: Discrete Mathematics and Theoretical Computer Science, pp. 137–156 (2007)
9. Geravand, S., Ahmadi, M.: Bloom filter applications in network security: a state-of-the-art survey. Comput. Netw. **57**(18), 4047–4064 (2013)
10. Guo, H., Tang, R., Ye, Y., Li, Z., He, X.: DeepFM: a factorization-machine based neural network for CTR prediction. In: IJCAI, p. 1725–1731 (2017)
11. Hochreiter, S., Schmidhuber, J.: Long short-term memory. Neural Comput. **9**(8), 1735–1780 (1997)
12. Kim, K., Ji, B., Yoon, D., Hwang, S.: Self-knowledge distillation with progressive refinement of targets. In: ICCV, pp. 6567–6576 (2021)
13. Kraska, T., Beutel, A., Chi, E.H., Dean, J., Polyzotis, N.: The case for learned index structures. In: SIGMOD, pp. 489–504 (2018)
14. LeCun, Y., et al.: Handwritten digit recognition: applications of neural network chips and automatic learning. IEEE Commun. Mag. **27**, 41–46 (1989)
15. Lin, T.Y., Goyal, P., Girshick, R., He, K., Dollár, P.: Focal loss for dense object detection. In: ICCV, pp. 2980–2988 (2017)
16. Liu, Q., Zheng, L., Shen, Y., Chen, L.: Stable learned bloom filters for data streams. PVLDB **13**(12), 2355–2367 (2020)
17. Mitzenmacher, M.: Compressed bloom filters. Trans. Netw. **10**(5), 604–612 (2002)
18. Mitzenmacher, M.: A model for learned bloom filters, and optimizing by sandwiching. In: NIPS, pp. 462–471 (2018)
19. Montgomery, D.C., Peck, E.A.: Introduction to Linear Regression Analysis (2001)
20. Nakamoto, S.: Bitcoin: a peer-to-peer electronic cash system. Decentralized Bus. Rev. 21260 (2008)

21. Putze, F., Sanders, P., Singler, J.: Cache-, hash-, and space-efficient bloom filters. JEA **14**, 4 (2010)
22. Rae, J., Bartunov, S., Lillicrap, T.: Meta-learning neural bloom filters. In: ICML, pp. 5271–5280 (2019)
23. Rigatti, S.J.: Random forest. J. Insur. Med. **47**(1), 31–39 (2017)
24. Vaswani, A., et al.: Attention is all you need. In: NIPS, pp. 5998–6008 (2017)

ACR-Tree: Constructing R-Trees Using Deep Reinforcement Learning

Shuai Huang, Yong Wang, and Guoliang Li$^{(\boxtimes)}$

Tsinghua University, Beijing, China
{huang-s19,wangy18}@mails.tsinghua.edu.cn, liguoliang@tsinghua.edu.cn

Abstract. The performance of an R-tree mostly depends on how it is built (how to pack tree nodes), which is an NP-hard problem. The existing R-tree building algorithms use either heuristic or greedy strategy to perform node packing and mainly have 2 limitations: (1) They greedily optimize the short-term but not the overall tree costs. (2) They enforce full-packing of each node. These both limit the built tree structure. To address these limitations, we propose ACR-tree, an R-tree building algorithm based on deep reinforcement learning. To optimize the long-term tree costs, we design a tree Markov decision process to model the R-tree construction. To effectively explore the huge searching space of non-full R-tree packing, we utilize the Actor-Critic algorithm and design a deep neural network model to capture spatial data distribution for estimating the long-term tree costs and making node packing decisions. We also propose a bottom-up method to efficiently train the model. Extensive experiments on real-world datasets show that the ACR-tree significantly outperforms existing R-trees.

Keywords: R-tree · Reinforcement learning · Spatial index

1 Introduction

Querying spatial objects is fundamental in location-based applications such as map services and social networking. For example, an urban resident may search points of interest (POIs) such as restaurants in some region (range queries) specified on Google Maps or sometimes query the k-nearest POIs (kNN queries). R-trees [5] are adopted to index these spatial objects and speed up spatial queries, especially when there are hundreds of millions of objects or a huge number of queries.

Given a set of spatial objects, an R-tree can be constructed by incrementally inserting each object [2,3,5,16]. However, this approach is not efficient for building an R-tree from scratch and often leads to poor R-tree structure with bad query performance. Therefore, there are approaches attempting to build more efficient R-trees in a bulk-loading manner. There are two types of strategies: (1) Bottom-up methods [1,6,8,12,14] pack objects into parent nodes recursively, based on hand-crafted heuristics. (2) Top-down method [4] partitions a node into

© The Author(s), under exclusive license to Springer Nature Switzerland AG 2023
X. Wang et al. (Eds.): DASFAA 2023, LNCS 13943, pp. 80–96, 2023.
https://doi.org/10.1007/978-3-031-30637-2_6

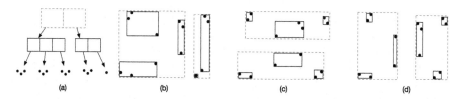

Fig. 1. R-tree constructed under various settings (the MBR of nodes in the higher level is represented by rectangles with dashed edges and lower level with solid edges). (a) (b): under full-node-packing constraint, (c): allowing packing of non-full nodes, (d): considering long-term tree costs.

child nodes recursively, greedily optimizing each tree construction step. These approaches try to achieve evenly distributed objects in different subtrees and full packing of entries of each subtree.

Limitations of the Existing Methods. (**L1**) They omit the long-term results (the whole node partition or the overall tree construction) when making the decision in each step. It may lead to a bad overall result. (**L2**) They impose an unnecessary full-packing constraint, *i.e.,* full-filling each tree node with entries to its capacity, on the tree construction. It limits the resulting tree structure.

For example, Fig. 1 shows 3 cases when we build a 3-level R-tree on a set of spatial object data, and we want to minimize the total area of minimum bounding rectangle (MBR) of tree nodes which mainly affects the query performance. If we build it by [4] under full-packing constraint, we will get the R-tree shown in (a) (b). We can observe that data points which are far from each other are packed together, resulting in nodes with big MBRs. As a comparison, if we allow a variable number of entries in each node, we can build an R-tree as shown in (c), where the MBR areas are significantly smaller. Furthermore, since (c) greedily optimizes the areas in the current level, we can further improve it by considering the areas in lower levels. In this way, we can build an R-tree in (d) of which the upper-level nodes are a bit larger than (c) but nodes in the lower level are significantly smaller thus the overall result is better.

Challenges and Our Proposed Solutions. To address these limitations and build R-trees with better query performance, we need to search in a much larger space, optimizing the long-term tree costs, and the traditional methods are impractical, *i.e.,* unable to enumerate all the possibilities. Therefore, we propose to build R-trees using deep reinforcement learning (DRL), which has been proven successful in applications of database systems such as configuration parameters tuning [13] and join order selection [18]. RL can learn to optimize the overall benefits of a complex task through trial-and-error explorations. It typically incorporates the generalization ability of deep learning (DRL), which allows us to effectively learn a strategy from large search space with limited explorations.

However, there are several challenges in utilizing DRL in R-tree building: (**C1**) How to model the R-tree building process and optimize the long-term tree costs (**L1**)? We propose a tree Markov decision process (tree-MDP), which allows us to optimize the overall tree costs (*i.e.*, considering the rest steps and rest levels). (**C2**) The searching space to build an R-tree is huge, which becomes much larger when we remove the full-packing constraint (**L2**). How to effectively explore good solutions? We utilize Actor-Critic, a DRL algorithm using neural networks to automatically learn the R-tree building strategy from limited explored samples. We design a grid-based model to encode the spatial distribution of data and use a hierarchical convolutional network to embed it into a vector, for effectively estimating the long-term tree costs (Critic) and making R-tree building decisions (Actor). (**C3**) It's time-consuming to collect feedback (*i.e.*, R-tree building experiences) and train the multi-level Actor-Critic models. We design a bottom-up model training framework where we use a training-sharing method to reduce training rounds and a shortcut method to efficiently access feedback in each round. Our main contributions are summarized as follows.

- We propose to utilize DRL to address 2 limitations in the existing R-tree batch-building algorithms, *i.e.*, greedy strategy and full-packing constraint.
- To model the top-down R-tree construction, we propose tree-MDP which well fits the recursive R-tree building process and models the long-term tree costs.
- We use Actor-Critic networks to effectively explore the huge searching space of R-tree construction. We design a grid-based representation model and a hierarchical network to embed the spatial distribution of data. Then we can effectively estimate the long-term tree costs and make building decisions.
- We propose a bottom-up model training framework, where we use training-sharing and shortcut methods to accelerate model learning.
- We have conducted extensive experiments on real-world datasets and the results show that our method outperforms existing approaches by 20%–30%.

2   Related Work

2.1   Spatial Queries and R-Tree

We study spatial queries on multi-dimensional objects such as points, rectangles and polygons. Given a set of n d-dimensional objects, we consider two common types of spatial queries: *range (or window) query* and *k-nearest neighbor (kNN) query*. A range query retrieves all the objects that are included by or intersect a rectangular range. A kNN query retrieves k closest objects to a given coordinate p in Euclidean space.

To avoid scanning all objects to answer a query, spatial indices especially R-trees are proposed for efficient searching.

R-Tree. The R-tree [5] is a balanced tree structure for indexing spatial objects and is widely used. In an R-tree, Each tree node N contains up to B entries, where B is determined by the disk block size. Therefore, the height H of an R-tree indexing n objects is lower bounded by $O(\log_B n)$. Each entry in a node contains a pointer to an child node (or object) and the minimum bounding rectangle (MBR) that surrounds it.

An R-tree greatly speeds up spatial queries by traversing tree nodes in a top-down manner, during which subtrees whose MBRs do not intersect a range query rectangle or are not promising to be in results of a kNN query can be safely pruned.

Extensive studies [2,3,5,7,16] design heuristics to optimize the query performance during dynamic data insertion. For example, the R*-tree [2] minimizes the areas, overlaps and perimeters of node MBRs when choosing the node to insert a new object, so that less tree nodes are expected to be accessed in queries.

R-Tree Building. When building an R-tree from scratch, instead of one-by-one insertion, batch building (packing) methods are designed for reducing construction costs and achieving better tree structure. Existing methods consider full packing, i.e., filling each node to its capacity if possible, with a 100% space utilization.

Most of these methods take a bottom-up manner, i.e., packing every B objects into one parent node, recursively until reaching the root (less than B child nodes). For example, [8] applies Z-order and Hilbert-order on multi-dimensional spatial objects, then sequentially packs each B ones. [14] adds a further step, i.e., mapping the spatial points into a rank space, before sorting and packing them, which guarantees a theoretical worst case bound on the query performance. STR [12] first divides the objects into $\sqrt{n/B}$ groups by the order of their x-coordinates, then packs each B objects in each group by y-coordinates. PR-tree [1] provides a worst-case query cost, which is asymptotically optimal, by a recursive 6-partitioning process. These heuristic-based methods rely on the uniformity of data distribution and can not achieve good performance on many real-world datasets with skewness [4].

There are also top-down methods, i.e., partitioning a parent node into up to B child nodes, recursively until no more than B objects in each as the leaf nodes. TGS [4] uses a greedy partitioning strategy: repeatedly splitting the objects into 2 subsets, by a cut orthogonal to one of the d axes. TGS optimizes the area sum of the 2 MBRs of resulting subsets, by enumerating all the candidate cuts. TGS has better performance than others on most datasets. However, the greedy strategy omits the long-term tree costs and may have a bad overall result.

Moreover, all these methods are limited by the full-packing constraint which reduces the problem space but limits the built R-tree structure. Different from them, we remove this constraint and utilize deep reinforcement learning to effectively optimize the overall R-tree costs.

2.2 Deep Reinforcement Learning

Reinforcement learning (RL) is a powerful algorithm that can learn to make decisions in a complex task (e.g., chess [17]), maximizing specific benefits, through trial-and-error exploration. It has also been successfully applied to problems in databases such as knob tuning [13] and join order selection [18].

The task (e.g., building an R-tree) is usually divided into multiple steps (e.g., node packing). In each step, the decision maker, a.k.a. agent, based on the current state (e.g., the spatial distribution of data) takes an action (e.g., packing a node), then gets a *reward* (*e.g.*, cost of the node) and moves to the next step. This is the Markov decision process (MDP) into which we need to formulate the task in RL. The *agent* learns a *policy*, *i.e.*, action selection strategy, that maximizes the long-term *rewards* (*e.g.*, the total R-tree cost), through the exploration experiences. When the task has a large searching space, RL usually incorporates deep learning (DRL) of which the generalization ability enables it to effectively learn the *policy* from limited exploration.

Actor-Critic is a class of DRL methods which is wildly used. It uses a deep neural network (DNN) model to represent the *policy* (Actor) which can make decisions (*action*). It uses another DNN model to estimate the long-term *rewards* (Critic) which can help improving the Actor. PPO [15] is a popular policy gradient method, which is a default choice at OpenAI[1], that updates the policy (*i.e.*, Actor) through a "surrogate" objective function. Our method utilizes the Actor-Critic framework and uses PPO to learn the Actor model.

3   Framework

We build an R-tree in a top-down manner because tree nodes closer to the root have larger impact on query performance, which are better to be considered first [4]. Specifically, first, we divide the n objects into $x \leq B$ groups, and each group corresponds to a child node of the root node, *i.e.*, partitioning the root node. Next, it recursively partitions these child nodes until every node has no more than B objects which become the leaf nodes. The key point is how to partition each tree node. We first split the object set into 2 subsets by a cut orthogonal to an axis, then recursively split the resulting subsets and finally get x child nodes.

The main challenge in this R-tree building process is that the search space of node partition (by multiple splitting operations) is huge. To make it practical and efficient to implement, the traditional top-down method TGS [4] takes 2 settings:

Full-Packing Constraint. TGS enforces packing full-filled child nodes, *i.e.*, a node of abundant objects will be partitioned into the least possible nodes, with each node except the last one filled up to its capacity. Under this constraint, the possible choices of each split are limited to $O(B)$, instead of $O(n)$ when without this constraint.

Greedy Split. TGS only considers how to optimize the current split operation, *i.e.*, from the $O(B)$ candidates choosing the "best" split that minimizes the area sum of the MBRs of 2 resulting subsets, which may be further split.

However, these 2 settings limit the R-tree building results as Sect. 1 and Fig. 1 show. To overcome these 2 limitations and search a better R-tree structure from the larger space, we utilize Actor-Critic [9], a DRL algorithm and propose **ACR-tree** (Actor-Critic R-tree), of which the framework is shown in Fig. 2. We use tree-MDP (**M1**, Sect. 4.1, 4.2) to model the top-down R-tree building process, which provides a framework to optimize the long-term tree costs. We design the

---

[1] https://openai.com/.

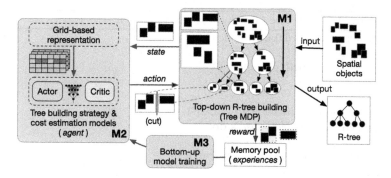

Fig. 2. ACR-tree building framework

Actor-Critic networks as the *agent* (**M2**, Sect. 4.3) to make decisions in each building step in the tree-MDP framework.

Now we introduce the workflow of building an ACR-tree. Given a set of spatial objects, we build an R-tree through top-down node partitioning, beginning at the root node containing all the objects. Each node partitioning is achieved by a recursive object set splitting process consisting of multiple steps. In each step, to split an object set, the spatial distribution of the objects is passed as the *state* to the Actor-Critic *agent*. The *agent* first uses a grid-based model to encode the spatial *state*, then uses a neural network model to embed the spatial features and generates an *action*. The *action* can be a splitting operation, *i.e.*, using a cut orthogonal to an axis to split the objects into 2 subsets. It can also be a packing operation, *i.e.*, stopping splitting the current set which can fit in one child node and packing it, and then getting a *reward* (*e.g.*, area of the node MBR). When all the nodes need no more partition, we built an R-tree.

To train the *agent* model, we repeatedly perform the above R-tree construction process and store the *rewards* into a memory pool, and use them as the feedback to update the parameters of both Actor and Critic models. We need to train several Actor-Critic pairs, one for each level. The training is time-consuming and we propose a bottom-up framework (**M3**, Sect. 5) for efficient model training.

4 Actor-Critic R-Tree

In this section, we first specifically show the R-tree building process without full-packing constraint, which consists of multiple splitting operations (Sect. 4.1). We then introduce the tree-MDP to model it (Sect. 4.2). Finally, we introduce the Actor-Critic networks, which make the splitting decisions in the process (Sect. 4.3).

4.1 Top-Down R-Tree Building Process

Algorithm 1 shows the top-down R-tree building framework, which is similar to TGS [4]. Partitioning the objects $\mathcal{O}$ contained in a node into at most B subsets (*i.e.*, child nodes) (line 5) is the key point in this framework, which decides the

---

Algorithm 1: CreateTree (Top-down R-tree Building)

Input: $\mathcal{O}$: n objects data, B: node capacity

1 **if** $n \leq B$ **then**
2 | $entries = \{\langle MBR(\{o\}), o\rangle | o \in \mathcal{O}\}$
3 **else**
4 | $entries = \{\}$
5 | $objectSubsets = NodePartition(\mathcal{O}, B)$
6 | **foreach** $\mathcal{O}_{ch} \in objectSubsets$ **do**
7 | | $N_{ch} = CreateTree(\mathcal{O}_{ch}, B)$
8 | | Add $\langle MBR(\mathcal{O}_{ch}), N_{ch}\rangle$ into $entries$
9 **return** *A pointer to entries*

Algorithm 2: NodePartition

Input: $\mathcal{O}$: n objects, P: max nodes, H: node height ($\lceil \log_B n \rceil$ by default)
Output: Object subsets contained by each child nodes

1 **if** $n \leq B^{H-1}$ *and* $(P = 1$ *or* $Pack(\mathcal{O}, P, H))$ **then**
2 | **return** $\{\mathcal{O}\}$
3 $dim, pos, \alpha = Cut(\mathcal{O}, P, H)$
4 sort $\mathcal{O}$ by the upper coordinates on dim, n_l of which are below pos
5 Adjust n_l to satisfy $0 < n_l < n$ and $\lceil \frac{n_l}{B^{H-1}} \rceil + \lceil \frac{n-n_l}{B^{H-1}} \rceil \leq P$
6 Take the first n_l ones from $\mathcal{O}$ as $\mathcal{O}_l$, the rest as $\mathcal{O}_r$
7 $P_l^{min} = \lceil \frac{|\mathcal{O}_l|}{B^{H-1}} \rceil, P_r^{min} = \lceil \frac{|\mathcal{O}_r|}{B^{H-1}} \rceil, P_{extra} = P - P_l^{min} - P_r^{min}$
8 $P_l = P_l^{min} + \lfloor \alpha P_{extra} \rfloor, P_r = P - P_l$
9 **return** $NodePartition(\mathcal{O}_l, P_l, H) \cup NodePartition(\mathcal{O}_r, P_r, H)$

---

final tree structure. Different from TGS, in our ACR-tree, *NodePartition* allows the child nodes to be not full packed, *i.e.*, for any $\mathcal{O}_{ch} \in objectSubsets$, $|\mathcal{O}_{ch}|$ can be arbitrary but not exceeding B^{H-1} where $H = \lceil \log_B |\mathcal{O}| \rceil$.

Algorithm 2 show the recursive *NodePartition* based on set splitting. P is the maximum number of nodes to be finally packed ($P = B$ at the beginning). We do not enumerate all possible splits but choose a cut orthogonal to 1 of the d dimensions to divide the object set (line 3). *e.g.*, When choose the X-*dim*, it's a vertical line $x = pos$. We also specify a ratio α to divide the available nodes P into P_l, P_r (line 7, 8) for the 2 resulting subsets $\mathcal{O}_l$ and $\mathcal{O}_r$ respectively. Then we call *NodePartition* on $\mathcal{O}_l$ and $\mathcal{O}_r$ recursively to further split them, and merge the results as the final set of child nodes (line 9). When the objects can fit in one child node ($n \leq B^{H-1}$), we are forced to stop splitting and pack them if the maximum node number $P = 1$, else we can choose whether to pack (line 1).

Example 1. Figure 3 shows the construction of a 3-level R-tree when $B = 3$. We first partition the objects $\{o_1 - o_{11}\}$ inside the root node. We first use a cut $x = c_0$ (yellow dash line) to split the objects into 2 subsets $\{o_6 - o_{11}\}, \{o_1 - o_5\}$, and use ratio $\alpha = 0.5$ to divide the maximum node number $P_0 = 3$ into $P_1 = 1, P2 = 2$. Then we are forced to pack $\{o_6 - o_{11}\}$ as N_1 since $P_1 = 1$. We also choose to pack $\{o_1 - o_5\}$ as N_2. Similarly, we can partition N_1, N_2 in level 2 and get $N_3 - N_7$, all of which contain no more than 3 objects, and we finally get an R-tree.

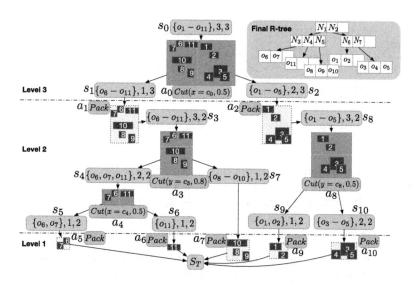

Fig. 3. An example of top-down R-tree construction when $B = 3$

The selection of *Cut* and *Pack* in each step will influence the future splitting and the final partitioning result. We utilize DRL to learn to make these splitting decisions that minimize the long-term R-tree costs. We first need to model the R-tree building process as an MDP.

4.2   Tree MDP Model

An MDP is usually sequential and chain-like, where we move from one state to another each time we make a decision. For example, a state can be the current board in a chess game [17]. However, the construction process of an R-tree is tree-like and the general MDP is hard to model it. Therefore, we propose tree-MDP, where a state (*i.e.*, an object set to be split) can have several succeeding states (*i.e.*, the resulting 2 subsets), and the long-term rewards of a state also depend on the succeeding actions and states only. Next, we specifically define the tree-MDP model (also use Fig. 3 for explanation).

State. A state s_t is a set $\mathcal{O}$ of rectangles, with a maximum node number P and a node height H. *e.g.*, the construction begins at $s_0 = \langle \{o_1 - o_{11}\}, 3, 3 \rangle$.

Action. There are 2 types of action:

- *Pack*: Packing the current objects into a new child node, then (a) if $H = 2$ (leaf node), move to the terminal state S_T, (b) else move to a new state $s_{t'}$ of a non-leaf node with $P' = B, H' = H - 1$, which need to be further partitioned.
- *Cut*: Using a cut (dim, pos, α) to split $\mathcal{O}$ and P into 2 succeeding states s_{t_l}, s_{t_r}.

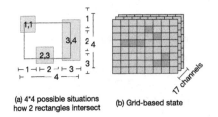

(a) 4*4 possible situations how 2 rectangles intersect   (b) Grid-based state

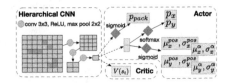

Fig. 4. Grid-based representation model   **Fig. 5.** Hierarchical Actor-Critic network

For example, the action a_0 on s_0 specifies a cut ($x = c_0$, the yellow dash line) to divide $\mathcal{O}$ into $\{o_6 - o_{11}\}$ and $\{o_1 - o_5\}$, and a ratio $\alpha_0 = 0.5$ to divide $P_0 = 3$ into $P_1 = 1, P_2 = 2$, into 2 succeeding states s_1, s_2 respectively. The action a_1 on s_1 packs the current object set where $|\{o_6 - o_{11}\}| \leq 3^2$ as a child node N_1.

A policy π is a mapping from states to the action distribution, *i.e.*, $a_t \sim \pi(s_t)$.

Reward. To optimize the overall R-tree cost (*i.e.*, node areas), we define the reward $r_t = r(s_t, a_t)$ as: (1) If a_t is *Cut*, then $r_t = 0$; (2) If a_t is *Pack* yielding a new child node N, then $r_t = area(N)$. e.g., $r_0 = 0$, $r_1 = -area(N_1)$.

We use a **value function** $V_\pi(s_t)$ to evaluate the long-term reward, *i.e.*, rewards of all the successors (subtree) of state s_t, under specific policy π. For terminal state $V_\pi(S_T) = 0$, else the value can be recursively defined as:

$$V_\pi(s_t) = \mathbb{E}_{a_t \sim \pi(s_t)} \begin{cases} r(s_t, a_t) + \gamma_1 \left[V_\pi(s_{t_l}) + V_\pi(s_{t_r}) \right] & , a_t \text{ is } Cut \\ r(s_t, a_t) + \gamma_2 V_\pi(s_{t'}) & , a_t \text{ is } Pack \end{cases} \quad (1)$$

where γ_1, γ_2 are 2 discounting factors (*e.g.*, 0.99) for better convergence.

We can observe that $V_\pi(s_t)$ is the discounted sum of node areas in the subtree from s_t. Therefore, we need to choose an action maximizing $V_\pi(s_t)$. However, in this tree construction process without full-packing constraint, the search space is huge and we can only explore a small part. Therefore a traditional search-based method (*e.g.*, Monte Carlo tree search) may yield a bad result. Therefore, we utilize the Actor-Critic algorithm, which uses neural network models to learn from limited samples and generalizes to the unexplored space, thus is effective for large searching space.

4.3 Actor-Critic Model

In order to make good decisions in the tree-MDP, the agent learns 2 models, *i.e.*, Actor to generate partitioning actions and Critic to estimate the long-term R-tree costs. Note that only Actor is not enough and we need Critic to assist in improving it. Moreover, a trained Critic model can also be utilized to accelerate the training process, which will be introduced in Sect. 5.2.

Since the policies and value functions in various tree levels are different, we learn $H-1$ models ($h = 2, ..., H$) for partitioning nodes of various object set sizes, *i.e.*, model h for partitioning $n \in (B^{h-1}, B^h]$ objects. Both of Actor and Critic in level h take a state $s_t = \langle \mathcal{O}_t, P_t, H_t \rangle$ (where $H_t = h$ is constant) as input, then Actor generates $a_t \sim \pi(s_t)$ and Critic estimates $V_\pi(s_t)$. The performance strongly depends on the effectiveness of embedding the spatial distribution of the rectangle set $\mathcal{O}_t$. First, we use a grid-based model to represent $\mathcal{O}_t$, which can be better captured by a neural network model. Then, we use a hierarchical convolutional network to aggregate the information into a hidden vector. Finally, we use the vector to generate the outputs of Actor and Critic.

Grid-Based Representation Model. It's hard for a model to learn the features of spatial distribution from raw input $\mathcal{O}_t$, i.e., coordinates of rectangle data of variable $O(n)$ size. Therefore, we propose to use a grid-based model for representing each state, of which the spatial object distribution is importance for partitioning decisions. Specifically, we divide the universal region of an object set into $W \times W$ equal-size grids and use the statics of rectangles intersecting each grid to represent a state.

If each object is a point, which is either contained by a grid or not, we can simply use 1 integer for each grid to represent the number of points contained by it. However, the possible cases of a rectangle intersecting a grid is not only 2. Therefore we need to use more channels to represent the intersection of a grid with the rectangle data. Specifically, the intersection of 2 rectangles is equivalent to the intersection of their coordinate ranges on each dimension. As Fig. 4a shows, the possible situations in how 2 coordinate ranges intersect is 4, thus there are 4^d cases for a rectangle data to intersect a grid in a d-dimensional space. Therefore, when $d = 2$, we use a 16-dimensional vector for each grid, concatenating 1 extra dimension representing P_t. As a result, as Fig. 4b shows, we can represent s_t by a $W \times W \times 17$ tensor $\mathbf{s}_t$, which is similar to an image ($H \times W \times 3$ with 3 channels of RGB).

Hierarchical Convolutional Network. Next, we aggregate the information from all the grids of $\mathbf{s}_t$ into a hidden vector. We leverage the convolutional neural network (CNN [11]) from the image processing field, which utilizes local perception to effectively capture the spatial features and is usually a basic unit to construct deep network structures such as AlexNet [10].

The network structure is shown in Fig. 5. We use $L = \log_2 W$ layers of CNN and Pooling modules to progressively reduce the $W \times W \times C$ input feature matrix into a $1 \times 1 \times d^h$ hidden embedding $\mathbf{Z}$, aggregating the spatial information from each location. Specifically, in each step i from L down to 1, we use a CNN $Conv_i^{3,3}$ with 3×3 receptive field to extract spatially local correlation from the feature map, followed by a Rectified Linear Unit ($ReLU(x) = max(x, 0)$) and a MaxPooling layer (reducing each 2×2 grids into 1 of the maximum value).

$$\mathbf{Z}_{i-1} = MaxPool^{2,2}(ReLU(Conv_i^{3,3}(\mathbf{Z}_i))), i = L, ..., 1 \qquad (2)$$

where $\mathbf{Z}_L = \mathbf{s}_t, \mathbf{Z}_i \in \mathbb{R}^{2^i \times 2^i \times d_i}$ and d_i increases as the matrix size shrinks since it needs to embed more information. Note that $\mathbf{Z} = \mathbf{Z}_0 \in \mathbb{R}^{d^h}$ embeds the global information, extracted by the local perception level by level, which can be used to generate outputs of Actor and Critic. Also note that the size of the parameter set is small since it does not depend on the number of grids $W \times W$.

Critic Output. The Critic needs to predict the state value (*a.k.a.* return) as $\hat{V}^{\pi}(s_t) \in \mathbb{R}_0^+$. We pass the embedding $\mathbf{Z}$ through a fully connected layer (FC) followed by a ReLU to generate $\hat{V}^{\pi}(s_t)$.

$$\hat{V}^{\pi}(s_t) = ReLU(FC_1(\mathbf{Z})) \in \mathbb{R}_0^+ \tag{3}$$

Actor Output. The Actor network needs to generate an action a_t. We first generate the probability p_{pack} of whether to *Pack*, by an FC layer and the Sigmoid function σ, where $\sigma(x) = \frac{1}{1+e^{-x}} \in (0,1)$.

$$z_{pack} = FC_2(\mathbf{Z}) \in \mathbb{R}^1, p_{pack} = \sigma(z_{pack}/\tau) \in \mathbb{R} \tag{4}$$

where τ is a temperature parameter controlling the trade-off between exploration and exploitation.

Similarly, we use another FC and the Softmax function (a high dimensional extension of Sigmoid, also denoted as σ) to generate the probabilities $\mathbf{p}_{dim}$ of each of the d dimensions to be chosen as the cutting dimension (Fig. 5 shows a 2-dimensional case that $\mathbf{p}_{dim} = \{p_x, p_y\}$).

$$\mathbf{z}_{dim} = FC_3(\mathbf{Z}) \in \mathbb{R}^d, \mathbf{p}_{dim} = \sigma(\mathbf{z}_{dim}/\tau) \in \mathbb{R}^d \tag{5}$$

Finally, we generate the cut position *pos* and the ratio α of each dimension for when it is chosen as the cut dimension. Since *pos* and α are both continuous values, we use Gaussian distribution $\mathcal{N}(\mu, \sigma^2)$ to model them. In the 2-dimensional case (*i.e.*, X and Y), $\mu_x^{pos}, \sigma_x^{pos}$ are generated by:

$$\mathbf{z}_x^{pos} = FC_x^{pos}(\mathbf{Z}) \in \mathbb{R}^2, \mu_x^{pos}, \sigma_x^{pos} = Sigmoid(\mathbf{z}_x^{pos}) \tag{6}$$

Similarly, we can use another 3 FC layers to generate the parameters of Gaussian distributions for Y-axis and for α.

For each step t in the R-tree construction process (Algorithm 2), if the object set can fit in one child node (line 1), we first generate p_{pack} from s_t and sample from the Bernoulli distribution $Bern(p_{pack})$ the decision whether to pack. If not (line 3), we choose a cut dimension from a categorical distribution of $\mathbf{p}_{dim}$, with the cut position *pos* and the node assigning ratio α sampled from the associating Gaussian distributions of that dimension.

5   Model Training

For building R-trees of H levels, we need to train $H-1$ models. A naive way is repeatedly building R-trees by these models, using the experiences (states, actions, and rewards) to update the Critic model by minimizing Mean squared error (MSE) loss and update the Actor model through PPO [15], a policy gradient method. However, each time we build an R-tree, the experiences provide unbalanced training samples (one for each decision step) for models of different levels. For example, the model in level H uses $O(B)$ (*e.g.*, several dozens) steps to partition the root node, while the model in level 2 takes $O(B^{H-1})$ (*e.g.*, dozens of thousands) steps to pack all the leaf nodes. Thus it's time-consuming to collect enough training samples for upper levels.

To accelerate this process, we propose to train the models level by level from bottom to up, during which: (1) We can **reuse the parameters** of lower-level models as a relatively good initialization of an upper-level model, and then fine-tune it with fewer training rounds (Sect. 5.1). (2) In each level, we do not build a whole subtree to get all the rewards. Instead, we only partition the current node and use a **short-cut** method to access the rest rewards, *i.e.*, not actually partitioning the child nodes but estimating their rewards by the trained Critic models (Sect. 5.2).

5.1   Bottom-Up Training Sharing

Intuitively, the tasks, *i.e.*, partitioning objects into $x \leq B$ subsets, for Actor-Critic models in various levels are related but different in some details (*e.g.*, child node capacity B^{h-1} and long-term subtree costs). Thus we can reuse a trained model in lower levels as a relatively good initialization of one in the next level.

As shown in Algorithm 3, we train the leaf-level (*i.e.*, $h = 2$) model at first. The training includes repeated exploration and updating rounds (line 3). In each round, we first randomly take an object set $\mathcal{O}$ with $|\mathcal{O}| \in (B^1, B^2]$ and perform *NodePartition* on it, on the policy represented by the current Actor θ_2 (line 4). During this process we collect experiences of all these steps. Then we calculate the long-term costs $V_\pi(s_t)$ (line 5), as a label of the Critic model and an evaluation of the action generated by the Actor model. We use the experiences and all the $V_\pi(s_t)$ to update these models by PPO [15] (line 6). After we finish training the model in level 2, we use its parameters as an initialization of the model in level 3 (line 7), which can be fine-tuned in fewer rounds to converge. Proceeding upwards until the root level, we can have all $H - 1$ models trained.

The calculation of long-term rewards costs most of the time in this process.

5.2   Shortcut Long-Term Reward Calculation

According to Eq. 1, the long-term reward $V_\pi(s_t)$ of a state s_t can be calculated as following: (1) In leaf level, we can simply cumulate the rewards of all the succeeding steps from the collected experiences. *e.g.*, in Level 1 of Fig. 3, $V_\pi(s_4) = r_4 + \gamma_1(r_5 + r_6)$. (2) In non-leaf levels, we need to further partition the child nodes to calculate $V_\pi(s_{t'})$. *e.g.*, in Level 2 of Fig. 3, to calculate $V_\pi(s_1) = r_1 + \gamma_2\{r_3 + \gamma_1[r_4 + \gamma_1(r_5 + r_6) + r_7]\}$, we need to build the whole subtree from s_3. This process is slow, especially for levels near to root, and we propose a short-cut method to avoid further partitions and directly approximate the long-term rewards.

Intuitively, since we have trained the Critic models of lower levels, we can use them to approximate the long-term reward by $V_\theta(s_{t'})$ for each child node $\mathcal{O}_{t'}$ in

---

Algorithm 3: Bottom-Up Model Training

---

1 Random initialize Actor-Critic parameters θ_2
2 **for** $h = 2, ..., H$ **do**
3   **for** $k = 0, 1, ...$ **do**
4   Choose an objects set $\mathcal{O}$ where $|\mathcal{O}| \in (B^{h-1}, B^h]$, perform *NodePartition*
         on policy $\pi(\theta_h)$, get experiences $\{s_t, r_t, a_t, \pi(a_t|s_t)\}$
5   Calculate $V_t = V_\pi(s_t)$ for each t according to Equation 1
6   *UpdatePPO*$(\theta_h, \{s_t, V_t, a_t, \pi(a_t|s_t)\})$
7   Copy parameters $\theta_{h+1} = \theta_h$

$O(1)$ time. We use these estimation results as a shortcut for calculating $V_\pi(s_t)$. For example, in Level 2 of Fig. 3, to calculate $V_\pi(s_1) = r_1 + \gamma_2 V_\pi(s_3)$, we use the trained Critic model θ_2 to approximate a $V_{\theta_2}(s_3)$, instead of further partitioning O_3. In this way, we can significantly accelerate the process of collecting training samples while the training quality does not decay too much, since the estimated long-term costs are relatively accurate.

6 Experiments

In this section, we conduct extensive experiments on real-world datasets to evaluate our method **ACR** and compare with the state-of-the-arts including **R*** [2], **HR** [8], **H4R** [6], **STR** [12], **TGS** [4] and **PR** [1].

6.1 Experiment Setup

Implementation Details. We use the codes[2] in [1] including the implementations of **R***, **HR**, **H4R**, **TGS**, **PR** and we also implement **STR**. The proposed R-tree building models are trained on NVIDIA GeForce RTX 3090 GPU, with PyTorch 1.8. The query tests of the R-tree indices are implemented in C++, on 3.10 GHz Intel(R) Xeon(R) Gold 6242R CPU, with 256 GB RAM.

Dataset	Region	#Objects
SD	South Dakota	53,771
WY	Wyoming	100,683
UT	Utah	323,636
AZ	Arizona	1,464,257
TX	Texas	4,081,258
CAL	California	6,321,254

Fig. 6. Datasets

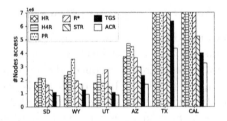

Fig. 7. Range query performance on various datasets

---

[2] https://www.cse.ust.hk/~yike/prtree/.

Datasets. We use several real-world datasets from OSM[3] including the buildings, represented by rectangles with 2-dimensional coordinates (*i.e.*, longitudes and latitudes), in various regions, as Fig. 6 shows.

R-Tree Parameters. In all the R-trees, we use 40 bytes for each entry in a node, *i.e.*, 32 B for 4 coordinates of an MBR and 8 B for a pointer referencing the corresponding child node (or the ID of an object data in a leaf node). The block size is 4 KB, thus the maximum entry number B of a node is 102. For the grid-based representation (Sect. 4.3), we observe $W = 64$ has the best performance, so we take it as the configuration through all the experiments.

Evaluation. We mainly focus on the number of node access, which is strongly correlated to the I/O costs and processing time, for answering queries. For window query, we generate rectangular windows at random locations, with the height and width randomly sampled from uniform distribution $U(0, 2L)$ with various scale L, and report the objects that intersect each query window. For kNN query, we generate the query points at random locations.

6.2   Results

Window Query Processing. We generate window queries on various datasets where $L = 0.04°$ (around 400 m in ground-distance). The number of queries on each dataset is the region area dividing the query window area. In this setting, the expected sum of the query result (*i.e.*, accessed object data) sizes is equal to the size of each dataset. Figure 7 shows the node access numbers of R-tree built by different methods on various datasets. We make the following observations:

(1) **TGS** and **STR** perform better than the other traditional methods. This is because (a) Sort-based methods **HR** and **H4R** can not well preserve the spatial proximity into 1-dimensional order; (b) **R*** builds an R-tree through one-by-one (*i.e.*, online) insertion, which omits the global structure, thus performs worse than the batch (*i.e.*, offline) building algorithms; (c) **PR** optimizes the worst-case performance and thus may not be the best on most datasets. The node accesses of **TGS** are less than **STR** on all the datasets, because **TGS** greedily optimizes the areas of nodes that are closer to the root, which has a larger impact on query performance.

(2) **ACR** has the best performance on all the datasets, whose node accesses are around 20% to 30% less than those of **TGS**. This is because the DRL-based method (a) considers the long-term tree costs and thus performs better than the greedy strategy, and (b) removes the full-packing constraint that limits the R-tree structure and uses Actor-Critic model to effectively explore a good result.

Varying Window Size and k. We evaluate the performance for range and kNN queries on dataset AZ. The range queries are generated in the same way

---

[3] http://download.geofabrik.de/.

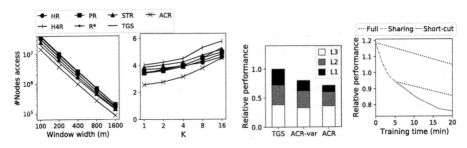

Fig. 8. Window que-ries of various scales   **Fig. 9.** kNN queries of various k **Fig. 10.** Performance under various settings   **Fig. 11.** Effect of bottom-up training

as above, with the window size L varying from 100 m to 1.6 km. The k of kNN queries varies from 1 to 16.

The results are shown in Fig. 8 and 9 and we make the following observations: (1) On all the query sets, the performance of **ACR** is significantly better than all other methods. (2) The node accesses of all the methods decrease as the query window size increases, because the number of queries decreases while the total number of reported objects fixes, thus the total node accesses becomes fewer. (3) For kNN queries, the node accesses increase when k increases, because we need to traverse more nodes to find more nearby objects.

Index Size and Building Time. As Table 1 shows, the index size of **ACR** is slightly larger than other methods. This is because the index size is proportional to the node number. Traditional full-packing methods use the fewest nodes to index all the objects while **ACR** allows nodes to not be full-filled and thus use more nodes. **ACR** also spends more time to build an R-tree because (a) it is implemented in Python which is much slower than C and (b) it passes the state into a neural network to make the decision in each step. However, the time cost is acceptable even for the largest dataset (*i.e.,* CAL of over 6M objects) and the space overhead is small thus **ACR** is practical for usage.

6.3   Self Studies

Effect of Removing Full-Packing Constraint and Long-Term Optimization. To evaluate the effect of overcoming the 2 limitations of **TGS** as introduced in Sect. 3, we train another model **ACR-Greedy** that builds R-trees without the full-packing constraint but only optimizes the area sum of the child nodes but not all the descendants (*i.e.,* the whole subtree), by setting the level discount factor $\gamma_2 = 0$ (in Eq. 1). And we show the numbers of access of nodes relative to TGS in various tree levels which depends on the node areas.

We can observe from Fig. 10 that: (1) **ACR-Greedy** has the better performance than **TGS**. This shows the effect of removing the full-packing constraint, which provides much more space for finding a better tree structure. (2) The node access in the top level (Level 3) has **ACR** more than **ACR-Greedy** because **ACR-Greedy** greedily optimizes the node partitioning of the current level.

Table 1. Index size and building time

Dataset	Index size (MB)		Building time (s)						
	Others	**ACR**	**R***	**HR**	**H4R**	**STR**	**TGS**	**PR**	**ACR**
SD	2	3	0.1	<0.1	<0.1	<0.1	<0.1	<0.1	21
WY	4	5	0.2	<0.1	<0.1	<0.1	0.1	0.1	85
UT	13	13	0.6	0.3	0.4	0.4	0.6	0.7	102
AZ	57	60	3	1	1	1	3	4	447
TX	158	168	7	3	3	3	11	13	1,075
CAL	244	254	12	5	5	5	18	21	1,601

However, **ACR** has much less node access in the rest 2 levels and the total performance is better than **ACR-Greedy**. This is because **ACR** optimizes the long-term tree costs, which can build an overall better R-tree.

Effect of Bottom-Up Training. To evaluate the effect of the bottom-up training framework along with the training sharing and the short-cut strategies, we train models of 2 levels on WY in different ways and report the query performance relative to TGS in Fig. 11. If we directly build the whole tree to train the 2 models (**Full**), the time cost is high. If we first train the model of level 1, of which the time cost is low, in the meanwhile sharing its parameters with the level 2 model (**Sharing**), it achieves good performance in much less time. Then we partition the tree nodes in level 2 and train the associating model. If we get the long-term rewards $V_\pi(s_t)$ by further partitioning the child nodes (**Full**), the speed is slow, while using the short-cut method to estimate $V_\pi(s_t)$ (**Short-cut**) costs much less time.

7   Conclusion

In this paper, we propose ACR-tree a DRL-based R-tree building algorithm, which overcomes 2 limitations of the existing methods. First, we propose tree-MDP to model the long-term tree costs, which is omitted by the traditional methods. Second, we remove the full-packing constraint and use Actor-Critic models to effectively explore the resulting huge search space, with a hierarchical CNN structure for embedding the spatial distribution of objects to effectively estimate the long-term costs and make the node partitioning decisions. We also propose a bottom-up framework with 2 strategies, *i.e.*, training-sharing and shortcut, to efficiently train the models. Extensive experiments on real-world datasets show that the ACR-tree significantly outperforms existing R-trees.

Acknowledgement. This paper was supported by National Natural Science Foundation of China (61925205, 62232009), Huawei, TAL education, and Beijing National Research Center for Information Science and Technology.

References

1. Arge, L., Berg, M.D., Haverkort, H., Yi, K.: The priority R-tree: a practically efficient and worst-case optimal R-tree. ACM Trans. Algorithms (TALG) **4**(1), 1–30 (2008)
2. Beckmann, N., Kriegel, H.P., Schneider, R., Seeger, B.: The R*-tree: an efficient and robust access method for points and rectangles. In: Proceedings of the 1990 ACM SIGMOD International Conference on Management of Data, pp. 322–331 (1990)
3. Beckmann, N., Seeger, B.: A revised R*-tree in comparison with related index structures. In: Proceedings of the 2009 ACM SIGMOD International Conference on Management of Data, pp. 799–812 (2009)
4. García R, Y.J., López, M.A., Leutenegger, S.T.: A greedy algorithm for bulk loading R-trees. In: Proceedings of the 6th ACM International Symposium on Advances in geoGraphic Information Systems, pp. 163–164 (1998)
5. Guttman, A.: R-trees: a dynamic index structure for spatial searching. In: Proceedings of the 1984 ACM SIGMOD International Conference on Management of Data, pp. 47–57 (1984)
6. Haverkort, H., Walderveen, F.V.: Four-dimensional Hilbert curves for R-trees. J. Exp. Algorithmics (JEA) **16**, 3-1 (2008)
7. Kamel, I., Faloutsos, C.: Hilbert R-tree: an improved R-tree using fractals. Technical report (1993)
8. Kamel, I., Faloutsos, C.: On packing R-trees. In: Proceedings of the Second International Conference on Information and Knowledge Management, pp. 490–499 (1993)
9. Konda, V., Tsitsiklis, J.: Actor-critic algorithms. Adv. Neural Inf. Process. Syst. **12** (1999)
10. Krizhevsky, A., Sutskever, I., Hinton, G.E.: ImageNet classification with deep convolutional neural networks. Adv. Neural. Inf. Process. Syst. **25**, 1097–1105 (2012)
11. LeCun, Y., Bottou, L., Bengio, Y., Haffner, P.: Gradient-based learning applied to document recognition. Proc. IEEE **86**(11), 2278–2324 (1998)
12. Leutenegger, S.T., Lopez, M.A., Edgington, J.: STR: a simple and efficient algorithm for R-tree packing. In: Proceedings 13th International Conference on Data Engineering, pp. 497–506. IEEE (1997)
13. Li, G., Zhou, X., Li, S., Gao, B.: QTune: a query-aware database tuning system with deep reinforcement learning. Proc. VLDB Endow. **12**(12), 2118–2130 (2019)
14. Qi, J., Tao, Y., Chang, Y., Zhang, R.: Theoretically optimal and empirically efficient R-trees with strong parallelizability. Proc. VLDB Endow. **11**(5), 621–634 (2018)
15. Schulman, J., Wolski, F., Dhariwal, P., Radford, A., Klimov, O.: Proximal policy optimization algorithms. arXiv preprint arXiv:1707.06347 (2017)
16. Sellis, T., Roussopoulos, N., Faloutsos, C.: The R+-tree: a dynamic index for multidimensional objects. Technical report (1987)
17. Silver, D., et al.: Mastering the game of go with deep neural networks and tree search. Nature **529**(7587), 484–489 (2016)
18. Yu, X., Li, G., Chai, C., Tang, N.: Reinforcement learning with tree-LSTM for join order selection. In: 2020 IEEE 36th International Conference on Data Engineering (ICDE), pp. 1297–1308. IEEE (2020)

A Scalable Query Pricing Framework for Incomplete Graph Data

Huiwen Hou[1], Lianpeng Qiao[2], Ye Yuan[1(✉)], Chen Chen[1], and Guoren Wang[1]

[1] Beijing Institute of Technology, Beijing, China
{houhw,yuan-ye}@bit.edu.cn
[2] Northeastern University, Shenyang, China
qiaolp@stumail.neu.edu.cn

Abstract. With the rapid growth of data, how to make full use of their value becomes a critical issue. In the past few years, it has been a popular method to buy and sell data through the data market. Meanwhile, a variety of data pricing mechanisms have been proposed. However, since most of them concentrate on relational data, little is known about graph data pricing, particularly incomplete graph data. In this paper, we mainly focus on the pricing problem for queries over incomplete graph data. We take *data provenance* as the key idea behind our pricing mechanism and assign a base price to each edge in the graph. Considering the arbitrage-free property of query price, and the lack of some potential answers due to data incompleteness, we propose two practical pricing functions for incomplete graph query respectively. Furthermore, we design feasible pricing algorithms based on subgraph matching to derive each type of query price. Extensive experiments on real graph datasets demonstrate the effectiveness and efficiency of our solutions.

Keywords: Data pricing · Incomplete graph data · Arbitrage-free

1 Introduction

Nowadays, due to the widespread use of various information technologies, data is increasing on a large scale every day. Rich and complex data has significant economic value in many areas. However, if data resources are only separately stored in their holders and not widely shared, it is impossible to maximize the value of data. This phenomenon is usually called *data island*. To solve this problem, an available solution is to regard data as a kind of digital commodity, and develop a data market platform that supports efficient data trading [1–3]. A number of data markets have been launched, such as GBDEx[1], GNIP[2], Factual[3], Windows Azure Marketplace[4], etc.

---

[1] https://www.gbdex.com/.
[2] https://www.support.gnip.com/.
[3] https://www.factual.com/.
[4] https://www.datamarket.azure.com/.

X. Wang et al. (Eds.): DASFAA 2023, LNCS 13943, pp. 97–113, 2023.
https://doi.org/10.1007/978-3-031-30637-2_7

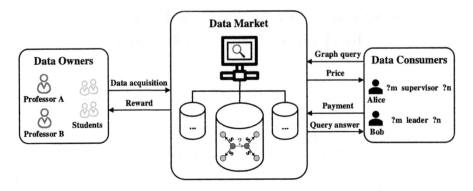

Fig. 1. A general framework of data trading

It is noteworthy that existing data markets mainly focus on trading for relational data. However, there are various kinds of data in our daily life, which could be introduced into the data market in theory. Especially, increasing amounts of data is modeled as graphs in order to overcome the limitations imposed by traditional relational databases with respect to capturing the complex data relevance. And it is common to mine the potential value of graph data in many applications, including bioinformatics [4], social networks [5], and Web analysis [6]. Therefore, we study the problem of query pricing in the scenario of graph data trading. Specifically, the data market should design a reasonable pricing mechanism when it intends to provide a data service called graph query. In addition, we further consider the situation that graph data stored in the data market is incomplete, which is common in practice.

Figure 1 shows a basic framework for data trading based on incomplete graph query. The framework consists of three major components, including data owners, data consumers, and a data market. Data owners are a group of entities with specific relationships. They provide raw data for the data market and receive appropriate rewards. Data consumers want to purchase some valuable information for research purposes. They send specific query requests to the data market and pay for the corresponding answers. The data market is an intermediary between data owners and data consumers. It makes efforts to collect and integrate data in order to construct underlying graph databases to be sold. Notice that some information of graph may be incomplete due to different reasons in data acquisition. On the other hand, it should efficiently execute the graph queries sent by data consumers. At the same time, it also needs to return a reasonable price for each query, which is our research emphasis in this paper.

We illustrate the motivation with a specific example in social network application. As shown in Fig. 1, we assume that Professor A and Professor B are supervisors of two students respectively, and Professor A is the leader of Professor B as well. Relationships among them are collected as a part of the underlying graph database by the data market. However, the detailed "leader" relationship from Professor A to Professor B is not obtained since it may involve their privacy. As a data consumer, Alice wants to purchase the statistics about supervisors and their corresponding students, and sends her query request to the data market. For the data market, how much should it charge for the query answer? Besides, we consider another data consumer Bob, whose query is about the "leader" relationship. As we have analyzed, data incompleteness is bound to have a definite impact on the query answer. Thus, how to reflect such impact on the price?

Although a variety of trading mechanisms in the data market tend to be mature in recent years, there are few researches about taking graph data as a digital commodity, which has great practical significance for the data market. Since the structure of graph data is more complex than other types of data, it becomes a troublesome problem to formulate an appropriate standard for pricing graph queries. Especially, pricing queries over incomplete graph data is more challenging, even though the incompleteness of data is inevitable.

According to the above analysis, we make the following contributions.

- We adopt data provenance as the key idea, and assign a base price to each edge according to the scenario of graph query pricing.
- We propose both an arbitrage-free pricing function and a discount pricing function for queries over incomplete graph data.
- We design efficient pricing algorithms to derive two types of query prices based on the idea of subgraph matching.
- We evaluate the effectiveness and efficiency of our proposed algorithms with extensive experiments on real graph datasets.

2   Problem Definition

In this section, we mainly define basic concepts for the query pricing on incomplete graph data. Table 1 summarizes the notations used in this paper.

Table 1. List of notations

Notation	Description
$\mathbb{G}$	Incomplete graph data
Q	Graph query
$l(r, \mathbb{G})$	Lineage set of result r
$l(Q, \mathbb{G})$	Lineage set of query Q
$\rho(e)$	Base price of edge e
$\epsilon(G)$	Quality of graph G
$C_{\mathbb{G}}(Q)$	Certainty coefficient
$p_{\mathbb{G}}(Q)$	Basic pricing function
$p_{\mathbb{G}}^a(Q)$	Arbitrage-free pricing function
$p_{\mathbb{G}}^d(Q)$	Discount pricing function

Definition 1 (Graph data). An edge-labeled graph is formally defined as $G = (V, E, \Sigma, L)$, where (1) V is a finite set of nodes; (2) $E \subseteq V \times V$ is a set of labeled edges; (3) Σ is a set of labels; and (4) $L : E \rightarrow \Sigma$ is a labeling function which maps each edge e in E to a label $L(e)$ in Σ. Normally, a node represents an object, and an edge from v to v' labeled a represents that there is a binary relation a between the two nodes, denoted by a triple (v, a, v').

Graph pattern matching is a common query method for graph data, which aims to find all subgraphs matching a given graph pattern. We usually regard a graph pattern GP as a set of triple patterns, i.e., $(?x, a, ?y)$ involving both node variables and edge labels. For the convenience of description, we get accustomed to defining a graph query Q in a form similar to SQL.

Definition 2 (Graph query). Given a graph G and a graph pattern GP, the graph query Q is defined as "*Select* RD *Where* GP", where the result description RD contains a subset of node variables in GP, similar to the projection operation.

Incomplete data is widespread in real life due to information loss, privacy preservation, or other reasons [7], which of course includes graph data. The incompleteness of graph is demonstrated in many aspects, and we mainly pay attention to the situation that the edge label is unknown in this paper. In other words, for an incomplete graph $\mathbb{G}$, there is $e \in E$ whose label cannot be obtained by the mapping function L. An incomplete graph $\mathbb{G}$ usually corresponds to a group of complete graph data $\{G_1, G_2, ..., G_m\}$, which is called *possible worlds* of $\mathbb{G}$. Each possible world G_i is a complete graph by replacing missing edge labels with specific labels in Σ. And for a query Q over $\mathbb{G}$, its output only contains the shared answers from all $Q(G_i)$ [8].

Definition 3 (Incomplete graph query). Given a query Q over an incomplete graph $\mathbb{G}$, $\{G_1, G_2, ..., G_m\}$ is the possible worlds of $\mathbb{G}$. The query answer $Q(\mathbb{G})$ is defined as:

$$Q(\mathbb{G}) = \bigcap_{1 \leq i \leq m} Q(G_i) \qquad (1)$$

Example 1. We take the incomplete graph data in Fig. 2 as data graph $\mathbb{G}$, and consider the following three queries. According to the above analysis, we can easily bind pattern variables to nodes in $\mathbb{G}$, and get the query answers where $Q_1(\mathbb{G}) = \{(v_1, v_5), (v_3, v_5)\}$ and $Q_2(\mathbb{G}) = \{(v_5, v_7)\}$. Especially for Q_3, it is equivalent to performing join operation on $Q_1(\mathbb{G})$ and $Q_2(\mathbb{G})$, thus the query answer is $Q_3(\mathbb{G}) = \{(v_1, v_5, v_7), (v_3, v_5, v_7)\}$.

$Q_1 := Select\ ?x_1,\ ?x_3\ Where\ \{?x_1\ a\ ?x_2,\ ?x_2\ b\ ?x_3\}.$

$Q_2 := Select\ ?x_1,\ ?x_3\ Where\ \{?x_1\ c\ ?x_2,\ ?x_2\ d\ ?x_3\}.$

$Q_3 := Select\ ?x_1,\ ?x_3,\ ?x_5\ Where\ \{?x_1\ a\ ?x_2,\ ?x_2\ b\ ?x_3,\ ?x_3\ c\ ?x_4,\ ?x_4\ d\ ?x_5\}.$

Actually, data markets should support trading on various types of data. Therefore, we regard the query request to graph data as a data service, and the data market can require buyers to pay for the service.

Definition 4 (Query pricing). Given a graph $\mathbb{G}$, the query price is determined by a pricing function $p_{\mathbb{G}} : \mathbb{Q} \rightarrow \mathbb{R}^+$, where $\mathbb{Q}$ is the set of queries and $\mathbb{R}^+ = [0, \infty)$.

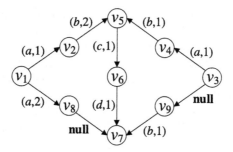

Fig. 2. The incomplete graph data $\mathbb{G}$ for the running example

3 Pricing Mechanism

In this section, we design specific pricing functions for incomplete graph query from several perspectives. Meanwhile, we propose corresponding algorithms for the computation of query prices.

3.1 Basic Pricing Function

The key idea of our pricing mechanism is based on data provenance. First, we refer to the concept of *data lineage* in relational databases [9], in order to evaluate the contribution of graph data to the query answer. Each result r in the query answer corresponds to at least one subgraph, which matches the graph pattern GP. In this paper, we call the subgraph that directly contributes to result r as the lineage set of r, and edges in such a subgraph as lineage elements of r.

Given an incomplete graph $\mathbb{G}$ and a query Q, the lineage set of one query result $r \in Q(\mathbb{G})$, denoted by $l(r, \mathbb{G})$, is defined as $l(r, \mathbb{G}) = G_r$ where (1) $Q(G_r) = \{r\}$ and (2) $\forall G'_r \subset G_r, Q(G'_r) = \varnothing$. These two properties indicate that the result r can be retrieved exactly by the lineage set G_r, and each edge in G_r is indispensable to r. For example, Fig. 3 shows the result lineage sets of queries illustrated in Example 1. For query Q_1, its two result lineage sets correspond to $l((v_1, v_5), \mathbb{G}) = G_1$ and $l((v_3, v_5), \mathbb{G}) = G_2$. Similarly, we can get $l((v_5, v_7), \mathbb{G}) = G_3$ for query Q_2 and $l((v_1, v_5, v_7), \mathbb{G}) = G_4$, $l((v_3, v_5, v_7), \mathbb{G}) = G_5$ for query Q_3.

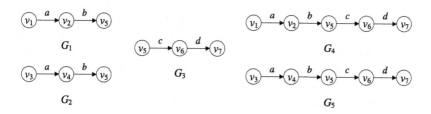

Fig. 3. Result lineage sets of Q_1, Q_2 and Q_3

In the practical data market, assigning a base price to each tuple is a common method to price relational data. And for graph data, we also choose a base price function ρ to generate a price for each edge in the graph. The base price $\rho(e)$ can be determined by data owners or the data market, in order to reflect the data value or collection cost. Meanwhile, the base price of an edge with unknown label is usually set to 0.

Therefore, considering the previous description of lineage set and base price, we formulate a basic pricing function for incomplete graph query.

Definition 5 (Basic pricing function). Given an incomplete graph $\mathbb{G}$, the price of query Q, denoted by $p_{\mathbb{G}}(Q)$, is defined as the sum of the price of each query result's lineage set.

$$p_{\mathbb{G}}(Q) = \sum_{r \in Q(\mathbb{G})} p(G_r) \tag{2}$$

where G_r is the lineage set of query result r, and $p(G_r) = \sum_{e \in G_r} \rho(e)$ is the price of G_r, which represents the total base prices of edges in G_r.

Example 2. As shown in Fig. 2, each labeled edge in $\mathbb{G}$ is assigned a base price marked in red. We can calculate the price of each query over $\mathbb{G}$ based on Definition 5, where $p_{\mathbb{G}}(Q_1) = p(G_1) + p(G_2) = 3 + 2 = 5$, $p_{\mathbb{G}}(Q_2) = p(G_3) = 2$ and $p_{\mathbb{G}}(Q_3) = p(G_4) + p(G_5) = 5 + 4 = 9$.

3.2 Arbitrage-Free Pricing Function

Although the pricing function based on lineage set can meet the basic requirements of incomplete graph query pricing, it also has some shortcomings that need to be improved. One is the *arbitrage-free* property, which is essential for data pricing [10]. This is because if there is an arbitrage opportunity in a pricing mechanism, some shrewd buyers may benefit from it. Formally, the arbitrage-free property is defined based on a notion of *query determinacy*.

Given two query bundles $\mathbf{Q_1}$ and $\mathbf{Q_2}$, which may contain multiple queries. If the answers of $\mathbf{Q_2}$ can be computed by applying a function f to the answers of $\mathbf{Q_1}$ without having access to graph data $\mathbb{G}$, such that $\mathbf{Q_2}(\mathbb{G}) = f(\mathbf{Q_1}(\mathbb{G}))$, we say that $\mathbf{Q_1}$ determines $\mathbf{Q_2}$ under $\mathbb{G}$, denoted by $\mathbb{G} \vdash \mathbf{Q_1} \twoheadrightarrow \mathbf{Q_2}$. Therefore, it is necessary for the pricing function to ensure that $p_{\mathbb{G}}(\mathbf{Q_1}) \geq p_{\mathbb{G}}(\mathbf{Q_2})$ to avoid arbitrage. Otherwise, the buyer will choose to purchase $\mathbf{Q_1}$ at a lower price instead of purchasing $\mathbf{Q_2}$. As mentioned above, the answer of Q_3 can be computed by performing join operation on $Q_1(\mathbb{G})$ and $Q_2(\mathbb{G})$. Thus, the query determinacy relation among the three queries is $\mathbb{G} \vdash Q_1, Q_2 \twoheadrightarrow Q_3$. According to Example 2, we notice that $p_{\mathbb{G}}(Q_1) + p_{\mathbb{G}}(Q_2) < p_{\mathbb{G}}(Q_3)$, which means the basic pricing function provides an arbitrage opportunity for buyers.

Therefore, we consider improving the basic pricing function proposed in the last subsection to make it arbitrage-free. We observe that the lineage sets of different query results may have the same edges. The price of such an edge will be recounted while calculating the query price, resulting in arbitrage opportunities.

In order to remove the duplicate edges, we further define the lineage set of query Q, denoted by $l(Q, \mathbb{G})$, where $l(Q, \mathbb{G}) = G_Q = \bigcup_{r \in Q(\mathbb{G})} G_r$. Based on the query lineage set, we propose an arbitrage-free pricing function.

Definition 6 (Arbitrage-free pricing function). Given an incomplete graph $\mathbb{G}$, the arbitrage-free price of query Q, denoted by $p_{\mathbb{G}}^a(Q)$, is defined as the total base prices of edges in G_Q.

$$p_{\mathbb{G}}^a(Q) = \sum_{e \in G_Q} \rho(e) \tag{3}$$

After the preceding analysis, the reason why $p_{\mathbb{G}}(Q_3)$ has an arbitrage opportunity is that the base prices of edge (v_5, v_6) and (v_6, v_7) are counted twice. According to Definition 6, we can get $p_{\mathbb{G}}^a(Q_3) = 9 - 1 - 1 = 7$, while $p_{\mathbb{G}}^a(Q_1)$ and $p_{\mathbb{G}}^a(Q_2)$ remain the same. Thus, they satisfy the arbitrage-free condition of $p_{\mathbb{G}}^a(Q_1) + p_{\mathbb{G}}^a(Q_2) \geq p_{\mathbb{G}}^a(Q_3)$.

Theorem 1. *The pricing function $p_{\mathbb{G}}^a(Q)$ defined by Eq. (3) is arbitrage-free.*

Proof. According to the definition of query determinacy, if $\mathbf{Q_1}$ determines $\mathbf{Q_2}$ under $\mathbb{G}$, there is a function f such that $\mathbf{Q_2}(\mathbb{G}) = f(\mathbf{Q_1}(\mathbb{G}))$. It is obvious that the query lineage sets of $\mathbf{Q_1}$ must cover the query lineage sets of $\mathbf{Q_2}$. Meanwhile, the pricing function $p_{\mathbb{G}}^a(Q)$ depends on the total base prices of edges in the query lineage set, which means we can derive $p_{\mathbb{G}}(\mathbf{Q_1}) \geq p_{\mathbb{G}}(\mathbf{Q_2})$. Therefore, the pricing function $p_{\mathbb{G}}^a(Q)$ is arbitrage-free. $\square$

Finding subgraphs matching the given graph pattern is a NP-hard problem, and many methods have been proposed to solve it based on a generic subgraph matching solution [11]. Thus, we propose an arbitrage-free pricing algorithm in order to derive the match set and price effectively. The details of our algorithm are shown in Algorithm 1. First, we retrieve a set of candidate nodes $\phi(u)$ for each node variable u in graph pattern GP according to specific labels of edges connected with u (lines 1–2). Next, we can use a greedy strategy to determine the query processing order, which gives priority to choosing a node variable with smaller candidate size (line 3). Then, we recursively call a *Search* subroutine to retrieve the set of matches M (line 5). After that, we use the query result r corresponding to each match m to update $Q(\mathbb{G})$. Meanwhile, we need to check whether the edges of lineage set G_r exist in query lineage set G_Q. If not exist, we also update $p_{\mathbb{G}}^a(Q)$ (lines 7–13).

For the subroutine *Search*, the recursion ends when each node variable matches a specific data node (lines 16–18). Otherwise, we get the node variable u_i to be matched. For each candidate node $v \in \phi(u_i)$ where v is not matched yet, we need to check whether the edges between u_i and already matched node variables in GP have corresponding edges between v and already matched data nodes in $\mathbb{G}$. If v is qualified, we update the matching information and continue to process the remaining node variables by recursively calling *Search*. Then, we restore the matching information to the status before updating (lines 20–25). The subroutine *Search* terminates when all correct matches are found.

Algorithm 1: Arbitrage-free Pricing

Input: data graph $\mathbb{G}$, and graph query Q.
Output: query answer $Q(\mathbb{G})$, and arbitrage-free price $p_{\mathbb{G}}^a(Q)$.

1 **for** each node variable u in GP **do**
2 $\quad$ generate candidate node set $\phi(u)$;

3 optimize search order $U \leftarrow \{u_1, u_2, ..., u_n\}$;
4 $M \leftarrow \emptyset$;
5 $Search(1, \emptyset, M)$;
6 $G_Q \leftarrow \emptyset$, $Q(\mathbb{G}) \leftarrow \emptyset$, $p_{\mathbb{G}}^a(Q) \leftarrow 0$;
7 **for** each match m in M **do**
8 $\quad$ generate query result r and its lineage set G_r;
9 $\quad$ $Q(\mathbb{G}) \leftarrow Q(\mathbb{G}) \cup \{r\}$;
10 $\quad$ **for** each edge $e \in G_r$ **do**
11 $\quad\quad$ **if** $e \notin G_Q$ **then**
12 $\quad\quad\quad$ $G_Q \leftarrow G_Q \cup \{e\}$;
13 $\quad\quad\quad$ $p_{\mathbb{G}}^a(Q) \leftarrow p_{\mathbb{G}}^a(Q) + \rho(e)$;

14 **return** $Q(\mathbb{G})$, $p_{\mathbb{G}}^a(Q)$;

15 **void** $Search(i, m, M)$
16 **if** $|m| = |U|$ **then**
17 $\quad$ $M \leftarrow M \cup \{m\}$;
18 $\quad$ **return**;
19 **else**
20 $\quad$ get current node variable u_i;
21 $\quad$ **for** each candidate node $v \in \phi(u_i)$ which is not yet matched **do**
22 $\quad\quad$ **if** v is joinable **then**
23 $\quad\quad\quad$ $m \leftarrow m \cup \{(u_i, v)\}$;
24 $\quad\quad\quad$ $Search(i + 1, m, M)$;
25 $\quad\quad\quad$ $m \leftarrow m/\{(u_i, v)\}$;

For arbitrage-free pricing algorithm, the computation phase of query price $p_{\mathbb{G}}^a(Q)$ has a time complexity of $O(|M||E_Q|)$, where M is the match set and E_Q is the edge set of Q.

Example 3. Let's take query Q_1 as an example. Table 2 shows the preorder node list and postorder node list based on edge labels, which help us to retrieve candidate nodes. For example, the candidate set for node variable x_2 is equivalent to the common part between postorder nodes of label a and preorder nodes of label b, i.e., $\phi(x_2) = \{v_2, v_4, v_8\} \cap \{v_2, v_4, v_9\} = \{v_2, v_4\}$. Similarly, we can get $\phi(x_1) = \{v_1, v_3\}$ and $\phi(x_3) = \{v_5, v_7\}$. Next, we optimize the search order by choosing a node variable with smaller candidate size preferentially, i.e., $\{x_1, x_2, x_3\}$. After that, we collect all possible matches by recursively calling $Search$, i.e., $M = \{\{(x_1, v_1), (x_2, v_2), (x_3, v_5)\}, \{(x_1, v_3), (x_2, v_4), (x_3, v_5)\}\}$. Finally, we can easily derive both query answer $Q(\mathbb{G})$ and arbitrage-free price $p_{\mathbb{G}}^a(Q)$ according to match set M, i.e., $Q_1(\mathbb{G}) = \{(v_1, v_5), (v_3, v_5)\}$ and $p_{\mathbb{G}}^a(Q_1) = 5$.

Table 2. The inverted edge label index of $\mathbb{G}$

Label	Preorder node	Postorder node
a	$\{v_1, v_3\}$	$\{v_2, v_4, v_8\}$
b	$\{v_2, v_4, v_9\}$	$\{v_5, v_7\}$
c	$\{v_5\}$	$\{v_6\}$
d	$\{v_6\}$	$\{v_7\}$
null	$\{v_3, v_8\}$	$\{v_7, v_9\}$

3.3 Discount Pricing Function

In addition, we further discuss the impact of query quality on the price. Specifically, it means that the information loss of graph data may lead to the lack of results in the query answer. Let's take query Q_1 as an example. As shown in Fig. 2, the label of edge (v_1, v_8) is a, and the label of edge (v_8, v_7) is unknown. For each complete graph data G_i with edge (v_8, v_7) labeled b, $Q_1(G_i)$ will contain result (v_1, v_7). However, $Q_1(\mathbb{G})$ only takes intersection of answers from possible worlds according to Eq. (1). Therefore, we regard (v_1, v_7) as the uncertain result of Q_1. Similarly, (v_3, v_7) is also an uncertain result.

We define the subgraph that contributes to an uncertain result as the *uncertain subgraph* of the query. For the convenience of discussion, we define the subgraph that contributes to a query result as the *certain subgraph* as well. In order to compare the difference between two types of subgraphs, we propose a metric to quantify the subgraph's quality. For a subgraph $\mathbb{G}'$ of $\mathbb{G}$, let α be the number of unknown labels in $\mathbb{G}'$. The quality of subgraph $\mathbb{G}'$, denoted by $\epsilon(\mathbb{G}')$, is defined as $\epsilon(\mathbb{G}') = q^\alpha$ where constant q $(0 < q < 1)$ represents the sensitivity of subgraph quality to α, which can be specified by the data market. Notice that the smaller the value of q, the quickly $\epsilon(\mathbb{G}')$ decreases with α, and the uncertain subgraph with more unknown labels will have a lower quality. Based on the notion of subgraph quality, we further quantify the impact of incomplete graph data on query quality and propose a novel metric, namely *certainty coefficient*.

Definition 7 (Certainty coefficient). Given an incomplete graph $\mathbb{G}$ and a query Q. Let $\mathbf{G_c}$ be the set of Q's certain subgraphs and $\mathbf{G_u}$ be the set of Q's uncertain subgraphs, the certainty coefficient of query Q, denoted by $C_{\mathbb{G}}(Q)$, is defined as:

$$C_{\mathbb{G}}(Q) = \frac{\sum_{G \in \mathbf{G_c}} \epsilon(G) + \sum_{G \in \mathbf{G_u}} \epsilon(G)}{|\mathbf{G_c}| + |\mathbf{G_u}|} \tag{4}$$

Example 4. We still take query Q_1 as an example, and assume that $q = \frac{1}{2}$. First, Q_1 has two certain subgraphs, corresponding to its two result lineage sets G_1 and G_2 respectively. Then, Q_1 also has two uncertain subgraphs, both of which contain one edge with unknown label, and their subgraph quality is equal to $\frac{1}{2}$. Finally, according to Eq. (4), the certainty coefficient of query Q_1 is $C_{\mathbb{G}}(Q_1) = \frac{1+1+\frac{1}{2}+\frac{1}{2}}{2+2} = 0.75$.

Algorithm 2: Discount Pricing

Input: data graph $\mathbb{G}$, and graph query Q.
Output: query answer $Q(\mathbb{G})$, and discount price $p_{\mathbb{G}}^d(Q)$.

1 **for** each node variable u in GP **do**
2   generate extended candidate node set $\phi^*(u)$;

3 optimize search order $U \leftarrow \{u_1, u_2, ..., u_n\}$;
4 $M_c \leftarrow \emptyset, M_u \leftarrow \emptyset$;
5 $Search^*(1, \emptyset, M_c, M_u)$;
6 derive $Q(\mathbb{G})$ and $p_{\mathbb{G}}^a(Q)$ based on M_c;
7 calculate $C_{\mathbb{G}}(Q)$ based on Eq. (4);
8 $p_{\mathbb{G}}^d(Q) \leftarrow C_{\mathbb{G}}(Q) \cdot p_{\mathbb{G}}^a(Q)$;
9 **return** $Q(\mathbb{G}), p_{\mathbb{G}}^d(Q)$;

10 **void** $Search^*(i, m, M_c, M_u)$
11 **if** $|m| = |U|$ **then**
12   **if** m corresponds to a subgraph without unknown labels **then**
13        $M_c \leftarrow M_c \cup \{m\}$;
14   **else**
15        $M_u \leftarrow M_u \cup \{m\}$;
16   **return**;
17 **else**
18   get current node variable u_i;
19   **for** each candidate node $v \in \phi^*(u_i)$ which is not yet matched **do**
20   **if** v is joinable **then**
21           $m \leftarrow m \cup \{(u_i, v)\}$;
22           $Search^*(i+1, m, M_c, M_u)$;
23           $m \leftarrow m/\{(u_i, v)\}$;

Actually, $C_{\mathbb{G}}(Q)$ represents the average quality of both certain and uncertain subgraphs. The more uncertain subgraphs, the smaller the value of $C_{\mathbb{G}}(Q)$, indicating that the query involves more uncertain results. As mentioned above, we consider taking $C_{\mathbb{G}}(Q)$ as a discount factor to make up for the buyer's loss caused by the lack of some potential results in the query answer, and propose a discount pricing function.

Definition 8 (Discount pricing function). Given an incomplete graph $\mathbb{G}$, the discount price of query Q, denoted by $p_{\mathbb{G}}^d(Q)$, is defined as:

$$p_{\mathbb{G}}^d(Q) = C_{\mathbb{G}}(Q) \cdot p_{\mathbb{G}}^a(Q) \tag{5}$$

Different from the arbitrage-free price, the discount price requires a discount factor $C_{\mathbb{G}}(Q)$. The key point of calculating $C_{\mathbb{G}}(Q)$ is to derive Q's uncertain subgraphs additionally. Therefore, we propose a discount pricing algorithm based on Algorithm 1. The details are shown in Algorithm 2. For each node variable u in graph pattern GP, we need to expand its candidate node set based on all possible uncertain subgraph structures (lines 1–2). Next, we recursively call another subroutine $Search^*$ to retrieve match set M_c corresponding to certain subgraphs and match set M_u corresponding to uncertain subgraphs (lines 12–15). After that, we can derive $Q(\mathbb{G})$ and $p^a_{\mathbb{G}}(Q)$ according to the process of Algorithm 1, and calculate $C_{\mathbb{G}}(Q)$ to adjust the query price (lines 6–8).

For discount pricing algorithm, the computation phase of certainty coefficient $C_{\mathbb{G}}(Q)$ has a time complexity of $O(|M_c| + |M_u|)$, where M_c is the match set of certain subgraphs and M_u is the match set of uncertain subgraphs.

Example 5. Since Table 2 also gives node lists of unknown label, we expand the candidate node set of each node variable to derive query's uncertain subgraphs. For example, query Q_1 corresponds to three possible uncertain subgraph structures. For node variable x_2, the candidate set needs to be appended: (1) the common part between postorder nodes of label a and preorder nodes of **null**, i.e., $\{v_8\}$, (2) the common part between postorder nodes of **null** and preorder nodes of label b, i.e., $\{v_9\}$ and (3) the common part between postorder nodes of **null** and preorder nodes of **null**, i.e., $\emptyset$. Therefore, the extended candidate node set $\phi^*(x_2)$ is $\{v_2, v_4, \boldsymbol{v_8}, \boldsymbol{v_9}\}$. Similarly, we can get $\phi^*(x_1) = \phi(x_1) \cup \{v_3, v_8\} = \{v_1, v_3, \boldsymbol{v_8}\}$ and $\phi^*(x_3) = \phi(x_3) \cup \{v_7, v_9\} = \{v_5, v_7, \boldsymbol{v_9}\}$. Next, we recursively call $Search^*$, and get match sets $M_c = \{\{(x_1, v_1), (x_2, v_2), (x_3, v_5)\}, \{(x_1, v_3), (x_2, v_4), (x_3, v_5)\}\}$, $M_u = \{\{(x_1, v_1), (x_2, v_8), (x_3, v_7)\}, \{(x_1, v_3), (x_2, v_9), (x_3, v_7)\}\}$. Finally, we calculate each uncertain subgraph's quality to get certainty coefficient $C_{\mathbb{G}}(Q_1)$, and derive the discount price such that $p^d_{\mathbb{G}}(Q_1) = C_{\mathbb{G}}(Q_1) \cdot p^a_{\mathbb{G}}(Q_1) = 0.75 \cdot 5 = 3.75$.

4   Experiments

In this section, we give an overview of details in our experimental settings, and present experimental results to evaluate our proposed methods.

4.1   Experimental Settings

Datasets. We choose four real knowledge graph datasets: FB15k-237 [12], WN18RR [12], YAGO3-10 [13] and DBpedia500k [14]. Detailed information of datasets is shown in Table 3. These four datasets are convenient for us to validate the scalability of our approach by changing the size of graphs. Meanwhile, label-dense and label-sparse graphs can also be used to compare the effect of label size on the performance.

Table 3. Statistics of datasets in the experiments

Dataset	Nodes	Edges	Labels
FB15k-237	14541	272115	237
WN18RR	40943	86835	11
YAGO3-10	123182	1079040	37
DBpedia500k	517475	3102677	654

Configurations. For each dataset, we randomly select a certain proportion of edges and set their labels to **null**. The missing rate, which represents the ratio of the number of missing labels to the number of edges in the dataset, is set to 0.1% with no special instructions. Meanwhile, the base price of each edge is set to 1, and parameter q in the subgraph quality function is set to 0.5 by default. Without loss of generality, we select six types of graph queries with different scales. Each type of graph query corresponds to five specific query requests, which are run 10 times to report the average result. The algorithms are implemented in Python 3.6. The experiments are conducted on a computer with Intel(R) Core(TM) i5-9500 CPU @ 3.00 GHz and 16 GB main memory.

4.2 Experimental Results

4.2.1 Efficiency of Pricing Algorithms

We evaluate the running time of three pricing algorithms for queries with different scales. Specifically, the query scale is expressed as each pair $(|V_Q|, |E_Q|)$ on the x-axis. As shown in Fig. 4, the efficiency of basic pricing algorithm is similar to that of arbitrage-free pricing algorithm. Discount pricing algorithm needs to retrieve uncertain subgraphs to calculate the certainty coefficient, and thus it will take longer to get the query price. It is noteworthy that with the increasing number of edges in Q, the efficiency gap between discount pricing algorithm and arbitrage-free algorithm is gradually widening. The direct reason we analyzed is that the more edges involved in a graph query, the more potential structures of uncertain subgraphs. Accordingly, each query node variable will correspond to more situations while expanding its candidate set. Therefore, discount pricing algorithm will consume a large amount of time in retrieving candidate nodes, rather than recursively calling the search subroutine. Especially, since both WN18RR and YAGO3-10 have a small number of labels, it is easy to cause candidate sets with large scale and affect the overall running time.

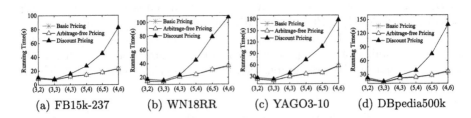

(a) FB15k-237   (b) WN18RR   (c) YAGO3-10   (d) DBpedia500k

Fig. 4. Running time for queries with different scales

4.2.2 Effect of Missing Rate on Efficiency

Next, we evaluate the effect of label missing rate on pricing algorithms' efficiency. In this part, we only select one triangle query whose scale is $(3,3)$ for each dataset to test the running time. As shown in Fig. 5, when the initial missing

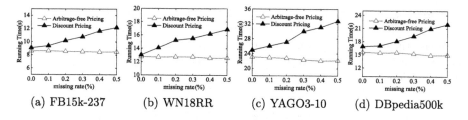

(a) FB15k-237   (b) WN18RR   (c) YAGO3-10   (d) DBpedia500k

Fig. 5. Running time under different missing rates

rate is 0, the running time of arbitrage-free and discount pricing algorithms are similar. This is because both of them are run on the complete graph data. With the increasing missing rate, the running time of discount pricing algorithm is gradually growing compared with arbitrage-free pricing algorithm. It is obvious that a higher missing rate is likely to produce more uncertain subgraphs for a query, thus affecting the efficiency of discount pricing algorithm.

4.2.3 Effect of Missing Rate on Prices

Furthermore, we continue to analyze the effect of missing rate on query prices. The triangle query for each dataset is consistent with the previous part. When the initial missing rate is 0, the prices derived by two algorithms must be the same. There will be no potential uncertain results for queries over complete graph data, and the certainty coefficient is equal to 1. Since we randomly erase a certain number of labels, any missing rate may reduce the size of query results, which leads to a cheaper arbitrage-free price. In Fig. 6(a) and Fig. 6(b), the experimental results of FB15k-237 and WN18RR prove such conclusion. For discount pricing algorithm, more uncertain subgraphs due to higher missing rate directly lower the value of certainty coefficient. Therefore, the smaller certainty coefficient makes the discount price gradually cheaper.

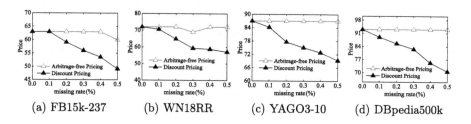

(a) FB15k-237   (b) WN18RR   (c) YAGO3-10   (d) DBpedia500k

Fig. 6. Query prices under different missing rates

4.2.4 Effect of Mechanism Parameter

Finally, we analyze the effect of parameter q in the subgraph quality function on certainty coefficient $C_{\mathbb{G}}(Q)$. The triangle query for each dataset is consistent with the previous part. The change trend of the certainty coefficient with q varying from 0.1 to 0.9 is shown in Fig. 7. As expected, a smaller value of q will directly decrease the quality of each uncertain subgraph. More importantly, for the same query of one dataset, the certainty coefficient increases linearly with q according to Definition 7.

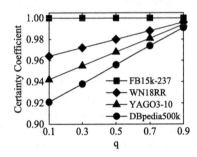

Fig. 7. Certainty coefficient under different q

5   Related Work

Data Pricing. The development of data markets has led to efficient data trading [2,15,16]. Especially, the query-based pricing model is a typical pricing solution in data markets, which can customize the data purchase according to the consumer's needs. Koutris et al. [17] proposed a query-based framework where a seller only needs to specify the prices of a few views, and then the price of any query can be derived automatically. Subsequently, Koutris et al. [18] further developed an integer linear programming formulation for computing prices of a large number of queries. Lin et al. [19] designed three types of arbitrage-free pricing functions for general data queries. Deep et al. [20] designed a novel pricing system, called QIRANA, which employs the possible world semantic to price a large class of SQL queries (including aggregation) in real time. Specific to data quality, Miao et al. [21] proposed two novel pricing functions for queries over incomplete relational data. And specific to data privacy, Ghosh et al. [22] considered selling privacy at auction when consumers want to purchase statistics. The follow-up work [23–25] further developed frameworks of pricing linear aggregate queries under noise perturbation.

Subgraph Isomorphism. Subgraph isomorphism has been an important problem in various applications, and many researches on subgraph matching algorithms are proposed to improve the search efficiency [11,26]. Ullmann [27] is the first practical algorithm to retrieve isomorphic subgraphs, which is based on the depth-first search. VF2 [28] selects the matching order by connection sequence, and exploits three pruning rules to prune out candidate nodes. GraphQL [29] proposed a new pattern-based graph query language to allow for flexible compositions of graph structures. QuickSI [30] introduced a novel tree-based matching sequence called QI-Sequence, which is used to prune out candidate nodes. GADDI [31] defines the concept of neighboring discriminating substructure (NDS) distance to reduce the candidate sets. SPath [32] takes a different approach, which matches a query path rather than a query node each time. [33–35] proposed approximate subgraph matching methods to address the limitation of exact matching, which allow some of nodes or edges not being matched exactly.

6 Conclusion

In this paper, we study the problem of query pricing on incomplete graph data. We design a basic pricing function based on the idea of data provenance. In order to satisfy the essential arbitrage-free property of data pricing, we propose an arbitrage-free pricing function based on the notion of query lineage set. Meanwhile, considering the query quality of incomplete graph data, we design a novel metric called certainty coefficient to adjust the query price. According to the generic subgraph matching solution, we design feasible algorithms to derive the query answer and price. Extensive experiments on real graph datasets demonstrate the effectiveness and efficiency of our pricing mechanism. In the future, we will study the query pricing theory for other types of incomplete graph data, such as the scenario of node information loss. Furthermore, we plan to design an acceptable compensation mechanism for data owners.

Acknowledgements. Ye Yuan is supported by the National Key Research and Development Program of China (Grant No. 2022YFB2702100) and the NSFC (Grant Nos. 61932004, 62225203, U21A20516).

References

1. Muschalle, A., Stahl, F., Löser, A., Vossen, G.: Pricing approaches for data markets. In: Castellanos, M., Dayal, U., Rundensteiner, E.A. (eds.) BIRTE 2012. LNBIP, vol. 154, pp. 129–144. Springer, Heidelberg (2013). https://doi.org/10.1007/978-3-642-39872-8_10
2. Balazinska, M., Howe, B., Suciu, D.: Data markets in the cloud: an opportunity for the database community. Proc. VLDB Endow. **4**(12), 1482–1485 (2011)
3. Fernandez, R.C., Subramaniam, P., Franklin, M.J.: Data market platforms: trading data assets to solve data problems. Proc. VLDB Endow. **13**(12), 1933–1947 (2020)
4. Leser, U.: A query language for biological networks. Bioinformatics **21**(2), 33–39 (2005)
5. Ronen, R., Shmueli, O.: SoQL: a language for querying and creating data in social networks. In: ICDE, pp. 1595–1602. IEEE (2009)
6. Gutierrez, C., Hurtado, C.A., Mendelzon, A.O., Pérez, J.: Foundations of semantic web databases. J. Comput. Syst. Sci. **77**(3), 520–541 (2011)
7. Miao, X., Gao, Y., Guo, S., Liu, W.: Incomplete data management: a survey. Front. Comp. Sci. **12**(1), 4–25 (2018). https://doi.org/10.1007/s11704-016-6195-x

8. Barceló, P., Libkin, L., Reutter, J.L.: Querying graph patterns. In: PODS, pp. 199–210. ACM (2011)
9. Cui, Y., Widom, J.: Practical lineage tracing in data warehouses. In: ICDE, pp. 367–378. IEEE (2000)
10. Deep, S., Koutris, P.: The design of arbitrage-free data pricing schemes. arXiv preprint arXiv:1606.09376 (2016)
11. Lee, J., Han, W.S., Kasperovics, R., Lee, J.H.: An in-depth comparison of subgraph isomorphism algorithms in graph databases. Proc. VLDB Endow. 6(2), 133–144 (2012)
12. Bordes, A., Usunier, N., Garcia-Duran, A., Weston, J., Yakhnenko, O.: Translating embeddings for modeling multi-relational data. In: NIPS, pp. 2787–2795 (2013)
13. Mahdisoltani, F., Biega, J., Suchanek, F.M.: YAGO3: a knowledge base from multilingual Wikipedias. In: CIDR (2015)
14. Lehmann, J., et al.: DBpedia - a large-scale, multilingual knowledge base extracted from Wikipedia. Semant. Web 6(2), 167–195 (2015)
15. Liang, F., Yu, W., An, D., Yang, Q., Fu, X., Zhao, W.: A survey on big data market: pricing, trading and protection. IEEE Access 6, 15132–15154 (2018)
16. Pei, J.: Data pricing - from economics to data science. In: SIGKDD, pp. 3553–3554. ACM (2020)
17. Koutris, P., Upadhyaya, P., Balazinska, M., Howe, B., Suciu, D.: Query-based data pricing. In: PODS, pp. 167–178. ACM (2012)
18. Koutris, P., Upadhyaya, P., Balazinska, M., Howe, B., Suciu, D.: Toward practical query pricing with querymarket. In: SIGMOD, pp. 613–624. ACM (2013)
19. Lin, B.R., Kifer, D.: On arbitrage-free pricing for general data queries. Proc. VLDB Endow. 7(9), 757–768 (2014)
20. Deep, S., Koutris, P.: QIRANA: a framework for scalable query pricing. In: SIGMOD, pp. 699–713. ACM (2017)
21. Miao, X., Gao, Y., Chen, L., Peng, H., Yin, J., Li, Q.: Towards query pricing on incomplete data. TKDE 34(8), 4024–4036 (2022)
22. Ghosh, A., Roth, A.: Selling privacy at auction. In: EC, pp. 199–208. ACM (2011)
23. Li, C., Li, D.Y., Miklau, G., Suciu, D.: A theory of pricing private data. TODS 39(4), 1–28 (2014)
24. Niu, C., Zheng, Z., Wu, F., Tang, S., Gao, X., Chen, G.: Unlocking the value of privacy: trading aggregate statistics over private correlated data. In: SIGKDD, pp. 2031–2040. ACM (2018)
25. Niu, C., Zheng, Z., Tang, S., Gao, X., Wu, F.: Making big money from small sensors: trading time-series data under pufferfish privacy. In: INFOCOM, pp. 568–576 (2019)
26. Fan, W., Li, J., Ma, S., Tang, N., Wu, Y., Wu, Y.: Graph pattern matching: from intractable to polynomial time. Proc. VLDB Endow. 3(1), 264–275 (2010)
27. Ullmann, J.R.: An algorithm for subgraph isomorphism. J. ACM 23(1), 31–42 (1976)
28. Cordella, L., Foggia, P., Sansone, C., Vento, M.: A (sub)graph isomorphism algorithm for matching large graphs. IEEE Trans. Pattern Anal. Mach. Intell. 26(10), 1367–1372 (2004)
29. He, H., Singh, A.K.: Graphs-at-a-time: query language and access methods for graph databases. In: SIGMOD, pp. 405–418. ACM (2008)
30. Shang, H., Zhang, Y., Lin, X., Yu, J.X.: Taming verification hardness: an efficient algorithm for testing subgraph isomorphism. Proc. VLDB Endow. 1(1), 364–375 (2008)

31. Zhang, S., Li, S., Yang, J.: GADDI: distance index based subgraph matching in biological networks. In: EDBT, pp. 192–203 (2009)
32. Zhao, P., Han, J.: On graph query optimization in large networks. Proc. VLDB Endow. **3**(1), 340–351 (2010)
33. Tian, Y., Patel, J.M.: TALE: a tool for approximate large graph matching. In: ICDE, pp. 963–972. IEEE (2008)
34. Zhang, S., Yang, J., Jin, W.: SAPPER: subgraph indexing and approximate matching in large graphs. Proc. VLDB Endow. **3**(1), 1185–1194 (2010)
35. Khan, A., Li, N., Yan, X., Guan, Z., Chakraborty, S., Tao, S.: Neighborhood based fast graph search in large networks. In: SIGMOD, pp. 901–912. ACM (2011)

Answering Label-Constrained Reachability Queries via Reduction Techniques

Yuzheng Cai and Weiguo Zheng[(✉)]

School of Data Science, Fudan University, Shanghai, China
yuzhengcai21@m.fudan.edu.cn, zhengweiguo@fudan.edu.cn

Abstract. Many real-world graphs contain edge labels for representing different types of relations between nodes, such as social networks, biological networks and knowledge graphs. As a fundamental task, the label-constrained reachability (LCR) query asks whether a given vertex s can reach another vertex t, only using a restricted set of given edge labels. However, existing works build a heavy index while taking too much time for answering queries online, exhibiting a poor performance on large graphs. To further reduce the index size and accelerate both index construction and query processing, we introduce two novel pruning techniques including degree-one reduction and unreachable query filter. Extensive experiments demonstrate that our proposed techniques significantly boost state-of-the-art methods.

Keywords: Label-constrained reachability · degree-one reduction · unreachable query filter

1 Introduction

To model different types of relationships between nodes, edges in many real-world graphs are usually labeled, such as social networks, biological networks and knowledge graphs [3,8,9,14,24]. For example, a social network might contain several edge labels such as "follow", "friendOf", and "relativeOf". As one of the most fundamental tasks for graph mining, the traditional reachability query asks whether a vertex can reach another vertex [12,15,19,22]. However, in many real-world scenarios, users may only care about the reachability under certain edge types, i.e., asking whether a vertex s can reach another vertex t just using a subset of edge labels $\mathcal{L}$. Such query is called *label-constrained reachability* (LCR) [3], denoted by $s \xrightarrow{?\mathcal{L}} t$. Take the graph G in Fig. 1 as an example, given a query $0 \xrightarrow{?\mathcal{L}} 12$ where $\mathcal{L} = \{a, c\}$, the answer is Yes since there is an s-t path $0 \xrightarrow{c} 1 \xrightarrow{a} 2 \xrightarrow{c} 12$. But when $\mathcal{L} = \{a, b\}$, the answer to the query $0 \xrightarrow{?\mathcal{L}} 12$ is No.

LCR queries have a wide range of applications, such as counter-terrorism on social networks, checking chain of protein interactions on biological networks,

X. Wang et al. (Eds.): DASFAA 2023, LNCS 13943, pp. 114–131, 2023.
https://doi.org/10.1007/978-3-031-30637-2_8

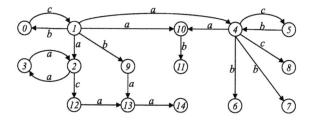

Fig. 1. An example of a directed edge-labeled graph G

regular path query on knowledge graphs and so on [3,9,14,24]. However, building an effective index for LCR query is much more complicated than the traditional reachability query, since there are $2^{|L|}$ possible combinations for the label constraint of a query, where $|L|$ is the number of edge labels in graph. Among the existing approaches [3,8,9,14,24], Pruned 2-hop index (P2H$^+$) [8,9] is the state-of-the-art method, which encodes the exact transitive closure between all pairs of vertices under all combinations of edge labels. Specifically, for each vertex v, P2H$^+$ materializes hop nodes u which can reach v or can be reached by v with corresponding label sets. For query $s \xrightarrow{?\mathcal{L}} t$, it returns Yes iff there exists a hop node u s.t. s can reach u and u can reach t under label constraint $\mathcal{L}$.

Though P2H$^+$ significantly outperforms all the previous approaches, in this paper we find that unnecessary costs are wasted in both its online query and offline index phases. Based on our discoveries, the following two reduction techniques are proposed to further improve the time efficiency for processing LCR queries. In summary, we make the following contributions in this paper.

Degree-One Reduction. Many real-world graphs are scale-free networks following the power-law degree distribution [7], and there may be lots of vertices whose in-degree or out-degree is 1. In P2H$^+$ index, we notice that if vertex v has in-degree (out-degree) one, all its in-coming (out-coming) hop nodes and label sets can be inferred from its unique parent (child). Such redundant entries should be removed to reduce index size (e.g., Table 1), and we further design new index algorithms to directly generate the reduced index instead of removing entries in the built P2H$^+$ index. Though it seems to be intuitive, the devil is in discussing all possible cases to make such index *sound, complete* and also *minimal*.

Unreachable Query Filter. The answer to $s \xrightarrow{?\mathcal{L}} t$ is definitely No if vertex s cannot reach t even without any label constraints. However, P2H$^+$ needs to iterate all index entries of s and t before returning No, which can be accelerated by introducing another light-weight index to rule out most of such queries in $\mathcal{O}(1)$ time. It contains three coordinates from FELINE index [15], along with two new coordinates designed for enhancing the pruning ability.

Extensive experiments demonstrate that equipped with above reduction techniques, answering online queries is much faster than P2H$^+$. Also, both indexing time and index size can be greatly reduced thanks to the degree-one reduction.

2   Problem Definition and Preliminary

We first define the problem of label-constrained reachability (LCR), and then introduce the state-of-the-art Pruned 2-hop method (P2H$^+$) [8,9].

2.1   Problem Definition

Let $G(V, E, L)$ denote a directed edge-labeled graph, where V is the vertex set, L is the label set, and $E \subseteq V \times V \times L$ is the edge set. Let $e(u, v, l) \in E$ represent a directed edge from vertex u to v with label l. The number of vertices, edges, and labels are denoted as $|V|$, $|E|$, and $|L|$, respectively. We denote a function $\lambda : E \to L$ mapping an edge to its label, i.e., $\lambda(e(u, v, l)) = l$. For any vertex $v \in V$, we denote its in-neighbor set under a specific label l as $N_{in}(v, l) = \{u \mid e(u, v, l) \in E\}$. Similarly, the out-neighbor set of v under label l is denoted by $N_{out}(v, l) = \{u \mid e(v, u, l) \in E\}$. And the in-degree of vertex v is denoted by $d_{in}(v) = \sum_{l \in L} |N_{in}(v, l)|$, while its out-degree $d_{out}(v) = \sum_{l \in L} |N_{out}(v, l)|$.

Given the source vertex s and target vertex t, an s-t path $p(s, t) = \{s = v_0, e_0, v_1, e_1, \cdots, v_{k-1}, e_{k-1}, v_k = t\}$ denotes a directed sequence $(k > 0)$, where $v_i \in V$ and $e_i \in E$ for every i. Given a label set $\mathcal{L} \subseteq L$, we say that $p(s, t)$ is an $\mathcal{L}$-constrained s-t path if and only if $\forall e_i \in p(s, t)$, $\lambda(e_i) \in \mathcal{L}$.

Definition 1 *(Label-constrained Reachability Query, shorted as LCR [14]). For a directed edge-labeled graph $G(V, E, L)$, given a query $s \xrightarrow{?\mathcal{L}} t$, where s and t are two distinct vertices in G and label set $\mathcal{L} \subseteq L$, the task is to determine whether there exists an $\mathcal{L}$-constrained s-t path in G.*

An LCR query asks whether vertex s can reach t only using labels in $\mathcal{L}$. For ease of presentation, we use $s \xrightarrow{\mathcal{L}} t$ to represent that s can reach t only using labels in $\mathcal{L}$, while $s \xmapsto{\mathcal{L}} t$ means that s cannot reach t only using labels in $\mathcal{L}$.

2.2   Pruned 2-Hop Index

The 2-hop cover framework has been widely used for processing graph queries, such as distance querying [1,6], reachability querying [2,19], and route planning [16,17]. Recently, pruned 2-hop method (P2H$^+$) [8,9] first applies it to answering LCR queries, in which each vertex $v \in V$ has an in-entry set $I_{in}[v] = \{\langle u, \mathcal{L} \rangle \mid u \xrightarrow{\mathcal{L}} v\}$ and an out-entry set $I_{out}[v] = \{\langle u, \mathcal{L} \rangle \mid v \xrightarrow{\mathcal{L}} u\}$. Intuitively, the answer

---

Algorithm 1. Query2HopIndex(s, t, $\mathcal{L}$)

1: **for** each entry $\langle u, \mathcal{L}_1 \rangle$ in $I_{out}[s] \cup \{\langle s, \emptyset \rangle\}$ **do**
2:     **if** $\mathcal{L}_1 \subseteq \mathcal{L}$ **then**
3:         **for** each entry $\langle u, \mathcal{L}_2 \rangle$ in $I_{in}[t] \cup \{\langle t, \emptyset \rangle\}$ **do**
4:             **if** $\mathcal{L}_2 \subseteq \mathcal{L}$ **then return** *Yes*
5: **return** *No*

Table 1. The built 2-hop index for example grpah G

(a) P2H$^+$ index				(b) Reduced 2-hop index		
v	$\mathbf{I_{in}[v]}$	$\mathbf{I_{out}[v]}$		**v**	$\mathbf{I_{in}[v]}$	$\mathbf{I_{out}[v]}$
0	$\langle 1, b\rangle$	$\langle 4, ac\rangle\langle 1, c\rangle$		0	-	-
1	-	$\langle 4, a\rangle$		1	-	$\langle 4, a\rangle$
2	$\langle 1, a\rangle$	-		2	$\langle 1, a\rangle$	-
3	$\langle 1, a\rangle\langle 2, a\rangle$	$\langle 2, a\rangle$		3	-	-
4	-	-		4	-	-
5	$\langle 4, c\rangle$	$\langle 4, b\rangle$		5	-	-
6	$\langle 4, b\rangle$	-		6	-	-
7	$\langle 4, b\rangle$	-		7	-	-
8	$\langle 4, c\rangle$	-		8	-	-
9	$\langle 1, b\rangle$	$\langle 13, a\rangle$		9	-	-
10	$\langle 4, a\rangle$	-		10	$\langle 4, a\rangle$	-
11	$\langle 4, ab\rangle\langle 10, b\rangle$	-		11	-	-
12	$\langle 1, ac\rangle\langle 2, c\rangle$	$\langle 13, a\rangle$		12	-	-
13	$\langle 1, ab\rangle\langle 1, ac\rangle\langle 2, ac\rangle$	-		13	$\langle 1, ab\rangle\langle 1, ac\rangle\langle 2, ac\rangle$	-
14	$\langle 1, ab\rangle\langle 1, ac\rangle\langle 2, ac\rangle\langle 13, a\rangle$	-		14	-	-

to $s \xrightarrow{?\mathcal{L}} t$ is Yes if there is a hop node u s.t. $s \xrightarrow{\mathcal{L}} u \xrightarrow{\mathcal{L}} t$. Thus, Algorithm 1 returns Yes if $\exists \langle u, \mathcal{L}_1\rangle \in I_{out}[s]$, $\langle u, \mathcal{L}_2\rangle \in I_{in}[t]$ s.t. $\mathcal{L}_1 \subseteq \mathcal{L} \wedge \mathcal{L}_2 \subseteq \mathcal{L}$.

P2H$^+$ regards all vertices in G as hop nodes, and conducts both forward and backward BFS from each hop node u. During forward BFS, it maintains a search queue Q, in which each element $(v, \mathcal{L})$ represents a vertex v accompanied by a label set $\mathcal{L}$. It always de-queues an element $(v, \mathcal{L})$ from Q having the smallest label set $\mathcal{L}$, then inserts index entry $\langle u, \mathcal{L}\rangle$ into $I_{in}[v]$ if query $u \xrightarrow{?\mathcal{L}} v$ cannot be answered by existing index. Otherwise, it skips exploring vertex v, since the reachability information from u to v has already been indexed. The skip rules include (1) checking if v has been processed before, and (2) querying $u \xrightarrow{?\mathcal{L}} v$ by Algorithm 1. Similarly, it conducts backward BFS for indexing I_{out}.

Example 1. Assume vertices in graph G (Fig. 1) are processed in descending order of vertex degree, i.e., $4, 1, 2, 10, 13, \cdots$. Next, we assume that the first vertex 4 has been processed and show how to process the following vertex 1. By iterating out-neighbors of 1, $I_{in}[0] = I_{in}[9] = \{\langle 1, b\rangle\}$ and $I_{in}[2] = \{\langle 1, a\rangle\}$. Note that we skip exploration for $(4, a)$ since $1 \xrightarrow{?a} 4$ can be answered by existing index $\langle 4, a\rangle \in I_{out}[1]$. Also, exploration for $(10, a)$ is skipped since $\langle 4, a\rangle \in I_{out}[1]$ and $\langle 4, a\rangle \in I_{in}[10]$ indicate that 1 can reach 10 via 4 under label set $\{a\}$. By continuing BFS, $I_{in}[3] = \{\langle 1, a\rangle\}$, $I_{in}[12] = \{\langle 1, ac\rangle\}$, $I_{in}[13] = I_{in}[14] = \{\langle 1, ab\rangle\langle 1, ac\rangle\}$. Similarly, after performing backward BFS from vertex 1, $I_{out}[0] = \{\langle 1, c\rangle\}$. The built P2H$^+$ index is shown in Table 1(a).

3   Degree-One Reduction

When a vertex v only has one in-neighbor or out-neighbor u, the reachability information of v can be easily inferred from u. Such property helps to reduce workloads for both index construction and online querying.

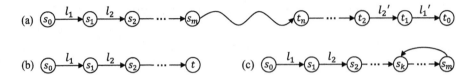

Fig. 2. Illustration for querying with degree-one reduction

Index Construction. If vertex v only has an in-neighbor (or out-neighbor) u, its in-entries $I_{in}[v]$ (out-entries $I_{out}[v]$) can be inferred from $I_{in}[u]$ ($I_{out}[u]$) in 2-hop index. We call such in-entries $I_{in}[v]$ (out-entries $I_{out}[v]$) as *redundant entries* (formally defined in Definition 2). Given the built P2H$^+$ index (e.g., Table 1(a)), all redundant entries can be removed to reduce index size (e.g., Table 1(b)). For example, $I_{in}[14]$ in Table 1(a) can be removed since vertex 14 has only one in-neighbor 13. Section 3.2 proves that the reduced index is sound, complete, and minimal. Furthermore, instead of deleting redundant entries on the built P2H$^+$ index, we propose a new index algorithm, which directly generates the reduced index without maintaining any redundant entries during index construction.

Online Querying. Intuitively, query process (Algorithm 1) becomes faster since there are less entries in the reduced index. Moreover, it can be stopped earlier if the label of the unique in-edge (out-edge) does not satisfy label constraint. Take query $0 \xrightarrow{?\{bc\}} 14$ as an example, since the unique in-edge of vertex 14 has label $a \notin \{bc\}$, we can directly return No without checking index entries.

In this section, we first formally define the *redundant entries* and *reduced 2-hop index*, then introduce a new query function utilizing the reduced index, followed by the new index algorithm with theoretical analysis.

Definition 2 *(Redundant Entries). Given a 2-hop index, for any vertex v s.t. $d_{in}(v) = 1$, all its in-entries $I_{in}[v]$ are redundant entries. For any vertex v s.t. $d_{out}(v) = 1$, all its out-entries $I_{out}[v]$ are also redundant entries.*

Definition 3 *(Reduced 2-hop Index). Removing all redundant entries in P2H$^+$ index leads to the reduced 2-hop index.*

3.1   Online Querying

As illustrated in Fig. 2(a), for query $s_0 \xrightarrow{?\mathcal{L}} t_0$, if s_0 only has one out-edge with label $l_1 \in \mathcal{L}$, intuitively we can move to its unique successor s_1. If s_1 also has only one out-neighbor, we keep moving to its unique successor s_2 when $l_2 \in \mathcal{L}$. Such movement continues for $s_2, s_3, \cdots$, until there is a vertex s_m s.t. $d_{out}(v) \neq 1$, or $d_{out}(v) = 1$ but the label of its unique out-edge is not in $\mathcal{L}$. Based on the above intuitions, the new query function is shown in Algorithm 2. Lines 1–6 iteratively trace the unique child of source vertex s. As illustrated in Fig. 2(b), when meeting the target vertex t, it directly returns Yes in line 5, since we have found a valid $\mathcal{L}$-constrained path. Note that we may fall into a cycle when

---

Algorithm 2. QueryWithDegreeOneReduction(s, t, $\mathcal{L}$)

1: $visited \leftarrow \{s\}$
2: **while** $d_{out}(s) = 1$ **do**
3: $\quad$ $s, e \leftarrow$ the unique out-neighbor and out-edge of s
4: $\quad$ **if** label of edge e is not in $\mathcal{L}$, or $s \in visited$ **then return** No
5: $\quad$ **if** $s = t$ **then return** Yes
6: $\quad$ $visited \leftarrow visited \cup \{s\}$
7: $visited \leftarrow \{t\}$
8: **while** $d_{in}(t) = 1$ **do**
9: $\quad$ $t, e \leftarrow$ the unique in-neighbor and in-edge of t
10: $\quad$ **if** label of edge e is not in $\mathcal{L}$, or $t \in visited$ **then return** No
11: $\quad$ **if** $s = t$ **then return** Yes
12: $\quad$ $visited \leftarrow visited \cup \{t\}$
13: **return** Query2HopIndex(s, t, $\mathcal{L}$) $\qquad \qquad \qquad \qquad \triangleright$ Algorithm 1

---

moving to a visited vertex s_k, as illustrated in Fig. 2(c). Thus, we introduce the *visited* set in line 1. If vertex s has been visited or the unique out-edge e does not satisfy label constraint, it directly returns No in line 4. Similarly, lines 7–12 trace the unique parent of target vertex t. Finally, line 13 returns the result by querying the reduced 2-hop index with Algorithm 1.

Theorem 1. *Algorithm 2 gives correct answers only using reduced 2-hop index.*

Proof. As discussed above, it is trivial for proving the answer returned in lines 4, 5, 10, or 11 is correct. When line 13 is invoked, $d_{out}(s) \neq 1$ and $d_{in}(t) \neq 1$. Thus, the redundant entries in P2H$^+$ index will never be used in Algorithm 2, which are removed in the reduced 2-hop index. Since all queries can be correctly answered by P2H$^+$ index [8,9], Algorithm 2 always returns the correct answer.

Time and Space Complexity. Algorithm 1 costs $\mathcal{O}\left((|I_{out}[s]| + |I_{in}[t]|)\,|L|\right)$ time and $\mathcal{O}(1)$ space [9]. For Algorithm 2, assume that we totally visit δ_s (δ_t) vertices in line 3 (line 9), which is much smaller than $|V|$ in practice. Thus, Algorithm 2 costs $\mathcal{O}\left((\delta_s + \delta_t + |I_{out}[s]| + |I_{in}[t]|)\,|L|\right)$ time and $\mathcal{O}(\delta_s + \delta_t)$ space.

3.2 Index Construction

By Definition 3, we can first build the P2H$^+$ index, then remove all redundant entries to obtain a reduced 2-hop index. Since unnecessary time and space are spent, we present a new index algorithm to generate the reduced 2-hop index directly without maintaining any redundant entries. Algorithm 3 conducts both forward and backward BFS from each hop node $u \in V$. Next, we focus on forward BFS (Algorithm 4) and omit backward BFS, since they work in a similar manner.

$\quad$ During forward BFS from current processing hop node u, intuitively redundant entries should not be inserted into $I_{in}[v]$ when $d_{in}(v) = 1$, as shown in line 5 of Algorithm 4. Assume that we are exploring vertex v under label set $\mathcal{L}$, i.e., indexing reachability information for $u \xrightarrow{\mathcal{L}} v$. If it has already indexed $u \xrightarrow{\mathcal{L}'} v$

Algorithm 3. Reduced 2-hop index generation

1: $marked \leftarrow \emptyset$
2: **for** each u in G with a specific order **do**
3:     $marked \leftarrow marked \cup \{u\}$
4:     ReducedForwardBFS(u)
5:     ReducedBackwardBFS(u)

Algorithm 4. ReducedForwardBFS(u) for building reduced 2-hop index

---

1: $Q \leftarrow \{(u, \emptyset)\}$
2: **while** $Q \neq \emptyset$ **do**
3:     pop $(v, \mathcal{L})$ having the smallest label set $\mathcal{L}$ from Q
4:     **if** SkipOrNot($u, v, \mathcal{L}$) **then** continue   ▷ new skip rules
5:     **if** $d_{in}(v) \neq 1 \wedge \mathcal{L} \neq \emptyset$ **then** insert entry $\langle u, \mathcal{L} \rangle$ into $I_{in}[v]$
6:     **for** each label l **do**
7:         **for** each out-neighbor $w \in N_{out}(v, l)$ **do**
8:             push $(w, \mathcal{L} \cup \{l\})$ into Q
9: **procedure** SKIPORNOT($u, v, \mathcal{L}$)
10:     **if** $v \in marked$ **then return** Yes ▷ skip rule 1
11:     **if** $d_{in}(v) = 1$ **then return** No ▷ skip rule 2 begins
12:     **for** each entry $\langle u, \mathcal{L}' \rangle$ in $I_{in}[v]$ **do**
13:         **if** $\mathcal{L}' \subseteq \mathcal{L}$ **then return** Yes ▷ skip rule 2 ends
14:     $u' \leftarrow u$ ▷ skip rule 3 begins
15:     **while** $d_{out}(u') = 1$ **do**
16:         $u' \leftarrow$ the unique out-neighbor of u'
17:         **if** $u' = v$ **then return** No
18:     **return** Query2HopIndex($u', v, \mathcal{L}$)   ▷ skip rule 3 ends with Algorithm 1

---

where $\mathcal{L}' \subseteq \mathcal{L}$, such exploration should be skipped. However, the skip rules of P2H$^+$ cannot work properly due to the absence of redundant entries, i.e., some unnecessary explorations have not been successfully skipped. Thus, we carefully design the following skip rules tailored for the reduced 2-hop index.

Rule 1. We skip vertex v if it is current processing or processed hop node, as $u \xrightarrow{\mathcal{L}} v$ has been indexed by hop node v. We implement it using $marked$ set.

Rule 2. Vertex v should be skipped if v has been explored from u under label set $\mathcal{L}' \subseteq \mathcal{L}$. Consider the following cases.

Case 1. When $d_{in}(v) \neq 1$ *(Fig. 3(a))*, if $\exists$ an in-entry $\langle u, \mathcal{L}' \rangle \in I_{in}[v]$ s.t. $\mathcal{L}' \subseteq \mathcal{L}$, we skip exploring v, as shown in line 13 of Algorithm 4.

Case 2. When $d_{in}(v) = 1$ *(Fig. 3(b))*, v cannot be explored from u under any label set $\mathcal{L}' \subseteq \mathcal{L}$, i.e., we cannot skip exploring v under label set $\mathcal{L}$. When tracing the unique parent of v iteratively, (1) if u is met, it is the first time to visit v from u since there is only one path from u to v. (2) If u is not met, there

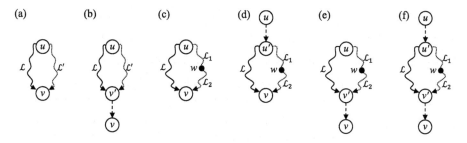

Fig. 3. Illustration for skipping explorations during reduced 2-hop index construction

must be a predecessor v' of v s.t. $d_{in}(v') \neq 1$, and no in-entry $\langle u, \mathcal{L}' \rangle \in I_{in}[v']$ satisfying $\mathcal{L}' \subseteq \mathcal{L}$, otherwise we have skipped v' in case 1 and it is impossible to visit v.

Rule 3. Vertex v should be skipped if $u \xrightarrow{?\mathcal{L}} v$ can be answered by any processed hop node w. Consider the following cases.

Case 1. When $d_{out}(u) \neq 1 \land d_{in}(v) \neq 1$ *(Fig. 3(c))*, since $I_{out}[u]$ and $I_{in}[v]$ are preserved, we skip exploring v if existing 2-hop index returns Yes for query $u \xrightarrow{?\mathcal{L}} v$ in line 18 of Algorithm 4.

Case 2. When $d_{out}(u) = 1 \land d_{in}(v) \neq 1$ *(Fig. 3(d))*, Algorithm 1 always returns No since $I_{out}[u] = \emptyset$. However, $u \xrightarrow{\mathcal{L}} v$ may have been indexed by a processed hop node w in $I_{out}[u']$ where u' is a unique child of u. Thus, we need to trace the unique child of u iteratively, during which (1) if v is met, we cannot skip exploring v (line 17 of Algorithm 4), since w is not met on the unique path from u to v, otherwise we have skipped it by Rule 1. (2) And if we do not meet u, there must be a successor u' of u s.t. $d_{out}(u') \neq 1$, and we skip exploring v if Algorithm 1 returns Yes for query $u' \xrightarrow{?\mathcal{L}} v$ in line 18 of Algorithm 4.

Case 3. When $d_{in}(v) = 1$ *(Figs. 3(e)–(f))*, $u \xrightarrow{\mathcal{L}} v$ has not been indexed, and we cannot skip exploring v under label set $\mathcal{L}$. Specifically, when tracing the unique parent of v iteratively, (1) if u is met, w is not in the unique path from u to v, otherwise we have skipped exploring w by Rule 1 and it is impossible to visit v. (2) And if we do not meet u, there must be a predecessor v' of v s.t. $d_{in}(v') \neq 1$, and existing index will not return Yes for query $u \xrightarrow{?\mathcal{L}} v$, otherwise we have skipped exploring v' in Case 1 or 2 and it is also impossible to visit v.

Example 2. Assume that we process all vertices in the same order as Example 1. After processing the first hop node 4, index entries related to it have been stored except redundant entries in Table 1(b). During forward BFS from hop node 1, we explore all its out-neighbors but skip vertex 4 by *Rule 1*, after which we only insert $\langle 1, a \rangle$ into $I_{in}[2]$ since other neighbors only have one in-edge. Next, we visit vertex 3 without inserting entry, after which we visit vertex 2 again but skip exploration since $\langle 1, a \rangle \in I_{in}[2]$ according to *Rule 2*. We also skip vertex 10 by *Rule 3*, since $1 \xrightarrow{\{a\}} 10$ has been indexed by $\langle 4, a \rangle \in I_{out}[1]$ and $\langle 4, a \rangle \in I_{in}[10]$.

Theorem 2. *Algorithm 3 builds a sound, complete, and minimal 2-hop index.*

Proof. **Soundness.** Every entry is added by finding an $\mathcal{L}$-constrained path during forward or backward BFS. Thus, it never produces false-positive answers.

Completeness. During forward and backward BFS from every vertex u, we only skip exploration if current reachability has been indexed. Thus, it covers all label-constrained reachability and no false-negative answer can be produced.

Minimality. Assume that Algorithm 3 process vertices $u_1, u_2, \cdots, u_{|V|}$ in order. We focus on discussing in-entries I_{in} and omit out-entries I_{out} due to symmetry. Removing any entry $\langle u_i, \mathcal{L} \rangle \in I_{in}[u_j]$ leads to a false-negative answer to $u_i \xrightarrow{?\mathcal{L}} u_j$. Specifically, during forward BFS from u_i in index construction, we do not skip u_j since $u_i \xrightarrow{\mathcal{L}} u_j$ has not been covered by processed hop nodes $u_1, u_2, \cdots, u_{i-1}$. And the modified BFS order [9] ensures that there is no entry $\langle u_i, \mathcal{L}' \rangle \in I_{in}[u_j]$ s.t. $\mathcal{L}' \subset \mathcal{L}$. Also, $u_i \xrightarrow{\mathcal{L}} u_j$ cannot be covered by the other hop nodes $u_{i+1}, u_{i+2}, \cdots, u_{|V|}$, since we skip exploring u_i during BFS from them.

Time Complexity. Algorithm 4 visits each edge at most $\mathcal{O}(2^{|L|})$ times. In function *SkipOrNot*, assume that line 16 visits δ vertices, much smaller than $|V|$ in practice. Since Algorithm 1 takes $\mathcal{O}\left((|I_{out}[s]| + |I_{in}[t]|) |L|\right) = \mathcal{O}(|V|2^{|L|})$ time [9], the index construction takes $\mathcal{O}(|V|^2|E|2^{2|L|})$ time in the worst.

Space Complexity. For each vertex v, there are at most $(|V| - 1) \cdot 2^{|L|}$ entries in $I_{in}[v]$ ($I_{out}[v]$). Thus, it consumes $\mathcal{O}(|V|^2 \cdot 2^{|L|})$ space in the worst.

4   Unreachable Query Filter

The answer to an LCR query $s \xrightarrow{?\mathcal{L}} t$ is definitely *No* if vertex s cannot reach t even without any label constraints. In this section, we propose another lightweight index named *Unreachable Query Filter* to rule out such unreachable queries on the fly. Specifically, by discarding all edge labels, we can condense all strongly connected components in graph G to obtain a DAG G_A. Take graph G in Fig. 1 as an example, Figs. 4(a)–(b) present the converted DAG G_A with a map M_{ID} for mapping vertex IDs in G to IDs in G_A. Next, we focus on reachability queries $\hat{s} \xrightarrow{?} \hat{t}$ in G_A, where $\hat{s} = M_{ID}[s]$ and $\hat{t} = M_{ID}[t]$.

4.1   FELINE Index

Topological Orders. In DAG G_A, denote X_s and X_t as the topological order of vertex s and t, respectively. If $X_s > X_t$, then we directly return *No* since s cannot reach t. Based on the first topological order X, [15] further generates another topological order Y by heuristic decision. Let *Roots* (R for short) be a priority queue storing all vertices which have no in-edges. Specifically, FELINE iteratively runs the following 3 steps until R is empty.

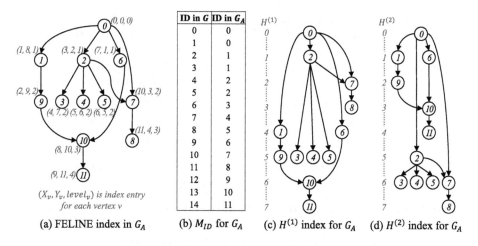

ID in G	ID in G_A
0	0
1	0
2	1
3	1
4	2
5	2
6	3
7	4
8	5
9	6
10	7
11	8
12	9
13	10
14	11

$(X_v, Y_v, level_v)$ is index entry
for each vertex v

(a) FELINE index in G_A (b) M_{ID} for G_A (c) $H^{(1)}$ index for G_A (d) $H^{(2)}$ index for G_A

Fig. 4. Illustration for False-Positive Reduction

- *Step 1.* Chooses the vertex v with largest X_v value from R to assign Y_v.
- *Step 2.* Remove all out-edges of v, then its children w may have no in-edges.
- *Step 3.* Add each w without in-edges to R, then repeat *Step 1*.

Level Filter. Each vertex v also has a level index $level_v$. If vertex v has no in-edges in G_A, its $level_v = 0$. Otherwise, $level_v = \max_u\{level_u\} + 1$, where u is the parent of v. Similarly, we directly return *No* when $level_s \geq level_t$.

Example 3. The built FELINE index for DAG G_A is shown in Fig. 4(a).

4.2 False-Positive Reduction

Given a query $s \xrightarrow{?} t$, if $X_s < X_t \wedge Y_s < Y_t \wedge level_s < level_t$ but actually s cannot reach t in G_A, FELINE index produces a false-positive answer. For example, vertex 3 cannot reach 10 in G_A (Fig. 4(a)), but it fails to return *No* since $X_3 < X_{10} \wedge Y_3 < Y_{10} \wedge level_3 < level_{10}$. Note that all key vertices contributing to such false-positive answers can be found [12], and we call them *False-Positive Contributors* (*FPC* for short). To reduce false-positive answers in FELINE index, we generate two more coordinates H^α and H^β based on *FPC*.

Finding False-Positive Contributors (FPC). After obtaining the first topological order X of FELINE index, FPC can be found when generating coordinate Y [12]. Specifically, vertex v is processed if it has been assigned Y coordinate, otherwise v is unprocessed. Recall that priority queue R contains all vertices which have no in-edges in G_A. And we can maintain a vertex set called *Candidate Free Nodes* (*CFN* for short), containing all unprocessed vertices v s.t. $v \notin R$ and at least one of v's parent has been processed. When popping vertex u from R, another vertex $v \in FPC$ iff $v \in CFN \wedge X_w < X_u$, where vertex w has the largest X_w value among all unprocessed parents of v [12].

Example 4. For G_A in Fig. 4(a), we first pop vertex 0 from R. Then, we pop vertex 6 from R, and $\exists$ vertex $7 \in CFN$ s.t. $X_2 < X_6$, indicating that vertex $7 \in FPC$. Similarly, when popping vertex 2, $\exists$ vertex $10 \in CFN$ s.t. $X_9 < X_2$ and vertex $10 \in FPC$ too. Finally, $FPC = \{7, 10\}$. Note that all false-positive queries in FELINE index can be attributed to at least one vertex in FPC [12]. The first false-positive contributor 7 makes it cannot return No for queries $6 \overset{?}{\to} 7$ and $6 \overset{?}{\to} 8$, while vertex 10 contributes to the false-positive answers for queries $2 \overset{?}{\to} 10$, $2 \overset{?}{\to} 11$, $3 \overset{?}{\to} 10$, $3 \overset{?}{\to} 11$, $4 \overset{?}{\to} 10$, $4 \overset{?}{\to} 11$, $5 \overset{?}{\to} 10$, and $5 \overset{?}{\to} 11$.

Given the computed FPC, we can generate new coordinates to return No for those false-positive query pairs. The naive ideas are presented as follows.

Naive Ideas of New Coordinates. Intuitively, for each false-positive contributor $v_i \in FPC$, we can add a new coordinate $H^{(i)}$ for vertices in G_A. Specifically, $H_{v_i}^{(i)} = level_{v_i}$, $H_u^{(i)} = level_u$ for all predecessors and descendants u of v_i. For the other vertices w, we have $H_w^{(i)} = level_w + \max\{H_u^{(i)}\}$. Given a query $s \overset{?}{\to} t$, we can return No if $\exists$ an $H^{(i)}$ coordinate s.t. $H_s^{(i)} \geq H_t^{(i)}$.

Example 5. For G_A in Fig. 4(a), $FPC = \{7, 10\}$ according to Example 4. Figure 4(c) shows the first new coordinate $H^{(1)}$ for avoiding false-positive answers caused by vertex 7. For queries $6 \overset{?}{\to} 7$ and $6 \overset{?}{\to} 8$ having false-positive answers from FELINE index, new coordinate $H^{(1)}$ helps to return No since $H_6^{(1)} > H_7^{(1)}$ and $H_6^{(1)} > H_8^{(1)}$. Figure 4(d) shows the new coordinate $H^{(2)}$, built for the second false-positive contributor 10 to avoid other mistakes made by FELINE.

Though all false-positive answers produced by FELINE index can be avoided, the size of FPC may be large and examining all new coordinates online is expensive. Next, we propose to generate only two new coordinates H^α and H^β based on FPC, s.t. if $H_s^\alpha \geq H_t^\alpha$ or $H_s^\beta \geq H_t^\beta$, we can safely return No for the query $s \overset{?}{\to} t$, avoiding as many false-positive answers as possible.

New Coordinates Generation. To generate the first coordinate H^α, we process each $v_i \in FPC$ in ascending order of $level_v$, while we process them in descending order when generating H^β. Since the generation of H^α and H^β only differs in process order, next we focus on generating H^α.

Definition 4. *(Affiliate Nodes of v, AN_v for short). When processing $v \in FPC$, $u \in AN_v$ iff: (1) H_u^α has not been assigned, (2) u and v are connected ignoring edge directions, and (3) u cannot be reached by any unprocessed $w \in FPC$.*

Take DAG G^τ in Fig. 5(a) as an example, where $FPC = \{v_1, v_2, v_3, v_4\}$ and C_i represents a subset of vertices with their inner edges. Each directed dashed line from v_i to C_j (from C_j to v_i) means that v_i can reach (be reached by) vertices in C_j. To generate H^α coordinate, we process v_1, v_2, v_3, and v_4 in order. Consider the following steps as illustrated in Fig. 5(b).

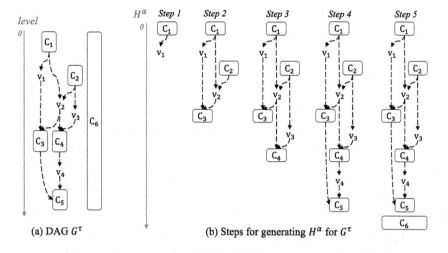

(a) DAG G^τ

(b) Steps for generating H^α for G^τ

Fig. 5. Illustration for generating new coordinate H^α

Table 2. Statistics of Datasets

Dataset	$|V|$	$|E|$	$|L|$	Synthetic Labels	Dataset	$|V|$	$|E|$	$|L|$	Synthetic Labels
Youtube	15K	14M	5		Epinions	132K	841K	64	✓
ArXiv	35K	422K	8	✓	NotreDame	326K	1.5M	64	✓
WebStanford	282K	2.3M	8	✓	WebGoogle	916K	5.1M	64	✓
Citeseer	384K	1.7M	8	✓	NELL	2.2M	2.8M	833	
Tech	5.8M	148M	8	✓	Flixster	2.5M	9.2M	64	✓
Indochina	7M	192M	8	✓	Twitter	11M	85M	64	✓
WikiLinkEn	27M	599M	8	✓	YAGO	37M	95M	96	

- *Step 1.* We process the first false-positive contributor v_1, and let $H_u^\alpha = level_u$ for all vertices in $\{v_1\} \cup AN_{v_1}$, where $AN_{v_1} = C_1$. Note that although v_1 can reach C_3, $C_3 \not\subset AN_{v_1}$ since the unprocessed $v_2 \in FPC$ can reach C_3 too.
- *Step 2.* When processing v_2, $H_u^\alpha = level_u + \Omega$ for vertices in $\{v_2\} \cup AN_{v_2}$, where $AN_{v_2} = C_2 \cup C_3$. Note that Ω is an offset, making sure that all H^α value assigned current step is larger than any H^α assigned in previous steps.
- *Steps 3–5.* Each step is shown in Fig. 5(b).

Before generating H^α, for each vertex u in DAG, we pre-compute a set $\Gamma_u = \{w \in FPC \mid w$ can reach u without passing any $w' \in FPC\}$. When processing $v \in FPC$, to assign H^α value for vertices in AN_v, we explore from v and visit vertices u satisfying (1) H_u^α has not been assigned, and (2) all vertices in $\Gamma_u \backslash \{v\}$ have been processed. For example, for each $u \in C_3$ in G^τ, $\Gamma_u = \{v_1, v_2\}$.

Time Complexity. Generating FELINE index costs $\mathcal{O}(|V| \log |V| + |E|)$. Computing Γ_u for each vertex u costs $\mathcal{O}(\gamma |E|)$, where γ is the largest size of Γ_u and is much smaller than $|V|$ in practice. Generating H^α and H^β costs $\mathcal{O}(|V| + |E|)$. Thus, building the *Unreachable Query Filter* costs $\mathcal{O}(|V| \log |V| + \gamma |E|)$.

Space Complexity. In *Unreachable Query Filter*, each vertex u has five coordinates $(X_u, Y_u, level_u, H_u^\alpha, H_u^\beta)$ and it totally needs $\mathcal{O}(|V|)$ space.

5 Experiments

5.1 Experimental Setup

Dataset. As shown in Table 2, we use 14 real graphs from a variety of domains [4,5,10,21]. For those without edge labels, we generate synthetic labels in the same way as previous works [8,9,14], which are exponentially distributed with $\lambda = |L|/1.7$. For each dataset, we have three query sets using different numbers of query labels $|\mathcal{L}|$, where $|\mathcal{L}| = \{|L|/4, |L|/2, |L| - 2\}$ for $|L| \leq 8$, while $|\mathcal{L}| = \{2, 4, 6\}$ for larger $|L|$. Each query set contains 1000 randomly generated queries.

Implementation. All algorithms use descending order of degree for processing hop nodes during index construction. Following P2H$^+$ method, for graphs with $|L| > 12$, 6 most frequent labels are selected as primary labels, while the others are evenly divided into 6 virtual labels for building index. We conduct experiments on a Linux server with Intel(R) E5-2596v4 CPU @2.2 GHz and 128G RAM. Programs are implemented in C++ and compiled with -O3 Optimization.

5.2 Online Query

The efficiency of online querying is crucial for LCR queries, and different sizes of query label sets also affect the performance. Table 3 reports the total querying time of 1000 queries by using all our reduction techniques, including degree-one reduction and unreachable query filter. By varying different query label size $|\mathcal{L}|$, reduction techniques consistently improve the query efficiency on all datasets, compared with state-of-the-art method P2H$^+$. Benefiting from the reduced search space, it runs even 3X faster over graphs *ArXiv*, *Citeseer*, and *Flixster* when $|\mathcal{L}| = 6$, which indicates that the improvement becomes more significant for queries with large-size label sets.

Table 3. Online querying time in microseconds (μs)

Dataset	2 or $|L|/4$		4 or $|L|/2$		6 or $|L|-2$	
	P2H$^+$	Ours	P2H$^+$	Ours	P2H$^+$	Ours
Youtube	417	**313**	436	**184**	536	**118**
ArXiv	2496	**1534**	3663	**984**	2544	**609**
WebStanford	319	**265**	334	**236**	344	**235**
Citeseer	1182	**377**	1618	**425**	3119	**461**
Tech	325	**200**	342	**167**	354	**143**
Indochina	776	**527**	811	**484**	804	**503**
WikiLinkEn	862	**435**	749	**397**	675	**359**
Epinions	54.2 ms	**50.6 ms**	192 ms	**134 ms**	434 ms	**244 ms**
NotreDame	389	**196**	1106	**785**	954	**607**
WebGoogle	681	**546**	1243	**1024**	10315	**5839**
NELL	383	**185**	390	**131**	399	**128**
Flixster	322 ms	**246 ms**	1.79 s	**941 ms**	2.34 s	**761 ms**
Twitter	7.19 s	**4.44 s**	43.1 s	**19.3 s**	86.7 s	**39.7 s**
YAGO	512	**329**	608	**280**	634	**291**

Table 4. Evaluating index construction time, number of index entries and index size

Dataset	Index Time (s)		Number of Entries		Index Size (MB)	
	P2H$^+$	Ours	P2H$^+$	Ours	P2H$^+$	Ours
Youtube	49	**43**	1816K	**1811K**	**13.85**	14.16
ArXiv	302	**147**	31.4M	**31M**	239.5	**237.2**
WebStanford	11	**10**	9.9M	**9.15M**	75.6	**71.5**
Citeseer	1195	**587**	174M	**169M**	1328	**1298**
Tech	**218**	243	45.4M	**39.2M**	**347**	423
Indochina	1168	**630**	380M	**342M**	2903	**2671**
WikiLinkEn	7199	**4492**	1092M	**989M**	8331	**8039**
Epinions	2.74	**2.67**	1103K	**728K**	8.42	**7.75**
NotreDame	29	**22**	15.3M	**11.6M**	116	**93**
WebGoogle	195	**190**	66M	**60M**	504	**473**
NELL	85	**52**	19M	**11M**	146	**137**
Flixster	33	**30**	16M	**9M**	123	**121**
Twitter	427	**401**	77M	**44M**	590	**521**
YAGO	509	**414**	293M	**203M**	**2239**	2380

Table 5. Index and query performance only using degree-one reduction

Dataset	Index Performance				Query Performance (µs)					
	Time (s)		Size (MB)		2 or $|L|/4$		4 or $|L|/2$		6 or $|L|$-2	
	P2H$^+$	DOR	P2H$^+$	DOR	P2H$^+$	DOR	P2H$^+$	DOR	P2H$^+$	DOR
Youtube	49	42	13.85	13.82	417	385	436	392	536	431
ArXiv	302	145	239.5	236.7	2496	2493	3663	2651	2544	738
WebStanford	11	9	75.6	69.8	319	273	334	314	344	336
Citeseer	1195	533	1328	1290	1182	634	1618	751	3119	799
Tech	218	217	347	299	325	272	342	248	354	262
Indochina	1168	610	2903	2609	776	673	811	785	804	691
WikiLinkEn	7199	4114	8331	7545	862	602	749	600	675	595
Epinions	2.74	2.6	8.42	5.55	54.2 ms	52.8 ms	192 ms	134 ms	434 ms	227 ms
NotreDame	29	20	116	88	389	228	1106	847	954	649
WebGoogle	195	174	504	461	681	599	1243	1122	10315	5932
NELL	85	51	146	87	383	260	390	255	399	258
Flixster	33	29	123	65	322 ms	289 ms	1.79 s	1.1 s	2.34 s	855 ms
Twitter	427	369	590	339	7.19 s	4.63 s	43.1s	20.9s	86.7s	40.1s
YAGO	509	393	2239	1549	512	471	608	462	634	505

Table 6. Index and query performance only using unreachable query filter

Dataset	Index Performance				Query Performance (µs)					
	Time (s)		Size (MB)		2 or $|L|/4$		4 or $|L|/2$		6 or $|L|$-2	
	P2H$^+$	UQF	P2H$^+$	UQF	P2H$^+$	UQF	P2H$^+$	UQF	P2H$^+$	UQF
Youtube	49	50	13.85	14.2	417	357	436	208	536	124
ArXiv	302	304	239.5	240.1	2496	1930	3663	1803	2544	1454
WebStanford	11	12	75.6	77.2	319	253	334	249	344	256
Citeseer	1195	1241	1328	1336	1182	1474	1618	834	3119	690
Tech	218	241	347	471	325	253	342	199	354	152
Indochina	1168	1195	2903	2965	776	631	811	579	804	565
WikiLinkEn	7199	7462	8331	8825	862	589	749	515	675	442
Epinions	2.74	2.83	8.42	10.61	54.2 ms	51.6 ms	192 ms	181 ms	434 ms	540 ms
NotreDame	29	31	116	122	389	265	1106	1019	954	733
WebGoogle	195	217	504	515	681	593	1243	1157	10315	9377
NELL	85	86	146	195	383	195	390	153	399	149
Flixster	33	34	123	179	322 ms	250 ms	1.79s	1.29s	2.34s	1.7s
Twitter	427	463	590	772	7.19 s	7.02 s	43.1 s	42.9 s	86.7 s	85.3 s
YAGO	509	528	2239	3070	512	413	608	377	634	360

5.3   Index Construction

We also report indexing performance (including index time and index size) in Table 4, comparing the proposed reduction techniques with P2H$^+$ method. As discussed in Sect. 3, degree-one reduction reduce index size by removing all redundant entries. Since unreachable query filter in Sect. 4 builds a light-weight index, the overall index size and time may be slightly larger than P2H$^+$ index for a few graphs, which is still acceptable since it improves query efficiency.

5.4 Effects of Degree-One Reduction

To figure out how single reduction technique affect both index and query performance, in this section we only adopt degree-one reduction (DOR for short) to perform reduction. As shown in Table 5, it achieves a smaller index size on all datasets as expected. By using degree-one reduction, building the reduced 2-hop index requires less index time, since we need to query the existing 2-hop index for ensuring minimality during index construction, which is also faster benefiting from a smaller index size.

5.5 Effects of Unreachable Query Filter

We also investigate the effect of the proposed unreachable query filter (UQF for short). Table 6 shows the overall performance when only using UQF. In terms of index time and index size, we find that its additional cost is marginal compared with P2H$^+$ method. It can greatly reduce online query time for most datasets since it only costs $\mathcal{O}(1)$ time for answering most of unreachable queries, while P2H$^+$ need to iterate all index entries of query vertices. Note that for certain query label size on a few datasets, it may not be paid-off for cases that is fast enough to return *No* even if iterating all 2-hop index entries.

6 Related Work

6.1 Reachability Queries

Existing works on answering reachability queries can be divided into two categories, i.e., *Index-only* and *Index-search*. *Index-only* approaches (e.g., [12,19,23]) compress the full transitive closure of input graph G, only examining index for online query vertex pairs without graph traversal. *Index-search* methods (e.g., [11,13,15,18,20]) construct much smaller index compared with *Index-only* approaches, but cost more time in answering queries due to online search. Since condensing all strongly connected components (SCCs) in graph G leads to a DAG which preserves reachability information, transitive reduction, and equivalence reduction for DAG [22] can be used to further accelerate query processing.

6.2 Label-Constrained Reachability Queries

Jin et al. [3] introduce LCR queries and propose a tree-based index consisting of both spanning tree and partial transitive closure. By decomposing the input graph G into SCCs, Zou et al. [24] build a complete index which cannot effectively work when G contains large SCCs. LI$^+$ index [14] contains both landmark-based and non-landmarks index for answering LCR queries, which cannot scale to large graphs due to large index size. The state-of-the-art method for LCR problem is P2H$^+$, which builds a 2-hop cover to preserve all label-constrained reachability information. However, as shown in the experiments, it still suffers unnecessary time and space costs in both index construction and query processing.

7   Conclusion

In this paper, we study the problem of answering label-constrained reachability queries, and introduce novel reduction techniques for improving both index construction and online query efficiency. *Degree-one Reduction* directly builds a reduced 2-hop index, removing redundant index entries for vertices, which is proved to be sound, complete, and minimal. *Unreachable Query Filter* utilizes five coordinates to avoid iterating all entries for unreachable queries. Extensive experiments confirm that our proposed reduction strategies achieve significantly better performance compared with the state-of-the-art method.

Acknowledgement. This work was supported by National Natural Science Foundation of China (Grant No. 61902074).

References

1. Akiba, T., Iwata, Y., Yoshida, Y.: Fast exact shortest-path distance queries on large networks by pruned landmark labeling. In: SIGMOD 2013, pp. 349–360 (2013)
2. Cohen, E., Halperin, E., Kaplan, H., Zwick, U.: Reachability and distance queries via 2-hop labels. SIAM J. Comput. **32**(5), 1338–1355 (2003)
3. Jin, R., Hong, H., Wang, H., Ruan, N., Xiang, Y.: Computing label-constraint reachability in graph databases. In: SIGMOD 2010, pp. 123–134 (2010)
4. Kunegis, J.: KONECT: the Koblenz network collection. In: WWW 2013, pp. 1343–1350 (2013)
5. Leskovec, J., Krevl, A.: SNAP datasets: Stanford large network dataset collection (2014)
6. Li, W., Qiao, M., Qin, L., Zhang, Y., Chang, L., Lin, X.: Scaling distance labeling on small-world networks. In: SIGMOD 2019, pp. 1060–1077 (2019)
7. Pastor-Satorras, R., Rubí, M., Diaz-Guilera, A.: Statistical Mechanics of Complex Networks. Lecture Notes in Physics (2003)
8. Peng, Y., Lin, X., Zhang, Y., Zhang, W., Qin, L.: Answering reachability and k-reach queries on large graphs with label constraints. VLDB J. (2021)
9. Peng, Y., Zhang, Y., Lin, X., Qin, L., Zhang, W.: Answering billion-scale label-constrained reachability queries within microsecond. VLDB **13**(6), 812–825 (2020)
10. Rossi, R.A., Ahmed, N.K.: The network data repository with interactive graph analytics and visualization. In: AAAI 2015, pp. 4292–4293 (2015)
11. Su, J., Zhu, Q., Wei, H., Yu, J.X.: Reachability querying: can it be even faster? TKDE **29**(3), 683–697 (2017)
12. Tang, X., Chen, Z., Zhang, H., Liu, X., Shi, Y., Shahzadi, A.: An optimized labeling scheme for reachability queries. Comput. Mater. Continua **55**, 267–283 (2018)
13. Trißl, S., Leser, U.: Fast and practical indexing and querying of very large graphs. In: SIGMOD 2007, pp. 845–856 (2007)
14. Valstar, L.D., Fletcher, G.H., Yoshida, Y.: Landmark indexing for evaluation of label-constrained reachability queries. In: SIGMOD 2017, pp. 345–358 (2017)
15. Veloso, R.R., Cerf, L., Meira Jr, W., Zaki, M.J.: Reachability queries in very large graphs: a fast refined online search approach. In: EDBT, pp. 511–522 (2014)
16. Wang, S., Lin, W., Yang, Y., Xiao, X., Zhou, S.: Efficient route planning on public transportation networks: a labelling approach. In: SIGMOD 2015, pp. 967–982 (2015)

17. Wang, S., Xiao, X., Yang, Y., Lin, W.: Effective indexing for approximate constrained shortest path queries on large road networks. VLDB **10**(2), 61–72 (2016)
18. Wei, H., Yu, J.X., Lu, C., Jin, R.: Reachability querying: an independent permutation labeling approach. VLDB J. **27**(1), 1–26 (2018)
19. Yano, Y., Akiba, T., Iwata, Y., Yoshida, Y.: Fast and scalable reachability queries on graphs by pruned labeling with landmarks and paths. In: CIKM 2013, pp. 1601–1606 (2013)
20. Yildirim, H., Chaoji, V., Zaki, M.J.: GRAIL: a scalable index for reachability queries in very large graphs. VLDB J. **21**(4), 509–534 (2012)
21. Zafarani, R., Liu, H.: Social computing data repository at ASU (2009)
22. Zhou, J., Yu, J.X., Li, N., Wei, H., Chen, Z., Tang, X.: Accelerating reachability query processing based on DAG reduction. VLDB J. **27**(2), 271–296 (2018)
23. Zhu, A.D., Lin, W., Wang, S., Xiao, X.: Reachability queries on large dynamic graphs: a total order approach. In: SIGMOD 2014, pp. 1323–1334 (2014)
24. Zou, L., Xu, K., Yu, J.X., Chen, L., Xiao, Y., Zhao, D.: Efficient processing of label-constraint reachability queries in large graphs. Inf. Syst. **40**, 47–66 (2014)

JG2Time: A Learned Time Estimator for Join Operators Based on Heterogeneous Join-Graphs

Hao Miao[1], Jiazun Chen[1], Yang Lin[2], Mo Xu[2], Yinjun Han[2], and Jun Gao[1(✉)]

[1] School of Computer Science, Peking University, Beijing, China
{miaohao,chenjiazun}@stu.pku.edu.cn, gaojun@pku.edu.cn
[2] ZTE Corporation, Beijing, China
{lin.yang,xu.mo1,han.yinjun}@zte.com.cn

Abstract. The join operator is one of the key operators in RDBMS, and estimating its evaluation time is a fundamental task in query optimization, scheduling, etc. However, it is hard to make a precise estimation, which is not only related with the physical join implementations (hash, sort, loop) but also with the corresponding parameters, like the size of the data, the number of partitions, the number of threads in a modern hash join. Existing works rely on the time complexity analysis but yield rough results, or employ machine learning techniques to build a predictive model but require many training instances. In this paper, we propose a method, named JG2Time, to estimate the running time using the join-graphs constructed from the source codes. Specifically, we construct a heterogonous join-graph by annotating parameter nodes to a call-graph generated by running time analysis tools, and propose ReGAT, a heterogonous graph neural network, to fully capture the edge weights (the number of function calls) in the join-graph. The embeddings learned from ReGAT can be used to predict the running time. In addition, we optimize JG2Time with a multi-task model that also predicts the times of function calls, and an unsupervised code learning method to enhance its generalization. The experimental results illustrate the effectiveness of JG2Time and its optimization strategies.

Keywords: running time estimation of join operator · call-graph · heterogonous join-graph · heterogeneous graph neural network

1 Introduction

The join operator is one of the key operators in RDBMS. For a query with join operators, the time consumed by join operators takes a major part of the total query time. To predict the running time of the join operator precisely is highly needed in query optimization, scheduling, etc. For example, in query optimization, DBMS needs to estimate the cost of candidate query plans and select the one with the lowest estimated cost. The precise estimation of the join

X. Wang et al. (Eds.): DASFAA 2023, LNCS 13943, pp. 132–147, 2023.
https://doi.org/10.1007/978-3-031-30637-2_9

operator (along with other operators) is no doubt helpful to the optimizer. We also notice that the overall query evaluation time serves as an important training signal for learned query optimizers [29]. The query scheduling task also needs to estimate the query cost, based on which different scheduling strategies can be applied to improve the system throughput [24].

However, predicting the time cost of a join operator is not trivial, as it is a function of many factors, including different physical join implementations, input parameters, the interactions among them, and hardware resources [21]. There are different physical join implementations (like nested loop join [16,17], hash join [6,8], and sort-merge join [2,4]) with various time complexities. Even for the same physical join implementation, its evaluation time is affected by different parameters. Take a radix join as an example, its parameters include the size of the input data, the number of passes in the radix join, the number of partitions, etc. The interactions among these parameters are complex and are related to the cache miss, TLB miss, etc. The time complexity analysis only produces very coarse results. Some work [23] has performed extensive experimental studies to analyze the absolute time under different parameters.

With the development of deep learning techniques, it is promising to learn a model to capture the relationships between parameters and the evaluation time of a join operator. We can run the join operator using different query workloads and parameters, and train a model with the collected latency. The model can then predict the running time with given parameters. Such an approach can ease the burden of developing estimation rules. However, due to the lack of knowledge of the underlying interactions among the parameters, the deep learning techniques require many training instances if one wishes to achieve relatively satisfactory results.

From the viewpoint of database vendors, the source codes of the join operator are actually accessible. Therefore, we are interested in whether the source codes can provide more hints for learning the interactions between parameters. We notice that source code analysis using deep learning methods is an important research direction in the field of software engineering, and most related works attempt to understand the source codes for code summarization [26], code classification [30], duplication detection [27], etc. These works, although bring some inspiration, cannot be migrated to our context directly, as they focus on the fine-granularity codes analysis. For example, some works learn code representation for summarization by using Transformer on code tokens, or modeling [9,20,27] the ASTs (abstract syntax trees) parsing from the source codes. Considering thousands of lines of source codes for a physical join implementation, these fine-granularity analysis works will incur a huge cost in modeling, and may introduce irrelevant features in estimating time cost.

In this paper, we select the call-graph to represent the source codes, which is constructed from the source code analysis tools, like *Valgrind*[1]. Compared with ASTs or code tokens for source codes, the call-graph has the following advantages in our context. First, one goal of *Valgrind* is to locate the performance bottleneck

---

[1] https://valgrind.org/.

in the codes, and the call-graph produced is then directly related with our time estimation task. Second, the call-graph is expressive and space-efficient. The nodes in the call-graphs represent functions, and the edges between the nodes are function call relationships. The size of the call-graphs can be controlled by only adding the functions that consume more time than a given threshold. Specifically, the contributions of this paper are as follows:

- We propose JG2Time, a running time estimation model based on join-graphs. A join-graph is constructed by annotating parameter nodes to function nodes in the call-graph, on which we propose ReGAT to capture the edge weights (the number of function calls) in code representation learning. The learned representation for the join-graph can be used to predict the running time. To the best of our knowledge, this is the first work to predict the running time of the join operator by utilizing its source codes and parameters.
- We design two optimization strategies on JG2Time. First, we build a multi-task model that contains a sub-model to predict the number of function calls. Then, in the inference phase, the sub-model can yield important hints for time estimation. Second, we borrow the idea of existing unsupervised source codes learning to initialize the embeddings of function nodes, with the expectation of easily applying JG2Time to other join implementations with new functions that have never seen during training.
- We perform experimental studies on JG2Time, and the results illustrate the effectiveness of our design choices. a) JG2Time can predict the running time precisely but with much fewer training instances, compared with deep predictive models using parameters only, which reflects the roles of source codes. b) The join-graph is more suitable in terms of time estimation compared with code tokens or ASTs. c) Two optimization strategies are helpful to JG2Time in terms of prediction precision and generalization, respectively.

2 Related Work

Join Algorithms. The join algorithm receives continuous attention from the database research field. With the advances of hardware, many join algorithms [2,13,19] are developed to fully exploit the capability of the new hardware. For example, considering the features of SIMD and NUMA, massive parallel sort-merge join methods are developed in [4]. The parallel radix hash join algorithm, a representative hash join in a multi-core computing environment, has been extensively studied with different optimization strategies [5]. A space-efficient hash join method is proposed in [7]. As it is hard to tell which kind of join algorithms work better, an extensive experimental study is performed to evaluate 13 physical join implementations over different data sets and query sets along with different parameters [23]. Such a work actually inspires us to design a model to predict the running time of the join operator.

Code Representation. Code representation [15], a hot research topic in the field of software engineering, is the process of capturing the major features of the source code snippets into continuous vectors (or embeddings). These embeddings are then fed into downstream tasks such as code classification, clone detection, code search, and so on. Code2vec [3] is an early work to learn code embeddings by sampling paths in the AST for a code snippet. ASTNN [30], attempting to fully explore local constraints in ASTs, splits the large AST of one code snippet into a set of small sub-trees at the statement level and adopts a bidirectional GRU network to learn embeddings on all sub-trees. When viewing the source code as a sequence of code tokens, a transformer model is used to capture the long-term dependencies in codes [1]. InferCode [10] is a recent work that mainly develops a self-supervised learning technique by predicting the same syntax subtrees among different source codes.

Application Performance Prediction. It is a challenging work to predict the performance of an application, like the execution time, the response time and throughput. Ramadan et al. [21] simplify the problem by converting it into a classification problem, and propose a tree-based LSTM model to predict whether the execution time becomes longer or shorter if the source codes change. DeepPerf [14] proposes to use a deep sparse MLP, modeling binary and numeric configuration options of software systems, to predict the response time or throughput of software systems. Samoaa et al. [22] have also shown that given sufficient training data the execution time of an application for different parameterizations is highly predictable. In this work, we demonstrate that predicting absolute performance values is possible in another context, namely for the running time of the join operator.

Heterogeneous Graph Neural Networks. The join-graph annotated with the parameter nodes can be viewed as a heterogeneous graph, which contains at least two kinds of nodes and edges. Heterogeneous graph neural networks are powerful models for learning the representation of heterogeneous graphs. Among models emerged recently [28], RGCN [12] introduces different graph convolutional layers, one for each type of relationships. Based on RGCN, RGAT [11]extends graph attention mechanisms to the relational graph domain. HAN [25] proposes a node-level attention to aggregate neighbor information and semantic-level attention to aggregate predefined meta-path information. HGCN [31] extends GCN [18] model to solve collective classification. However, we cannot directly adopt these models in join-graphs, as there is no built-in support for edge weights in these models, while the edge weights, representing the number of function calls, play important roles in the time estimation.

3   Methodology

In this section, we first formulate the problem and provide an overview of JG2Time. Then, we focus on the major components of JG2Time, including the construction of the join graph, and a ReGAT model to estimate running time on the join graph. Finally, we propose two optimization strategies for JG2time,

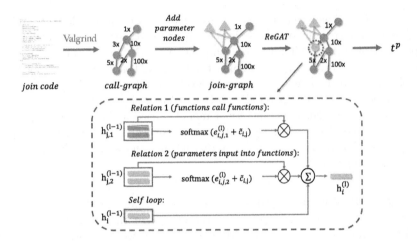

Fig. 1. A schematic overview of the main phases of JG2Time. A join-graph is generated from source codes by invoking *Valgrind* first, and then augmented with parameter nodes and edges. The result join-graph serves as an input into proposed ReGAT to generate embedding for the time estimation.

a multi-task model which contains a sub-model to learn the number of function calls, and an unsupervised function node representation learning inspired by InferCode [10] to enhance the generalization.

3.1 Problem Definition

Let O_i be the source codes for a physical join implementation. Suppose that we run O_i under different parameters set $P_j^i = \{p_1, p_2, p_3...p_k\}$ and get the running time t_j^i. We can construct a dataset $S = ((O_i, P_j^i), t_j^i)$. Our goal is to train a deep learning model $M = (O_i, P_j^i)$ to minimize the loss between $M = (O_i, P_j^i)$ and t_j^i.

3.2 Model Framework

Our goal is to predict the running time of join operators based on the source codes and related parameters. The general procedure of our approach is shown in Fig. 1. Given the source codes of a physical join implementation, we first invoke running time code analysis tools like *Valgrind* to get the call-graph for the source codes, in which the nodes are for the functions, and the edges are for the function call relationships with weights representing the number of function calls. And then we get the join-graph by augmenting the call-graph with parameter nodes and edges. We propose ReGAT by extending RGAT [11] to handle the edge weights in two sub-graphs generated from two relationships. The hidden representations learned from two sub-graphs are then combined into one, which is fed into a regression model to predict the running time.

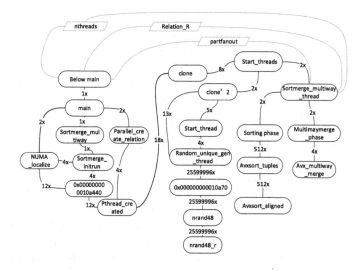

Fig. 2. An example of program heterogeneous join-graph.

3.3   Building Heterogeneous Join-Graphs

Code representation learning is a hot research topic in software engineering, and many existing works [9,20,27] parse code snippets into ASTs and design different models to learn embeddings of the ASTs. However, we claim ASTs are not suitable in our context due to the following reasons: First, ASTs model the computation flow on variables, which results in a very complex graph even for a code snippet with 100 lines. As a physical join implementation usually includes thousands of lines, the corresponding AST is too large. Second, we are not particularly concerned about the role of each variable. Instead, we are interested in expensive functions and their relationships. Therefore, we adopt a coarse-grained join-graph to represent source codes for a join operator.

We use the m-way sort merge join described in [4] to show the process of building a heterogeneous join-graph. First, we invoke *Valgrind* to monitor the running of the m-way sort merge join, and obtain its call-graph. Each black-lined node in the call-graph is for a function, and the edges represent their call relationships. For example, in Fig. 2, the "sorting phase" function node points to the "avxsort tuples" function node, indicating that the former function calls the latter, and "512x" on the edge indicates that the number of function calls equals 512.

Second, we annotate parameters nodes to function nodes to produce the join-graph, and attempt to capture the interactions between parameters. Ideally, we would like to connect parameters to functions associated with them one by one, but it is not easy to locate the functions precisely in practice, as the call-graph contains many built-in standard functions, and some functions are associated with the intermediate variables of parameters. To simplify the problem, we have three choices in this step. First, we can associate all parameters with all functions, but this yields a large number of unnecessary edges, which will contain noise for the learned model. Second, we can associate the parameters with the entry function, like the "below main" function in Fig. 2. Although the strategy is simple, it is not easy to propagate parameters to other functions far from the entry. Third, we can connect the parameters nodes with functions whose code snippets can be found in the source codes. We mainly adopt the third approach to construct the heterogeneous join-graph.

A generated join-graph containing black function nodes and orange parameter nodes is illustrated in Fig. 2. The heterogeneous join-graph $G = (V, E, R, C)$ with nodes $v_i \in V$, labeled edges (relations) $(v_i, v_j, r) \in E$, and relation types $r \in R$. The join-graph G contains two types of nodes V_f and V_p, representing the function nodes and parameter nodes, respectively, and contains two types of edges E_{ff} and E_{fp}, where E_{ff} is for the function call relationship between two nodes in V_f and E_{fp} contains edge between parameter node in V_p and function node V_f. The edges in E_{ff} and E_{fp} are undirected, as we expect that the features of different nodes can be propagated efficiently by the graph neural network in the following. For nodes i and j, we express the number of times node i calls node j by $c_{ij} \in C$, and we define $c_{ij} = 1$ if there exists an edge between a parameter node and a function node. Moreover, we use one-hot encodings to represent different function names, and we concatenate the values of the parameters as the features for parameter nodes.

3.4   ReGAT Model for Join-Graph Representation and Running Time Prediction

We propose ReGAT, a heterogeneous graph neural network, to learn the embedding of the join-graph. The join-graph is a special kind of heterogeneous graph containing two kinds of nodes and edges, and the weights are on one kind of edge. It seems that we can apply existing heterogeneous graph neural networks, such as RGCN [12], HAN [25], over the join-graph. Although these models can implicitly learn the dependencies between different nodes, they cannot fully incorporate the explicit edge weights (the number of function calls), which plays an important role in time estimating. Therefore, we extend the existing heterogeneous graph neural network to handle the edge weights, and devise ReGAT.

Specifically, to facilitate the calculation, we first normalize the times of function calls in a graph following the form:

$$\bar{c}_{ij} = \frac{c_{ij}}{\sum_{j \in N_i^r} c_{ij}}, \tag{1}$$

where N_i^r denotes the set of neighbor indices of node i under relation $r \in R$.

We see $\bar{c}_{ij}$ as a scalar weight on the edge from node i to node j and then compute a Gaussian bias:

$$\tilde{c}_{ij} = -\frac{(1 - \bar{c}_{ij})^2}{2\delta^2}, \tag{2}$$

where δ denotes the standard deviation, serving as a hyperparameter to control the significance of the number of function calls.

We combine this Gaussian bias with the previous attention in RGAT [11] to form a new attention mechanism in Eq. 3, where we adopt as a substitute for the attention in RGAT to learn the explicit and implicit importance between each node and its neighbors in the heterogeneous join-graph:

$$\alpha_{ijr} = \text{softmax}\left(e^l_{ijr} + \tilde{c}_{ij}\right),$$ (3)

$$e^l_{ijr} = \text{LeakyReLU}\left(a^T \left[W_0^{(l)} h_i \| W_r^{(l)} h_j\right]\right).$$ (4)

The hidden representation of nodes in $(l+1)^{th}$ layer in ReGAT can be formulated as the following equation:

$$h_i^{l+1} = \sigma\left(W_0^{(l)} h_i^{(l)} + \sum_{r \in R} \sum_{j \in N_i^r} \alpha_{ijr} W_r^{(l)} h_j^{(l)}\right),$$ (5)

where α_{ijr} is the attention score with edge weights between node i and node j. The final predicted time of join algorithm execution t^p is as follows:

$$t^p = MLP_t(h_{root}^L \| mean(\sum_{v \in V} (h_v^L))),$$ (6)

in which MLP_t is a Multi-Layer Perceptron model, h_{root}^L is the final embedding of the entry function node of join-graph, and h_v^L is the final embedding of nodes in the heterogeneous graph. As the life-time of the entry function equals the running time of the join operator, the learned embedding of the entry function should be a major part of the features for the time estimator. Besides, we consider the mean pooled embedding of all other nodes as another part. These two parts are concatenated as the input of the time prediction.

Mean Squared Error(MSE) $loss_t$ is then used to form the learning objective:

$$loss_t = \frac{1}{|S|} \sum_{G \in S} \left(t^l - t^p\right)^2,$$ (7)

where t^l and t^p are the labels and prediction time of the join operator.

3.5   Optimization

In this subsection, we propose two optimization strategies, one is to predict the number of function calls in the inference phase, another is to initialize the function node embeddings to enhance the generalization capability of the model.

A Multi-task Model to Predict the Times of Function Calls Simultaneously. The times of function calls should be captured from the join-graph so as to predict the running time more precisely. However, unlike the fixed structure of the join-graph, the times of function calls are ad-hoc and related with the input parameters. In the inference phase with different parameters, the times of function calls are not available actually unless we run the join implementation in the context of *Valgrind*. One possible solution is to set times of function calls to a constant, e.g., 1. We can see that such a method degrades the performance of the learned model due to the change in values between the training and inference phases.

In order to remedy the issue above, we leverage multi-task models and attempt to train a sub-model to predict times of function calls in the training phase simultaneously. The sub-model also requires a combination of content features and structural features of the nodes on the join-graph. Thus, we apply the similar ReGAT sub-model to the join-graph with the default edge weights (1 in our paper) to predict the times of function calls between any two function nodes in the sub-task:

$$\dot{H}^L = ReGAT_c(V, E, R, 1), \tag{8}$$

where $\dot{H}^L$ denotes the embedding of the function nodes learned by ReGAT sub-model.

The times of function calls prediction can be seen as a regression task, for which the call time c_{ij}^p between function node i and node j can be calculated as follows:

$$c_{ij}^p = MLP_c(\dot{h}_i^L || \dot{h}_j^L). \tag{9}$$

For supervised the times of function calls prediction, we can minimize the Mean Squared Error over all function call edges between the ground-truth and the prediction:

$$loss_c = \frac{1}{|S|} \sum_{G \in S} \sum_{i,j \in E_{call}} \frac{\left(\tilde{c}_{ij} - c_{ij}^p\right)^2}{|E_{call}|}, \tag{10}$$

where E_{call} includes all edges for function call relationships. Then, the final loss can be written as:

$$loss_f = \sigma * loss_c + (1 - \sigma) * loss_t, \tag{11}$$

σ is used to adjust the importance between the two tasks, by default we set it to 0.5.

InferCode for Function Nodes Features. The call-graph is a concise structure for source codes. Although providing the explicit function call relationships, the call-graph preserves limited features (only function name) over the function nodes. In the basic method, we have to represent different function names with one-hot encodings. However, that method will not only lose the rich semantics in the code snippets for the function nodes, but also make it challenging for JG2Time to predict the time when a new function name is met.

We plan to combine the advances of the AST-based learning into call-graph-based learning. The AST-based method runs at the granularity of a code snippet for a function. The function embedding with rich semantics learned from the AST-based method can enhance the function node features in the call-graph.

In this paper, we adopt InferCode [10], a self-supervised learning method, to obtain more features of the function node from the code snippets. InferCode does not require any extra human labeling, but relies on the shared common sub-trees in ASTs as the training signals, which enables two functions with similar code snippets to have close representation. In addition, InferCode runs on the code snippet for a function instead of the entire source codes for the join operator. The limited size of code snippets enables InferCode to run efficiently.

4   Experiment

4.1   Experiment Setting

Dataset. We get 6 join implementations from previous works, including 2 sort-merge join implementations (m-way sort-merge join and massively parallel sort-merge join) in [4], and 4 hash join implementations in [5,7]. The underlying data are generated with the same method described in [23]. We extract the 11 most important parameters P for the source codes of join implementations, such as the number of threads, the size of the input relation, the number of radix bits for partitioning, etc. As described in Sect. 3.1, we run O_i with different parameters P_j^i, and collect its running time t_j^i, resulting in (O_i, P_j^i, t_j^i) as one instance. We collect 1800 instances and split them into 600, 200, and 1000 instances as training set, validation set, and test set, respectively.

Hardware and Implementation. All our experiments are performed on a server with 256 GB RAM, two Intel (R) Xeon (R) Silver 4210R CPUs @ 2.40 GHz, one 40 MB L2 data cache, one 55 MB L3 data cache, two NUMA nodes-each with 20 CPUs and a 24 GB GeForce RTX 3090 GPU. The 6 join implementations used to create dataset are all implemented in C/C++ and compiled by GCC version 9.4.0. JG2Time and other competitors are implemented in PyTorch.

Competitors. To our best knowledge, there is no work to utilize the call-graph to predict the running time. Then, we select four representative related methods as our competitors.

- Multi Layer Perceptron(MLP). MLP relies on the parameters to predict the running time of the join operator without considering the source codes. We want to compare the MLP method with JG2Time to study whether the source codes really contribute to the prediction performance. In our experiments, MLP contains four multi-layer perceptions with ReLU activation functions.
- Transformer. Transformer [1] achieves significant progress in many NLP tasks. The source codes can be viewed as a sequence of tokens, fed into the transformer which can capture the long-distance relationships between tokens. We are interested in whether the token based source code representation can also predict the running time or not.

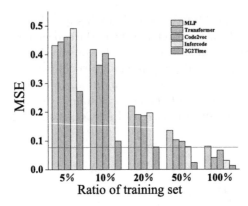

Fig. 3. Performance comparisons of different ratios of training set.

– Code2vec. Code2vec [3] learns representation from the AST for source codes. It extracts paths of any two leaf nodes in the AST, uses a neural network to learn the representation of each path, and aggregates paths through an attention mechanism to form the embedding of the code snippet.
– InferCode. InferCode [10] also works on the AST form for the source code. It learns the representation of the source code in a self-supervised way, by predicting whether a frequent sub-tree exists or not in the AST for a code snippet.

4.2   Results

Here, we report experimental results of competitors and JG2Time, study the performance of JG2Time with varying parameters, and investigate the generalization of different models to the join implementations with new functions.

Comparison of JG2Time Against Competitors. We adopt 5%, 10%, 20%, 50% and 100% ratios of the training set to train JG2Time and competitors. It can be seen intuitively from Fig. 3 that JG2Time always achieves the best performance under the same training ratios. In other words, JG2Time requires fewer training instances for a similar MSE loss. For example, the red line in Fig. 3 shows that JG2Time using 20% training instances achieves a MSE loss less than 0.1, while the competitors require even more than 50% training instances to reach such a loss.

Among all methods, MLP performances the worst, which shows that the source codes (no matter whether in the form of tokens, AST, or call-graph) can benefit from understanding the interaction between parameters, and consequently the running time prediction. We also notice that JG2Time significantly outperforms Code2Vec, InferCode, and Transformer, which illustrates that the coarse granularity join-graph is concise and effective in capturing the relationships between parameters and predicting running time.

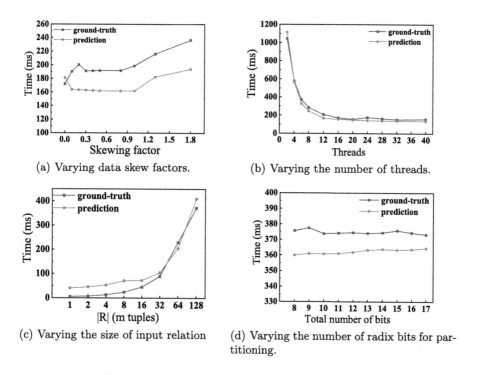

Fig. 4. Accuracy of JG2Time Varying Parameters.

Accuracy of JG2Time Varying Parameters. We study the performance of JG2Time using different parameters, and report the absolute predicted results with the ground-truth. Figure 4 depicts results for a partition-based hash join described in [5] varying data skew factors (zipf factors)(Fig. 4(a)), the number of threads (Fig. 4(b)), the size of input data (Fig. 4(c)), and the number of bits used in hash (Fig. 4(d)). We can see that the predicted results can match the ground-truth well.

Generalization to Join Implementations with New Functions. We also perform studies on whether the model learned can be applied to other join implementations with new functions. Specifically, we use four hash join implementations to train a new model, trying to predict the running time of the other two sort-merge join implementations that have never seen during training. The results are shown in Table 1. It can be seen that JG2Time achieves a much lower MSE loss than its competitors. The improvement comes from two aspects. First, JG2Time makes use of the rich information learned by InferCode on the code snippet for a function. Second, JG2Time captures the function call relationships and interactions between parameters and functions in one join-graph. We also notice from the results that using InfoCode alone is not sufficient.

Table 1. Performance of generalization to join implementations with new functions.

Method	MSE
MLP	0.823
Transformer	0.794
Code2vec	0.782
InferCode	0.771
JG2Time	0.278

Table 2. Ablation study.

Method	MSE
JG2Time$_{-infercode}$	0.0148
JG2Time$_{-function\ call\ times}$	0.0159
JG2Time$_{-annotating\ parameter\ nodes}$	0.0286
JG2Time	0.0122

4.3   Ablation Study

In this section, we evaluate three different variants of JG2Time:

- JG2Time$_{-InferCode}$. It uses one-hot encodings for different function names rather than embeddings learned by InferCode on code snippets for functions.
- JG2Time$_{-function\ call\ times}$. It does not include the sub-model to predict the number of function calls, and sets weights to 1 for all edges in the join graph in the inference phase.
- JG2Time$_{-annotating\ parameter\ nodes}$. Such a method associates all parameter nodes with all function nodes, instead of partial functions whose code snippets can be found in the source codes.

We use 100% training set for the ablation experiments. The results shown in Table 2 demonstrate that the design choices selected and optimization strategies are effective. The performance of JG2Time$_{-annotating\ parameter\ nodes}$ decreases the most, which reveals that the interaction between parameter nodes and function nodes is important to in the running time estimation. Node representations learned from InferCode can improve both generalization and the quality of predicted results, as they provide meaningful features to function nodes. The number of function calls is also helpful for time estimation. Note that even without predicting the number of function calls, the parameter nodes in the join graph can still provide explicit hints, and this may be the reason why JG2Time$_{-function\ call\ times}$ performs relatively well.

5   Conclusion and Future Work

In this paper, we propose JG2Time, an effective method for predicting the running time of join operators. JG2Time is based on a heterogeneous join-graph, which incorporates the parameters and call-graphs generated by running time analysis tools. We propose ReGAT to learn the representation of the join-graph, which can handle edge weights (the number of function calls). We also devise two optimization strategies, a multi-task model to learn the edge weights simultaneously, and an unsupervised learning method to obtain the features of function nodes to improve the generalization. Experimental results illustrate the effectiveness of JG2Time.

In the future, we plan to leverage the pre-trained model trained in massive source codes, which may capture more syntactic and semantic code features of function nodes, to further improve the generalization of JG2Time. In addition, we plan to enhance JG2Time with context information to predict the running time of concurrent join operations.

Acknowledgement. This work was partially supported by NSFC under Grant No. 62272008 and 61832001, and ZTE-PKU Joint Program.

References

1. Ahmad, W.U., Chakraborty, S., Ray, B., Chang, K.W.: A transformer-based approach for source code summarization. arXiv preprint arXiv:2005.00653 (2020)
2. Albutiu, M.C., Kemper, A., Neumann, T.: Massively parallel sort-merge joins in main memory multi-core database systems. arXiv preprint arXiv:1207.0145 (2012)
3. Alon, U., Zilberstein, M., Levy, O., Yahav, E.: code2vec: learning distributed representations of code. Proc. ACM Program. Lang. **3**(POPL), 1–29 (2019)
4. Balkesen, C., Alonso, G., Teubner, J., Özsu, M.T.: Multi-core, main-memory joins: sort vs. hash revisited. Proc. VLDB Endowment **7**(1), 85–96 (2013)
5. Balkesen, C., Teubner, J., Alonso, G., Özsu, M.T.: Main-memory hash joins on multi-core CPUs: tuning to the underlying hardware. In: 2013 IEEE 29th International Conference on Data Engineering (ICDE), pp. 362–373. IEEE (2013)
6. Bandle, M., Giceva, J., Neumann, T.: To partition, or not to partition, that is the join question in a real system. In: Proceedings of the 2021 International Conference on Management of Data, pp. 168–180 (2021)
7. Barber, R., et al.: Memory-efficient hash joins. Proc. VLDB Endowment **8**(4), 353–364 (2014)
8. Blanas, S., Li, Y., Patel, J.M.: Design and evaluation of main memory hash join algorithms for multi-core CPUs. In: Proceedings of the 2011 ACM SIGMOD International Conference on Management of data, pp. 37–48 (2011)
9. Büch, L., Andrzejak, A.: Learning-based recursive aggregation of abstract syntax trees for code clone detection. In: 2019 IEEE 26th International Conference on Software Analysis, Evolution and Reengineering (SANER), pp. 95–104. IEEE (2019)
10. Bui, N.D., Yu, Y., Jiang, L.: Infercode: self-supervised learning of code representations by predicting subtrees. In: 2021 IEEE/ACM 43rd International Conference on Software Engineering (ICSE), pp. 1186–1197. IEEE (2021)

11. Busbridge, D., Sherburn, D., Cavallo, P., Hammerla, N.Y.: Relational graph attention networks. arXiv preprint arXiv:1904.05811 (2019)
12. Chen, J., Hou, H., Gao, J., Ji, Y., Bai, T.: RGCN: recurrent graph convolutional networks for target-dependent sentiment analysis. In: Douligeris, C., Karagiannis, D., Apostolou, D. (eds.) KSEM 2019. LNCS (LNAI), vol. 11775, pp. 667–675. Springer, Cham (2019). https://doi.org/10.1007/978-3-030-29551-6_59
13. Fang, Z., He, Z., Chu, J., Weng, C.: SIMD accelerates the probe phase of star joins in main memory databases. In: Li, G., Yang, J., Gama, J., Natwichai, J., Tong, Y. (eds.) DASFAA 2019. LNCS, vol. 11448, pp. 476–480. Springer, Cham (2019). https://doi.org/10.1007/978-3-030-18590-9_70
14. Ha, H., Zhang, H.: DeepPerf: performance prediction for configurable software with deep sparse neural network. In: 2019 IEEE/ACM 41st International Conference on Software Engineering (ICSE), pp. 1095–1106. IEEE (2019)
15. Han, S., Wang, D., Li, W., Lu, X.: A comparison of code embeddings and beyond. arXiv preprint arXiv:2109.07173 (2021)
16. He, B., Luo, Q.: Cache-oblivious databases: limitations and opportunities. ACM Trans. Database Syst. (TODS) 33(2), 1–42 (2008)
17. He, B., et al.: Relational joins on graphics processors. In: Proceedings of the 2008 ACM SIGMOD International Conference on Management of Data, pp. 511–524 (2008)
18. Kipf, T.N., Welling, M.: Semi-supervised classification with graph convolutional networks. arXiv preprint arXiv:1609.02907 (2016)
19. Lang, H., Leis, V., Albutiu, M.-C., Neumann, T., Kemper, A.: Massively parallel NUMA-aware hash joins. In: Jagatheesan, A., Levandoski, J., Neumann, T., Pavlo, A. (eds.) IMDM 2013-2014. LNCS, vol. 8921, pp. 3–14. Springer, Cham (2015). https://doi.org/10.1007/978-3-319-13960-9_1
20. Mou, L., Li, G., Zhang, L., Wang, T., Jin, Z.: Convolutional neural networks over tree structures for programming language processing. In: Thirtieth AAAI Conference on Artificial Intelligence (2016)
21. :Ramadan, T., Islam, T.Z., Phelps, C., Pinnow, N., Thiagarajan, J.J.: Comparative code structure analysis using deep learning for performance prediction. In: 2021 IEEE International Symposium on Performance Analysis of Systems and Software (ISPASS), pp. 151–161. IEEE (2021)
22. Samoaa, H., Leitner, P.: An exploratory study of the impact of parameterization on JMH measurement results in open-source projects. In: Proceedings of the ACM/SPEC International Conference on Performance Engineering, pp. 213–224 (2021)
23. Schuh, S., Chen, X., Dittrich, J.: An experimental comparison of thirteen relational equi-joins in main memory. In: Proceedings of the 2016 International Conference on Management of Data, pp. 1961–1976 (2016)
24. Wagner, B., Kohn, A., Neumann, T.: Self-tuning query scheduling for analytical workloads. In: Proceedings of the 2021 International Conference on Management of Data, pp. 1879–1891 (2021)
25. Wang, X., et al.: Heterogeneous graph attention network. In: The World Wide Web Conference, pp. 2022–2032 (2019)
26. Wei, B., Li, G., Xia, X., Fu, Z., Jin, Z.: Code generation as a dual task of code summarization. In: Advances in Neural Information Processing Systems, vol. 32 (2019)
27. White, M., Tufano, M., Vendome, C., Poshyvanyk, D.: Deep learning code fragments for code clone detection. In: 2016 31st IEEE/ACM International Conference on Automated Software Engineering (ASE), pp. 87–98. IEEE (2016)

28. Yang, C., Xiao, Y., Zhang, Y., Sun, Y., Han, J.: Survey, benchmark, evaluation, and beyond, Heterogeneous network representation learning (2020)
29. Yu, X., Li, G., Chai, C., Tang, N.: Reinforcement learning with tree-lstm for join order selection. In: 2020 IEEE 36th International Conference on Data Engineering (ICDE), pp. 1297–1308. IEEE (2020)
30. Zhang, J., Wang, X., Zhang, H., Sun, H., Wang, K., Liu, X.: A novel neural source code representation based on abstract syntax tree. In: 2019 IEEE/ACM 41st International Conference on Software Engineering (ICSE), pp. 783–794. IEEE (2019)
31. Zhu, Z., Fan, X., Chu, X., Bi, J.: HGCN: a heterogeneous graph convolutional network-based deep learning model toward collective classification. In: Proceedings of the 26th ACM SIGKDD International Conference on Knowledge Discovery & Data Mining, pp. 1161–1171 (2020)

Time Series Data

Long-Tailed Time Series Classification via Feature Space Rebalancing

Pengkun Wang[1], Xu Wang[1], Binwu Wang[1], Yudong Zhang[1], Lei Bai[2(✉)], and Yang Wang[1(✉)]

[1] University of Science and Technology of China, Hefei 230000, China
{pengkun,wx309,wbw1995,zyd2020}@mail.ustc.edu.cn, angyan@ustc.edu.cn
[2] Shanghai AI Laboratory, Shanghai 200000, China
baisanshi@gmail.com

Abstract. Learning unbiased decision boundaries is crucial for time series classification. Real-world datasets typically exhibit long-tailed natures of class distributions, which results in an imbalanced feature space after training, i.e., decision boundaries will be easily biased towards dominant classes that dominate the feature space. However, existing methods mostly train models from artificially balanced datasets, making it still unclear how to deal with the long-tailed natures of time series data in real-world scenarios. Motivated by this question, we analyze the similarities and differences between long-tailed time series classification and general long-tailed recognition, and propose a Feature Space Rebalancing (FSR) strategy for time series classification, which works jointly from both representation and data perspectives. Specifically, from the representation perspective, we design Balanced Contrastive Learning (BCL), which avoids excessive intra-class compaction of tail classes by introducing a balanced supervised contrastive loss with hierarchical prototypes, resulting in a balanced feature space and better generalization. From the data perspective, we explore the effectiveness of traditional data augmentation on long-tailed distributions and propose an Adaptive Temporal Augmentation (ATA) to rebalance the potential feature space at the temporal level. Extensive experiments on multiple long-tailed time series datasets demonstrate its superiority, including different class distributions and imbalance ratios.

Keywords: Time series classification · Long-tailed recognition · Contrastive learning

1 Introduction

Time series classification (TSC) has been widely explored as it is associated with massive real-world applications and has a significant impact on human life [1,4,8,9,12,17,18,26,28]. For instance, some recent TSC methods [4,26] have spared no effort to introduce deep neural networks into the medical field, and

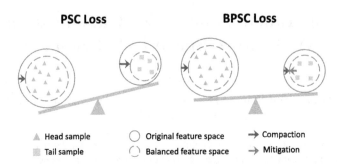

Fig. 1. BPSC loss dynamically rebalances the feature space with a class-dependent compaction factor. It is 'forgiving' to tail classes and avoids the final feature spaces of tail classes becoming over-compacted and biased.

realize intelligent disease diagnosis by automatically classifying the ECG signals of patients. Meanwhile, massive real data make it possible to train more complex deep models. Therefore, researchers are focusing on designing exquisite network structures or extracting semantically precise high-dimensional representations to further improve the discriminativeness of models.

Despite significant progress, most existing TSC methods [8, 26] focus on learning unbiased decision boundaries from artificially balanced datasets (i.e. all the classes have similar sample sizes). However, in the real world, class distributions of time series data typically exhibit long-tailed nature, which makes the decision boundaries easily biased towards the dominant classes with massive training data and thus decreases the classification accuracy. Although a few methods have considered class imbalance, they either only explored simple tasks with two categories [9,12] or seriously ignore the long-tailed nature [30]. Thus, several issues are worth pondering, e.g., 1) *What are the challenges of long-tailed time series classification relative to the other domain (especially vision)?* 2) *Are general long-tailed recognition methods applicable to the time series domain?* 3) *How to realize efficient long-tailed time series classification?* These issues are closely related to the scalability and generalization ability of long-tailed learning but have not been well-explored in the TSC area.

Motivated by the above problems, we analyze the **similarities** and **differences** between long-tailed time series classification (Long-tailed TSC) and general long-tailed recognition (GLR). When ignoring the difference in the data dimension, Long-tailed TSC and GLR can be regarded as homogeneous problems. For example, we can extract label-dependent information based on the overall class distribution of the dataset to achieve supervised inter-class balance or apply the idea of clustering to realize weakly-supervised/unsupervised inter-class balance. Therefore, existing methods of resampling or improving general cost-sensitive functions have certain generalizations to Long-tailed TSC, e.g., the improved Softmax for multi-class probability calculation [22]. However, unlike images, time series have unique temporal properties, which determines that we cannot solve Long-tailed TSC from a sample perspective alone. From the perspective of the data dimension, existing methods alleviate the long-tailed nature by designing complex network structures or semantically related processing modules, but they cannot be directly transferred to time series due to the inability to model temporal information. Moreover, the correlation information between variables is also impossible to model by existing long-tailed methods. The above inspires us to consider this peculiar property when solving Long-tailed TSC.

Motivated by this, based on a comprehensive consideration of temporal properties, we propose a novel Feature Space Rebalancing (FSR) strategy for long-tailed time series classification that works jointly from both representation and data perspectives. From the representation perspective, we design a Balanced Contrastive Learning (BCL) that consists of two key parts: balanced prototypical supervised contrastive (BPSC) loss and hierarchical prototypes. The former adjusts the degree of intra-class compaction by a class-dependent compaction factor when computing the intra-class prototype similarity, thus avoiding an imbalanced feature space. The latter considers the unique temporal properties of time series and introduces additional temporal prototypes when computing the contrastive loss, which are extracted by a simple temporal module. These hierarchical prototypes characterize time series more comprehensively, bringing the learned feature space closer to the essence of time series. These hierarchical prototypes comprehensively characterize time series, bringing the learned feature space closer to the essence of time series.

We also rebalance the feature space from the data perspective. It is well known that traditional data augmentation (e.g. jittering) can expand the feature space and improve the intra-class diversity, thereby making the class distribution closer to the true distribution [25]. It is a common practice to apply the same degree of data augmentation to all classes. However, there is a fact that is easily neglected in long-tailed datasets, that is, the feature distribution of the head class is relatively close to the true distribution because of the massive samples, while the feature distribution of the tail class may be biased because of the sparse samples and poor intra-class diversity. We propose Adaptive Temporal Augmentation (ATA) to alleviate this problem. The core idea is to assign different degrees of temporal augmentation (e.g. jittering) to each class according to the sample size, thus improving the intra-class diversity of the tail class and balancing the augmented feature space. Specifically, for a multivariate time series, we set independent degree-consistent augmentation for each variable, and the degree is determined by its label information.

Our contributions are summarized as follows:

- We comprehensively discuss the long-tailed time series classification learning and construct three corresponding long-tailed datasets. To the best of our knowledge, this is the first long-tailed time series classification work, which fills a gap in the field.
- To address the above Long-tailed TSC, we propose a novel Feature Space Rebalancing (FSR) strategy. First, we design a Balanced Contrastive Learning (BCL) from the representation perspective, which avoids imbalanced fea-

ture spaces by introducing compaction factors and hierarchical prototypes in the supervised contrastive loss. Second, from the data perspective, we rethink traditional data augmentation and propose an Adaptive Temporal Augmentation (ATA) to balance the augmented feature space.
- We conduct extensive experiments on the three proposed datasets and demonstrate that the proposed FSR is more suitable for long-tailed time series classification than existing methods.

2   Related Works

Time Series Classification. In recent years, extensive studies have been made on time series classification with deep neural networks [1,8,17,18,26]. These methods aim to achieve better model performance by designing exquisite network structures [1,18,26] or improving plug-and-play modules [8,17]. However, state-of-the-art methods mainly experiment with balanced datasets to demonstrate their capacity, ignoring the imbalance problem of real-world datasets. To this end, several studies propose diverse strategies to address imbalanced time series classification [9,12,30]. But these methods mainly focus on scenarios with a small number of classes (e.g. 2), while real-world scenarios usually have a large number of classes. In comparison, we construct three time series datasets with different class distributions and imbalance ratios for long-tailed natures, and propose FSR to model complex data distributions.

Long-Tailed Recognition. In real-world scenarios, class distributions typically exhibit long-tailed natures, which makes the trained model easily biased toward head classes with massive data [29]. Many methods have made efforts to address this class imbalance and they can be grouped into three categories: class re-balancing [3,6,20,22], information augmentation [27], and module improvement [14,19,24]. In the field of class re-balancing, a mainstream strategy is to design cost-sensitive loss functions to adjust loss values for different classes during training, e.g., CB loss [6] or balanced softmax loss [22]. Further, some methods combine improved cost-sensitive loss functions with well-designed network structures to achieve efficient long-tailed learning, e.g., decoupled training [15] or ensemble learning [24]. Most recent studies focus on general long-tailed recognition, however, limited effort has been made for long-tailed time series classification due to the lack of proper benchmarks. Inspired by this, we construct three benchmarks to fill the gap and design hierarchical prototypes based on temporal properties to ensure semantic consistency in contrastive learning.

Contrastive Learning. Contrastive learning has achieved outstanding success in self-supervised representation learning, which has profound implications for a variety of downstream tasks [5]. The basic principle of contrastive learning is to learn a high-dimensional semantic feature space by constructing positive and negative sample pairs, and attract the positive sample pairs and repulsing the negative sample pairs. Recent works also found that using contrastive loss in long-tailed learning can obtain representation models generating a better feature space [13,14,19,24]. It is worth noting that Hybrid [24] proposes a hybrid network structure with a prototypical supervised contrastive loss, which resolves the memory bottleneck resulting from standard supervised contrastive learning. However, this loss is not friendly to feature space balancedness, since imposing the same degree of intra-class constraints on all the classes would result in excessive intra-class compaction of tail classes. In our work, we focus on the adaptability in representing distance computations and propose a balanced prototypical supervised contrastive loss to avoid excessive intra-class compaction of tail classes.

3   Long-Tailed Time Series Classification

Conventional time series classification cannot cope with the long-tailed natures in real-world applications, resulting in poor performance of the trained model on tail classes. However, tail classes are critical for tasks such as abnormal activities in behavior recognition and rare conditions in disease diagnosis. Considering that no research has been explicitly investigated in this direction so far, we give a detailed problem definition and corresponding datasets to fill this gap.

3.1   Problem Definition

Conventional time series classification methods mostly train models on balanced datasets. Differently, long-tailed time series classification focuses on training a robust deep neural network from a time series dataset with a long-tailed class distribution. This long-tailed nature can be understood as the fact that a small number of classes have massive time series samples while other classes are only related to a few samples. More formally, let $\{x_i, y_i\}_{i=1}^{N}$ be the long-tailed time series training set, where each time series x_i corresponds to a class label y_i. Assuming that a dataset contains $\mathcal{C}$ classes, the sample number of class c is n_c, and the total number of the entire dataset is $N = \sum_{k=1}^{\mathcal{C}} n_c$, the imbalance ratio of the time series dataset can be defined as n_{max}/n_{min}, where n_{max} and n_{min} denote the sample size of the class containing the most and least samples, respectively. Without loss of generality, the class distribution in the training set exhibits a long-tailed trend when sorted by cardinality in decreasing order. Additionally, a time series dataset may be univariate or multivariate.

3.2   Proposed Datasets

Based on existing datasets, we construct three derived long-tailed time series classification datasets to fill the gaps in this field. Referring to mainstream datasets, we divide the classes of each dataset into head classes, medium classes, and tail classes according to the sample size. The visualized class distributions and statistics are shown in Fig. 2 and Table 1 respectively.

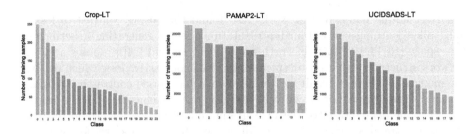

Fig. 2. Class distributions of proposed long-tailed time series classification datasets.

Table 1. Statistics of proposed datasets ($\star$ and $\#$ denote the "number" and "sample size" respectively).

Dataset	Crop-LT	PAMAP2-LT	UCIDSADS-LT
Variable$\star$	1	36	45
Class$\star$	24	12	19
Length	46	20	20
Training set$\star$	2,145	173,309	42,692
Validation set$\star$	1,200	6,000	9,500
Test set$\star$	16,800	12,000	19,000
Head classes$\star$	2 ($\# >200$)	2 ($\# >20000$)	4 ($\# >3000$)
Medium classes$\star$	5 ($200 \geq \# \geq 100$)	6 ($20000 \geq \# \geq 11000$)	9 ($3000 \geq \# \geq 1700$)
Tail classes$\star$	17 ($\# <100$)	4 ($\# <11000$)	6 ($\# <1700$)
Imbalance ratio	17	9	5

Crop-LT. Crop is a univariate dataset from the well-known UCR time series archive [7]. It consists of 24 classes with the same sample size and the length of each sample is 46. To better evaluate long-tailed time series classification methods, we resample a long-tailed training set, i.e., Crop-LT, which has 2 head classes, 7 medium classes, and 17 tail classes. The imbalance ratio is 17. The overall training set size is 2,145.

PAMAP2-LT. PAMAP2 is a multivariate benchmark for daily physical activity classification, and its data is collected with three IMUs placed on the subject's chest, dominant wrist, and dominant ankle, respectively, under the sampling frequency of 100 Hz [21]. The sampled dataset PAMAP2-LT contains 12 classes with a total of 173,309 training data and an imbalance ratio of 9. In PAMAP2-LT, each time series are acquired by 36 sensors, and the time step length of each stream is 20. Similar to the Crop-LT dataset, we define 2 head classes, 6 medium classes, and 4 tail classes.

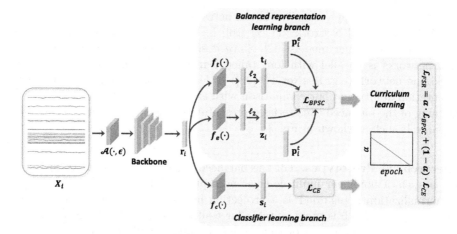

Fig. 3. Overview of the proposed feature space rebalancing based network structure. The network is hybrid, and it contains a balanced representation learning branch and a classifier learning branch. The former utilizes balanced contrastive learning to learn a balanced and unbiased feature space, while the latter is a traditional classification strategy. During training, the curriculum learning is used to achieve a smooth transition from the balanced representation learning branch to the classifier learning branch.

UCIDSADS-LT. UCIDSADS is a multivariate benchmark specially devised for daily and sports activities, which comprises the motion sensor data of 19 daily and sports activities [2]. The samples in this benchmark are acquired by 45 sensors at the sampling frequency of 25 Hz. We sample 42,692 time series from the original dataset to form the UCIDSADS-LT dataset with an imbalance ratio of 5, and the length of each time series is 20. Like the PAMAP2-LT dataset, we define 4 head classes, 9 medium classes, and 6 tail classes.

4 Methodology

Based on the above problem definition, we try to solve the problem: *How to realize efficient long-tailed time series classification?* To this end, we propose a feature space rebalancing (FSR) strategy that consists of two parts: balanced contrastive learning (BCL) and adaptive temporal augmentation (ATA). Motivated by [24], we design an improved hybrid network for Long-tailed TSC as shown in Fig. 3. Furthermore, in the initialization phase, we incorporate the ATA module into the framework.

4.1 Balanced Contrastive Learning

Balanced contrastive learning aims to learn a balanced feature space that balances head and tail classes while achieving intra-class compactness and inter-class separability, thereby helping the classifier learn unbiased decision boundaries. For a time series (x_i, y_i), its corresponding high-dimensional representation

r_i can be generated by various backbone networks, such as the widely used Temporal Convolutional Network (TCN) [16], LSTM [11], and more powerful and advanced Transformer models [32]. Since designing more powerful representation networks is parallel with long-tailed learning, we simply utilize a shared backbone network (e.g. ResNet-TS [28]) to learn r_i considering its stable representation ability and fair comparison with existing methods, which drives the learning of the balanced time series representation learning branch and classifier learning branch as shown in Fig. 3.

Hierarchical Prototypes. For the balanced time series representation learning branch, a nonlinear multiple-layer perceptron (MLP) $f_e(\cdot)$ combined with ℓ_2 normalization is regarded as a projection head to map r_i into a vector representation z_i, which is more suitable for contrastive learning. Considering the unique temporal properties of time series, we additionally use an LSTM $f_t(\cdot)$ to extract the corresponding temporal representation t_i from r_i. $f_t(\cdot)$ aggregates high-dimensional representations into compact representations, which also combines with ℓ_2 normalization.

$$\mathbf{z}_i = \ell_2(f_e(\mathbf{r}_i)), \quad \mathbf{t}_i = \ell_2(f_t(\mathbf{r}_i)). \tag{1}$$

For these two different perspectives of representation, we utilize their average representation as prototype representations, which are defined as $\mathbf{p}_i^e$ and $\mathbf{p}_i^t$.

Balanced Prototypical Supervised Contrastive Loss. As we mentioned, the prototypical supervised contrastive (PSC) loss can resolve the memory bottleneck issue by learning a prototype for each class [24]. For a long-tailed dataset with $\mathcal{C}$ classes, the goal of PSC is to learn a prototype feature for each class during training and guide the vector representation to be closer to the prototype of their class and far away from the prototypes of other classes. The formulation of PSC loss can be written as

$$\mathcal{L}_{PSC}(\mathbf{z}_i) = -log\frac{e^{(\mathbf{z}_i \cdot \mathbf{p}_{y_i}/\tau)}}{\sum_{j=1, j \neq y_i}^{\mathcal{C}} e^{(\mathbf{z}_i \cdot \mathbf{p}_j/\tau)}}, \tag{2}$$

where $\tau > 0$ is a scalar temperature parameter, and $\mathbf{p}_{y_i}$ is the prototype representation for class y_i, which is normalized to the unit hypersphere.

Although reducing memory consumption, PSC imposes the same degree of intra-class constraints on all the classes. As shown in Fig. 1, for the head classes, this constraint enforces a more compact feature space, thus mitigating their intrusion into tail classes. For the tail classes, their feature spaces also become more compact. However, the prototype representations of these tail classes learned by the model may be biased due to the small sample size, which leads to the final feature spaces of these classes being over-compacted and biased. Overall, this balanced constraint will exacerbate the feature space imbalancedness and impair the generalization of the model on the tail classes.

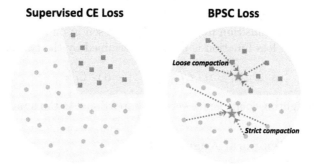

Fig. 4. Feature spaces learnd by our BPSC loss. Compared with the supervised cross-entropy (CE), our BPSC loss learns a balanced feature space. The *star* indicates the prototype of each class, and the *shadow area* indicates the decision boundary.

To alleviate the above-mentioned problem, we propose to impose different degrees of constraints on different classes and design a balanced prototypical supervised contrastive (BPSC) loss as

$$\mathcal{L}_{BPSC}(\mathbf{z}_i) = -log \frac{\omega_i \cdot e^{(\mathbf{z}_i \cdot \mathbf{P}^e_{y_i}/\tau)}}{\sum_{j=1,j\neq y_i}^{C} e^{(\mathbf{z}_i \cdot \mathbf{P}^e_j/\tau)}}$$
$$+ \beta \cdot \left(-log \frac{\omega_i \cdot e^{(\mathbf{z}_i \cdot \mathbf{P}^t_{y_i}/\tau)}}{\sum_{j=1,j\neq y_i}^{C} e^{(\mathbf{z}_i \cdot \mathbf{P}^t_j/\tau)}}\right) \qquad (3)$$
$$with \quad \omega_i = \frac{n_{max}}{n_i},$$

where β is a weighting coefficient of the temporal property, and ω_i represents the compaction factor, which leads the contrastive loss to be more 'forgiving', i.e., maintaining strict intra-class compaction for head classes and relatively loose intra-class compaction for tail classes. According to the theory that a balanced feature space helps to learn high-quality representations, the compaction factor can avoid excessive intra-class compaction of tail classes to a certain extent, resulting in a balanced inter-class feature space, as illustrated in Fig. 4.

Nevertheless, we found that the compaction factor of the tail class tends to be 0 when the imbalance rate of the dataset is high, which leads to overly loose feature space and thus fails to learn high-quality representations. For this mixed blessing, we propose a mitigative compaction factor

$$\omega_i = e^{-n_{max}/n_i \cdot \rho_b}. \qquad (4)$$

By adjusting the mitigation coefficient ρ_b, we can obtain the optimal intra-class compaction which is beneficial to balancing the feature space and learning high-quality representations.

Moreover, the classifier learning branch is simpler which applies a single linear layer $f_c(\cdot)$ to $\mathbf{z}_i$ to predict the class-wise logits $\mathbf{s}_i$. During the training process, we also employ a curriculum [31] to adjust the weightings of these two branches to realize a smooth transition from balanced representation learning to classifier learning. The final loss function is jointly determined by the two branches:

$$\mathcal{L}_{FSR} = \alpha \cdot \mathcal{L}_{BPSC} + (1 - \alpha) \cdot \mathcal{L}_{CE}, \tag{5}$$

where α is a weighting coefficient inversely proportional to the number of epochs.

4.2 Adaptive Temporal Augmentation

Mainstream long-tailed learning methods focus on model structure and representation, while data augmentation has received little attention. It has been proven that traditional data augmentation can enlarge the feature space and increase the intra-class diversity, and is beneficial to improving the generalization of the model [25]. A consensus approach is to apply the same degree of data augmentation to all the classes, which is not suitable for long-tailed datasets due to high imbalance ratios. Specifically, the feature space of the tail class is usually smaller than that of the head class for the long-tailed dataset. Adding the same degree of augmentation will expand the feature space of both the head and tail classes, leaving the feature space imbalancedness still existing. If we want to get a balanced feature space, it is necessary to balance the inter-class diversity as much as possible, which is similar to the idea of resampling.

To achieve this goal, we propose an Adaptive Temporal Augmentation (ATA) to assign different degrees of augmentation to each class according to the sample size. We define a parametric temporal augmentation method (e.g. jittering) as $\mathcal{A}(\cdot, \epsilon)$, where ϵ is the augmentation factor. Then, the augmented sample is

$$\hat{x}_i = \mathcal{A}(x_i, \epsilon). \tag{6}$$

For example, when using temporal jittering augmentation, we append an independent degree-consistent noise sequence to each variable in a multivariate time series.

In traditional data augmentation, the augmentation factor ϵ is a constant for all classes. In adaptive temporal augmentation, our goal is to balance the inter-class diversity, so the augmentation factor of the head class will be smaller than that of the tail class, thus maintaining the stability of the head class and the diversity of the tail class. Similar to BCL, it is better to make the model have better generalization performance on the head class, so we propose a mitigative augmentation factor

$$\epsilon_i' = \epsilon \cdot e^{-\frac{n_i}{n_{max}} \cdot \rho_a}, \tag{7}$$

where ρ_a is the mitigation factor. If $n_i > n_j$, then $\epsilon_i' < \epsilon_j'$. The mitigative augmentation factor forces the inter-class diversity to be closer, and ultimately, the trained model not only learns a more balanced feature space but also recognizes tail classes generalized.

Table 2. Performance on univariate Crop-LT dataset and multivariate PAMAP2-LT and UCIDSADS-LT datasets.

Method	Crop-LT				PAMAP2-LT				UCIDSADS-LT			
	Head	Medium	Tail	All	Head	Medium	Tail	All	Head	Medium	Tail	All
CE [10]	89.65	46.77	57.93	58.25	89.25	71.30	74.25	75.28	82.15	72.87	70.51	74.08
Focal [20]	90.15	46.37	53.42	55.01	93.90	72.90	73.83	76.71	85.45	70.06	65.45	71.84
CB [6]	79.43	38.63	63.19	59.43	90.75	70.87	76.73	76.13	79.85	72.67	72.77	74.21
LDAM [3]	85.93	45.11	62.14	60.58	85.65	71.17	73.38	74.32	77.95	71.23	66.82	71.25
BS [22]	81.14	44.60	63.71	61.18	89.35	69.80	78.30	75.89	78.08	76.28	67.35	73.84
Seesaw [23]	85.29	47.69	61.53	60.62	92.35	72.12	75.03	76.46	79.70	72.14	80.53	76.38
Hybrid [24]	87.57	42.14	59.96	58.55	92.35	69.40	73.40	74.56	80.22	72.04	67.53	72.34
KCL [14]	88.21	47.98	62.84	61.85	92.15	70.82	75.27	75.86	80.33	71.57	71.95	73.53
TSC [19]	88.01	48.10	63.27	62.17	91.80	71.03	77.24	76.56	81.59	74.37	72.79	75.39
FSR	89.75	50.29	66.31	**64.93**	93.10	73.70	80.55	**79.05**	82.55	77.48	78.57	**78.89**

5   Experiments

Implementation Details. For all three datasets, for a fair comparison with existing long-tailed learning methods, we use the general time series feature extraction network ResNet-TS [28] as the backbone network. For FSR, $f_e(\cdot)$ is a nonlinear MLP with one hidden layer, $f_t(\cdot)$ is a two-layer LSTM, and $f_c(\cdot)$ is a single linear layer. We use Pytorch to implement all neural networks and train the model on 8 NVIDIA Tesla V100 GPUs. The networks are trained for 200 epochs by the Adam optimizer with a learning rate of 10^{-4} and weight decay of 4×10^{-3}. For Crop-LT dataset, the batch size is 128, the weighting coefficient α is $1 - (Epoch_{now}/Epoch_{max})$, and β is 0.5. For PAMAP2-LT and UCIDSADS-LT datasets, the batch size is 256, the weighting coefficient α is $1 - (Epoch_{now}/Epoch_{max})^2$, and β is 0.9. In BCL, we set the mitigation coefficient ρ_b to be 0.5. In ATA, we use a jittering with an augmentation factor ϵ of $(0, 0.1)$ and mitigation factor ρ_a of 1 as augmentation.

Compare with State-of-the-Art Methods. The comparison between the proposed FSR and state-of-the-art methods on three long-tailed datasets is presented in Table 2. Based on the partitioning of the datasets above, we show the average accuracy on all classes, and also on each subset. For a comprehensive comparison, in addition to the cross-entropy (CE) loss, we select a variety of long-tailed recognition methods as baselines, which are based on *different theoretical ideas*, including class-level re-weighting [6,20,23], class-level re-margining [3],class-balanced re-sampling [22], metric learning [24], and decoupled training [14,19]. And our proposed FSR can be classified into metric learning or a new class-balanced augmentation.

As can be seen from the table, on the univariate Crop-LT dataset, the existing long-tailed learning methods can generally improve the overall classification accuracy compared to CE. FSR is no exception, it outperforms the compared methods on the medium and tail subsets. In particular, it outperforms CE by 3.52% and 8.38%, respectively. However, the performance of the compared meth-

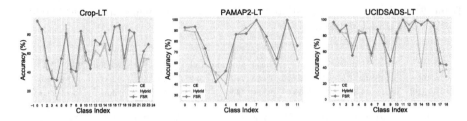

Fig. 5. Visualization of the accuracy of each class on three proposed long-tailed time series classification datasets.

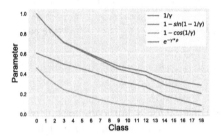

Fig. 6. Visualization trends of the proposed mitigation compaction factor and its variants.

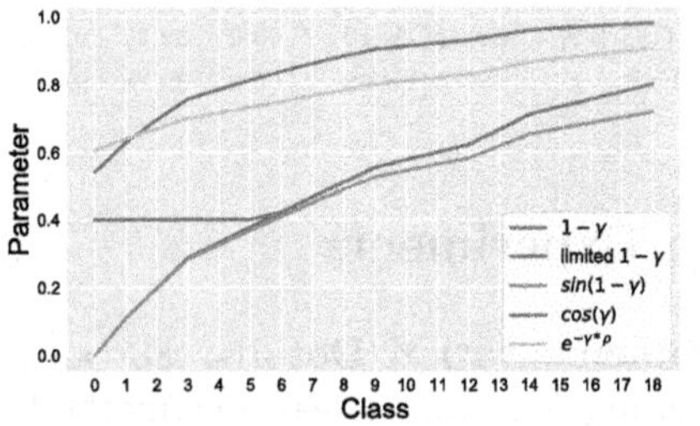

Fig. 7. Visualization trends of the proposed mitigative augmentation factor and its variants.

ods is unstable on the multivariate PAMAP2-LT and UCIDSADS-LT datasets. It is obvious that the method based on the improved loss function cannot effectively improve the performance on the tail class, due to the imbalanced feature space learned by the single-stage classifier learning. The recent mainstream methods based on contrastive learning have a steady improvement on the tail class, but the overall accuracy is not excellent. The reason is that although a balanced feature space is learned, the temporal information is ignored. Our methods address such limitations in that: 1) Hierarchical prototypes that consider temporal information; 2) Class-dependent intra-class compaction that balances the feature space; 3) Adaptive augmentation that improves tail class diversity.

Visualization of Accuracy on Each Class. To more intuitively observe the superiority of FSR, we visualize the accuracy of each class in Fig. 5. It can be found that compared with baselines, the accuracy of FSR is more stable, which means that it balances all classes as much as possible, especially the difficult ones. In the head class, the accuracy of FSR is slightly better than Hybrid, because the mitigative augmentation factor also improves the diversity of the head class. In the medium and tail classes, FSR significantly outperforms baselines, which benefits from the 'forgiving' of balanced contrastive learning for these classes.

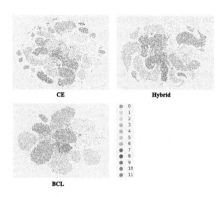

Fig. 8. T-SNE visualizations of representations of CE, Hybrid, and our proposed BCL on the PAMAP2-LT dataset. Different colors represent different classes.

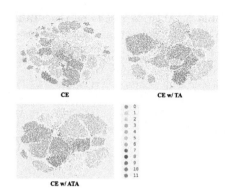

Fig. 9. T-SNE visualizations of representations of CE, CE with TA, and CE with ATA on the PAMAP2-LT dataset. Different colors represent different classes.

Table 3. Comparison of the proposed mitigation compaction factor and its variants on the UCIDSADS-LT dataset.

δ	Head	Medium	Tail	All
Hybrid	80.22	72.04	67.53	72.34
$1/\gamma$	80.17	74.39	72.82	75.11
$1 - sin(1 - 1/\gamma)$	80.30	74.87	73.62	75.62
$1 - cos(1/\gamma)$	80.89	76.78	74.05	76.78
$e^{-\gamma \cdot \rho}$	81.75	77.50	76.43	78.06

Table 4. Comparison of the proposed mitigative augmentation factor and its variants on the UCIDSADS-LT dataset.

δ	Head	Medium	Tail	All
CE	82.15	72.87	70.51	74.08
$1 - \gamma$	79.50	73.43	75.33	75.30
$limited \; 1 - \gamma$	81.14	73.60	76.21	76.01
$sin(1 - \gamma)$	80.16	73.57	75.92	75.70
$cos(\gamma)$	81.50	73.81	76.47	76.27
$e^{-\gamma \cdot \rho}$	82.55	74.00	77.30	76.84

Mitigative Compaction Factor for BCL. To illustrate the rationality of the proposed mitigative compaction factor in BCL, we compare different variants on the UCIDSADS-LT dataset. Here, we use γ to denote n_{max}/n_i. Figure 6 shows the changing trend of different variants. From the results in Table 3, the proposed compaction factor can significantly alleviate the imbalance problem caused by traditional compaction. In addition, the mitigative compaction factor ensures that the tail classes can also be reasonably compacted.

Mitigative Augmentation Factor in ATA. In ATA, the mitigative augmentation factor is crucial for adaptive augmentation. Here, we use γ to denote n_i/n_{max}, and δ to denote the dynamic coefficient of ϵ'_i, then

$$\epsilon'_i = \epsilon \cdot \delta = \epsilon \cdot e^{-\gamma \cdot \rho_a}. \tag{8}$$

To illustrate the rationality of the proposed dynamic coefficient, we compare different variants of δ on the UCIDSADS-LT dataset. As shown in Fig. 7, these

Table 5. Ablation results on the PAMAP2-LT dataset.

Method	Head	Medium	Tail	All
BCL	92.40	71.45	78.23	77.20
w/o BPSC & HP	92.35	69.40	73.40	74.56
w/o BPSC	92.48	70.75	76.22	76.20
w/o HP	92.27	71.24	77.87	76.96

Table 6. Comparison between baselines, baselines with TA, and baselines with ATA.

Method	Crop-LT				UCIDSADS-LT			
	Head	Medium	Tail	All	Head	Medium	Tail	All
CE	89.65	46.77	57.93	58.25	82.15	72.87	70.51	74.08
w/ TA	90.28	46.06	58.57	58.61	77.07	73.40	75.77	74.92
w/ ATA	89.72	42.08	60.97	59.43	82.55	74.00	77.30	76.84
BCL	87.86	49.29	63.96	62.90	81.75	77.50	76.43	78.06
w/ TA	86.07	45.14	64.29	62.11	81.33	75.57	74.95	76.59
w/ ATA	89.75	50.29	66.31	64.93	82.55	77.48	78.57	78.89

variants have different trends, some increase rapidly from 0, and some increase slowly from a lower limit. Here, we apply these variants to CE. From the results in Table 4, we have two intuitive findings: 1) the class-dependent augmentation factor can improve the generalization of the model; 2) the fixed lower limit or mitigative increase from a lower limit is a better choice because it can increase the diversity of all classes and not just the tail class.

T-SNE Visualization. To demonstrate that the representations learned by BCL and ATA can distinguish different classes in the latent space, we visualize the representations of different methods on the PAMAP2-LT dataset by T-SNE. As shown in Fig. 8, compared with Hybrid, our proposed BCL based on the compaction factor learns a more balanced feature space, so that the model can better recognize the tail classes. The effect of ATA is also significant, in Fig. 9, we can observe that the feature space of the tail class is expanded without being suppressed by the head class.

Ablation Study for Balanced Contrastive Learning. In balanced contrastive learning, balanced prototypical supervised contrastive (BPSC) loss and hierarchical prototypes (HP) are the core of recognizing long-tailed time series and rebalancing feature space. Without using ATA, we compare BCL and its variants on the PAMAP2-LT dataset. As shown in Table 5, when only BPSC is used, the accuracy of the model on the medium and tail classes is significantly improved, proving that BPSC can obtain a more balanced feature space than PCS. When only HP is used, both prototypes accurately model the time series and correct the bias of the model, resulting in an overall improvement in accuracy on each subset. And when neither is used, it is obvious that the performance of the model degrades significantly.

Ablation Study for Adaptive Temporal Augmentation. To demonstrate the effectiveness of adaptive temporal augmentation (ATA), we apply ATA and traditional temporal augmentation (TA) to CE and BCL, respectively. The experimental results on Crop-LT and UCIDSADS-LT datasets are shown in Table 6. For CE, although TA improves the generalization of the model, ATA achieves a more significant improvement, especially for tail classes. This phenomenon proves that ATA helps to improve the diversity of tail classes, thereby making the feature space more balanced. For BCL, using TA leads to a decrease in the accuracy of the model, while using ATA still improves performance steadily. From a representation perspective, we argue that a balanced and unbiased feature space helps learn accurate prototypes, while the prototypes learned by TA are biased, which leads to a significant decrease in model performance.

6   Conclusion

In this work, we construct three long-tailed time series classification datasets and propose a feature space rebalancing strategy, FSR. To the best of our knowledge, this is the first long-tailed time series classification method, which fills a gap in the field. We rebalance the feature space from two perspectives, including representation-based balanced contrastive learning and data-based adaptive temporal augmentation. Experiments on the three proposed datasets demonstrate the superiority of FSR.

Acknowledgements. This paper is partially supported by the National Natural Science Foundation of China (No. 62072427, No. 12227901), the Project of Stable Support for Youth Team in Basic Research Field, CAS (No. YSBR-005), Academic Leaders Cultivation Program, USTC.

References

1. Bai, L., Yao, L., Wang, X., Kanhere, S.S., Guo, B., Yu, Z.: Adversarial multi-view networks for activity recognition. Proc. ACM Interact. Mob. Wearable Ubiquit. Technol. **4**, 1–22 (2020)
2. Barshan, B., Yüksek, M.C.: Recognizing daily and sports activities in two open source machine learning environments using body-worn sensor units. Comput. J. **57**, 1649–1667 (2014)
3. Cao, K., Wei, C., Gaidon, A., Aréchiga, N., Ma, T.: Learning imbalanced datasets with label-distribution-aware margin loss. In: Proceedings of NeurIPS (2019)
4. Chen, H., Huang, C., Huang, Q., Zhang, Q., Wang, W.: ECGadv: generating adversarial electrocardiogram to misguide arrhythmia classification system. In: Proceedings of AAAI (2020)
5. Chen, T., Kornblith, S., Norouzi, M., Hinton, G.E.: A simple framework for contrastive learning of visual representations. In: Proceedings of ICML (2020)
6. Cui, Y., Jia, M., Lin, T., Song, Y., Belongie, S.J.: Class-balanced loss based on effective number of samples. In: Proceedings of CVPR (2019)
7. Dau, H.A., et al.: The UCR time series archive. IEEE CAA J. Autom. Sinica **6**, 1293–1305 (2019)
8. Dempster, A., Schmidt, D.F., Webb, G.I.: MiniRocket: a very fast (almost) deterministic transform for time series classification. In: Proceedings of KDD (2021)
9. Deng, G., Han, C., Dreossi, T., Lee, C., Matteson, D.S.: IB-GAN: a unified approach for multivariate time series classification under class imbalance. In: Proceedings of SDM (2022)

10. He, K., Zhang, X., Ren, S., Sun, J.: Deep residual learning for image recognition. In: Proceedings of CVPR (2016)
11. Hochreiter, S., Schmidhuber, J.: Long short-term memory. Neural Comput. **9**, 1735–1780 (1997)
12. Huang, H., Xu, C., Yoo, S., Yan, W., Wang, T., Xue, F.: Imbalanced time series classification for flight data analyzing with nonlinear granger causality learning. In: Proceedings of CIKM (2020)
13. Jiang, Z., Chen, T., Chen, T., Wang, Z.: Improving contrastive learning on imbalanced data via open-world sampling. In: Proceedings of NeurIPS (2021)
14. Kang, B., Li, Y., Xie, S., Yuan, Z., Feng, J.: Exploring balanced feature spaces for representation learning. In: Proceedings of ICLR (2021)
15. Kang, B., et al.: Decoupling representation and classifier for long-tailed recognition. In: Proceedings of ICLR (2020)
16. Lea, C., Flynn, M.D., Vidal, R., Reiter, A., Hager, G.D.: Temporal convolutional networks for action segmentation and detection. In: Proceedings of CVPR (2017)
17. Lee, D., Lee, S., Yu, H.: Learnable dynamic temporal pooling for time series classification. In: Proceedings of AAAI (2021)
18. Li, G., Choi, B., Xu, J., Bhowmick, S.S., Chun, K., Wong, G.L.: ShapeNet: a shapelet-neural network approach for multivariate time series classification. In: Proceedings of AAAI (2021)
19. Li, T., et al.: Targeted supervised contrastive learning for long-tailed recognition. In: Proceedings of CVPR (2022)
20. Lin, T., Goyal, P., Girshick, R.B., He, K., Dollár, P.: Focal loss for dense object detection. In: Proceedings of ICCV (2017)
21. Reiss, A., Stricker, D.: Introducing a new benchmarked dataset for activity monitoring. In: 16th International Symposium on Wearable Computers, ISWC 2012, Newcastle, United Kingdom, 18–22 June 2012 (2012)
22. Ren, J., et al.: Balanced meta-softmax for long-tailed visual recognition. In: Proceedings of NeurIPS (2020)
23. Wang, J., et al.: Seesaw loss for long-tailed instance segmentation. In: Proceedings of CVPR (2021)
24. Wang, P., Han, K., Wei, X., Zhang, L., Wang, L.: Contrastive learning based hybrid networks for long-tailed image classification. In: Proceedings of CVPR (2021)
25. Wen, Q., et al.: Time series data augmentation for deep learning: a survey. In: Proceedings of IJCAI (2021)
26. Yue, Z., et al.: TS2Vec: towards universal representation of time series. In: Proceedings of AAAI (2022)
27. Zang, Y., Huang, C., Loy, C.C.: FASA: feature augmentation and sampling adaptation for long-tailed instance segmentation. In: Proceedings of ICCV (2021)
28. Zha, D., Lai, K.H., Zhou, K., Hu, X.: Towards similarity-aware time-series classification. In: Proceedings of SDM (2022)
29. Zhang, Y., Kang, B., Hooi, B., Yan, S., Feng, J.: Deep long-tailed learning: a survey. CoRR (2021)
30. Zhao, P., et al.: T-SMOTE: temporal-oriented synthetic minority oversampling technique for imbalanced time series classification. In: Proceedings of IJCAI (2022)
31. Zhou, B., Cui, Q., Wei, X., Chen, Z.: BBN: bilateral-branch network with cumulative learning for long-tailed visual recognition. In: Proceedings of CVPR (2020)
32. Zhou, H., et al.: Informer: beyond efficient transformer for long sequence time-series forecasting. In: Proceedings of AAAI (2021)

Flow-Based End-to-End Model for Hierarchical Time Series Forecasting via Trainable Attentive-Reconciliation

Shiyu Wang[✉], Yinbo Sun, Yan Wang, Fan Zhou, Lin-Tao Ma, James Zhang, and YangFei Zheng

Antgroup, Hangzhou 310000, China
{weiming.wsy,yinbo.syb,luli.wy,hanlian.zf,lintao.mlt,james.z,
yangfei.zyf}@antgroup.com

Abstract. Time Series (TS) is one of the most common data formats in modern world, which often takes hierarchical structures, and is normally complicated with non-Gaussian and non-linear properties. Many businesses rely on accurate TS forecasting, under these complications, to help with operational efficiencies. In this paper, we present a novel approach for Hierarchical Time Series (HTS) prediction via trainable attentive reconciliation and Normalizing Flow (NF), which is used to approximate the complex (normally non-Gaussian) data distribution for multivariate TS forecasting. To reconcile the HTS data, we also propose a new flexible reconciliation strategy via the attention-based encoder-decoder neural network, unlike the existing methods with strong assumptions, such as unbiased forecasts and Gaussian noises. In addition, by using the reparameterization trick, we are able to combine forecasting and reconciliation into a trainable end-to-end model. Extensive empirical evaluations are conducted on real-world hierarchical datasets and the preliminary results demonstrate the efficacy of our proposed method.

Keywords: hierarchical time series · reconciliation · normalizing flow · attention · neural networks

1 Introduction

Multivariate *time series* (TS) forecasting with hierarchical structure has become increasingly more important in real-world applications [2,10], e.g., commercial organizations often want to forecast logistics demands/sales simultaneously at store, city, and state levels [16]. Smart grid for electricity also forms a natural hierarchy with different levels of aggregations [18]. Typically, forecasting tasks at different levels are for distinct purposes, i.e., bottom-level forecasts are often about the specific item to help with the micro-tactical decision, while upper-level forecasts look at the whole system to assist macro-strategic decisions, e.g., in the sales forecast application [1], predicting the bottom-level store can help to manage the inventory, while predicting the upper-level cities or states can facilitate

X. Wang et al. (Eds.): DASFAA 2023, LNCS 13943, pp. 167–176, 2023.
https://doi.org/10.1007/978-3-031-30637-2_11

mid- and long-term strategies. These applications demand better predictions by exploiting the information available across the hierarchy while maintaining the coherence that upper-level forecasts are the sum of bottom-level forecasts [7]. The classic approaches for this goal by reconciling forecasts can be found in [8].

The reconciliation method, i.e., producing predictions for all variables and making a subsequent adjustment to ensure these predictions adhere to known linear constraints [11], has recently motivated active research on hierarchical forecasting. Previous works first generate independently the base forecast from each TS in the hierarchy, and then revises the base forecast (i.e., reconciliation) to ensure coherency [17], e.g. MinT [20] is the general reconciliation procedure that revises unbiased independent univariate base forecasts, minimizing variances of forecast errors. However, these existing works face the following challenges: **(i)** The base forecast is obtained independently for each TS in the hierarchy, without exploiting the information available across all levels. **(ii)** Data in *Hierarchical Time Series* (HTS) is often non-Gaussian/non-linear due to the nature of data and the statistical discrepancies in TS at each level. **(iii)** The existing methods are based on strong assumptions, e.g., all forecasts are unbiased (assuming both the base forecasts and the revised forecasts are unbiased [6]), and noise distributions are Gaussian. **(iv)** The state-of-the-art methods mostly follow the two-stage approach mentioned above, and in the reconciliation stage, the base forecasts are revised without any regard to the learned model parameters.

In this paper, we propose a novel trainable end-to-end approach for hierarchical forecast that combines two independent components, i.e., *Flow-Based Multivariate Probabilistic Time-series Forecasting* and *Trainable Attentive Reconciliation* (employing the reparameterization trick [9]). **First**, we generate the base forecast by the Multivariate Probabilistic Forecasting [14] to exploit the information available across all levels, and employ *Normalizing Flow* (NF) [12] to handle the complex non-Gaussian/non-linear data distributions. **Second**, to address the problems of the current reconciliation methods mentioned above, by incorporating end-to-end model [13], we ensure the coherence of the forecasts at all-levels using the attentive reconciliation without any strong assumption, where coherence constraints are ensured, while the discrepancy between the reconciled forecast and the ground truth are dynamically minimized (via attentive-network with learning parameters). **Third**, we further take advantage of the end-to-end model to improve the forecasting performance by combining two separate stages, i.e., the generation and the reconciliation, of our base forecasts.

Our Contributions. We summarize our contributions as follows:

- A multivariate probabilistic framework employing NF for data with complex statistical characteristics
- A general and flexible attentive reconciliation method without any strong assumption
- An end-to-end model optimizing the overall performance of all components

2   Background and Related Work

In this paper, we focus on *Hierarchical Time Series* (HTS) with tree structures, which can be regarded as a multivariate TS with linear aggregation constraints. We can thus use an aggregation matrix $S \in \mathbb{R}^{n \times m}$ (m is number of bottom-level nodes, n is total number of nodes) to represent HTS, where each node represents one TS, to be predicted over a horizon.

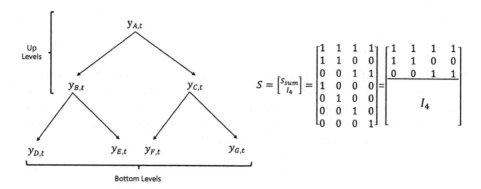

Fig. 1. The tree and the matrix representation of an HTS with 4 bottom-level series and 3 upper-level series.

2.1   Preliminaries

Let us denote the values of a multivariate HTS by $y_{t,i} \in \mathbb{R}$, where $i \in \{1, 2, ..., n\}$ indexes the individual univariate component TS, and t indexes time on a time horizon $t \in \{1, 2, ..., T\}$. Here we assume that the index i of the individual TS is given by the level-order traversal of the hierarchical tree going from left to right at each level. We define $x_{t,i}$ to be time-varying covariate vectors associated with each univariate TS i at time step t.

In the tree structure, TS' at the leaf nodes of the hierarchy are the bottom-level series $b_t \in R^m$ and those of the remaining nodes are the upper-level series $u_t \in R^r$. Clearly, the total number of nodes $n = r + m$. Each node in the tree has at most one parent node and may have zero or more child nodes. Then $y_t = [u_t, b_t]^T \in R^n$ contains observations at time t for all levels [5], which satisfy (by the aggregation matrix S):

$$y_t = [u_t, b_t]^T \quad \Leftrightarrow \quad y_t = Sb_t \quad \Leftrightarrow \quad [u_t b_t] = [S_{sum} I_m] b_t, \qquad (1)$$

where for every time step t, $S_{sum} \in \{0, 1\}^{r \times m}$ is a summation matrix and I_m is the $m \times m$ identity. For example in Fig. 1, $u_t = [y_{A,t}, y_{B,t}, y_{C,t}] \in R^3$ and $b_t = [y_{D,t}, y_{E,t}, y_{F,t}, y_{G,t}] \in R^4$, the aggregation matrix $S = [S_{sum}, I_4]^T$.

For given multivariate HTS, we define base forecast from any TS estimator that does not consider structural information as $\hat{y}_t$. We also define reconciliation forecast which enforces hierarchical constraints on the predictions as $\tilde{y}_t$. Typically, base and reconciliation forecasts can be linked by the following equation which is also called reconciliation process:

$$\tilde{y}_t = SP\hat{y}_t, \tag{2}$$

where $P \in \mathbb{R}^{m \times n}$ is a matrix that maps the base forecast (of dimension n) into the bottom-forecast (of dimension m) of leaf nodes. The aggregation matrix S sums bottom-forecasts using the aggregation structure to produce the reconciliation forecast $\tilde{y}_t$ under the coherence constraints of the hierarchical structure.

2.2   Related Work

Reconciliation of Hierarchical Forecasting. Existing hierarchical forecasting methods mostly follow the two-stage approach: (i) Predicting each TS independently to generate the base h-period-ahead forecasts $\hat{y}_{T+h}$; (ii) Then revise the base forecasts by the reconciliation to ensure coherency to obtain the reconciliation forecasts $\tilde{y}_{T+h}$. In [20], Mint obtains the reconciliation forecasts

$$\tilde{y_{T+h}} = SP\hat{y_{T+h}}, \tag{3}$$

which reconciles forecasts using a matrix $P = (S^T W_h^{-1} S)^{-1}(S^T W_h^{-1})$, minimizing $tr[SPW_h P^T S^T]$ with constraint $SPS = S$, where W_h is the covariance matrix of the h-period-ahead forecast errors $\hat{\varepsilon}_{T+h} = y_{T+h} - \hat{y_{T+h}}$ (S and P are matrices defined in Eq. (2)).

This approach has multiple advantages: (1) The forecasts are coherent by construction; (2) A forecast combination from all levels is applied through the matrix P, but with the strong assumptions that the base forecasts are unbiased and the error covariance W_h is hard to derive for general h.

The **Regularized Regression for Hierarchical Forecasting** [3] is also a two-stage approach, but relaxing unbiased assumptions, and seeks the revised forecasts with the best trade-off between bias and forecast variance [3]. The hierarchical forecasting problem is formulated as an empirical risk minimization (ERM) problem which directly minimizes the mean squared forecast errors.

The current work in **end-to-end model** [13] of hierarchical forecasting uses DeepVAR [15] to obtain the base forecast, assuming that the covariance matrix captures the correlations imposed by the hierarchy and the relationships among the bottom-level TS' is a diagonal matrix. The reconciliation is achieved using a closed-form formulation of an optimization problem, minimizing the errors of base forecast $\hat{y}$ and reconciling forecast $\tilde{y}$ subject to coherent constraints of hierarchical structure:

$$\tilde{y}_t = \underset{y \in \mathbb{R}^n}{\arg\min} \quad \|y - \hat{y}_t\|_2 \quad \text{s.t. } Ay = 0,$$

where $A := [I_r | - S_{sum}] \in \{0, 1\}^{r \times n}$, 0 is an r-vector of zeros, and I_r is the $r \times r$ identity matrix. By using Lagrangian relaxation, a closed-form solution can be obtained:

$$\tilde{y}_t = M\hat{y}_t \text{ , where } M := I - A^T(AA^T)^{-1}A.$$

M is a fixed matrix, which is time-invariant and can be computed offline once, prior to the training. Unlike Mint, there is no relationship with the predicted value y, and this kind of reconciliation is essentially a fine-tuning of the base forecast under the coherence constraint, i.e., the learned model parameters are not used to revise the base forecast in the reconciliation stage.

3 Methodology

In this section, we detail our end-to-end method combining base TS estimator at all-level and the reconciliation module. Our base estimator is a multivariate probabilistic TS forecasting model, characterizing non-Gaussian/ nonlinear data using NF [12]. We parameterize the reconciliation module using a trainable attention-based [19] neural-network [4] to minimize reconciliation error under coherence constraints. By using the reparameterization trick [9], we combine the above independent components into a trainable end-to-end model.

3.1 Model

We now introduce each component with a schematic view in Fig. 2.

Multivariate Probabilistic Time Series (TS) Forecasting via NF
We first use Recurrent Neural Network (RNN) to capture patterns inside each individual TS, e.g., trends and cycles. Note that RNN is unrolled for each TS separately during training and testing, but its parameters are shared across all TS' to exploit common patterns in the entire history.

With the patterns encoded by RNN, we are able to generate probabilistic forecasts, feeding hidden states of RNN to a fully connected dense layer to create a parameterized multivariate Gaussian distribution $\Theta_t = \{\mu_t, \Sigma_t\}$:

$$\Theta_t = \Psi(\Phi(x_t, y_{t-1}, h_{t-1})), \tag{4}$$

where Ψ is the dense layer for learning distribution parameters, Φ is the recurrent function of RNN whose parameters are globally shared across all TS', covariate vector x_t and lags y_{t-1} are the inputs, and h_{t-1} is the hidden state of RNN. We use a diagonal covariance Σ_t matrix to represent the base distribution.

For many real-world HTS, bottom-level data usually are very sparse, worsened by the non-Gaussian/nonlinear distribution, making it hard to characterize with Gaussian distribution of hidden states generated from RNN module. Equipped with the powerful density approximator, NF, we are able to tackle these challenges, capturing the nonlinear relationships among all levels in hierarchy. Specifically, we employ Real-NVP with K layers of flow modules to map data sampled from Gaussian distribution to target distribution.

By using Real-NVP, we can get nonparametric distribution $p(y_t|x_t, y_{1:t-1})$ transformed from the base distribution $\Theta_t = \{\mu_t, \Sigma_t\}$ via the flow module Λ. The covariate vector $x_{1:t}$ and past target multivariate lags $y_{1:t-1}$ are encoded in

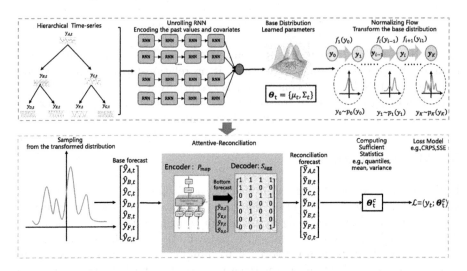

Fig. 2. Model Architecture. Red dashed lines represent Multivariate Probabilistic Time-series Forecasting via NF (Sect. 3.1) and blue dashed lines highlight Sampling and Attentive-Reconciliation (Sect. 3.1). The HTS is encoded by the multivariate forecasting model via NF to obtain the complex target distribution. By using the reparameterization trick, samples from the target distribution are reconciled to revise base forecasts for coherency. We can compute the sufficient statistics from the samples via the empirical distribution to adapt to various loss functions.

the hidden state h_{t-1} which will serve as input of distribution learning function Ψ.

$$p(y_t|x_{1:t}, y_{1:t-1}) = p(\Theta_t; \Lambda) = p(f^{-1}(y_t)|x_t, y_{t-1}, h_{t-1}; \Phi)|det([Jac_{y_t}(f^{-1})])|, \tag{5}$$

where the first term in $p(\Theta_t)$ is the density of the multivariate gaussian distribution from Eq. (4), and the second is the absolute value of the determinant of the Jacobian of $f_{y_t}^{-1}$ given the flow module Λ, evaluated at y_t.

Sampling and Attentive-Reconciliation. Next we produce the reconciliation forecast given the predicted nonparametric distribution $p(y_t|x_t, y_{1:t-1})$ from NF. First, we generate a set of Monte Carlo samples from the predicted distribution $p(y_t|x_t, y_{1:t-1})$, but since it is a nonparametric distribution, we need to ensure that the sampling step is differentiable, for which we apply the reparameterization trick with $\Theta_t = \{\mu_t, \Sigma_t\}$. Specifically, we can inverse-transform the samples via NF

$$\hat{y}_t = f^{-1}(\mu_t + \Sigma_t^{1/2} z), \tag{6}$$

with $z \sim \mathcal{N}(0, I)$, and f^{-1} is the inverse transform of NF.

After producing the samples of the base forecast $\hat{y}_t$, we reconcile $\hat{y}_t$ to get the reconciliation forecast $\tilde{y}_t$ under the coherence constraints of the hierarchical structure. According to Eq. (2) ($\tilde{y}_t = SP\hat{y}_t$), we have to generate the mapping matrix P and the aggregation matrix S. With the powerful expressiveness of the attention-based encoder-decoder framework, we are able to hit two birds with one stone. The mapping matrix P is to generate the bottom forecast by integrating the information at all levels while capturing the nonlinear relationships between all-TS', which can be achieved using the attention mechanism.

Specifically, we parameterize the mapping matrix P by an attentive-encoder neural network. The aggregation matrix S representing the (fixed) decoder, which is time-invariant, can be computed offline once, prior to the training process. We denote number of all nodes by da, the number of upper-level nodes by du, the number of bottom-level nodes by db, and the length of prediction time steps by dh. We define the query matrix $\boldsymbol{Q}$, the key matrix $\boldsymbol{K}$, and the value matrix $\boldsymbol{V}$ as follows:

$$Q = FC(\hat{y}_{bottom-level}),\ K = FC(\hat{y}_{all-level}),\ V = FC(\hat{y}_{all-level}), \qquad (7)$$

where $\boldsymbol{Q} \in R^{db \times dh}$, $\boldsymbol{K} \in R^{da \times dh}$, and $\boldsymbol{V} \in R^{da \times dh}$.

According to Eq. (2), we define the Attentive-Reconciliation process as

$$\text{Attentive-Reconciliation}_{\Psi}(Q, K, V) = \text{softmax}\left(\frac{QK^T}{\sqrt{d_K}}\right) VS, \qquad (8)$$

where the reconciled bottom-forecast $\tilde{b}_t$ is obtained from softmax $\left(QK^T/\sqrt{d_K}\right)$, Ψ is the learned parameters of Attentive-Reconciliation, d_K is the dimension of keys, and the aggregation matrix S enforces coherence constraints. Our reconciliation method not only ensures hierarchical coherence constraints but also dynamically revises forecasts via attentive-network with learned parameters that minimize the discrepancy between the reconciliation forecast $\tilde{y}_t$ and the ground truth y_t. The overall performance is optimized in our end-to-end approach that trains both stages at the same time. Please also note that our objective can accommodate different loss functions in the reconciliation stage (Fig. 3).

We repeat the above procedure in the h-period-ahead forecasting to obtain a set of coherent forecasts $\{\tilde{y_T}, y_{\tilde{T}+1}, ..., y_{\tilde{T}+h}\}$. Please note that since we can

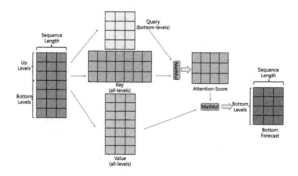

Fig. 3. In the attentive-reconciliation stage, the bottom-level forecast is the query, and the all-level forecast is the key and value. The information of the all-level forecast is used to generate the reconciled bottom-level forecast.

obtain sufficiently many samples from NF to compute any appropriate statistics, our method can generate both point estimates and probabilistic forecasts.

4   Experiment

4.1   Evaluation

Evaluation Metrics. Considering that our method generates probabilistic forecasts instead of point forecasts, it is necessary to evaluate the corresponding probabilistic accuracy with proper metrics, but commonly-used *Mean Absolute Error* (MAE) or *Mean Absolute Percent Error* (MAPE) cannot be directly used for this purpose. On the other hand, *Continuous Probability Ranked Score* (CRPS) generalizes the MAE for probabilistic measurement, making it one of the most widely used accuracy metrics for probabilistic forecasts. We therefore introduce *Continuous Ranked Probability Score* (CRPS), measuring the compatibility of a cumulative distribution function $\hat{F}_{t,i}^{-1}$ for TS i against the ground-truth observation $y_{t,i}$, for our evaluation, defined as

$$CRPS(\hat{F}_t, y_y) := \sum_i \int_0^1 QS_q(\hat{F}_{t,i}^{-1}(q), y_{t,i}) \, dq, \tag{9}$$

where QS_q is the quantile score for the q-th quantiles:

$$QS_q = 2(\mathbb{1}\{y_{t,i} \le \hat{F}_{t,i}^{-1}(q)\} - q)(\hat{F}_{t,i}^{-1}(q) - y). \tag{10}$$

Table 1. We use CRPS (see reason in (Sect. 4.1)) to evaluate performances. The table shows CRPS values (lower is better) averaged over 5 runs. We conduct an ablation study in Multivariate Forecast (Diagonal Gaussian) and Multivariate Forecast (Flow-based). State-of-the-art methods except for DeepVAR-lowrank-Copula and Hier-E2E produce consistent results over multiple runs.

Method	Labour	Traffic	Tourism	Tourism-L
ARIMA-NaiveBU	0.0453	0.0808	0.1138	0.1741
ETS-NaiveBU	0.0432	0.0665	0.1008	0.1690
ARIMA-MinT-shr	0.0467	0.0770	0.1171	0.1609
ARIMA-MinT-ols	0.0463	0.1116	0.1195	0.1729
ETS-MinT-shr	0.0455	0.0963	0.1013	0.1627
ETS-MinT-ols	0.0459	0.1110	0.1002	0.1668
ARIMA-ERM	0.0399	0.0466	0.5887	0.5635
ETS-ERM	0.0456	0.1027	2.3755	0.5080
N-BEATS-NaiveBU	0.0703	0.1062	0.1912	0.1213
Transformer-NaiveBU	0.0621	0.1102	0.1877	0.1121
Informer-NaiveBU	0.0571	0.0911	0.1601	0.0701
DeepAR-NaiveBU	0.0574 ± 0.0026	0.1023 ± 0.0019	0.1816 ± 0.0088	0.0636 ± 0.0073
DeepVAR-lowrank-Copula	0.0583 ± 0.0071	0.0991 ± 0.0083	0.1782 ± 0.0093	0.1125 ± 0.0041
Hier-E2E	0.0340 ± 0.0088	0.0376 ± 0.0060	0.0834 ± 0.0052	0.1520 ± 0.0032
Hier-Flow-Attention(*Ours*)	**0.0304 ± 0.0076**	**0.0339 ± 0.0081**	**0.0752 ± 0.0091**	**0.1401 ± 0.0089**
Multivariate Forecast (Diagonal Gaussian)	0.0376 ± 0.078	0.0418 ± 0.0067	0.0926 ± 0.0048	0.1589 ± 0.0062
Multivariate Forecast (Flow-based)	0.0351 ± 0.067	0.0378 ± 0.0075	0.0854 ± 0.0089	0.1525 ± 0.0093

Experiment Results. Our method obtains the base forecasts via NF and then uses the attentive-reconciliation to revise the base forecast for coherency. For evaluation, we generate 200 samples from the learned parameters to create an empirical predictive distribution. We run our method 5 times and report the mean and standard deviation of the CRPS scores. The results are shown in Table 1, where we can see our Hier-Flow-Attention model provides significant improvements on most datasets.

Ablation Study. In the ablation study, we compare the two variants of our model to analyze the sources of improvements. The first experiment is the Multivariate Forecast via Gaussian distribution with the diagonal covariance Matrix. The second experiment is Multivariate Forecast via NF. These experiments do not include any reconciliation steps. By comparison, using NF to fit the hierarchical data distribution generates better results than directly applying Gaussian distribution, as expected. In the meantime, Multivariate Forecast also predicts better than state-of-the-art hierarchical methods. This demonstrates that learning the joint distribution of all the TS' improves the forecasting performance. Moreover, using attentive-reconciliation as a flexible reconciliation strategy, our method improves over unreconciled Multivariate Forecasts over coherency.

As in the above table, benefiting from powerful NF as density approximator to better characterize complex non-Gaussian/nonlinear distributions of actual data, our Multivariate Forecast significantly outperforms DeepVAR-lowrank-Copula and Multivariate Forecast with Diagonal Gaussians at all levels. Moreover, using the flexible reconciliation strategy via attention-based encoder-decoder mechanism, the accuracy at the upper level is further improved compared to other methods.

5 Conclusion

In this paper, we proposed a novel approach for HTS prediction via NF and trainable attentive reconciliation, where non-Gaussian/nonlinear data at all levels of the hierarchy are better modeled, and by using the attentive-reconciliation, we enforced the coherent probabilistic forecasts. We were able to combine independent components into a trainable end-to-end model with the help of the reparameterization trick. The comparison against other state-of-the-art methods in real-world data demonstrates the competitiveness of our method.

References

1. Anderer, M., Li, F.: Forecasting reconciliation with a top-down alignment of independent level forecasts. arXiv preprint arXiv:2103.08250 (2021)
2. Athanasopoulos, G., Ahmed, R.A., Hyndman, R.J.: Hierarchical forecasts for Australian domestic tourism. Int. J. Forecast. **25**(1), 146–166 (2009)
3. Ben Taieb, S., Koo, B.: Regularized regression for hierarchical forecasting without unbiasedness conditions. In: Proceedings of the 25th ACM SIGKDD International Conference on Knowledge Discovery & Data Mining, pp. 1337–1347 (2019)

4. Burba, D., Chen, T.: A trainable reconciliation method for hierarchical time-series. arXiv preprint arXiv:2101.01329 (2021)
5. Corani, G., Azzimonti, D., Augusto, J.P.S.C., Zaffalon, M.: Probabilistic reconciliation of hierarchical forecast via Bayes' rule. In: Hutter, F., Kersting, K., Lijffijt, J., Valera, I. (eds.) ECML PKDD 2020. LNCS (LNAI), vol. 12459, pp. 211–226. Springer, Cham (2021). https://doi.org/10.1007/978-3-030-67664-3_13
6. Han, X., Dasgupta, S., Ghosh, J.: Simultaneously reconciled quantile forecasting of hierarchically related time series. In: International Conference on Artificial Intelligence and Statistics, pp. 190–198. PMLR (2021)
7. Hollyman, R., Petropoulos, F., Tipping, M.E.: Understanding forecast reconciliation. Eur. J. Oper. Res. **294**(1), 149–160 (2021)
8. Hyndman, R.J., Ahmed, R.A., Athanasopoulos, G., Shang, H.L.: Optimal combination forecasts for hierarchical time series. Comput. Stat. Data Anal. **55**(9), 2579–2589 (2011)
9. Kingma, D.P., Welling, M.: Auto-encoding variational Bayes. arXiv preprint arXiv:1312.6114 (2013)
10. Liu, Z., Yan, Y., Hauskrecht, M.: A flexible forecasting framework for hierarchical time series with seasonal patterns: a case study of web traffic. In: The 41st International ACM SIGIR Conference on Research & Development in Information Retrieval, pp. 889–892 (2018)
11. Panagiotelis, A., Gamakumara, P., Athanasopoulos, G., Hyndman, R.J., et al.: Probabilistic forecast reconciliation: properties, evaluation and score optimisation. In: Monash Econometrics and Business Statistics Working Paper Series, vol. 26, p. 20 (2020)
12. Papamakarios, G., Nalisnick, E., Rezende, D.J., Mohamed, S., Lakshminarayanan, B.: Normalizing flows for probabilistic modeling and inference. arXiv preprint arXiv:1912.02762 (2019)
13. Rangapuram, S.S., Werner, L.D., Benidis, K., Mercado, P., Gasthaus, J., Januschowski, T.: End-to-end learning of coherent probabilistic forecasts for hierarchical time series. In: International Conference on Machine Learning, pp. 8832–8843. PMLR (2021)
14. Rasul, K., Sheikh, A.S., Schuster, I., Bergmann, U., Vollgraf, R.: Multivariate probabilistic time series forecasting via conditioned normalizing flows. arXiv preprint arXiv:2002.06103 (2020)
15. Salinas, D., Bohlke-Schneider, M., Callot, L., Medico, R., Gasthaus, J.: High-dimensional multivariate forecasting with low-rank gaussian copula processes. arXiv preprint arXiv:1910.03002 (2019)
16. Seeger, M., et al.: Approximate Bayesian inference in linear state space models for intermittent demand forecasting at scale. arXiv preprint arXiv:1709.07638 (2017)
17. Taieb, S.B., Taylor, J.W., Hyndman, R.J.: Coherent probabilistic forecasts for hierarchical time series. In: International Conference on Machine Learning, pp. 3348–3357. PMLR (2017)
18. Taieb, S.B., Taylor, J.W., Hyndman, R.J.: Hierarchical probabilistic forecasting of electricity demand with smart meter data. J. Am. Stat. Assoc. **116**(533), 27–43 (2021)
19. Vaswani, A., et al.: Attention is all you need. In: Advances in Neural Information Processing Systems, pp. 5998–6008 (2017)
20. Wickramasuriya, S.L., Athanasopoulos, G., Hyndman, R.J.: Optimal forecast reconciliation for hierarchical and grouped time series through trace minimization. J. Am. Stat. Assoc. **114**(526), 804–819 (2019)

Orthrus: A Dual-Branch Model for Time Series Forecasting with Multiple Exogenous Series

Ziang Yang[1,2], Biyu Zhou[1(✉)], Xuehai Tang[1], Ruixuan Li[1], and Songlin Hu[1,2]

[1] Institute of Information Engineering, Chinese Academy of Sciences, Beijing, China
{yangziang,zhoubiyu,tangxuehai,liruixuan,husonglin}@iie.ac.cn
[2] School of Cyber Security, University of Chinese Academy of Sciences, Beijing, China

Abstract. Extending the forecasting horizon is a crucial demand for real applications in time series forecasting with multiple exogenous series (TFME). Previous studies adopt Transformer to effectively capture long-term dependency coupling between output and input in a sequence. However, the potential entanglement in multi-dimensional feature space still precludes the application in TFME tasks. In this paper, we propose a dual-branch network named *Orthrus* to solve this issue by differentiating the processing of target and exogenous sequences. *Orthrus* takes the long target sequence as input and uses multi-head self-attention mechanism to capture long-term cyclical patterns. Concurrently, it applies the Local Mutual Dependency Analysis module to extract the sub-sequences of the exogenous sequences with the maximum expected information and adopts the Multi-scale Convolutional Neural Network module to capture the dependencies among the sub-sequences and align with the target sequence. In this way, the feature entanglement issue is largely alleviated. Extensive experiments on four real-world datasets verify that *Orthrus* is superior to all baselines in prediction accuracy, inference efficiency, and memory usage, providing an effective solution to the TFME task.

Keywords: Time Series · Forecasting · Neural Network

1 Introduction

Time series forecasting, as a significant branch of dynamic data analysis, plays a fundamental guiding role in many real-world applications, such as bio-surveillance, financial analytics, and smart city solutions [14,19,25]. Time series forecasting with multiple exogenous series (TFME) task is to study how to accurately predict future values of target series based on historical observations and a set of associated exogenous series. It has been widely applied in various domains. For example, forecasting river flow based on historical flow, precipitation, temperature, etc., can assist in issuing early river flow warnings to minimize loss [2].

Prior works for TFME have made a substantial effort in single-step or short-term forecasting primarily based on nonlinear fitting and RNNs [10,12,18,20].

© The Author(s), under exclusive license to Springer Nature Switzerland AG 2023
X. Wang et al. (Eds.): DASFAA 2023, LNCS 13943, pp. 177–187, 2023.
https://doi.org/10.1007/978-3-031-30637-2_12

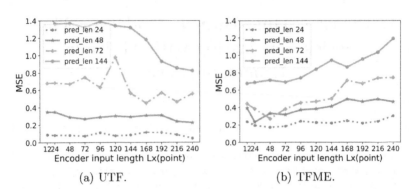

Fig. 1. Forecasting results on a real-world dataset.

However, in real-world scenarios, one pressing demand is to prolong the prediction horizon into the far future. This demands a high prediction capacity of the model, which is the ability to capture precise long-term dependency coupling between output and input efficiently.

Transformer [24] as the state-of-the-art solution to long sequence modeling has shown superior capacity of capturing the recurring patterns with long-term dependency [16,26,27]. As an empirical example, Fig. 1(a) presents the results of univariate long-term series forecasting (UTF) on a real-world dataset, where Transformer accurately predicts the temperature from the short-term period (24 points, 6 h) to the long-term period (144 points, 36 h). MSE (i.e., Mean Squared Error) drops smoothly and slowly within the growing input length.

Nevertheless, the direct application of Transformer to TFME tasks is not as effective as expected. This is because when multiple exogenous sequences are provided, Transformer lacks the ability to align target sequence and multiple exogenous sequences, leading to feature entanglement. In Fig. 1(b), MSE firstly decreases slightly and then rises sharply. The performance of TFME is even worse than that of the prediction in Fig. 1(a), which only uses the target sequence as the input when the input length increases to a certain extent.

Inspired by the observation that exogenous series is indeed beneficial, while excessive exogenous series play a negative role in accurately modeling, we propose a novel architecture *Orthrus*, which deals with the target sequence and the exogenous sequences differently. Long target sequence is processed by multi-head self-attention mechanism to capture long-term cyclical patterns, while short exogenous sequences are processed by Multi-scale Convolutional Neural Network (MSCNN) to comprehensively capture the dependencies among multi-source sequences and align with the target sequence. *Orthrus* adopts the Local Mutual Dependency Analysis module (LocMDA) to find the sub-sequence with the largest expected information among each exogenous sequence to reduce the learning difficulty of the model and thus alleviate the feature entanglement problem. The contributions of this paper are summarized as follows:

- We conduct extensive experiments and find that more historical input of exogenous series is not always better in the TFME task, revealing that the hypothesis of most previous studies is not applicable.
- We propose a novel solution *Orthrus*, with the differentiated process of target series and exogenous series in terms of both feature extraction and input sequence length, thus greatly improving the prediction effect of TFME task.
- We propose LocMDA and MSCNN modules to reduce the inter-series feature entanglement and enhance the model's alignment capacity comprehensively.
- Extensive experiments based on real-world datasets verify that *Orthrus* is superior to all baselines in prediction accuracy, inference efficiency, and memory usage. Specifically, the MSE of *Orthrus* is 31% lower than that of the state-of-the-art Transformer-based solution on average, while the running time is saved by 59% and the runtime memory usage is saved by 17%.

2   Related Work

Time series forecasting methods start from the classic tools. ARIMAX [1,6], considers more exogenous variables and transforms the non-stationary process to stationary through differencing for linear prediction. Nonlinear autoregressive exogenous (NARX) models [10,12,18] learn a nonlinear mapping using kernel methods [7], ensemble methods [5], and Gaussian processes [11], but they employ predefined nonlinear forms, failing to reflect genuine underlying relations.

Deep learning methods have been proposed to flexibly detect nonlinear relations across related time series. Temporal Convolution Networks (TCN) [3] attempt to model the temporal causality with the causal convolution. RNNs and their variants [4,9,13] demonstrate great advantages and evolve into a sequence-to-sequence prediction paradigm [17,20,22]. However, suffering from inherent recurrent structure, vanishing and exploding gradient problems cannot be avoided. LSTNet [15] combines convolutional layers and recurrent layers to capture short-term and long-term temporal patterns. Recently, LogTrans [16] and TST [26] based on self-attention mechanism [24] have made strides in UTF but perform mediocrely in TFME tasks. Although the Probsparse self-attention mechanism proposed by Informer [27] partially eliminates interference and improves alignment ability by finding the more crucial Query, anisotropy in feature dimensions' prediction capacity still precludes the application in TFME tasks.

3   Methodology

First, we formulate the TFME problem. Given a target series $y_{1:t_0}$, where $y_{1:t_0} := [y_1, y_2, \cdots, y_{t_0}]$ and $y_t \in \mathbb{R}$ denotes the value of time series y at time t. Besides let $\{z_{i,1:t_0}\}_{i=1}^{N}$ be a collection of N related exogenous series that are assumed to be known over the past time period. TFME is to predict the next τ time steps for target series $y_{t_0+1:t_0+\tau}$ and a longer output's length τ is encouraged.

$$y_{t_0+1:t_0+\tau} = F(y_{1:t_0}, \{z_{i,1:t_0}\}_{i=1}^{N}; \Phi), \qquad (1)$$

where Φ denotes the learnable parameters shared by all time series in the set. An appropriate model is explored to learn the nonlinear mapping function $F(\cdot)$.

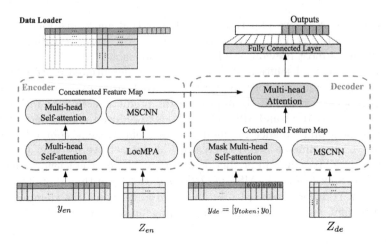

Fig. 2. An overview of *Orthrus*.

The overview of *Orthrus* is presented in Fig. 2. It holds the encoder-decoder architecture [8]. The target sequence input and exogenous sequence input are processed differentially. Please refer to the following sections for details.

3.1   Target Input Processing: Capturing Long-Term Cyclical Patterns

Target input processing uses multi-head self-attention to handle the longer sequence input. Multi-head self-attention allows accessing any part of the history regardless of distance, making it potentially more suitable for grasping the recurring patterns with long-term dependency for long-term prediction.

First, we give the input representation to enhance the global positional context and local temporal context of time series inputs. Assuming we have p types of global time stamps (like weeks, months, holidays). Each global timestamp is employed by a learnable stamp embedding SE with limited vocab size. The local context is preserved by using the canonical position embedding PE. We project the target sequence scalar input y into d_{model}-dim vector $y^{\dagger}$ with Conv1D filters. Thus, we have the feeding vector:

$$y_{en[i]} = \alpha y_i^{\dagger} + PE_i + \sum_p [SE_i]_p, \tag{2}$$

where $i = 1, \cdots, L_y$, L_y is the length of scalar input y and α is the factor balancing the magnitude between the scalar projection and local/global embeddings.

Then, a multi-head self-attention layer receives long input y_{en} and computes output.

3.2 Exogenous Input Processing: Extracting Intra-series and Inter-series Relations

Exogenous input processing includes two key modules: (1) LocMDA to select more reasonable input and (2) MSCNN to capture intra-series and inter-series relations flexibly. Each module is presented in the following sections.

Fine-Grained Input Filtering. In the TFME scenario, since the long-term cyclical pattern of the target sequence has been extracted, it is thus possible to use and process only the exogenous inputs that have the greatest impact on the current prediction target. This can avoid interference with model judgments caused by excessively long exogenous sequence. Moreover, it can lessen the difficulty and computation of multi-source series learning.

To this end, the "amount of information" obtained from exogenous sequences should be evaluated. Compact and appropriate input should be chosen to learn the mapping. We propose a general LocMDA module to analyze local mutual dependency between target and exogenous sequences at each time step. LocMDA adopts the maximal information coefficient (MIC) [21] since it is more robust to outliers, noise, and abrupt trend shifts than mutual information [23]. Furthermore, as our purpose is to evaluate the mutual dependency of each time step, we consider measuring the local MIC of the sub-sequence via sliding window.

Firstly, we introduce the calculation of MIC. Given a finite set D of ordered pairs, the x-values of D can be partitioned into a bins and the y-values of D can be partitioned into b bins. This creates an a−by−b grid G in the finite two-dimensional space. Let $D|_G$ be the distribution induced by the points in D on the cells of G. For $D \subset \mathbb{R}^2$ and $a, b \in N^*$, define

$$I^*(D, a, b) = \max_G I(D|_G), \tag{3}$$

where the maximum is over all grids G with a columns and b rows, and $I(D|_G)$ is the MI of $D|_G$. The characteristic matrix $M(D)$) is defined as follows:

$$M(D)_{a,b} = \frac{I^*(D, a, b)}{\log \min\{a, b\}}. \tag{4}$$

The MIC of a finite set D with sample size $|D|$ and grid size less than $B(n)$ is:

$$MIC(D)_{a,b} = \max_{ab < B(|D|)} \{M(D)_{a,b}\}, \tag{5}$$

where $B(\cdot)$ is a function of sample size.

We set a fixed window ω_{tg} for the target sequence $\omega_{tg} = y_{t_0+1:t_0+\tau}$, which is the segment to be predicted. Starting from t_0 time step and the prediction horizon is τ. As the motivation experiments show that the recent short-term exogenous series is more effective, we define a sliding window in each exogenous sequence to begin at the most recent time and slide to an earlier time in steps of j: $\omega_{i,j} = z_{i,t_0-\tau-j*s:t_0-j*s}, j = 0, 1, 2, \cdots, \lfloor (l - \tau)/s \rfloor$, where s is the stride of sliding window and l denotes the input length of exogenous sequence.

LocMDA repeats the process of calculating MIC in the set W which consists of all the point in the sliding window $\omega_{i,j}$ and target window ω_{tg}:

$$mic_{i,j} = MIC(W)_{a,b}, \tag{6}$$

until local maximum value $|M|$ is found. At this time, the sub-sequence in the sliding window is the most relevant segment to the prediction sequence. As the window moves forward, the calculated MIC value forms a correlation curve. We adaptively select the most relevant sub-sequences for each exogenous sequence. In concrete terms, we take the most relevant segment above as the center, expand the sub-sequence chosen to the left and right, and stop expanding once the window's correlation is less than a threshold M'. The threshold M' is as follows:

$$M' = \lambda \cdot |M|, \tag{7}$$

where $\lambda \in (0, 1]$ is the scaling factor that controls sub-sequence length macroscopically. If λ is larger, the selected exogenous sub-sequences are more compact. Note that when $\lambda = 1$, the dynamic length selection degrades to fixed length selection, thus it can be seen as a generalization.

Intra-series and Inter-series Relations Extraction. Despite the interesting insights provided by input filtering, capturing patterns from multi-source sequences to further increase the model's alignment ability remains challenging. The key to resolving this problem is identifying intra-series relations (relations within a series) and inter-series relations (relations between different series).

Convolution operation, hierarchically collecting local features as powerful data representations, is suitable for feature extraction of high-dimensional and short-term time series. However, the optimal filter differs depending on the dataset. MSCNN with various sizes of filters are employed to capture various lengths of temporal patterns. As different self-attention weights are learned, the proposed model learns diverse dependency structures among time series.

3.3   Encoder-Decoder and Model Training

In the encoder, we aggregate the outputs from the target and exogenous input processing. Then, the output of encoder is sent to the decoder for interaction. We feed the decoder with the target sequence vector as $y_{de} = [y_{token}; y_0] \in \mathbb{R}^{(L_{token}+L_{y_0}) \times d_{model}}$, where $y_{token} \in \mathbb{R}^{L_{token} \times d_{model}}$ is the start token, an earlier slice before the output sequence. $y_0 \in \mathbb{R}^{L_y \times d_{model}}$ is a placeholder for the target sequence. Masked multi-head self-attention is applied in the computing by setting masked dot-products to $-\infty$, preventing each position from attending to coming positions. The decoder predicts all the outputs by one forward procedure, eliminating the time-consuming "decoding step-by-step" transaction.

4   Experiment

4.1   Datasets and Metrics

We first describe four datasets for empirical studies. *SML* is collected by a domestic monitor system. The target series is "room temperature". *ETT* collects hourly electric power data, comprising the target series "oil temperature" and 6 power load feature series. *ECL* is converted into hourly electricity consumption of 321 clients (coding "MT000" to "MT320") by Informer [27]. We set *"MT320"* as the target series. *Weather* contains hourly climatologic data of almost 1,600 locations, and comprises the target series "wet-bulb" and 11 climate features series. The division of train/val/test datasets refers to related works.

We evaluate the model regarding prediction accuracy and overhead. We use MSE (mean squared error) and MAE (mean average error) on prediction windows as accuracy metrics. In terms of overhead, we compare the inference efficiency (i.e., time to complete a prediction) and memory consumption at runtime.

Table 1. Forecasting results on TFME.

Models		*Orthrus*		*Orthrus°*		Informer [27]		TST [26]		LSTNet [15]		DA-RNN [20]		ARIMAX [1]	
Metrics		MSE	MAE	MSE	MAE	MSE	MAE	MSE	MAE	MSE	MAE	MSE	MAE	MSE	MAE
SML	24	**0.115**	**0.245**	0.131	0.263	0.202	0.342	0.172	0.274	0.125	0.275	0.221	0.269	0.203	0.351
	48	**0.171**	**0.306**	0.193	0.321	0.260	0.396	0.233	0.343	0.296	0.389	0.438	0.393	0.556	0.647
	168	**0.461**	**0.514**	0.512	0.579	0.651	0.644	0.678	0.686	0.523	0.610	0.538	0.612	0.529	0.659
	720	**0.587**	**0.610**	0.599	0.628	0.696	0.694	0.786	0.759	0.680	0.733	0.671	0.747	-	-
ETT	24	**0.047**	**0.166**	0.095	0.253	0.187	0.361	0.167	0.336	1.175	0.793	0.065	0.178	0.106	0.176
	48	**0.081**	**0.232**	0.140	0.314	0.550	0.615	0.556	0.607	1.344	0.864	0.105	0.250	0.143	0.295
	168	0.194	0.372	**0.168**	**0.337**	0.430	0.599	0.794	0.899	1.865	1.092	0.275	0.543	0.385	0.572
	720	0.274	0.481	**0.255**	**0.454**	0.841	0.668	0.916	0.760	1.916	1.173	0.368	0.619	-	-
ECL	24	**0.177**	**0.310**	0.195	0.330	0.256	0.381	0.243	0.367	0.184	0.348	0.840	0.539	0.602	0.458
	48	0.268	0.383	**0.245**	**0.373**	0.326	0.422	0.326	0.425	0.279	0.437	1.030	0.616	0.861	0.528
	168	0.409	**0.466**	0.433	0.498	0.436	0.467	0.411	0.585	**0.378**	0.498	1.147	0.676	1.101	0.610
	720	**0.448**	**0.501**	0.451	0.507	0.466	0.509	0.553	0.690	0.494	0.619	1.586	0.927	-	-
Weather	24	0.190	0.324	0.199	0.338	0.173	0.315	**0.171**	**0.310**	0.575	0.507	0.210	0.320	0.244	0.394
	48	0.270	0.394	0.262	0.393	**0.236**	**0.372**	0.239	0.380	0.622	0.553	0.284	0.427	0.381	0.520
	168	**0.381**	**0.483**	0.408	0.507	0.432	0.518	0.474	0.561	0.676	0.585	0.491	0.722	0.521	0.726
	720	**0.417**	**0.513**	0.426	0.522	0.685	0.573	0.682	0.746	0.731	0.625	0.760	0.901	-	-
wining-count	21		6		2		2		1		0		0		

The '-' indicates failure for the unacceptable results.

4.2   Experimental Details

Baselines: We compare with several different types of baselines, including ARIMAX [1]: a classical method, DA-RNN [20]: an RNN-based auto-regressive model, LSTnet [15]: a hybrid model combined Conv1D and recurrent layers , TST [26]: a vanilla Transformer-based model , Informer [27]: the SOTA solution based on sparse Transformer. Besides, *Orthrus* is more of a framework innovation and can benefit directly from Transformer's variants. *Orthrus°* with Informer's lighter *ProbSparse* self-attention is also tested. For the sake of fairness, *Orthrus* should be compared to TST and *Orthrus°* to Informer.

Setup: The input of each dataset is zero-mean normalized. We prolong the prediction windows size τ progressively to cover both short-term and long-term series forecasting, i.e., {24 points, 48 points, 168 points, 720 points}. The learning rate starts from $1e^{-4}$, decaying 10 times smaller every 2 epochs, and the total epoch is 10. The batch size is 32. All models are training/testing on single NVIDIA Tesla V100 GPUs with 32G memory.

4.3 Results and Analysis

Prediction Accuracy. Table 1 summarizes the evaluation results of all the methods on 4 datasets. We steadily extend the prediction horizon τ as a higher requirement of prediction capacity. The best results are highlighted in boldface.

We observe that: **(1)** *Orthrus* obtains consistently state-of-the-art performance across all 4 datasets in wining-counts, and prediction errors climb smoothly and gently within the ever-increasing prediction horizon. This indicates *Orthrus*'s success in enhancing the prediction capacity in TFME problem. **(2)** The proposed *Orthrus* beats DA-RNN and LSTnet, which employ RNN as the backbone, with the MSE decreasing 40.5% (at 24), 43.2% (at 48), 40.0% (at 168) and 31.7% (at 720) on average, which thoroughly verifies *Orthrus*'s potential to predict long-term series in TFME task. **(3)** The Transformer-based TST outperforms ARIMAX, DA-RNN, and LSTnet. TST's performance is still limited due to feature entanglement in multivariable time series. Our proposed *Orthrus* decreases the MSE by 30.3% (at 24), 29.2% (at 48), 30.7% (at 168), and 33.6% (at 720). **(4)** *Orthrus*$^\circ$ beats Informer which employs the identical *ProbSparse* self-attention setting. This proves the strategy of differentiating the exogenous sequence from the target sequence is wise for TFME problem. It can also be noticed that *Orthrus* is slightly better than *Orthrus*$^\circ$. The *ProbSparse* self-attention adopted by *Orthrus*$^\circ$ may not be as stable as the canonical self-attention, although consuming less memory (as illustrated in Fig. 3(b)).

Inference Efficiency. With the same setting and each method's current finest implement, we perform a rigorous inference runtime comparison in Fig. 3(a). During the inference phase, the *Orthrus* (purple line) achieves the highest inference efficiency among all the methods with the generative style decoding.

Memory Consumption. *Orthrus* is a hybrid model, which is hard to have a closed-form of memory usage, so we show the actual running memory consumption in Fig. 3(b). As can be observed, *Orthrus* reduces memory consumption by roughly 17%. It confirms that *Orthrus* alleviates Transformer's infamous memory bottleneck. The memory saving of *Orthrus*$^\circ$ reaches up to 26%. This implies that if a sparse self-attention is adopted, the memory can be further lowered.

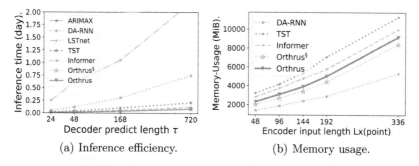

(a) Inference efficiency.   (b) Memory usage.

Fig. 3. Comparisons of inference efficiency and memory usage.

4.4 Ablation Study

The Performance of LocMDA. As shown in Table 2, firstly, the "w/o LocMDA", which foolishly shortens the length of the exogenous sequence, cannot improve performance by leaps and bounds. Secondly, "w/o LocMDA°" replaces the part that sets a scaling threshold to adaptively determine the sub-sequence length with fixed-length. Its suboptimal results demonstrate that dynamic selection is beneficial to improving model prediction capability.

Table 2. Ablation of LocMPA and MSCNN

Methods	MSE			
	24	48	168	720
w/o both	0.097	0.266	0.495	0.549
w/o LocMDA	0.061	0.174	0.299	0.356
w/o LocMDA°	0.052	0.093	0.228	0.298
w/o MSCNN	0.058	0.121	0.277	0.304
w/o MSCNN°	0.050	0.095	0.209	0.287
Orthrus	**0.047**	**0.081**	**0.194**	**0.274**

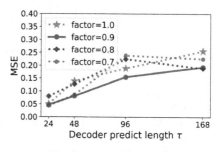

Fig. 4. Scaling Factor Sensitivity

The Performance of MSCNN. The model of "w/o MSCNN" is much more inferior than *Orthrus* which manifests that CNN can extract the salient local features of the shorter neighboring exogenous sequence than Transformer. Then, we explore different single kernel sizes $k \in \{1, 2, 3, 6\}$ on the model and fix all other settings. The results of "w/o MSCNN°" indicate that even the optimal single convolution kernel cannot be comparable to MSCNN, showing that MSCNN fully captures intra-series and inter-series relationships.

Scaling Factor Sensitivity. The scaling factor $\lambda \in (0, 1]$ controls the selected sub-sequence length of LocMDA in Eq. 7. Figure 4 shows the results on ETT. We start from the small factor ($\lambda = 0.7$) to large ones. The general MSE drops to the bottom at $\lambda = 0.9$ and remains stable, as shown in Fig. 4.

5   Conclusion

In this paper, we studied the TFME problem and proposed *Orthrus* to prolong the prediction horizon. *Orthrus* differentiates the processing of target and exogenous sequences. We designed the LocMPA module to pick out the concise segment of exogenous sequence, eliminating the interference of excessive long exogenous series and reducing computation costs. The carefully-designed MSCNN module with multi-scale filters can enhance the alignment ability of the model. Extensive experiments on four real-world datasets demonstrated the effectiveness of *Orthrus* in prediction accuracy, inference efficiency, and memory usage, providing an effective solution to the TFME task.

References

1. Asteriou, D., Hall, S.G.: Arima models and the Box-Jenkins methodology. Appl. Econometrics **2**(2), 265–286 (2011)
2. Atiya, A.F., El-Shoura, S.M., Shaheen, S.I., El-Sherif, M.S.: A comparison between neural-network forecasting techniques-case study: river flow forecasting. IEEE Trans. Neural Netw. **10**(2), 402–409 (1999)
3. Bai, S., Kolter, J.Z., Koltun, V.: An empirical evaluation of generic convolutional and recurrent networks for sequence modeling. arXiv:1803.01271 (2018)
4. Bengio, Y., Simard, P.Y., Frasconi, P.: Learning long-term dependencies with gradient descent is difficult. IEEE Trans. Neural Netw. **5**, 157–166 (1994)
5. Bouchachia, A., et al.: Ensemble learning for time series prediction. In: The 1st International Workshop on Nonlinear Dynamics and Synchronization (2008)
6. Box, G.E., Jenkins, G.M.: Some recent advances in forecasting and control. J. R. Stat. Soc. Ser. C Appl. Stat. **17**(2), 91–109 (1968)
7. Chen, S., Wang, X., Harris, C.: NARX-based nonlinear system identification using orthogonal least squares basis hunting. IEEE Trans. Control Syst. Technol. **16**, 78–84 (2008)
8. Cho, K., et al.: Learning phrase representations using RNN encoder-decoder for statistical machine translation. In: EMNLP (2014)
9. Chung, J., et al.: Empirical evaluation of gated recurrent neural networks on sequence modeling. arXiv:1412.3555 (2014)
10. Diaconescu, E.: The use of NARX neural networks to predict chaotic time series. WSEAS Trans. Comput. Arch. **3**, 182–191 (2008)
11. Frigola, R., et al.: Integrated pre-processing for Bayesian nonlinear system identification with gaussian processes. In: Conference on Decision and Control (2013)
12. Gao, Y., Er, M.J.: NARMAX time series model prediction: feedforward and recurrent fuzzy neural network approaches. Fuzzy Sets Syst. **150**(2), 331–350 (2005)
13. Hochreiter, S., et al.: Long short-term memory. Neural Comput. **9**, 1735–1780 (1997)
14. Huang, X., et al.: TEALED: a multi-step workload forecasting approach using time-sensitive EMD and auto LSTM encoder-decoder. In: DASFAA 2022. Lecture Notes in Computer Science, vol. 13246, pp. 706–713. Springer, Cham (2022). https://doi.org/10.1007/978-3-031-00126-0_55
15. Lai, G., Chang, W.C., Yang, Y., Liu, H.: Modeling long- and short-term temporal patterns with deep neural networks. In: International ACM SIGIR Conference on Research and Development in Information Retrieval (2018)

16. Li, S., et al.: Enhancing the locality and breaking the memory bottleneck of transformer on time series forecasting. In: Advances in Neural Information Processing Systems, vol. 32 (2019)

17. Liang, Y., Ke, S., Zhang, J., Yi, X., Zheng, Y.: GeoMAN: multi-level attention networks for geo-sensory time series prediction. In: IJCAI (2018)

18. Lin, T., Horne, B.G., Tino, P., Giles, C.L.: Learning long-term dependencies in NARX recurrent neural networks. Neural Netw. **7**, 1329–1338 (1996)

19. Matsubara, Y., Sakurai, Y., Van Panhuis, W.G., Faloutsos, C.: FUNNEL: automatic mining of spatially coevolving epidemics. In: ACM SIGKDD, pp. 105–114 (2014)

20. Qin, Y., Song, D., Cheng, H., Cheng, W., Jiang, G., Cottrell, G.W.: A dual-stage attention-based recurrent neural network for time series prediction. In: International Joint Conference on Artificial Intelligence (2017)

21. Reshef, D.N., et al.: Detecting novel associations in large data sets. Science **334**(6062), 1518–1524 (2011)

22. Song, H., Rajan, D., Thiagarajan, J., Spanias, A.: Attend and diagnose: clinical time series analysis using attention models. In: AAAI. vol. 32 (2018)

23. Thomas, M., Joy, A.T.: Elements of Information Theory. Wiley-Interscience, Hoboken (2006)

24. Vaswani, A., et al.: Attention is all you need. In: Advances in neural information processing systems, pp. 5998–6008 (2017)

25. Xu, C., et al.: Graph attention networks for new product sales forecasting in e-commerce. In: Jensen, C.S., et al. (eds.) DASFAA 2021. LNCS, vol. 12683, pp. 553–565. Springer, Cham (2021). https://doi.org/10.1007/978-3-030-73200-4_39

26. Zerveas, G., et al.: A transformer-based framework for multivariate time series representation learning. In: ACM SIGKDD (2021)

27. Zhou, H., et al.: Informer: beyond efficient transformer for long sequence time-series forecasting. In: AAAI (2021)

A Predictive Coding Approach to Multivariate Time Series Anomaly Detection

Zhi Qi[1], Hong Xie[2(✉)], and Mingsheng Shang[2]

[1] College of Computer Science, Chongqing University, Chongqing, China
okjamesqi@gmail.com
[2] Chongqing Institute of Green and Intelligent Technology,
Chinese Academy of Sciences, Chongqing, China
{xiehong,msshang}@cigit.ac.cn

Abstract. This paper proposes `LPC-AD`, a fast and accurate multivariate time series (MTS) anomaly detection method. `LPC-AD` is motivated by the ever-increasing needs for fast and accurate MTS anomaly detection methods to support fast troubleshooting in cloud computing, micro-service systems, etc. `LPC-AD` is fast in the sense that it reduces the training time by as high as 38.2% compared to the state-of-the-art (SOTA) deep learning methods that focus on training speed. `LPC-AD` is accurate in the sense that it improves the detection accuracy by as high as 18.9% compared to SOTA sophisticated deep learning methods that focus on enhancing detection accuracy. Methodologically, `LPC-AD` contributes a generic architecture `LPC-Reconstruct` for one to attain different trade-offs between training speed and detection accuracy. More specifically, `LPC-Reconstruct` is built on ideas from autoencoder for reducing redundancy in time series, latent predictive coding for capturing temporal dependence in MTS, and randomized perturbation for avoiding overfitting of anomalous dependence in the training data. We present simple instantiations of `LPC-Reconstruct` to attain fast training speed, where we propose a simple randomized perturbation method. The superior performance of `LPC-AD` over SOTA methods is validated by extensive experiments on four large real-world datasets. Experiment results also show the necessity and benefit of each component of the `LPC-Reconstruct` architecture and that `LPC-AD` is robust to hyper parameters.

Keywords: MultiTime Series · Anomaly Detection · Predictive coding

1 Introduction

Real-world applications like cloud computing, micro-service systems, etc., generate large amounts of high dimensional time series data and they need to be processed by fast and accurate MTS anomaly detection methods. Although many machine learning algorithms were proposed to detect anomalies in MTS [3,12,18, 26,31], how to achieve a fast training speed while retaining fairly high detection accuracy is underexplored. Classical methods like [15,28,30] have a fast training speed, but their detection accuracy is not high, due to the low expressiveness capability of their model. Using deep learning models, methods like [5,13,21,25] break the records of classical methods with a surprisingly high detection accuracy thanks to the high expressiveness of these models. In the research line of modern methods, most efforts were spent on developing sophisticated models to improve detection accuracy [5,13,21,25]. As a consequence, these sophisticated models become more and more time-consuming for training. Recently, the USAD [3] and TranAD [26] improve the training speed, while retaining a fairly high accuracy. These two algorithms follow the same architecture of adversarial training (refer to Sect. 2 for details) and their detection accuracy can be low in some datasets (refer to Sect. 6 for details). This paper explores faster training speed and more accurate MTS anomaly detection methods.

It is challenging to design fast and accurate MTS anomaly detection algorithms. First, the MTS training data is usually not associated with anomalous labels and there is no specification on the pattern of an anomaly in general [3,26]. Second, the anomalous data points in MTS training data can mislead the trained model [13]. Third, the MTS data is usually of high dimension and it has complex spatial and temporal dependence [3,26], where the spatial dependence refers to the dependence among different time series. This makes it difficult to learn the normal and abnormal patterns from the MTS. Fourth, the MTS data is usually non-stationary [14], making it difficult to learn stable patterns. In this paper, we propose an algorithm called LPC-AD to address these challenges.

The LPC-AD has a shorter training time than SOTA deep learning methods that focus on reducing training time, and it has a higher detection accuracy than SOTA sophisticated deep learning based methods that focus on enhancing detection accuracy. These merits of LPC-AD are supported by novelty in the design and extensive empirical evaluation on four large real-world datasets.

Novelty in the Design. Though the LPC-AD falls into the research line of autoencoder based methods [3,12,26,31] that learn the normal spatial and temporal dependence in the MTS data to detect anomalies (refer to Sect. 2 for details on this research line), it contributes several new ideas. Unlike previous works [3,12,26,31] which use autoencoder to capture spatial and temporal dependence in many MTS, we use autoencoder to reduce redundancy in the time series data. Through this, we obtain a low dimensional latent representation of the data. Evidence of redundancy in MTS can be found in [14,17]. As our purpose is to reduce redundancy, simple autoencoders such as vanilla LSTM based autoencoder suffice. To capture temporal dependence in MTS, we apply ideas from predictive coding [2], which have been successfully applied to word representation [19], visual representation learning [7], etc. Note that we are the first to explore predictive coding for MTS anomaly detection. More specifically, we use a predictor to capture temporal dependence between two consecutive sliding windows of the MTS. The predictor aims to learn the normal dependence via predicting future latent variables from its preceding latent variables in the sliding window. Here, we do not need sophisticated prediction networks, as they may overfit the anomalous patterns in the training data. We then use randomized perturbation to inject some noise on the output of the predictor. Finally, we use the decoder to decode this perturbed prediction and use the absolute error between the decoded data and the ground truth time series data to measure the anomalous score of a data point. The purpose of this perturbation is to avoid overfitting to the patterns of anomaly data points, as the decoder aims to learn the normal patterns. This randomized perturbation approach has a theoretical foundation in online learning in the presence of adversarial data [1]. As an analogy, the anomalous data points in the time series are equivalent to the adversarial data in online learning. Note that this randomized perturbation only incurs a negligible computational cost. Combining them all, we contribute a generic architecture called `LPC-Reconstruct` in the sense that one can select different instances of the above mentioned autoencoder network, the predictor network and the randomized perturbation method to attain different tradeoffs between the training speed and detection accuracy.

Extensive Empirical Validation. We instantiate `LPC-Reconstruct` with the vanilla LSTM autoencoder, three simple predictors (i.e., linear predictor, LSTM enabled predictor and attention enabled predictor) and propose a simple randomized perturbation method based on Gaussian noise. We conduct extensive experiments on four public datasets that are widely used for evaluating MTS anomaly detection algorithms. Experiment findings are summarized into four folds. First, the `LPC-AD` reduces the training time of SOTA deep learning methods that focus on fast training by as high as 38.2%. It improves the detection accuracy of SOTA sophisticated deep learning based methods that focus on high accuracy by as high as 18.9%. Second, `LPC-AD` has a high sample efficiency, i.e., reducing the training data points by 75%, its detection accuracy only drops by less than 2%. Third, `LPC-AD` is robust to its hyper parameters in terms of accuracy, i.e., the variation of detection accuracy is at most three percent when the hyper parameters vary in a fairly large range. Fourth, the ablation study shows the necessity of each component and the benefit of each component.

2   Related Work

Our work is closely related to reconstruction based methods, which have two lines. *(1) Autoencoder (AE) based methods.* The EncDec-AD [18] instantiates the autoencoder architecture with the vanilla LSTM for MTS anomaly detection. The reconstruction error of a data point quantifies the anomaly score, and EncDec-AD reports an anomaly whenever the anomaly score exceeds a given threshold. The MSCRED [31] improves the expressiveness of EncDec-AD. It uses a convolutional network to capture spatial dependence and a convolutional LSTM to capture temporal dependence. However, it is time-consuming to train the model, and it requires a large amount of training data to fit the model well. Recent autoencoder based methods [3,12,26] contribute simple models with a fast training speed, while retaining a fairly high detection accuracy. The open-Gauss [12] instantiates the autoencoder architecture with a tree-based LSTM for MTS anomaly detection. It has a fast training speed and low memory requirement. USAD [3] further improves the training speed via adversarial training. It has one simple encoder and two decoders. Two decoders share the same encoder, and the purpose of two decoders is to enable adversarial training, which amplifies the temporal anomaly. The encoder and two decoders are composed of several linear layers, enabling a fast training speed. However, simple linear layers cannot capture temporal dependence well because they can not process longer input. TranAD [26] replaced the linear layer of USAD with transformer [27] network to capture temporal dependence better. TranAD contributes a slightly more sophisticated model than USAD. It is shown to have a higher detection accuracy than USAD, while retaining a fast training speed. Compared to these works, our work presents a new architecture, which combines autoencoder, predictive coding and randomized perturbation. Furthermore, our work outperforms them in both training speed and detection accuracy. *(2) Variational autoencoder (VAE) based methods.* The Donut [29] extends the vanilla VAE by changing the linear layers to represent the variance of latent variables to a layer with the soft plus operation. The LSTM-VAE [21] instantiates the VAE architecture by replacing the feed-forward network of VAE with a vanilla LSTM network to capture the temporal dependence in MTS. To improve the expressiveness of Donut and LSTM-VAE, OmniAnomaly [25] changes the latent variables' distribution of LSTM-VAE from standard Gaussian distribution to a more sophisticated distribution. The sophisticated distribution in OmniAnomaly is a combination of a linear Gaussian state space model [10] and a normalizing flow [24]. OmniAnomaly further uses latent variable linking to enhance the temporal dependence between latent variables. InterFusion [13] improves OmniAnomaly by using an anomaly pre-filtering algorithm to filter out potential anomalous data in the training set. It also compresses the multi-time series via two-view embedding. To better capture spatial dependence in MTS, the SDFVAE [5] instantiates the VAE architecture with a convolutional neural network, a BiLSTM, and recurrent VAE. Although Omni-Anomaly, InterFusion and SDFVAE have high expressiveness because of their sophisticated models, their training processes are very resource-intensive and time-consuming. Our work is an autoencoder based method, and it outperforms them in both training speed and detection accuracy.

Koopman operator is a classical method in control theory [16]. It has a similar idea of latent predictive coding. Its deep learning variants have been applied to do time series forecasting [20]. Theoretically, in an infinite-dimensional latent state space, the next state can be a linear function of the states in a recent time window [16]. In particular, when we instantiate the predictor in our LPC-AD architecture as a linear function, we get a deep Koopman operator [16]. For practical applications where we cannot have finite-dimensional states, non-linear predictor function is needed for state transitions in latent space. This is validated by our experiment in Sect. 6.4. Due to page limit, a more comprehensive literature review can be found in our technical report [8].

3   Problem Formulation

We consider an $M \in \mathbb{N}_+$ dimensional time series indexed by time stamps $t \in \mathbb{N}_+$. Let $\boldsymbol{x}_t \triangleq (x_{t,1}, \ldots, x_{t,M}) \in \mathbb{R}^M$ denote the t-th observation or sample point of the time series. For example, $\boldsymbol{x}_t$ can be the values of M KPIs of a cloud computing system measured at time stamp t. Note that in practice, the observations of a time series are sampled at a fixed rate or time-varying rate. In this paper, we do not make any assumptions about the sampling rate. We consider the setting that we are given a training dataset of $T \in \mathbb{N}_+$ data points of the time series denoted by $\mathcal{T} = \{\boldsymbol{x}_1, \ldots, \boldsymbol{x}_T\}$. Each data point is not associated with a label on whether it is anomalous or not. Our objective is to design and train an anomaly detection algorithm from the training dataset $\mathcal{T}$. Following previous works [3,13,23,25,26], we normalized the time series data to the range of $[0, 1]$.

4   The LPC-AD Algorithm

Denote a sliding window of $\ell_h \in \mathbb{N}_+$ latest historical data points up to time stamp t as $\boldsymbol{W}_t^h \triangleq [\boldsymbol{x}_{t-\ell_h+1}, \cdots, \boldsymbol{x}_t]$, where $t \geq \ell_h$. Denote a sliding window of $\ell \in \mathbb{N}_+$ future data points starting from time stamp $t+1$ as $\boldsymbol{W}_{t+1} = [\boldsymbol{x}_{t+1}, \ldots, \boldsymbol{x}_{t+\ell}]$. The window size ℓ_h and ℓ are two hyperparameters, and in general, they may be different, i.e., $\ell_h \neq \ell$. We aim to design and train an algorithm to learn the dependence between two consecutive windows, i.e., $\boldsymbol{W}_t^h$ and $\boldsymbol{W}_{t+1}$, from $\mathcal{T}$. And then, we utilize it to decide whether a data point $\boldsymbol{x}_t$ outside the training dataset, where $t > T$, is an anomaly or not.

Algorithm 1. LPC-AD Algorithm

Input: $(\boldsymbol{W}_t^h, \boldsymbol{W}_{t+1}), \forall t \geq T$, alert threshold λ
1: $\epsilon \sim \mathcal{N}(\mathbf{0}, \boldsymbol{\Sigma})$
2: $[\widetilde{\boldsymbol{x}}_{t+1}, \ldots, \widetilde{\boldsymbol{x}}_{t+\ell}] \leftarrow \texttt{LPC-Reconstruct}(\boldsymbol{W}_t^h, \boldsymbol{W}_{t+1}, \epsilon; \boldsymbol{\Theta})$
3: $\text{AnomalyScore}_i \leftarrow \|\boldsymbol{x}_{t+i} - \widetilde{\boldsymbol{x}}_{t+i}\|_2, \forall i = 1, \ldots, \ell$
4: $a_i \leftarrow \mathbb{I}_{\{\text{AnomalyScore}_i \geq \lambda\}}, \forall i = 1, \ldots, \ell.$
5: **Return** a.

Formally, we design an anomaly detection algorithm based on the latent predictive coding. Algorithm 1 outlines our LPC-AD (Latent Predictive Coding for Anomaly Detection) algorithm. The LPC-AD takes two consecutive sliding windows, i.e., $\boldsymbol{W}_t^h$ and $\boldsymbol{W}_{t+1}$, and an alert threshold $\lambda \in \mathbb{R}_+$ as input. Note that the sliding window $\boldsymbol{W}_{t+1}$ is outside the training dataset, i.e., $t \geq T$. It outputs an ℓ-dimensional binary vector to indicate where a data point in the window $\boldsymbol{W}_{t+1}$ is anomalous or not. In Step 1 to 3, an anomaly score for each data point in $\boldsymbol{W}_{t+1}$ is computed, which quantifies the likelihood for a data point to be an anomaly. More specifically, the anomaly score is built on two key ideas. The one is to utilize latent predictive coding to capture the normal dependence between two consecutive sliding windows. This is achieved by the $\texttt{LPC-Reconstruct}(\boldsymbol{W}_t^h, \boldsymbol{W}_{t+1}, \epsilon; \boldsymbol{\Theta})$ algorithm in Step 2. The details of $\texttt{LPC-Reconstruct}$ are deferred to Sect. 5. The other idea is random perturbation, which is achieved by generating a stochastic noise in Step 1 and taking it as an input to the $\texttt{LPC-Reconstruct}$ algorithm in Step 2. The purpose of this random perturbation is to avoid overfitting the patterns of anomaly data points. Based on the anomaly score, in the remaining steps, we use the alert threshold λ to report the anomaly. In particular, when the anomaly score exceeds the alert threshold λ, we report an anomaly; otherwise, we report no anomaly. Using an alert threshold to report anomaly is widely used in previous works [3,13,23,25,26]. Exhaustive search (with a proper discretization of search space) is widely adopted [3,13,23,25,26].

5   Design and Training of LPC-Reconstruct

5.1   General Model Architecture

We parameterize $\texttt{LPC-Reconstruct}(W_t^h, W_{t+1}, \epsilon; \Theta)$ via a neural network as shown in Fig. 1. Four building blocks of this architecture are: a sequence encoder $\texttt{SeqEnc}(\cdot; \Theta_{\mathrm{SE}})$ with parameter Θ_{SE}, a sequence decoder $\texttt{SeqDec}(\cdot; \Theta_{\mathrm{SD}})$ with parameter Θ_{SD}, a predictor $\texttt{Predic}(\cdot; \Theta_{\mathrm{PD}})$ with parameter Θ_{PD} and a randomized data perturbation operator $\texttt{RandPerturb}(\cdot; \Theta_{\mathrm{RP}})$ with parameter Θ_{RP}. We defer the design details of each of these building blocks to Sect. 5.2, and let us state some properties of them first. The $\texttt{SeqEnc}(\cdot; \Theta_{\mathrm{SE}})$ takes a consecutive sequence of data $x_{t_1}, x_{t_1+1}, \ldots, x_{t_2}$ as the input, where $t_1, t_2 \in \{1, \ldots, T\}$ and $t_1 \leq t_2$, and outputs an encoded low dimensional data sequence denoted by $z_{t_1}, z_{t_1+1}, \ldots, z_{t_2}$, where $z_t \in \mathbb{R}^N$, $t \in \{t_1, \ldots, t_2\}$, $N \in \mathbb{N}_+$ and $N < M$. We call z_t the latent variable and it is a compression of the original data x_t. The $\texttt{SeqDec}(\cdot; \Theta_{\mathrm{SD}})$ takes a low dimensional latent variable sequence $z_{t_1}, \ldots, z_{t_2}$ as the input, and outputs a high dimensional decoded sequence denoted by $\widehat{x}_{t_1}, \ldots, \widehat{x}_{t_2}$, where $\widehat{x}_t \in \mathbb{R}^M$, $t \in \{t_1, \ldots, t_2\}$. The $\widehat{x}_t$ is a reconstruction of the original data x_t. The $\texttt{Predic}(\cdot; \Theta_{\mathrm{PD}})$ takes a sequence of ℓ_h latent variables $z_{t-\ell_h+1}, \ldots, z_t$ as the input, and outputs a prediction on the next ℓ latent variables $z_{t+1}, \ldots, z_{t+\ell}$. The $\texttt{RandPerturb}(\cdot, \epsilon; \Theta_{\mathrm{RP}})$ takes the latent variable sequence predicted by $\texttt{Predic}(\cdot; \Theta_{\mathrm{PD}})$ and the ground truth latent variable sequence as the input, and generates a zero mean stochastic noise vector ϵ to perturb the predictions.

In $\texttt{LPC-Reconstruct}$, we first input the data points of two consecutive sliding window, i.e., W_t^h, W_{t+1}, into the encoder to obtain a low dimensional representation Z_t^h, Z_{t+1}, where $[Z_t^h, Z_{t+1}] = \texttt{SeqEnc}(W_t^h, W_{t+1}; \Theta_{\mathrm{SE}})$, and $Z_t^h = [z_{t-\ell_h 1+1}, \cdots, z_t]$ and $Z_t^2 = [z_{t+1}, \cdots, z_{t+\ell}]$. The purpose of this low dimensional representation is to eliminate the redundancy in the time series data. Then we use the decoder to reconstruct the original time series from the encoded low dimensional data, i.e., Z_t^h and Z_{t+1}, $[\widehat{W}_t^h, \widehat{W}_{t+1}] = \texttt{SeqDec}(Z_t^h, Z_{t+1}; \Theta_{\mathrm{SD}})$, where $\widehat{W}_t^h = [\widehat{x}_{t-\ell_h+1}, \cdots, \widehat{x}_t]$ and $\widehat{W}_{t+1} = [\widehat{x}_{t+1}, \cdots, \widehat{x}_{t+\ell}]$. To capture the dependence in the low dimensional representation, we use a prediction operator, which use the encoded data of the historical window to predict the data associated with the future window, formally $\widehat{Z}_{t+1} = \texttt{Predic}(Z_t^h; \Theta_{\mathrm{PD}})$. Then we use a randomized perturbation operator to perturb the above prediction. $\widetilde{Z}_{t+1} = \texttt{RandPerturb}(Z_{t+1}, \widehat{Z}_{t+1}, \epsilon; \Theta_{\mathrm{RP}})$, where $\widetilde{Z}_{t+1} = [\widetilde{z}_{t+1}, \cdots, \widetilde{z}_{t+\ell}]$ and ϵ denotes an ℓ-dimensional zero mean stochastic noise. The purpose of this perturbation is to avoid overfitting the patterns of anomaly data points, as there may be some anomaly data points in the training data. This operation also makes $\widetilde{Z}_{t+1}$ more robust to small outliers in training data. It has a theoretical foundation in online learning in the presence of adversarial data [1]. Finally, we obtain the reconstructed data as: $[\widehat{W}_t^h, \widehat{W}_{t+1}] = \texttt{SeqDec}(Z_t^h, \widetilde{Z}_{t+1}; \Theta_{\text{SD}})$, where $\widetilde{W}_{t+1} = \texttt{LPC-Reconstruct}(W_t^h, W_{t+1}, \epsilon; \Theta)$. Then, we have $\Theta = [\Theta_{\text{SE}}, \Theta_{\text{SD}}, \Theta_{\text{PD}}, \Theta_{\text{RP}}]$.

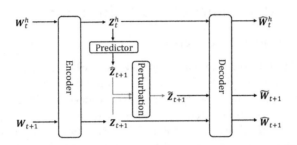

Fig. 1. Architecture of $\texttt{LPC-Reconstruct}(W_t^h, W_{t+1}, \epsilon; \Theta)$.

To learn the function $\texttt{LPC-Reconstruct}(W_t^h, W_{t+1}, \epsilon; \Theta)$ from the training data $\mathcal{T}$, we consider a loss function associated with time stamp t defined as:

$$L_t(\Theta) = \underbrace{\|W_t^h - \widehat{W}_t^h\|_2 + \|W_{t+1} - \widehat{W}_{t+1}\|_2}_{\text{redundancy reduction loss}} + \underbrace{\mathbb{E}_\epsilon[\|W_{t+1} - \widetilde{W}_{t+1}\|_2]}_{\text{dependency capturing loss}},$$

where the expectation notation $\mathbb{E}_\epsilon$ means take expectation with respect to the random vector ϵ. Given a training dataset $\mathcal{T}$, the total loss is $L(\Theta) = \sum_{t=\ell_h}^{T-\ell} L_t(\Theta)$. The physical meaning of the above loss is to find the parameters Θ that best capture the dependence between consecutive sliding windows. Note that in the above loss, we start from $t = \ell_h$ instead of $t = 1$, to avoid the corner case where the historical window W_t^h contains less than ℓ_h data points. Similarly, we end at $t = T - \ell - 1$ instead of $t = T$, to avoid the corner case that the future window W_{t+1} contains less than ℓ data points.

5.2   Instantiation on the Model

Instantiation of $\texttt{SeqEnc}(\cdot; \Theta_{\text{SE}})$ and $\texttt{SeqDec}(\cdot; \Theta_{\text{SD}})$. We instantiate the sequence encoder $\texttt{SeqEnc}(\cdot; \Theta_{\text{SE}})$ and the sequence decoder $\texttt{SeqDec}(\cdot; \Theta_{\text{SD}})$ by adding a linear layer to the classical LSTM. The figure of this architecture is in our technical report [8]. The purpose of adding a linear layer to the classical LSTM is to reduce the number of dimensions and eliminate redundant information. In particular, a more sophisticated sequence encoder and sequence decoder can have better performance in dimension reduction which may, in turn, improve the final detection accuracy of the architecture. Since we are processing a multivariate time series data with complex temporal dependency, a simple encoder composed of linear layers is oversimplified to encode the original data well [3]. Recurrent neural networks are more sophisticated than this oversimplified linear encoder, and they can capture temporal dependency in the time series data. However, simple RNNs suffer from the gradient vanishing issue, which makes them unsuitable to capture long dependence in the data. One simple variant of RNNs, which is LSTM, can solve this gradient vanishing issue.

Instantiation of `Predic(·; ΘₚD)`. Again we consider simple instantiations of the predictor `Predic(·; Θ_PD)`, to better demonstrate the power of our proposed architecture. We consider three instances of `Predic(·; Θ_PD)`. (1) **Instantiating** `Predic(·; Θ_PD)` **using linear transformation.** This is the simplest instance of `Predic(·; Θ_PD)`. This instance enables us to compare our architecture with the methods based on the Koopman operator theory [16], which uses a linear transformation to capture dependence between latent embeddings. Formally, for a latent embedding sequence $Z_t = [z_{t-\ell_h+1}, \cdots, z_t]$, where z_t is a N-dimensional vector, the linear transformation predicts the latent embedding vector in the next time slot as

$$\texttt{Predic}(Z_t^h; \Theta_{\text{PD}}) = PZ_tQ, \tag{1}$$

where parameters $P \in \mathbb{R}^{N \times N}$, $Z_t \in \mathbb{R}^{N \times \ell_h}$ and $Q \in \mathbb{R}^{\ell_h}$. (2) **Instantiating** `Predic(·; Θ_PD)` **via LSTM enabled Seq2Seq.** We consider a simple nonlinear instantiation of the predictor. This instantiation enables us to understand the impact of nonlinearity on the final detection accuracy of our proposed architecture. Formally, we instantiate `Predic(·; Θ_PD)` with a basic seq2seq neural network complemented with a single LSTM layer. Details of this instantiation are in our technical report [8]. (3) **Instantiating** `Predic(·; Θ_PD)` **via attention enabled Seq2Seq.** It was shown that seq2seq with attention mechanism is good at capturing temporal dependency and can achieve more accurate prediction [22]. Hence, the purpose of this instance is to understand the impact of the prediction accuracy of `Predic` on the accuracy of anomaly detectiono in our LPC-AD framework. Formally, this instance of `Predic(·; Θ_PD)` is developed in [22]. Details of this instantiation are in our technical report [8]. The attention enabled Seq2Seq can be trained in an end-to-end manner.

Instantiation of Randomized Perturbation `RandPerturb(·, ε; Θ_RP)`. Again we consider simple instantiations on `RandPerturb(·, ε; Θ_RP)`, to better demonstrate the power of our proposed architecture. Note that ϵ is a zero mean random vector generated from the Gaussian distribution $\mathcal{N}(0, \Sigma)$. Formally, we use the following formula to instantiate `RandPerturb(·, ε; Θ_RP)`:

$$\texttt{RandPerturb}(Z_{t+1}, \widehat{Z}_{t+1}, \epsilon; \Theta_{\text{RP}}) = Z_{t+1} + \epsilon \odot (Z_{t+1} - \widehat{Z}_{t+1})^{\text{abs}},$$

where $\odot$ denotes component-wise multiplication of vectors and $(Z_{t+1} - \widehat{Z}_{t+1})^{\text{abs}}$ means taking the absolute of each component of the vector $Z_{t+1} - \widehat{Z}_{t+1}$.

5.3 Offline Training

We generate $K \in \mathbb{N}_+$ samples of the perturbation noise denoted by $\epsilon_1, \ldots, \epsilon_K$. Let $\widetilde{W}_{t+1,k}$ denote the reconstructed data associated with ϵ_k. Formally, we approximate the expectation $\mathbb{E}_\epsilon[\|\widetilde{W}_{t+1} - W_{t+1}\|_2]$ via $\mathbb{E}[\|\widetilde{W}_{t+1} - W_{t+1}\|_2] \approx \frac{1}{K}\sum_{k=1}^{K}\|\widetilde{W}_{t+1,k} - W_{t+1}\|_2$. We use the following approximation on the per time slot loss to train the model:

$$L_t(\Theta) \approx \|W_t^h - \widehat{W}_t^h\|_2 + \|W_{t+1} - \widehat{W}_{t+1}\|_2 + \frac{1}{K}\sum_{k=1}^{K}\|\widetilde{W}_{t+1,k} - W_{t+1}\|_2. \quad (2)$$

Algorithm 2 outlines our model training procedure.

Algorithm 2. The Generic Training algorithm of LPC-AD

Input: $\{(W_{\ell_h}^h, W_{\ell_h+1}), (W_{\ell_h+1}^h, W_{\ell_h+2}), \cdots, (W_{T-\ell}^h, W_{T-\ell+1})\}$, $MaxEpoch$.
Output: Trained $\Theta_{SE}, \Theta_{SD}, \Theta_{RP}, \Theta_{PD}$
 1: Initialize $\Theta_{SE}, \Theta_{SD}, \Theta_{RP}, \Theta_{PD}, e \leftarrow 1$
 2: **repeat**
 3:    **for** $t = \ell_h$ to $T - \ell + 1$ **do**
 4:       $[Z_t^h, Z_{t+1}] \leftarrow \texttt{SeqEnc}(W_t^h, W_{t+1}; \Theta_{SE})$, $\widehat{Z}_{t+1} \leftarrow \texttt{Predic}(Z_t^h; \Theta_{PD})$
 5:       $[\widehat{W}_t^h, \widehat{W}_{t+1}] \leftarrow \texttt{SeqDec}(Z_t^h, Z_{t+1}; \Theta_{SD})$
 6:       **for** $k = 1$ to K **do**
 7:          $\epsilon_k \sim \mathcal{N}(0, \Sigma)$, $\widetilde{Z}_{t+1,k} \leftarrow \texttt{RandPerturb}(Z_{t+1}, \widehat{Z}_{t+1}, \epsilon_k; \Theta_{RP})$
 8:          $[*, \widehat{W}_{t+1,k}] \leftarrow \texttt{SeqDec}(Z_t^h, \widetilde{Z}_{t+1,k}; \Theta_{SD})$
 9:       **end for**
10:       Update $\Theta_{SE}, \Theta_{SD}, \Theta_{PD}$ and Θ_{RP} using Eq. (2)
11:    **end for**
12:    $e \leftarrow e + 1$
13: **until** $e = MaxEpoch$

---

6   Experiments on Real-World Datasets

6.1   Experiment Setting

Device and Datasets. We run algorithms on a server, which has a CPU (Intel(R) Xeon(R) Platinum 8151 CPU @ 3.40 GHz, 96 GB memory) and a GPU (Quadro GV100, 32 GB memory). We use four public datasets which are extensively used in previous works [3,5,13,25,26,29,31]. More details of datasets are in our technical report [8].

Comparison Baselines. OmniAnomaly is built from a variational autoencoder. Its sophisticated model to capture temporal dependence of time series enhances the detection accuracy, but leads to a slow training speed. **Inter-Fusion** is an improved variant of OmniAnomaly. The improvement is on the detection accuracy of OmniAnomaly. The model is still sophisticated and the training speed is slow. **USAD** achieves a fast training speed by trading-off the detection accuracy. It is built on autoencoder and uses adversarial training. **TranAD** focuses on a fast training speed by trading-off detection accuracy. It is built on the transformer architecture. We do not compare with a number of other notable baselines such as [6,11,32,33], due to that they are shown inferior to TranAD and we omit them for brevity and simplicity of presentation. **LPC-AD-SA** sets the predictor function Predic$(\cdot; \Theta_{PD})$ to be an attention enabled Seq2Seq. **LPC-AD-S** sets the predictor function Predic$(\cdot; \Theta_{PD})$ to be an LSTM enabled Seq2Seq. **LPC-AD-L** sets the predictor function Predic$(\cdot; \Theta_{PD})$ as a linear function shown in Eq. (1).

Evaluation Metrics. Following previous works [3,5,13,25,26,29,31], we use precision (P), recall (R), area under the receiver operating characteristic curve (AUROC), and the micro F1 score (F_1) macro F1 score (F_1^*) to evaluate the detection accuracy of all algorithms. We use a point-adjustment strategy proposed by [29] to calculate the performance metrics. This point-adjustment strategy was widely recognized and applied in previous works [3,5,13,25,26,31]. We are aware of the latest alternative of point-adjustment strategy [9], which is shown to have its own defficiencies [4].

Hyperparameter Setting & Implementation Details. For the baseline algorithms (OmniAnomaly, InterFusion, USAD and TranAD), we set their hyperparameters such as window size, embedding size, etc., according to their paper or their open-source code (when it is not stated in the paper). For three instances of LPC-AD, i.e., LPC-AD-SA, LPC-AD-S and LPC-L, we use the following default hyper parameters: historical window size $\ell_h = 10$, future window size $\ell = 2$, hidden layer dimension of LSTM $= M/2$, embedding dimension $N = 8$ (16 for WADI dataset), covariance matrix $\Sigma = I$, learning rate $= 0.001$, maximum training epoch $MaxEpoch = 40$ (25 for WADI dataset), training batch size $= 64$. We also vary the hyperparameters to study their impact on the detection accuracy and training time. By default, we use 100% training dataset for all the algorithms. We implemented instances of LPC-AD with PyTorch-1.9.0 library, trained with Adam optimizer.

6.2 Comparison with SOTA Baselines

Q1: Can LPC-AD Improve the Detection Accuracy and Training Time Compared to the SOTA Baselines?

Table 1 omits the precision metric due to page limit. Consider two baselines with sophisticated models with SOTA detection accuracy. From Table 1 one can observe that compared with OmniAnomaly, our LPC-AD-SA improves its F_1 and F_1^* by at least 11.2% over the ASD, WADI and SWaT datasets and by 2.2% over the SMD dataset. It is worth noting that on the WADI dataset, our LPC-AD-SA improves the F_1 and F_1^* of OmniAnomaly by around 57.4%. This drastic improvement is due to the fact that the WADI dataset is highly non-smooth, over which the sophisticated model of OmniAnomaly overfits the data. Compared with InterFusion, which is an improved variant of OmniAnomaly, our LPC-AD-SA improves its F_1 and F_1^* by at most 18.9% over the WADI and SWaT datasets and by at least 2.1% over the SMD and ASD dataset. Namely, LPC-AD-SA significantly outperforms the OmniAnomaly and InterFusion in terms of F_1 and F_1^* For the AUROC accuracy measure, LPC-AD-SA also significantly outperforms the OmniAnomaly and InterFusion.

Table 1. Comparison of detection accuracy.

Method	SMD				ASD			
	R	AUROC	F_1	F_1^*	R	AUROC	F_1	F_1^*
OmniAnomaly	0.9091	0.9539	0.9275	0.9338	0.8593	0.9801	0.8348	0.8531
InterFusion	0.9394	0.9939	0.9287	0.9328	**0.9734**	0.997	0.9342	0.936
USAD	0.9054	0.9499	0.8874	0.9011	0.8742	0.9906	0.8989	0.9093
TranAD	0.9465	0.9954	0.9414	0.947	0.9027	0.9846	0.8813	0.8942
LPC-AD-SA	**0.967**	**0.9973**	**0.9483**	**0.9533**	0.9539	**0.998**	**0.935**	**0.9367**
Method	WADI				SWaT			
	R	AUROC	F_1	F_1^*	R	AUROC	F_1	F_1^*
OmniAnomaly	0.4587	0.7225	0.537	0.5461	0.756	0.8519	0.7256	0.7397
InterFusion	0.6589	0.9303	0.7114	0.7217	0.8292	0.9363	0.814	0.8392
USAD	0.535	0.8966	0.6649	0.6791	0.8625	0.9686	0.9074	0.9078
TranAD	0.2405	0.6864	0.3865	0.3865	0.6963	0.9338	0.8138	0.8138
LPC-AD-SA	**0.7745**	**0.9755**	**0.8455**	0.961	**0.9376**	**0.9926**	**0.9489**	**0.9492**

Furthermore, our LPC-AD-SA can significantly outperform these baselines in terms of training speed as well. In particular, Table 2 shows that the per epoch training time of our LPC-AD-SA is less than 10% of that of both OmniAnomaly and InterFusion over four datasets. In other words, the training time of Omni-Anomaly and InterFusion are roughly an order of magnitude larger than our LPC-AD-SA.

Answer 1.1: *LPC-AD-SA improves the detection accuracy of both Omni-Anomaly and InterFusion significantly, and it reduces the training time of both OmniAnomaly and InterFusion drastically.*

Table 2. Comparison of per epoch training time (in seconds).

	Omni	Inter	TranAD	USAD	LPC		Omni	Inter	TranAD	USAD	LPC
SMD	2945	2891	304.8	229.7	**188.1**	WADI	3937	2432	275.2	252.3	**209.7**
ASD	293.4	397.1	397.1	31.2	**26.4**	SWaT	2956	2076	168.5	161.2	**118.2**

Consider two baselines with simple models and SOTA training speed, i.e., USAD and TranAD. Table 2 shows that LPC-AD-SA has a shorter per epoch training time than both TranAD and USAD. More specifically, compared with TranAD, our LPC-AD-SA reduces its per epoch running time by at most 38.2% and by at least 24% over the SMD, WADI and SWaT datasets. Over the ASD dataset, LPC-AD-SA reduces the per epoch training time of TranAD by 3.7%. One reason to have this small reduction is that the number of dimension of the ASD dataset is small, i.e., 19, thus TranAD is already fast enough, leaving a small room for further improvement. Compared with USAD, our LPC-AD-SA reduces its per epoch running time by 16%–27% over four datasets. One reason to have this large reduction in training time over all datasets is that the USAD is not fast enough even when the dimension of a dataset is small, leaving a large room for further improvement. These results show that our LPC-AD-SA significantly reduces the running time of TranAD and USAD. Moreover, as shown in Table 1, our LPC-AD-SA can also significantly improve the detection accuracy of TranAd and USAD. In particular, consider the WADI dataset, our LPC-AD-SA improves the F_1 and F_1^* of both TranAD and USAD by 24.9%–119.42%. On the WADI dataset, LPC-AD-SA improves the AUROC of USAD and TranAD by 8.7% and 42.1%, respectively. One reason to have this drastic improvement is that the WADI is highly non-smooth, over which simple models of TranAD and USAD may under-fit the data, leading to low detection accuracy and leaving a large room for further improvement. For the other three datasets, the time series are smoother than WADI, over which LPC-AD-SA has a smaller improvement (i.e., several percent) of F_1, F_1^* and AUROC over the TranAD and USAD than the WADI dataset. The reason is that over these datasets, the accuracy of TranAD and USAD is already not low, leaving limited room for further improvement.

Answer 1.2: *LPC-AD-SA improves the detection accuracy of both USAD and TranAD significantly, especially when the time series data is highly non-smooth. It also reduces the training time.*

Q2: Is the Improvement Robust to Training Dataset Size? Due to page limit, we present experiment details in our technical report, but we present the observations as follows.

Answer 2: *For different training dataset size, LPC-AD-SA robustly improves the detection accuracy and training time over the baselines.*

6.3 Parameter Sensitivity Analysis

Similar with [3,5,13,25,26,31], we focus on SMD for the brevity of presentation. One reason for selecting SMD is that the SMD dataset has a relatively larger number of time series and has rich temporal patterns in the datasets.

Q3: Is LPC-AD Robust Under Different Parameter Settings?

- **Impact of history window size.** Figure 2 shows the impact of historical window size ℓ_h on the detection accuracy and training time on the AutoEncoder baseline (AE) and three instances of LPC-AD. Figure 2(a) shows that the F_1 curves of LPC-AD-SA, LPC-AD-S and AE are roughly flat as the historical window size ℓ_h increases from 2 to 3. The same findings can be observed on both the F_1^* and AUROC metrics, as shown in Fig. 2(b) and 2(c). Namely, the detection accuracy of LPC-AD-SA, LPC-AD-S and AE is not sensitive to the historical window size. Moreover, from Fig. 2(b)–2(c), we observe that the detection accuracy of the linear variant LPC-AD-L has a larger variation as the historical window size ℓ_h changes. Note that the detection accuracy of each algorithm is not monotone in the historical window size. This phenomena that the detection accuracy is not monotone in window size is also observed in the experiments of previous works [3,13,26]. One reason of this non-monotonicity is that increasing the historical window size increases the parameters of the model and it does not increase the volume of training data. Figure 2(d) shows that the training time of LPC-AD-SA, LPC-AD-S, LPC-AD-L and AE increases slightly in the historical window size ℓ_h nearly at a linear rate. This implies that these algorithms scale well with respect to the historical window size ℓ_h. Furthermore, as shown in Fig. 2(c), the gap between F_1 (or AUROC) curves of LPC-AD-SA and LPC-AD-L is quite large. This means that capturing the nonlinear dependence in the time series data can improve the anomaly detection accuracy significantly. Figure 2(d) shows that this improvement in the detection accuracy is achieved at the cost of slowing down the training speed.

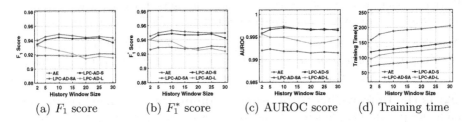

(a) F_1 score   (b) F_1^* score   (c) AUROC score   (d) Training time

Fig. 2. Impact of historical window size ℓ_h on the detection accuracy and training time.

- **Impact of future window size.** Figure 3 shows the impact of future window size ℓ on the detection accuracy and training time of three instances of LPC-AD and AE. From Fig. 3(a) one can observe that the F_1 curves of LPC-AD-SA, LPC-AD-S and AE are flat as the historical window size ℓ_h increases from 1 to 8. Figure 3(b) and 3(c) show similar findings on the F_1^* and AUROC metric. In summary, the detection accuracy of LPC-AD-SA, LPC-AD-S and AE is not sensitive to the historical window size. Figure 3(a)–3(c) show that the detection accuracy of LPC-AD-L decreases in future window size. The reason that the linear predictor of LPC-AD-L under fits the data, making it less accurate for larger future window size. Figure 3(d) shows that the training time of LPC-AD-SA, LPC-AD-S and LPC-AD-L increase nearly linearly in the future window size ℓ. Meanwhile, the training time of AE is almost unchanged, because the AE does not have the predictor component. These results show that three instances of LPC-AD scale well with respect to the future window size ℓ. Lastly, Fig. 3 shows similar accuracy vs. training speed tradeoffs as Fig. 2, which is caused by capturing the nonlinear temporal dependence in the time series data.

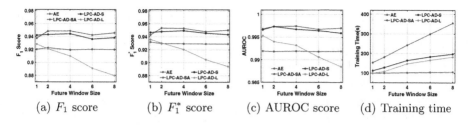

(a) F_1 score   (b) F_1^* score   (c) AUROC score   (d) Training time

Fig. 3. Impact of future window size on the detection accuracy and training time.

- **Impact of latent variable dimension, variance of random perturbation.** Due to page limit, we present experiment details in our technical report.

Answer 3: *Our LPC-AD-SA algorithm is robust to parameters of history window size, future window size, latent variable dimension and variance of random perturbation.*

6.4 Ablation Study

We conduct an ablation study to reveal a fundamental understanding on how two key components of our proposed algorithm, i.e., the predictor and the randomized perturbation operator, improve the anomaly detection accuracy.

Q4: Are the Random Perturbations and Predictor Necessary?

- **Impact of randomized perturbation.** To study the impact of randomized perturbation, we consider a variant of LPC-AD without randomized perturbation denoted by LPC-AD-N, where N refers to no randomized perturbation. In particular, LPC-AD-N is obtained by setting the randomized perturbation operator in the following deterministic form $\texttt{RandPerturb}(Z_{t+1}, \widehat{Z}_{t+1}, \epsilon; \Theta_{\mathrm{RP}}) = \widehat{Z}_{t+1}$. Figure 4(a) shows that LPC-AD-SA always have a higher F_1 than LPC-AD-N, where the improvement reaches a significant number of 5%. Figure 4(b) shows that LPC-AD-SA always has a

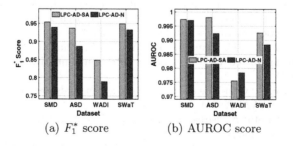

(a) F_1^* score   (b) AUROC score

Fig. 4. Impact of randomized perturbation on the detection accuracy.

higher AUROC than LPC-AD-N, except on the WADI dataset. On the WADI dataset, the AUROC of LPC-AD-SA is around 0.5% lower than LPC-AD-N.
- **Impact of the predictor.** Due to page limit, we present the experiment details in our technical report.

Answer 4: *Adding the predictor can increase the F1 score by as high as 9.16%. Adding the randomized perturbation can increase the F1 score by as high as 5%.*

7   Conclusion

This paper presents LPC-AD, which has shorter training time than SOTA deep learning methods that focus on reducing training time and it has a higher detection accuracy than SOTA sophisticated deep learning methods that focus on enhancing detection accuracy. These merits of LPC-AD are supported by novelty in the design and extensive empirical evaluations. In particular, LPC-AD contributes a generic architecture LPC-Reconstruct, which is a novel combination of ideas from autoencoder, predictive coding and randomized perturbation. We also contribute a new randomized perturbation method to avoid overfitting of anomalous dependence patterns. Extensive experiments validate superior performance and reveal fundamental understating of our method.

Acknowledgment. This work was supported in part by the Alibaba Innovative Research grant (ATA50DHZ4210003), National Natural Science Foundation of China (62072429), the Chinese Academy of Sciences "Light of West China" Program, the Key Cooperation Project of Chongqing Municipal Education Commission (HZ2021008, HZ2021017), and the "Fertilizer Robot" project of Chongqing Committee on Agriculture and Rural Affairs.

References

1. Abernethy, J., Lee, C., Tewari, A.: Perturbation techniques in online learning and optimization. Perturbat. Optim. Stat. 223 (2016)
2. Atal, B.S., Schroeder, M.R.: Adaptive predictive coding of speech signals. Bell Syst. Tech. J. **49**(8), 1973–1986 (1970)
3. Audibert, J., Michiardi, P., Guyard, F., Marti, S., Zuluaga, M.A.: USAD: unsupervised anomaly detection on multivariate time series. In: ACM KDD (2020)
4. Carmona, C.U., Aubet, F.X., Flunkert, V., Gasthaus, J.: Neural contextual anomaly detection for time series. In: IJCAI (2022)
5. Dai, L., Lin, T., Liu, C., Jiang, B., Liu, Y., Xu, Z., Zhang, Z.L.: SDFVAE: static and dynamic factorized VAE for anomaly detection of multivariate CDN KPIs. In: WWW (2021)
6. Deng, A., Hooi, B.: Graph neural network-based anomaly detection in multivariate time series. In: AAAI (2021)
7. Doersch, C., Gupta, A., Efros, A.A.: Unsupervised visual representation learning by context prediction. In: IEEE ICCV (2015)
8. Full Paper. https://1drv.ms/b/s!AkqQNKuLPUbEi1enh5C0RrTsDVL2?e=tvPAnY

9. Kim, S., Choi, K., Choi, H.S., Lee, B., Yoon, S.: Towards a rigorous evaluation of time-series anomaly detection. In: AAAI (2022)
10. Kitagawa, G., Gersch, W.: Linear gaussian state space modeling. In: Kitagawa, G., Gersch, W. (eds.) Smoothness Priors Analysis of Time Series, vol. 116, pp. 55–65. Springer, New York (1996). https://doi.org/10.1007/978-1-4612-0761-0_5
11. Li, D., Chen, D., Jin, B., Shi, L., Goh, J., Ng, S.-K.: MAD-GAN: multivariate anomaly detection for time series data with generative adversarial networks. In: Tetko, I.V., Kůrková, V., Karpov, P., Theis, F. (eds.) ICANN 2019. LNCS, vol. 11730, pp. 703–716. Springer, Cham (2019). https://doi.org/10.1007/978-3-030-30490-4_56
12. Li, G., et al.: openGauss: an autonomous database system. VLDB **14**, 3028–3042 (2021)
13. Li, Z., et al.: Multivariate time series anomaly detection and interpretation using hierarchical inter-metric and temporal embedding. In: ACM KDD (2021)
14. Liu, D., et al.: MicroHECL: high-efficient root cause localization in large-scale microservice systems. In: IEEE ICSE-SEIP (2021)
15. Liu, F.T., Ting, K.M., Zhou, Z.H.: Isolation forest. In: IEEE ICDM (2008)
16. Lusch, B., Kutz, J.N., Brunton, S.L.: Deep learning for universal linear embeddings of nonlinear dynamics. Nat. Commun. **9**(1), 1–10 (2018)
17. Ma, M., et al.: Diagnosing root causes of intermittent slow queries in cloud databases. VLDB **13**, 1176–1189 (2020)
18. Malhotra, P., Ramakrishnan, A., Anand, G., Vig, L., Agarwal, P., Shroff, G.: LSTM-based encoder-decoder for multi-sensor anomaly detection. arXiv preprint arXiv:1607.00148 (2016)
19. Mikolov, T., Chen, K., Corrado, G., Dean, J.: Efficient estimation of word representations in vector space. arXiv preprint arXiv:1301.3781 (2013)
20. Nguyen, N., Quanz, B.: Temporal latent auto-encoder: a method for probabilistic multivariate time series forecasting. In: AAAI (2021)
21. Park, D., Hoshi, Y., Kemp, C.C.: A multimodal anomaly detector for robot-assisted feeding using an LSTM-based variational autoencoder. IEEE Robot. Autom. Lett. **3**(3), 1544–1551 (2018)
22. Qin, Y., Song, D., Chen, H., Cheng, W., Jiang, G., Cottrell, G.: A dual-stage attention-based recurrent neural network for time series prediction. arXiv preprint arXiv:1704.02971 (2017)
23. Reynolds, D.A.: Gaussian mixture models. Encycl. Biometr. **741**(659–663) (2009)
24. Rezende, D., Mohamed, S.: Variational inference with normalizing flows. In: ICML (2015)
25. Su, Y., Zhao, Y., Niu, C., Liu, R., Sun, W., Pei, D.: Robust anomaly detection for multivariate time series through stochastic recurrent neural network. In: ACM KDD (2019)
26. Tuli, S., Casale, G., Jennings, N.R.: TranAD: deep transformer networks for anomaly detection in multivariate time series data. arXiv preprint arXiv:2201.07284 (2022)
27. Vaswani, A., et al.: Attention is all you need. In: NIPS (2017)
28. Wang, Y., Masoud, N., Khojandi, A.: Real-time sensor anomaly detection and recovery in connected automated vehicle sensors. IEEE TITS **22**(3), 1411–1421 (2020)
29. Xu, H., et al.: Unsupervised anomaly detection via variational auto-encoder for seasonal KPIs in web applications. In: WWW (2018)
30. Yaacob, A.H., Tan, I.K., Chien, S.F., Tan, H.K.: Arima based network anomaly detection. In: CCSN (2010)

31. Zhang, C., et al.: A deep neural network for unsupervised anomaly detection and diagnosis in multivariate time series data. In: AAAI (2019)
32. Zhao, H., et al.: Multivariate time-series anomaly detection via graph attention network. In: IEEE ICDM (2020)
33. Zong, B., et al.: Deep autoencoding gaussian mixture model for unsupervised anomaly detection. In: ICLR (2018)

RpDelta: Supporting UCR-Suite on Multi-versioning Time Series Data

Xiaoyu Han[1], Fei Ye[1], Zhenying He[1,2(✉)], X. Sean Wang[1], Yingze Song[3], and Clement Liu[4]

[1] School of Computer Science, Fudan University, Shanghai, China
{xyhan22,fye21}@m.fudan.edu.cn, {zhenying,xywangCS}@fudan.edu.cn
[2] Shanghai Key Laboratory of Data Science, Shanghai, China
[3] University of Liverpool, Liverpool, UK
sgyson16@liverpool.ac.uk
[4] Chase Grammar School, Cannock, UK
clement.liu@chasegrammar.com

Abstract. In real applications, various cleaning strategies are adopted to repair a specific time series several times for better effects. These multiple versions of the repaired time series, along with the raw time series, are often stored directly in the system for the users. However, as the scale of data explodes, high storage cost becomes a non-negligible problem. To address this problem, we propose RpDelta, a repaired time series storage strategy, under which a repaired time series can be represented as the combination of the raw time series and a differential file to use the storage space more efficiently. Meanwhile, we design a sequential reading strategy based on a finite state machine to make RpDelta adaptive to practical uses, which will almost not introduce additional time and space overheads. We also take the UCR-Suite algorithm as an example to introduce our optimizations on a simultaneous-operation circumstance with the help of RpDelta's properties. The extensive experiments show the effectiveness and efficiency of our work.

Keywords: Repaired time series · Multiple versions · Differential file

1 Introduction

Data quality for time series is critical in time series analysis and forecasting. The credibility of the results depends on assumptions that collected time series are reliable, which only sometimes hold in reality. For example, in finance, the correct rate of stock information on Yahoo Finance is 93% [11]. Moreover, in manufacturing, the collected data may be partially noisy or missing due to the unreliability of the physical sensor devices [4]. Therefore, before operating on time series, to avoid the degradation of analysis and forecasting caused by poor data quality, the data quality of time series should be checked from several aspects, such as data validity, completeness, and consistency [9].

X. Wang et al. (Eds.): DASFAA 2023, LNCS 13943, pp. 205–220, 2023.
https://doi.org/10.1007/978-3-031-30637-2_14

Recent works have proposed many data-cleaning algorithms to repair time series data to improve data quality. However, as we discuss below, it is difficult for a single data-cleaning algorithm to meet the requirements of real-world applications. Therefore, we will adopt different cleaning algorithms or parameters under one specific cleaning algorithm for better effects according to our comprehension of the collected data as time passes. In this way, we cannot overwrite the raw time series after we get a repaired one, which means that the raw one is non-tamperable [8]. Actually, Apache IoTDB integrates data-cleaning algorithms as functions into SQL operations, which keeps the raw time series unchanged unless we use update operation. So, a better way is to store multiple repaired time series separately from the original time series. Figure 1(a) shows the different versions of repaired time series derived from the same raw time series.

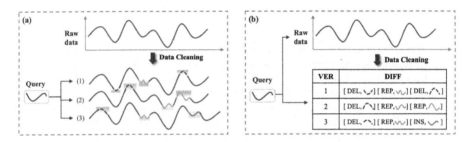

Fig. 1. (a) Subsequence matching on traditional multiple versions of repaired time series.(b) An optimization of subsequence matching under RpDelta storage strategy.

Under the above constraint of storing both the original time series and multiple repaired time series, there are several fundamental problems:

1. **Unnecessary Storage Use.** What is required to be repaired only accounts for a small part of the raw time series, generally at most 15%, because anomalies are often rare [2] or the collected data is untrusted. In other words, the repaired time series stores a lot of data identical to the raw one, which causes an enormous waste of space resources.
2. **High Disk I/O.** Sometimes we need to conduct the same operation(e.g., subsequence match) on all the repaired time series to analyze the effects of the data-cleaning algorithms and ensure robustness. To achieve this, we must read all the repaired time series, thus leading to a great deal of disk I/O, including many redundancies and jeopardizing the life of the disks.
3. **Extra Operation Time Overheads.** Considering the similarities between different versions of repaired time series, we will perform many repetitive calculations during the operation, which incurs extra time overheads.

To solve the above problem, we design a new storage strategy named RpDelta for multiple versions of repaired time series. Besides the raw time series, RpDelta stores all the repair operations from the raw to a repaired time series in a differential file called delta file. The number of delta files corresponds to the number of repaired versions. In this way, we can lower the storage cost from a wholly repaired time series to a delta file containing only necessary repairs. Since we can quickly restore the repaired time series according to its corresponding delta file and the raw time series, this will not harm the data integrity. Moreover, we redesign the sequential reading strategy under RpDelta based on the finite state machine approach and use the subsequence matching algorithm UCR-Suite [6] as an example to introduce the way to lower the disk I/O and repetitive calculations. Figure 1(b). shows the basic idea of RpDelta. It can be seen as extracting the unique part of each repaired time series to organize different delta files to replace the complete series. Meanwhile, the database engine can encapsulate all operations modifications under the RpDelta structure. Users do not need to do additional operations by themselves and can directly call the given interface.

To sum up, the main contributions of our paper are summarized as follows:

- We propose RpDelta, a storage strategy aimed at multiple versions of the repaired time series, which can significantly lower storage cost while maintaining data integrity.
- We further propose a sequential reading strategy under RpDelta based on the finite state machine approach. This approach possesses the same time complexity $O(n)$ and space complexity $O(1)$ as reading the complete time series directly, without prior restoration of the complete repaired time series.
- We take the UCR-Suite subsequence searching algorithm as an example to introduce the optimization of the circumstance about performing the same operations on all repaired time series. This optimization can efficiently reduce the disk I/O and unnecessary time overhead.
- We conduct extensive experiments on the UCR Time Series Archive. They demonstrate our optimizations of storage use, disk I/O and time overhead.

2 Preliminaries

2.1 Basic Concepts

Definition 1 (Time Series). A *time series* is a sequence of data points ordered by the time that can be denoted as the form $T = (t_1, t_2, \ldots, t_n)$, where $n = |T|$ stands for the length of time series T. A subsequence of T with length k is part of the whole time series that can be denoted as the form $T_{i,k}$, where i stands for its start position and satisfies $1 \leq i \leq n - k + 1$. For q time series, we use $T^1, T^2, \ldots, T^q$ to distinguish them and the same for subsequences $T_{i,k}^1, T_{i,k}^2, \ldots T_{i,k}^q$.

Definition 2 (Distance Metrics). Here we give the definition of two existing distance metrics used for time series calculation in our work, *Euclidean distance* (ED) and *Dynamic Time Warping* (DTW).

Given two time series T and T' with length n:

$$ED(T,T') = \sqrt{\sum_{i=1}^{n}(t_i - t'_i)^2} \tag{1}$$

$$DTW(O,O) = 0, DTW(O,T) = DTW(T,O) = +\infty$$

$$DTW(T,T') = \sqrt{min \begin{cases} DTW(T_{1,n-1}, T'_{1,n-1}) \\ DTW(T, T'_{1,n-1}) \\ DTW(T_{1,n-1}, T') \end{cases} + (t_n - t'_n)^2} \tag{2}$$

where O stands for empty series, $T_{1,0} = T'_{1,0} = O$.

We use $dist(T, T')$ to signal the distance between T and T' in this paper.

2.2 Problem Statement

Definition 3 (Subsequence Matching Problem). Given a time series T with length n and a query time series Q with length m(usually $m \ll n$), find a subsequence of T starting from position i which minimizes $dist(T_{i,m}, Q)$.

3 Repaired Time Series Storage

3.1 RpDelta Construction

The design for the new storage strategy RpDelta is inspired by the properties of the repaired time series. We investigated that the content needed to be repaired in a time series often accounts for only about 3% to 5% of the whole, at most no more than 15%, so most of the time series data after the repair remains the same as the raw series. In this case, considering the immutability of the raw series, we can use a differential file to record all the repair operations needed from the raw series to the repaired time series, which is similar to the binlog in MySQL. Here we denoted this differential file as a delta file. In Fig. 2, the delta file records three different operations that are above the arrow. We can quickly obtain a complete repaired time series by sequentially implementing those operations on the raw time series, as the Fig. 2 shows. This strategy helps us to ensure data integrity while lowering the storage costs. Our later experiments prove this point.

3.2 Delta File StorageFormat

We next define the exact form of the delta files. Here how to represent a repair operation is the foremost thing. After we research the current data cleaning algorithms, we classify the operations into three basic types: insert, delete and replace. We may use insertion in interpolation, deletion in removing repetitive samplings, and replacement in fixing the noises. The other complicated operations can be decomposed into these three types.

Moreover, we involve three parameters: start position, operation length, and data points. All the operations need start position and operation length to describe their scopes. The start position of insertion means we will insert a segment before the position in the raw series. The start position of deletion means we will delete a segment with this position in the raw series as the first data point. Only insertion and replacement need data points to determine their operation contents. In this way, we can define an operation as *"Type Length Position Data(Array)"*. Take "INS 2 1 [9, 6]" as an example, it means that we will insert an array [9, 6] whose length is 2 into the position before the time point 1. We can see this example and other operations in Fig. 2.

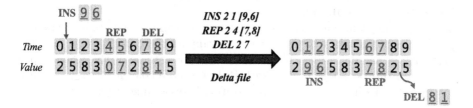

Fig. 2. Using a raw time series and a delta file to restore the repaired time series.

Besides, the data points in different time series have distinct types for various uses. Suppose in a situation requiring high precision, treating double data as float in the delta files is a disaster. So we should be informed of the data points' type when recording the repaired operations. At the beginning of the delta file, we can include the type information matching with the data points in the time series. Of course, we can add other information as well if necessary.

In short, a delta file involves two parts, the first part is some basic information about the repair operations and the corresponding time series, and the second part is a set of operations stored in the above format.

3.3 Delta File Constraints

To avoid ambiguity in different organizational forms in delta files, here we make some reasonable constraints to make our processing and explanation clearer and more convenient.

Constraint 1 (Position Constraint). A repair operation's start position refers to the raw series's corresponding position

For insertion, it may cause the index position in the raw series to change, so we need to make this constraint to ensure a delta file is free from duality of position. Sometimes the repair positions are off the raw series, such as inserting first and then restricting the speed of the inserted segment. However, we can merge them into one operation of inserting an already speed-restricted segment, performed on the raw series and equivalent to the previous two operations. The above shows the rationality of this constraint.

Constraint 2 (Overlap Constraint). There is no overlap between the two repair operations

According to constraint 1, we will not perform any operation on the inserted or deleted segments. If two insertion or deletion operations are performed in the same position, we can merge them according to their order. If an insertion operation shares the same position with a deletion or replacement operation, that has no problem because the inserted segment will not be affected. Some illegal situations, such as replacing or deleting some parts of a deleted subsequence, will not be allowed in our delta files. In this way, this constraint is reasonable.

Constraint 3 (Order Constraint). The repair operations are sorted by the position order in the delta file

Combining constraint one with constraint two, we can conclude that each repair operation is independent of the other. So swapping the order between any two operations in the delta file will not affect the correctness of the whole repair process. On this occasion, to deal with RpDelta more conveniently, we sort the repair operations in the delta files by position in ascending order.

4   Operations and Optimizations on RpDelta

When implementing analysis on time series, we are likely to perform operations on them. Sometimes we need to find the best-matching subsequence, while other times, we need to cluster multiple different time series. The existing algorithms are based on the input of a complete time series, however, under RpDelta each time series consists of two parts, thus making us fail to perform the algorithms directly. We will take the subsequence searching algorithm UCR-Suite as an example to introduce how to support some basic operations under RpDelta. Moreover, we will also utilize RpDelta's properties to accelerate some circumstances.

4.1   Sequential Reading

Sequential reading (reading the data points one by one) is the most basic operation on time series, and we can find it in many prevalent time series algorithms. Most of the algorithms read the data points of a time series one by one for calculations, even if in a batch process(data will be put into the buffer at first). On this occasion, we need to implement it on RpDelta first for further implementation.

A simple and intuitive approach is to restore the repaired time series according to the raw series and the corresponding delta file and then read it sequentially. Nevertheless, we will encounter two problems using this method. First, to restore, we need to put the entire raw series into the memory, which is a large burden. Second, no matter what data structure we use, the time complexity to restore a time series with length n and m operations in its delta file will be $O(mn)$, since insertion and deletion need $O(n)$ time to complete. The restoration process itself will likely take longer than the time series algorithms. Under this circumstance, we need to design a more efficient and clever way to read sequentially in real time on RpDelta. Previously, we read the raw series from start to end directly, so our approach reads both the raw series and the delta file only once.

Problem Simplification. Here, we refer to where we can get the data as the **data source**. In traditional complete storage, the data source is the raw time series, and it is unique, so it is easy to conduct each read. While under RpDelta, all the data points in repaired time series can be obtained from either the raw series or the corresponding delta file, which means that we have two data sources. So considering both data sources at the same time is necessary. In this way, we can reduce the sequential reading problem to the strategy of deciding the data source of the next read according to the current reading status.

FSM (Finite State Machine) Method. We use FSM to solve the problem of data source choice. FSM is a computational model with a finite number of states, which can transfer from one to another in response to some inputs.

Here we use ST_Raw, ST_Ins, ST_Del, and ST_Rep to stand for the four different reading states, and we use ST_Cmp to stand for a checking state.

While dealing with a repaired time series, we first check the position of the following repair operation in ST_Cmp. Before arriving at this position, the repaired time series is the same as the raw. In this way, we can choose the raw series as the data source in ST_Raw and then return to the ST_Cmp state for the next check. While encountering such a position, we move to different states according to their corresponding operation types and return to ST_Cmp when the operation is over after several cycles. Table 1 shows the change between different states.

Table 1. Transfer between different states

Now	ST_Cmp		$ST\_Ins/Del/Rep$		ST_Raw
Input	(Pos) matched	not matched	End of read	Not end of read	End of read
Next	$ST\_Ins/Del/Rep$	ST_Raw	ST_Cmp	Remain unchanged	ST_Cmp

In the insertion state, the delta file is the data source, and we will directly return to ST_Cmp when finished. In the deletion state, we take raw series as the data source and skip a specific length of raw series before reading the point we want. Finally, in the replace state, the delta file is the data source, and we also need to skip the corresponding subsequence replaced in the raw series. This method allows us to read the repaired time series on RpDelta sequentially, almost without extra time or space complexity. Figure 3 exemplifies the reading process.

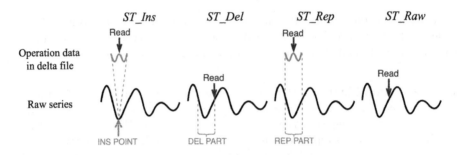

Fig. 3. Reading process in different reading states.

With the help of the above state machine, the space complexity can reach $O(1)$, and we only need to read the raw series and Delta File once to obtain the repaired time series sequentially on RpDelta.

4.2 Fundamental Subsequence Search

Here we take the UCR-Suite algorithm, a prevalent subsequence searching algorithm, as an example to illustrate how to implement such kind of operations on RpDelta.

The time series addressed by UCR-Suite are stored as .txt format files in the file system. The algorithm uses file pointers to read them sequentially, reflecting in the code as *fscanf*. Under the RpDelta strategy, we only need to replace it with a reading interface implemented according to the FSM method, and the rest of the algorithm remains unchanged.

The introductions of delta files and FSM reading strategy will bring about extra time overhead in the read-in of the UCR-Suite. However, it is small enough to be ignored, especially when the cost of the calculations is high. Our experiments displayed in Sect. 5.3 prove this point in detail.

Through the example of UCR-Suite, we can see that for all operations similar to it, in which we need to read the data points in time series one by one, the previous reading method(e.g., file pointer or others under different storage form) can be modified to the interface based on FSM method without any pain. This also proves the practical feasibility of RpDelta.

4.3 Subsequence Search on Multiple Repaired Time Series

Multi-versioning time series data is used to provide robustness for downstream tasks. To avoid potential errors caused by a single cleaning algorithm, we provide multi-versioning results or select the most reliable version for a given task. In this process, comparison and analysis between different versions of time series will bring unnecessary I/Os and repetitive calculations. For example, in UCR-Suite, we will calculate the distance between the query series and a repaired time series subsequence to find the best match. We are likely to read and take the same subsequence to calculate many times because this part does not need to be repaired and thus appears in many repaired time series, which causes much unnecessary extra time overhead.

Given that our RpDelta solves the repetitive storage by recording duplicates in one raw series and storing the exclusive parts in corresponding delta files, we will use this property to eliminate the repetitive calculations. Here we still take UCR-Suite as an example to illustrate how to optimize such occasion. Our optimization includes three steps.

Parallel Read Processing. We use a parallel approach to read multiple series at the same time. We should note that this approach is parallel abstractly and does not mean multiple cores or threads here(though it can be realized in such ways). In the past, we read in a data point, calculated the metrics, compared the distance with the best-so-far, and then read in the following data point and repeated this process. Now we read in an array directly, which consists of many data points from different time series. After that, we take the same procedure above on each of them. Our core idea of optimization is to reuse the calculations. If we deal with the time series one by one in this situation, we need to pay a lot for storing multiplexing value. It is because we must record the calculations of all the possible repetitive parts during the first repaired time series process. When reading them parallel, we just need to store them temporarily because they will soon be consumed by the others and can be thrown.

Moreover, we only need to access the raw series on the disk once in parallel reading, lowering the I/O cost. This is because we will put a data point into the buffer in the memory when first met, and while it has been used by all the sequential readings of the repaired time series(FSM method), it will be removed from the buffer. In this way, we can also lower time overhead since there is no need to access the disk for the raw series frequently.

Repetitive Subsequence Judgment. Moreover, we must decide on which circumstance we should reuse the calculations. As mentioned above, to the identical subsequences appearing many times, we just need to calculate once and store the value for later use. Therefore, one of the most important things is to propose a strategy that allows us to judge repetitive subsequences in different repaired time series. Here, we propose a coarse-grained method to determine them: if a subsequence has not been repaired by any data-cleaning algorithm, it is our target because it will appear in all of the repaired time series. In this method, we deliberately abandon many possible repetitive subsequences, such as those that appear $l - 1$ times in the total l repaired time series. This makes the optimization more manageable and less costly in terms of implementation, maintenance and judgment. On the other hand, if we insist on finding all of the repetitive subsequences, chances are that its time cost will outweigh what we save on the calculation reuse.

In the concrete implementation, we scan all the delta files and mark all the operation positions during preprocessing. After this, we can get an array B in which $B[i]$ stands for the total number of operations on raw series $T_{i-m+1,m}$(m is the query length). Here we should notice that the influence of an operation is in the range of its length, not the single start data point. By examining $B[i] = 0$ or not, we can quickly learn whether a subsequence is our target during runtime with the time cost of $O(1)$.

Algorithm 1: Optimizations in UCR-Suite under DTW metric.

Input: $n = |T^k|$, $m = |Q|$, q is the series number
Output: The best-match subsequence in each time series

 1 **for** $p \leftarrow 1$ **to** q **do**
 2   **for** $i \leftarrow 1$ **to** $n - m + 1$ **do**
 3   **if** $T^p_{i,m}$ *belongs to repetitive subsequence* **then**
 4   **if** *meet* $T^p_{i,m}$ *first time* **then**
 5 | Calculate *lb_kim* and put into cache(may early abandon);
 6   **else**
 7   Get *lb_kim* from cache;
 8   **if** *lb_kim* $< bsf^p$ *and is not completely calculated* **then**
 9 | Calculate *lb_kim* more exactly and refresh it in cache;
10   **end**
11   **end**
12   **else**
13 | Calculate *lb_kim*;
14   **end**
15   **if** *lb_kim* $< bsf^p$ **then**
16 | Continue to calculate *lb_k*, *lb_k2*, *dtw* in order like above;
17   **else**
18 | Early abandon;
19   **end**
20   **end**
21 **end**

Repetitive Calculation Elimination. In UCR-Suite under DTW metric, the main computations are *lb_kim* lower bound generated by the LB_KimFL algorithm, *lb_k* lower bound generated by the LB_KeoghEQ algorithm, *lb_k2* lower bound generated by the LB_KeoghEC algorithm and the DTW itself. All the above lower bound are served for early abandoning in UCR-Suite. We should point out that what we get in the calculation may be different from the exact value of them because of early abandoning. When first meeting a repetitive subsequence, we will manage it as usual and save the main computations in a buffer. Next time we process it again, all the saved computations will be reused. Suppose they are enough for early abandoning or distance calculation. In that case, we can eliminate repetition and adopt them; if not, we will calculate again for more exact values and update them in a buffer. Algorithm 1 shows part of this procedure related to *lb_kim*, and other computations are similar, so we omit them here. In this way, most calculations of repetitive subsequences can be combined into nearly one or, at most several times, thus significantly improving the subsequence-searching efficiency on multi-versioning time series.

5   Experiments

5.1   Experimental Setup

Datasets. We conduct extensive experiments on the latest UCR Time Series Classification Archive Dataset [3], which contains plenty of real-world datasets, such as ECG and Electric Devices data.

Baselines. Our baseline stores every repaired time series entirely on the disk for the storage experiment. In addition, to the operation experiment, since our optimization is introduced with the example of the UCR-Suite algorithm, here we use the UCR-Suite algorithm under ED and DTW respectively as our baselines to show our optimization. We get the implementation of UCR-Suite from its official website. Moreover, we identify that UCR-Suite uses the idea of early abandoning to accelerate the matching process, which makes our optimization of repetitive calculation less influential, but this is not always the case. Therefore, to make the results of the experiments more apparent to us, we modify UCR-Suite under the ED metric by removing the early abandoning strategy and use it as a baseline as well.

Our experiments use **ED** and **DTW** to stand for the raw UCR-Suite algorithm under the corresponding distance metric. ED^- stands for UCR-Suite under ED without early abandoning. ED_p uses parallel read processing under ED, and **ED-M** uses repetitive subsequences elimination.

The reason why we do not choose the latest subsequence searching algorithms, such as KV-Match [12] or ULISSE [5], as the baselines is that what we have done is not a searching algorithm but an idea of optimization under RpDelta. Meanwhile, such algorithms need many times the space we require because they use indexes to accelerate.

Implementation Details. We use a 10^7 long time series with the data type of double as the raw time series. This time series is concatenated by the time series in UCR Archive. We obtain query series of the desired length from the subsequence of the raw time series. In terms of repaired time series, we generate them artificially under the guidance of the Pareto principle. All of our data-cleaning strategies randomly put 80% of the repairs on 20% of the raw series and 20% of the repairs on 80% of the raw series. Replacement operations account for 60%, while insertion and deletion account for 20% respectively. The average length of each repair operation is 10, and the value of data points in operation is between $[\mu - \sigma, \mu + \sigma]$, where μ is the mean and σ is the standard deviation of the raw time series.

In the storage experiment, all the time series are stored as .txt files in the file system. For convenience in observing, we directly use the space of the raw time series to approximate the space of the repaired time series, which is reasonable in terms of mathematical expectations and will not influence our conclusions.

In the operation experiment, we have 3 key parameters: series number, repair rate, and query length. Here the series number reflects the number of data-cleaning strategies used to achieve multi-versioning time series. Each data-cleaning strategy corresponds to a version. We set the standard value of the series number to 6, the repair rate to 4%, and the query length to 128. We take the standard value if we do not specify the parameters' values. In the different experiments, we will adjust the value of the corresponding parameter. The Sakoe-Chiba Band [7] used on DTW calculations is 0.05(ratio to query series).

Testbed. Our algorithms are implemented in C++ and compiled by g++ 7.5.0 on Ubuntu 18.04 system with a 4.15.0-189-generic Linux kernel. We conduct experiments on an Intel(R) Xeon(R) Silver 4208 CPU @ 2.10 GHz machine with 64 GB RAM.

5.2   Storage Performance

We compare the storage use of RpDelta with the complete storage baseline. Table 2 shows RpDelta's storage performance. We find out from Table 2 that the storage use ascends as the repair rate and series number increases, but is much slower than the baseline, with less than 20% of the baseline at 8 repaired time series.

Table 2. Storage use of multi-versioning repaired time series under different repair rates and series numbers and the value of θ to each repair rate.

Rate	Number					
	1	2	4	6	8	θ
Baseline	182.02M	273.03M	455.04M	637.06M	819.08M	1
1%	91.92M	92.83M	94.67M	96.48M	98.30M	0.01001
2%	92.84M	94.67M	98.30M	101.95M	105.61M	0.02006
4%	94.66M	98.30M	105.56M	112.86M	120.12M	0.03999
8%	98.32M	105.60M	120.18M	134.77M	149.29M	0.08005

Experiments show that as the number of time series increases, the ratio of space to the baseline will decrease, but it will not converge to 0. Here we assume that the size of the delta file under a specific repair rate is a fixed value γ, and the size of a repaired time series or the raw one is a fixed value λ. The total number of repaired time series is n. We find that the ratio will converge to the value θ:

$$\theta = \lim_{n \to +\infty} \frac{n \cdot \gamma + \lambda}{(n+1)\lambda} = \frac{\gamma}{\lambda} \tag{3}$$

The last column of the Table 2 shows our experiments' value of θ at different repair rates. It illustrates the limits of RpDelta's space-saving capabilities.

This experiment demonstrates that in the situation of multi-versioning time series, RpDelta can effectively lower the space cost to a low level and solve the problem of unnecessary disk storage occupation.

5.3 Operation and Optimization Performance

We conduct extensive experiments to demonstrate the effectiveness of our optimization under different circumstances. Experiments show that we usually have a 60%–100% speed improvement under the ED metric, and under the DTW metric, a 30%–50% one. Moreover, they also display the good scalability of our method. The specific experiments are as follows.

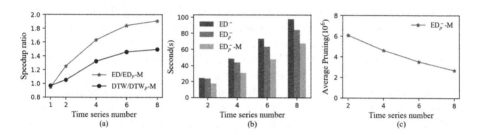

Fig. 4. (a) The speedup ratio of post-optimized algorithm to the baseline under different series numbers. (b) The running time of $\mathbf{ED}^-$, $\mathbf{ED_p^-}$, $\mathbf{ED_p^-}$-M under different series numbers. (c) The average number of eliminated repetitive subsequences in a repaired time series under different series numbers (divide by series number - 1 when calculating).

Effects of Series Number. In this experiment, we explore the impact of the series number on our optimization. Figure 4(a) shows the speedup ratio under different series numbers. When the series number is 1, we only apply the FSM method described in Sect. 4.1. At this point, the sequential reading time under the RpDelta strategy is almost the same as that of the baseline. Only maintaining the state machine to select the data source makes it slightly slower, but it does not matter.

As the series number increases, the speedup ratio gradually ascends, but the growth rate becomes smaller. We can get the reason from Fig. 4(b) Experiments on UCR-Suite without early abandoning show that despite the more I/O time saved by parallel read processing in the situation of more series, the effect of repetitive calculation elimination also descends. This is because the random distribution of the repair part leads to a decreasing number of repetitive subsequences per series, as Fig. 4(c) exhibits. These two factors together give rise to the results in Fig. 4(a).

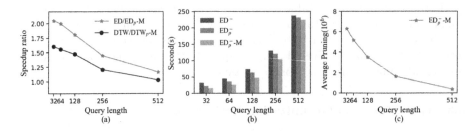

Fig. 5. (a) The speedup ratio of post-optimized algorithm to the baseline under different query length. (b) The running time of $\mathbf{ED^-}$, $\mathbf{ED_p^-}$, $\mathbf{ED_p^-}$-$\mathbf{M}$ under different query length. (c) The average number of eliminated repetitive subsequences in a repaired time series under different query length (divide by series number - 1 when calculating).

Effects of Query Length. In this experiment, we explore the impact of the query length on our optimization. Figure 5(a) shows the speedup ratio under different query lengths. The speedup ratio remains at a high level of more than 1.8 under ED and 1.5 under DTW when the query length is less than 128 and gradually descends to a comparatively low level (nearly 1) as the query length increases. This is because it is more difficult to find such long repetitive subsequences on this occasion, as Fig. 5(c) exhibits. Meanwhile, the increase in query length leads to a longer running time, thus making the optimization of parallel read processing inconspicuous, which is displayed in Fig. 5(b).

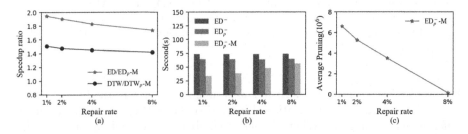

Fig. 6. (a) The speedup ratio of post-optimized algorithm to the baseline under different repair rates. (b) The running time of $\mathbf{ED^-}$, $\mathbf{ED_p^-}$, $\mathbf{ED_p^-}$-$\mathbf{M}$ under different repair rates. (c) The average number of eliminated repetitive subsequences in a repaired time series under different repair rates (divide by series number - 1 when calculating).

Effects of Repair Rate. In this experiment, we explore the impact of the repair rate on our optimization. Figure 6(a) shows the speedup ratio under different repair rates. The speedup ratio declines slowly when the repair rate ascends but still maintains a high level. Generally speaking, the higher the repair rate, the fewer the repetitive subsequences, but the speedup ratio is not seriously affected here. We can get the reason for this phenomenon from the experiment on ED^-. Through Fig. 6(b), we can find that the running time of ED and ED_p is nearly unchanged as the repair rate ascends. Nevertheless, repetitive subsequence elimination has little effect at a high repair rate because few repetitive subsequences exist, as Fig. 6(c) exhibits. Parallel read processing plays a more significant role in speedup when the query length is 128, and the series number is 6, which explains Fig. 6(a).

To sum up, our optimization of operations on RpDelta owns a significant effect. The parallel read processing can provide stable improvement, and the repetitive subsequences elimination is considerably helpful, especially when the series number, query length, and repair rate are relatively small and still has sound effects when they increase.

6   Related Work

There are many studies on time series data cleaning and time series operations. Xi Wang pointed out that the current data-cleaning algorithms can be divided into three categories [11]. The first is a smoothing-based cleaning algorithm, such as the interpolation method used by S. Xu [13]. The second is a constraint-based cleaning algorithm. For example, Shaoxu Song proposed SCREEN [10], which uses the speed restrictions on value changes in a given interval. The third is a statistics-based cleaning algorithm, such as a method based on the HMM for RFID data cleaning put forward by Baba et al. [1]. However, all the above algorithms just focus on how to repair the time series more effectively and ignore the high storage costs of repaired time series.

In time series operations, there exist many kinds, like time series subsequence searching, time series clustering, and so on. Our works focus on time series subsequence searching. UCR-Suite is a state-of-the-art approach to solving the normalized subsequence searching problem, while KV-Match [12] and ULISSE [5] are the latest searching algorithms based on indexes. However, current operations are performed on the complete time series and fail to deal with the circumstances of many repetitive subsequences between sequential input time series.

7   Conclusion

In this paper, we propose a repaired time series storage strategy called RpDelta for efficient use of storage space. We first introduce the basic idea and the storage format of RpDelta. Then, we impose three constraints on the delta files in RpDelta to avoid possible ambiguities. In order to put RpDelta into practical use, we design a sequential reading strategy based on a finite state machine and take UCR-Suite as an example to illustrate how it works in a concrete operation of time series. Moreover, we also optimize the simultaneous-operation circumstance with the help of the RpDelta's properties. The experiments performed on

UCR 2018 archive show that we can save more than 80% of the storage space by using RpDelta when the series number is 8. Meanwhile, we can usually obtain a 60%–100% speed improvement under the ED metric and 30%–50% under the DTW metric, which shows the effectiveness and efficiency of our approach.

Acknowledgements. The authors would like to thank all the anonymous reviewers for their insightful comments and suggestions. This work was supported by the National Key R&D Program of China (No. 2021YFB3300502).

References

1. Baba, A.I., Jaeger, M., Lu, H., et al.: Learning-based cleansing for indoor RFID data. In: SIGMOD Conference, pp. 925–936. ACM (2016)
2. Benkő, Z., Bábel, T., Somogyvári, Z.: Model-free detection of unique events in time series. Sci. Rep. **12**(1), 227 (2022)
3. Dau, H.A., Keogh, E., Kamgar, K., et al.: The UCR time series classification archive (2018). https://www.cs.ucr.edu/~eamonn/time_series_data_2018/
4. Jeffery, S.R., Alonso, G., Franklin, M.J., Hong, W., Widom, J.: Declarative support for sensor data cleaning. In: Fishkin, K.P., Schiele, B., Nixon, P., Quigley, A. (eds.) Pervasive 2006. LNCS, vol. 3968, pp. 83–100. Springer, Heidelberg (2006). https://doi.org/10.1007/11748625_6
5. Linardi, M., Palpanas, T.: Scalable data series subsequence matching with ULISSE. VLDB J. **29**(6), 1449–1474 (2020). https://doi.org/10.1007/s00778-020-00619-4
6. Rakthanmanon, T., Campana, B.J.L., Mueen, A., et al.: Searching and mining trillions of time series subsequences under dynamic time warping. In: KDD, pp. 262–270. ACM (2012)
7. Sakoe, H., Chiba, S.: Dynamic programming algorithm optimization for spoken word recognition. IEEE Trans. Acoust. Speech Signal Process. **26**(1), 43–49 (1978)
8. Sathe, S., Papaioannou, T.G., Jeung, H., Aberer, K.: A survey of model-based sensor data acquisition and management. In: Aggarwal, C. (ed.) Managing and Mining Sensor Data, pp. 9–50. Springer, Boston (2013). https://doi.org/10.1007/978-1-4614-6309-2_2
9. Song, S., Zhang, A.: IoT data quality. In: CIKM, pp. 3517–3518. ACM (2020)
10. Song, S., Zhang, A., Wang, J., Yu, P.S.: SCREEN: stream data cleaning under speed constraints. In: SIGMOD Conference, pp. 827–841. ACM (2015)
11. Wang, X., Wang, C.: Time series data cleaning: a survey. IEEE Access **8**, 1866–1881 (2020)
12. Wu, J., Wang, P., Pan, N., et al.: KV-match: a subsequence matching approach supporting normalization and time warping. In: ICDE, pp. 866–877. IEEE (2019)
13. Xu, S., Lu, B., Baldea, M., et al.: Data cleaning in the process industries. Rev. Chem. Eng. **31**, 453–490 (2015)

GP-HLS: Gaussian Process-Based Unsupervised High-Level Semantics Representation Learning of Multivariate Time Series

Chengyang Ye[✉][ID] and Qiang Ma[ID]

Graduate School of Informatics, Kyoto University, Kyoto, Japan
ye@db.soc.i.kyoto-u.ac.jp, qiang@i.kyoto-u.ac.jp

Abstract. Representation learning of multivariate time series is a significant and challenging task, which is helpful in various tasks such as time series data search, trend analysis, and forecasting. In practice, unsupervised learning is strongly preferred owing to sparse labeling. Most existing studies focus on the representation of independent subseries and do not take into consideration the relationships among different subseries. In certain situations, this may lead to failure of downstream tasks. This study proposes an unsupervised representation learning model for multivariate time series by considering high-level semantics. Specifically, we introduce the covariance calculated by the Gaussian process to the self-attention mechanism to reveal the high-level semantics features of the subseries. Additionally, we design a novel unsupervised method to learn the representation of multivariate time series. Moreover, to deal with the challenge of variable lengths of input subseries of multivariate time series, a temporal pyramid pooling (TPP) method is applied to construct input vectors with equal length. The experimental results show that our model has substantial advantages compared with other semantic-based representation learning models and can be well applied in various downstream tasks.

Keywords: Unsupervised representation learning · Gaussian process · High-level semantics · Multivariate Time Series

1 Introduction

The significant progress of the Internet and widespread use of sensors has driven the remarkable development of multivariate time series, such as electrocardiograms [1] and daily stock prices [2]. This plays an important role in the field of data engineering. As the application scenarios and downstream tasks of multivariate time series become increasingly complex, representation learning can advance the analysis of multivariate time series and become a universal tool for feature detection and preprocessing of raw data. Representation learning

X. Wang et al. (Eds.): DASFAA 2023, LNCS 13943, pp. 221–236, 2023.
https://doi.org/10.1007/978-3-031-30637-2_15

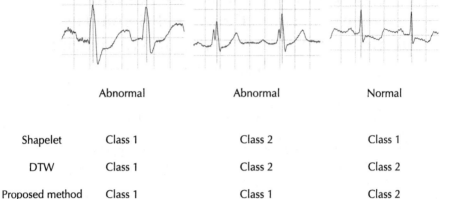

	Abnormal	Abnormal	Normal
Shapelet	Class 1	Class 2	Class 1
DTW	Class 1	Class 2	Class 2
Proposed method	Class 1	Class 1	Class 2

Fig. 1. Example of the issue in semantic-based time subseries. These figures represent different states of the heart. Class 1 and class 2 represents abnormal state and normal state respectively.

replaces manual feature engineering and enables the learning and use of features to perform a specific task.

Conversely, several semantic-based methods and algorithms have recently demonstrated good performance in the areas of natural language processing (NLP) and computer vision (CV). Extracting semantic-based features is the first step of almost all CV models [3]. Inspired by this progress, an increasing number of algorithms for multivariate time series have used semantic information in a wide variety of tasks, especially in a data search of time series [4]. These algorithms convert multivariate time series data into several subseries based on semantic information. Semantic information can be revealed by the shape of the curve of time series data (e.g., the shapelet learning method [5]) or other statistical information, such as the mean and maximum values of time series data [6]. Semantic-based methods have a natural advantage, i.e., they can convert a time series into several subseries according to their semantic information; this characteristic can be beneficial for data storage and search.

These algorithms have shown good results in information retrieval and classification tasks. However, there are still some limitations and weaknesses in previous studies. Most of these semantic-based algorithms focus on obtaining accurate semantic subseries rather than the relationships among different subseries. They require another algorithm to learn the relationships among subseries, which may increase the computational cost of downstream algorithms and affect the performance. This can be illustrated by analyzing the incorrect results in the experiments with these semantic-based algorithms. Figure 1 shows a typical example of this issue in electrocardiogram (ECG) classification tasks.

As can be seen in the Fig. 1, the three curves of the ECG time series have similar shapes, although they represent different states of the heart. In such a situation, the traditional semantic-based algorithms classified them into incorrect classes. This is a common phenomenon in the real world, which is mainly caused by disregarding the relationships between different subseries. We refer to these semantic relationships as high-level semantics. High-level semantics is a fundamental concept in CV [7] that can distinguish an object in an image by considering the surrounding information of the neighbors of the target object. Given this definition, real-world time series also have high-level semantics, i.e., the relation between neighbor subseries and target subseries that can enhance the performance of semantic-based time series methods. Thus, our motivation is to design a representation learning model by representing subseries of multivariate time series with high-level semantics.

In addition, there are other problems with traditional time series algorithms that are challenging for various reasons. First, most real-life time series are unlabeled. Therefore, unsupervised algorithms are strongly preferred because of their broader application scenarios, i.e., unlabeled time series data can be used, and more adaptive features can be learned. Second, the methods should deliver compatible representations while allowing the input time series to have unequal lengths. Given that the algorithm divides the entire time series into several subseries according to semantic information, the length of each subseries may differ.

In this study, we propose a novel unsupervised learning framework to learn the representation of semantic-based subseries of multivariate time series. The proposed model represents the subseries by considering the covariance calculated by the Gaussian process (GP) to reveal their high-level semantics (HLS) and is named GP-HLS. First, a Gaussian process-based attention mechanism is introduced to the encoder of the transformer [8] as the representation learning model. It uses the covariance calculated by the GP as the external information to consider the high-level semantics of each subseries of the multivariate time series. Subsequently, a Gaussian drop-based triplet network is designed for multivariate time series to construct the positive and negative sample pairs of unsupervised training. In addition, we use an advanced segmentation algorithm named greedy Gaussian segmentation (GGS) [9] to generate several subseries of multivariate time series. And a widely used input regularization method, named temporal pyramid pooling (TPP) [10], is considered to generate regular inputs for time series subseries with unequal lengths.

In summary, the main contributions of our work are as follows:

- We propose a transformer encoder-based architecture with the GP in the self-attention mechanism (Sect. 3.2) that uses covariance information to learn high-level semantic features in subseries inputs.
- We develop an unsupervised training method (Sect. 3.3). Triplet sample pairs for multivariate time series data based on the Gaussian drop are also designed to construct the unsupervised sample pairs of multivariate time series.
- We conduct extensive experiments on several datasets from different fields (Sect. 4). In comparison to other baseline algorithms, the proposed GP-HLS model achieves better results and is applicable different tasks.

The remainder of this paper is organized as follows. Section 2 outlines previous studies on representation learning for multivariate time series and self-

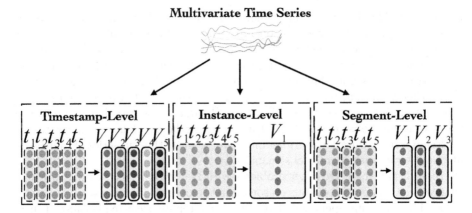

Fig. 2. Main differences among three types of representation learning of multivariate time series.

attention mechanisms from existing literature. Section 3 describes the architecture of the proposed model in detail. Finally, Sect. 4 presents the experimental results, and the study conclusions are summarized in Sect. 5.

2   Related Work

2.1   Representation Learning of Time Series

The representation learning of time series data has become a topic of considerable research interest. Most models aim to discover spatial-temporal dependencies in data. According to the representation granularity, there are three types of representation learning for time series: the timestamp-level, the instance-level, and the segment-level. The differences between these three types are shown in Fig. 2.

In timestamp-level learning, the model represents each timestamp for time series; it is the most traditional idea for representation learning of time series and very complex. It focuses more on the relationship between the different dimensions of time series; an example of a model that uses such type of representation learning is TS2Vec [11]. TS2Vec has been recently proposed as a universal framework for learning time series representations by hierarchically performing contrastive learning over augmented contextual information. Although timestamp-level representation learning can achieve superior results in time series forecasting and anomaly detection tasks, such algorithms still have limitations. Particularly, they are not intended to represent the state of subseries and cannot be applied to certain tasks like data retrieval.

Many studies have also focused on learning instance-level representations, which describe the entire segment of the input time series and have shown excellent performance in clustering and classification tasks [12]. In addition, recent works have employed contrastive loss to learn the inherent structure of time

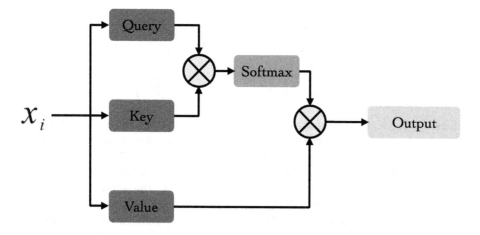

Fig. 3. Schematic of original self-attention of transformer.

series. Nevertheless, they also have certain limitations. Instance-level representations may not be suitable for fine-grained forecasting models, which must infer the target in a specific subseries.

The segment-level representation of time series combines the advantages of timestamp-level and instance-level representation learning. It is somewhere in the middle of the timestamp-level and instance-level representation at the level of granularity, such as a scalable time series pre-training model SETP [13]. This model learns temporal patterns from long-term multivariate time series to generate segment-level representations, which help to develop dependencies between multivariate time series. A problem with these algorithms is the segmentation rule; this model divides multivariate time series data into subseries using a regular sliding window. In this manner, the subseries is random without any semantic information and relationship between different subseries. This may also lead to confusion in the representation of results. The representation of these subseries cannot be used in tasks that require semantic information, such as data retrieval. In other words, none of these segment-based representation learning methods can learn high-level semantic information in time series. This is the main issue that the proposed model can address in this study.

2.2 Self-attention Mechanism

Self-attention (also called intra-attention [14]) is an attention mechanism that constructs attention models using the relationship between the input samples themselves. It is useful in a wide field of machine learning, such as image processing, text representation and data prediction. The most well-known application of self-attention is the transformer [8] proposed for NLP tasks.

Assuming that x_i represents a certain training batch consisting of several subseries of multivariate time series, the original self-attention can be described as shown in Fig. 3.

The function of self-attention is expressed as

$$Q = x_i W_i^Q; K = x_i W_i^K; V = x_i W_i^V \tag{1}$$

$$Attention(Q, K, V) = softmax\left(\frac{QK^T}{\sqrt{d_k}}\right) V \tag{2}$$

where Q, K, V are the matrices of queries, keys of dimension d_k, and values of dimension d_k, respectively. As shown in Eq. (1), queries, keys, and values are projected by the linear transformations $W_i^Q \in \mathbb{R}^{d_m \times d_q}$, $W_i^K \in \mathbb{R}^{d_m \times d_k}$ and $W_i^V \in \mathbb{R}^{d_m \times d_v}$, respectively, where d_m is the dimension of the input.

As shown in the paper on ordinary transformers [8], the self-attention mechanism represents inputs by calculating their similarity, which can generate a suitable representation for words. However, this is not sufficient for representing subseries with high-level semantics. It not only requires similarity information among different subseries, but also the correlation among each subseries that plays a significant role in the representation learning of time series.

3 Methodology

3.1 Overview

In this section, the proposed GP-HLS model structure and the relevant algorithms are described. The structure of GP-HLS is shown in Fig. 4. First, an input regularization method of one-dimensional data is considered to generate regular inputs for time series subseries with unequal lengths. Subsequently, a GP-based attention mechanism is introduced to the encoder of a transformer as

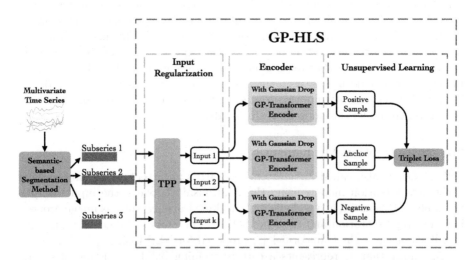

Fig. 4. Structure of unsupervised representation learning for time series with high-level semantic features.

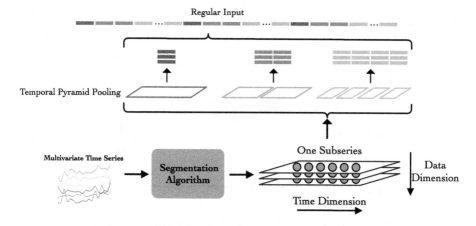

Fig. 5. Detailed diagram of input regularization method.

a representation learning model. It uses covariance calculated by the GP as the external information to consider the high-level semantic features of each sub-series of the multivariate time series. Then, a Gaussian drop-based triplet loss function is designed for multivariate time series to construct the positive and negative sample pairs of unsupervised training.

Because most semantic-based segmentation methods divide the entire time series into several subseries with varying lengths, we must reshape them with unequal lengths. Then, the model learns their representation. We apply TPP [10] to regularize the subseries input generated by the segmentation method, which was proposed to deal with the varying length issue of the input for one-dimensional data. The TPP method is illustrated in Fig. 5.

3.2 Gaussian Process-Based Self-attention Mechanism

As introduced earlier, the original self-attention mechanism is not sufficient to represent subseries with high-level semantics. The correlation among each sub-series is necessary for the representation learning of time series, and especially for revealing the high-level semantics in time series.

Based on this concept, we propose a GP-based self-attention mechanism in the encoder of the transformer architecture to add correlation information to the representation learning of multivariate time series. The diagram of the proposed model is shown in Fig. 6, where X represents the entire sequence of the multivariate time series. In our model, the covariance function is learned from the GP and the covariance matrix is then generated according to the subseries in the batch. The covariance matrix can reveal the correlation among each subseries in the input batch, which can be used as the correlation matrix for subsequent calculations.

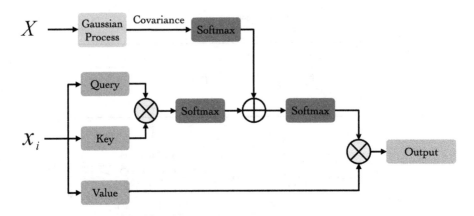

Fig. 6. Schematic of Gaussian process-based self-attention mechanism.

After adding the GP part to the self-attention mechanism, the function of self-attention described in Eq. (2) can be rewritten as:

$$Attention(Q, K, V) = softmax \left(softmax \left(\frac{QK^T}{\sqrt{d_k}} \right) + softmax \left(Cov(x_i) \right) \right) V$$

(3)

where $Cov(x_i)$ represents the covariance matrix of subseries in the input batch. In Eq. (3), he first component $\frac{QK^T}{\sqrt{d_k}}$ represents the similarity relationship of the input, and the second component $Cov(x_i)$ represents the correlation relationship of the input.

For covariance, a fundamental fact of GP is that it can be defined entirely by second-order statistics [15]. Thus, if a GP is assumed to have a mean of zero, the covariance function ultimately defines the behavior of the process.

Covariance is the core of GP, which can be determined by the different kernel functions. This also expands the scope of the application of our model. For other types of data, we can choose different kernel functions to obtain a better representation of data correlations. In this study, we chose a radial basis function kernel (RBF). It is also known as the squared-exponential kernel.

The RBF kernel is stationary and parameterized by a length scale $l > 0$, which can be either a scalar (an isotropic variant of the kernel) or a vector with the same dimensions as the inputs x (an anisotropic variant of the kernel). The kernel is expressed as

$$k(x_i, x_j) = \sigma^2 exp \left(-\frac{d(x_i, x_i)^2}{2l^2} \right)$$

(4)

where σ^2 is a hyperparameter, l is the kernel length scale, and $d(\cdot, \cdot)$ is the Euclidean distance.

3.3   Unsupervised Training

The triplet network was developed from the Siamese network [16], which is an artificial neural network that uses the same weights while working in tandem on two different input vectors to compute comparable output vectors. In comparison with the Siamese network, the triplet network uses both positive and negative samples. This joint training of positive and negative pairs could help the model easily distinguish the input from the same class and different classes. To use the triplet network, labeled data are necessary. However, most real-life time series are unlabeled. Therefore, unsupervised representation learning was suitable for training.

In this section, we introduce our simple unsupervised training method. The key point of unsupervised representation learning is to ensure that similar time series obtain similar representations with no supervision to learn such similarities. While there are some unsupervised methods for time series representation learning, most of them require manual training pair design. This not only increases the complexity of the algorithm but also makes the training pairs rely on the precision of manual methods, which cannot generate universal training pairs for most training models. Hence, we design an unsupervised method for time series to select pairs of similar time series inspired by the recent development of unsupervised methods and contrast learning in CV [17] and NLP [18]. This is a sample method that can be added to most training models.

Dropout is a relatively general and straightforward method for machine learning models. Owing to the random characteristic of dropout, one input will have two different eigenvectors when going through a model with a dropout layer. We develop an unsupervised method for converting the Siamese network into a triplet network, which achieves data enhancement without changing the original high-level semantic features and information of the data. While some contrast learning models in CV and NLP use the standard dropout layer to generate positive pairs, we choose the Gaussian dropout for representation learning of multivariate time series. A diagram of the generation of the training pairs (anchor, positive, and negative samples) for the triplet network of representation learning is shown in Fig. 7.

In comparison with the standard dropout layer, Gaussian dropout discards neurons using a probability that fits a Gaussian distribution. This is equivalent to adding multiplicative noise to the input signal that obeys a Gaussian distribution. This Gaussian noise does not change the original distribution of the multivariate time series and can maintain the consistency of the data distribution in the model.

The original training object of the triplet loss is calculated by the distance between the positive, anchor, and negative samples expressed as:

$$max(d(x, x^+) - d(x, x^-) + margin, 0) \qquad (5)$$

where x^+ is the positive sample, and x^- is the negative sample; $d(\cdot)$ is the distance between the input pairs, and margin is a hyperparameter to control the distances. Considering the dropout unsupervised method in NLP [19] and the

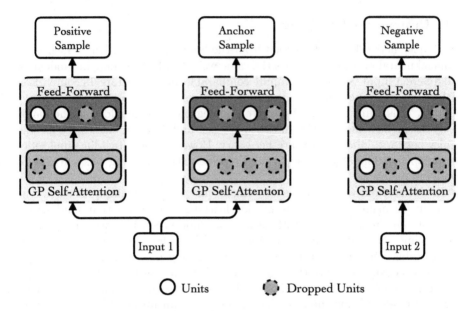

Fig. 7. Schematic of generating training pairs for triplet network of representation learning.

process of generating the sample pairs in our model, the training objective can be defined as follows:

$$
- \log \frac{e^{\cos(x_i, x_i^+)/\tau}}{\Sigma_{j=1}^{N} \left(e^{\cos(x_i, x_j^+)/\tau} + e^{\cos(x_i, x_j^-)/\tau} \right)} \tag{6}
$$

where $\cos(\cdot, \cdot)$ is the cosine distance, N is the mini-batch size, and τ is a temperature hyperparameter. Based on this principle, the convergence speed of the triplet network can be improved.

4   Experiments

In this section, we test the effectiveness of our model by analyzing its performance on different tasks. Classification and retrieval tasks are used as downstream tasks to prove the effectiveness of high-level semantic information in representation learning. In addition, a case study is conducted to recall the example introduced in Sect. 1.

4.1   Classification

In the classification task, the output vector of our model was passed through a softmax function to obtain a distribution over classes, and its cross-entropy

Table 1. Summary of UEA multivariate datasets.

Dataset	Train Size	Test Size	Length	Classes	Dimensions
EthanolConcentration	261	263	1751	4	3
Handwriting	150	850	152	26	3
Heartbeat	204	205	405	2	61
PEMS-SF	267	173	144	7	983
SpokenArabicDigits	6599	2199	93	10	13
HJapaneseVowels	270	370	29	9	12

with the categorical ground truth labels was considered as the sample loss. In this task, we show that our model performs better than other unsupervised methods.

We used the following six multivariate datasets from the UEA time series classification archives [20], which provide multiple datasets from different domains with varying dimensions, unequal lengths, and missing values. A summary of these datasets is presented in Table 1.

Meanwhile, the UEA archives provide an initial benchmark for existing models that provided accurate baseline information. Based on the performance metrics provided by the UEA archives, we chose the following three models as our baseline:

- Dimension-dependent dynamic time warping (DTW_D) [21]: it uses a weighted combination of raw series and first-order differences for neural network classification with either Euclidean distance or full-window dynamic time warping (DTW). It combines two distances, i.e., the DTW distance between two series and two different series, using a weighting parameter. It develops the traditional DTW method and suits every series of data.
- ROCKET [22]: it is based on a random convolutional kernel similar to a shallow convolutional neural network. It can achieve fast and accurate time series classification using random convolutional kernels.
- Time series transformer (TST) model [23]: it largely fills the gap in the application of the transformer model to the representation learning of time series. This model achieves a better learning performance by introducing a transformer-based pre-training model.

Table 2 presents the classification results for the multivariate time series, where bold indicates best values. As shown in Table 2, the proposed model demonstrated the best performance among the four datasets. From the data presented in the table, it can be concluded that the effectiveness of our model is significantly enhanced as the amount of data increases.

However, our model is relatively more advantageous for small datasets than baselines. The results for the Heartbeat datasets revealed that binary classification is more likely to exploit contrastive learning. Conversely, our model yielded better results for datasets with trend changes. In general, the results of the SpokenArabicDigits data indicate a relative weakness of our model, i.e., it has no

Table 2. Accuracy results of proposed and other methods.

Dataset	GP-HLS	DTW_D	ROCKET	TST
EthanolConcentration	**0.467**	0.452	0.326	0.326
Handwriting	0.312	0.286	**0.588**	0.309
Heartbeat	**0.781**	0.717	0.756	0.776
PEMS-SF	**0.919**	0.711	0.751	0.896
SpokenArabicDigits	0.968	0.963	0.712	**0.993**
HJapaneseVowels	**0.997**	0.949	0.962	0.994

significant advantage when dealing with large scale data. And for handwriting, our proposed method and TST both have undesirable results. We can draw a conclusion that attention mechanism has a weak ability in dealing with the low-dimensional data, especially when the training data is obviously less than test data. To mitigate these issues, we intend to set new feature parameters and try other mechanism to increase the sensitivity of the model to such data in our future work.

4.2 Retrieval

For the time series retrieval task, we evaluate the effectiveness of the proposed model for unsupervised time series retrieval tasks based on two different datasets. Table 3 The statistics of two multivariate time series datasets in experiment. N.A. denotes not available.

The EEG Eye State dataset was collected from one continuous EEG measurement using the Emotiv EEG Neuroheadset [24]. All data has 117 s duration of the measurement. The eye state was detected using a camera during the EEG measurement. '1' represents a closed eye, and '0' the eye-open state. In this experiment, we generate 6012 segments by GGS algorithm.

The Twitter dataset was collected to predict Buzz from the Buzz in social media Dataset [25]. It contains examples of buzz events from Twitter. And it does not have any label of class information. In this experiment, we generate 49803 segments by GGS algorithm.

The details of two datasets are shown in Table 3. For those segments in these two datasets, we select 50% as the training data, next 10% as the validation data, and the last 40% as the test data.

Table 3. The details of two multivariate time series datasets in experiment. N.A. denotes not available.

Dataset	Number of Attributes	Number of Instances	Classes
EEG Eye State	15	14980	2
Twitter	77	583250	N.A

Table 4. Unsupervised multivariate time series retrieval performance (MAP).

Dataset	EEG Eye State			Twitter		
Hidden size	64	128	256	64	128	256
GP-HLS	**0.282**	**0.336**	**0.395**	**0.108**	**0.144**	**0.171**
DeepBit	0.225	0.284	0.325	0.040	0.089	0.102
HashGAN	0.206	0.299	0.320	0.051	0.101	0.101
LSTM-ED	0.245	0.325	0.357	0.077	0.113	0.143

We compared our model with three typical baseline methods in time series retrieval. All these methods are unsupervised. DeepBit [26] is an unsupervised deep learning approach. It can learn binary descriptors in an unsupervised manner. HashGAN [27] is a deep unsupervised hashing function, which is also designed for image retrieval. The last baseline is Long Short-Term Memory (LSTM) encoder-decoder (LSTM-ED) [28]. It uses an encoder LSTM to map an input sequence into a fixed length representation.

To evaluate the performance of proposed model and baseline models in the task of unsupervised multivariate time series retrieval, we calculate the K nearest neighbors (KNN) based on Euclidean distance (ED). For each query segment, we first calculate its KNN as the ground truth (KNN = 100 for EEG Eye State and KNN = 500 for Twitter dataset). Then, we search the representation of similar segments based on the Hamming distance. Finally, the mean average precision (MAP) is reported for comparison purposes. Meanwhile, to evaluate the performance of each model more comprehensively, we use three different hidden size of each model, 64, 128 and 256. The MAP results of each model are shown in Table 4. We notice that our model has a strong advantage compared to baseline models. It is mainly owing to the use of semantic-based segments and our high-level semantics representation learning algorithm. Meanwhile, we observed that LSTM-ED consistently outperformed DeepBit and HashGAN. The reason may be the DeepBit and HashGAN are specifically designed for images and cannot represent the temporal information in the input segment. These two algorithms could need more necessary improvement for use in time series tasks.

4.3   Case Study

In this case study, we revisit the example in Sect. 1 at greater depth to explain the motivation for this work. As introduced in Sect. 1, by analyzing the wrong results in experiments of some semantic-based algorithms, we conclude that the relationship among subseries plays a significant role in representation learning of multi-variate time series. In this section, we design an experiment to further address this issue.

Most time series datasets are carefully designed and selected with a perfect distribution or measure precision. However, time series from the real world may have many problems, such as noise, loss, or measurement errors. Additionally,

Table 5. Summary of ECG200 and TwoLeadECG.

Dataset	Train Size	Test Size	Length	Classes	Dimensions
ECG200	100	100	96	2	1
TwoLeadECG	23	1139	82	2	1
Combined ECG	123	1239	82	2	1

Table 6. Accuracy results of proposed and other methods.

Dataset	GP-HLS	DTW_D	ST
ECG200	**0.902**	0.880	0.840
TwoLeadECG	**0.991**	0.868	0.984
Combined ECG	**0.752**	0.442	0.510

measuring data from different equipment or sources interferes with each other. These issues can significantly affect the performance of models. Therefore, we combined two other datasets from the same subject that were obtained from various sources. Specifically, we used ECG 200 and TwoLeadECG as the datasets from UCR time series classification archive [29]. Both datasets trace the recorded electrical activity and contain two classes: normal heartbeat and myocardial infarction (MI). We randomly combined these two datasets and reshaped the length of the combined ECG dataset to obtain a regular length of time series. The details of these datasets are listed in Table 5.

We chose the DTW_D and Shapelet Transform as the baseline algorithms. The Shapelet Transform (ST) [30] is based on the shapelet method, which separates the shapelet discovery from the classifier by finding the top k shapelets in a single run. Shapelets were used to transform the data, and each attribute in the new dataset represented the distance of a series to one of the shapelets. This is a semantic-based method for time series. First, we conducted experiments on these two datasets separately. Then, we conducted an experiment using the combined dataset. The results of the experiments in the case study are listed in Table 6.

5   Conclusion

High-level semantics is essential for representation learning of time series data. This is particularly true in data search of time series. Our high-level semantic methods can represent time series by converting them into several subseries according to their high-level semantics information. This characteristic is beneficial for data storage and search. In this study, we propose a novel unsupervised representation learning model with high-level semantic features of multivariate time series. A Gaussian process-based self-attention mechanism was introduced to the encoder of the transformer as the representation learning model. In addition, a Gaussian drop-based triplet net-work was designed for multivariate time series to construct positive and negative sample pairs of unsupervised training. The experiments show that the proposed model demonstrates significant improvement in multivariate time series representation learning and can be used in various downstream tasks such as classification and retrieval. In future research, our efforts will be devoted to the design of the triplet loss function. So far, many different loss functions have been designed for various applications. Thus, we believe that the loss function may improve the performance of our model.

References

1. Tseng, K., Li, J., Tang, Y., Yang, C., Lin, F.: Healthcare knowledge of relationship between time series electrocardiogram and cigarette smoking using clinical records. BMC Med. Inf. Decis. Making. **20**, 1–11 (2020). https://doi.org/10.1186/s12911-020-1107-2
2. Lawi, A., Mesra, H., Amir, S.: Implementation of long short-term memory and gated recurrent units on grouped time-series data to predict stock prices accurately. J. Big Data **9**, 1–19 (2022). https://doi.org/10.1186/s40537-022-00597-0
3. Cao, J., Pang, Y., Zhao, S., Li, X.: High-level semantic networks for multi-scale object detection. IEEE Trans. Circ. Syst. Video Technol. **30**, 3372–3386 (2019)
4. Lu, Y.,et al.. STARDOM: semantic aware deep hierarchical forecasting model for search traffic prediction. In: Proceedings of the 31st ACM International Conference On Information & Knowledge Management, pp. 3352–3360 (2022)
5. Li, G., Choi, B., Xu, J., Bhowmick, S., Chun, K., Wong, G.: ShapeNet: a Shapelet-neural network approach for multivariate time series classification. In: Proceedings of the AAAI Conference on Artificial Intelligence, vol. 35, pp. 8375–8383 (2021)
6. Middlehurst, M., Vickers, W., Bagnall, A.: Scalable dictionary classifiers for time series classification. In: Yin, H., Camacho, D., Tino, P., Tallón-Ballesteros, A.J., Menezes, R., Allmendinger, R. (eds.) IDEAL 2019. LNCS, vol. 11871, pp. 11–19. Springer, Cham (2019). https://doi.org/10.1007/978-3-030-33607-3_2
7. Liu, W., Liao, S., Ren, W., Hu, W., Yu, Y.: High-level semantic feature detection: a new perspective for pedestrian detection. In: Proceedings of the IEEE/CVF Conference on Computer Vision and Pattern Recognition, pp. 5187–5196 (2019)
8. Vaswani, A., et al.: Attention is all you need. In: Advances in Neural Information Processing Systems, vol. 30 (2017)
9. Hallac, D., Nystrup, P., Boyd, S.: Greedy Gaussian segmentation of multivariate time series. Adv. Data Anal. Classif. **13**, 727–751 (2019). https://doi.org/10.1007/s11634-018-0335-0
10. Chen, Y., Fang, W., Dai, S., Lu, C.: Skeleton moving pose-based human fall detection with sparse coding and temporal pyramid pooling. In: 2021 7th International Conference on Applied System Innovation (ICASI), pp. 91–96 (2021)
11. Yue, Z., et al.: TS2Vec: towards universal representation of time series. In: Proceedings of the AAAI Conference on Artificial Intelligence, vol. 36, pp. 8980–8987 (2022)
12. Tonekaboni, S., Eytan, D., Goldenberg, A.: Unsupervised representation learning for time series with temporal neighborhood coding. ArXiv Preprint ArXiv:2106.00750 (2021)
13. Shao, Z., Zhang, Z., Wang, F., Xu, Y.: Pre-training enhanced spatial-temporal graph neural network for multivariate time series forecasting. In: Proceedings of the 28th ACM SIGKDD Conference on Knowledge Discovery and Data Mining, pp. 1567–1577 (2022)

14. Zhao, H., Jia, J., Koltun, V.: Exploring self-attention for image recognition. Proceedings of the IEEE/CVF Conference on Computer Vision and Pattern Recognition, pp. 10076–10085 (2020)
15. Hadji, A., Szabó, B.: Can we trust Bayesian uncertainty quantification from Gaussian process priors with squared exponential covariance kernel? SIAM/ASA J. Uncertainty Quantification **9**, 185–230 (2021)
16. He, A., Luo, C., Tian, X., Zeng, W.: A twofold Siamese network for real-time object tracking. In: Proceedings of the IEEE Conference on Computer Vision and Pattern Recognition, pp. 4834–4843 (2018)
17. Wu, L., et al.: R-drop: regularized dropout for neural networks. Adv. Neural Inf. Process. Syst. **34**, 10890–10905 (2021)
18. Gao, T., Yao, X., Chen, D.: SimCSE: simple contrastive learning of sentence embeddings. ArXiv Preprint ArXiv:2104.08821 (2021)
19. Ha, C., Tran, V., Van, L., Than, K.: Eliminating overfitting of probabilistic topic models on short and noisy text: the role of dropout. Int. J. Approximate Reasoning **112**, 85–104 (2019)
20. Bagnall, A., et al.: The UEA multivariate time series classification archive. ArXiv Preprint ArXiv:1811.00075 (2018)
21. Chen, Y., Hu, B., Keogh, E., Batista, G. DTW-D: time series semi-supervised learning from a single example. In: Proceedings of the 19th ACM SIGKDD International Conference on Knowledge Discovery and Data Mining, pp. 383–391 (2013)
22. Dempster, A., Petitjean, F., Webb, G.I.: ROCKET: exceptionally fast and accurate time series classification using random convolutional kernels. Data Min. Knowl. Disc. **34**(5), 1454–1495 (2020). https://doi.org/10.1007/s10618-020-00701-z
23. Zerveas, G., Jayaraman, S., Patel, D., Bhamidipaty, A., Eickhoff, C.: A transformer-based framework for multivariate time series representation learning. In: Proceedings of the 27th ACM SIGKDD Conference on Knowledge Discovery & Data Mining, pp. 2114–2124 (2021)
24. Wang, T., Guan, S., Man, K., Ting, T.: EEG eye state identification using incremental attribute learning with time-series classification. Math. Probl. Eng. **2014** (2014)
25. Stattner, E., Collard, M.: Modèles et l'analyse des réseaux: approches mathématiques et informatiques (MARAMI). In: Conférence Sur Les Modèles Et L'analyse Des Réseaux: Approches Mathématiques Et Informatiques (MARAMI), vol. 4, p. 40 (2013)
26. Lin, K., Lu, J., Chen, C., Zhou, J.: Learning compact binary descriptors with unsupervised deep neural networks. In: Proceedings of the IEEE Conference on Computer Vision and Pattern Recognition, pp. 1183–1192 (2016)
27. Dizaji, K., Zheng, F., Sadoughi, N., Yang, Y., Deng, C., Huang, H.: Unsupervised deep generative adversarial hashing network. In: Proceedings of the IEEE Conference on Computer Vision and Pattern Recognition, pp. 3664–3673 (2018)
28. Srivastava, N., Mansimov, E., Salakhudinov, R.: Unsupervised learning of video representations using LSTMs. in: International Conference on Machine Learning, pp. 843–852 (2015)
29. Dau, H., et al.: The UCR time series archive. IEEE/CAA J. Autom. Sinica **6**, 1293–1305 (2019)
30. Arul, M., Kareem, A.: Applications of Shapelet transform to time series classification of earthquake, wind and wave data. Eng. Struct. **228**, 111564 (2021)

Towards Time-Series Key Points Detection Through Self-supervised Learning and Probability Compensation

Mingxu Yuan[1,2], Xin Bi[1,2(✉)], Xuechun Huang[3], Wei Zhang[1,2], Lei Hu[1,2],
George Y. Yuan[4], Xiangguo Zhao[5], and Yongjiao Sun[3]

[1] Key Laboratory of Ministry of Education on Safe Mining of Deep Metal Mines,
Northeastern University, Shenyang 110819, China
[2] Key Laboratory of Liaoning Province on Deep Engineering and Intelligent
Technology, Northeastern University, Shenyang 110819, China
`bixin@mail.neu.edu.cn`
[3] School of Computer Science and Engineering, Northeastern University,
Shenyang, China
[4] Thinvent Digital Technology Co., Ltd., Nanchang, India
[5] College of Software, Northeastern University, Shenyang, China

Abstract. Key points detection is crucial for signal analysis by marking the identification points of specific events. Deep learning methods have been introduced into key points detection tasks due to their significant representation learning ability. However, in contrast to common time series classification and prediction tasks, the target key points correspond to significantly different time-series patterns and account for an extremely small proportion in a whole sample. Consequently, existing end-to-end methods for key points detection encounter two major problems: specificity and sparsity. Thus, in this work, we address these issues by proposing a probability compensated self-supervised learning framework named ProCSS. Our ProCSS consists of two major components: 1) a pretext task module pretraining an encoder based on self-supervised learning to capture effective time-series representations with a higher generalization ability; 2) a joint loss function providing both dynamic focal adaptation and probability compensation by extreme value theory. Extensive experiments using both real-world and benchmark datasets are conducted. The results indicate that our method outperforms our rival methods for time-series key points detection.

Keywords: time series learning · key points detection · self-supervised learning · extreme value theory

1 Introduction

Sensors continuously collect signals that reflect the real-time status and metrics, which cannot be obtained by human senses. As crucial components, sensors have been widely deployed in various fields, such as engineering monitoring [7], environmental protection [26], and medical diagnosis [38].

X. Wang et al. (Eds.): DASFAA 2023, LNCS 13943, pp. 237–252, 2023.
https://doi.org/10.1007/978-3-031-30637-2_16

The sensor signal is typical time-series data. The processing and analysis using various intelligent methods have great significance for realizing early prediction and decision-making [6]. Key points detection is crucial for signal analysis by marking the identification points of specific events, e.g., anomalous patterns [33] and process transitions [2].

In traditional time-series classification and prediction tasks, the deep learning methods can effectively capture the key features of time series information for accurate classification [23] and also use the time relationship of feature extraction to predict the future trend [22]. Thus, deep learning methods have been introduced into key points detection tasks due to their significant representation learning ability.

However, in contrast to common time series classification and prediction tasks, the features of key points detection tasks have two characteristics: 1) *Specificity*. The features of the time-series patterns to identify different key points have enormous variations. Directly applying common feature learning models may decrease the generalization and robustness. 2) *Sparsity*. The target key points account for only a small proportion of the total number of sampling points. Effective time-series features often cannot be fully learned and focused on. Thus, the key points detection tasks encounter two major problems: how to learn effective features to improve generalization and how to boost the training process that is diluted by the sparsity in this task.

To address these issues, we aim to learn high-quality time-series representations and improve generalization by decreasing the dependency of the learned time-series features on the original downstream task labels. This learning process is realized by designing a pretext task to pretrain a powerful encoder in a self-supervised learning manner. On the other hand, to alleviate the impact of sparse key points, we consider two critical factors in our loss function: the imbalance between point types and the extreme value distribution obeyed by the key points.

Therefore, in this paper, we propose a Probability Compensated Self-Supervised learning framework named ProCSS. Our ProCSS consists of two major components: a pretext task module that captures effective time-series representations with a higher generalization ability and a joint loss function considering both imbalance and probability compensation.

To summarize, the contributions of this paper are as follows:

(1) We design a pretext task module based on self-supervised learning to tackle the specificity issue;
(2) We design a joint loss function consisting of a self-adaptive focal loss and a probability compensated loss to tackle the sparsity issue;
(3) We conduct extensive experiments and ablation studies on both real-world and benchmark datasets to evaluate the performance of our method.

The remainder of this paper is organized as follows. Section 2 reports the summary of related works. The details of our method are presented in Sect. 3. Section 4 evaluates the performance of our method. Finally, Sect. 5 concludes this paper.

2   Related Work

Over the years, many researchers have addressed the key points detection problem in the related fields. We studied traditional methods, deep learning methods, and self-supervised learning methods of the key points detection tasks.

2.1   Traditional Methods

The short-term to long-term average ratio (STA/LTA) [3] has been widely used in studies on microseismic signals. STA/LTA utilizes the difference in amplitudes of seismic signals and noise, and the first arrival time of each seismic wave is determined by calculating the ratio of two time windows with different lengths. STA/LTA depends on a manually set threshold and a high signal-to-noise ratio.

Several studies address the key points detection problem by clustering analysis. Clustering analysis is a classification algorithm that can divide similar data into one cluster and dissimilar data into a different cluster. Spectral multi-manifold clustering [24] is proposed to select P-wave arrivals, which can extract the low-dimensional manifold features from a suitable affinity matrix. Ma et al. [21] proposed a method for microseismic data based on locally linear embedding (LLE) and an improved particle swarm optimization (PSO) clustering algorithm. TSSC [42] was proposed to cluster time series segmentations based on the K-means algorithm to select the S-phase of a microseismic signal.

With the development of neural networks and deep learning, the shortcomings of traditional methods have been gradually revealed. Traditional methods often rely on experience to establish models and select parameters. The results are greatly affected by subjective factors, so they are not sufficiently flexible to be applied in actual scenarios.

2.2   Deep Learning Methods

Wang et al. [31] took the long-short window segment and the spectrum, which corresponded to the window as the input of the model and compared the two output results to obtain the detection results. Ross et al. [29] proposed a convolutional neural network structure with high robustness, so that the results could not be masked by high-frequency noise. Ince et al. [17] proposed a 1D convolutional neural network-based real-time motor condition monitoring and early fault-detection system. Zahid et al. [34] designed an encoder-decoder model for R-peak detection in low-quality Holter ECGs. Based on the recurrent neural network, Zheng et al. [39] proposed a method to simultaneously identify multiple events at different signal-to-noise ratios. Laitala et al. [19] proposed a method based on LSTM to detect the R-peak of ECG signals. Chen et al. [8] utilized an unsupervised learning algorithm to cluster the sampling points into waveform point clusters and nonwaveform point clusters. PhaseNet [41] first used the U-Net model, which performs well in the field of medical image segmentation to realize the first break picking of microseismic waves. The network used unfiltered waveforms as the input, and the performance was greatly improved compared with traditional methods. He et al. [15] presented PickCapsNet, which is a highly scalable capsule network for P-wave arrival picking of microseismic signals from a single waveform without feature extraction.

Although deep learning methods have significant representation learning ability, most existing methods cannot handle the before mentioned issues of specificity and sparsity in the time-series key points detection tasks.

2.3   Self-supervised Learning

In order to reduce the dependence on labeled data and improve the generalization ability of models, self-supervised learning (SSL) [27] has been proposed. In self-supervised learning, models are trained by solving a series of handcrafted pretext tasks, in which the supervision information is acquired from data itself automatically without the need for manual annotation.

Self-supervised learning has been utilized in various fields. Guizilini et al. [13] proposed a method to realizing automatically segmenting dynamic objects in an urban environment in a self-supervised learning manner based on Gaussian Processes. Wang et al. [32] designed a Siamese-triplet network to learn the visual representations using videos. Lee et al. [20] leveraged the temporal coherence as a supervisory signal by formulating representation learning as a sequence sorting task. Jenni et al. [18] trained a discriminator network to distinguish real images from images with synthetic artifacts for feature learning. Zhang et al. [37] proposed a graph-based distillation framework for video classification. Sterzentsenko et al. [30] designed an autoencoder for deep depth denoising without clean data as ground truth. PsyNet [5] addressed the object localization problem by implementing point symmetric transformation and utilizing its parameters as an artificial label for self-supervised learning. Zeng et al. [36] created augmented graphs out of the original graphs to perform contrastive self-supervised learning for graph classification. Gong et al. [12] pretrained the audio spectrogram transformer model with joint discriminative and generative masked spectrogram patch modeling using unlabeled audio.

However, for the time-series key points detection tasks, existing self-supervised learning models cannot handle the specificity and sparsity issues as well. How to learn effective time-series representations with a higher generalization and alleviate the impact caused by extreme values remains a challenge.

3   Our Method

3.1   Overview

We propose a probability compensated self-supervised learning framework ProCSS for time-series key points detection. Our ProCSS consists of two major modules, namely, a *pretext task module* for learning the high-quality representations of time series in the self-supervised learning manner, and a *detection module* that outputs the key points detection results. The structure of the proposed method is shown in Fig. 1.

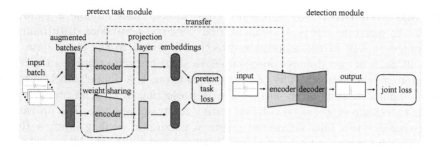

Fig. 1. Structure of ProCSS.

In the pretext task module, we pretrain a pretext task encoder in a self-supervised manner. Two different versions of the same time-series sample batch are obtained using different augmentation settings. These two augmented batches are fed into different encoding channels to generate corresponding embeddings. Then, we deploy a pretext task loss function that is capable of measuring embedding similarities of each augmented sample pair. In this way, the high-quality time-series representations can be obtained without dependence on the downstream detection labels.

In the detection module, we design an encoder-decoder structure to complete the detection task. The encoder is initialized by transferring the encoder that are pretrained from our pretext task. Then the encoder is fine-tuned and the decoder trained by designing a joint loss function. The output of the detection module is a sequence of probabilities of all points in the input sample being positive. However, the key points account for an extremely small proportion in a whole sample. In addition, the difference between the output probabilities of positive and negative points is expected to be as large as possible. To tackle this issue consistent with extreme value theory, we compensate for the probability distribution of the key points to enhance the learning performance of our module. We design a loss function based on probability compensation that considers both the point-type imbalance and the extreme value distribution obeyed by the key points.

3.2 Pretext Task Module

Since the features of the time-series patterns to identify different key points have enormous variations, directly applying straight-forward feature learning modules has relatively poor generalization and robustness.

To tackle this issue of specificity, we design a pretext task module to obtain the high-quality time-series representations by learning the correlation of different augmented versions of an input sample batch. It can provide more reasonable initialization weights to obtain higher generalized features and make full use of the prior information to improve the generalization and robustness of our network.

Module Structure. The structure of the pretext task module is shown in Fig. 2. To generate different augmentation versions, a random disturbance in the range of ±10 in the amplitude direction is added to the input data. We take all pairs of two different augmented versions of each sample as inputs to the pretext task network. The pretext task network is a dual-channel network. Each channel consists of an encoder and a projection layer. The two encoders in the two channels are weight-shared during the training phase. The projection layer contains two fully connected layers, a maximum pooling layer, a ReLU activation function, and a batch normalization layer. The embeddings obtained by the projection layer are then used to calculate our designed pretext task loss.

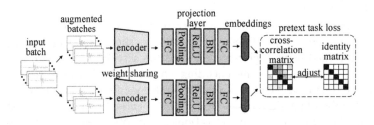

Fig. 2. Structure of the pretext task module.

Pretext Task Loss. To realize the tuning of the network under self-supervision, we design our pretext task loss based on redundancy reduction [35]. This loss can be automatically calculated without dependence on the downstream detection labels. We generate a cross-correlation matrix $\mathbf{C}$ within a training batch. The diagonal elements $\mathbf{C}_{ii}$ indicates the correlation of the same vector components of the embeddings calculated through all the samples in a training batch, whereas the off-diagonal elements $\mathbf{C}_{ij}$ indicates the correlation between different vector components of the embeddings.

The intuition is that we intend to enhance both the *invariance* of the embeddings to the disturbance during augmentation and the *decorrelation* among vector components in the same embeddings. Thus, we expect the cross-correlation matrix $\mathbf{C}$ to be similar to an identity matrix $\mathbf{I}$.

Given a batch of sample sequence X, the embeddings obtained from different augmentations in different channels are denoted as Z^1 and Z^2. The cross-correlation matrix is calculated as [35]:

$$\mathbf{C}_{ij} = \frac{\sum_{k=0}^{m} Z_{ki}^1 Z_{kj}^2}{\sqrt{\sum_{k=0}^{m} (Z_{ki}^1)^2} \sqrt{\sum_{k=0}^{m} (Z_{kj}^2)^2}} \tag{1}$$

where m is the batch length. The value of $\mathbf{C}_{ij}$ is between -1 and 1. A closer value to 1 indicates a stronger correlation between the two variables, a closer value to 0 indicates a weaker correlation between the two variables is, and negative values represent a negative correlation.

To generate a $\mathbf{C}$ more similar to the identity matrix, the pretext task loss function is as:

$$L = \sum_i (1 - \mathbf{C}_{ii})^2 + \lambda \sum_i \sum_{j \neq i} \mathbf{C}_{ij}^2 \qquad (2)$$

where λ is a tradeoff hyperparameter between invariance and redundancy reduction of the embeddings.

3.3   Detection Module

To provide a higher detection performance, the composite convolution blocks that obtain the perception information of different scales are fused to enhance the learning ability of the network composed of convolution kernels of different sizes. Then, the residual learning blocks are combined to reduce the number of parameters in the network and improve the convergence speed of the model. For skip connections, we deploy a feature fusion structure of a bidirectional cross-layer, which uses the outputs of different nodes as the input to the node. We only retain intermediate nodes with at least two inputs to improve the feature fusion ability and efficiency.

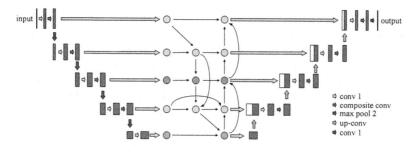

Fig. 3. Structure of the detection module.

The structure is shown in Fig. 3. The detection module consists of an encoder-decoder structure, which is connected by skip connections. The encoder structure contains four encoder layers. Each encoder layer contains a one-dimensional convolutional layer, a composite convolutional layer, and a max-pooling layer. The decoder structure also contains four decoder layers, and each decoder layer is symmetric with the corresponding encoder layer.

3.4   Joint Loss Function

The key points detection tasks can be considered a binary classification problem of key points and background points. However, the learning process may face the following problems. 1) The classification categories are extremely imbalanced. 2) Since the key points in a sample is extremely sparse, the distribution of key points is similar to the long-tailed distribution containing extreme values.

To tackles these issues, we propose a joint loss function that not only provides dynamic focal adaptation, but also compensate the distribution of key points by extreme value theory.

Self-adaptive Focal Loss. The basic cross-entropy loss is calculated as:

$$L = -\sum_{i=1}^{n} y_i \log \hat{y}_i \tag{3}$$

where y_i is the ground-truth label, $\hat{y}_i$ is the predicted value, and n is the number of samples. The cross-entropy loss treats all samples the same, does not distinguish the difficulty of sample classification, and calculates the loss after all samples are given identical weights. Thus, in the scenario of unbalanced data, using cross-entropy may lead to insufficient learning of key points.

To solve this problem, the focal loss introduces two hyperparameters to adjust sample weights, which is calculated as:

$$L_{fl} = \begin{cases} -\alpha(1 - \hat{y})^{\gamma} \log \hat{y}, & y = 1 \\ -(1 - \alpha)\hat{y}^{\gamma} \log(1 - \hat{y}), & y = 0 \end{cases} \tag{4}$$

where α is a hyperparameter to alleviate the imbalance between classes, γ is a hyperparameter to focus on samples that are more likely to be misclassified. However, these hyperparameters must be manually set.

To improve the adaptation ability of our network, we design a self-adaptive focal loss with dynamic parameters. Compared with Eq. 4, the parameter γ can be automatically adapted according to the predicted value and loss during the learning process. The loss of positive and negative samples is calculated as:

$$L = \begin{cases} -(1 - \hat{y})^{\gamma \times e^{\hat{y}}} \log \hat{y}, & y = 1 \\ -\hat{y}^{\gamma \times e^{1-\hat{y}}} \log(1 - \hat{y}), & y = 0 \end{cases} \tag{5}$$

where $L_{(y=1)}$ is the loss of positive samples, and $L_{(y=0)}$ is the loss of negative samples. With the result of $\hat{y}$ from 0 to 1, the coefficients $\gamma \times e^{\hat{y}}$ and $\gamma \times e^{1-\hat{y}}$ changes from γ to $e\gamma$.

The parameter α is limited within the range of $[0.5, 0.9]$ to ensure that sparse key points are effectively learned, which is calculated as:

$$\alpha = \frac{0.4 \times sum(L_{(y=0)})}{sum(L_{(y=1)}) + sum(L_{(y=0)})} + 0.5 \tag{6}$$

where $sum(L_{(y=1)})$ is the sum of losses of positive samples, and $sum(L_{(y=0)})$ is the sum of losses of negative samples.

Finally, the self-adaptive focal loss is calculated as:

$$L_{saf} = \alpha \times L_{(y=1)} + (1 - \alpha) \times L_{(y=0)} \tag{7}$$

Probability Compensated Loss. Key points are regarded as extreme events in a long tail distribution. The imbalance of the proportion of head data and tail data in the long-tailed distribution will make the model lack adequate learning of tail data. Thus, we apply the probability compensated loss function based on extreme value theory [9] to better protect the extreme value distribution obeyed by the key points.

Let $\mathbf{x} = \{x_1, x_2, \ldots, x_t, \ldots, x_m\}$ be a time series with length of m, $\hat{\mathbf{y}} = \{\hat{y}_1, \hat{y}_2, \ldots, \hat{y}_t, \ldots, \hat{y}_m\}$ be corresponding predicted values sequence. Let $\mathbf{v} = \{v_1, v_2, \ldots, v_t, \ldots, \ldots, v_m\}$ be an auxiliary sequence with the same length as $\mathbf{x}$. Each v_t is calculated as:

$$v_t = \begin{cases} 1, \hat{y}_t > \varepsilon \\ 0, \hat{y}_t \in [0, \varepsilon] \end{cases} \tag{8}$$

where $\varepsilon > 0$ is a manually set threshold. The points whose $v_t = 0$ are normal points, while whose $v_t = 1$ represents extreme events. Extreme value theory studies the distribution of maximum in observed samples. The form of the generalized extreme value distribution [11] is as follows:

$$G(\hat{y}) = \begin{cases} exp(-(1 - \frac{1}{\beta}\hat{y})^\beta), \beta \neq 0, 1 - \frac{1}{\beta}\hat{y} > 0 \\ exp(-e^{-\hat{y}}), \qquad\quad \beta = 0 \end{cases} \tag{9}$$

where β is the extreme value index. Previous work [10] extend the extreme value to model the tail distribution when $\beta \neq 0$, which is calculated as follows:

$$1 - F(\hat{y}) \approx P(v_t = 1)(1 - \log G(\frac{\hat{y} - \varepsilon}{f(\varepsilon)})) \tag{10}$$

where F is the distribution of $\hat{y}$, $f(*)$ is a scale function, $\frac{y - \varepsilon}{f(\varepsilon)}$ can approximately correspond to the probability prediction value [9]. For binary classification tasks, tail estimation is added to each item of the binary classification cross entropy loss function as weight, and the calculation is as follows:

$$\begin{aligned} L_{pc} &= -P(v_t = 1)\lfloor(1 - \log G(\hat{y}))\rfloor y \log(\hat{y}) \\ &\quad - P(v_t = 0)\lfloor(1 - \log G(1 - \hat{y}))\rfloor(1 - y)\log(1 - \hat{y}) \\ &= -\alpha_1 \left[1 + (1 - \frac{\hat{y}}{\beta})^\beta\right] y \log(\hat{y}) - \alpha_0 \left[1 + (1 - \frac{1 - \hat{y}}{\beta})^\beta\right] (1 - y)\log(1 - \hat{y}) \end{aligned} \tag{11}$$

where α_1 is the proportion of extreme events in the sample, i.e., the ratio of the number of key points to the length of the sequence. α_0 is the ratio of the number of mormal points to the length of the sequence. The term $\left[1 + (1 - \frac{\hat{y}}{\beta})^\beta\right]$ could increase the penalty when the model recognize the extreme event with little confidence.

Joint Loss. To take both self-adaptive focal loss and probability compensated loss, the joint loss function is as:

$$L_{total} = L_{saf} + L_{pc} \tag{12}$$

where L_{saf} is the self-adaptive focal loss function that provides dynamic focal adaptation for different point types, L_{pc} is the probability compensated loss function that compensate for the extreme values by considering the ratio of key points and the distribution of long-tailed samples.

4   Experiments

In this section, we evaluate the performance of ProCSS on two time-series datasets: microseisms and ECGs. Extensive experiments of both performance evaluation and ablation studies are conducted.

4.1   Evaluation Metrics

To evaluate the results of the experiment, we use accuracy (ACC), mean absolute error (MAE), and mean square error (MSE) as our evaluation metrics. We define the absolute error of prediction as the absolute value of the number of sampling points between the detection result and the real label. We set a threshold as an acceptable error range. If the absolute error is less than the threshold value, the detection result is considered correct.

4.2   Datasets

Two datasets collected by sensors are used in our experiments:

- Microseism: A dataset containing microseism signals collected from a project in southwest China. It contains 5350 signals. The amplitude of the microseism signal lies in the range of $10^{-6} - 10^{-5}$ m/s. The sampling frequency is 6000 Hz, and each signal has 3000 sample points. Each signal contains a P-wave arrival event and an S-wave arrival event.
- ECG: A dataset of ECG signals that record the heartbeat rhythm. The ECG recordings are from the MIT-BIH Normal Sinus Rhythm Database of Physionet. It contains 18 long-term ECG recordings of subjects referred to the Arrhythmia Laboratory Beth Israel Hospital. The bandwidth of the original ECG lies in the range of $0.05 - 100$ Hz [1].

In the microseism dataset, we evaluate P-wave arrival event (P-wave) and S-wave arrival event (S-wave). In the ECG dataset, we only evaluate R-peak of the signals (R-peak) in the experiments. Each sample in the two datasets is a segmented as a complete waveform with a length of 2048 from the original time series.

4.3 Rival Methods

We compare the performance of our method with six rival methods:

- U-Net [28]: A simple and efficient detection model that uses a symmetric encoder-decoder structure and a skip connection to fuse different features.
- ResNet [14]: A variation of the U-Net model prevents the optimization effect from becoming worse after deepening the network when the model is saturated. Compared with the U-Net model, ResNet replaces some convolutional blocks with residual learning modules.
- Att-UNet [25]: A variation of the U-Net model. It introduces an attention mechanism into the U-Net model, monitors the feature information of the upper level through the feature mapping of the lower level, suppresses the parts irrelevant to the target task, and emphasizes the features of interest in learning.
- Att-Res-UNet [4]: It combines the residual module with the attention mechanism and introduces the advantages of the two structures.
- U-Net++ [40]: It improves the skip-connection part of the U-Net model by using nested and dense skip-connections and a more flexible feature fusion method, which decreases the difference between features during fusion and can adjust the complexity of the model by pruning.
- U-Net3+ [16]: It uses a full-scale skip connection to fuse features of different scales. Each decoder layer combines features of all coding layers.

4.4 Results

We first present the settings of the hyperparameters. Then, we present the accuracy of ProCSS under different thresholds. With the determined network structure, the performance of the ProCSS is compared with rival methods. Finally, ablation studies are presented to demonstrate the effectiveness of each module in our framework.

Hyperparameters. In the process of detection module training, each convolution layer uses the ReLU function as the activation function. We use a normal distribution to initialize the weights, and the standard deviation is 0.1. We use the Adam optimization algorithm to optimize the model, and the learning rate is set to 0.001. We use batch training to improve the memory efficiency and reduce the training time, the batch size is set to 32, and the epoch is set to 100.

To determine the error threshold, we evaluate the detection performance under different error thresholds. For the microseismic dataset, we set the error threshold from 1 to 10; for the ECG dataset, we set the error threshold from 1 to 5. The experimental results are presented in Fig. 4.

Based on the above results, the error threshold is set to 4 on the microseismic dataset and 2 on the ECG dataset.

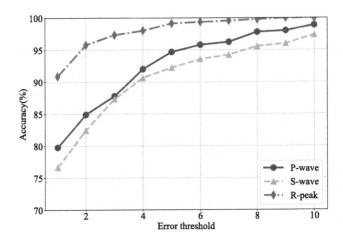

Fig. 4. Accuracies of detection with varied error thresholds.

Performance Comparison. We compare the performance of three evaluation metrics before the error threshold is set. Table 1 presents the results on the two datasets. We highlight the highest performance of each case in bold.

Table 1. Key points detection performance of microseismic and ECG datasets (%)

	P-wave			S-wave			R-peak		
	ACC	MAE	MSE	ACC	MAE	MSE	ACC	MAE	MSE
U-Net	84.75	2.54	2.95	79.32	3.04	2.97	95.66	0.94	1.97
ResNet	93.32	1.68	2.17	85.56	2.96	2.45	96.79	0.73	1.89
Att-UNet	94.95	1.57	2.11	86.22	2.89	2.37	98.15	0.61	1.40
Att-Res-UNet	95.05	1.55	2.10	87.17	2.65	2.34	98.07	0.52	1.33
U-Net++	93.86	1.96	2.35	88.79	2.57	2.33	97.90	0.34	1.65
U-Net3+	94.17	1.87	2.27	89.15	2.54	2.32	98.13	0.26	1.12
ProCSS	**96.34**	**1.42**	**2.08**	**90.99**	**2.48**	**2.30**	**98.25**	**0.17**	**0.99**

From the results, we observe that our ProCSS method achieves the highest accuracy value, the lowest MAE value and the lowest MSE value in all cases. The results demonstrate that our method outperforms other rival methods in various key points detection scenarios.

The detection performance of P-wave in the microseismic dataset is better than that of S-wave because the S-wave makes it more difficult to determine the location of the label in actual situations. The detection performance of R-peak in the ECG dataset is relatively easier than the detection of the microseismic signal, and there is no large gap between the detection results of each model.

However, the key points detection performance of our method also shows better performance, which can also prove the effectiveness and progressiveness of the ProCSS.

Ablation Study. To verify the effectiveness of the pretext module in our proposed framework, we conduct ablation studies evaluating the performance of an ProCSS without a self-supervised pretrain module and an ProCSS with a self-supervised pretrain module. The experimental results are presented in Table 2. ProCSS- represents the ProCSS without the pretext module.

Table 2. Performance without/with pretext module on accuracy (%).

	P-wave	S-wave	R-peak
U-Net	84.75	79.32	95.66
ProCSS-	93.17	86.45	97.89
ProCSS	**96.34**	**90.99**	**98.25**

The results demonstrate that the detection performance decreases without the pretext module. U-Net cannot fully capture the high-quality representations of sensor data when the key points vary enormously in the time-series detection tasks. Thus, the key points detection performance is not ideal.

To verify the effectiveness of the proposed compensational joint loss function, we conduct ablation studies to evaluate different loss functions, including the entropy loss, focal loss with different γ values and α values, self-adaptive focal loss, and probability compensation loss. The experimental results are presented in Table 3.

Table 3. Performance with different loss function on accuracy (%)

	P-wave	S-wave	R-peak
cross entropy loss	94.35	87.53	96.62
Focal loss ($\gamma=2,\alpha=0.75$)	94.92	89.49	97.16
Focal loss ($\gamma=1,\alpha=0.75$)	94.44	86.25	97.35
Focal loss ($\gamma=2,\alpha=0.5$)	93.87	85.66	97.69
Focal loss ($\gamma=3,\alpha=0.75$)	95.21	88.17	98.04
Self-adaptive focal loss	95.47	89.15	**98.51**
Probability compensated loss	95.91	90.70	97.93
Joint loss	**96.34**	**90.99**	98.25

The experimental results demonstrate that the focal loss function can effectively improve the model performance, and the probability compensation loss function can play a positive role in improving the model detection performance. The accuracy on the microseismic dataset is improved more than that on the ECG dataset. Analysis shows that the problem of sample imbalance between point types and the extreme value distribution obeyed by the key points in the microseismic dataset is more serious than that in the ECG dataset. Therefore, the experimental results prove that the proposed compensational joint loss function can compensate for the impact of the sample imbalance of point types and extreme value distribution on key points detection tasks.

5   Conclusions

To tackle the problems of specificity and sparsity in the time-series key points detection tasks, in this paper, we proposed a probability compensated self-supervised learning framework ProCSS. Our ProCSS consists of a pretext task module that tackles the specificity issue and a detection module that tackles the sparsity issue. The pretext task module is capable of learning high-quality representations with a higher generalization by pretraining an encoder in the self-supervised learning manner. The detection module utilizes a designed joint loss function, which realizes dynamic focal adaptation and probability compensation by extreme value theory. Experimental results and ablation studies on both real-world and benchmark datasets indicate an excellent performance of our framework in the time-series key points detection tasks.

Acknowledgments. The work is supported by National Key R&D Program of China (Grant No. 2022YFB3304302), the National Natural Science Foundation of China (Grant No. 62072087, 61972077, 52109116), LiaoNing Revitalization Talents Program (Grant No. XLYC2007079).

References

1. Aditi, B., Sanjay, K.: A robust approach to denoise ECG signals based on fractional Stockwell transform. Biomed. Signal Process. Control **62**, 102090 (2020)
2. Ahad, N., Davenport, M.A.: Semi-supervised sequence classification through change point detection. CoRR abs/2009.11829 arXiv:2009.11829 (2020)
3. Allen, R.V.: Automatic earthquake recognition and timing from single traces. Bull. Seismol. Soc. Am. **68**(5), 1521–1532 (1978)
4. Alom, Z., Hasan, M., Yakopcic, C.: Recurrent residual convolutional neural network based on u-net (r2u-net) for medical image segmentation. CoRR abs/1802.06955 (2018)
5. Baek, K., Lee, M., Shim, H.: PsyNet: self-supervised approach to object localization using point symmetric transformation. In: The Thirty-Fourth AAAI Conference on Artificial Intelligence, pp. 10451–10459 (2020)
6. Cauteruccio, F., et al.: A framework for anomaly detection and classification in multiple IoT scenarios. Futur. Gener. Comput. Syst. **114**, 322–335 (2021)
7. Chen, J., Liu, Y., Carey, S.J., Dudek, P.: Proximity estimation using vision features computed on sensor. In: 2020 IEEE International Conference on Robotics and Automation (ICRA), pp. 2689–2695 (2020)

8. Chen, Y.: Automatic microseismic event picking via unsupervised machine learning. Geophys. J. Int. **222**(3), 1750–1764 (2020)
9. Ding, D., Zhang, M., Pan, X., Yang, M., He, X.: Modeling extreme events in time series prediction. In: Proceedings of the 25th ACM SIGKDD International Conference on Knowledge Discovery & Data Mining, pp. 1114–1122 (2019)
10. Dwass, M.: The asymptotic theory of extreme order statistics (janos galambos). SIAM Rev. **22**(3), 379 (1980). https://doi.org/10.1137/1022076
11. Fisher, R.A., Tippett, L.H.C.: Limiting forms of the frequency distribution of the largest or smallest member of a sample. Math. Proc. Camb. Philos. Soc. **24**(2), 180–190 (1928)
12. Gong, Y., Lai, C.J., Chung, Y., Glass, J.R.: SSAST: self-supervised audio spectrogram transformer. CoRR abs/2110.09784 arXiv:2110.09784 (2021)
13. Guizilini, V., Ramos, F.: Online self-supervised segmentation of dynamic objects. In: 2013 IEEE International Conference on Robotics and Automation, pp. 4720–4727 (2013)
14. He, K., Zhang, X., Ren, S., Sun, J.: Deep residual learning for image recognition. In: 2016 IEEE Conference on Computer Vision and Pattern Recognition (CVPR), pp. 770–778 (2016)
15. He, Z., Peng, P., Wang, L., Jiang, Y.: Pickcapsnet: capsule network for automatic p-wave arrival picking. IEEE Geosci. Remote Sens. Lett. **18**(4), 617–621 (2021)
16. Huang, H., Lin, L., Tong, R., Hu, H.: UNet 3+: a full-scale connected UNet for medical image segmentation. In: ICASSP 2020–2020 IEEE International Conference on Acoustics, Speech and Signal Processing (ICASSP), pp. 1055–1059 (2020)
17. Ince, T., Kiranyaz, S., Eren, L., Askar, M., Gabbouj, M.: Real-time motor fault detection by 1-d convolutional neural networks. IEEE Trans. Ind. Electron. **63**(11), 7067–7075 (2016)
18. Jenni, S., Favaro, P.: Self-supervised feature learning by learning to spot artifacts. In: Proceedings of the IEEE Conference on Computer Vision and Pattern Recognition (CVPR), June 2018
19. Laitala, J., et al.: Robust ECG R-peak detection using LSTM. In: Proceedings of the 35th Annual ACM Symposium on Applied Computing, pp. 1104–1111 (2020)
20. Lee, H.Y., Huang, J.B., Singh, M., Yang, M.H.: Unsupervised representation learning by sorting sequences. In: 2017 IEEE International Conference on Computer Vision (ICCV), pp. 667–676 (2017)
21. Ma, H., Wang, T., Li, Y., Meng, Y.: A time picking method for microseismic data based on LLE and improved PSO clustering algorithm. IEEE Geosci. Remote Sens. Lett. **15**(11), 1677–1681 (2018)
22. Ma, K., Leung, H.: A novel lstm approach for asynchronous multivariate time series prediction. In: 2019 International Joint Conference on Neural Networks (IJCNN), pp. 1–7 (2019)
23. Ma, Q., Zheng, Z., Zheng, J., Li, S., Zhuang, W., Cottrell, G.W.: Joint-label learning by dual augmentation for time series classification. In: Thirty-Fifth AAAI Conference on Artificial Intelligence, pp. 8847–8855. AAAI Press (2021)
24. Meng, Y., Li, Y., Zhang, C., Zhao, H.: A time picking method based on spectral multimanifold clustering in microseismic data. IEEE Geosci. Remote Sens. Lett. **14**(8), 1273–1277 (2017)
25. Ozan, O., Jo, S., Loïc, L.: Attention u-net: learning where to look for the pancreas. CoRR abs/1804.03999 (2018)
26. Qin, X., et al.: Street-level air quality inference based on geographically context-aware random forest using opportunistic mobile sensor network. In: 2021 the

5th International Conference on Innovation in Artificial Intelligence, pp. 221–227 (2021)

27. Raina, R., Battle, A., Lee, H., Packer, B., Ng, A.Y.: Self-taught learning: transfer learning from unlabeled data. In: Proceedings of the 24th International Conference on Machine Learning, pp. 759–766 (2007)

28. Ronneberger, O., Fischer, P., Brox, T.: U-Net: convolutional networks for biomedical image segmentation. In: Navab, N., Hornegger, J., Wells, W.M., Frangi, A.F. (eds.) MICCAI 2015. LNCS, vol. 9351, pp. 234–241. Springer, Cham (2015). https://doi.org/10.1007/978-3-319-24574-4_28

29. Ross, Z.: Generalized seismic phase detection with deep learning. Bull. Seismol. Soc. Am. **108**(5A), 2894–2901 (2018)

30. Sterzentsenko, V., et al.: Self-supervised deep depth denoising. In: 2019 IEEE/CVF International Conference on Computer Vision (ICCV), pp. 1242–1251 (2019)

31. Wang, J., Teng, T.: Artificial neural network-based seismic detector. Bull. Seismol. Soc. Am. **85**(1), 308–319 (1995)

32. Wang, X., Gupta, A.: Unsupervised learning of visual representations using videos. In: 2015 IEEE International Conference on Computer Vision (ICCV), pp. 2794–2802 (2015)

33. Xue, F., Yan, W.: Multivariate time series anomaly detection with few positive samples. In: 2022 International Joint Conference on Neural Networks (IJCNN), pp. 1–7 (2022)

34. Zahid, M.U., et al.: Robust r-peak detection in low-quality Holter ECGs using 1D convolutional neural network. IEEE Trans. Biomed. Eng. **69**(1), 119–128 (2022)

35. Zbontar, J., Jing, L., Misra, I., LeCun, Y., Deny, S.: Barlow twins: self-supervised learning via redundancy reduction. In: Proceedings of the 38th International Conference on Machine Learning, pp. 12310–12320 (2021)

36. Zeng, J., Xie, P.: Contrastive self-supervised learning for graph classification. CoRR abs/2009.05923 arXiv:2009.05923 (2020)

37. Zhang, C., Peng, Y.: Better and faster: knowledge transfer from multiple self-supervised learning tasks via graph distillation for video classification. In: Proceedings of the Twenty-Seventh International Joint Conference on Artificial Intelligence, IJCAI-18, pp. 1135–1141 (2018)

38. Zhao, Y., Shang, Z., Lian, Y.: A 13.34 μw event-driven patient-specific ANN cardiac arrhythmia classifier for wearable ECG sensors. IEEE Trans. Biomed. Circuits Syst. **14**(2), 186–197 (2020)

39. Zheng, J.: An automatic microseismic or acoustic emission arrival identification scheme with deep recurrent neural networks. Geophys. J. Int. **212**(2), 1389–1397 (2017)

40. Zhou, Z., Siddiquee, M.M.R., Tajbakhsh, N., Liang, J.: UNet++: redesigning skip connections to exploit multiscale features in image segmentation. IEEE Trans. Med. Imaging **39**(6), 1856–1867 (2020)

41. Zhu, W., Beroza, G.: PhaseNet: a deep-neural-network-based seismic arrival-time picking method. Geophys. J. Int. **216**(1), 261–273 (2018)

42. Zhu, X., Chen, B., Wang, X., Li, T.: Time series segmentation clustering: a new method for s-phase picking in microseismic data. IEEE Geosci. Remote Sens. Lett. **19**, 1–5 (2022)

SNN-AAD: Active Anomaly Detection Method for Multivariate Time Series with Sparse Neural Network

Xiaoou Ding, Yida Liu, Hongzhi Wang$^{(\boxtimes)}$, Donghua Yang, and Yichen Song

School of Computer Science and Technology, Harbin Institute of Technology, Harbin, China
{dingxiaoou,wangzh}@hit.edu.cn

Abstract. Anomaly detection of time series data is an important and popular problem in both research and application fields. Kinds of solutions have been developed to uncover the anomaly instances from data. However, the labelled data is always limited and costly for real applications, which adds to the difficulty of identifying various anomalies in multivariate time series. In this paper, we propose a novel active anomaly detection method with sparse neural network (SNN-AAD) to improve the accuracy and efficiency in anomaly detection for time series with limited labels. SNN-AAD is designed for two objectives: (1) to achieve sufficient generalization capacity of the model with small-size labels, and (2) to effectively reduce the human cost in active learning process. We introduce sparse neural network in training which minimizes the detection loss caused by the sparsity of labels. We improve the active anomaly detection with the design of sample selection strategy and abnormal feature order generation algorithm, which extracts the important features of instances and reduce the cost of human intelligence. Experimental results on four real-life datasets show SNN-AAD has good detection performance with limited labels, and improves the detection accuracy by 10.2%–18.2% from benchmark methods, while it achieves 12.1% human cost reduction.

Keywords: time series data mining · anomaly detection · active learning · sparse neural network

1 Introduction

The rapid development of sensor technologies and the widespread use of sensor devices witness the flowering of data mining technologies in time series data. However, anomaly problems are seriously prevalent in time series [3,20], where anomalies are summarized as any unusual change in a value or a pattern, which do not conform to specific expectations [4,18]. To identify anomalies is one of the most important and exciting tasks for time series data mining [17,18]. On one hand, anomaly detection techniques are applied to identify the unexpected

© The Author(s), under exclusive license to Springer Nature Switzerland AG 2023
X. Wang et al. (Eds.): DASFAA 2023, LNCS 13943, pp. 253–269, 2023.
https://doi.org/10.1007/978-3-031-30637-2_17

events that the data reflects. On the other hand, it also assists to guarantee that the data analysis result would not suffer from outliers in the later data management pipeline.

Researchers have gone a long way in anomaly detection studies. Anomaly detection techniques with different methods such as Density-, Window- and Constraint-based approaches have been developed and applied in various real scenarios (see [7,12,20] as surveys). In recent years, deep learning models have also demonstrated good performance in anomaly detection tasks, and classical models include autoencoders, recurrent neural networks (RNNs), convolutional neural networks (CNNs) and generative adversarial networks (GANs), etc. [12]. As the labelled data are always limited and costly for real applications [10], research has been conducted on how to train models with limited labels to ensure the sensitivity of detection methods. In addition, active learning is introduced into anomaly detection techniques (e.g., Active-MTSAD [19], HOD [2]), which involves artificial help to improve the detection effectiveness of the model when the labelled data is inadequate, or even unavailable. Although the existing methods perform well in different scenarios, it is still challenged to achieve high accuracy and efficiency in anomaly detection for multivariate time series with limited labels. Challenges lie in the following aspects.

First, *how to train an accurate model with limited labels to achieve sufficient generalization capacity*? Detection results likely to contain more FPs and FNs when labels are limited. The main reason is that features contained in the labels are not complete enough, and some features often lead to *redundant* calculation of the abnormal representation. This results in low efficiency of feature utilization and overfitting phenomenon. Although some semi-supervised methods [15] change the cost function to forcibly minimize the reconstruction error of normal data and maximize the reconstruction error of abnormal data to strengthen the reconstruction ability of the training network. It is still difficult to effectively avoid overfitting and achieve high generalization ability.

Second, *how to select the data for manual labeling and reduce the human cost while ensuring detection accuracy*? As it is always challenged to define what are anomalies compared to normal patterns, one will likely obtain some fuzzy data from the detection model. Methods are expected to be carefully designed for selecting samples from such fuzzy data so that they cover more anomaly instances while minimizing the human efforts. To achieve this, it is necessary to sort and select appropriate features feedback to human annotation. Otherwise, it will be quite expensive and difficult to leverage human intelligence.

Contributions. Motivated by the above, we study the anomaly detection problem for multivariate time series data with limited labels. Our contributions are summarized as follows:

(1). We propose a novel active anomaly detection method with sparse neural network (SNN-AAD) to solve the anomaly detection problem for time series data when the label data are limited, which also effectively involves human intelligence for the detection process (in Sect. 2.2).

(2). We innovatively introduce sparse neural network into the studied anomaly detection, which effectively reduces the overfitting phenomenon caused by limited labels (in Sect. 3). We propose back propagation functions to prune *less-important* connections in SNN, which achieves to minimize the detection loss caused by the sparsity of labels. The proposed SNN-AAD achieves efficient training process with the feedback from humans.

(3). We design the anomaly sample selection strategy to choose samples for manual labelling in order to reduce human cost in active anomaly detection (in Sect. 4.1). We formalize the criterion of fuzzy data in time series anomaly detection and propose the abnormal feature order generation algorithm (AFOGSNN) to extract the importance of various features of anomalies (in Sect. 4.2). The proposed SNN-AAD improves the learning ability of more valuable anomaly information with low human effort.

(4). We conduct extensive experiments on three public datasets and a real IoT dataset (in Sect. 5). Experimental results show that the proposed SNN-AAD has a good performance in anomaly detection tasks with limited labels. SNN-AAD improves the detection accuracy by 10.2%-18.2% from benchmark methods, while it achieves 12.1% human cost reduction with the proposed active learning process.

2   Problem Overview

2.1   Problem Statement

We define $\mathcal{S} = \{S_1, ..., S_M\} \in \mathbb{R}^{N \times M}$ to be a M-dimensional time series, where M is the total number of dimensions, where $S = \langle s_1, ..., s_N \rangle$ is a sequence, and $N = |S|$ is the length of S, i.e., the total number of elements in S. $s_n = \langle x_n, t_n \rangle, (n \in [1, N])$, where x_n is a real-valued number with a time point t_n. A sequence tuple in a M-dimensional $\mathcal{S}$ is the set of all data points recorded at time t_i, denoted by $\mathcal{S}(t_i) = \langle s_{i1}, s_{i2}, ..., s_{iM} \rangle$, i.e., the i-th row of $\mathcal{S}$.

As mentioned above, one critical problem for active anomaly detection is how to maximize the improvement of the model with a small size of samples for manual labeling. The key problem is to label the most valuable instances with a data selection mechanism. For anomaly detection of time series data, the data with different prediction results from multiple classifiers are expected to be selected, because such instances are believed to have greater uncertainty. We formalize such data as X_{fuzzy} in Definition 1. Then, we apply the concept of information entropy to measure the "uncertainty" property of the data to be selected.

Definition 1. *Fuzzy Data* *refers to the data points that represent the maximum prediction probability of the current prediction result of $\mathcal{S}(t_i)$ which has*

$$X_{\text{fuzzy}} = \arg \min_x \Pr \left(l_{\max p} | \mathcal{S}(t_i) \right)$$

Here, $\Pr$ represents the prediction of the trained model for data points, and $l_{\max p}$ represents the category with the highest probability for data $\mathcal{S}(t_i)$.

In order to measure the efficiency of our current anomaly data detection, we propose a measurement algorithm for the cost of manual anomaly detection to facilitate the calculation of the optimization degree of this method for the manual annotation part.

Definition 2 (*Manual Annotation Cost*). *Given the time series to be detected,* $\mathcal{S} = \{S_1, ..., S_M\} \in \mathbb{R}^{N \times M}$, *where* $t_i \in R^K$. *Suppose* D *is the sample set expected to be labelled, the number of features involved in marking data points is* $Cost(s), s \in D$. *The cost of manual labeling can be expressed as*

$$MCost = \frac{\sum_{s \in D} Cost(s)}{N \times M}$$

where $Cost(s)$ *indicates that the corresponding node needs to refer to several dimensions to determine its label.*

Accordingly, $MCost$ expresses the cost of manual annotation as the ratio between the data requiring manual inspection for annotation and the total data amount.

Problem Definition. Given two set of time series data $\mathcal{S}_l$ and $\mathcal{S}_{ul}$, while only $\mathcal{S}_l$ has limited labels $L = \{l_i | l_i \in [0, 1]\}$, where the total number of labels $|L|$ is quite smaller than the data amount in $\mathcal{S}_l$. The active anomaly detection problem is to train multiple regression models $\mathcal{I} = \{I_1, I_2, ..., I_p\}$, and involve human to label the fuzzy data identified by Definition 1. Our goal is to design the learning exception scoring function and obtain the anomaly score $\phi(x_i)$ for data. The score of real anomalies is expected to be close to 1 as possible, while the score of normal data is as close as possible to 0. In addition, the human cost is minimized by reduction of the size of set D and $Cost(s)$ w.r.t s according to Definition 2.

Intractability Analysis. We first analyze the solution space of the problem. For the binary classification problem, where the abnormal value is marked as 1, and the normal value is marked as 0, suppose that the original dataset S has a total of M dimensions, and the time series size is N, where the number of K is labeled data. Without losing generality, it can be considered that the labeled data is the first K data $\{x_1, x_2, ...x_k\}$. Any dimension of one data instance can be abnormal, which means that there are $f = \{2^M - (2^{M-1} - 1) - (2^{M-2} - 1) - ...\} = \{M - 1\}$ possibilities. So for a dataset of size N, there are M^N possible rows in total, of which the first N numbers, namely M^N possibilities, have been determined. Therefore, there is a solution space with size M^K in total. Since the anomaly score is a continuous value, the solution space for finding the anomaly score is larger than the solution space for 01 classification of anomaly data. It can be seen that the size of the solution space of the problem is exponential. To reduce the solution space of the problem, we need to consider the following two aspects in our solution: (1) to reduce the number of data that require accurate solutions, and (2) to propose the internal relationship between data characteristics and solutions in order to reduce the solution space.

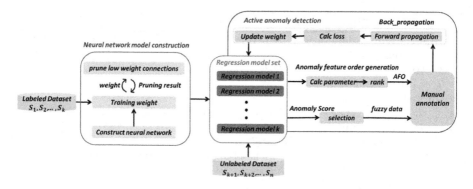

Fig. 1. Method Framework of SNN-AAD

2.2   Method Overview

The whole solution SNN-AAD is outlined in Figure 1, including two main phases: sparse neural network construction and active anomaly detection.

For the sparse neural network construction phase, the proposed SNN-AAD first constructs multiple learning models $\mathcal{I} = \{I_1, I_2, ..., I_p\}$ with the labeled dataset. Note that we generate multiple SNNs with different structures to ensure the diversity and robustness of the learning models. During the process, we initialize the weights of models and construct the corresponding weight matrix. SNN-AAD computes the matrices with lower absolute value of weight in order to identifies the connections of less importance in the networks. It prunes the learning models $\mathcal{I}$ to remove such connections. We iteratively process the training and pruning until the connection weights in the model reached a certain absolute value. Finally, we get multiple regression models $\mathcal{I} = \{I_1, I_2, ..., I_p\}$.

In the active anomaly detection step, SNN-AAD will first trains the input data $\mathcal{S}$ and obtains the corresponding labels of $\mathcal{S}$. We then introduces the active learning model to help improve the detection accuracy. The main objective of this step is to select more appropriate samples for humans and update the training model with human feedback. Specifically, SNN-AAD executes the sample data selection function to uncover fuzzy data which are identified to have much high uncertainty degree. We then apply the algorithm AFOGSNN to realize the quantitative judgment of important features with high contribution to the detection results and determine the data that needs to be handed over to manual labeling. The manual feedback results will be returned to the regression model set $\mathcal{I}$, and SNN-AAD will update the training model with backpropagation functions. The pseudo-code of SNN-AAD process is formalized in Algorithm 1.

3   Sparse Neural Network Construction

In this section, we introduce how to construct sparse neural network and train the model for the proposed SNN-AAD solution. Multi-Layer Perceptrons (MLP)

Algorithm 1: SNN-AAD

    Input: Training data S_l with limited labels L, test data S, learning λ, iteration
            time *epoch*
    Output: set of S with *Label* $\{l_1, l_2, ..., l_m\}$
1 $\mathcal{I} \leftarrow$ **Initialization**(S_l, L, λ);
2 **for** *i=1 to epoch* **do**
3 $L = \{l_1, l_2, ..., l_m\} \leftarrow$ the annotation result of $\mathcal{I}$ on $T_{ul} = \{t_1, t_2, ..., t_m\}$;
4 $U \leftarrow$ **Selection**(X_{test}, L);
5 $AFO \leftarrow$ **AFOGSNN**$(datalist, \mathcal{I})$;
6 $L = \{l_1, l_2, ..., l_m\} \leftarrow$ Manually label data points in the set U according to
        AFO;
7   **for** I *in* $\mathcal{I}$ **do**
8 | **Back_propagate**(U, L, I, λ);
9 *Label* $\leftarrow$ the annotation result of $\mathcal{I}$ on $T_{ul} = \{t_1, t_2, ..., t_m\}$;
10 **return** S with *Label*

---

based on feedforward neural networks has demonstrated to be suitable for types of learning tasks, including classification, regression, and even unsupervised learning. However, MLP would not perform well when faced with limited labels which always fail to provide adequate representations for identifying various anomaly instances. In this case, we propose the sparse neural network to train the detection model, and realize drops of connections between the input layer and the hidden layer through calculation, so as to improve the training effect of the model. The improved sparse neural network has the following advantages:

1. Diversity. Since the hidden layer can abstract each feature of the original data into another dimension space to facilitate a better division of the original data, SNN could change the "attention" to some unimportant features, and make the hidden layer representation of the multi-learner have greater diversity, when removing part of the links from the input layer to the hidden layer.

2. Avoid Overfitting. In the limited label cases, the sample data is always not enough to provide comprehensive features of anomalies. Compared with MLP, SNN can effectively avoid overfitting of the detection model and maintain sufficient generalization capacity for identifying novel anomalies.

3. Efficiency. When some links in the network are reduced, the redundant calculation of unimportant features is effectively avoided. SNN has a smaller time cost than traditional neural networks, and has great potential to be applied to ensemble learning models such as Bagging to further improve the performance of the model.

3.1 SNNs Generation

In the step of SNN model generation, we mainly consider how to generate the structure of SNN model and how to prune SNNs. Specifically, we first need to determine the number of nodes contained in the input layer, hidden layer and

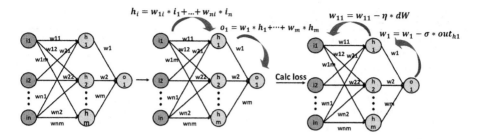

Fig. 2. The process of back propagation

output layer. In the proposed method, we assume that the three values are M, $\frac{M}{2}$ and 1, respectively. We use full connection to establish connection between layers. For example, if there are l_1 and l_2 nodes between the adjacent two layers, we would establish a total of $l_1 * l_2$ connections between the two layers.

Then, based on the traditional MLP method, we use the sparse neural network model to prune the model. We reduce the number of connections to achieve unstructured pruning. We need to determine the sparsity for the model, where sparsity is expressed as $\frac{\#Connections\ after\ pruning}{\#Full\ connections}$. Let the number of connections between two layers be $l_1 * l_2$, we are able to reduce $l_1 * l_2 * (1 - Sparsity)$ connections to avoid redundant computing on unrelated connections. For connection selection, we define the importance of connection as $W = Abs(weight)$, and we select connections to prune according to W in the ascending order. In order to ensure the reliability of the model, we adopt the method of generating multiple neural networks with different structures in SNN-AAD.

3.2 Back Propagation of SNNs

Different from the traditional neural network, we note that SNNs do not need to update the elements that are zero in the parameter matrix when the weights are updated by back propagation functions. We preform the back propagation function according to the following steps, and present the process in Fig. 2.

1. First, we conduct a forward propagation. To take h_1 node as an example, it has

$$net_{h1} = w_{11} * i_1 + w_{21} * i_2 + ... + w_{n1} * i_n, \tag{3.1}$$

$$out_{h1} = \frac{1}{1 + e^{(-net_{h1})}}. \tag{3.2}$$

2. Next, we calculate the total error with

$$E_{total} = \sum \frac{1}{2} * (target - output)^2. \tag{3.3}$$

3. Finally, we update the connection weight on the overall error. To take w_1 connection as an example, it has

$$new(w_1) = w_1 - \lambda * \frac{\partial E_{total}}{\partial w_1} = w_1 - \sigma * out_{h1}. \tag{3.4}$$

Algorithm 2: Back_propagation

Input: input data $T = \{t_1, t_2, ...t_k\}$, label $L = \{l_1, l_2, ...l_k\}$, model $\mathcal{I}$, learning rate λ

Output: Updated Model $\mathcal{I}_{new}$

1 Forward propagation;

2 **for** $i=1$ *to* k **do**

3 $\quad\mid\quad E_i \leftarrow sigmoid(t_i) * (l_i - t_i)$;

4 **for** $j=1$ *to* h **do**

5 $\quad\mid\quad H_i \leftarrow sigmoid(E_j) * \sigma_{l=1}^{d} E_j W_{jl}$;

6 **for** $i=1$ *to* h **do**

7 $\quad\mid\quad$ **for** $j=1$ *to* m **do**

8 $\quad\mid\quad\mid\quad W_{ij} \leftarrow W_{ij} + \lambda E_j O_i$;

9 **for** $i=1$ *to* h **do**

10 $\quad\mid\quad$ **for** $j=1$ *to* m **do**

11 $\quad\mid\quad\mid\quad$ **if** $W_{ij} \neq 0$ **then**

12 $\quad\mid\quad\mid\quad\mid\quad W_{ij} \leftarrow W_{ij} + \lambda H_j O_i$;

13 **return** $\mathcal{I}_{new}$

Algorithm 2 shows the back propagation process of SNN. It first performs a forward propagation and calculates the values of each node. Then the derivative of the SIGMOID function is applied to calculate the error from the hidden layer to the output layer. It then calculates the error from the input layer to the hidden layer. We highlight that we only need to update the value of non-zero elements in the parameter matrix, which avoids redundant calculation for unimportant features.

4 Active Anomaly Detection

In this section, we introduce how to efficiently involve human intelligence in identifying anomalies and improve the training model when faced with limited labels. We apply active learning to select the fuzzy data that cannot be classified well by the learning model set $\mathcal{I}$, and determine the more important features for identifying anomaly instances. During the process, we aim to solve two critical issues for active anomaly detection, i.e., the annotation data selection and manual labelling cost optimization, respectively.

4.1 Annotation Data Selection

Since the selection of a batch of the most valuable samples for manual process is the crucial point in active anomaly detection techniques, we consider to design selection mechanism in SNN-AAD. The general idea is to let human deal with the instances difficult to distinguish for the multiple models, which are formalized as the data with the smallest prediction probability in the current prediction result, i.e., $X_{\mathsf{fuzzy}} = \arg\min_x \Pr\left(l_{\max p}|\mathcal{S}(t_i)\right)$ according to Definition 1.

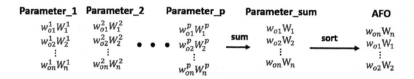

Fig. 3. Demonstration of AFO generation

We treat the uncertainty in anomalies detected from the training process from the perspective of information entropy. Uncertain sampling refers to selecting the sample points that are difficult to distinguish for the model in order to quickly improve the accuracy of the model. We use the method of information entropy to calculate the statistical average of uncertainty $\log_2[\Pr(y_i = w|x_i)]$ and measure the value of data for manual work with Eq. (4.1).

$$H^{BAG}(x_i) = \sum \Pr(y_i = w|x_i)|\log_2[\Pr(y_i = w|x_i)]| \qquad (4.1)$$

As shown in Eq. (4.1), for p regression models, $\Pr(y_i = w|x_i)$ denotes the probability that the sample x predicted to be w by p models, that is, the number of votes that sample x is predicted to be w divided by p. A lower value of $H^{BAG}(x)$ represents higher confidence of the detection result with model $\mathcal{I}$. Specifically, for the binary classification problem, the threshold setting of selection methods which could be regarded as query functions is generally considered as a 2–1 divergence phenomenon. That is, the ratio of abnormal and normal labels in the sample points is 2: 1. In this case, we have $H^{BAG} = 0.92$. Thus, it generally makes sense to select samples with H^{BAG} higher than 0.92 to ask for human feedback in practical applications.

4.2   Manual Cost Optimization with AFO

We highlight that it is not enough to only implement a selection of "valuable" samples for a good active anomaly detection model. It is expected to be able to return the instances with such determinative and important features that contribute more to the detection results. In this case, human can well use these data "profiles" to quickly understand the labeling tasks at hand, and provide reliable feedback to the model. Accordingly, we propose the concept of *anomaly feature order* (AFO), and design sorting algorithm for the important features w.r.t anomaly instances.

AFO Generation. In order to generate AFO with the propose SNN, we calculate the corresponding parameter for each feature. The larger absolute value of the parameter represents higher importance of the feature for the current classification result. Note that the SIGMOID function applied in SNN-AAD is a monotone function, the activation function will not cause the absolute value of each feature parameter to affect the feature importance. Given a total of d-dimensions input layer, and h neurons in the hidden layer, we can construct AFO with the representation space of the hidden layer R_{ih} as follows.

Algorithm 3: AFOGSNN

Input: Manually annotate data $\mathcal{S}_{ano} = \{S_1, S_2, ..., S_M\}$, the model set
$\quad\quad \mathcal{I} = \{I_1, I_2, ..., I_n\}$
Output: anomaly feature order AFO
1 initialize list $parameter_{sum}$;
2 **foreach** I_i *in* $\mathcal{I}$ **do**
3 $\quad\quad$ $parameter_i \leftarrow$ compute $parameter_x$ according to Equation (4.5);
4 $\quad\quad$ $parameter_{sum} \leftarrow l \cup parameter_i$;
5 $AFO \leftarrow$ Sort $\mathcal{S}_{ano}$ by ($parameter_{sum}$);
6 **return** AFO

$$R_{ih} = \begin{bmatrix} x_1 \cdots x_n \end{bmatrix} \begin{bmatrix} w_{11} & \cdots & w_{n1} \\ \vdots & \ddots & \vdots \\ w_{1n} & \cdots & w_{nn} \end{bmatrix} = \begin{bmatrix} W_1 X^T & \cdots & W_n X^T \end{bmatrix}, \quad\quad (4.2)$$

$$\theta_1 = \begin{bmatrix} w_{11} & \cdots & w_{12} \\ \vdots & \ddots & \vdots \\ w_{n1} & \cdots & w_{nn} \end{bmatrix} \quad\quad (4.3)$$

where θ_1 denotes the weight matrix from the input layer to the hidden layer, $W_i^T = \begin{bmatrix} w_{i1} \cdots w_{in} \end{bmatrix}$, $X^T = \begin{bmatrix} x_1 \cdots x_n \end{bmatrix}$.

Accordingly, the representation space R_{h0} of the output layer is expressed as

$$R_{h0} = \begin{bmatrix} W_1 X^T & \cdots & W_n X^T \end{bmatrix} \begin{bmatrix} w_{o1} \\ \vdots \\ w_{on} \end{bmatrix} = (w_{o1} W_1 + \cdots + w_{on} W_n) X^T, \quad\quad (4.4)$$

where $\theta_2^T = \begin{bmatrix} w_{o1} \cdots w_{on} \end{bmatrix}$ is the weight matrix from the hidden layer to the output layer. From the above, we obtain the parameter matrix of the anomaly features in Eq. (4.5).

$$parameter_x = (w_{o1} W_1 + \cdots + w_{on} W_n) \quad\quad (4.5)$$

As shown in Fig. 3, $parameter_x$ is a $1 \times t$-size matrix. Each column of the matrix corresponds to a dimension of the data. Considering the model set from the above step, we summarize the anomaly feature sequences corresponding to all sub learning models as the average anomaly feature sequence, and obtain the sorted AFOs.

We present our AFO generation process in Algorithm 3. After initializing the model set $\mathcal{I}$, we begin to calculate the abnormal feature order corresponding to each network. During the process, we compute the AFO matrix with the above equations, and maintain a mapping from attribute sets to $parameter_x$. We sort the elements in the mapping set according to the value in AFO matrix. We finally integrate the multiple neural networks and sort all neural network attributes to obtain the final AFO.

Table 1. Dataset Content

Dataset	Dimension	Data amount	Proportion of anomalies
Har	561	6771	322 (4.76%)
Covtype	54	232933	21093 (9.06%)
Oil chromatogram	10	12380	1049 (8.47%)
SWaT	51	449919	44992 (10%)

The selected samples with AFO will be return for human feeback, and SNN-ADD is able to learn the incremental information iteratively with back propagation mechanism on the proposed SNNs to achieve high detection performance.

5   Experimental Evaluation

5.1   Experimental Settings

Datasets. We employ three real-life public datasets, namely *HAR*, *Covtype*, and *SwaT*, and one real IoT business dataset *Oil chromatogram* in our experiments. Table 1 summarizes the four datasets. The HAR dataset records human activity data, such as x, y and z accelerometer data from smart phones and gyroscope data, with a sampling frequency 50 Hz. Covtype dataset represents different characteristics of vegetation. The data are taken from four wilderness areas of Roosevelt National Forest. The oil chromatogram data set provides the content of dissolved gas and free gas in the transformer, which is used to judge whether the transformer has faults and the type of faults. SWaT data comes from a water treatment test platform for research in the field of network security, including data of corresponding sensors, actuators and network data.

Implementation. We have implemented the whole process of SNN-AAD method. In each iteration, we choose the sample points with H^{BAG} greater than 0.92. We set the number of base learners to be 100, the number of training rounds 2000, and the learning rate 0.01. Besides the proposed SNN-AAD, we also implement three classical detection strategies for comparative evaluation.

1. **Isolated forest** [11] is a classical baseline anomaly detection algorithm. In isolated forests, small data sets can often achieve better results. A large number of samples will reduce the ability to isolate outliers in isolated forests, because normal samples will interfere with the isolation process and reduce the ability to isolate anomalies.
2. **GBDT** [9] (Gradient enhanced decision tree) is a popular tree ensemble model, which uses the regression tree to predict a continuous weight on one leaf, and accumulate prediction of all trees to get the final result.
3. **MLP** has multiple neuron layers, and it always has "full connection" between layers of multilayer perceptron. We combine MLP and active learning in our experiments.

Table 2. Performance Comparison of various anomaly detection methods

Method	Har				Covtype				Oil chromatogram				SWaT			
	Prec	Rec	F1	T(s)	Prec	Rec	F1	T(s)	Prec	Rec	F1	T(s)	Prec	Rec	F1	T(s)
SNN-AAD	**0.82**	**0.79**	**0.80**	14.57	**0.87**	0.82	0.84	176	**0.95**	**0.96**	**0.95**	2.88	**0.78**	**0.79**	**0.78**	420
MLP	0.6	0.62	0.60	15.03	0.75	0.76	0.75	168	0.87	0.82	0.84	1.57	0.64	0.62	0.62	478
GBDT	0.72	0.7	0.70	**10.07**	0.84	**0.88**	**0.85**	121	0.94	0.90	0.91	2.84	0.71	0.69	0.69	820
Iso Forests	0.52	0.44	0.47	10.10	0.56	0.52	0.53	137	0.78	0.69	0.73	**1.34**	0.43	0.34	0.27	**356**

Metrics. Since SNN-AAD outputs anomaly score ϕ for data, the test data with $\phi > 0.5$ are treated as anomalies, and the algorithm performances are measured by Precision (P), Recall (R), and F1: $P = \frac{TP}{TP+FP}$, $R = \frac{TP}{TP+FN}$, $F1 = \frac{2 \cdot P \cdot R}{P+R}$, which are widely applied in anomaly detection performance evaluation.

5.2   General Performance Comparison

Table 2 shows the general detection performance of algorithms on the four datasets. It shows that the proposed SNN-AAD has the highest F1 performance in most cases. The GBDT method obtains the second performance, and at some times, its detection results would be higher than the proposed SNN-AAD. This is because GBDT can effectively combines the initial training set and is robust to outliers. For MLP, it performs worse than either SNN-AAD and GBDT under the same conditions. The result confirms that SNNs constructed in this paper can effectively train the data with limited labels.

It shows that the performance difference between the proposed method and MLP, GBDT is small on Oil chromatography data, while the difference is larger on the other three data sets. We explain that a small-scale of attributes in oil chromatography results in less gap in the detection performance of the four algorithms. However, in the other cases, the proposed SNN-AAD shows stronger advantage in the detection performance. It reflects the superiority of the proposed sparse neural network model, which achieves good model generalization ability by effectively eliminating irrelevant connections.

5.3   Performance Comparison with Varying Parameters

We next evaluate the performance of algorithms with four kinds of critical parameters of the studied problem.

Exp1: Performance Comparison with Varying Iterations. We use 2–20 iterations for the performance test of the training method for the number of iterations. As shown in Fig. 4, with the increase of the number of iterations, the precision of the proposed SNN-AAD is significantly improved to 0.9 after the iteration times reaches 16. The performance of the isolation forest and MLP methods remain at a low level, and the Recall of MLP even shows a small decrease. This is because that adopting a fully connected network is not good enough to *capture* the feature information from limited labels. This also confirms

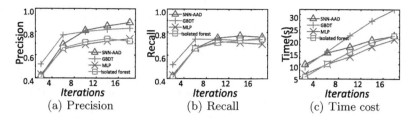

Fig. 4. Performance comparison with varying iterations on *Har*

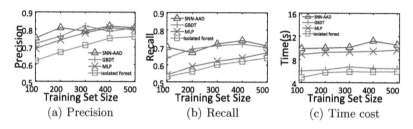

Fig. 5. Performance comparison with varying initial training set size on *Har*

that the proposed SNN-AAD embodies the effectiveness in multiple iterations of training and beats GBDT after a higher number of iterations. Although there is little difference between the Recall of SNN-AAD and GBDT algorithms, the Precision of SNN-AAD shows the potential to improve the advantage. In addition, by training the SNNs, SNN-AAD also achieves high detection efficiency, which effectively saves time compared with GBDT (Fig. 6).

Exp2: Performance Comparison with Varying Initial Training Set Size. To evaluate the generalization capacity of methods with limited labels, we use 5 different training sets $[100, 200, 300, 400, 500]$ in Exp2. As shown in Fig. 5, the proposed SNN-AAD shows the highest detection performance even with fewer training sets, which confirms the effectiveness of the proposed model. And the performance of SNN-AAD remains stable with the increase of the training set size. In general, the Precision of SNN-AAD remains at a higher level compared to the Recall. Although MLP and Isolated Forest improve in both P and R with the growth of the training set size, under the same circumstances, these two methods never outperform SNN-AAD, and have lower performance when labels are limited. This confirms again the necessity of model optimization for limited labels in anomaly detection tasks.

Exp3: Performance Comparison with Varying Manual Annotation Ratios. We report the performance comparison with 5 different manual annotation ratios $[0.05, 0.10, 0.15, 0.20, 0.25]$ for SNN-AAD and MLP with active learning process. When the manual labeling ratio is 0.05, SNN-AAD achieves high recall but low precision, however, when the ratio reaches 0.1, the performance of SNN-AAD is significantly improved. Compared with MLP, the advantage of

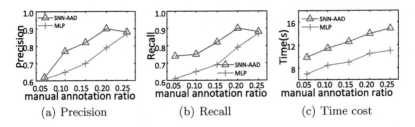

Fig. 6. Performance comparison on *Har*

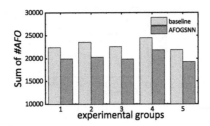

Fig. 7. Cost optimization with AFOGSNN

SNN-AAD is quite obvious, which confirms that the proposed AFO generation function effectively contribute to the identification of the more important features of anomalies. The experimental results confirm the effectiveness of the proposed active anomaly detection process.

Exp4: Cost Optimization of Manual Annotation. We conduct five groups of experiments to evaluate the optimization effectiveness of the proposed AFOGSNN in Sect. 4, compared with the baseline method without processing the features of anomalies. Figure 7 shows the results of five experiments, in which the ordinate represents the number of attributes manually labeled and the abscissa corresponds to five groups of experiments. The average value of the number of attributes manually labeled before optimization is 22978, that is the number of attributes that need to be viewed manually to execute the annotation task. After optimization, the number of attributes manually labeled decreases to 20229. It reports that the number of attributes decreased by 12.1% with the cost optimization algorithm AFOGSNN.

6   Related Work

Anomaly detection (see [7,12,20] as surveys) is a important step in time series management process [3,20], which aims to discover unexpected changes in patterns or data values in time series. Applying density-based methods is one of the earliest known anomaly detection solutions. The basic idea of the methods is that anomalies usually occur in low-density areas, while normal data usually occur in dense areas. The local outlier factor method (LOF), which uses K nearest neighbor is proposed in [1]. The algorithm mainly judges whether a point is an outlier by comparing the density of each point and its neighborhood points. The lower the density of the point, the more likely it is to be recognized as an outlier. [16] proposes a clustering based framework to detect outliers in changing data streams, They assign weight to features according to their correlation. Weighted features restrain the influence of noise in the calculation process, which is of great significance. The migration ability of clustering based anomaly detection methods is poor. In the cases of large changes in data requirement, if more data needs to be added or modified, it is necessary to retrain the anomaly detection model based on clustering.

The methods combined with neural network have also been introduced into anomaly detection, such as CNN [13], ResNet [8] and Inception Time [5]. Deep-ConvLSTM [14] is a deep learning framework composed of convolution layer and LSTM loop layer. It can automatically learn feature representation and model the time dependence between their activations.Compared with other algorithms, the method combined with neural network has higher mobility, and continues training based on the original data after adding new data. However, they still require a long training time and need require a lot of storage space to store over parameterized models. One way to reduce the complexity of neural network (NN) is to delete "unnecessary" connections to obtain a smaller network with similar performance than the original network. An important discovery in this field is the Lottery Hypothesis (LTH) [6]. This hypothesis indicates that most neural networks contain smaller subnets that can be trained independently to achieve similar or even higher accuracy. By training the sub network, the resources needed in the neural network can be reduced.

7   Conclusions

In this paper, we study the problem of active anomaly detection with sparse neural network for multivariate time series with limited label data. We develop SNN-AAD method to solve the problem, which consists of sparse neural network construction and active anomaly detection. We design SNNs to achieve high diversity of learners and high efficiency for training. We propose the anomaly sample selection strategy and propose the abnormal feature order generation algorithm to extract the importance of various features of anomalies. Experimental results on four datasets show that SNN-AAD has a good performance in anomaly detection tasks with limited labels. SNN-AAD improves the detection accuracy by 10.2%–18.2% from benchmark methods, while it achieves 12.1% human cost reduction with the proposed active learning process.

Acknowledgements. This work was supported by the National Key Research and Development Program of China (2021YFB3300502); National Natural Science Foundation of China (NSFC) (62232005, 62202126, U1866602); China Postdoctoral Science Foundation (2022M720957); Heilongjiang Postdoctoral Financial Assistance (LBH-Z21137); and Sichuan Science and Technology Program (2020YFSY0069).

References

1. Breunig, M.M., Kriegel, H., Ng, R.T., Sander, J.: LOF: identifying density-based local outliers. In: Chen, W., Naughton, J.F., Bernstein, P.A. (eds.) Proceedings of the 2000 ACM SIGMOD International Conference on Management of Data, 16–18 May 2000, Dallas, Texas, USA, pp. 93–104. ACM (2000)
2. Chai, C., Cao, L., Li, G., Li, J., Luo, Y., Madden, S.: Human-in-the-loop outlier detection. In: Proceedings of the 2020 International Conference on Management of Data, SIGMOD Conference 2020, online conference, 14–19 June 2020, pp. 19–33. ACM (2020)
3. Dasu, T., Duan, R., Srivastava, D.: Data quality for temporal streams. IEEE Data Eng. Bull. **39**(2), 78–92 (2016)
4. Dasu, T., Loh, J.M., Srivastava, D.: Empirical glitch explanations. In: Proceedings of the 20th ACM SIGKDD International Conference on Knowledge Discovery and Data Mining, KDD 2014, New York, NY, USA, 24–27 August 2014, pp. 572–581 (2014)
5. Fawaz, H.I., et al.: InceptionTime: finding AlexNet for time series classification. Data Min. Knowl. Discov. **34**(6), 1936–1962 (2020)
6. Frankle, J., Carbin, M.: The lottery ticket hypothesis: finding sparse, trainable neural networks. In: 7th International Conference on Learning Representations, ICLR 2019, New Orleans, LA, USA, 6–9 May 2019. OpenReview.net (2019)
7. Gupta, M., Gao, J., Aggarwal, C.C., Han, J.: Outlier Detection for Temporal Data. Synthesis Lectures on Data Mining and Knowledge Discovery. Morgan & Claypool Publishers, San Rafael (2014)
8. He, K., Zhang, X., Ren, S., Sun, J.: Deep residual learning for image recognition. In: 2016 IEEE Conference on Computer Vision and Pattern Recognition, CVPR 2016, Las Vegas, NV, USA, 27–30 June 2016, pp. 770–778. IEEE Computer Society (2016)
9. Jiang, J., Cui, B., Zhang, C., Fu, F.: DimBoost: boosting gradient boosting decision tree to higher dimensions. In: Proceedings of the 2018 International Conference on Management of Data, SIGMOD Conference, pp. 1363–1376. ACM (2018)
10. Le, K., Papotti, P.: User-driven error detection for time series with events. In: 36th IEEE International Conference on Data Engineering, ICDE 2020, Dallas, TX, USA, 20–24 April 2020, pp. 745–757. IEEE (2020)
11. Liu, F.T., Ting, K.M., Zhou, Z.: Isolation-based anomaly detection. ACM Trans. Knowl. Discov. Data **6**(1), 3:1–3:39 (2012)
12. Lundström, A., O'Nils, M., Qureshi, F.Z., Jantsch, A.: Improving deep learning based anomaly detection on multivariate time series through separated anomaly scoring. IEEE Access **10**, 108194–108204 (2022)
13. Martínez, H.P., Bengio, Y., Yannakakis, G.N.: Learning deep physiological models of affect. IEEE Comput. Intell. Mag. **8**(2), 20–33 (2013)
14. Morales, F.J.O., Roggen, D.: Deep convolutional and LSTM recurrent neural networks for multimodal wearable activity recognition. Sensors **16**(1), 115 (2016)
15. Munawar, A., Vinayavekhin, P., Magistris, G.D.: Limiting the reconstruction capability of generative neural network using negative learning. In: 27th IEEE International Workshop on Machine Learning for Signal Processing, MLSP, pp. 1–6. IEEE (2017)
16. Qin, X., Cao, L., Rundensteiner, E.A., Madden, S.: Scalable kernel density estimation-based local outlier detection over large data streams. In: Advances in Database Technology - 22nd International Conference on Extending Database

Technology, EDBT 2019, Lisbon, Portugal, 26–29 March 2019, pp. 421–432. Open-Proceedings.org (2019)

17. Takeuchi, J., Yamanishi, K.: A unifying framework for detecting outliers and change points from time series. IEEE Trans. Knowl. Data Eng. **18**(4), 482–492 (2006)

18. Toledano, M., Cohen, I., Ben-Simhon, Y., Tadeski, I.: Real-time anomaly detection system for time series at scale. In: Proceedings of the KDD Workshop on Anomaly Detection, pp. 56–65 (2017)

19. Wang, W., Chen, P., Xu, Y., He, Z.: Active-MTSAD: multivariate time series anomaly detection with active learning. In: 52nd Annual IEEE/IFIP International Conference on Dependable Systems and Networks, DSN 2022, Baltimore, MD, USA, 27–30 June 2022, pp. 263–274. IEEE (2022)

20. Wang, X., Wang, C.: Time series data cleaning: a survey. IEEE Access **8**, 1866–1881 (2020)

Spatial Data

Continuous k-Similarity Trajectories Search over Data Stream

Rui Zhu[1]([✉]), Meichun Xiao[1], Bin Wang[2], Xiaochun Yang[2], Xiufeng Xia[1],
Chuanyu Zong[1], and Tao Qiu[1]

[1] Shenyang Aerospace University, Shenyang, China
{zhurui,xiaomeichun,xiufengxia,zongcy,qiutao}@mail.sau.edu.cn
[2] Northeastern University, Shenyang, China
{binwang,yangxc}@mail.neu.edu.cn

Abstract. Continuous k-similarity trajectories search (short for CKST) over data stream is a fundamental problem in the domain of spatio-temporal database. Let $\mathcal{T}$ be a set of trajectories, T_q be the query trajectory. T_q monitors elements in $\mathcal{T}$, retrieves k trajectories that are the most similar to T_q whenever trajectories in $\mathcal{T}$ are updated. Some existing works study k-similarity trajectories search over historical trajectory data. Few efforts could support continuous k-similarity trajectories search over data stream, but they cannot accurately measure similarity among trajectories.

In this paper, we propose a novel framework named SLBP (Score Lower-Bound-based Prediction) to support CKST. It is based on the following observation, that is, given T_q and one trajectory $T_i \in \mathcal{T}$, if the distance between their corresponding last generated GPS point is large, the distance between T_q and T_i also may be large, and it cannot become a query result trajectory for a long time. In this way, we can predict the earliest time T_i could become a query result. We develop a group of algorithms to support CKST via using the above property.

Keywords: Trajectory stream · k-similarity trajectories search · Prediction · Continuous query

1 Introduction

k-similarity trajectories search is a fundamental problem in the domain of spatio-temporal database. It has many applications, such as route planning [5], trajectory compression [6,8], car pooling [4], and clustering of moving objects [7].

In this paper, we study the problem of continuous k-similarity trajectories search (short for CKST) over data stream, i.e., k-similarity trajectories search under streaming data environment. Let T be the trajectory of a moving object o. It contains a set of n GPS points $\{p_1, p_2, \cdots, p_n\}$ generated by o during the last n *time units*. We use time-based window [10] to model these points. Whenever the window slides(one-time unit is passed), the first generated point p_1 is removed from T, one newly generated point is inserted into T. Let T_q be

© The Author(s), under exclusive license to Springer Nature Switzerland AG 2023
X. Wang et al. (Eds.): DASFAA 2023, LNCS 13943, pp. 273–282, 2023.
https://doi.org/10.1007/978-3-031-30637-2_18

a query trajectory, and $\mathcal{T}$ be the trajectory set. T_q monitors trajectories in $\mathcal{T}$, returns k trajectories with the lowest scores to the system whenever one time unit is passed. Here, the score of a trajectory $T_i \in \mathcal{T}$, i.e., denoted as $\mathsf{D}(T_i)$, equals the distance between T_i and the query trajectory. For simplicity, in this paper, the distance between two trajectories is measured by the sum of Euclidean Distance at corresponding GPS points.

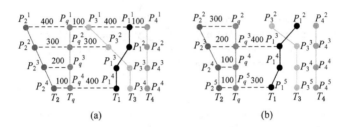

(a)   (b)

Fig. 1. Continuous k-Similarity Trajectory Search($k = 2$)

Take an example in Fig. 1(a). There are 4 trajectories $\{T_1, T_2, T_3, T_4\}$ contained in $\mathcal{T}$, each trajectory T_i contains 4 GPS points generated by the moving object o_i during the last 4 time units, i.e., T_2 contains 4 GPS points generated by the moving object o_2, which are $\{p_2^1, p_2^2, p_2^3, p_2^4\}$. T_q is the query trajectory. The distance between T_2 and T_q equals $1000(=400+300+200+100)$. The distances among T_q and these 4 trajectories are $\{1800, 1000, 1400, 2400\}$ respectively. As $k = 2$, query results are $\{T_2, T_3\}$. As shown in Fig. 1(b), after one time unit is passed, points in T_2 are updated to $\{p_2^2, p_2^3, p_2^4, p_2^5\}$, the distance between T_q and these 4 trajectories are updated to $\{1600, 700, 1800, 2400\}$ respectively, query results are updated to $\{T_2, T_1\}$.

In this paper, we propose a novel framework named SLBP (Score Lower-Bound-based Prediction) to support CKST. It is based on the following observation, that is, let T_i be a non-query result trajectory, T_k be the query result trajectory with the k-th lowest score, and T_q is the query trajectory. As the maximal speed of moving objects is usually bounded, we can evaluate the distance lower-bound between T_i and T_q, i.e., denoted as $\mathsf{LB}(T_i)$, based on a few points contained in them. If $\mathsf{D}(T_k) \ll \mathsf{LB}(T_i)$, $\mathsf{D}(T_k)$ must be smaller than $\mathsf{D}(T_i)$ for a long time. In this way, we can predict the earliest moment, i.e., denoted as $T_i.t$, T_i has the chance to become a query result trajectory(*prediction moment of T_i*). In other words, before $T_i.t$, T_i cannot become a query result trajectory, and we need not monitor T_i before $T_i.t$. Based on this, we can support CKST via accessing, as few as, GPS points contained in a small number of trajectories whenever the window slides. Above all, our contributions can be summarized as follows.

Firstly, We propose a tree-based structure named PDS (short for Partition-Distance Sketch) to summarize distance distribution among GPS points corresponding to each trajectory $T_i \in \mathcal{T}$ and T_q. In this way, we can more accurately evaluate the score lower-bound of trajectories.

Secondly, We propose a novel algorithm named PDSP (short for <u>P</u>D<u>S</u>-based <u>P</u>rediction) to predict the earliest moment each trajectory T_i has a chance to become a query result trajectory via adaptively forming their corresponding PDS structure. In this way, we can effectively reduce the frequency of access trajectories. In addition, we propose a PDSP-based incremental maintenance algorithm named INC-PDSP to support CKST under the data stream.

2 Preliminary

2.1 Related Works

The state of the arts efforts could be divided into two categories: *Ad-hoc k-similarity trajectories query* and *continuous k-similarity trajectories query*.

Ad-hoc k-Similarity Trajectories Query. These efforts mainly focus on designing a reasonable similarity function so as to improve the accuracy of query results. Among all efforts, Lei Chen et al. [1] introduced a new similarity measure function called EDR (<u>E</u>dit <u>D</u>istance on <u>R</u>eal sequence), which evaluates trajectory similarity via counting the number of insertions, deletions, and replacements required to transform one trajectory into the query trajectory. This problem is similar to that of minimal edit distance. Accordingly, they proposed a dynamic programming-based algorithm to support query processing. Gajanan Gawde [2] proposed a PBD-based method for comparing similarity among trajectories. They considered the polygon shape of trajectories, which is more robust for processing trajectories with noise.

Continuous k-Similarity Trajectories Query. The state of the arts efforts could be further divided into: *historical trajectory data-based* and *streaming trajectory data-based*. For the former one, Güting et al. [3] proposed a Spatio-temporal k-nearest neighbor search algorithm. They regarded each trajectory as a sequence of *units*. They build a Spatio-temporal(3D) R-tree for supporting query processing. For the latter one, Sacharidis et al. [9] studied the problem of continuous k-similarity trajectories query over data stream. The distance between the query object and other moving objects in a time window is determined by their maximum or minimum distance within all timestamps. Accordingly, this algorithm uses a few GPS points to represent the distance between two trajectories, which greatly reduces the query accuracy. Zhang et al. [11,12] proposed a distributed top-k similarity search algorithm over big trajectory streams. They utilized the multi-resolution property of Haar wavelet to compress trajectory data and determined a similarity bound for the compressed data.

2.2 Problem Definition

In this section, we first introduce the concept of the time-based sliding window. Formally, a time-based sliding window $W\langle n, \triangle_t\rangle$ has a fixed window size n and a fixed slide $\triangle_t$. It contains objects generated in the last n time units, and $\triangle_t$

represents the duration in terms of time units between two adjacent windows sliding. In this paper, we apply time-based sliding windows to model trajectory data as stated in Definition 1.

Definition 1. Trajectory under Sliding Window. *A trajectory T of an object o, i.e., expressed by the tuple $T\langle o, n, \triangle_t, s, e, P \rangle$, is modeled as a time-ordered sequence of GPS points generated in the last n time units with P being the GPS point set, s and e being the sequence Id of the earliest and last generated GPS point in P, and $\triangle_t$ being the duration in terms of time units between two adjacent generated GPS points.*

To be more specific, given a current trajectory T_i in the trajectory set $\mathcal{T}$, $T_i.P$ contains the set of n GPS points. Their sequence Ids are $\{T.s, T.s + 1, \cdots, T.e\}$, and they are expressed as $\{p_i^1, p_i^2, \cdots, p_i^n\}$ respectively. Whenever one time unit $\triangle_t$ is passed, the earliest generated point in $T.P$ is removed from $T.P$, and a new point is inserted into $T.P$. $T.s$ and $T.e$ are updated to $T.s + 1$ and $T.e + 1$. Besides, let p_1^j and p_2^j be two GPS points contained in $T_1.P$ and $T_2.P$, the distance between p_1^j and p_2^j, i.e., denoted as $\mathsf{D}(p_1^j, p_2^j)$, equals to the Euclidean distance between p_1^j and p_2^j. Accordingly, the distance between T_1 and T_2 is calculated via Eq. 2.

$$D(T_1.P, T_2.P) = \sum_{j=s}^{j=e} D(p_1^j, p_2^j) \tag{1}$$

Definition 2. CKST query. *Let $\mathcal{T} = \{T_1, T_2, ..., T_N\}$ be the set of N trajectories, $q\langle k, T_q \rangle$ be a query with T_q being a query trajectory and k being a query parameter. A CKST (short for $\underline{C}$ontinuous $\underline{k}-\underline{S}$imilarity $\underline{T}$rajectory Search) monitors these N trajectories, returns k trajectories with the smallest scores to the system whenever one time unit is passed. Here, the score of each trajectory T_i, i.e., denoted as $D(T_i)$, equals the distance between T_i and T_q.*

3   The Prediction-Based Framework

In this section, we propose a novel framework named SLBP(short for $\underline{S}$core $\underline{L}$ower $\underline{B}$ound-based $\underline{P}$rediction) for supporting CKST over data stream. In the following, we first explain the structure of PDS.

3.1   The Partition-Based Distance Sketch

Our solution starts from the following observation, that is, given the query trajectory T_q and a trajectory $T_i \in \mathcal{T}$, if $\mathsf{D}(p_i^n, p_q^n)$ is large, $\mathsf{D}(T_i)$ still may be large. As stated in Lemma 1, we can evaluate scores lower-bound and upper-bound of T_i, i.e., denoted as $\mathsf{LB}(T_i)$ and $\mathsf{UB}(T_i)$, via the last generated GPS points p_i^n and p_q^n corresponding to T_i and T_q respectively. Here, U_{max} refers to objects' maximal traveling speed per time unit.

Lemma 1. *Given T_i and T_q, if $D(p_i^n, p_q^n) = d$, UB(T_i) is bounded by $nd + n(n-1)U_{\max}$. In addition, LB(T_i) is bounded by $nd - n(n-1)U_{\max}$ when $d \geq 2(n-1)U_{max}$, and $\frac{d}{2U_{max}}(\frac{d}{2} + U_{max})$ when $d < 2(n-1)U_{max}$.*

We use the following example in Fig. 2(a)–(b) to explain it. p_2^8 is the last GPS points of T_2 generated by the object o_2 with $D(p_2^8, p_q^8)$ being 3.2 km($=32 \cdot 100 > 2 \cdot (8-1) \cdot 100$). Accordingly, $D(p_2^7, p_q^7)$ is no less than 3200-100·2($=3000$), but no large than 3200+100·2($=3400$). Similarly, $D(p_2^6, p_q^6)$ range from 2800m to 3600m, etc. Accordingly, LB(T_2) and UB(T_2) are 20 km($=3200+3000+...=8 \cdot 3200 - 8(8-1) \cdot 100$) and 31.2 km($=3200+3400+...=8 \cdot 3200 + 8(8-1) \cdot 100$).

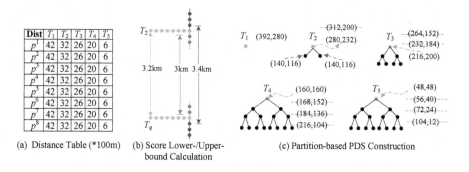

(a) Distance Table (*100m) (b) Score Lower-/Upper-   (c) Partition-based PDS Construction
bound Calculation

Fig. 2. The PDS Construction and Maintenance($k = 2$)

If $D(p_i^n, p_q^n)$ is large, $D(T_k)$ must be smaller than $D(T_i)$ for a long time. Under this case, it is unnecessary to timely monitor $D(T_i)$ whenever the window slides, and Lemma 1 helps us save a lot of running costs in monitoring scores of all trajectories. Otherwise($D(p_i^n, p_q^n)$ is not large), LB(T_i) and UB(T_i) cannot be used for evaluation as they are loose in many cases.

Based on the above observation, we propose a partition-based method to tighten the score lower/upper-bound of trajectories. Formally, let $T_i \in \mathcal{T}$ be equally partitioned into a group of sub-trajectories $\{T_i(1, m_i), T_i(2, m_i), ..., T_i(n_i, m_i)\}$. $T_i(j, m_i)$ refers to the j-th sub-trajectory of T_i with scale being m_i. It contains a group of GPS points that are the $\{(j-1) \cdot m_i + 1, (j-1) \cdot m_i + 2, ...j \cdot m_i\}$-th generated GPS points in T_i, and n_i refers to the number of sub-trajectories. Based on the partition result, LB(T_i) and UB(T_i) is updated to $\sum_{j=1}^{j=n_i} LB(T_i(j, m_i))$ and $\sum_{j=1}^{j=n_i} UB(T_i(j, m_i))$ respectively. LB($T_i(j, m_i)$) and UB($T_i(j, m_i)$) could be calculated based on Lemma 1.

We now formally discuss the structure of PDS (Partition-based Distance Sketch). It is a tree-based structure, which summarizes the distance distribution among GPS points in each trajectory T_i and T_q. Let ST_i be the PDS structure corresponding to T_i. The root node R_i of ST_i corresponds to the whole trajectory T_i, its two child node e and e' corresponds to sub-trajectories $T_i(1, \frac{n}{2})$ and $T_i(2, \frac{n}{2})$. Similarly, each child node of e(also e') corresponds to a sub-trajectory with the length being $\frac{n}{4}$, etc. Each node $e'' \in ST_i$ is associated with the tuple $\langle u, l \rangle$. They record the score upper-bound and score lower-bound of the corresponding sub-trajectory between T_i and T_q. If e'' is a leaf node, $e''.u$ and $e''.l$ is computed based on Lemma 1.

3.2 Sketch-Based Prediction Algorithm

The Self-adaptively Partition. We first scan each trajectory $T_i \in \mathcal{T}$, compute $\mathsf{LB}(T_i)$ and $\mathsf{UB}(T_i)$ via Lemma 1. After scanning, if existing k trajectories have score upper-bound lower than $\mathsf{LB}(T_i)$, it is removed from $\mathcal{T}$ to the set T'. Next, we partition each $T_i' \in \mathcal{T}$ into 2 sub-trajectories with equal scale, update $\mathsf{LB}(T_i')$ and $\mathsf{UB}(T_i')$ accordingly. Again, after scanning, if existing k trajectories have score upper-bound lower than $\mathsf{LB}(T_i')$, it is removed from $\mathcal{T}$ to the set T'. From then on, we repeat the above operations until the number of trajectories in $\mathcal{T}$ reduces to k. At that moment, we can use these k trajectories as query result trajectories in the current window. Note, during the partition, we can form the corresponding PDS structure for each trajectory.

Back to the example in Fig. 2(c). We first scan these 5 trajectories and compute their score lower-/upper-bound based on their last generated GPS points. We find that $\mathsf{LB}(T_1) = 28$km, which is larger than both $\mathsf{UB}(T_4)$ and $\mathsf{UB}(T_5)$. We then remove it from $\mathcal{T}$ to T'. At that moment, the corresponding PDS structure of T_1 contains a root node. We then partition each reminding trajectory into 2 sub-trajectories and update its score lower-/upper-bound. After re-computing, we find that $\mathsf{LB}(T_2)$ is updated to 23.2 km. In addition, both $\mathsf{UB}(T_4)$ and $\mathsf{UB}(T_5)$ are updated to 18.4 km and 7.2 km. We then remove it into T'. In the third round, we further partition T_3, T_4 and T_5 into 4 sub-trajectories respectively. At that moment, we find that $\mathsf{LB}(T_3)$ is updated to 20 km, and both $\mathsf{UB}(T_4)$ and $\mathsf{UB}(T_5)$ are updated to 16.8 km and 5.6 km. We then remove it into T'. Accordingly, T_4 and T_5 could be regarded as query results.

The PDS-Based Prediction. It is to predict the earliest moment, i.e., denoted as $T_i.t$, each trajectory T_i has a chance to become the query result trajectory. The reason we calculate it is: (i) we can avoid monitoring T_i until $T_i.t$; (ii) the corresponding GPS points generated within $[T_i.s, T_i.t - 1]$ could be deleted. In this paper, $T_i.t$ is calculated based on Theorem 1. Here, $\mathsf{LB}(T_i, \delta_i)$ and $\mathsf{UB}(T_i, \delta_i)$ refer to the score lower-bound and upper-bound of T_i after δ_i time units.

Theorem 1. *Given T_i, T_k and T_q, $D(T_i)$ is no smaller than $D(T_k)$ after δ_i time units with δ_i being the maximal value satisfying $LB(T_i, \delta_i) \leq UB(T_k, \delta_i)$.*

$$LB(T_i, \delta_i) = \begin{cases} \sum_{j=\frac{\delta_i}{m_i}+1}^{j=\frac{n}{m_i}} LB(T_i(j, m_i)) + \delta_i D(p_i^e, p_q^e) - \delta_i(1 + \delta_i)U_{max} \\ \delta_i = \frac{D(p_i^e, p_q^e)}{2U_{max}}, \text{ if } \delta_i > \frac{D(p_i^e, p_q^e)}{2U_{max}} \end{cases} \quad (2)$$

$$UB(T_i, \delta_i) = \sum_{j=\frac{\delta_i}{m_i}+1}^{j=\frac{n}{m_i}} UB(T_i(j, m_i)) + \delta_i D(p_i^e, p_q^e) + \delta_i(1 + \delta_i)U_{max} \quad (3)$$

In the implementation, $T_i.t$ could be computed via *Binary Search*. Specially, we first check whether $\mathsf{LB}(T_i, n) \geq \mathsf{UB}(T_k, n)$ is held. If the answer is yes, the algorithm is terminated, and $T_i.t$ is set to $T_i.e + n$. Otherwise, we check whether $\mathsf{LB}(T_i, \frac{n}{2}) \geq \mathsf{UB}(T_k, \frac{n}{2})$ is held. If the answer is yes, we further check whether $\mathsf{LB}(T_i, \frac{3n}{4}) \geq \mathsf{UB}(T_k, \frac{3n}{4})$. Otherwise, we check whether $\mathsf{LB}(T_i, \frac{n}{4}) \geq \mathsf{UB}(T_k, \frac{n}{4})$ is held. From then on, we repeat the above operations until meeting a leaf node. In particular, $T_i.t$ still cannot be computed after searching, we should go on further partitioning T_i into a set of sub-trajectories with length $\frac{m_i}{2}$, update ST_i based on the partition result(form new child nodes for the original leaf nodes), and re-compute $T_i.t$ accordingly. If $T_i.t$ could be computed based on the new partition, the algorithm is terminated. Otherwise, we should repeat the above operations. For the limitation of space, we skip the details. After PDS structure of all trajectories are accessed, we use an inverted-list to maintain these trajectories in descendent order based on their predicted moment.

Algorithm 1: The INC-PDSP Algorithm

Input: The Query Trajectory T_q, The Inverted-list I_c, I_q,
Output: Query result set $\mathcal{Q}$

1 **for** *i from 1 to k* **do**
2    $D(I_q[i]) \leftarrow \mathsf{updateScore}(I_q[i])$;

3 **while do** $I_q[i].t = T_{cur}$
4   Trajectory $T \leftarrow I_q[i]$, $t \leftarrow I_q[i].t$, $t' \leftarrow I_q[i].t'$;
5   The partition $P' \leftarrow \mathsf{partition}(T(t', t), I_q[i].m)$;
6   PDS $ST' \leftarrow \mathsf{formSubTree}(P')$;
7    $ST \leftarrow \mathsf{combine}(ST, ST')$;
8    $T.t \leftarrow \mathsf{prediction}(ST, I_q[k])$;
9   **if** $T.t = T.cur$ **then**
10      $I_q \leftarrow I_q \cup T\text{-}I_q[k]$;
11   **else**
12      $\mathsf{re\text{-}insertion}(T)$;

13 **return** $\mathcal{Q}$;

The Incremental Maintenance Algorithm. In the following, we propose the algorithm INC-PDSP, which explains how PDSP is applied for supporting CKST under data stream. Here, I_c and I_q are two inverted-list that maintains non query result trajectories and query result trajectories respectively. Whenever the window slides, we first update scores of k query result trajectories(lines 1–2). Let the current time unit be T_{cur}. Next, we access each trajectory with the predicted moment being T_{cur} and check whether it can become a query result trajectory. If the answer is yes, we insert it into the query result set I_q. Otherwise, we update its predicted moment and re-insert it into the inverted-list I_c.

Specially, let T_i be a trajectory with the predicted moment being $T_i.t$, $T_i.t'$ be the last time unit T_i is accessed. We first partition GPS points generated within $[T_i'.t, T_i.t]$ into a group of sub-trajectories with length m_i, form a PDS structure ST' based on these sub-trajectories, and then combine into its original PDS structure ST(line 5–7). Then, we re-calculate the predicted moment of T_i in the manner discussed in the initialization algorithm(lines 5–8). After calculation, if it has a chance to become a query result, it is inserted into I_q. Otherwise, we re-insert it into I_c based on its new predicted moment(lines 10–12).

4   Performance Evaluation

4.1   Experiment Settings

Datasets. In total, four datasets are used in our experiments, including three real datasets, namely BEIJING, PORTO, and NYC, and a synthetic dataset namely NORMAL. BEIJING comes from the Microsoft T-Drive project, which contains GPS trajectories of 10,357 taxis from Feb.2 to Feb.8, 2008. PORTO includes 1,710,671 trajectories, which describes the trajectory of 442 taxis from Jan.7, 2013, to Jun. 30, 2014, in the city of Porto. NYC comes from New York City Taxi and Limousine Commission. It contains 2.36 GB of trip records. Normal is a synthesized trajectory dataset by simulating the trajectories of moving objects on urban roads.

Table 1. Parameter Settings

Parameter	value
N	200KB, 400KB, **600KB** ,800K, 1M
n	200, 400, 600,**800**, 1K
k	20, **40**, 60, 80, 100
U_{max}	20km/h, 30km/h, **40km/h**, 50km/h, 60km/h

Parameters. In our study, we evaluate the performance of different algorithms under four parameters, including N, n, k, and U_{max}. Here, N refers to the number of trajectories in the trajectory set $\mathcal{T}$; n refers to the number of GPS points contained in each trajectory; k is the input of **CKST**; U_{max} refers to objects' maximal traveling speed per time unit. The parameter settings are listed in Table 1 with the default values bolded.

Competitors. In addition to the algorithms included in the SLBP framework: INC-PDSP, we also implement the algorithm developed by Sacharidis named HRZ and a baseline algorithm named BASE.

4.2   Performance Comparison

We first report the running time of all algorithms under different k values in Fig. 3(a)–(d). Among all k values evaluated, we can observe that INC-PDSP outperforms BASE, and HRZ consistently, for all four datasets. For example, the running time of INC-PDSP is 2% of BASE and 16.7% of HRZ. The significant improvement lies in the fact that INC-PDSP considers spatio-temporal correlation among GPS points in each trajectory. The corresponding virtual trajectory is closer to the real trajectory. Thus, it could provide trajectories with tight score bounds by accessing a small number of GPS points in them.

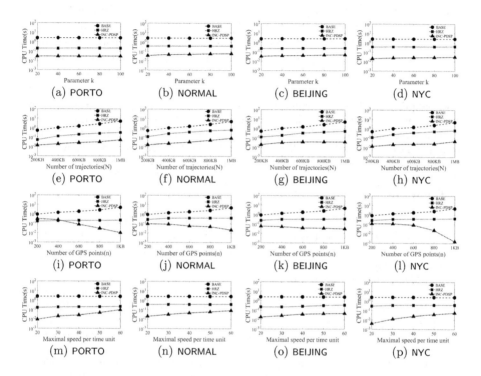

Fig. 3. Running time comparison of different algorithms under different data sets

We also report the running time of different algorithms under different N values in Fig. 3(e)–(h). Similar to the above observations, INC-PDSP consistently performs best. We also observe that, as N increases, the running time of BASE and HRZ goes up rapidly. In addition, INC-PDSP shows a much more stable performance under various parameter settings, as compared with its competitors. It demonstrates that INC-PDSP is resilient to multiple parameters.

The running time of different algorithms under different n is reported in Fig. 3(i)–(l). We find that INC-PDSP performs the best, BASE's running time gradually increasing. HRZ's running time is also increased with n. The reason is the larger than n, the more GPS points it has to access whenever trajectories' prediction moment should be updated. By contrast, when n increases, INC-PDSP's running time drops. This is because the larger the n, the larger the distance from starting points of objects' trajectories to their end points. The running time of different algorithms under different U_{max} is reported in

Fig 3(m)–(p). We find that INC-PDSP performed the best of all again. In addition, BASE is not sensitive to U_{max}.

5   Conclusion

In this paper, we propose a novel framework named SLBP to support continuous k-similarity trajectories search over data stream. It is able to return the query result trajectories via accessing as few as GPS points of a small number of trajectories in the whole trajectory set. We have conducted extensive experiments to evaluate the performance of our proposed algorithms on several datasets. The results demonstrate the superior performance of our proposed algorithms.

Acknowledgements. This paper is partly supported by the National Key Research and Development Program of China (2020YFB1707901), the National Natural Science Foundation of Liao Ning (2022-MS-303, 2022-MS-302,and 2022-BS-218), the National Natural Science Foundation of China (62102271, 62072088, Nos. U22A2025, 62072088, 62232007, 61991404), and Ten Thousand Talent Program (No. ZX20200035).

References

1. Chen, L., Özsu, M.T., Oria, V.: Robust and fast similarity search for moving object trajectories. In: Proceedings of the 2005 ACM SIGMOD, pp. 491–502 (2005)
2. Gawde, G., Pawar, J.: Similarity search of time series trajectories based on shape. In: Proceedings of the ACM India Joint International Conference on Data Science and Management of Data, pp. 340–343 (2018)
3. Güting, R.H., Behr, T., Xu, J.: Efficient k-nearest neighbor search on moving object trajectories. VLDB J. **19**(5), 687–714 (2010). https://doi.org/10.1007/s00778-010-0185-7
4. Hsieh, F.S.: Car pooling based on trajectories of drivers and requirements of passengers. In: 2017 IEEE 31st AINA, pp. 972–978. IEEE (2017)
5. Jiang, J., Xu, C., Xu, J., Xu, M.: Route planning for locations based on trajectory segments. In: Proceedings of the 2nd ACM SIGSPATIAL, pp. 1–8 (2016)
6. Li, T., Chen, L., Jensen, C.S., Pedersen, T.B.: TRACE: real-time compression of streaming trajectories in road networks. Proc. VLDB Endow. **14**(7), 1175–1187 (2021)
7. Li, T., Chen, L., Jensen, C.S., Pedersen, T.B., Gao, Y., Hu, J.: Evolutionary clustering of moving objects. In: 2022 IEEE 38th ICDE, pp. 2399–2411. IEEE (2022)
8. Li, T., Huang, R., Chen, L., Jensen, C.S.: Compression of uncertain trajectories in road networks. Proc. VLDB Endow. **13**(7), 1050–1063 (2020)
9. Sacharidis, D., Skoutas, D., Skoumas, G.: Continuous monitoring of nearest trajectories. In: Proceedings of the 22nd ACM SIGSPATIAL, pp. 361–370 (2014)
10. Tang, L.A., et al.: On discovery of traveling companions from streaming trajectories. In: 2012 IEEE 28th ICDE, pp. 186–197. IEEE (2012)
11. Zhang, Z., Qi, X., Wang, Y., Jin, C., Mao, J., Zhou, A.: Distributed top-k similarity query on big trajectory streams. Front. Comp. Sci. **13**, 647–664 (2019). https://doi.org/10.1007/s11704-018-7234-6
12. Zhang, Z., Wang, Y., Mao, J., Qiao, S., Jin, C., Zhou, A.: DT-KST: distributed top-k similarity query on big trajectory streams. In: Candan, S., Chen, L., Pedersen, T.B., Chang, L., Hua, W. (eds.) DASFAA 2017. LNCS, vol. 10177, pp. 199–214. Springer, Cham (2017). https://doi.org/10.1007/978-3-319-55753-3_13

Towards Effective Trajectory Similarity Measure in Linear Time

Yuanjun Liu[1], An Liu[1(✉)], Guanfeng Liu[2], Zhixu Li[3], and Lei Zhao[1]

[1] School of Computer Science and Technology, Soochow University, Suzhou, China
20214227045@stu.suda.edu.cn, {anliu,zhaol}@suda.edu.cn
[2] School of Computing, Macquarie University, Macquarie Park, Australia
guanfeng.liu@mq.edu.au
[3] School of Computer Science and Technology, Fudan University, Shanghai, China
zhixuli@fudan.edu.cn

Abstract. With the utilization of GPS devices and the development of location-based services, a massive amount of trajectory data has been collected and mined for many applications. Trajectory similarity computing, which identifies the similarity of given trajectories, is the fundamental functionality of trajectory data mining. The challenge in trajectory similarity computing comes from the noise in trajectories. Moreover, processing such a myriad of data also demands efficiency. However, existing trajectory similarity measures can hardly keep both accuracy and efficiency. In this paper, we propose a novel trajectory similarity measure termed ITS, which is robust to noise and can be evaluated in linear time. ITS converts trajectories into fixed-length vectors and compares them based on their respective vectors' distance. Furthermore, ITS utilizes interpolation to get fixed-length vectors in linear time. The robustness of ITS owes to the interpolation, which makes trajectories aligned and points in trajectories evenly distributed. Experiments with 12 baselines on four real-world datasets show that ITS has the best overall performance on five representative downstream tasks in trajectory computing.

Keywords: Trajectory similarity measure · Trajectory distance measure · Trajectory data mining · GPS data mining

1 Introduction

The last decade has witnessed unprecedented growth in the availability of location-based services. Large amounts of mobile data described as sequences of locations, known as trajectories, are generated and mined for many applications. Typical examples include collecting GPS location histories for management purposes, such as tracking typhoons for better precautions, providing ordinary users better routes, planning for trucks and carpooling, and tracking migration of animals in biological sciences [15,23]. Trajectory offers unprecedented information to help understand the behaviors of moving objects, resulting in a growing

X. Wang et al. (Eds.): DASFAA 2023, LNCS 13943, pp. 283–299, 2023.
https://doi.org/10.1007/978-3-031-30637-2_19

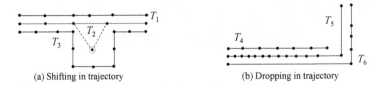

(a) Shifting in trajectory   (b) Dropping in trajectory

Fig. 1. Noise in trajectories. (a): The real trajectory of T_1 and T_2 are more similar, since they are both straight. Unfortunately, a shifted point brings a fake corner to T_2, so T_2 may be more similar to T_3 than T_1. (b): The real trajectory of T_5 and T_6 are more similar. Nevertheless, most points of T_5 are closer to points of T_4, then T_5 may be more similar to T_4 than T_6.

interest in trajectory data mining. An essential problem in mining is designing measures for identifying similar trajectories. Many data analysis tasks can use trajectory similarity measures, including trajectory clustering, classification, and k-nearest neighbor (k-nn) search, which have a broad range of real applications. For example, in the urban construction and planning bureau, it is helpful to locate popular routes by comparing similarities between vehicle trajectories and deciding on new transportation routes. During COVID-19, governments can track epidemics and warn citizens by searching other trajectories similar to those of attacked person [14]. Navigation systems search similar trajectories close to origin and destination and recommend better routes to users [3].

Several challenges lie in trajectory similarity computing. GPS devices collect trajectories while objects are moving, but noises may be caused during collection. Two kinds of noises, shifting and dropping, exist in trajectories and bring error. Firstly, the positions of points may be shifted since the precision of GPS devices, as reported in [20]: "Typically, the GPS nominal accuracy is about 15 m". As shown in Fig. 1(a), there is a shifted point in T_2. Technically, the actual trajectories of T_1 and T_2 are more similar since they are both straight. Unfortunately, the shifted point brings a fake corner to T_2 so that T_2 may be more similar to T_3 than T_1.

Secondly, objects move at various speeds, but GPS devices sample at consistent intervals. A fast speed leads to a high dropping rate, and a slow speed leads to a low dropping rate. On the one hand, the speeds between objects are typically different, so the dropping rates between trajectories are uncertain. On the other hand, the speed of an object may change over time, so the dropping rate inside a trajectory changes over time, also causing unfairness when counting matched points. Figure 1(b) shows T_5 has a low dropping rate at the left half and a high dropping rate at the right half. Meanwhile, the dropping rate of T_6 is higher than T_5. The actual trajectories of T_5 and T_6 are similar. However, most points of T_5 are closer to points of T_4 than points of T_6, leading T_5 to be more similar to T_4 than T_6. Shifting and dropping make it challenging to keep accuracy while measuring trajectory similarity.

The third challenge is efficiency. With the ubiquity of positioning techniques, various trajectory data are generated constantly. For example, there are about

Table 1. Summary of trajectory similarity measures. n is the number of points in a trajectory. A good measure should have a short running time and high robustness.

Measures	Time	Robustness	Parameter
DTW [1]	$O(n^2)$	High	Free
PDTW [9]	$O(n^2)$	High	Need
FastDTW [21]	$O(n)$	Low	Need
BDS [24]	$O(n^2)$	Medium	Free
SSPD [2]	$O(n^2)$	Medium	Free
FD [15]	$O(n^2)$	High	Free
LIP [18]	$O(nlogn)$	Medium	Free
OWD [13]	$O(n^2)$	High	Need
STS [10]	$O(n^2)$	Medium	Need
LCSS [26]	$O(n^2)$	Low	Need
EDR [5]	$O(n^2)$	High	Need
ERP [4]	$O(n^2)$	Medium	Free
MD [7]	$O(n^2)$	Medium	Free
CATS [6]	$O(n^2)$	Medium	Need
STED [17]	$O(n^2)$	Medium	Free
STLC [22]	$O(n^2)$	Low	Need
STLCSS [26]	$O(n^2)$	Low	Need
RID [25]	$O(n)$	Low	Free
EDwP [19]	$O(n^2)$	Medium	Free
ITS	$O(n)$	High	Free

1 TB GPS data generated each day in JD [11]. Suppose there is a set of trajectories and each trajectory has $n = 1000$ points. While an $O(n)$ method takes 1 s to process such a dataset, an $O(n^2)$ method takes about 2 min. Therefore, a fast trajectory similarity measure is a must to mine such a massive amount of trajectory data.

Table 1 summarizes existing trajectory similarity measures. An important observation from the table is that all existing measures can hardly be accurate and efficient. Matching-based measures [1,2,21,24] are generally robust to variable dropping rates but sensitive to shifting. Sequence-based measures [4,5,7,26] are sensitive to dropping rates as the number of points in trajectories changes with the dropping rate. Shape-based measures [13,15,18] are robust to the dropping rate, but most need to improve at handling shifting. The running time of time-based measures [6,17,22] is $O(n^2)$, and they also generally need parameters that are set manually. EDwP [19] employs projection to align trajectories but takes $O(n^2)$ time to find the best projection positions. RID [25] is $O(n)$ time, but it aims to identify trajectories after the rotation operation. FastDTW [21] is also $O(n)$ time but is not robust. Apart from the above methods, machine learning has recently been adopted to calculate the similarity of trajectories. These machine learning-based methods [12,28] can calculate similarity in $O(n)$ time but need training and parameter-tuning. As they are not measures, we do not include them in Table 1.

To overcome the above limitations of existing measures, we propose a novel trajectory distance measure termed ITS, which is robust to shifting and variable dropping rates and can be evaluated in linear time. The basic idea behind it is to generate a fixed-length vector for each trajectory and then compare two trajectories based on the distance of their respective vectors. Furthermore, unlike machine learning-based methods [12,28] in which trajectories are also represented by vectors, ITS employs a fast interpolation strategy to convert trajectories into vectors, thus avoiding expensive training and parameter-tuning. During interpolation, two trajectories are aligned, and points in a trajectory are evenly distributed, so ITS is robust to variable dropping rates. Shifting can only disturb interpolation points around the shifted area, so ITS is also robust to shifting.

In summary, the paper makes the following contributions:

- We propose a novel trajectory measure termed ITS, which has linear time complexity and space complexity.
- ITS is robust to dropping and shifting in trajectory data and satisfies the triangle inequality.
- We compare ITS with 12 baselines on five important downstream tasks in trajectory computing using four real-world datasets. Experimental results show that ITS has the best overall performance.

The rest of the paper is organized as follows. Section 2 discusses the related work. The definition and problem statement are given in Sect. 3. Section 4 presents the details of our method. The experimental results are presented in Sect. 5. Finally, we summarize our work in Sect. 6.

2   Related Work

Existing trajectory similarity measures can be classified into several groups according to the technologies they use. We summarize them in Table 1 and discuss their details below.

Matching is one of the most popular technology to measure trajectory similarity. Dynamic Time Warping (DTW) [1] matches points to the nearest point and measures the distances of matched point-pairs. DTW warps trajectories in a non-linear way while allowing two trajectories have different dropping rates. However, DTW matches every point and ignores point distribution, so it cannot handle changing dropping rates inside the trajectory. The running time of DTW is $O(n^2)$ as it compares every point pair. Piecewise DTW (PDTW) [9] takes every several points into a piece, replaces them with their average, and then applies DTW to measure the similarity of averaged trajectories. Averaging makes PDTW robust to shifting, but the number of points to be averaged should be determined artificially. PDTW reduces the number of points to speed up DTW but also consumes $O(n^2)$ time. FastDTW [21] restricts the matching window size to get $O(n)$ computation complexity but at the cost of accuracy. The window size of FastDTW needs to be set manually. Symmetrized Segment-Path Distance (SSPD) [2] matches points to the nearest segment, and Bi-Directional Similarity (BDS) [24] matches points to every segment. They measure the distances of matched point-segment pairs. They are robust to variable dropping rates but not robust to shifting.

In the case of sequence based measures, Longest Common Subsequence (LCSS) [26], Edit Distance on Real sequence (EDR) [5] and Edit distance with Real Penalty (ERP) [4] treat points to be the same as long as the distance between two points is less than a threshold. The threshold of ERP is the distance between a point to the original point. Others should be set manually. They are robust to the shifting that is smaller than the threshold. Merge Distance (MD) [7] finds the shortest trajectory merged by two trajectories. It is robust to variable dropping rates since it considers the length of segments. They all compare the distance of every point pair, which yields $O(n^2)$ time.

Regarding shape-based measures, Fréchet Distance (FD) [15] between two curves is defined as the minimum length of a leash required to connect two separate paths. It regards trajectories as curves, thus being robust to dropping. It compares all point pairs, which yields $O(n^2)$ time. Locality In-between Polylines distance (LIP) [18] counts the area where two trajectories are enclosed. LIP can be transformed into the red-blue intersection problem so that it can be solved in $O(n \log n)$ time. One Way Distance (OWD) [13] grids trajectories and calculates the distance of grids. Its running time is $O(n^2)$, and the grid size needs to be set manually. OWD is robust to shifting that does not exceed the grid size. LIP and OWD fill the gap between adjacent points, which normalizes the distribution of points, thus being robust to variable dropping rates.

Time is also taken into consideration for trajectory similarity computation. Spatiotemporal Euclidean Distance (STED) [17] compares positions for each time instant, but it requires a common temporal domain for trajectories, which is a hard requirement. It excludes subsequence matching and shifting that tries to align trajectories. It has $O(n^2)$ running time. The time instant interpolates the unsampled area; thus, STED is robust to variable dropping rates. Spatiotemporal Linear Combine distance (STLC) [22] measures similarities from both spatio and temporal aspects and needs parameters to control the relative importance of the spatial and temporal similarities. Clue-Aware Trajectory Similarity (CATS) [6] and SpatioTemporal LCSS (STLCSS) [26] match points if the time interval is less than a threshold. Just like LCSS, they are robust to shifting that is smaller than the threshold. They compare every point pair, yielding $O(n^2)$ time.

Spatial-Temporal Similarity (STS) [10] estimates the probabilities of points and measures the similarity of probabilities. Its running time is $O(n^2)$. Probability brings robustness of shifting. Rotation Invariant Distance (RID) [25] records polar radius at each polar angle and utilizes FastDTW to measure the similarity of radius. Its running time is $O(n)$, but the problem it solves is the similarity of trajectories after rotation. Edit Distance with Projections (EDwP) [19] aligns trajectories by projecting points onto another trajectory, thus being robust to dropping. However, finding the best projection positions takes $O(n^2)$ time.

ML-based methods [12,28] generate representation vectors for trajectories based on neural networks. They are robust to variable dropping rates because they drop points while training. However, training and parameter-tuning are needed in ML-based methods, making them data-dependent and resource-consuming. Besides, it is hard to use them in applications where indexing and pruning are essential. What is more, they are not measures and are generally unexplainable.

3   Problem Formulation

Definition 1 (Original Trajectory). *An original trajectory is a sequence of points $T = [p_1, p_2, \cdots, p_n]$ where $p_i = (x_i, y_i)$ is a spatial point in 2D space, x_i and y_i are latitude and longitude, respectively. The length of T is defined as the number of points in T, that is, n.*

Definition 2 (Sample Point). *Given a point $p = (x, y)$, its sample point is given by $\phi(p) = (x + \Delta(x), y + \Delta(y))$ where ϕ is a sample function, $\Delta(x)$ and $\Delta(y)$ are the noise introduced during sampling.*

In the above definition, we consider shifting, one noise resulting from physical devices that generate trajectory data. Next, we will consider another kind of noise.

Definition 3 (Sample trajectory). *Given an original trajectory T, its sample trajectory is also a sequence of points $T' = \varphi(T) = [\phi(p_{s_1}), \phi(p_{s_2}), \cdots, \phi(p_{s_m})]$ where $1 \leq s_1 < s_2 < \cdots < s_m \leq n, \varphi$ samples the trajectory.*

The above definition of sample trajectory indicates that the length of a sample trajectory T' is less than or equal to the original trajectory T. In addition, this definition simulates another noise called dropping in practice; that is, not all points in a trajectory can be sampled due to the continuity of movement and the discretization of sampling.

Problem Statement. *The problem to be addressed is to accurately evaluate the similarity of two trajectories in the presence of noise. Specifically, given two sample trajectories T_i' and T_j', we need to design a function $f(\cdot)$ to compute their similarity, with a goal of the similarity of two sample trajectories $f(T_i', T_j')$ should be as close as possible to the similarity of their respective original trajectories $f(T_i, T_j)$, that is, $f(T_i', T_j') \approx f(T_i, T_j)$.*

4   Interpolation Based Trajectory Similarity

As mentioned earlier, lots of works try to transform trajectories into fixed-length vectors so that the similarity of trajectories can be evaluated efficiently based on the similarity of their respective vectors. Following this idea, we propose in this paper an interpolation-based trajectory similarity measure ITS, which adopts interpolation to generate a fixed-length vector representation for trajectories. Interpolation is a lightweight operation, so ITS is efficient. It can be done without the time-consuming training and parameter-tuning involved in the ML-based methods or the complex computation (e.g., typically quadratic time) to find the best matching between points needed in traditional methods such as DTW and EDR.

Generally speaking, ITS interpolates a predefined number of points, say h points, to a trajectory despite its length. Therefore, given two trajectories T_1 and T_2 that may have different lengths, ITS generates two interpolated trajectories $\hat{T}_1 = [\hat{p}_1, \hat{p}_2, \cdots, \hat{p}_h]$ and $\hat{T}_2 = [\hat{q}_1, \hat{q}_2, \cdots, \hat{q}_h]$. The numbers of points in the interpolated trajectories are the same, so they can be matched efficiently by taking $(\hat{p}_i, \hat{q}_i)$ as a pair.

The interpolation should be designed delicately to keep the accuracy of similarity computing. Given a trajectory $T = [p_1, p_2, \cdots, p_n]$, let $\lambda(p_i) \in [0, 1]$ be the normalized arc length from p_1 to p_i, that is, $\lambda(p_i) = \sum_{j=2}^{i} d(p_{j-1}, p_j)/L(T)$, where $L(T) = \sum_{j=2}^{n} d(p_{j-1}, p_j)$ is the arc length of trajectory T, d is the Euclidean distance. Let $P(l)$ be an interpolated point with the normalized arc length l, that is, $\lambda(P(l)) = l$. ITS interpolates points evenly over the arc length of the whole trajectory. Figure 2 illustrates the interpolation procedure. First, the trajectory T is considered to be a set of consecutive segments, as shown in Fig. 2(a). Then ITS straightens these segments so that the trajectory becomes one large segment shown in Fig. 2(b). After that, this large segment is divided into $h - 1$ parts evenly. Finally, as shown in Fig. 2(c), the h endpoints of these parts are reported as the interpolation points. Formally, given a trajectory T and an integer h, ITS generates an interpolated trajectory $\hat{T} = [\hat{p}_i \mid 1 \leq i \leq h]$, where $\lambda(\hat{p}_i) = \frac{i-1}{h-1}$.

Algorithm 1 shows the details of interpolation. Given a trajectory T and an integer h, the algorithm generates h points and returns the interpolated trajectory in the form of vector v. The algorithm walks on the trajectory with a constant step ε (Line 2), and yields $(v_j, v_{j+h}) = P(\frac{j-1}{h-1})$ at each interpolation position.

The above interpolation strategy is efficient since the trajectory needs to be traveled once. It is also effective as it tries to keep the semantics of the original trajectory. In particular, the shape of the trajectory mostly stays the same. Moreover, the points in a trajectory are evenly distributed, which is very useful when estimating the similarity of two trajectories.

Algorithm 2 shows how ITS works. To compute the similarity of two trajectories T_1 and T_2, ITS first generates two interpolated trajectories $\hat{T}_1 = [\hat{p}_j \mid 1 \leq j \leq h]$ where $\hat{p}_j = P_{\hat{T}_1}(\frac{j-1}{h-1})$ and $\hat{T}_2 = [\hat{q}_j \mid 1 \leq j \leq h]$ where $\hat{q}_j = P_{\hat{T}_2}(\frac{j-1}{h-1})$. Then it takes the square root of the average sum of squared Euclidean distances between every pair $(\hat{p}_j, \hat{q}_j)$ as the distance of T_1 and T_2:

$$\text{ITS}(T_1, T_2) = \sqrt{\frac{1}{h} \sum_{j=1}^{h} d(\hat{p}_j, \hat{q}_j)^2} \tag{1}$$

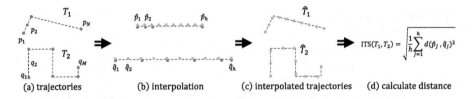

Fig. 2. Illustration of ITS. Blue points are points in the original trajectories, and red points are interpolation points generated by ITS. (Color figure online)

Finally, the similarity of trajectories T_1 and T_2 can be set to their negative distance, that is, $-\text{ITS}(T_1, T_2)$.

Algorithm 1: Interpolation

Input: trajectory T, integer h
Output: interpolated vector v

1   $v_j \leftarrow 0, 1 \leq j \leq 2h$
2   $\varepsilon \leftarrow \frac{L(T)}{h-1}$
3   $a, b \leftarrow 0, 0$
4   $j \leftarrow 0$
5   **for** $i \leftarrow 2$ **to** n **do**
6      $w \leftarrow d(p_{i-1}, p_i)$
7      $b \leftarrow b + w$
8   **while** $b - a > \varepsilon$ **do**
9          $v_j \leftarrow (x_i - x_{i-1})(1 - \frac{b-a}{w}) + x_{i-1}$
10         $v_{j+h} \leftarrow (y_i - y_{i-1})(1 - \frac{b-a}{w}) + y_{i-1}$
11         $j \leftarrow j + 1$
12         $a \leftarrow a + \varepsilon$

Algorithm 2: ITS

Input: trajectory T_1, trajectory T_2, integer h
Output: distance its

1   $v_1, v_2 \leftarrow \text{Interpolation}(T_1, h), \text{Interpolation}(T_2, h)$
2   $its \leftarrow d(v_1, v_2)/\sqrt{h}$

Theoretical Analysis. ITS can be evaluated in $O(n)$ time and $O(h)$ space. During interpolation (Algorithm 1), calculating $L(T)$ costs n arithmetical operations. The outer loop traverses n points, and the inner loop steps h interpolation positions. Thus, interpolation needs $2n + h$ operations and $2h$ space. During ITS (Algorithm 2), interpolating trajectories costs $4n + 2h$ operations, and calculating the vectors' distance costs h operations. Therefore, ITS needs $4n + 3h$ operations and $4h$ space. Moreover, ITS has some desirable properties, given by the following theorem.

Theorem 1. *Given three trajectories T_1, T_2, T_3, ITS satisfies:*

1. *Non-negativity:* ITS$(T_1, T_2) \geq 0$.
2. *Semi-identity:* $T_1 = T_2 \Rightarrow$ ITS$(T_1, T_2) = 0$.
3. *Symmetry:* ITS$(T_1, T_2) =$ ITS(T_2, T_1).
4. *Triangle inequality:* ITS$(T_1, T_2) +$ ITS$(T_2, T_3) \geq$ ITS(T_1, T_3).

Proof. Given two trajectories T_1, T_2, $\hat{T}_1 = [\hat{p}_j \,|\, 1 \leq j \leq h]$ where $\hat{p}_j = P_{\hat{T}_1}(\frac{j-1}{h-1})$ and $\hat{T}_2 = [\hat{q}_j \,|\, 1 \leq j \leq h]$ where $\hat{q}_j = P_{\hat{T}_2}(\frac{j-1}{h-1})$ are interpolated trajectories,

$$d(\cdot) \geq 0 \Rightarrow \sqrt{\frac{1}{h}\sum_{j=1}^{h} d(\hat{p}_j, \hat{q}_j)^2} \geq 0 \Rightarrow \text{ITS}(T_1, T_2) \geq 0,$$

$$T_1 = T_2 \Rightarrow d(P_{\hat{T}_1}(l), P_{\hat{T}_1}(l)) = 0 \Rightarrow \sqrt{\frac{1}{h}\sum_{j=1}^{h} d(\hat{p}_j, \hat{q}_j)^2} = 0 \Rightarrow \text{ITS}(T_1, T_2) = 0,$$

$$\text{ITS}(T_1, T_2) = \sqrt{\frac{1}{h}\sum_{j=1}^{h} d(\hat{p}_j, \hat{q}_j)^2} = \sqrt{\frac{1}{h}\sum_{j=1}^{h} d(\hat{q}_j, \hat{p}_j)^2} = \text{ITS}(T_2, T_1).$$

Given three trajectories T_1, T_2, T_3,

$$\text{ITS}(T_1, T_2) + \text{ITS}(T_2, T_3) = d(v_1, v_2)/\sqrt{h} + d(v_2, v_3)/\sqrt{h} \geq d(v_1, v_3)/\sqrt{h}$$
$$= \text{ITS}(T_1, T_3)$$

where $v_1 = \text{Interpolation}(T_1, h), v_2 = \text{Interpolation}(T_2, h), v_3 = \text{Interpolation}(T_3, h)$, and v_1, v_2, v_3 are viewed as points in $2h$ dimensional space.

As ITS satisfies the triangle inequality, generic indexing structures and pruning strategies can be applied when dealing with a huge amount of trajectories.

The performance of ITS depends on the value of h. Generally, a larger h means an interpolated trajectory has more points, which improves the accuracy of ITS but increases its running time. On the contrary, the evaluation of ITS with a smaller h is more efficient. However, its accuracy will decrease as it may need to capture more semantic information about trajectories. Therefore, it is necessary to strike a balance between accuracy and efficiency. As discussed in Sect. 5.3, a suitable h depends on the length of trajectories. When evaluating the similarity of two trajectories, h can be set to their average length. Therefore, we consider ITS to be parameter-free. As $h = O(n)$, ITS has linear time and space complexity.

Table 2. Dataset statistics

Dataset	Split gap	Mean length	# Trajectory
Porto [16]	–	59	1,243,663
T-drive [27]	12 mins	111	146,761
GeoLife [29]	10 secs	131	1,392
ASL [8]	–	57	2,565

5 Experiment

We evaluate ITS by comparing it with 12 baselines on four real-world datasets. Section 5.1 gives the details of the selected baselines and datasets. In Sect. 5.2, we evaluate different measures using five downstream tasks, including rank, precision, cross-similarity, k-nn queries, and clustering. Finally, we analyze the effect of the parameter of ITS in Sect. 5.3.

5.1 Experimental Setup

Datasets. Table 2 shows four real-world datasets used in our experiments. For ASL, we use it in clustering tasks only and take the x and y dimensions only. Followed by [12,19], for T-drive and GeoLife, we split the trajectories if the time gap between consecutive points is too large. We remove the trajectories with points less than 30 since they can hardly contain enough information after dropping. We divide the trajectories with points more than 300 since they have sufficient information, and more points cost too much time for $O(n^2)$ methods. Meanwhile, the mean lengths of datasets are far away from 300. We sample 100 trajectories randomly and yield 100^2 pairs in each dataset. Due to the limited space, we show the average results on all datasets in Sect. 5.2.

Baselines. We select 12 representative baselines for experiments: FD [15], DTW [1], FastDTW [21], LCSS [26], EDR [5], ERP [4], OWD [13], CATS [6], BDS [24], SSPD [2], EDwP [19] and STS [10]. All of them are introduced in Sect. 2. We implement them using Python. Following [10,23], we do not include ML-based methods since they are not measures. First, we set the spatial threshold of LCSS, EDR, and CATS to 15 m [24] on other tasks, and to 0.01 m on the clustering task duo to the nature of ASL dataset. Next, we set the temporal threshold of CATS to infinity, the radius of FastDTW to 10, and the granularity of OWD to 20. Finally, we convert the global grid size of STS into the co-location grid number, which is set to 16.

Downstream Tasks. Inspired by [2,10,12], we evaluate different measures using rank and precision, cross-similarity, k-nn queries, and clustering. The noises on trajectories are shifting and dropping. Let $\mathbb{T} = (T_1, T_2, \ldots, T_{|\mathbb{T}|})$ be original

Table 3. Average MR of measures with dropping and shifting

	r_1				r_2				r_3				r_4				Summary	
	0.2	0.4	0.6	0.8	0.2	0.4	0.6	0.8	0.2	0.4	0.6	0.8	0.2	0.4	0.6	0.8	Mean	Rank
FD	1.04	1.10	1.36	2.84	1.02	1.14	1.42	2.25	1.20	1.25	1.20	1.20	1.18	1.40	1.40	1.74	1.42	5
DTW	1.00	1.08	1.70	5.66	1.00	1.01	1.08	1.22	1.05	1.21	1.26	1.37	1.31	1.71	2.12	2.38	1.64	6
FastDTW	20.38	24.36	18.98	28.56	20.51	23.21	16.66	22.89	42.99	42.02	41.61	41.02	42.26	41.93	41.27	41.27	31.87	11
LCSS	9.54	11.64	8.42	20.84	17.31	8.31	16.60	21.93	21.10	13.51	14.03	11.37	11.48	10.50	12.37	6.09	13.44	8
EDR	1.00	1.01	1.03	1.05	1.00	1.00	1.00	1.00	1.00	1.00	1.00	1.00	1.03	1.26	2.05	3.00	1.21	1
ERP	21.29	36.11	44.31	42.02	17.46	21.51	31.19	36.04	12.39	12.39	12.39	12.39	12.39	12.39	12.40	12.39	21.82	9
OWD	1.47	1.56	1.48	1.56	1.47	1.55	1.59	1.47	1.00	1.01	1.03	1.04	1.01	1.07	1.10	1.15	1.28	3
CATS	1.02	1.02	1.02	1.04	1.02	1.03	1.04	1.07	1.02	1.02	1.06	1.10	1.22	1.75	2.47	2.91	1.30	4
BDS	73.71	72.53	73.10	70.97	73.75	73.85	73.36	73.46	75.68	76.02	76.71	77.36	78.86	77.84	76.18	74.79	74.89	13
SSPD	5.41	5.66	6.01	9.32	4.64	5.51	5.14	5.99	5.19	6.14	7.20	7.05	9.48	8.17	7.43	7.48	6.61	7
EDwP	31.43	32.95	33.64	37.29	25.54	28.35	30.43	31.46	17.85	22.34	21.29	18.99	22.24	23.07	22.69	23.64	26.45	10
STS	32.99	32.71	32.51	31.44	33.25	32.97	32.64	32.21	32.90	32.53	32.47	32.19	32.34	32.06	31.71	31.41	32.40	12
ITS	1.00	1.30	1.96	2.19	1.00	1.00	1.42	1.21	1.04	1.04	1.04	1.04	1.03	1.06	1.13	1.16	1.23	2

trajectories, $\mathbb{T}' = \{T'|T' = \varphi(T), T \in \mathbb{T}\}$ be sampled trajectories with noises. We take results on $\mathbb{T}$ as ground truth to evaluate the performances on $\mathbb{T}'$. To study the robustness of measures with noises, we perform four kinds of transformations. First, we drop the points at the whole range of T with a rate r_1, representing various dropping rates between trajectories. Next, we drop the points at the first half or the second half of T with a rate r_2, representing changing dropping rates inside trajectories. Furthermore, we introduce shifting by adding Gaussian noise with radius β meters. First, we set $\beta = 30$ m for points sampled by a rate r_3, representing shifting for some points. Next, we set $\beta = r_4 \cdot 100$ m for all points, representing shifting for all points. Finally, we apply one kind of transform each time and set other rates to 0.

$$x_i' = x_i + \text{Gaussian}(0, 1) \cdot \beta$$
$$y_i' = y_i + \text{Gaussian}(0, 1) \cdot \beta$$

Rank and Precision. We denote $s_{ij'} = f(T_i, T_j')$ as the similarity of $T_i \in \mathbb{T}$ and $T_j' \in \mathbb{T}'$, f is a trajectory similarity measure function, γ_i as the rank of $s_{ij'}$ in $(s_{i1'}, s_{i2'}, \ldots, s_{i|\mathbb{T}'|})$, χ_i as 1 if $\gamma_i = 1$ else 0, mean rank MR $= \frac{1}{|\mathbb{T}|} \sum_{i=1}^{|\mathbb{T}|} \gamma_i$ and mean precision MP $= \frac{1}{|\mathbb{T}|} \sum_{i=1}^{|\mathbb{T}|} \chi_i$. A smaller MR and a bigger MP indicate that measures can better recognize trajectories with noises.

Cross-similarity. We denote $s_{i'j'} = f(T_i', T_j')$ as the similarity of $T_i' \in \mathbb{T}'$ and $T_j' \in \mathbb{T}'$, cross similarity CS $= \frac{1}{|\mathbb{T}|^2} \sum_{i=1}^{|\mathbb{T}|} \sum_{j=1}^{|\mathbb{T}|} \frac{\min(|s_{ij}|, |s_{i'j'}|)}{\max(|s_{ij}|, |s_{i'j'}|)}$. The bigger CP is, the more robust measures recognize trajectory variants.

k-nn Queries. We denote $\mathcal{N}_i$ as the k-nearest-neighbors (k-nn) of T_i in $\mathbb{T}$, $\mathcal{N}_i'$ as the k-nn of T_i' in $\mathbb{T}'$, k-nn precision KP $= \frac{1}{|\mathbb{T}|} \sum_{i=1}^{|\mathbb{T}|} \frac{|\mathcal{N}_i \cap \mathcal{N}_i'|}{k}$. A bigger KP indicates that measures can find similar neighbors with noises. k is ten here.

Clustering. We execute K-Medoids on the ASL dataset and count how many trajectory pairs are classified into correct clusters. Unlike the above tasks evaluate the robustness of measures with dropping and shifting, clustering evaluates

Table 4. Average MP of measures with dropping and shifting

	r_1				r_2				r_3				r_4				Summary	
	0.2	0.4	0.6	0.8	0.2	0.4	0.6	0.8	0.2	0.4	0.6	0.8	0.2	0.4	0.6	0.8	Mean	Rank
FD	0.93	0.85	0.79	0.67	0.95	0.92	0.86	0.79	0.94	0.93	0.91	0.91	0.94	0.89	0.88	0.84	0.87	6
DTW	1.00	0.95	0.83	0.48	1.00	0.99	0.95	0.89	0.97	0.92	0.90	0.89	0.89	0.83	0.79	0.77	0.88	5
FastDTW	0.71	0.58	0.70	0.52	0.71	0.64	0.73	0.62	0.50	0.52	0.53	0.54	0.53	0.53	0.53	0.54	0.59	8
LCSS	0.25	0.24	0.26	0.07	0.20	0.31	0.11	0.24	0.06	0.13	0.15	0.17	0.37	0.21	0.20	0.51	0.22	11
EDR	1.00	0.99	0.97	0.96	1.00	1.00	1.00	1.00	1.00	1.00	1.00	1.00	0.98	0.87	0.71	0.62	0.94	3
ERP	0.03	0.01	0.00	0.01	0.15	0.03	0.02	0.01	0.41	0.41	0.41	0.41	0.41	0.41	0.41	0.41	0.22	10
OWD	0.99	0.98	**0.98**	0.94	0.99	0.98	0.99	0.99	1.00	0.99	0.98	0.96	**0.99**	0.95	**0.94**	0.91	**0.97**	1
CATS	0.99	0.98	**0.98**	**0.97**	0.98	0.97	0.97	0.95	0.98	0.99	0.96	0.93	0.89	0.76	0.64	0.57	0.91	4
BDS	0.01	0.01	0.01	0.01	0.01	0.01	0.01	0.01	0.00	0.00	0.00	0.00	0.00	0.00	0.00	0.00	0.00	13
SSPD	0.57	0.61	0.54	0.39	0.61	0.59	0.58	0.51	0.59	0.50	0.45	0.40	0.35	0.34	0.29	0.29	0.48	9
EDwP	0.11	0.09	0.08	0.01	0.18	0.11	0.12	0.12	0.21	0.21	0.21	0.25	0.26	0.22	0.24	0.20	0.16	12
STS	0.67	0.67	0.67	0.67	0.67	0.67	0.67	0.67	0.67	0.67	0.67	0.67	0.67	0.67	0.67	0.67	0.67	7
ITS	**1.00**	**0.99**	**0.98**	0.90	**1.00**	**1.00**	0.98	0.94	0.98	0.97	0.97	0.97	0.98	**0.97**	0.93	**0.92**	**0.97**	1

Table 5. Average CS of measures with dropping and shifting

	r_1				r_2				r_3				r_4				Summary	
	0.2	0.4	0.6	0.8	0.2	0.4	0.6	0.8	0.2	0.4	0.6	0.8	0.2	0.4	0.6	0.8	Mean	Rank
FD	**1.00**	**0.99**	**0.98**	**0.95**	**1.00**	**1.00**	**0.99**	**0.98**	0.99	0.99	0.99	0.99	0.99	0.98	0.97	0.97	**0.98**	1
DTW	0.98	0.97	0.95	0.91	0.97	0.94	0.91	0.86	0.99	0.99	0.99	0.99	0.99	0.98	0.97	0.96	0.96	3
FastDTW	0.03	0.02	0.01	0.00	0.05	0.03	0.03	0.02	0.03	0.02	0.03	0.03	0.02	0.03	0.00	0.01	0.02	13
LCSS	0.69	0.62	0.60	0.28	0.67	0.54	0.61	0.62	0.89	0.90	0.75	0.66	0.68	0.62	0.64	0.68	0.65	6
EDR	0.68	0.54	0.42	0.29	0.77	0.61	0.48	0.37	0.73	0.57	0.47	0.39	0.41	0.33	0.29	0.26	0.48	9
ERP	0.82	0.82	0.81	0.79	0.85	0.82	0.81	0.81	1.00	1.00	1.00	1.00	1.00	1.00	1.00	1.00	0.91	5
OWD	0.98	0.96	0.93	0.89	0.99	0.98	0.96	0.94	0.95	0.94	0.93	0.93	0.94	0.91	0.89	0.87	0.94	4
CATS	0.69	0.46	0.30	0.14	0.78	0.64	0.50	0.36	0.63	0.44	0.31	0.24	0.26	0.18	0.15	0.13	0.39	10
BDS	0.76	0.53	0.33	0.13	0.88	0.77	0.66	0.59	0.77	0.62	0.51	0.45	0.46	0.41	0.38	0.37	0.54	8
SSPD	0.80	0.67	0.56	0.38	0.87	0.80	0.73	0.67	0.68	0.60	0.57	0.55	0.55	0.51	0.49	0.47	0.62	7
EDwP	0.32	0.21	0.14	0.10	0.40	0.31	0.24	0.21	0.29	0.24	0.20	0.18	0.19	0.15	0.15	0.12	0.21	12
STS	0.33	0.33	0.33	0.33	0.33	0.33	0.33	0.33	0.33	0.33	0.33	0.33	0.33	0.33	0.33	0.33	0.33	11
ITS	**1.00**	**0.99**	0.96	0.93	**1.00**	0.99	0.98	0.95	0.98	0.98	0.98	0.97	0.98	0.97	0.96	0.95	0.97	2

Table 6. Average KP of measures with dropping and shifting

	r_1				r_2				r_3				r_4				Summary	
	0.2	0.4	0.6	0.8	0.2	0.4	0.6	0.8	0.2	0.4	0.6	0.8	0.2	0.4	0.6	0.8	Mean	Rank
FD	0.98	0.96	0.91	0.84	0.99	0.98	0.96	0.91	0.97	0.96	0.96	0.95	0.96	0.94	0.92	0.91	0.94	2
DTW	0.96	0.95	0.92	0.84	0.96	0.92	0.87	0.82	0.99	0.98	0.98	0.98	0.98	0.97	0.94	0.93	0.94	2
FastDTW	0.73	0.66	0.71	0.64	0.56	0.75	0.61	0.64	0.98	0.98	0.98	0.98	0.98	0.98	0.98	0.98	0.82	6
LCSS	0.23	0.28	0.47	0.19	0.40	0.29	0.26	0.22	0.67	0.81	0.54	0.23	0.38	0.33	0.29	0.26	0.37	13
EDR	0.91	0.86	0.79	0.70	0.94	0.88	0.83	0.76	0.92	0.87	0.82	0.79	0.80	0.75	0.71	0.68	0.81	7
ERP	0.56	0.54	0.53	0.53	0.59	0.56	0.53	0.54	1.00	1.00	1.00	1.00	1.00	1.00	1.00	1.00	0.77	9
OWD	0.94	0.91	0.86	0.80	0.95	0.93	0.90	0.85	0.93	0.91	0.91	0.91	0.91	0.88	0.86	0.84	0.89	5
CATS	0.92	0.84	0.75	0.63	0.93	0.90	0.84	0.77	0.90	0.82	0.78	0.72	0.74	0.67	0.63	0.61	0.78	8
BDS	0.89	0.78	0.68	0.61	0.91	0.88	0.80	0.75	0.84	0.75	0.71	0.62	0.59	0.61	0.58	0.56	0.72	10
SSPD	0.69	0.57	0.49	0.40	0.78	0.69	0.60	0.57	0.56	0.44	0.41	0.34	0.35	0.32	0.34	0.32	0.49	11
EDwP	0.67	0.61	0.54	0.46	0.68	0.61	0.56	0.52	0.51	0.42	0.38	0.37	0.38	0.34	0.33	0.33	0.48	12
STS	**1.00**	**1.00**	**1.00**	0.98	**1.00**	**1.00**	**1.00**	0.96	**1.00**	**1.00**	**1.00**	**1.00**	**1.00**	**1.00**	0.99	**1.00**	**0.99**	1
ITS	0.98	0.96	0.92	0.87	0.99	0.97	0.94	0.91	0.96	0.95	0.95	0.94	0.95	0.93	0.92	0.90	0.94	2

measures at a semantic scene. There are 95 clusters in the ASL dataset, and each cluster contains 27 trajectories. We select clusters randomly and evaluate measures with the various numbers of clusters. We denote ς_i as standard label of trajectory T_i, ς'_j as predicted label of trajectory T'_j, $\zeta_{ij} = 1$ if $\varsigma_i = \varsigma_j$ and $\varsigma'_i = \varsigma'_j$ else 0, clustering precision CP $= \frac{1}{|T|^2} \sum_{i=1}^{|T|} \sum_{j=1}^{|T|} \zeta_{ij}$. A bigger clustering precision indicates that trajectories are classified correctly.

5.2   Performance

Rank. Table 3 shows the average results of the Rank task on three datasets. MR values increase as noises generally increase. EDR performs the best, yet it is not good at handling shifting on all points. ITS is just next to EDR and only 0.02 worse than EDR. OWD is robust to shifting thanks to the grid but not to changing dropping rates inside the trajectory. ITS and FastDTW are $O(n)$ time complexity, but FastDTW is less robust than ITS.

Precision. Table 4 shows the average results of the Precision task on three datasets. MP values descend as noises generally increase. OWD and ITS perform best. Similar to the results of the Rank task, EDR is not good at handling shifting on all points, and ITS performs better than FastDTW.

Cross-Similarity. Table 5 shows the average results of the Cross-similarity task on three datasets. CS values descend as noises increase generally. FD is the best because dropping and shifting can hardly change the cross similarity of the shape itself, and shape-based OWD also performs well. ITS is just next to FD and only 0.01 worse than FD. ERP is robust to shifting when shifting is less than the threshold, which is the distance from points to the origin. However, ERP is not doing well with dropping.

k-nn Queries. Table 6 shows the average k-nn precision task results on three datasets. KP values descend as noises generally increase. STS performs the best, and ITS is tied for second with FD and DTW. FastDTW restricts the matching window size, so it cannot match points out of the window when the drop rate gets bigger. Just like the results of the Cross-similarity task, ERP is robust to shifting but not good at dropping.

Clustering. Table 7 shows the results of the Clustering task on ASL. CP values increases as the number of clusters increase for some measures since the more clusters there are, the more semantic information in trajectories such measures can identify. Conversely, CP values descend for others since they cannot identify semantic information. ITS performs best and performs better as the number of clusters increases. SSPD is the second since it aims at clustering tasks.

Runtime. Table 8 shows the running time of measures among increasing n. The time complexity of ITS is $O(4n + 3h)$, and FastDTW is $O(n(8r + 14))$ where r is the radius. We set h and r fixed here. ITS is the fastest measure. OWD grows slowly since the number of girds reaches almost max at the very first.

Table 7. CP of measures

#Clusters	4	5	6	7	Rank
FD	0.60	0.53	0.65	0.73	8
DTW	0.62	0.68	0.70	0.73	5
FastDTW	0.26	0.20	0.17	0.14	13
LCSS	0.26	0.21	0.36	0.14	11
EDR	0.28	0.24	0.22	0.20	12
ERP	**0.72**	0.73	0.77	0.77	3
OWD	0.63	0.59	0.68	0.78	7
CATS	0.30	0.28	0.29	0.31	10
BDS	0.58	0.66	0.69	0.73	6
SSPD	**0.72**	0.77	0.76	0.78	2
EDwP	0.63	0.71	0.73	0.74	4
STS	0.53	0.59	0.53	0.63	9
ITS	**0.72**	**0.79**	**0.83**	**0.83**	1

Table 8. Runtime(s) of measures

n	400	600	800	1000	1200	Rank
FD	0.779	1.906	3.055	4.617	6.576	9
DTW	0.707	1.601	2.941	4.517	6.389	8
FastDTW	0.023	0.030	0.045	0.052	0.061	2
LCSS	0.558	1.238	2.216	3.467	5.903	6
EDR	0.674	1.651	3.445	4.575	6.756	10
ERP	1.970	4.826	10.647	16.530	22.533	11
OWD	1.255	1.391	1.601	1.558	1.350	5
CATS	0.641	1.770	3.020	4.252	5.943	7
BDS	3.486	8.310	15.316	22.713	31.491	12
SSPD	4.668	11.725	20.697	30.231	42.594	13
EDwP	0.034	0.050	0.066	0.081	0.102	3
STS	0.509	0.702	0.958	1.203	1.508	4
ITS	**0.019**	**0.026**	**0.035**	**0.041**	**0.047**	1

ERP calculates the threshold dynamically, so it is slower than EDR. SSPD and BDS calculate distance twice to get symmetry, so they are the slowest.

Summary. To evaluate the overall performance of measures on both effectiveness of five downstream tasks and the efficiency of running time, we sum the rank of mean results on five tasks and the rank of running time. The smaller the sum is, the better the measure can keep both accuracy and efficiency. Table 9 shows that ITS has the best overall performance on the five downstream tasks and the running time. ITS is best on the Precision and Clustering tasks. Moreover,

Table 9. Summary of measures on downstream tasks

	Rank of Tasks					Rank of Runtime	Summary	
	MR	MP	CS	KP	CP		Sum	Rank
FD	5	6	1	2	8	9	31	4
DTW	6	5	3	2	5	8	29	3
FastDTW	11	8	13	6	13	2	53	10
LCSS	8	11	6	13	11	6	55	12
EDR	1	3	9	7	12	10	42	5
ERP	9	10	5	9	3	11	47	8
OWD	3	1	4	5	7	5	25	2
CATS	4	4	10	8	10	7	43	6
BDS	13	13	8	10	6	12	62	13
SSPD	7	9	7	11	2	13	49	9
EDwP	10	12	12	12	4	3	53	10
STS	12	7	11	1	9	4	44	7
ITS	2	1	2	2	1	1	9	1

it is second on the Rank, Cross-similarity, and k-nn precision tasks. ITS is not the best on all tasks but is in the top 2 for all tasks, and other measures cannot achieve it. What is more, ITS is the fastest measure. ITS and FastDTW are both $O(n)$ time complexity, but FastDTW is less robust than ITS, and FastDTW is not as fast as ITS. Thus, ITS can keep both accuracy and efficiency.

5.3   Effect of h

Table 10 shows the running time in seconds(s) of ITS grows slowly as the h increases while n is fixed. However, the running times of $h = 10$ and $h = 100$ are close since the running time is dominated by n when $h < n$.

Figure 3 shows that ITS performs better as h grows generally, but performances grow slower as h increases. The Y-axis indicate the results of downstream tasks. Let $m = \frac{\sum_{T \in \mathbb{T}} |T|}{|\mathbb{T}|}$ be the mean numbers of points of trajectories. Setting $h = m$ can obtain almost the best results within a very short time.

Table 10. Runtime(s) of ITS with various h

h	10	10^2	10^3	10^4	10^5
Runtime(s)	0.035	0.035	0.041	0.083	0.514

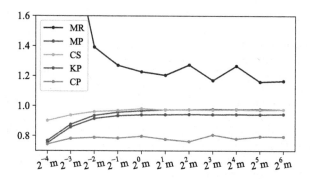

Fig. 3. Performances of ITS with various h

6   Conclusion

In this paper, we revisited the problem of measuring trajectory similarity effectively and efficiently. The problem is challenging due to the ubiquitous shifting in trajectory data and variable dropping rates of different objects. In addition, most existing methods have a quadratic running time, which is good but provides enough room for improvement. Motivated by these observations, we proposed ITS, an effective and efficient trajectory measure. The key idea of the measure is a fast interpolation that transforms trajectories into vectors without losing the semantics. The effectiveness of our measure results from the well-defined properties of interpolation: trajectories are aligned, the points in a trajectory are distributed evenly, and the shifting can only disturb surrounding interpolated points. Furthermore, the efficiency of our measure is guaranteed by fast interpolation, which is linear time, and fast computation of the distance of vectors, which is linear time too. We conducted extensive experiments with 12 baselines on four real-world datasets. The results showed that ITS is the fastest measure and has the best overall performance on a series of downstream tasks that are important in trajectory computing.

Acknowledgements. This work is supported by Natural Science Foundation of Jiangsu Province (Grant Nos. BK20211307), and by project Funded by the Priority Academic Program Development of Jiangsu Higher Education Institutions.

References

1. Berndt, D.J., Clifford, J.: Using dynamic time warping to find patterns in time series. In: KDD, pp. 359–370 (1994)
2. Besse, P.C., Guillouet, B., Loubes, J., Royer, F.: Review and perspective for distance-based clustering of vehicle trajectories. IEEE Trans. Intell. Transp. Syst. **17**(11), 3306–3317 (2016)
3. Ceikute, V., Jensen, C.S.: Vehicle routing with user-generated trajectory data. In: MDM, pp. 14–23 (2015)
4. Chen, L., Ng, R.T.: On the marriage of lp-norms and edit distance. In: VLDB, pp. 792–803 (2004)
5. Chen, L., Özsu, M.T., Oria, V.: Robust and fast similarity search for moving object trajectories. In: SIGMOD, pp. 491–502 (2005)
6. Hung, C., Peng, W., Lee, W.: Clustering and aggregating clues of trajectories for mining trajectory patterns and routes. VLDB J. **24**(2), 169–192 (2015)
7. Ismail, A., Vigneron, A.: A new trajectory similarity measure for GPS data. In: IWGS 2015, pp. 19–22 (2015)
8. Kadous, M.W.: Temporal Classification: Extending the Classification Paradigm to Multivariate Time Series. Ph.D. thesis (2002)
9. Keogh, E.J., Pazzani, M.J.: Scaling up dynamic time warping for datamining applications. In: KDD, pp. 285–289 (2000)
10. Li, G., Hung, C., Liu, M., Pan, L., Peng, W., Chan, S.G.: Spatial-temporal similarity for trajectories with location noise and sporadic sampling. In: ICDE, pp. 1224–1235 (2021)
11. Li, R., et al.: JUST: JD urban spatio-temporal data engine. In: ICDE, pp. 1558–1569 (2020)
12. Li, X., Zhao, K., Cong, G., Jensen, C.S., Wei, W.: Deep representation learning for trajectory similarity computation. In: ICDE 2018, pp. 617–628 (2018)
13. Lin, B., Su, J.: Shapes based trajectory queries for moving objects. In: ACM-GIS, pp. 21–30 (2005)
14. Luo, Y., et al.: Deeptrack: monitoring and exploring spatio-temporal data: a case of tracking COVID-19. Proc. VLDB Endow. **13**(12), 2841–2844 (2020)
15. Magdy, N., Sakr, M.A., Mostafa, T., El-Bahnasy, K.: Review on trajectory similarity measures. In: ICICIS, pp. 613–619 (2015)

16. Moreira-Matias, L., Gama, J., Ferreira, M., Mendes-Moreira, J., Damas, L.: Predicting taxi-passenger demand using streaming data. IEEE Trans. Intell. Transp. Syst. **14**(3), 1393–1402 (2013)
17. Nanni, M., Pedreschi, D.: Time-focused clustering of trajectories of moving objects. J. Intell. Inf. Syst. **27**(3), 267–289 (2006)
18. Pelekis, N., Kopanakis, I., Marketos, G., Ntoutsi, I., Andrienko, G.L., Theodoridis, Y.: Similarity search in trajectory databases. In: TIME 2007, pp. 129–140 (2007)
19. Ranu, S., P, D., Telang, A.D., Deshpande, P., Raghavan, S.: Indexing and matching trajectories under inconsistent sampling rates. In: ICDE 2015, pp. 999–1010 (2015)
20. Rohani, M., Gingras, D., Gruyer, D.: A novel approach for improved vehicular positioning using cooperative map matching and dynamic base station DGPS concept. IEEE Trans. Intell. Transp. Syst. **17**(1), 230–239 (2016)
21. Salvador, S., Chan, P.: Toward accurate dynamic time warping in linear time and space. Intell. Data Anal. **11**(5), 561–580 (2007)
22. Shang, S., Chen, L., Wei, Z., Jensen, C.S., Zheng, K., Kalnis, P.: Trajectory similarity join in spatial networks. Proc. VLDB Endow. **10**(11), 1178–1189 (2017)
23. Su, H., Liu, S., Zheng, B., Zhou, X., Zheng, K.: A survey of trajectory distance measures and performance evaluation. VLDB J. **29**(1), 3–32 (2020)
24. Ta, N., Li, G., Xie, Y., Li, C., Hao, S., Feng, J.: Signature-based trajectory similarity join. IEEE Trans. Knowl. Data Eng. **29**(4), 870–883 (2017)
25. Vlachos, M., Gunopulos, D., Das, G.: Rotation invariant distance measures for trajectories. In: KDD, pp. 707–712 (2004)
26. Vlachos, M., Gunopulos, D., Kollios, G.: Discovering similar multidimensional trajectories. In: ICDE 2002, pp. 673–684 (2002)
27. Yuan, J., et al.: T-drive: driving directions based on taxi trajectories. In: ACM-GIS 2010, pp. 99–108 (2010)
28. Zhang, H., et al.: Trajectory similarity learning with auxiliary supervision and optimal matching. In: IJCAI, pp. 3209–3215 (2020)
29. Zheng, Y., Xie, X., Ma, W.: Geolife: a collaborative social networking service among user, location and trajectory. IEEE Data Eng. Bull. **33**(2), 32–39 (2010)

Accurate and Efficient Trajectory-Based Contact Tracing with Secure Computation and Geo-Indistinguishability

Maocheng Li[1] , Yuxiang Zeng[1] , Libin Zheng[2]([✉]) , Lei Chen[1,4] ,
and Qing Li[3]

[1] The Hong Kong University of Science and Technology, Hong Kong, China
{csmichael,yzengal,leichen}@cse.ust.hk
[2] Sun Yat-sen University, Guangzhou, China
zhenglb6@mail.sysu.edu.cn
[3] The Hong Kong Polytechnic University, Hong Kong, China
csqli@comp.polyu.edu.hk
[4] The Hong Kong University of Science and Technology (Guangzhou),
Guangzhou, China

Abstract. Contact tracing has been considered as an effective measure to limit the transmission of infectious disease such as COVID-19. Trajectory-based contact tracing compares the trajectories of users with the patients, and allows the tracing of both direct contacts and indirect contacts. Although trajectory data is widely considered as sensitive and personal data, there is limited research on how to securely compare trajectories of users and patients to conduct contact tracing with excellent accuracy, high efficiency, and strong privacy guarantee. Traditional Secure Multiparty Computation (MPC) techniques suffer from prohibitive running time, which prevents their adoption in large cities with millions of users. In this work, we propose a technical framework called ContactGuard to achieve accurate, efficient, and privacy-preserving trajectory-based contact tracing. It improves the efficiency of the MPC-based baseline by selecting only a small subset of locations of users to compare against the locations of the patients, with the assist of Geo-Indistinguishability, a differential privacy notion for Location-based services (LBS) systems. Extensive experiments demonstrate that ContactGuard runs up to 2.6× faster than the MPC baseline, with no sacrifice in terms of the accuracy of contact tracing.

Keywords: Contact tracing · Differential privacy · Spatial database

1 Introduction

Contact tracing has been considered as one of the key epidemic control measures to limit the transmission of infectious disease such as COVID-19 and Ebola [14].

X. Wang et al. (Eds.): DASFAA 2023, LNCS 13943, pp. 300–316, 2023.
https://doi.org/10.1007/978-3-031-30637-2_20

In contrast to traditional contact tracing conducted normally by interviews and manual tracking, digital contact tracing nowadays rely on mobile devices to track the visited locations of users, and are considered as more accurate, efficient, and scalable [19,21]. Taking the contact tracing of COVID-19 as an example, more than 46 countries/regions have launched contact tracing applications [14], *e.g.*, TraceTogether in Singapore and LeaveHomeSafe in Hong Kong SAR.

As a complementary technique to the Bluetooth-based contact tracing applications such as the Exposure Notification developed by Apple and Google[1], *trajectory-based contact tracing* [6,13] compares the trajectories of users with the patients, and allows the tracing of both the direct contacts and the indirect contacts. The direct contacts happen when the users and the patients co-visit the same location at the *same* time. In contrast, the indirect contacts normally happen when users and patients visit the same location at *different* times, and the location (*e.g.*, environment, surface of objects) is contaminated and becomes the transmission medium of the virus. Because its capability of tracing both direct and indirect contacts, trajectory-based contact tracing has been widely deployed in countries/regions where there is a stricter policy of COVID-19 control. For example, in China, prior to entering high-risk locations such as restaurants, bars, and gyms, users need to check-in with a health code, which is linked with their personal identities.

Despite the usefulness of trajectory-based contact tracing, the privacy of trajectory data is an obvious issue. In fact, privacy concerns prevent the contact tracing application from being widely adopted in cities like Hong Kong, and there are extreme cases that citizens use two different mobile phones, one for their daily use, and the other one solely for the purpose of fulfilling the check-in requirement when entering the high-risk locations, in order to prevent any privacy breach. Existing privacy-preserving trajectory-based contact tracing research relies on specific hardware (Trusted Hardware such as Intel SGX [13]), but the security guarantee heavily depends on the hardware, and it leads to poor portability. In fact, Intel plans to stop its support for SGX from its 11th and 12th generation processors. Thus, in this work, we aim to develop a hardware-independent software-based solution, which can provide high accuracy of contact tracing, excellent execution efficiency, and strong privacy guarantee.

In this paper, we formulate the *Privacy-Preserving Contact Tracing (PPCT)* problem, which is illustrated with the simplified example in Fig. 1(a). We aim to correctly identify whether each user u_i is a close contact or not. If a user u_i co-visited some location with some patient under some spatial and temporal constraints (*e.g.*, co-visit the same location with 5 m apart, and within a time window of 2 days), then our proposed solution should correctly identify the user u_i as a close contact. The key challenge of PPCT is that there is a strong privacy requirement: during the entire process of contact tracing, the private information (the visited locations of users, together with the timestamps) is strictly protected from the access of other parties.

---

[1] https://covid19.apple.com/contacttracing.

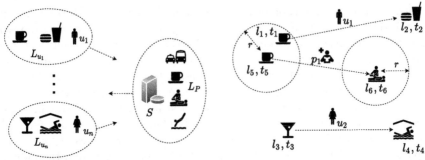

(a) A simplified example of our problem.   (b) An example of the close-contact.

Fig. 1. (a) A simplified example of our problem. L_{u_i} denotes the visited location of a user u_i. L_P denotes the collection of all visited locations of all patients. (b) An example of the close-contact. u_1 is a close contact, while u_2 is not.

The challenge of PPCT goes beyond the strong privacy requirement. On one hand, the contact tracing application requires a high level of accuracy. Taking COVID-19 as an example, some countries spend tremendous amount of resources to identify potential close contacts of patients, and sometimes a whole region of citizens need to go through mandated testing after only one case of patient was found in the area. Thus, our proposed solution needs to avoid false negatives as far as possible, maintaining a reasonably high recall, if not close to 100%. On the other hand, we aim to offer an efficient solution to the PPCT problem, as traditional Secure Multiparty Computation (MPC) techniques usually induce unpractical running time. As our experimental results show, a naïve MPC solution requires more than 5 days for 1 million users, which is the population of a modern city. Such long running time prohibits its daily execution, because it could not even terminate within 24 h.

To tackle the aforementioned challenges, we propose a novel technical framework ContactGuard. ContactGuard improves the efficiency of MPC operations by selecting only small subsets of users' locations to compare with the patients. The subset selection process is assisted by an efficient privacy-preserving mechanism – Geo-Indistinguishability (Geo-I) for Location-based services (LBS) systems. The basic idea is that each user perturbs his/her true visited locations with Geo-I and submits only the perturbed locations to the server, where the patients' locations are stored. Using the perturbed locations of the user, the server compares them with the patients' locations, and then informs the user about which locations are more "risky", because they are closer to the patients. After that, the user could use MPC only on the high-risk locations, to largely reduce the computational overhead.

To summarize, we make the following contributions in this paper:

- We formally define an important problem, the Privacy-Preserving Contact Tracing (PPCT) problem in Sect. 2, which addresses the privacy issue in trajectory-based contact tracing.

- We propose a novel solution ContactGuard in Sect. 3. It combines the strengths of two different privacy-preserving paradigms: differential privacy and secure multiparty computation. The running time is improved significantly because it selects only a small subset of locations to compare with the patients during the MPC operation.
- We conduct extensive experiments to validate the effectiveness and efficiency of ContactGuard in Sect. 4. It runs up to 2.6× faster than the MPC baseline, with no sacrifice in terms of the accuracy of contact tracing. For moderate privacy budget settings for each user, ContactGuard obtains close to 100% recall and 100% precision.

In addition, we review related works in Sect. 5 and conclude in Sect. 6.

2   Problem Definition

In this section, we introduce some basic concepts, the adversary model, and a formal definition of the Privacy-Preserving Contact Tracing (PPCT) problem.

2.1   Basic Concepts

Definition 1 (Location). *A location $l = (x, y)$ represents a 2-dimensional spatial point with the coordinates (x, y) on an Euclidean space.*

Definition 2 (Trajectory). *A trajectory is a set of (location, timestamp) tuples indicating the visited location and time that the location is visited. A trajectory $L = \{(l_1, t_1), \ldots, (l_{|L|}, t_{|L|})\}$, where $|L|$ is the number of visited locations in the trajectory.*

Examples: let location $l_1 = (300, 500)$ and l_2 be another location $l_2 = (200, 200)$. An example of trajectory $L_{example} = \{(l_1,$ 2020-06-10 12:28:46$), (l_2,$ 2020-06-10 17:03:24$)\}$ indicates that location l_1 is visited at timestamp 2021-06-10 12:28:46 and location l_2 is visited at timestamp 2021-06-10 17:03:24. The size of $L_{example}$ is 2, i.e., $|L_{example}| = 2$.

Here, the definition of *trajectory* includes both short sequences (*e.g.*, a user leaves home in the morning, moves to the working place, and comes back home at night) and longer sequences across multiple days. In our problem setting, in order to trace contacts, the trajectory of a user contains the union of all visited locations of a particular user in the incubation period of the disease (*e.g.*, past 14 days for COVID-19 [10,12]).

Definition 3 (Users). *A set U denotes all n users. Each user $u \in U$ is associated with a trajectory $L_u = \{(l_1, t_1), \ldots, (l_{|L_u|}, t_{|L_u|})\}$, following Definition 2, indicating the locations the user u visited and the time of the visits.*

Each user needs to be checked against the patients to see whether the user is a contact of some patient or not. The definition of a patient is similar to the one of the user, and we define them separately for easier illustration in the contact tracing problem.

Definition 4 (Patients). *A set P denotes all m patients. Each patient $p \in P$ has a trajectory $L_p = \{(l_1, t_1), \ldots, (l_{|L_p|}, t_{|L_p|})\}$, following Definition 2, indicating the locations the patient p visited and the time of the visits. The set $L_P = L_{p_1} \cup L_{p_2} \cup \ldots \cup L_{p_m}$ is a union of trajectories of all patients $p_1, \ldots, p_m$.*

In our problem setting, we use L_P to denote the aggregated set of all trajectories of patients. It is an important notion because a user $u \in U$ is defined as a *contact* (see a more formal definition in Definition 5) if the user u's trajectory L_u overlaps with some location in L_P. It is not necessary to identify which specific patient $p \in P$ leads to the contact. Indeed, as we will explain further in our adversary model and system settings in Sect. 2.2, it suffices to store the union of all trajectories L_P of the patients, without distinguishing each patient's individual trajectories, and it also enhances privacy protection.

Definition 5 (Contacts). *Given a distance threshold r, and a time difference threshold δ, a user $u \in U$ is called the **contact** if the user u has visited a location l_u that is within r distance to some visited location l_p of some patient $p \in P$, and the time difference between the two visits is within δ, i.e.,*

$$\exists (l_u, t_u) \in L_u \; \exists (l_p, t_p) \in L_P \; ((d_s(l_u, l_p) \leq r) \wedge (d_t(t_p, t_u) \leq \delta)) \tag{1}$$

where the function $d_s(l_u, l_p)$ represents the spatial distance – the Euclidean distance between locations l_u and l_p. $d_t(t_u, t_p)$ represents the temporal difference – the time difference in seconds from an earlier timestamp t_p to a later timestamp t_u.

We give a toy example in Example 1 to better illustrate the concept of contacts.

Example 1. As shown in Fig. 1b, there is one patient p_1 and two users u_1-u_2. User u_1 has a trajectory $L_{u_1} = \{(l_1, t_1), (l_2, t_2)\}$. User u_2 has a trajectory $L_{u_2} = \{(l_3, t_3), (l_4, t_4)\}$. The set of patients P (only contains one single patient p) has the trajectory $L_P = \{(l_5, t_5), (l_6, t_6)\}$. The locations are shown on the figure, and let us further specify the time. Let $t_1 = $ 2021-06-10 11:00:00 and $t_5 = $ 2021-06-10 10:00:00, indicating that patient p visits the location l_5 an hour earlier than the time when user u visits l_1.

Based on Definition 5, let r be the distance shown in the figure, and the time difference threshold be $\delta = 2$ h, then the user u_1 is a contact, since $d_s(l_1, l_5) \leq r$, i.e., u_1 has visited l_1 and l_1 is within distance r to patient p_1's visited location l_5, and $d_t(t_1, t_5) = 1$ h $\leq \delta$, i.e., the time difference between the two visits is within δ ($\delta = 2$ h).

On the other hand, user u_2 is not a close-contact, because her visited locations, l_3 and l_4, are not in proximity of any patient's visited locations.

2.2 Adversary Model

There are two major roles in our application: the users (the clients) and the government (the server). All trajectories by the patients (represented as L_P as in

Definition 4) are aggregated and stored at the server. The setting follows the real-world situation: governments often collect whereabouts of confirmed patients in order to minimize any further transmission, usually starting with tracing the close contacts of the patients. For example, this is enforced by legislation in Hong Kong[2].

We adopt a semi-honest model as the adversary model in our problem. Adversary could exist on both the client (the user) and the server side (the government). We assume both the client and the server are curious about other parties' private information, but they are not malicious and follow our designed system protocols. The semi-honest model is a commonly adopted setting in recent privacy-preserving LBS related applications [23,24].

2.3 Privacy-Preserving Contact Tracing Problem

Based on the concepts and adversary model introduced previously, we now define the Privacy-Preserving Contact Tracing (PPCT) problem as follows.

Definition 6 (Privacy-Preserving Contact Tracing (PPCT) problem).
Given a set U of n users , the trajectory L_u for each user $u \in U$, a set P of m patients, a set L_P containing the union of trajectories for all patients $p_1, \ldots, p_m \in P$, a distance threshold r, and a time difference threshold δ, the PPCT problem needs to correctly identify each user as a contact (see Definition 5) or not.

In addition, PPCT has the following privacy requirements:

- *The computation process needs to be differentially private/confidential w.r.t. each user's trajectory L_u.*
- *The computation process needs to be differentially private/confidential w.r.t. L_P, the union of trajectories of the patients.*

3 Methodology

In this section, we first introduce two baselines, one based on Secure Multiparty Computation (MPC) and the other one based on Geo-Indistinguishability (Geo-I). Then, we introduce our proposed method – ContactGuard.

3.1 Baselines

MPC and Geo-I are two different paradigms to satisfy the privacy requirements in the PPCT problem, but they differ greatly in terms of efficiency and accuracy. MPC allows the server and the client to use cryptographic primitives to securely compare each visited location without revealing the exact locations of one party to the other parties. However, it induces significant computation overhead.

---

[2] https://www.elegislation.gov.hk/hk/cap599D!en?INDEX_CS=N.

Geo-I [2] is an extended notion of differential privacy into the spatial domain (strictly speaking, Geo-I is a variant of local differential privacy notion [4,8]). It is an efficient mechanism to be applied to the location inputs, however Geo-I injects noise into the inputs (protecting the input location by perturbing it to an obfuscated location). Using the perturbed locations to compare trajectories would then lead to errors and reduce the accuracy of contact tracing.

MPC Baseline. The MPC baseline is to directly apply existing secure multi-party computation techniques to our problem. In our setting, each time, we treat a client (a user) and the server (which holds the patients' data) as two parties. Each of the party holds their own visited locations, which are private. Then, they use MPC operations to compare their visited locations and check whether the user is a contact or not, according to Definition 5. Note that the baseline computes the *exact* result, which means that it does not lose any accuracy.

Geo-I Baseline. In Geo-I baseline, each user perturbs his/her trajectory L_u to a protected noisy location set L'_u using Geo-I (Algorithm 1). Then, the user submits L'_u to the server. The server directly compares L'_u with the patients' locations L_P. If there exists some perturbed location $l' \in L'_u$ that is within distance r' (a system parameter), then u is identified as a contact.

Algorithm 1 is a generic method to obtain a set of perturbed locations L'_u, given a set of original locations L_u and a total privacy budget of ϵ. At Line 2, the total privacy budget ϵ is equally divided into $|L_u|$ shares. Thus, each share is $\epsilon' = \epsilon/|L_u|$. Then we perturb each location $l \in L_u$ into a new location l' by the Geo-I mechanism. All perturbed locations l' compose the set L'_u of the perturbed locations.

Algorithm 1: Perturb_Location_Set

Input: ϵ, L_u.
Output: L'_u.

1 $L'_u := \{\}$
2 $\epsilon' = \epsilon/|L_u|$
3 **foreach** $l \in L_u$ **do**
4 $\quad$ $l' = \text{Geo-I}(\epsilon', l)$
5 $\quad$ $L'_u.\text{insert}(l')$
6 **return** L'_u

Note that in the original definition for the trajectory L_u in Definition 2, each visited location l is associated with a timestamp that the location is visited (see Example 1). We omit the timestamp information for each visited location, and perturb *only* the locations to perturbed locations. We abuse the notation L_u and L'_u to denote only the spatial locations (Definition 1) and the generated perturbed locations.

Time Complexity. For Algorithm 1 (the client side), the time complexity is $\mathcal{O}(|L_u|)$, linear to the number of visited locations given in the input L_u. For the server side, because it compares each location of L'_u against every location of L_P, the time complexity is $\mathcal{O}(|L'_u||L_P|) \to \mathcal{O}(|L_u||L_P|)$. For the server S to finish

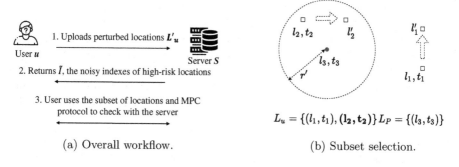

(a) Overall workflow.

(b) Subset selection.

Fig. 2. The overall workflow and the subset selection process in Step 3.

the processing of all users, the total time complexity is thus $\mathcal{O}(\sum_u |L_u||L_P|) \rightarrow \mathcal{O}(|U||L_P| \max_u |L_u|)$.

Privacy Analysis. We defer the privacy analysis of Algorithm 1 to Sect. 3.2, as it is the same as the privacy analysis of ContactGuard.

3.2 Our Solution ContactGuard

In this section, we propose a novel method called ContactGuard. As we have previously introduced, the MPC baseline provides excellent accuracy, because it invokes computationally heavy MPC operations to compare all visited locations of users against the visited locations of the patients. The weakness of the MPC baseline is its poor efficiency. On the other hand, the Geo-I baseline is efficient because each client (user) applies an efficient Geo-I mechanism to change his/her visited locations to perturbed ones. But the inaccurate perturbed locations could lead to false positives/negatives for the PPCT problem.

The ContactGuard aims to combine the advantages of both baselines. On one hand, it uses the exact locations and MPC to determine whether a user is a contact or not to ensure excellent accuracy while providing strong privacy protection. On the other hand, it accelerates the MPC operation by selecting only a subset of users' visited locations to invoke the heavy MPC operations. The selection is done with the help of Geo-I. The overall workflow of ContactGuard is illustrated in Fig. 2a. The details are explained next.

Step 1. Location Perturbation. At this step, the user perturbs his/her visited location set L_u to a perturbed visited location set L'_u, and submits L'_u to the server. The steps are similar to the Geo-I baseline (Sect. 3.1).

Given a privacy budget ϵ for each user, and the true location set L_u, the user generates a perturbed location set L'_u using Algorithm 1. Similar to the Geo-I baseline, the total privacy budget ϵ is equally divided into $|L_u|$ shares. Each location $l \in L_u$ is then perturbed to a noisy location l' by Geo-I mechanism with a privacy budget of $\epsilon' = \epsilon/|L_u|$.

Similar to the Geo-I baseline, we abuse notations L_u and L'_u to denote only the locations (without the timestamps) and the perturbed locations. This is slightly different from the original definition for the visited location set L_u in Definition 2, where each location is associated with a timestamp. We simply ignore all the temporal information during the perturbation process, as a way to provide strong privacy guarantee for the timestamps (as we do not use it at all). The temporal information will be used to check for the temporal constraints for determining a contact in later MPC steps (Step 3).

Privacy Analysis. To understand the level of privacy guarantee of Algorithm 1, we provide the theoretical privacy analysis for it. It extends Geo-I [2] from protecting a single location to a set of locations. The brief idea of the extension was mentioned in [2], we provide a concrete and formal analysis here.

First, we extend the privacy definition of Geo-I from a single location to a set of locations.

Definition 7. *(General-Geo-I) For two tuples of locations* $\mathbf{l} = (l_1, \ldots, l_n), \mathbf{l}' = (l'_1, \ldots, l'_n)$, *a privacy parameter* ϵ, *a mechanism* M *satisfies* ϵ-*General-Geo-I iff:*

$$d_\rho(M(\mathbf{l}), M(\mathbf{l}')) \le \epsilon d_\infty(\mathbf{l}, \mathbf{l}'),$$

where $d_\infty(\mathbf{l}, \mathbf{l}') = max_i d(l_i, l'_i)$, which is the largest Euclidean distance among all pairs of locations from $\mathbf{l}$ and $\mathbf{l}'$.

Different from the Geo-I definition for a single location, here the distance between two tuples of locations is defined as the largest distance at any dimensions. If a randomized mechanism M satisfies the General-Geo-I in Definition 7, then given the any perturbed location set output $\mathbf{o} = (o_1, \ldots, o_n)$, which is a sequence of perturbed locations generated by M, no adversary could distinguish whether the perturbed locations are generated from true location set $\mathbf{l}$ or from $\mathbf{l}'$. The ratio of the probabilities of producing the same output $\mathbf{o}$ from the true location set $\mathbf{l}$ or its neighboring input $\mathbf{l}'$ is bounded by $\epsilon d_\infty(\mathbf{l}, \mathbf{l}')$.

As a special case, when there is only one location in $\mathbf{l}$, Definition 7 becomes the original definition in Geo-I . We start the privacy analysis by showing the composition theorem for the General-Geo-I.

Lemma 1. *Let* K_0 *be a mechanism satisfying* ϵ_1-*Geo-I and* K_1 *be another mechanism satisfying* ϵ_2-*Geo-I. For a given 2-location tuple* $\mathbf{l} = (l_1, l_2)$, *the combination of* K_0 *and* K_1 *defined as* $K_{1,2}(\mathbf{l}) = (K_1(l_1), K_2(l_2))$ *satisfies* $(\epsilon_1 + \epsilon_2)$-*General-Geo-I (Definition 7).*

Proof. Let a potential output of the combined mechanism, $K_{1,2}(\mathbf{l})$, be $\mathbf{o} = (o_1, o_2)$, indicating a tuple of two perturbed locations. The probability of generating $\mathbf{o}$ from two input location sets $\mathbf{l} = (l_1, l_2)$ and $\mathbf{l}' = (l'_1, l'_2)$ are $\Pr[K_{1,2}(\mathbf{l}) = \mathbf{o}]$ and $\Pr[K_{1,2}(\mathbf{l}') = \mathbf{o}]$, respectively. Then, we measure the ratio of the twoprobabilities:

$$\frac{\Pr[K_{1,2}(\mathbf{l}) = \mathbf{o}]}{\Pr[K_{1,2}(\mathbf{l'}) = \mathbf{o}]} = \frac{\Pr[K_1(l_1) = o_1] \cdot \Pr[K_2(l_2) = o_2]}{\Pr[K_1(l'_1) = o_1] \cdot \Pr[K_2(l'_2) = o_2]} \tag{2}$$

$$= \frac{\Pr[K_1(l_1) = o_1]}{\Pr[K_1(l'_1) = o_1]} \cdot \frac{\Pr[K_2(l_2) = o_2]}{\cdot \Pr[K_2(l'_2) = o_2]} \tag{3}$$

$$\leq e^{\epsilon_1 \cdot d(l_1, l'_1)} \cdot e^{\epsilon_2 \cdot d(l_2, l'_2)} \tag{4}$$

$$\leq e^{\epsilon_1 \cdot d_\infty(\mathbf{l}, \mathbf{l'})} \cdot e^{\epsilon_2 \cdot d_\infty(\mathbf{l}, \mathbf{l'})} \tag{5}$$

$$= e^{(\epsilon_1 + \epsilon_2) d_\infty(\mathbf{l}, \mathbf{l'})} \tag{6}$$

The end result in Eq. (6) shows that $K_{1,2}$ satisfies $(\epsilon_1 + \epsilon_2)$-General-Geo-I (Definition 7).

Theorem 1. *Algorithm 1 satisfies ϵ-General-Geo-I (Definition 7).*

Proof. We use the composition theorem of Geo-I in Lemma 1 to show that Algorithm 1 composes linearly and consumes a total privacy budget of ϵ over the entire set of locations.

At Line 4 of Algorithm 1, for each location $l \in L_u$, it is perturbed to l' with a privacy budget $\epsilon' = \epsilon_u / |L_u|$. Then, for each location, we achieve ϵ'-Geo-I. If there are two locations in L_u, and because each location is perturbed independently, we achieve $\epsilon' + \epsilon' = 2\epsilon'$-General-Geo-I. Then, by induction, for $|L_u|$ locations, we achieve $\sum_{l \in L_u} \epsilon' = \epsilon' \cdot |L_u| = \frac{\epsilon}{|L_u|} \cdot |L_u| = \epsilon$-General-Geo-I.

Step 2. Subset Selection. After Step 1, each user has a set of perturbed locations L'_u and submits it to the server. At Step 2, subset selection, upon receiving L'_u, the server selects a subset of high-risk locations based on the perturbed locations. This step is done at the server side, where the patients' true locations are stored. The server returns to the client the index of the selected high-risk locations (*e.g.*, if the 2nd location and the 4th location are high-risk locations, then the server returns $\{2, 4\}$) to indicate which location out of L_u are close to the patients. In order to protect the privacy of the patients, randomized response is used to perturb the indexes of high-risk locations to noisy indexes.

On the server side, the server does not know the true locations nor the timestamps of the visits, which are stored in L_u on the client side (the user). However, the server has access to the true locations visited by the patients, which are denoted as L_P. L_P are the basis to judge whether a user location is high-risk or not.

Theorem 2. *The subset selection step satisfies ϵ_P-Local Differential Privacy .*

Due to space limit, we defer the proof to the full version [1]. We also provide a detailed running example therein for this step.

Step 3. Accelerated MPC. After Step 2, the user u receives the set $\tilde{I}$, which contains the indexes of high-risk locations as identified by the server. At Step 3,

Table 1. Comparison of baselines and ContactGuard

Methods	MPC baseline	Geo-I baseline	ContactGuard
Accuracy (Recall)	100% $\checkmark$	39.0–70% $\times$	86–100% $\checkmark$
Efficiency	# of Secure Operations: $\mathcal{O}(|U||L_P|\max_u|L_u|)$ $\times$	#of Plaintext Operations: $\mathcal{O}(|U||L_P|\max_u|L_u|)$ $\checkmark$	# of Secure Operations: $\mathcal{O}(|U||L_P|\max_u|I_u|)$ $\checkmark$
Privacy	Confidentiality $\checkmark$	Local-d DP $\checkmark$	Local-d DP $\checkmark$

accelerated MPC, the user uses MPC protocol to check with the server about whether the user is a contact or not, using only the locations indicated by the set $\tilde{I}$. In contrast, the MPC baseline uses every location in L_u to check with the server using MPC protocol. Here, the user uses only a small subset of locations to perform secure computations.

Our method runs the same secure computation procedure as the MPC baseline. However, it only provides the subset of locations as the inputs to the ProtocolIO object io. The timestamps are also included in the io object, and are used to compare with the patients' trajectories L_P. This is different from the Geo-I baseline or Step 1. location perturbation in ContactGuard, which only uses the spatial information (the locations).

Overall Analysis. Overall, for the server to finish processing all users, it takes $\mathcal{O}(\sum_u |I_u||L_P|) \to \mathcal{O}(|U||L_P|\max_u|I_u|)$, where I_u indicates the subset for each user u.

Privacy Analysis. ContactGuard offers end-to-end privacy guarantee $w.r.t.$ to both L_u and L_P. The detailed analysis is deferred to the full report [1].

Table 1 lists the comparison of the two baselines with our proposed Contact-Guard. While achieving the same privacy and accuracy, our proposed solution significantly improves the efficiency of the MPC solution.

4   Experimental Study

In this section, we first introduce the experimental setup in Sect. 4.1. Then, we present the detailed experimental results in Sect. 4.2.

4.1   Experimental Setup

Datasets. We use a real-world dataset Gowalla [3] and a randomly generated synthetic dataset in our experiments.

Baselines. We compare our proposed ContactGuard (short as **CG**) method with the MPC baseline (Sect. 3.1, short as **MPC**) and the Geo-I baseline (Sect. 3.1, short as **Geo-I**).

Metrics. We focus on the following metrics:

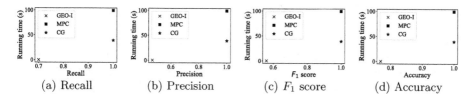

Fig. 3. Effectiveness (Recall/Precision/F_1/Accuracy) vs. Running time trade-off of different methods on the Gowalla dataset. $|U| = 200, \epsilon = 4.0$.

- Recall: the number of found true close-contacts divided by the total number of close-contacts.
- Precision: the number of found true close-contacts divided by the number of predicted close-contacts.
- F_1 score: the harmonic mean of the precision and recall. $F_1 = 2 \cdot \frac{\text{precision} \cdot \text{recall}}{\text{precision} + \text{recall}}$.
- Accuracy: the percentage of correctly classified users.
- Running time: the running time of the method in seconds.

Recall is considered as the most important metric in our experiments, as it measures the capability of the tested method for identifying potential close-contacts out of the true close-contacts, which is crucial in the application scenario of contact tracing of infectious disease.

Control Variables. We vary the following control variables in our experiments to test our method. Default parameters are in boldface. The privacy budget setting adopts the common setting with the real-world deployments (Apple and Google) [7].

- Number of users: $|U| \in [\mathbf{200}, 400, 800, 1600]$.
  Scalability test: $|U| \in [10K, 100K, 1M]$.
- Privacy budget for each user: $\epsilon \in [2.0, 3.0, \mathbf{4.0}, 5.0]$.
- Privacy budget for the patients: $\epsilon_P \in [2.0, 3.0, \mathbf{4.0}, 5.0]$.

4.2 Experimental Results

Our results show that the proposed solution ContactGuard achieves the best effectiveness/efficiency tradeoff, reducing the running time of the MPC baseline by up to 2.85×. The efficient Geo-I baseline, however, only provides poor effectiveness. Meanwhile, ContactGuard maintains the same level of effectiveness (measured by recall, precision, F_1 score and accuracy) as the one of the MPC baseline. The scalability tests also show that ContactGuard accelerates the MPC baseline significantly. When the number of users is large (*e.g.*, 500K), our ContactGuard supports daily execution, which is desirable in real-world application, while the MPC baseline fails to terminate within reasonable time (*e.g.*, 24 h).

Effectiveness vs. Efficiency Tradeoff. Figure 3 demonstrates the effectiveness and efficiency tradeoff of different methods. The y-axis are the running

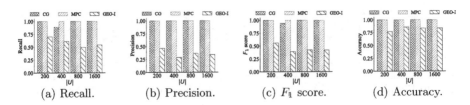

(a) Recall.   (b) Precision.   (c) F_1 score.   (d) Accuracy.

Fig. 4. Effectiveness (Recall/Precision/F_1/Accuracy) when varying the number of users ($|U|$) on the Gowalla dataset. $\epsilon = 4.0$.

time of different methods (MPC baseline, Geo-I baseline and our proposed ContactGuard), and the x-axis are the respective measures for effectiveness (recall, precision, F_1 and accuracy).

As we could see from Fig. 3, the MPC baseline is an *exact* method, which obtains 100% in all measures of effectiveness. However, the high accuracy comes with prohibitive running time. It runs for 99.35 s when the number of users is 200 ($|U| = 200$) and 194.99 s for 400 users ($|U| = 400$).

As a comparison, the Geo-I method is efficient, but comes with a significant decrease in effectiveness. For the setting $|U| = 200, \epsilon = 4.0$, Geo-I obtains about 70% recall and merely 46.7% precision. For the setting $|U| = 400, \epsilon = 3.0$, the recall drops to 39.0% and the precision drops to 42.0%.

Our proposed solution ContactGuard achieves the best effectiveness/efficiency tradeoff. For the setting $|U| = 200, \epsilon = 4.0$ (Fig. 3), CG obtains the same level of effectiveness as the one of the MPC baseline, achieving 100% in recall, precision and accuracy.

As compared to the Geo-I baseline, CG method obtains 2.28×, 2.38×, and 1.24× improvement in terms of recall, precision and accuracy, respectively. On the other hand, when we compare to the *exact* MPC baseline, the CG method is about 2.5× faster.

Effect of Control Variables. Then, we test the effectiveness across different input sizes ($|U|$) and different values of the privacy budget (ϵ).

Varying $|U|$: Fig. 4 shows measures of effectiveness (recall, precision, F_1, and accuracy) when we vary the number of users ($|U|$) for the Gowalla dataset. Across different input sizes, the CG method maintains the same level of recall/precision/F_1 score/accuracy as the *exact* MPC baseline across different $|U|$. This verifies that our CG method is highly effective across different input sizes.

On the other hand, the Geo-I baseline sacrifices significant effectiveness (in recall/precision/F_1/accuracy) across different $|U|$. The recall drops from 70% when $|U| = 200$ to 54.2% when $|U| = 1600$. The precision is constantly below 50%, and hovers around 35% when $|U|$ gets larger than 800. It shows that the pure differential privacy based solution injects noise to the perturbed locations, and the perturbed locations inevitably lead to poor effectiveness in the contact tracing applications.

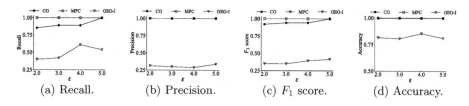

Fig. 5. Effectiveness (Recall/Precision/F_1/Accuracy) when varying the privacy budget (ϵ) on the Gowalla dataset. $|U| = 400$.

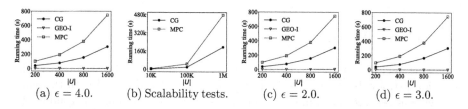

Fig. 6. Efficiency (running time in seconds) when varying the number of users ($|U|$) on the synthetic dataset.

Varying ϵ: Fig. 5 demonstrates the impact of the privacy budget on the effectiveness (recall/precision/F_1/accuracy). In terms of recall, CG obtains the same level of effectiveness (reaching 100%) as the MPC baseline when $\epsilon = 5.0$. The recall is 85.2% when the privacy budget is small ($\epsilon = 2.0$) and improves to 88.9% when $\epsilon = 3.0$. In these two settings, the improvement over the Geo-I baseline are about 2.12×.

In terms of precision, CG never introduces false positives, because positive results are returned only when there is indeed a true visited location of the user (included in the subset selected and verified with MPC operation) which overlaps with the patients. Thus, the precision is constantly at 100% over different ϵ.

The F_1 score reflects the similar pattern as the recall, as it is an average of recall and precision. The accuracy of CG stays at more than 99% across different ϵ.

Efficiency and Scalability. When compared to the MPC baseline, our solution CG achieves significant speed up. Figure 6 shows the running time of different methods across different input sizes, when the number of users scales up to 1 million.

When the number of users ($|U|$) increases from 200 to 1600, the running time of the MPC baseline grows from 99.35 s to 746.34 s. The running time grows linearly with $|U|$.

In comparison, the running time of CG grows from 39.31 s to 306.20 s, also with a linear growth with $|U|$. The speed up of CG over MPC is 2.52×, 2.50×, 2.48×, and 2.44× when $|U|$ increases from 200 to 1600. The results are shown in Fig. 6a.

Then, we use the synthetic dataset to compare the scalability of the MPC baseline and the CG method. The results are shown in Fig. 6b. The running time for MPC is 4754.7 s, 46,913 s (13 h) and 476,504.6 s (5.5 days) for $|U| = 10$ K, 100 K, and 1 M. As a comparison, the CG method runs in 1936.8 s, 19,087.2 s (5.3 h) and 195,060.7 s (2.25 days). It maintains a 2.5× speed up over the MPC baseline. This verifies that the subset selection in CG significantly reduces the running time of the MPC protocol. When the number of users reaches 1 million, the MPC baseline takes about 5.5 days to process all the users, while our proposed CG finishes in about 2 days.

5 Related Work

Our work is closely related to REACT [6], as it also uses Geo-I [2] to preserve the location privacy. In REACT, each location is applied with Geo-I with the same privacy budget. As shown in the experiments (both in [6] and in our results), the Geo-I baseline performs poorly in terms of both recall and precision, because it only uses perturbed locations. As a similar notion to Geo-I, local differential privacy has also been used to protect the trajectory data [5]. However, its main use is for aggregated statistics over relatively larger regions, where in our application, each user's contact tracing and fine-grained visited location is important.

Many other privacy-preserving techniques are used for contact-tracing, including blockchain based P^{2}B-Trace [18] and cryptographic primitives enabled BeeTrace [17] and Epione [26]. As opposed to using relative locations in these works, our setting uses absolute locations, which enables tracing indirect contacts (transmission via a contaminated environment) in addition to direct contacts.

On topics not related to contact tracing, there are many recent works on combining differential privacy and cryptographic techniques [27]. Secure shuffling and encryption-based techniques [9,20] could help reduce the error introduced by differential privacy, and differential privacy could enable more efficient secure computation Our solution, ContactGuard, is inspired by this line of ideas, and uses Geo-I to accelerate MPC computations. In addition, privacy issues are raised in many Location-based services (LBS) applications [11,22,23,25].

6 Conclusion

In this work, we study the Privacy-Preserving Contact Tracing (PPCT) problem, to identify the close-contacts of the patients while preserving the location privacy of the patients and the users. To address this problem, we propose an accurate and efficient solution named ContactGuard, which accelerates the Secure Multiparty Computation (MPC) with the help of Geo-Indistinguishability (Geo-I). Experimental results demonstrate that ContactGuard provides significant speed up over the MPC baseline, while maintains an excellent level of effectiveness (as measured by recall and precision). For future works, we look into more advanced ways to compress data [15,16] to achieve better privacy/utility tradeoff.

Acknowledgment. Libin Zheng is sponsored by the National Natural Science Foundation of China No. U22B2060 and 62102463, Natural Science Foundation of Guangdong Province of China No. 2022A1515011135 and the Basic and Applied basic Research Project of Guangzhou basic Research Program No. 202102080401. Lei Chen's work is partially supported by National Science Foundation of China (NSFC) under Grant No. U22B2060, the Hong Kong RGC GRF Project 16209519, RIF Project R6020-19, AOE Project AoE/E-603/18, Theme-based project TRS T41-603/20R, China NSFC No. 61729201, Guangdong Basic and Applied Basic Research Foundation 2019B151530001, Hong Kong ITC ITF grants MHX/078/21 and PRP/004/22FX, Microsoft Research Asia Collaborative Research Grant and HKUST-Webank joint research lab grants. This work has also been supported by the Hong Kong Research Grants Council through a General Research Fund (Project No. 11204919).

References

1. Full paper. https://github.com/csmichael/dpcovid/blob/main/full.pdf
2. Andres, M.E., Bordenabe, N.E., Cjhatzikokolakis, K., Palamidessi, C.: Geo-indistinguishability: differential privacy for location-based systems. In: CCS (2013)
3. Cho, E., Myers, S.A., Leskovec, J.: Friendship and mobility: user movement in location-based social networks. In: KDD (2011)
4. Cormode, G., Kulkarni, T., Srivastava, D.: Answering range queries under local differential privacy. In: VLDB (2019)
5. Cunningham, T., Cormode, G., Ferhatosmanoglu, H., Srivastava, D.: Real-world trajectory sharing with local differential privacy. VLDB (2021)
6. Da, Y., Ahuja, R., Xiong, L., Shahabi, C.: REACT: real-time contact tracing and risk monitoring via privacy-enhanced mobile tracking. In: ICDE (2021)
7. Domingo-Ferrer, J., Sánchez, D., Blanco-Justicia, A.: The limits of differential privacy (and its misuse in data release and machine learning). CACM (2021)
8. Duchi, J.C., Jordan, M.I., Wainwright, M.J.: Local privacy and statistical minimax rates. In: FOCS (2013)
9. Erlingsson, Ú., Feldman, V., Mironov, I., Raghunathan, A., Talwar, K., Thakurta, A.: Amplification by shuffling: From local to central differential privacy via anonymity. SODA (2019)
10. Ferretti, L., et al.: Quantifying SARS-CoV-2 transmission suggests epidemic control with digital contact tracing. Science **368**(6491) (2020)
11. Gao, D., Tong, Y., She, J., Song, T., Chen, L., Xu, K.: Top-k team recommendation in spatial crowdsourcing. In: Cui, B., Zhang, N., Xu, J., Lian, X., Liu, D. (eds.) WAIM 2016. LNCS, vol. 9658, pp. 191–204. Springer, Cham (2016). https://doi.org/10.1007/978-3-319-39937-9_15
12. Guan, W.J., Ni, Z.Y., Hu, Y., et al.: Clinical characteristics of coronavirus disease 2019 in China. N. Engl. J. Med. **382**(18), 1708–1720 (2020)
13. Kato, F., Cao, Y., Yoshikawa, M.: Secure and efficient trajectory-based contact tracing using trusted hardware. In: IEEE Big Data (2020)
14. Lewis, D.: Why many countries failed at COVID contact-tracing-but some got it right. Nature 384–387 (2020)
15. Liu, Q., Shen, Y., Chen, L.: LHist: towards learning multi-dimensional histogram for massive spatial data. In: ICDE (2021)
16. Liu, Q., Shen, Y., Chen, L.: HAP: an efficient hamming space index based on augmented pigeonhole principle. In: SIGMOD (2022)

17. Liu, X., Trieu, N., Kornaropoulos, E.M., Song, D.: BeeTrace: a unified platform for secure contact tracing that breaks data silos. IEEE Data Eng. Bull. (2020)
18. Peng, Z., Xu, C., Wang, H., Huang, J., Xu, J., Chu, X.: P^{2}b-trace: privacy-preserving blockchain-based contact tracing to combat pandemics. In: SIGMOD (2021)
19. Rodríguez, P., et al.: A population-based controlled experiment assessing the epidemiological impact of digital contact tracing. Nat. Commun. **12**(1), 1–6 (2021)
20. Roy Chowdhury, A., Wang, C., He, X., Machanavajjhala, A., Jha, S.: Crypte: crypto-assisted differential privacy on untrusted servers. In: SIGMOD (2020)
21. Salathé, M., et al.: Early evidence of effectiveness of digital contact tracing for SARS-CoV-2 in switzerland. medRxiv (2020)
22. She, J., Tong, Y., Chen, L., Song, T.: Feedback-aware social event-participant arrangement. In: SIGMOD (2017)
23. Tao, Q., Tong, Y., Zhou, Z., Shi, Y., Chen, L., Xu, K.: Differentially private online task assignment in spatial crowdsourcing: a tree-based approach. In: ICDE (2020)
24. Tong, Y., Yuan, Y., Cheng, Y., Chen, L., Wang, G.: Survey on spatiotemporal crowdsourced data management techniques. J. Softw. **28**(1), 35–58 (2017)
25. Tong, Y., Zeng, Y., Ding, B., Wang, L., Chen, L.: Two-sided online micro-task assignment in spatial crowdsourcing. TKDE (2021)
26. Trieu, N., Shehata, K., Saxena, P., Shokri, R., Song, D.: Epione: lightweight contact tracing with strong privacy. IEEE Data Eng. Bull. (2020)
27. Wagh, S., He, X., Machanavajjhala, A., Mittal, P.: DP-cryptography: marrying differential privacy and cryptography in emerging applications. Commun. ACM **64**(2), 84–93 (2021)

Efficient and Accurate Range Counting on Privacy-Preserving Spatial Data Federation

Maocheng Li[1], Yuxiang Zeng[1,2], and Lei Chen[1,3(✉)]

[1] The Hong Kong University of Science and Technology, Hong Kong, China
{csmichael,leichen}@cse.ust.hk
[2] School of Computer Science and Engineering, Beihang University, Beijing, China
turf1013@buaa.edu.cn
[3] The Hong Kong University of Science and Technology (Guangzhou),
Guangzhou, China

Abstract. A spatial data federation is a collection of data owners (*e.g.*, a consortium of taxi companies), and collectively it could provide better location-based services (LBS). For example, car-hailing services over a spatial data federation allow end users to easily pick the best offers. We focus on the range counting queries, which are primitive operations in spatial databases but received little attention in related research, especially considering the privacy requirements from data owners, who are reluctant to disclose their proprietary data. We propose a grouping-based technical framework named FedGroup, which groups data owners without compromising privacy, and achieves superior query accuracy (up to 50% improvement) as compared to directly applying existing privacy mechanisms achieving Differential Privacy (DP). Our experimental results also demonstrate that FedGroup runs orders-of-magnitude faster than traditional Secure Multiparty Computation (MPC) based method, and FedGroup even scales to millions of data owners, which is a common setting in the era of ubiquitous mobile devices.

1 Introduction

A federated database system is a collection of multiple cooperating but autonomous database systems [16]. It reflects the real-world situation that the entire database of customer records is usually divided among different companies, and each company owns a distinct share of the entire database. Federated database systems have drawn many research interests [3,4,17] and been deployed in real-world applications. For example, multiple hospitals participate in an alliance to collectively contribute their data for discovering new drugs [3].

In the specific domain of location-based services (LBS) systems, a *spatial data federation* considers a federation of *spatial data*, such as locations or trajectories. Similar to general-purpose federated databases, spatial data federation has also seen a wide range of real-world applications. For example, Alibaba's AMap [20] in

X. Wang et al. (Eds.): DASFAA 2023, LNCS 13943, pp. 317–333, 2023.
https://doi.org/10.1007/978-3-031-30637-2_21

China provides car-hailing services over a federation of different taxi companies, which enables users to have more flexibility and easily pick the best offers.

To support the aforementioned LBS applications, query processing techniques are needed for a wide range of spatial queries. Among them, *range counting queries* are one of the most important primitive operations. Range counting queries return the count of the spatial objects located within a given range. A car-hailing federated system may frequently issue the following range counting query: *"how many cars are within 500 m of the location of a customer?"*

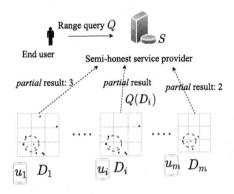

Fig. 1. FPRC problem. Each one of m data owners u_i has a spatial database D_i. An end user issues a range counting query Q over the spatial data federation: $D_1 \cup \ldots \cup D_m$.

In this paper, we target at the *Federated Privacy-preserving Range Counting (FPRC)* problem (Fig. 1), with a special focus on the large-scale spatial data federations, where the number of data owners is large (*e.g.*, more than 10 K). Such large-scale settings have recently drawn more attentions due to the wide adoption of mobile devices. For example, in a decentralized crowdsourcing platform [23], each data owner is a mobile device, and each user holds his own data and is not willing to disclose any unprotected private data to other parties. In such applications, the number of data owners could easily scale up to the order of millions.

It is challenging to support range counting queries over a spatial data federation, because each data owner is often reluctant to disclose its proprietary data. Directly applying traditional Secure Multiparty Computation (MPC) techniques or Differential Privacy (DP) solutions would either be too computationally expensive, or bring too much noise to outweigh any meaningful results for the LBS applications. As demonstrated by our experiments in Sect. 4, the MPC baseline is orders-of-magnitude slower than our solution, requiring more than 13 h when the input size is large, while the DP baseline loses more than 50% accuracy. Due to the interactive nature of the LBS applications, an ideal solution should offer excellent accuracy, practical efficiency, and proven privacy guarantee.

To tackle the challenges of FPRC problem, we propose a grouping-based technical framework named FedGroup. To reduce the large amount of noise injected by directly applying DP for each data owner, we assign similar data owners (mobile users in our setting) into groups. We argue that injecting DP noise in each group (as opposed to injecting noise for each data owner) suffices to provide privacy guarantee. The reason is due to the fact that certain users may have social relationships with other data owners (*e.g.*, they belong to the same family), and they share similar trajectories and whereabouts.

To achieve effective grouping, we adopt a commonly accepted concept named (k, r)-core in graph analysis [27] to ensure that the group has strong connections inside (here the nodes in the graph are data owners and the edges between nodes indicate that they have strong spatial similarities, *i.e.*, their trajectories are similar). We formulate an optimization problem to find the best way to put users into groups in order to minimize the error introduced by DP noise in the FPRC problem. We show that the problem is NP-hard and provide an efficient greedy-based algorithm with a performance guarantee.

Last but not least, the grouping of users relies on how to measure the spatial similarity between users (*e.g.*, their trajectories). It is a non-trivial task to construct the similarity graph between each pair of users, considering that the trajectories are private information of each data owner. We devise a novel hybrid solution to combine DP and MPC to achieve accurate similarity graph construction.

To summarize, we make the following contributions:

- We develop a novel technical framework FedGroup, to utilize an offline grouping step to largely reduce the amount of noise needed in the Federated Privacy-preserving Range Counting (FPRC) problem. We introduce the FPRC problem in Sect. 2.
- In order to achieve effective grouping in FedGroup, we show that it is an NP-hard problem to achieve optimal grouping to minimize the amount of noise needed in FPRC problem. Then, we propose an efficient greedy-based algorithm with a performance guarantee to achieve effective grouping.
- We also devise effective techniques to construct the spatial similarity graph between users by combining DP and MPC techniques, considering data owners' trajectories as private information. The details are presented in Sect. 3.
- We conduct extensive experiments to validate the effectiveness and efficiency of our proposed solution. The results are shown in Sect. 4.

In addition, we review related works in Sect. 5 and conclude in Sect. 6.

2 Problem Definition

In this section, we introduce some basic concepts, the adversary model, and the definition of the Federated Privacy-preserving Range Counting (FPRC) problem. Due to page limitations, a toy example is provided in the full report [1].

2.1   Basic Concepts

Definition 1 (Location). *A location $l = (x, y)$ represents a 2-dimensional spatial point with the coordinates (x, y) on an Euclidean space.*

Locations in an Euclidean space is commonly seen in existing work [10,15,19].

Definition 2 (Data owner). *There are m data owners $u_1, \ldots, u_m$. Each data owner u_i owns a spatial database D_i. Each spatial database D_i consists of multiple data records (locations as in Definition 1), i.e., $D_i = \{l_1, \ldots, l_{|D_i|}\}$.*

We also refer to each data owner's database D_i as a *data silo*. We let $D = \cup_{i=1}^m D_i$ denotes the collection of all data silos, *i.e.*, the union of all data records from each data silo.

Definition 3 (Range counting query). *A range counting query Q asks for how many data records are within distance r to the query location q_0, i.e.,*

$$Q(r, q_0) = \sum_{l \in D} \mathbb{I}(d(l, q_0) < r) \tag{1}$$

where $\mathbb{I}(\cdot)$ is the indicator function which equals to 1 if the predicate $d(l, q_0) < r$ is true, and 0 otherwise. $d(\cdot)$ is the Euclidean distance function. $l \in D$ is any record from the union of all data records from each data silo.

2.2   Privacy and Adversary Model

There are mainly three parties (roles) in our spatial data federation (see Fig. 1):
(1) the **end user** who issues the query; (2) the **service provider** who receives the query and coordinates the execution of the query; and (3) each **data owner** u_i who owns the private spatial database D_i.

We assume that all parties are *semi-honest*, meaning that they are *curious but not malicious*. They are *curious* about other parties' private information, but they honestly follow and execute system protocols. The setting has been widely adopted in recent privacy-preserving LBS related applications [18,21,24].

2.3   Federated Privacy-Preserving Range Counting Problem

Based on the previous concepts and the adversary model, we now define the Federated Privacy-preserving Range Counting (FPRC) problem as follows.

Definition 4 (Federated Privacy-preserving Range Counting problem). *Given a federation of m spatial databases $D_1, \ldots, D_m$, a range counting query $Q(q_0, r)$, and a privacy parameter (a.k.a., privacy budget) ϵ, the FPRC problem asks for a query answer $\tilde{Q}$, with the following privacy requirements:*

- *R1. The computed final result $\tilde{Q}$ satisfies ϵ-Differential Privacy (DP), where the definition of neighboring databases refers to changing one single record in any data silo D_i.*

- *R2. The intermediate result disclosed by any data silo D_i satisfies ϵ-DP.*
- *R3. The private inputs for each data silo D_i are confidential if there is any multiparty computation involved.*

Two baselines, each respectively based on MPC and DP, are proposed. The MPC baseline directly applies MPC technique to compute a secure summation over the query result from each data silo. The DP baseline injects an instance of Laplace noise to each data silo, and collects the aggregated noisy summation.

As mentioned in Sect. 1, the **key challenges** to solve the FPRC problem are threefold: 1) privacy: each data owner is reluctant to disclose its sensitive data; 2) efficiency: directly applying MPC results an impractical solution; 3) accuracy: since the number of data owners could reach the order of millions, the scale of total noise injected by the DP baseline is too large.

3   Our Solution FedGroup

In this section, we introduce the technical details of FedGroup, which addresses the challenges of the FPRC problem. We highlights its overall workflow first (Fig. 2 is an illustrative figure), and then introduce the details in each of the steps: 1) constructing the spatial similarity graph; 2) finding groups given the similarity graph; and 3) partial answers and aggregation. Due to page limitations, please refer to the full report for more examples and detailed proofs [1].

Key Idea and Intuition. In our setting of the Federated Privacy-preserving Range Counting (FPRC) problem (as in Sect. 2), the number of data owners (m) is potentially very large (*e.g.*, millions), as each data owner could be a mobile user. The key idea of FedGroup is based on the following observation: certain mobile users have strong connections (due to social ties or other collaboration relationships) with each other. If we consider such pairs of data owners to belong to the same group, then the privacy protection inside the group could be relaxed. For example, there is no need to consider privacy protection of users' trajectories within a family, as family members are very likely to know about other members' whereabouts during the day. Thus, DP noise is only needed cross groups, as opposed to the case in the DP baseline, where an instance of Laplace noise is injected for every mobile user. As a result, the overall noise injected for the query result for the FPRC problem could be greatly reduced.

3.1   Spatial Similarity Graph Construction

In this step, the goal is to construct an undirected graph $G_s = (V, E)$ between data owners. Each node in V of the graph represents a data owner u_i, and an edge in E between two nodes u_i and u_j indicates the similarity of the spatial databases D_i and D_j owned by u_i and u_j, respectively. The weight of an edge measures the strength of the connection between u_i and u_j, and we use cosine similarity as the weight of the edge.

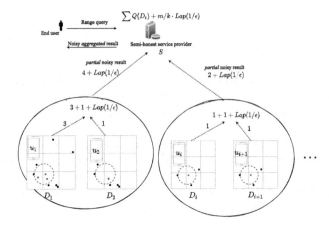

Fig. 2. FedGroup workflow.

We first introduce the details of similarity graph, including how to measure the spatial similarity between data owners. Then, we consider the privacy-preserving setting, where each data owner's spatial database is considered private, and present our solutions of computing the similarity function between data owners in a differentially private manner.

Spatial Similarity Graph. The key to construct the spatial similarity graph $G_s = (V, E)$ is measuring the weight of each undirected edge $e = (u_i, u_j)$ between two data owners $u_i, u_j \in V$. In the following section, we introduce how to compute the similarity function between the two data silos D_i and D_j.

For each data silo D_i, we use a grid structure T to decompose the spatial domain into a number of grids, where $|T|$ denotes the number of grids. Then, within each grid, a count is computed to denote the number of spatial points (records) of D_i falling inside the grid. Thus, we obtain a $|T|$-dimensional count vector for the data silo D_i.

Similarly, for data silo D_j, we could obtain another count vector $v_j = [c_{j,1}, c_{j,2}, \ldots, c_{j,|T|}]$. Since the two vectors v_i and v_j are with the same length $|T|$, we could use *cosine similarity* to measure their similarity as:

$$\text{sim}(D_i, D_j) = \cos(v_i, v_j) = v_i \cdot v_j / \|v_i\| \|v_j\| \tag{2}$$

The cosines similarity ranges from 0 to 1, and the larger the value is, the more similar the two count vectors are. In our application, the higher the cosine similarity between the two count vectors is, the more similar are the two associated spatial databases, owned by two different data owners.

Constructing the Graph. To construct the spatial similarity graph $G_s = (V, E)$, first we insert all the data owners as the node set V. Then, we iterate over all pairs of data owners $u_i, u_j \in V$, and measure the spatial similarity of their spatial databases D_i and D_j, according to Eq. (2). The weight of the corresponding edge $e = (u_i, u_j)$ is set as $\text{sim}(D_i, D_j)$, i.e., $w(e) = \text{sim}(D_i, D_j)$.

Without considering the privacy of each data silo D_i, constructing the spatial similarity graph G_s is straightforward. The method is running on the service provider S. First, we create the node set V of G_s by inserting all data owners $u_1, \ldots, u_m$ into V. Then, the method simply iterates over the pairs of data owners u_i, u_j for $i, j \in [m]$ and $i < j$, and calculates $\text{sim}(D_i, D_j)$. If the similarity is larger than the given threshold r, then an edge (u_i, u_j) will be inserted into the edge set E of G_s, with the edge weight set as $w(u_i, u_j) = \text{sim}(D_i, D_j)$.

The time complexity is $O(m^2)$, since we iterate over all possible pairs of data owners. The space complexity required is also $O(m^2)$, because we need to store G_s on S, including all the nodes and the edges (and their weights).

Privacy-Preserving Computation. Although it is straightforward to construct G_s without considering the privacy of each data silo D_i, it is crucial to provide a proven privacy guarantee when we measure the spatial similarity between each pairs of data silos. The first and foremost motivation of this work is to consider privacy-preservation in the spatial federation setting, where each data owner's data silo D_i is considered sensitive.

Obviously, the non-private way of constructing the graph fails to meet the privacy requirement R2 in our problem definition in Definition 4. In R2, it requires that the intermediate results shared by each data owner satisfy ϵ-DP. However, in the non-private version, the provided D_i (or its count vectors v_j) are directly published by each data owner to the service provider S. Clearly it is a privacy failure. Furthermore, R1 in Definition 4 may fail. It is not clear whether the constructed similarity graph G_s provides any formal privacy guarantee *w.r.t.* changing of one record in any data silo D_i.

Thus, we present a hybrid solution, which utilizes both Secure Multiparty Computation (MPC) and the standard Laplace mechanism in DP.

Hybrid Solution. The hybrid solution combines the advantages of both worlds – the high accuracy of MPC (because the computations are *exact*), and the high efficiency of DP (because each data silo performs the Laplace computation independently, and injecting Laplace noise itself is an efficient operation).

The key idea of the hybrid solution is that: we use the efficient Laplace mechanism in DP to quickly compute a noisy weight $\tilde{w}$. Instead of directly using this noisy weight as a surrogate weight for the edges on the ground-truth graph, we only use it as a filtering mechanism. If the noisy weight $\tilde{w}$ is too small as compared to the given threshold r, we could quick conclude that the edge should *not* be inserted to the graph. Or, if the noisy weight $\tilde{w}$ appears to be high, we quickly conclude that the edge should be inserted to the graph. If the noisy weight lies on the borderline, which shows uncertainty and may introduce errors, we invoke the computationally heavy MPC method to calculate the exact edge weight given the two private inputs from two data owners.

Algorithm 1: Construct G_s - Hybrid solution

Input: $u_1, \ldots, u_m, \tilde{v}_1, \ldots, \tilde{v}_m, r, r_l, r_u$.
Output: $\tilde{G}_s = (V, \tilde{E})$.

1   $V := \{u_1, \ldots, u_m\}$
2   $\tilde{E} := \{\}$
3   **foreach** $u_i \in V$ **do**
4   **foreach** $u_j \in V$ and $j > i$ **do**
5       $e := (u_i, u_j)$
6       $\tilde{w} := \cos(\tilde{v}_i, \tilde{v}_j)$
7   **if** $e.weight < r_l$ **then**
8   Continue
9   **else if** $\tilde{w} > r_u$ **then**
10        $e.\text{weight} := \tilde{w}$
11   Insert e to $\tilde{E}$
12   **else**
13         $e.\text{weight} := \text{Secure\_Sim}(D_i, D_j)$
14   **if** $e.weight > r$ **then**
15   Insert e to $\tilde{E}$

16   **return** $\tilde{G}_s = (V, \tilde{E})$

Algorithm 1 shows the detailed steps. The inputs $\tilde{v}_1, \ldots, \tilde{v}_m$ are the noisy counts vectors, obtained by injecting Laplace noise to the counts vectors $v_1, \ldots, v_m$.

Privacy Analysis. We show that our hybrid solution satisfies the privacy requirements as in Definition 4.

Theorem 1. *The hybrid solution in Algorithm 1 satisfies all privacy requirements (R1-R3) in Definition 4.*

Proof. Since the only released intermediate results are the noisy counts $\tilde{v}_i$, R2 is satisfied. R1 is satisfied by the post-processing and parallel composition theorem of DP, because the output graph $\tilde{G}_s$ only depends on the noisy counts. R3 is satisfied because the multi-party computation only happens at Line 12, and the confidentiality of the data is provided by the MPC protocol.

3.2 Finding Groups

(k, r)**-Core.** We adopt a commonly accepted concept (k, r)-core [27] from graph mining literature to define the groups. The concept of (k, r)-core considers two important criteria of determining closely related groups (*a.k.a.*, community detection): *engagement* and *similarity*, such that the group members not only have strong social connections with each other (strong engagement), but also share high similarities (*i.e.*, their trajectories are similar).

In our setting, as we focus on the spatial database of data owners, rather than the social graph, we require that the groups we form demonstrate strong similarities between data owners. Thus, we require each group to be a r-clique, meaning each group is a clique, and the edge weight (*i.e.*, the spatial similarity, introduced in Sect. 3.1) between the data owners is larger or equal to r.

Definition 5 (r-group). *A r-group g is a clique in the spatial similarity graph G_s, such that there exists an edge between any two data owners belonging to the group, and the edge weight is larger than the threshold r, i.e., for any two data owners $u_i, u_j \in g$,*

$$\exists e = (u_i, u_j) \in E \ , w(e) > r \tag{3}$$

Optimizing the Grouping. The concept of r-group gives meaning to grouping data owners together, because they are similar. However, there could be various ways of putting data owners into different groups. For example, a data owner could belong to multiple r-group, and which group should we choose to put the data owner in?

In this section, we formulate an optimization problem to connect the utility goal of the ultimate problem of this paper – the FPRC problem, with the grouping strategy. Then, we show that this optimization is an NP-hard problem, and we propose a greedy algorithm, which is effective and efficient.

We first define the Data Owner Grouping (DOG) problem.

Definition 6 (Data Owner Grouping (DOG) problem). *Given the inputs to the FPRC problem in Definition 4, the DOG problem asks for a way to assign data owners into λ disjoint groups: $g_1, g_2, \ldots, g_\lambda$, where each group g_i is a r-group (according to Definition 5), such that the aggregated amount of noise injected by FedGroup for the FPRC problem is minimized.*

Next, we show that to minimize the error (the total amount of noise injected) for the FPRC problem, it is equivalent to minimize the number of groups λ.

Lemma 1. *Minimizing the error in the FPRC problem is equivalent to minimizing λ, which is the total number of groups in the DOG problem.*

Proof. We review the error of our FedGroup solution in the FPRC problem. FedGroup assigns data owners into disjoint groups, and each group injects one single instance of Laplace noise to the query result. Then, the partial noisy result is collected from each group and aggregated as the final answer. Thus, the total error of our final answer $\tilde{Q}$, which is measured by its variance, is:

$$\text{Var}(\tilde{Q}) = \text{Var}(\sum_{i=1}^{\lambda} \text{Lap}(1/\epsilon)) = \lambda \cdot \text{Var}(\text{Lap}(1/\epsilon)) = 2\lambda/\epsilon^2. \tag{4}$$

Thus, to minimize the total error in Eq. (4), we need to minimize λ.

Algorithm 2: Greedy Find Groups

Input: $\tilde{G}_s = (V, \tilde{E})$.

Output: $g_1, g_2, \ldots, g_\lambda$.

1   $G_c := \text{ComplementGraph}(\tilde{G}_s)$

2   assigned := InitializeIntegerArray(size=$|V|$, initValue=0)

3   $\lambda := 1$; assigned[u_1] = 1 ; g_1.insert(u_1)

4   $u := \text{NextUnassignedVertex}(V, \text{assigned})$

5 **while** u *exists* **do**

6     $gid := \text{NextGroupId}(u.\text{neighbors}(), \text{assigned})$

7   assigned[u] = gid ; g_{gid}.insert(u)

8   **if** $gid > \lambda$ **then**

9       $\lambda = gid$

10     $u := \text{NextUnassignedVertex}(V, \text{assigned})$

11 **return** $g_1, g_2, \ldots, g_\lambda$

Now, our DOG problem becomes finding an optimal way of assigning data owners into disjoint groups, where each group is a r-clique, and the number of groups is minimized. In fact, the problem could be reduced from the Minimum Clique Partition (MCP) problem and the hardness result is shown as follows.

Theorem 2. *The DOG problem is NP-hard.*

Proof. We reduce the NP-hard problem, Minimum Clique Partition (MCP), to the DOG problem in the full report [1]. Thus, DOG shares the same hardness.

Greedy Algorithm. Since the DOG problem is an NP-hard problem, we devise an efficient greedy solution (Algorithm 2) to obtain effective grouping and provide a bounded noise scale in the returned solution. The solution is inspired by the greedy solution to the Minimum Graph Coloring problem, which is shown as an equivalent problem to the MCP problem.

Performance Guarantee. In Theorem 3, we show the greedy solution has a bounded number of groups and offers a bounded noise for the FPRC problem.

Theorem 3. *If d is the largest degree of a node in the complement graph G_c, then Algorithm 2 returns at most $d + 1$ groups.*

Proof. We focus on a data owner u with degree d (the maximum degree in the graph) as Algorithm 2 proceeds. There are at most d neighbors of u that we should avoid using the same group id. Since we are using the lowest-numbered group ids that have not been used by any of the neighbors, among group id $1, 2, \ldots, d + 1$, there is at least one id that could be used by u (*e.g.*, the first d group ids are used, and we now use $d + 1$). Thus, we conclude that at most $d + 1$ groups are returned by Algorithm 2.

Time Complexity. Each time when Algorithm 2 processes a vertex, it iterates over the neighbors of the vertex to find the lowest-numbered group id that is available. Thus, overall, it takes $O(|E|)$ time to run the algorithm. The graph complement step at the initialization also takes $O(|E|)$ time.

3.3   Partial Answers and Aggregation

Section 3.2 describes the most critical step of FedGroup, which is finding the groups for data owners. After the grouping, data owners belonging to the same group could avoid injecting separate instances of Laplace to their individual query answers. Instead, each group aggregates the partial answers from data owners and injects one instance of Laplace noise to the partial aggregated answer.

As the last step, the service provider S collects the partial noisy answers from all groups $g_1, \ldots, g_\lambda$. The noisy answers are aggregated (taking a summation) and then returned to the end user.

3.4   Extension to Range Aggregation Queries

In this paper, we focus on the most fundamental primitives for LBS systems, the range counting queries. However, the techniques we present could be easily extended to other aggregation queries, including range SUM() and AVG().

To extend FedGroup to SUM(), the important extension is the scale of noise injected for each group (the sensitivity). Since now each data record in a spatial database does not only affect the final query result by $+/-1$, we need to inject the worst-case scale to satisfy differential privacy. To avoid injecting unbounded Laplace noise, a truncation could be performed, *i.e.*, we truncate all data records to make a certain attribute smaller or equal to a truncation parameter θ.

The extension to AVG() is straightforward as the average could be calculated by SUM()/COUNT().

4   Experimental Study

In this section, we first introduce the experimental setup in Sect. 4.1 and present the detailed experimental results in Sect. 4.2.

4.1   Experimental Setup

Datasets. We use a real-world geo-social network datasets Gowalla [7] and a randomly generated synthetic dataset in our experiments.

Baselines. We compare our proposed FedGroup (short as **FG**) method with the DP baseline (short as **DP**) and the MPC baseline (short as **MPC**).

For FedGroup, we also implement a variant based on cliques sizes (a heuristic based solution). It lists the cliques in the graph in descending sizes, and keeps adding the largest clique into a new group until all nodes are assigned with a group. We name this variant as FedGroup-Exhaustive (short as **FG-Ex**). Note that finding the maximum clique in a graph is in general an NP-hard problem, so we expect that this variant is only tractable on small graphs.

Metrics. We focus on the following end-to-end metrics:

- Mean relative error (MRE): the ratio of error as compared to the true result, *i.e.*, |returned result - true result|/true result, averaged over repeated queries.
- Mean absolute error (MAE): the absolute error of the returned result as compared to the true results, *i.e.*, |returned result - true result|, averaged over repeated queries.
- Query evaluation time: the time to execute the spatial count queries in seconds, averaged over repeated queries.

Control variables.

- The number of data owners: $m \in [500, 1K, 2K, 3K]$. For the scalability test: $m \in [100K, 250K, 500K, 1M]$.
- Privacy budget: $\epsilon \in [0.2, \mathbf{0.3}, 0.4, 0.5]$. The privacy budget considered is relatively small, because it is a personal budget for each data owner, and repeated queries may result linear composition of the budget.

4.2 Experimental Results

Here, we present the results on query accuracy to first verify the motivation of our work – our proposed solution FedGroup offers superior query accuracy (up to 50% improvement) as compared to the DP baseline. We also compare the running time of query execution to show that the MPC baseline is not tractable even in moderate data sizes, while our solution FedGroup offers excellent scalability. Overall, we have a better utility/efficiency trade-off than the MPC baseline.

Query Accuracy. We test the query accuracy (using MRE and MAE) of different methods. The results are shown in Fig. 3 and Fig. 4. Over different privacy budget, FedGroup consistently outperforms the DP baseline by a large margin (30–50% improvement). Across different input sizes (the number of federations m), FedGroup also provides a significant accuracy improvement.

Varying ϵ. Figure 3 shows the query accuracy results when we test the compared methods over different privacy budget ϵ for a spatial federation of $m = 500$ data owners using the Gowalla dataset. FedGroup (FG) offers a clear improvement over the DP baseline. Specifically, as shown in Fig. 3a, the MRE for the DP baseline at $\epsilon = 0.3$ is 0.801 (80.1%), while FG reduces the error to 0.551 (55.1%), which provides a $(0.801 - 0.551)/0.801 * 100\% = 31.2\%$ improvement.

When ϵ gets larger, indicating a more relaxed privacy requirement, the query accuracy generally improves for all methods, reflected by a smaller MRE and MAE. At $\epsilon = 0.5$ (as shown in Fig. 3a and Fig. 3b), the MRE and MAE for the DP baseline is 0.618 and 58.069, respectively. In comparison, the MRE and MAE of FG improves to 0.271 and 25.448, respectively. The query accuracy

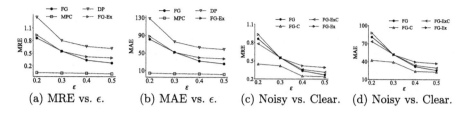

Fig. 3. End-to-end query accuracy of different methods vs. the privacy budget ϵ. (Gowalla dataset, $m = 500$.)

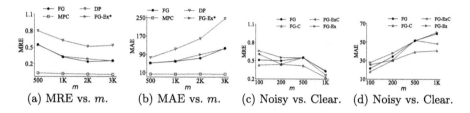

Fig. 4. End-to-end query accuracy of different methods vs. the number of federations m. (Gowalla dataset, $\epsilon = 0.3$.)

improvement of FG over DP is about 56.1%. The accuracy improvement is more significant than the one of $\epsilon = 0.3$ (31.2%), and the explanation is that when ϵ gets larger, more accurate spatial similarity graphs are constructed (leading to more correctly identified edges), and FG could assign data owners into smaller number of groups, which eventually leads to a smaller end-to-end query error.

It is worth noting that the computationally heavy MPC baseline offers the best query accuracy (due to the $O(1)$ noise) across all ϵ. Despite its impracticability in real-world data sizes, we include its results here for completeness.

Varying m. For a fixed privacy budget ϵ, we test the query accuracy of different methods across different m. The results (Fig. 4) verify that FG consistently outperforms the DP baseline by a significant margin. As m grows from 500 to 3 K, the MAE of the DP baseline grows from 75.3 to 246.5. In comparison, our FG provides a lower MAE, which grows from 51.8 to 117.4. From Fig. 4b, we could see that the MAE of FG grows more slowly than the DP baseline.

When m increases, the true query count also increases. This results a decreasing trend for MRE for all methods, as shown in Fig. 4a. When $m = 2$ K, the MRE of the DP baseline is 0.517, while our FG's MRE is 0.238, which provides a 53.97% improvement. For different m, the MRE of DP is about 2× as large (worse) as the one of our FG.

Query Efficiency. We then conduct experiments to test the query efficiency for different methods. The results are shown in Fig. 5. The results on the scalability tests on the synthetic dataset are shown in Fig. 6.

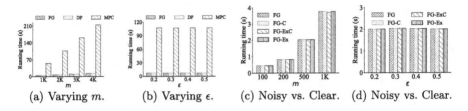

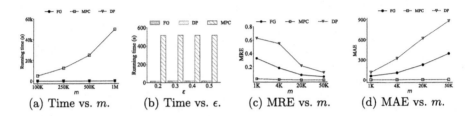

Fig. 5. Efficiency (running time) for different control variables. (Gowalla dataset)

Fig. 6. Scalability tests. (Synthetic dataset)

Gowalla Dataset. For smaller m in the range of 1 K–4 K, we show the query execution time of different methods in Fig. 5, using the Gowalla dataset. The query time using the MPC method grows linearly as m grows (Fig. 5a), and stays at the same level across different ϵ (Fig. 5b), but it is orders-of-magnitude larger (slower) than FG and DP.

Figure 5c and Fig. 5d show that the running time is not impacted much by whether we use the ground-truth graph or the noisy graph. The time grows roughly linearly when m increases, but all FG variants finish within 4 s when $m = 1K$, which is not comparable to the time-consuming MPC baseline.

Scalability Tests. We further verify the scalability of our FG method by testing on the synthetically generated dataset. Figure 6a shows the query execution time when we scale m to a million. Despite the linear growth of running time of MPC, it requires more than 13 h to execute a query when $m = 1M$. This is not acceptable in real-world LBS applications. In comparison, both the DP and our FG answers the query within 20 s at the largest input size.

For a given m, the running time of different methods do not vary much over different ϵ (Fig. 6b). In addition to the efficiency results, we also include some query accuracy results when we test the methods on extremely large inputs from the synthetic dataset (shown in Fig. 6c–Fig. 6d). These results further verify that our FG method is consistently outperforming the equally efficient DP method.

Summary. We conduct extensive experiments to examine both the effectiveness (in terms of query accuracy) and the efficiency of our proposed method FedGroup. In terms of query accuracy, FedGroup provides up to 50% improvement over the DP baseline. Though the computationally heavy MPC baseline delivers the best accuracy, its inefficiency prevents it from real-world application. The FedGroup runs orders-of-magnitude faster than the MPC baseline and executes a query within 20 s for a million data owners.

5   Related Work

In this section, we first review the works on general-purposed data federations, and then discuss specific works on spatial data federation, as well as other spatial data management techniques considering data privacy.

General-Purposed Data Federations. The concept of data federation dated back to the 90s when the need arises to manage autonomous but cooperating database systems [16]. In recent years, there are a growing number of works on optimizing and adopting secure multiparty computations to build privacy-preserving data federations [3–5]. Bater et al. first implemented the MPC-based data federation system called SMCQL [3], and then improved the system to Shrinkwrap [4] by using differential privacy to reduce the amount of dummy records inserted during query execution in order to make the computation *oblivious*. Its later effort, SAQE [5], further improves the tradeoff among utility, efficiency, and privacy by considering approximate query processing, as a by-product of injecting necessary noise for achieving differential privacy. Conclave [25] is another system which adopts secure query processing, *i.e.*, the private data are *confidential* during the execution of the queries.

These security-based systems offer strong privacy guarantee, however the efficiency issue is a bottle-neck. Most of the works are limited to only two parties in the secure computation, and experimental results [22] show that the current system is still far from being scalable in real-world data sizes. For example, joining a private table L (shared by multiple parties) and a public table R takes more than 25 min, even if R contains only 100 objects.

Specialized Spatial Data Federations. In an effort to develop specialized and more optimized spatial data federation, Shi et al. [17] first studied how to efficiently perform approximate range aggregation queries without considering privacy. Later, a MPC-based system HuFu [22] was built, with specialized optimization made for spatial operators. Though significant improvements are made over other existing works (*e.g.*, Conclave), the running time is still a major concern. In our experiments, we demonstrate even the most simplified MPC baseline runs orders-of-magnitude slower than our proposed solution.

Privacy-Preserving Spatial Data Management. In parallel, there are a fruitful amount works of spatial data management using DP or local DP (LDP). The DP model masks the existence of a single location record [8] or an entire trajectory [11]. There are also other efforts on protecting location data [18,26] or using LDP and its variants on offering theoretical guarantee to protect the location data [2,6,9] The DP and LDP model both differ from our privacy model, where a number of data owners exist.

6   Conclusion

In this paper, we target at the Federated Privacy-preserving Range Counting (FPRC) problem, where the number of federations (data owners) is large. By utilizing the social ties between data owners (such as family members), we propose a grouping-based framework called FedGroup. It reduces the amount of noise required by only injecting one instance of Laplace noise per group, while the DP baseline requires one instance of noise for each data owner, resulting an unacceptably large scale of noise, overweighing the true query answer. In addition, FedGroup is vastly more efficient than the MPC-based baseline, especially when we scale to a million data owners. Extensive experimental results verify the accuracy improvement and efficiency advantage of our proposed solution. For future works, we look into more advanced ways to compress data [12–14] to achieve better privacy/utility tradeoff.

Acknowledgements. Lei Chen's work is partially supported by National Science Foundation of China (NSFC) under Grant No. U22B2060, the Hong Kong RGC GRF Project 16213620, RIF Project R6020-19, AOE Project AoE/E-603/18, Theme-based project TRS T41-603/20R, China NSFC No. 61729201, Guangdong Basic and Applied Basic Research Foundation 2019B151530001, Hong Kong ITC ITF grants MHX/078/21 and PRP/004/22FX, Microsoft Research Asia Collaborative Research Grant and HKUST-Webank joint research lab grants.

References

1. Full paper (2022). https://github.com/csmichael/dpfed/blob/main/full.pdf
2. Andres, M.E., Bordenabe, N.E., Cjhatzikokolakis, K., Palamidessi, C.: Geo-indistinguishability: differential privacy for location-based systems. In: CCS (2013)
3. Bater, J., Elliott, G., Eggen, C., Goel, S., Kho, A.N., Rogers, J.: SMCQL: secure query processing for private data networks. VLDB (2017)
4. Bater, J., He, X., Ehrich, W., Machanavajjhala, A., Rogers, J.: Shrinkwrap: efficient SQL query processing in differentially private data federations. VLDB (2018)
5. Bater, J., Park, Y., He, X., Wang, X., Rogers, J.: SAQE: practical privacy-preserving approximate query processing for data federations. VLDB (2020)
6. Chen, R., Li, H., Qin, A.K., Kasiviswanathan, S.P., Jin, H.: Private spatial data aggregation in the local setting. In: ICDE (2016)
7. Cho, E., Myers, S.A., Leskovec, J.: Friendship and mobility: user movement in location-based social networks. In: KDD (2011)
8. Cormode, G., Procopiuc, C., Srivastava, D., Shen, E., Yu, T.: Differentially private spatial decompositions. In: ICDE (2012)
9. Cunningham, T., Cormode, G., Ferhatosmanoglu, H., Srivastava, D.: Real-world trajectory sharing with local differential privacy. VLDB (2021)
10. Gao, D., Tong, Y., She, J., Song, T., Chen, L., Xu, K.: Top-k team recommendation in spatial crowdsourcing. In: WAIM (2016)
11. He, X., Cormode, G., Machanavajjhala, A., Procopiuc, C.M., Srivastava, D.: DPT: differentially private trajectory synthesis using hierarchical reference systems. VLDB (2015)

12. Liu, Q., Shen, Y., Chen, L.: LHist: towards learning multi-dimensional histogram for massive spatial data. In: ICDE (2021)
13. Liu, Q., Shen, Y., Chen, L.: HAP: an efficient hamming space index based on augmented pigeonhole principle. In: SIGMOD (2022)
14. Liu, Q., Zheng, L., Shen, Y., Chen, L.: Stable learned bloom filters for data streams. VLDB (2020)
15. She, J., Tong, Y., Chen, L., Song, T.: Feedback-aware social event-participant arrangement. In: SIGMOD (2017)
16. Sheth, A.P., Larson, J.A.: Federated database systems for managing distributed, heterogeneous, and autonomous databases. CSUR (1990)
17. Shi, Y., Tong, Y., Zeng, Y., Zhou, Z., Ding, B., Chen, L.: Efficient approximate range aggregation over large-scale spatial data federation. TKDE (2021)
18. Tao, Q., Tong, Y., Zhou, Z., Shi, Y., Chen, L., Xu, K.: Differentially private online task assignment in spatial crowdsourcing: a tree-based approach. In: ICDE (2020)
19. Tao, Q., Zeng, Y., Zhou, Z., Tong, Y., Chen, L., Xu, K.: Multi-worker-aware task planning in real-time spatial crowdsourcing. In: DASFAA (2018)
20. Technode: Alibaba's amap launches taxi ride-hailing platform in beijing (2021)
21. To, H., Shahabi, C., Xiong, L.: Privacy-preserving online task assignment in spatial crowdsourcing with untrusted server. In: ICDE (2018)
22. Tong, Y., et al.: Hu-Fu: efficient and secure spatial queries over data federation. VLDB (2022)
23. Tong, Y., Yuan, Y., Cheng, Y., Chen, L., Wang, G.: Survey on spatiotemporal crowdsourced data management techniques. J. Softw. **28**(1), 35–58 (2017)
24. Tong, Y., Zhou, Z., Zeng, Y., Chen, L., Shahabi, C.: Spatial crowdsourcing: a survey. VLDBJ (2020)
25. Volgushev, N., Schwarzkopf, M., Getchell, B., Varia, M., Lapets, A., Bestavros, A.: Conclave: secure multi-party computation on big data. In: EuroSys (2019)
26. Xiao, Y., Xiong, L.: Protecting locations with differential privacy under temporal correlations. In: CCS (2015)
27. Zhang, F., Zhang, Y., Qin, L., Zhang, W., Lin, X.: When engagement meets similarity: efficient (k, r)-core computation on social networks. VLDB (2017)

Predicting Where You Visit in a Surrounding City: A Mobility Knowledge Transfer Framework Based on Cross-City Travelers

Shuai Xu[1,2(✉)], Jianqiu Xu[1], Bohan Li[1], and Xiaoming Fu[3]

[1] Nanjing University of Aeronautics and Astronautics, Nanjing, China
{xushuai7,jianqiu,bhli}@nuaa.edu.cn
[2] State Key Laboratory for Novel Software Technology, Nanjing University, Nanjing, China
[3] University of Göttingen, Göttingen, Germany
fu@cs.uni-goettingen.de

Abstract. The increasingly built intercity transportation enables people to visit surrounding cities conveniently. Hence it is becoming a hot research topic to predict where a traveler would visit in a surrounding city based on check-in data collected from location-based mobile Apps. However, as most users rarely travel out of hometown, there is a high skew of the quantity of check-in data between hometown and surrounding cities. Suffering from the severe sparsity of user mobility data in surrounding city, existing approaches do not perform well as they can hardly maintain travelers' intrinsic preference and meanwhile adapt to travelers' interest drift. To address these concerns, in this paper, taking cross-city travelers as the medium, we propose a novel framework called *CityTrans* to transfer traveler mobility knowledge from hometown city to surrounding city, which considers both the long-term preference in hometown city and short-term interest drift in surrounding city. Various attention mechanisms are leveraged to obtain traveler representation enriched by long-term and short-term preferences. Besides, we propose to portray POIs through GNN incorporating POI attributes and geographical information. Finally, the traveler and POI representations are combined for prediction. To train the framework, the transfer loss as well as the prediction loss are jointly optimized. Extensive experiments on real-world datasets validate the superiority of our framework over several state-of-the-art approaches.

Keywords: Location prediction · Mobility modeling · Knowledge transfer · Cross-city travelers · Attention mechanism

1 Introduction

Nowadays, the intercity transportation system has become an infrastructure to connect neighboring cities, which enables people to visit a surrounding city

X. Wang et al. (Eds.): DASFAA 2023, LNCS 13943, pp. 334–350, 2023.
https://doi.org/10.1007/978-3-031-30637-2_22

conveniently from their hometown city[1]. Meanwhile, mobile Apps that provide POI (Point-of-Interest) gathering services such as Yelp, Foursquare, and Wechat Moment are becoming widespread, which have accumulated massive user check-in data rapidly over time. Hence a more refined and intriguing research problem, i.e., predicting where a user would visit when they travel to a surrounding city, is coming into focus. However, a high skew of quantity and quality of user check-in data between hometown city and surrounding city makes the problem extremely tricky. For one thing, user-POI interactions in an out-of-town city are about one less order of magnitude than those in hometown city on average [3], which means there is a severe data sparsity problem in surrounding city. For another thing, due to the advent of various restrictions for personal data collection, a considerable proportion of high-quality user check-in data is discarded by the mobile Apps [6], which further exacerbates the data sparsity problem. As a consequence, although human mobility modeling inside a city has been extensively studied by recent works [2,13], most of them struggle in such data scarcity as it is far from enough to train a powerful prediction model for out-of-town travelers.

In recent years, approaches based on knowledge transfer have been proposed to solve the cross-city traveler visit location prediction problem, where three types of strategies are applied to relieve data sparsity. The first type is to transfer context-based knowledge such as the semantic features of POIs from the hometown city into out-of-town city [15], which does not transfer travelers' preference when they visit different cities. The second type directly introduces travelers' preference into out-of-town city [12], which ignores the phenomenon of traveler interest drift and thus can degrade the prediction performance. The third type leverages probabilistic generative models to incorporate traveler interest drift [24], which strengthens the learning effect of traveler preference by separating the city-dependent topics of each city from the common topics shared by all cities. We argue that travelers' check-in behavior is determined by not only the long-term hometown preference (which is relatively invariant) but also the short-term travel intention (which can be regarded as interest drift), thus existing approaches do not perform well as they can hardly maintain travelers' intrinsic preference and meanwhile adapt to travelers' interest drift when modeling travelers' visit preference in the surrounding city.

To address these concerns, in this paper, we conceive a novel mobility knowledge transfer framework called *CityTrans* for predicting a traveler's visit location in a surrounding city based on cross-city travelers. The motivation of our work is stated as follows (a visual description is shown in Fig. 1). Different from previous studies such as [1,9,17] which do not rely on cross-city travelers, we depend on similar crossing-city travelers to transfer mobility knowledge from the hometown city to the surrounding city, so as to alleviate the data sparsity issue. In previous studies, the geographical distance between source city and target city can be very far, thus the number of cross-city travelers is very small and the

---

[1] For clarity, in this paper, we refer to people's residential city or working city as the hometown city, and refer to others as the out-of-town city. Besides, if an out-of-town city is close to the hometown city, we refer to it as the surrounding city.

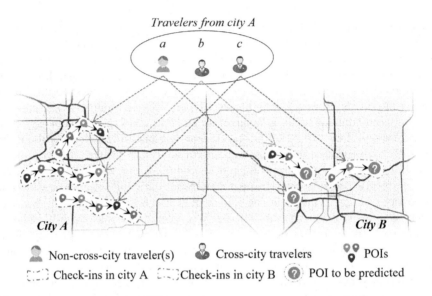

Fig. 1. A visual description of the cross-city traveler visit location prediction problem, where city A and city B are neighboring cities. Traveler b and c are the cross-city travelers who have more check-ins in city A and fewer check-ins in city B. Suppose travelers b and c are similar users in city A, we can predict where traveler b and traveler c would visit in city B by jointly considering the check-ins of b and c in city A and city B. For traveler a, which corresponds to the cold-start prediction, our framework still works with certain requirements.

amount of cross-city check-ins is very limited. Consequently, preference of cross-city travelers can not be directly transferred across cities. However, in this paper, we study human mobility between neighboring cities, where *most travelers have more check-ins in the hometown city and fewer check-ins in the surrounding city*. For each cross-city traveler, his/her check-in data in the surrounding city is not enough to build a reliable prediction model, so we can use the check-in data of similar cross-city travelers from the source city to strengthen the prediction model in the target city. Specifically, as his/her check-in data in the source city is relatively abundant, we use it to extract the long-term hometown preference based on self-attention mechanism. Meanwhile, following the idea of social collaborative filtering [4], we find his/her similar travelers in the source city, and sample their check-in data in the target city, based on which we can extract the short-term travel preference. To enable mutual interactions between the long-term preference and the short-term preference, we leverage the cross-attention mechanism for preference transfer across cities, based on which the ultimate traveler representation can be obtained. As for POI modeling, we incorporate geographical information as well as POI attributes, where the graph neural network (GNN) is leveraged to obtain the ultimate POI representation. Finally, the traveler and POI representations are combined for prediction via inner product.

In summary, this paper brings several unique contributions:

- First, we study an intriguing problem with a different scenario from existing studies, i.e., predicting where a traveler would visit in a surrounding city, given the relatively rich mobility data in the hometown city and the much sparser mobility data in the surrounding city. To relieve the severe data scarcity, with cross-city travelers as the medium, we propose a novel mobility knowledge transfer framework.
- Second, we propose to model travelers' long-term preference which is invariant in the source city, and model travelers' short-term preference with interest drift in the target city, respectively. Based on various attention mechanisms, cross-city traveler preference transfer can be realized. Besides, we propose to portray POIs in the target city through GNN, with consideration of geographical information and POI attributes.
- Third, we verify the performance of the proposed framework based on real-world LBSN datasets involving three source-target city pairs. Extensive experimental results demonstrate the effectiveness and superiority of our framework over existing state-of-the-art approaches. Furthermore, we validate that even for cold-start travelers, our approach still has good performance.

2   Related Work

2.1   User Visit Location Prediction

User mobility modeling is the groundwork for user future visit location prediction, where the key point is to model users' visit interest under different temporal-spatial contexts [20]. To accomplish this, related studies can be roughly divided into two types.

The first type applies the variants of RNNs to model sequential influence and temporal dynamics in user trajectories, so that a user's preference for the next mobility can be perceived. The difficulty lies in how to capture the correlation of successive check-ins and the dynamic changing rule of mobility preference when user trajectory extends [10]. Xu *et al.* [19] and Yang *et al.* [22] introduce time gates to control the influence of the hidden state of a previous RNN unit based on the time interval between successive check-ins, which encodes the temporal correlation for context-aware user location prediction. Feng *et al.* [5] and Yu *et al.* [25] divide a user's whole trajectory into several sub-trajectories and input them into the reformed RNN unit. With the matching ability of the self-attention mechanism, the informative check-in behavior in the past can be fused to improve the learning of up-to-date mobility preference. The second type leverages graph embedding models to obtain the embeddings of various kinds of nodes such as user, time, location, category and so on, based on which the prediction can be made by calculating the vector similarity [16,21,23].

2.2 Mobility Knowledge Transfer

Transfer learning has long been addressed for tasks like smart transportation [7], focusing on using traffic network and location images while ignoring the use of social network users and location dynamics. Comparing with user location prediction problem within a city, cross-city user location prediction is more intractable and specialized due to the cold start, interest drift, and other domain gap issues [14]. For cold start issues, the missing user-POI interactions in target city can be supplemented by transferring the content information of POIs (categories or tags) [6,9] or borrowing the POI preference from social friends [12]. For interest drift, which refers to the phenomenon that users' out-of-town check-ins are not aligned to their hometown preference, one effective but conventional way is based on probabilistic generative models [24], and the other up-to-date way is to transfer user interest which requires joint learning of users' previous visits in both two cities [18].

In the case of visit location prediction for cross-city travelers, it is more important to transfer users' intrinsic preference from source city to target city and meanwhile adapt to users' immediate interest in the target city. It requires joint learning of users' previous visits in both two cities, which can be hardly done by existing studies. Our work differentiates itself from previous works by a novel application scenario and a mobility knowledge transfer framework based on cross-city travelers, where a traveler's long-term preference in the source city and short-term interest drift in the target city can be effectively captured for cross-city preference transfer.

3   Problem Definition

Suppose we have collected traveler check-in data generated in two neighboring cities. We denote the set of travelers in the hometown city as $\mathcal{U}$, among which most travelers have check-in records both in hometown city and surrounding city, denoted by $\mathcal{D}^S$ and $\mathcal{D}^T$, where the superscripts S and T represent source city and target city, respectively. We denote the set of POIs in the hometown city and the surrounding city as $\mathcal{V}^S$ and $\mathcal{V}^T$. With a slight abuse of notation, the dataset for any city—either target or source—is assumed to consist of travelers $|\mathcal{U}| = M$, POIs $|\mathcal{V}| = N$, and an affinity matrix $\mathbf{R} = \{r_{uv}\}_{M \times N}$. We populate entries in $\mathbf{R}$ as row-normalized number of check-ins made by a traveler to a POI (i.e., multiple check-ins mean higher value). This can be further weighed by the traveler-POI ratings. A traveler u's check-in set is denoted as $\mathcal{C}^u = \{c_i^u | u \in \mathcal{U}, 1 \leq i \leq n\}$, where n is the number of u's check-ins, $c_i^u = (u, v_i, t_i, r_i)$ means traveler u visits POI v_i at time slot t_i with a numerical rating r_i. Last but not least, the attributes of a POI v include the geo-location, the category (such as Hotel, Cafe, etc.), and the number of visits on this POI.

Based on the notations above, without loss of generality, we formally define the cross-city traveler visit location prediction problem as follows:

Problem Formulation. Given a set of travelers $\mathcal{U}$ from the source city S, a set of POIs $\mathcal{V}^T$ in the target city T, as well as the check-in set $\mathcal{D}^S$ and $\mathcal{D}^T$, we aim to

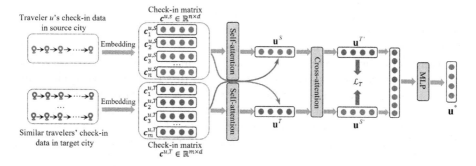

Fig. 2. Overview of the cross-city traveler preference transfer module.

predict which POI in the target city will be visited by a traveler from the source city. Specifically, we aim to produce a top-K POI list, so that the ground-truth POI v^o ($v^o \in \mathcal{V}^T$) where traveler u^o ($u^o \in \mathcal{U}$) would visit in the target city can be ranked as high as possible.

4   The CityTrans Framework

4.1   Cross-City Preference Transfer

For transferring travelers' preference from their hometown to the surrounding city effectively, enlightened by the idea of social collaborative filtering, we make full use of cross-city travelers' check-in data and conceive a dual-channel module with the transfer loss $\mathcal{L}_T$, so as to depict a traveler's preference comprehensively. The overview of cross-city preference transfer module is displayed in Fig. 2.

First of all, for each cross-city traveler u, his/her check-in data in the source city is taken as input for one channel to capture the long-term preference $\mathbf{u}^S$ in the hometown city. As cross-city traveler u only has a few check-in records in the target city, which is not enough to extract the short-term preference, we propose to find the most similar travelers from the hometown and sample m check-ins from their check-in records in the target city. The sampled check-in data in the target city is taken as input for the other channel to capture the short-term preference $\mathbf{u}^T$ in the surrounding city. To find similar travelers, we calculate the Pearson correlation coefficient between any pair of travelers in the hometown on top of the affinity matrix $\mathbf{R}$, based on which a symmetric matrix $\mathbf{P}$ containing similarity coefficients between travelers can be obtained. Next, for a cross-city traveler u, we sort other travelers according to the similarity coefficients to obtain an ordered traveler list. Then, we repeatedly sample a similar traveler's check-in records according to the ordered list until m check-in records are collected.

Second, for each check-in record c_i^u of cross-city traveler u (we take the upper channel in Fig. 2 as an example), we encode the discrete variables (POI ID, POI category, time slot and numerical rating). Then we concatenate these codes and embed it into the d-dimensional space, namely, $\mathbf{c}_i^{u,S} \in \mathbb{R}^d$. Afterwards, for all check-in records in $\mathcal{C}^u$, we obtain the check-in embedding matrix, $\mathbf{C}^{u,S} \in \mathbb{R}^{n \times d}$. In a similar way, we can also obtain the check-in embedding matrix for the sampled check-in records in the target city, namely, $\mathbf{C}^{u,T} \in \mathbb{R}^{m \times d}$.

Third, in order to infer traveler u's long-term preference in the hometown city, we apply the self-attention mechanism to discover informative check-ins in his/her trajectory, so that the indicative temporal-spatial information within his/her historical trajectory can be used to strengthen the preference modeling in the hometown city. Specifically, given the input $\mathbf{C}^{u,S} \in \mathbb{R}^{n \times d}$, we calculate the attention weights matrix $\mathbf{A} \in \mathbb{R}^{n \times n}$ based on the following equations:

$$
\begin{aligned}
\mathbf{Q} &= \mathbf{W}^Q \cdot \mathbf{C}^{u,S} \\
\mathbf{K} &= \mathbf{W}^K \cdot \mathbf{C}^{u,S} \\
\mathbf{V} &= \mathbf{W}^V \cdot \mathbf{C}^{u,S} \\
\mathbf{A} &= \mathrm{Softmax}(\mathbf{K} \cdot \mathbf{Q}^T)
\end{aligned}
\tag{1}
$$

where each element $a_{i,j} \in \mathbf{A}$ denotes the attention weight corresponding to check-in $c_i^{u,S}$ and check-in $c_j^{u,S}$. $\mathbf{W}^Q, \mathbf{W}^K, \mathbf{W}^V \in \mathbb{R}^{d \times d}$ are the weight matrices to be learned. Based on the attention weights matrix, we can obtain a new check-in embedding matrix $\tilde{\mathbf{C}}^{u,S}$:

$$
\tilde{\mathbf{C}}^{u,S} = \mathbf{A} \cdot \mathbf{V}
\tag{2}
$$

where $\tilde{\mathbf{C}}^{u,S} \in \mathbb{R}^{n \times d}$. In the end, we concatenate the embeddings in $\tilde{\mathbf{C}}^{u,S}$ and obtain traveler u's long-term preference $\mathbf{u}^S$ through dimension reduction:

$$
\mathbf{u}^S = \mathrm{Relu}(\mathbf{W}_1 \cdot (\tilde{\mathbf{c}}_1^{u,S} || \tilde{\mathbf{c}}_2^{u,S} || ... || \tilde{\mathbf{c}}_n^{u,S}) + \mathbf{b}_1)
\tag{3}
$$

where $\mathbf{u}^S \in \mathbb{R}^d$, $\mathbf{W}_1$ and $\mathbf{b}_1$ are the weight matrix and bias vector to be learned, and $||$ stands for the concatenation operation.

Similarly, we can obtain traveler u's short-term preference $\mathbf{u}^T \in \mathbb{R}^d$ according to the same operations above.

Next, we apply the cross-attention mechanism to enable mutual interactions between long-term preference in the hometown city and short-term preference in the surrounding city, so that traveler u's short-term preference in the surrounding city can have more hometown style. Specifically, given $\mathbf{u}^S$ and the check-in matrix $\mathbf{C}^{u,T}$ as input, we calculate the attention weight β_j between each check-in embedding $\mathbf{c}_j^{u,T}$ and $\mathbf{u}^S$ based on the following equation:

$$
\beta_j = \mathrm{Softmax}(\mathbf{c}_j^{u,T} \cdot \mathbf{W}^S \cdot \mathbf{u}^S)
\tag{4}
$$

where $\mathbf{W}^S \in \mathbb{R}^{d \times d}$ is the weight matrix. In the end, we obtain the updated short-term preference $\mathbf{u}^{T'} \in \mathbb{R}^d$ based on the attention weights:

$$
\mathbf{u}^{T'} = \sum_{j=1}^{m} \beta_j \mathbf{c}_j^{u,T}
\tag{5}
$$

Similarly, we can obtain the updated long-term preference $\mathbf{u}^{S'} \in \mathbb{R}^d$ of traveler u according to the same operations above.

Last, we concatenate the updated long and short-term traveler preference, and map it to the ultimate traveler representation $\mathbf{u}^*$ via the Relu activation:

$$\mathbf{u}^* = \text{Relu}(\mathbf{W}_2 \cdot (\mathbf{u}^{T'} || \mathbf{u}^{S'}) + \mathbf{b}_2) \tag{6}$$

where $\mathbf{u}^* \in \mathbb{R}^d$, $\mathbf{W}_2$ and $\mathbf{b}_2$ are the weight matrix and bias vector to be learned.

As shown in Fig. 2, in order to make the updated long and short-term traveler preference as consistent as possible, i.e., the distance between $\mathbf{u}^{T'}$ and $\mathbf{u}^{S'}$ in the shared vector space as close as possible, the transfer loss $\mathcal{L}_T$ is introduced. We consider Euclidean distance, KL divergence and cosine similarity for $\mathcal{L}_T$. Empirical analysis shows that transfer loss based on cosine similarity performs best, so we adopt cosine similarity as the loss function for this module.

4.2 Geo-Sensitive POI Modeling

For a cross-city traveler, his/her decision to visit a POI in an out-of-town city is not only affected by immediate travel intention, but also affected by how much he/she is interested in nearby POIs. In other words, nearby POIs can reinforce each other to attract travelers. In this paper, we hold that a traveler's mobility in surrounding city is bounded by geographical distance and thus check-ins to distant POIs are very unlikely. As a result, we propose to model POIs in the target city considering POI attributes as well as geographical information.

First, for each POI v_j^T in the target city, we encode POI ID, POI category and the number of visits (which represents the popularity of a POI) using one-hot encoding scheme. Then we concatenate these codes and embed it into the d-dimensional space, namely, $\mathbf{v}_j^T \in \mathbb{R}^d$. Afterwards, the initial embedding matrix $\mathbf{V}^T \in \mathbb{R}^{N \times d}$ for all POIs in the target city can be obtained.

Second, we build an undirected POI graph $\mathcal{G}^T = (\mathcal{V}^T, \mathcal{E}^T)$, where $\mathcal{V}^T$ stands for the POI set, and $\mathcal{E}^T$ represents the edge set. For each edge $e_{p,q} \in \mathcal{E}^T$, its weight can be calculated as:

$$e_{p,q} = \exp(-dist(v_p^T, v_q^T)) \tag{7}$$

where $dist(\cdot, \cdot)$ is the Haversine distance between two POIs. In this way, we can obtain an adjacent matrix $\mathbf{E}^T \in \mathbb{R}^{N \times N}$ based on the geographical distance between each pair of POIs in the target city.

Third, to capture the relations among POIs in a spatial perspective, we employ the graph neural network (GNN) to learn updated POI representations enriched by geographical distance, which prevents the need for additional handcrafting. Below is the equation to update the POI representations:

$$\mathbf{V}^{T*} = \text{Relu}(\mathbf{E}^T \cdot \mathbf{V}^T \cdot \mathbf{W}^T + \mathbf{b}^T) \tag{8}$$

where $\mathbf{V}^{T*} \in \mathbb{R}^{N \times d}$ is the ultimate POI representation matrix, $\mathbf{W}^T$ and $\mathbf{b}^T$ are the weight matrix and bias vector to be learned. Note that ultimate POI representations not only cover the structural features of the POI graph, but also contain popularity and geographical information. Correspondingly, we can obtain the ultimate representation of POI v_j^T in the target city, i.e., $\mathbf{v}_j^{T*} \in \mathbb{R}^d$.

4.3   Check-in Probability Estimation

Based on the two modules above, we can estimate the probability of a cross-city traveler u checking in at POI v in the target city. From the travelers' perspective, we adopt the idea of matrix factorization (MF) to explore the interactions between travelers and POIs in the target city, where the check-in probability $\hat{y}_{u,v}$ is regarded as the inner product of traveler u's representation and POI v's representation:

$$\hat{y}(u, v) = \mathbf{u}^* \cdot \mathbf{v}^{*\mathrm{T}} \tag{9}$$

where $\mathbf{u}^* \in \mathbb{R}^d$, $\mathbf{v}^* \in \mathbb{R}^d$ are the representations of the cross-city traveler u and the POI in the target city, respectively. $\mathbf{v}^{*\mathrm{T}}$ is the transpose of $\mathbf{v}^*$.

4.4   Model Training

We follow the assumption of Bayesian Personalized Ranking (BPR) [11], which is indeed a learning-to-rank strategy, to optimize the model parameters. To be concrete, BPR can make use of both observed and unobserved traveler-POI interactions by learning a pair-wise ranking loss in the training process, where the probability of an observed check-in should be ranked higher than that of an unobserved check-in. The prediction loss $\mathcal{L}_P$ with a regularization term is described below:

$$\mathcal{L}_P = -\sum_{u \in \mathcal{U}} \sum_{v^+ \in \mathcal{C}^{u,T}} \sum_{v^- \notin \mathcal{C}^{u,T}} \ln \delta(\hat{y}(u, v^+) - \hat{y}(u, v^-)) + \varepsilon ||\Theta||^2 \tag{10}$$

where $\mathcal{C}^{u,T}$ denotes cross-city traveler u's check-ins in the target city, and (u, v^+) and (u, v^-) represent an observed check-in and an unobserved check-in, respectively. Besides, $\delta(\cdot)$ is the sigmoid function, Θ denotes the parameter set in the model, and ε is the regularization coefficient.

As we consider the transfer loss $\mathcal{L}_T$ and the prediction loss $\mathcal{L}_P$ at the same time, we can minimize the following composite loss function to jointly train our model in an end-to-end manner:

$$\mathcal{L} = \lambda \mathcal{L}_T + (1 - \lambda)\mathcal{L}_P \tag{11}$$

where λ is a hyper-parameter controlling the contribution of each loss function.

Algorithm 1: Cross-city traveler visit location prediction algorithm

---

Input : u^o, $\mathcal{D}^S$, $\mathcal{D}^T$, $\mathcal{V}^T$, $\boldsymbol{\Theta}$, $\mathbf{P}$, m, K.
Output: the predicted top-K POI list L.
1 find the most similar travelers of u^o according to similarity matrix $\mathbf{P}$;
2 repeatedly sample similar travelers' check-ins until m check-in records are collected;
3 obtain the traveler representation $\mathbf{u}^{o*}$ according to Eq.(1) $\sim$ Eq.(6);
4 for *each candidate POI* $v' \in \mathcal{V}^T$ **do**
5 | obtain the POI representation $\mathbf{v}'^*$;
6 | compute $\hat{y}(u^o, v')$ according to Eq.(9);
7 end
8 sort POIs based on the computed probabilities in the descending order;
9 produce the prediction list L by selecting the top-K POIs from $\mathcal{V}^T$;
10 return L

Note that parameters to be learned include the set of weight matrices and the set of biases in the neural network layers. As the whole model is indeed a feed-forward neural network, where all parameters in the loss function are differentiable, we adopt Adam optimizer [8] with mini-batch strategy and negative sampling to minimize the composite loss automatically. For each positive sample (u, v^+), we randomly sample 10 negative samples (u, v^-). Besides, for each epoch during model training, we would re-sample negative samples for each positive sample, so that each negative sample only gives very weak negative signal in the training process.

4.5   Location Prediction in the Surrounding City

Given a traveler u^o $(u^o \in \mathcal{U})$, we can estimate the check-in probability $\hat{y}(u^o, v^o)$ for each POI v^o $(v^o \in \mathcal{V}^T)$ in the surrounding city. After check-in probabilities toward all candidate POIs are calculated, we select the POIs with top-K probabilities in the surrounding city as the result.

Algorithm 1 illustrates the process to predict a traveler's visit location when he/she travels to the surrounding city. We notice that the overall time complexity of Algorithm 1 is $\mathcal{O}(N \cdot n \cdot m) + \mathcal{O}(N \cdot \log(N))$, where N is the number of POIs, n is the length of traveler's trajectory, and m is a constant, representing the number of sampled check-ins for similar travelers.

5   Experiments

The experiments are implemented on a workstation with dual processors (Intel Xeon E5), four GPUs (Nvidia Titan Xp, 12 GB), and 188 GB RAM. Codes are written in Python 3.7 and the model architecture is built by Pytorch 1.3.1.

Table 1. Major statistics of the datasets.

Datasets & Properties	LV-HD	PH-MS	BS-CA
S→T	Las Vegas→Henderson	Phoenix→Mesa	Boston→Cambridge
# Travelers of city S	5,239	4,196	3,075
# Check-ins in city S	221,370	168,574	142,376
# Cross-city travelers	4,675	3,397	2,493
# POIs in city T	1,254	1,596	942
# Cross-city check-ins in city T	27,743	20,966	15,352

5.1 Datasets and Evaluation Methodology

Dataset. We evaluate the approach using real-world check-in data published by Yelp[2]. This large-scale dataset contains user check-in records in more than 100 North American cities. Each check-in record is formed as anonymous user-ID, POI-ID, POI coordinates, POI category, timestamp. Since hometown information is not provided in this dataset, we choose the city where a user has the highest number of check-ins as his/her hometown. We select three hometown-surrounding city pairs, i.e., Las Vegas-Henderson (LV-HD), Phoenix-Mesa (PH-MS) and Boston-Cambridge (BS-CA), where each pair of cities is within 20 miles of each other. For each cross-city pair, we first filter out inactive travelers with less than 10 check-ins and POIs with less than 10 visits. Then we keep travelers who have check-ins in both hometown city and surrounding city[3]. The major statistics of three selected datasets are shown in Table 1.

Evaluation Metrics. For evaluation, travelers' check-ins in hometown and their former 80% check-ins in surrounding city are used for training. Besides, their remaining 20% check-ins in the surrounding city are used for testing. To avoid information leakage, we also remove a part of each traveler's check-in records in hometown from the training data if these records are temporally after the earliest check-in record in the testing data of this traveler. Since we may have more than one out-of-town check-ins for each traveler, and we aim to predict the top-K ranking among all POIs, we employ Acc@K and NDCG@K as our evaluation metrics. The larger the values of the metrics are, the better the models perform.

5.2 Baselines

We compare our approach with the following baselines:

- **CCTP** [17]: This approach considers the impact of various city characteristics and travelers' preference, based on which a XGBoost model is built to predict which POI a traveler would visit in a new city.

---

[2] https://www.yelp.com/dataset/challenge.
[3] Please note that we still regard those travelers with only one check-in record in the surrounding city as non-cross-city travelers, because their only check-in record in the surrounding city needs to be used for cold-start prediction.

- **UIDT** [3]: This approach portrays traveler interest drift and transfer among different cities via matrix factorization, where representations of traveler and POI both contain city-independent feature and city-dependent feature.
- **CHAML** [1]: This is a meta-learning based transfer framework for cross-city location prediction. It is meticulously designed for travelers to a new city, which is regarded as the state-of-the-art approach for cold-start problem.
- **TRAINOR** [18]: This is a state-of-the-art knowledge transfer model that considers cross-city traveler preference, interest drifts, travel intention and out-of-town geographical influence at the same time. Traveler preference are transferred from hometown to target city by a MLP-based mapping function.
- **TransRec** [9]: This state-of-the-art approach integrates the deep neural network, transfer learning technique, and density-based resampling to support crossing-city traveler location prediction.
- **CitynoTrans**: It is a variant of our approach removing transfer loss $\mathcal{L}_T$ in the final loss function $\mathcal{L}$.

5.3   Experimental Settings

All the parameters in Θ are initialized according to the uniform distribution $U(-0.01, 0.01)$. We fix the batch size to 128, and set the initial learning rate of the Adam optimizer as 0.001 with the L2 regularization coefficient as $\varepsilon = 1e^{-5}$. The embedding size d is fixed to 128 for all latent representations. The number of layers in GNN is set to 1. To avoid over-fitting, we adopt dropout on the embedding layer and each hidden layer, where the dropout rate is set as 0.5. For the balancing coefficient λ and the sample size m, we conduct grid search in the range of [0.1,0.9] and [100,1000], respectively. Limited by pages, we directly report the optimal values, i.e., $\lambda = 0.1$, $m = 1000$.

5.4   Empirical Analysis

Performance Comparison. For cross-city travelers, we report the performance of various approaches across all datasets in Table 2. From the results, we have the following observations. First, the proposed approach CityTrans performs the best in terms of Acc@K metric when K = {1,5,10,20} among all datasets, and it outperforms other approaches in terms of NDCG@K metric when K = {5,10} among 4 out of 6 tasks, except for NDCG@10. With regard to Acc@10, which is widely used to evaluate a prediction model, CityTrans achieves 5.6%~35.1%, 3.5%~14.4%, 4.3%~28.2% performance improvements compared with three state-of-the-art approaches (i.e., CHAML, TRAINOR, TransRec) on LV-HD, PH-MS and BS-CA datasets, respectively. This indicates the effectiveness and superiority of our approach by considering cross-city travelers for mobility knowledge transfer. As a contrast, when we remove the preference transfer module (i.e., CitynoTrans), its performance decreases sharply due to the severe data scarcity in the surrounding city, which further verifies the value of the preference transfer module. As for the baselines, we observe that TransRec performs fairly good in terms of two metrics, being the best among all baselines.

Table 2. Performance comparison among various approaches. Numbers in bold font indicate the best performing approach. All results marked ¶ are statistically significant (i.e., two-sided Fisher's test with $p \leq 0.05$) over the second best one.

	Metric	CCTP	UIDT	CHAML	TRAINOR	TransRec	CitynoTrans	CityTrans
LV-HD	Acc@1	0.126	0.142	0.165	0.194	0.211	0.160	**0.224**¶
	Acc@5	0.134	0.152	0.181	0.208	0.224	0.173	**0.236**¶
	Acc@10	0.148	0.166	0.194	0.236	0.248	0.182	**0.262**¶
	Acc@20	0.203	0.248	0.275	0.334	0.350	0.268	**0.360**¶
	NDCG@5	0.188	0.234	0.362	0.381	0.386	0.315	**0.397**¶
	NDCG@10	0.206	0.268	0.379	0.402	**0.412**¶	0.338	0.406
PH-MS	Acc@1	0.107	0.134	0.156	0.188	0.202	0.151	**0.210**¶
	Acc@5	0.112	0.146	0.173	0.198	0.214	0.168	**0.221**¶
	Acc@10	0.127	0.169	0.208	0.214	0.230	0.179	**0.238**¶
	Acc@20	0.140	0.185	0.216	0.235	0.247	0.206	**0.252**¶
	NDCG@5	0.173	0.224	0.346	0.365	0.371	0.307	**0.375**¶
	NDCG@10	0.190	0.252	0.354	0.369	0.377	0.309	**0.379**¶
BS-CA	Acc@1	0.131	0.158	0.179	0.212	0.225	0.174	**0.238**¶
	Acc@5	0.142	0.171	0.193	0.228	0.237	0.188	**0.246**¶
	Acc@10	0.160	0.182	0.209	0.247	0.257	0.201	**0.268**¶
	Acc@20	0.216	0.260	0.288	0.342	0.360	0.271	**0.364**¶
	NDCG@5	0.194	0.245	0.371	0.389	0.397	0.328	**0.399**¶
	NDCG@10	0.215	0.273	0.385	0.408	**0.417**¶	0.343	0.415

This is due to the use of a complex deep neural network architecture which not only models the complex traveler-POI interactions enriched by textual contents, but also reduces the discrepancy between distributions over POIs caused by city-dependent features. In comparison with TransRec, CityTrans further considers travelers' long-term preference in hometown city and short-term interest drift in surrounding city, and promotes the mutual interaction between them via preference transfer, which ultimately contributes to the advantage. Second, we notice that as the list length K increases, the advantages of CityTrans over other approaches are decreasing, which indicates that there perhaps exist an upper bound of performance for traveler mobility prediction [26], and our approach performs better in the process of approximating this upper bound.

Moreover, to verify the value of cross-city check-ins in the proposed framework, we analyze how the trend of various metrics undergoes (Acc@10 and NDCG@10), with the percentage of cross-city check-ins used for training increasing from 10% to 100%. The results are shown in Fig. 3. As can be observed, the higher the proportion of cross-city check-ins used for training, the better the prediction performance of our approach, which is consistent with intuition.

Cold-Start Prediction. We further test different approaches in the extreme case where target travelers do not have any out-of-town check-ins, i.e., cold-start problem. It should be noted that this experiment is aimed at non-cross-city travelers, therefore to ensure the availability of the trained model, we remove those travelers whose visited POIs do not appear in the training data. The results are shown in Fig. 4. Notably, CHAML achieves the best performance among its

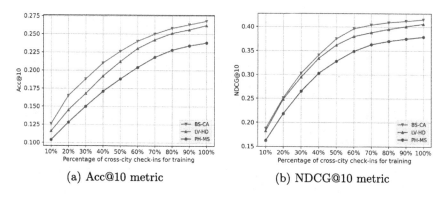

(a) Acc@10 metric   (b) NDCG@10 metric

Fig. 3. Impact of percentage of cross-city check-ins on prediction performance.

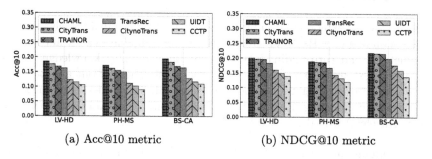

(a) Acc@10 metric   (b) NDCG@10 metric

Fig. 4. Performance of various approaches for cold-start predictions.

competitors for this task. The fundamental reason is that CHAML does not rely on cross-city check-ins for knowledge transfer. By incorporating meta-learning and non-uniform sampling strategies, CHAML can match a cold-start travelers' interest in the surrounding city more precisely. In contrast, CityTrans and other approaches depend on cross-city check-ins for knowledge transfer. In terms of CityTrans, cross-city travelers' mobility in both hometown and surrounding city has been portrayed in the trained model. Even though a cold-start traveler is not contained in the model, CityTrans is still able to achieve desirable performance due to similar check-in records in hometown city.

Ablation Study. In the end, we analyze the performance of multiple variants of CityTrans by removing key features one by one. To be specific, CityTrans-S denotes a variant by removing the long-term preference modeling in the hometown city; CityTrans-T denotes a variant by removing the short-term preference modeling in the surrounding city; CityTrans-Att denotes a variant by removing POI attributes; in the end, CityTrans-Geo denotes a variant by removing the geographical weights in POI graph $\mathcal{G}^T$. Results of the ablation experiments are shown in Fig. 5. Based on the gaps among different variants, we confirm that cross-city travelers' check-ins in the surrounding city are vital for our framework.

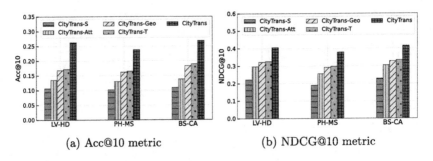

(a) Acc@10 metric   (b) NDCG@10 metric

Fig. 5. Performance of multiple variants of CityTrans in ablation study.

6   Conclusion

In this paper, we explore a novel problem of predicting where a traveler will visit in a surrounding city. Based on cross-city travelers who have check-in data both in hometown and surrounding city, we propose a mobility knowledge transfer framework. Extensive experiments verify the effectiveness of our approach. For future work, we consider to incorporate sequential features to strengthen traveler preference modeling. Besides, information from real social friends (not just similar travelers) will also be exploited.

Acknowledgements. This work was supported in part by the Natural Science Foundation of Jiangsu Province under Grant BK20210280, in part by the Fundamental Research Funds for the Central Universities under Grant NS2022089, and in part by the Jiangsu Provincial Innovation and Entrepreneurship Doctor Program under Grant JSSCBS20210185.

References

1. Chen, Y., Wang, X., Fan, M., Huang, J., Yang, S., Zhu, W.: Curriculum meta-learning for next poi recommendation. In: Proceedings of the 27th ACM SIGKDD Conference on Knowledge Discovery and Data Mining, pp. 2692–2702 (2021)
2. Dang, W., et al.: Predicting human mobility via graph convolutional dual-attentive networks. In: Proceedings of the 15th ACM International Conference on Web Search and Data Mining, pp. 192–200 (2022)
3. Ding, J., Yu, G., Li, Y., Jin, D., Gao, H.: Learning from hometown and current city: cross-city poi recommendation via interest drift and transfer learning. In: Proceedings of the ACM on Interactive, Mobile, Wearable and Ubiquitous Technologies, vol. 3, no. 4, pp. 1–28 (2019)
4. Fan, Z., Arai, A., Song, X., Witayangkurn, A., Kanasugi, H., Shibasaki, R.: A collaborative filtering approach to citywide human mobility completion from sparse call records. In: Proceedings of the 25th International Joint Conference on Artificial Intelligence, pp. 2500–2506 (2016)
5. Feng, J., et al.: Deepmove: predicting human mobility with attentional recurrent networks. In: Proceedings of the 2018 World Wide Web Conference, pp. 1459–1468 (2018)

6. Gupta, V., Bedathur, S.: Doing more with less: overcoming data scarcity for poi recommendation via cross-region transfer. ACM Trans. Intell. Syst. Technol. **13**(3), 1–24 (2022)

7. Jiang, R., et al.: Transfer urban human mobility via poi embedding over multiple cities. ACM Trans. Data Sci. **2**(1), 1–26 (2021)

8. Kingma, D.P., Ba, J.: Adam: a method for stochastic optimization. In: Proceedings of the 3rd International Conference on Learning Representation, pp. 1–15 (2015)

9. Li, D., Gong, Z.: A deep neural network for crossing-city poi recommendations. IEEE Trans. Knowl. Data Eng. **34**(8), 3536–3548 (2022)

10. Luca, M., Barlacchi, G., Lepri, B., Pappalardo, L.: A survey on deep learning for human mobility. ACM Comput. Surv. (CSUR) **55**(1), 1–44 (2021)

11. Rendle, S., Freudenthaler, C., Gantner, Z., Schmidt-Thieme, L.: BPR: bayesian personalized ranking from implicit feedback. arXiv preprint arXiv:1205.2618 (2012)

12. Wang, H., Fu, Y., Wang, Q., Yin, H., Du, C., Xiong, H.: A location-sentiment-aware recommender system for both home-town and out-of-town users. In: Proceedings of the 23rd ACM SIGKDD International Conference on Knowledge Discovery and Data Mining, pp. 1135–1143 (2017)

13. Wang, H., Li, Y., Jin, D., Han, Z.: Attentional markov model for human mobility prediction. IEEE J. Sel. Areas Commun. **39**(7), 2213–2225 (2021)

14. Wang, L., Geng, X., Ma, X., Liu, F., Yang, Q.: Cross-city transfer learning for deep spatio-temporal prediction. In: Proceedings of the 28th International Joint Conference on Artificial Intelligence, pp. 1893–1899 (2019)

15. Wei, Y., Zheng, Y., Yang, Q.: Transfer knowledge between cities. In: Proceedings of the 22nd ACM SIGKDD International Conference on Knowledge Discovery and Data Mining, pp. 1905–1914 (2016)

16. Xie, M., Yin, H., Wang, H., Xu, F., Chen, W., Wang, S.: Learning graph-based poi embedding for location-based recommendation. In: Proceedings of the 25th ACM International on Conference on Information and Knowledge Management, pp. 15–24 (2016)

17. Xie, R., Chen, Y., Xie, Q., Xiao, Y., Wang, X.: We know your preferences in new cities: mining and modeling the behavior of travelers. IEEE Commun. Mag. **56**(11), 28–35 (2018)

18. Xin, H., et al.: Out-of-town recommendation with travel intention modeling. In: Proceedings of the AAAI Conference on Artificial Intelligence, vol. 35, pp. 4529–4536 (2021)

19. Xu, J., Zhao, J., Zhou, R., Liu, C., Zhao, P., Zhao, L.: Predicting destinations by a deep learning based approach. IEEE Trans. Knowl. Data Eng. **33**(2), 651–666 (2021)

20. Xu, S., Fu, X., Cao, J., Liu, B., Wang, Z.: Survey on user location prediction based on geo-social networking data. World Wide Web **23**(3), 1621–1664 (2020). https://doi.org/10.1007/s11280-019-00777-8

21. Xu, S., Fu, X., Pi, D., Ma, Z.: Inferring individual human mobility from sparse check-in data: a temporal-context-aware approach. IEEE Trans. Comput. Soc. Syst. (2022). https://doi.org/10.1109/TCSS.2022.3231601

22. Yang, D., Fankhauser, B., Rosso, P., Cudre-Mauroux, P.: Location prediction over sparse user mobility traces using rnns: flashback in hidden states! In: Proceedings of the 29th International Joint Conference on Artificial Intelligence, pp. 2184–2190 (2020)

23. Yang, D., Qu, B., Yang, J., Cudré-Mauroux, P.: Lbsn2vec++: heterogeneous hypergraph embedding for location-based social networks. IEEE Trans. Knowl. Data Eng. **34**(4), 1843–1855 (2022)

24. Yin, H., Zhou, X., Cui, B., Wang, H., Zheng, K., Nguyen, Q.V.H.: Adapting to user interest drift for poi recommendation. IEEE Trans. Knowl. Data Eng. **28**(10), 2566–2581 (2016)
25. Yu, F., Cui, L., Guo, W., Lu, X., Li, Q., Lu, H.: A category-aware deep model for successive poi recommendation on sparse check-in data. In: Proceedings of the 2020 World Wide Web Conference, pp. 1264–1274 (2020)
26. Zhang, C., Zhao, K., Chen, M.: Beyond the limits of predictability in human mobility prediction: context-transition predictability. In: IEEE Transactions on Knowledge and Data Engineering (2022). https://doi.org/10.1109/TKDE.2022.3148300

Approximate k-Nearest Neighbor Query over Spatial Data Federation

Kaining Zhang[1], Yongxin Tong[1(✉)], Yexuan Shi[1], Yuxiang Zeng[1,2], Yi Xu[1],
Lei Chen[2,3], Zimu Zhou[4], Ke Xu[1], Weifeng Lv[1], and Zhiming Zheng[1]

[1] State Key Laboratory of Software Development Environment,
Beijing Advanced Innovation Center for Future Blockchain and Privacy Computing,
School of Computer Science and Engineering and Institute of Artificial Intelligence,
Beihang University, Beijing, China
{zhangkaining,yxtong,skyxuan,turf1013,xuy,kexu,lwf}@buaa.edu.cn,
zzheng@pku.edu.cn
[2] Department of Computer Science and Engineering, HKUST, Hong Kong, China
leichen@cse.ust.hk
[3] Data Science and Analytics Thrust, HKUST (GZ), Guangzhou, China
[4] City University of Hong Kong, Hong Kong, China
zimuzhou@cityu.edu.hk

Abstract. Approximate nearest neighbor query is a fundamental spatial query widely applied in many real-world applications. In the big data era, there is an increasing demand to scale these queries over a spatial data federation, which consists of multiple data owners, each holding a private, disjoint partition of the entire spatial dataset. However, it is non-trivial to enable approximate k-nearest neighbor query over a spatial data federation. This is because stringent security constraints are often imposed to protect the sensitive, privately owned data partitions, whereas naively extending prior secure query processing solutions leads to high inefficiency (*e.g.*, 100 s per query). In this paper, we propose two novel algorithms for efficient and secure approximate k-nearest neighbor query over a spatial data federation. We theoretically analyze their communication cost and time complexity, and further prove their security guarantees and approximation bounds. Extensive experiments show that our algorithms outperform the state-of-the-art solutions with respect to the query efficiency and often yield a higher accuracy.

Keywords: Approximate nearest neighbor · Spatial data federation

1 Introduction

k-Nearest Neighbor (kNN) query is one of the most fundamental queries in spatial databases, which aims to find k spatial objects that are closest to a given location. The approximate solutions to kNN queries (*a.k.a.*, approximate kNN or ANN) are of particular research interest since they are better suited for real-time response over large-scale spatial data than the exact counterparts [6,7,12,13,25].

© The Author(s), under exclusive license to Springer Nature Switzerland AG 2023
X. Wang et al. (Eds.): DASFAA 2023, LNCS 13943, pp. 351–368, 2023.
https://doi.org/10.1007/978-3-031-30637-2_23

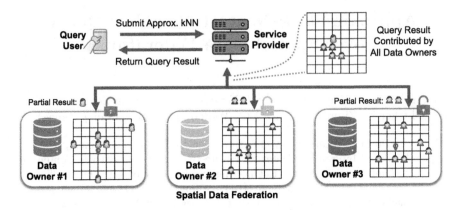

Fig. 1. Example of approximate kNN query over spatial data federation

There has been widespread adoption of approximate kNN in various application domains, such as transportation and map service, to name a few [12,13,21].

As the scale of real-world spatial applications continues to grow from region- to city- or even nation-wide, there has been a sharp urge to support approximate kNN queries over a *spatial data federation* [2,18,24]. A spatial data federation consists of multiple data owners who agree on the same schema and manage their own partition of the entire spatial dataset autonomously. Direct access to each raw data partition is prohibited, and secure queries over the federation are compulsory due to data sensitivity or commercial reasons. Take Amap [1], a major map service company in China, as an example. It offers a taxi-calling service via an integrated platform uniting dozens of taxi companies including Caocao, Shouqi, Yidao, *etc.* Prior to dispatching taxi orders, Amap may want to retrieve the k nearest drivers of a given passenger over the entire dataset of these companies via an approximate kNN query (see Fig. 1). The query should deliver high accuracy within a short time for satisfactory user experiences. It should also provide security guarantees for data partitions of the taxi companies by not leaking sensitive information such as locations during the query processing.

A naive solution is to extend general-purpose secure query schemes over data federations [2,24] to approximate kNN queries over a spatial data federation. However, they can be highly inefficient when processing approximate kNN queries. For instance, on the OpenStreetMap benchmark dataset [20,27] with 10k spatial objects and 6 data owners, the query delay of Conclave [24] is 100 s per query, which can hardly support real-time responses in taxi-calling applications. Other proposals (*e.g.*, SAQE [4] and Shrinkwrap [3]) are dedicated to specific relational queries and cannot be easily extended to spatial queries such as approximate kNN.

In this paper, we focus on efficient and secure solutions to approximate kNN query over spatial data federation (*"federated approximate kNN query"* for short). Specifically, we first design an approximation algorithm called *one-round*. The main idea is to (1) retrieve kNN over each data owner's local dataset in plaintext, (2) securely estimate how much each local kNN will contribute to the final result, and (3) securely collect the final result from all data owners based on the estimated contribution ratio. Due to the possible errors introduced in *one-round*, we further propose an improved algorithm called *multi-round* which optimizes the contribution ratio estimation procedure for a better trade-off between query accuracy and efficiency. To summarize, we have made following contributions.

- To the best of our knowledge, we are the first to study approximate kNN queries over a spatial data federation.
- We design two novel algorithms for federated approximate kNN queries. We theoretically analyze their communication cost and time complexity, and prove their security guarantees and approximation bounds.
- We conduct extensive experiments on both synthetic and real datasets and compare our algorithms with the state-of-the-art solutions. Results show that our method is always more efficient and often achieves higher accuracy. For instance, our solution can be 2–4 orders of magnitude faster than SMCQL [2] and Conclave [24] on the real dataset.

The rest of this paper is organized as follows. We first formally define the federated approximate kNN query in Sect. 2. Next, we introduce our algorithms and present our theoretical analysis in Sect. 3. Finally, we conduct experiments in Sect. 4, review related studies in Sect. 5, and conclude in Sect. 6.

2   Problem Statement

In this section, we first introduce some basic concepts and then present the formal definition of the approximate kNN query over spatial data federation.

Definition 1 (Spatial Object). *A spatial object d_i is denoted by a location $l_{d_i} = (x_{d_i}, y_{d_i})$ on a 2-dimensional Euclidean space.*

Based on Definition 1, the distance between spatial objects d_1 and d_2 is computed by the Euclidean distance function $dis(l_{d_1}, l_{d_2}) = \sqrt{(x_{d_2} - x_{d_1})^2 + (y_{d_2} - y_{d_1})^2}$.

Definition 2 (Data Owner). *A data owner S_i owns a set D_i of spatial objects $d_1, d_2, \cdots, d_{|D_i|}$, where $|D_i|$ is the number of spatial objects in D_i.*

In real-world applications, a data owner is often a company that owns a certain amount of spatial data [19,22,26]. Moreover, these data owners may want to run spatial analytics jointly (over the union of their datasets) but are unwilling/unable to share their raw data directly. As a result, a *spatial data federation* can be established to provide unified analytic services with security guarantees.

Definition 3 (Spatial Data Federation). *A spatial data federation $F = \{S_1, S_2, ..., S_n\}$ units n data owners to provide spatial query services. A set D is used to represent all the spatial objects, i.e., $D_1 \bigcup D_2 \bigcup ... \bigcup D_n$, where a subset D_i of spatial objects is held by the data owner S_i.*

The query processing over a spatial data federation usually requires a joint computation across the data owners. Therefore, this federation also needs to guarantee the security of each data owner. As a result, existing work [2–4,24] usually assumes that each data owner can be a *semi-honest* attacker, which is defined by the following threat model.

Definition 4 (Semi-honest Threat Model). *Referring to existing work [2, 24], the data owners over spatial data federation are presumed to be semi-honest. A semi-honest attacker follows the query execution plan (e.g., correctly executing spatial queries over their owned dataset), but is also curious about the sensitive data (e.g., the locations of spatial objects) of the other data owners.*

Our problem is defined upon the (exact) k nearest neighbor (kNN) query.

Definition 5 ((Exact) kNN [13]). *Given a set D of spatial objects, a query location $l_q = (x_q, y_q)$, and a positive integer k, kNN retrieves a set $res^* \subseteq D$ of k spatial objects that are closest to the query location l_q, i.e., $\forall d \in res^*$ and $d' \in D - res^*$, $dis(l_d, l_q) \leq dis(l_{d'}, l_q)$.*

Based on these concepts, we introduce the approximate kNN query over spatial data federation ("*federated approximate kNN query*" for short) as follows.

Definition 6 (Federated Approximate kNN Query). *Given a spatial data federation $F = \{S_1, S_2, ..., S_n\}$, a query location $l_q = (x_q, y_q)$, and a positive integer k, a federated approximate kNN query $q(F, l_q, k)$ aims to find a set res of k spatial objects over the whole dataset D such that the **result accuracy** δ of this approximate answer can be maximized as much as possible*

$$\delta = \frac{|res \cap res^*|}{k}, \text{ where } res^* \text{ is the exact } kNN \tag{1}$$

while satisfying the following security constraint.

– ***Security constraint.** Under the semi-honest threat model, the query processing algorithm should ensure a data owner cannot infer any sensitive information about other data owners except for the query result.*

In Definition 6, the result accuracy defined in Eq. (1) is a widely-used metric [13,25] to assess the quality of the retrieved results. In the security requirement, the extra sensitive information can be the locations of any other owner's spatial objects, the ownership of the spatial objects, and the cardinality of query results over another owner's local dataset ("local result" as short).

Example 1. Three data owners S_1, S_2, and S_3 constitute a spatial data federation F. The locations of the spatial objects owned by S_1, S_2, and S_3 are listed in Table 1. A query user submits a kNN query $q(F, l_q, k)$ to this federation F, where the query location $l_q = (3, 3)$ and $k = 5$. Suppose the exact kNN is $res^* = \{(2,3),(3,2),(3,2),(3,3),(3,4)\}$, and the result found by an approximation algorithm is $res = \{(3,3),(3,2),(2,3),(0,4),(6,4)\}$. The accuracy δ of this approximate result is $|res \cap res^*|/k = 3/5 = 60\%$.

3   Our Approximation Algorithms

In this section, we first introduce an overview of our solution in Sect. 3.1. Then, two basic and secure computation operations are presented in Sect. 3.2. Based on these basic operations, we propose two algorithms with different trade-offs for federated approximate kNN queries, *i.e.*, one-round algorithm (Sect. 3.3) and multi-round algorithm (Sect. 3.4). The former is faster than the latter, while the latter is more accurate than the former. Finally, we prove the approximation guarantees of both algorithms in Sect. 3.5.

Table 1. Locations of spatial objects owned by S_1, S_2, and S_3

S_1		S_2		S_3	
ID	Location	ID	Location	ID	Location
1	(0,4)	1	(1,6)	1	(1,4)
2	(3,0)	2	(3,2)	2	(2,3)
3	(3,4)	3	(3,3)	3	(3,2)
4	(3,5)	4	(3,6)	4	(4,2)
5	(4,4)	5	(4,5)	5	(4,6)
6	(7,6)			6	(6,4)
				7	(7,3)

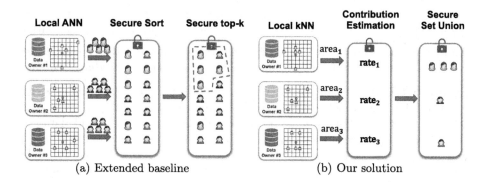

(a) Extended baseline   (b) Our solution

Fig. 2. Comparison of the extended baseline and our solution

3.1   Overview

Limitation of Extended Baseline. As shown in Fig. 2(a), existing general-purpose solutions (*e.g.*, Conclave [24]) can be extended as a baseline. In general, it first asks each data owner to perform an approximate kNN (*i.e.*, ANN) over its local dataset, and then uses secure sort and secure top-k to compute the final result. From experiments in Sect. 4, we have two observations on the limitation of baselines. (1) The *inefficiency* is mainly caused by the secure operations, which are complicated and time-consuming. (2) The errors of totally $O(n)$ ANNs may exacerbate the error of the final answer. Thus, it motivates us to design a new query processing solution that requires *more light-weight* secure operations and *fewer inaccurate approximations*.

Overview of Our Solution. To overcome the limitation, the *main idea* of our solution is illustrated in Fig. 2(b). Specifically, each silo first executes exact kNN (instead of ANN) to produce k candidate objects. Next, an accurate method is devised to approximately estimate the contribution ratio to the final answer in each silo. This method relies on secure summation, which is one of the simplest secure operations. Based on this ratio, a secure set union is used to collect the final answer from all silos. By contrast, our solution only involves *light-weight secure operations* and *one approximation operation*.

3.2 Preliminary

Our algorithm is designed based on two primitive operations as follows, *i.e.*, secure summation and secure set union.

Secure Summation Operation [5]. The secure summation operation gets the sum of the private values held by multiple data owners while satisfying the security constraint. For example, when data owners S_1-S_3 hold $value_1$-$value_3$ respectively, a secure summation operation works as follows. Each pair of S_i and S_j $(i < j)$ negotiates a random number $sc_{i,j}$ secretly in advance. When a secure summation request is submitted, each data owner S_i perturbs $value_i$ as $value'_i = value_i + \sum_{j \in [1,3], i < j} sc_{i,j} - \sum_{j \in [1,3], i > j} sc_{j,i}$. Next, S_i sends $value'_i$ to the spatial data federation. Finally, the spatial data federation adds up $value'_1, value'_2$, and $value'_3$ in plain text as the final result, which equals to $value_1 + value_2 + value_3$.

Secure Set Union Operation [10]. The secure set union operation gets the union of the private sets held by multiple data owners while satisfying the security constraint. For example, suppose 3 data owners S_1-S_3 hold datasets $set_1 - set_3$, respectively. First, S_1, S_2 and S_3 generate random sets $rset_1, rset_2$, and $rset_3$, respectively. Then, S_1 sends $tset_1 = set_1 \bigcup rset_1$ to S_2, S_2 sends $tset_2 = tset_1 \bigcup set_2 \bigcup rset_2$ to S_3, and S_3 sends $tset_3 = tset_2 \bigcup set_3 \bigcup rset_3$ to S_1. Next, S_1 sends $tset_1 = tset_3 - rset_1$ to S_2, S_2 sends $tset_2 = tset_1 - rset_2$ to S_3, and S_3 sends $tset_3 = tset_2 - rset_3$ to S_1. Finally, S_1 submits $tset_3$ to the spatial data federation, which equals the union of set_1, set_2, and set_3.

3.3 One-Round Algorithm

Main Idea. Each data owner first performs an exact kNN query over its local dataset to get the local kNN set. Then the spatial data federation estimates the contribution ratio of each data owner's local kNN to the final answer. The larger the ratio is, the more objects the data owner's local kNN will contribute to the final answer.

Algorithm Details. Algorithm 1 illustrates the detailed procedure. In lines 1–2, each data owner executes its local kNN. In lines 3–5, we use a radius r_i to denote the kth nearest distance to the query location l_q and $area_i$ to denote the area of a circle with the radius r_i to the center l_q. Then, the contribution ratio $rate_i$ of the data owner S_i's local kNN to the final result is inversely proportional to $area_i$. To compute $rate_i$, we first compute $sum = \sum_{i=1}^{n} \frac{1}{area_i}$ by the secure summation operation $SecureSum()$, and then each data owner S_i can calculate $rate_i$ as $\frac{1/area_i}{sum}$. Accordingly, S_i provides the top $(rate_i \times k)$NN as the partial result res_i. Finally, the spatial data federation performs the secure union $SecureUnion()$ among these partial results $res_1, res_2, \cdots, res_n$ to collect the final result.

Example 2. Back to Example 1. The values of some intermediate variables are shown in Table 2. Data owners calculate the kth nearest distance to l_q based on their local kNN first. For example, since the location of the kth nearest neighbor of S_1 is $(0,4)$, the kth nearest distance to l_q can be calculated as $r_1 = \sqrt{(0-3)^2 + (4-3)^2} = \sqrt{10}$. Similarly, we have $r_2 = \sqrt{13}$ and $r_3 = \sqrt{10}$. Therefore, $area_1 = 10\pi, area_2 = 13\pi$, and $area_3 = 10\pi$. By the secure summation, we have $sum = 18/65\pi$. Then, $rate_1$ can be computed as $area_1/sum = 13/36$. $rate_2$ and $rate_3$ can be calculated similarly. It indicates the data owners S_1-S_3 would offer their 2NN, NN and 2NN as the partial result. By using a secure union, the final result by Algorithm 1 is $res = \{(3,4), (4,4), (3,3), (3,2), (2,3)\}$.

Algorithm 1: One-round algorithm

Input: spatial object sets $D_1, D_2, \cdots, D_n$, the query request $q(F, l_q, k)$
Output: query result

1   **for** $i \in [1, n]$ **do**
2     $nn_i \leftarrow S_i$'s local kNN over D_i

3   **for** $i \in [1, n]$ **do**
4     $r_i \leftarrow \max_{j \in [1,k]} dis(l_{nn_i[j]}, l_q)$
5     $area_i \leftarrow \pi(r_i)^2$

6   $sum \leftarrow SecureSum(\frac{1}{area_1}, \frac{1}{area_2}, \cdots, \frac{1}{area_n})$
7   **for** $i \in [1, n]$ **do**
8     $rate_i \leftarrow \frac{1/area_i}{sum}$
9     $num_i \leftarrow rate_i \times k$
10   $res_i \leftarrow$ the top num_i-NN in nn_i

11 **return** $res \leftarrow SecureSetUnion(res_1, res_2, \cdots, res_n)$

Table 2. Values of some intermediate variables in Example 2

	S_1	S_2	S_3
ID of local kNN	{3,5,4,2,1}	{3,2,5,4,1}	{2,3,4,1,5}
r	$\sqrt{10}$	$\sqrt{13}$	$\sqrt{10}$
$area$	10π	13π	10π
$rate$	13/36	5/18	13/36
num	2	1	2
ID of res	{3,5}	{3}	{2,3}

Communication Complexity. In Algorithm 1, the communication mainly occurs in the secure summation and secure set union. Since each data owner needs to send information to other $n-1$ data owners, the communication complexity of secure summation is $O(n^2)$. In the secure set union, the communication complexity of interactions among these n data owners is $O(n)$. Thus, the communication complexity of Algorithm 1 is $O(n^2)$.

Time Complexity. The time complexity of lines 1–5 is $O(\log m)$, where m is the maximum number of objects owned by a data owner. In line 6, the time complexity of the secure summation is $O(n)$, where n is the number of data owners. The time complexity of lines 7–10 is $O(1)$. Note that although the total communication cost is $O(n^2)$, the time complexity of communication is still bounded by $O(n + logm)$ for each data owner. In line 11, the time complexity of the secure set union is $O(n)$. Thus, the time complexity of Algorithm 1 is $O(n + \log m)$.

Security Proof. We prove the security of Algorithm 1 in Lemma 1 by the composition lemma [8] in cryptography theory.

Fact 1 (Composition Lemma [8]). *Given a secure protocol $\phi(y|x)$ that can securely compute y based on plain-text query x, the operation y can be securely computed by executing protocol $\phi(y|x)$ but substitutes every plain-text query x with a secure protocol $\phi(x)$.*

Lemma 1. *Algorithm 1 is secure against the semi-honest threat model in Definition 4.*

Proof. Let $\phi(x)$ be the secure summation (line 6) and $\phi(y|x)$ be the calculation procedure in lines 1–10 based on the plain-text summation. It can be observed that lines 1–5 and lines 7–10 do not involve interactions across data owners, which means each data owner can execute lines 1–10 independently when the summation result is known. Based on the composition lemma [8], the computations in lines 1–10 are also secure. As the secure set union (line 11) is secure, the whole calculation procedure of Algorithm 1 is secure.

3.4   Multi-round Algorithm

Motivation. The one-round algorithm roughly estimates the contribution ratio of data owners' local kNN to the query result, which may result in unsatisfied accuracy (see our experiments in Sect. 4). Thus, we present a multi-round algorithm that slightly sacrifices efficiency for an improvement in the accuracy.

Main Idea. To further improve the query *accuracy*, we use a more fine-grained approach to estimate the contribution of each data owner to the final query result. Specifically, we divide the query processing procedure into multiple rounds and each round contributes a part of the final result. For example, when the number of rounds is W, k/W nearest neighbors (k/WNN) of kNN will be determined in each round. In this way, the estimation of the contribution ratio in each round can be more accurate than that in Algorithm 1. Besides, those objects, which have been determined, will not appear in subsequent rounds.

Algorithm Details. Algorithm 2 illustrates the detailed procedure. In lines 1–3, each data owner S_i maintains a set $UnadSet_i$ that contains candidate objects and a set $AdSet_i$ that contains selected objects. The local kNN of S_i is put into $UnadSet_i$ as initialization. Different from Algorithm 1, Algorithm 2 takes W ($W > 1$) rounds to compute the final results in lines 4–8 and each round decides k/W nearest neighbors (k/WNN). Specifically, for each round w, we estimate the contribution ratio of $UnadSet_1, UnadSet_2, \cdots, UnadSet_n$ to the k/WNN in this round, which is similar to the one-round algorithm. Note that in the w ($w \geq 2$)th round, we fine-tune the $area_i$ as $\pi[r_i^2 - (r_i')^2]$, where r_i is defined as line 4 of Algorithm 1 and r_i' is the cached value of r_i in the $(w-1)$th round. The selected objects of S_i are removed from $UnadSet_i$ to $AdSet_i$. In line 9, the final result is calculated by a secure set union over $AdSet_1, AdSet_2, \cdots, AdSet_n$.

Example 3. Back to Example 1. We aim to find 2NN and 3NN in the 1st and 2nd rounds, respectively. The values of the intermediate variables of the 1st and 2nd rounds are listed in Table 3 and Table 4, respectively. By merging $AdSet_1, \cdots, AdSet_n$, we have $res=\{(3,4),(3,3),(3,2),(2,3),(3,2)\}$, whose query accuracy is 100%.

Communication and Time Complexity. Algorithm 2 involves W contribution estimation and 1 secure set union operation. Since the number W of rounds is a constant parameter, the communication complexity and time complexity of Algorithm 2 are same as those of Algorithm 1, *i.e.*, $O(n^2)$ and $O(n + \log m)$, respectively.

Algorithm 2: Multi-round algorithm

Input: spatial object sets $D_1, D_2, \cdots, D_n$, the query request $q(F, l_q, k)$, the query round W

Output: query result

1 **for** $i \in [1, n]$ **do**
2 $\quad$ $UnadSet_i \leftarrow S_i$'s local kNN over D_i
3 $\quad$ $AdSet_i \leftarrow \phi$

4 **for** $w \in [1, W]$ **do**
5 $\quad$ **for** $i \in [1, n]$ **do**
6 $\quad\quad$ $NewadSet_i \leftarrow res_i$ found by Algorithm 1 whose input is $\{UnadSet_1, UnadSet_2, \cdots, UnadSet_n, q(F, l_q, k/W)\}$
7 $\quad\quad$ $AdSet_i \leftarrow AdSet_i \cup NewadSet_i$
8 $\quad\quad$ $UnadSet_i \leftarrow UnadSet_i - NewadSet_i$

9 **return** $res \leftarrow SecureUnion(AdSet_1, AdSet_2, \cdots, AdSet_n)$

Table 3. Values of some intermediate variables of the 1st round in Example 3

	S_1	S_2	S_3
ID of objects in $NewadSet$	$\emptyset$	$\{3\}$	$\{2\}$
ID of objects in $AdSet$	$\emptyset$	$\{3\}$	$\{2\}$
ID of objects in $UnadSet$	$\{1,2,3,4,5,6\}$	$\{1,2,4,5\}$	$\{1,3,4,5,6,7\}$

Security Proof. We prove the security of Algorithm 2 in Lemma 1 by the composition lemma [8] (*i.e.*, Fact 1 in Sect. 3.3).

Lemma 2. *Algorithm 2 is secure against the semi-honest threat model in Definition 4.*

Proof. Let $\phi(x)$ be the calculation procedure in line 6 and $\phi(y|x)$ be the calculation procedure in lines 1–8 based on the plain-text calculation procedure in line 6. It can be observed that lines 1–5 and lines 7–8 do not involve interactions across data owners, which means each data owner can execute lines 1–8 independently when the result of line 6 is known. Since the calculation procedure in line 6 is proved to be secure in Sect. 3.3, the computations in lines 1–8 are also secure based on the composition lemma [8]. As the secure set union (line 9) is secure, the whole calculation procedure of Algorithm 2 is secure.

3.5 Approximation Guarantees of both Algorithms

The following theorem proves the approximation guarantees of our one-round algorithm (when $W = 1$) and multi-round algorithm (when $W > 1$).

Theorem 1. *The accuracy δ of the approximate kNN query result satisfies*

$$\Pr(\delta < 1 - \varepsilon) \leq 2 \exp(\frac{-2W\varepsilon^2}{n}),$$

Table 4. Values of some intermediate variables of the 2nd round in Example 3

	S_1	S_2	S_3
ID of $NewadSet$	{3}	{2}	{3}
ID of $AdSet$	{3}	{3,2}	{2,3}
ID of $UnadoptedSet$	{1,2,4,5,6}	{1,4,5}	{1,4,5,6,7}

where W, n, and k represent the query round, the number of data owners, and the number of objects to be queried respectively.

Proof. Assume that the spatial objects owned by data owners $S_1, S_2, \cdots, S_n$ roughly follow the uniform distribution in each round. Denote the density of objects in S_i as ρ_i. We have $\mathbb{E}[\rho_i] = \frac{k/W}{area_i}$, where $area_i$ is the circle's area of local $\frac{k}{W}$NN in data owner S_i. Thus, the expected area of the circle of global $\frac{k}{W}$NN is

$$\mathbb{E}[area] = \frac{k/W}{\sum_{i=1}^{n} \mathbb{E}[\rho_i]} = \frac{1}{\sum_{i=1}^{n} \frac{1}{area_i}}.$$

Denote the contribution ratio of the data owner S_i as $rate_i$. We have

$$\mathbb{E}[rate_i] = \frac{\frac{1}{area_i}}{\sum_{i=1}^{n} \frac{1}{area_i}}.$$

Then, by applying the Hoeffding's inequality [23], we can derive that for the data federation F,

$$\Pr(\sum_{j=1}^{W} \sum_{i=1}^{n} |rate_i^j - \mathbb{E}[rate_i^j]| > \varepsilon) \leq 2 \exp(\frac{-2\epsilon^2}{Wn}).$$

And the accuracy δ of the approximate kNN query result is

$$\delta = 1 - \frac{(\sum_{j=1}^{W} \sum_{i=1}^{n} |rate_i^j - \mathbb{E}[rate_i^j]|) \cdot \frac{k}{W}}{k}.$$

Therefore, the accuracy δ of the approximate result satisfies

$$\Pr(\delta < 1 - \varepsilon) \leq 2 \exp(\frac{-2W\varepsilon^2}{n}).$$

4 Experimental Evaluation

This section presents the experimental setup in Sect. 4.1 and results in Sect. 4.2.

4.1 Experimental Setup

Datasets. Both real dataset (MBJ) and synthetic dataset (OSM) are used in our experimental evaluation.

- **Multi-company Spatial Data in Beijing (MBJ).** We randomly select 10^6 pieces of spatial data records from the original dataset comprising 1,029,081 pieces of records from 10 companies in Beijing. Each company is regarded as a data owner.

Table 5. Parameter settings

Parameter	Setting
#(Nearest neighbors) k	4, 8, **16**, 32, 64
#(Data onwers) n	2, 4, **6**, 8, 10
Data size $|D|$	10^4, 10^5, $\mathbf{10^6}$, 10^7, 10^8

- **OpenStreetMap (OSM).** This dataset is widely used in large-scale spatial data systems [20,27]. We randomly select 10^4, 10^5, 10^6, 10^7, and 10^8 pieces of location records and assign each record a random data owner number so that data owners have the same number of objects.

Parameter Settings. Referring to the parameter settings in [9,20], we vary k, $|D|$, and n from 4 to 64, 10^4 to 10^8, and 2 to 10, respectively. The parameter settings are summarized in Table 5, in which default parameters are in bold.

Compared Algorithms. We compare the performance of our algorithms and the following baselines (*i.e.*, SMCQL [2], which is developed by ObliVM [15], and Conclave [24], which is developed by MP-SPDZ [11]).

For our proposed algorithms, we use OR and MR to denote the one-round algorithm (*i.e.*, Algorithm 1) and multi-round algorithm (*i.e.*, Algorithm 2), respectively.

For the compared baselines, we make extensions on supporting approximate kNN queries as follows. Each data owner calculates its local approximate kNN by a seminal indexing approach, the ANN library [16]. Then the final kNN is derived by a secure sorting over the union of local kNN computed by data owners. Notice that SMCQL only supports queries over a federation of two data owners.

Evaluation Metrics. We use *running time* and *communication cost* to test the efficiency and result accuracy ("*accuracy*" as short) to test the effectiveness.

Experimental Environment. The experiments are conducted on 10 docker containers (*i.e.*, up to 10 data owners), each with 16 AMD Ryzen 3.4GHz CPU cores. Each docker container can be regarded as a data owner. The experimental results are the average of 50 repetitions.

4.2   Experimental Results

Note that the results of SMCQL are only available when n is 2, since its adopted security technique, ObliVM [15], is only applicable for 2 data owners.

Effect of k. Figure 3-4 illustrate the results when varying k on MBJ and OSM, respectively. We can first observe that our solution always achieves a better efficiency than Conclave. For example, on the real dataset (MBJ), Conclave can be 4 orders of magnitude slower than our one-round (OR) and multi-round (MR) algorithms, and its communication cost is at least 5 orders of magnitude higher than that of our solution. In terms of result accuracy, our solution is often better than Conclave. For instance, the accuracy of MR is up to 13% higher than that of Conclave. The error of Conclave is mainly caused by the accumulative errors of its local approximate kNN queries ($O(n)$ in total). By contrast, the error of our solution is only produced in one operation, $i.e.$, contribution estimation. Moreover, the results also show that MR can effectively improve the accuracy with only marginal sacrifice in the efficiency.

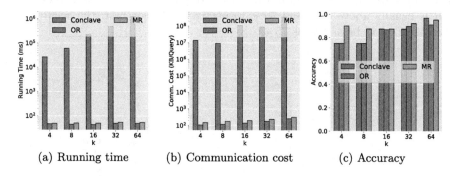

(a) Running time   (b) Communication cost   (c) Accuracy

Fig. 3. Results when varying k on the MBJ dataset

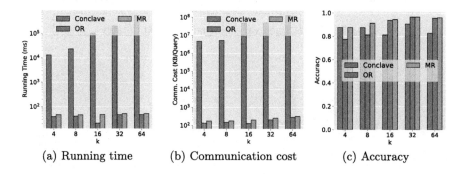

(a) Running time   (b) Communication cost   (c) Accuracy

Fig. 4. Results when varying k on the OSM dataset

Effect of n. Figure 5-6 present the results when varying n on MBJ and OSM, respectively. In terms of efficiency, the running time and communication cost of our algorithms are still significantly lower than those of SMCQL and Conclave. For instance, OR and MR are up to 4 orders of magnitude faster than the baselines. The communication cost of Conclave is up to 1.6 TB per query, while that of ours is below 0.27 MB. Besides, the running time and communication cost of all algorithms show an upward trend, when the number n of data owners is increasing. This is because (1) the secure operations are efficiency bottleneck of the baselines and (2) the efficiency of secure operations is sensitive to n. In terms of effectiveness, the result accuracy of our algorithms is higher than that of baselines in most cases. For example, the accuracy of OR and MR is up to 26.25% and 31.25% higher than that of SMCQL and Conclave, respectively.

Effect of $|D|$. Figure 7 shows the results when varying the data size $|D|$ on OSM. When varying the data size, the running time and communication cost of our algorithms are constantly lower than those of Conclave. For instance, OR is usually 3 orders of magnitude faster than Conclave. Between our algorithms, MR is slightly less efficient than OR. Overall, the results show that our solution is much more scalable than the state-of-the-art. As for the result accuracy, our algorithms are more stable than Conclave. Specifically, the accuracy of OR and MR is usually above 83% and 90%, respectively. By contrast, the accuracy of the baseline can be as low as 75%.

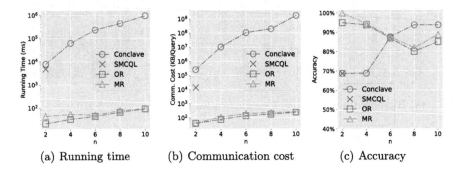

(a) Running time   (b) Communication cost   (c) Accuracy

Fig. 5. Results when varying n on the MBJ dataset

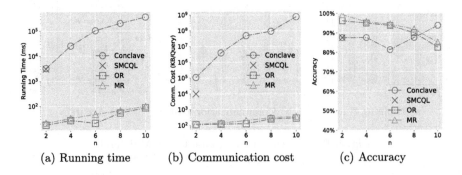

(a) Running time   (b) Communication cost   (c) Accuracy

Fig. 6. Results when varying n on the OSM dataset

Summary of results. The experimental findings are summarized as follows.

(1) Our solution is notably more efficient than the baselines in query efficiency. It can be 2–4 orders of magnitude faster and take 2–6 orders of magnitude lower communication cost than that of SMCQL and Conclave.
(2) The result accuracy of our solution is often higher than that of baselines. For example, the result accuracy of our solution can be up to 31.25% higher than the accuracy of SMCQL and Conclave.
(3) Our multi-round algorithm can effectively improve the accuracy of one-round algorithm by sacrificing only a little efficiency.

5   Related Work

Our work is related to the domains of *Querying over Data Federation* and *Approximate kNN Query*.

Querying over Data Federation. Existing algorithms perform queries over a data federation by utilizing the Secure Multiparty Computation (SMC) [8] technique, which protects the input and intermediate data from leaking. After receiving a query request submitted by the user, the data federation parses the query into a series of secure and plain-text operations and coordinates multiple data owners to get query results based on the union of their local datasets.

Hu-Fu [17,20] is a spatial data federation system that supports spatial queries with exact results. However, unlike SMCQL [2] and Conclave [24], it is non-trivial to extend Hu-Fu to support approximate kNN queries. This is because (1) the query rewriter of Hu-Fu decomposes a kNN query into a series of range counting queries and one range query (instead of several kNN queries on each local dataset) and (2) the results of range counting and range query here must be accurate to ensure the final answer has exactly k objects. Other data federation systems, *SAQE* [4] and *Shrinkwrap* [3], consider the trade-offs among result accuracy, query efficiency, and differential privacy. For example, *SAQE* improves the query efficiency by sampling and protects the query result by differential

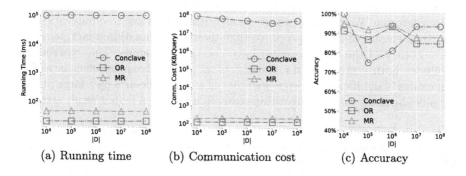

(a) Running time   (b) Communication cost   (c) Accuracy

Fig. 7. Results when varying $|D|$ on the OSM dataset

privacy, while *Shrinkwrap* protects the intermediate results by using differential privacy. However, since differential privacy is not the main concern of our work, their solution cannot be extended to our problem setting.

Some recent work also studies graph (data) federation [28]. A typical application of the graph federation is multi-modal route planning [14] over multiple transportation networks. To solve this problem, Li *et al.* [14] proposed a novel solution to achieve very high efficiency and low communication overhead.

Approximate kNN Query. Existing solutions to approximate kNN queries can be divided into three categories: *tree-based*, *graph-based*, and *hash-based* solutions. In *tree-based solutions*, the main idea is to divide the entire space into multiple disjoint areas, and kNN often falls in the area that contains or is adjacent to the query location [13]. For *graph-based solutions*, the core idea is to construct a proximity graph based on the neighbor relationship between objects and search for the "nearest neighbors" first [25]. As for *hash-based solutions*, the basic idea is to map data objects to lower-dimensional hash values. Then, the closer the distance between two objects is, the higher the probability of being mapped to the same hash value [6]. These algorithms do not consider data security/privacy and hence cannot be used as the solutions to our federated approximate kNN query problem. However, these algorithms can be used in querying the local dataset in each data owner.

Some existing work [7,12], which considers data privacy, is mostly studied under the scenario of outsourced databases. Although these solutions [7,12] can protect security by searchable asymmetric encryption schemes, their application scenario is quite different from ours (*i.e.*, a data federation). For example, an outsourced database assumes the raw data of one data owner is encrypted and stored in the third party (*e.g.*, a cloud server). By contrast, in a data federation, the local dataset of each data owner does not need to be encrypted and stored in a third party. Thus, queries can be securely executed in plaintext over each local dataset, which is quite different from the scenario of outsourced databases.

6   Conclusion

This paper investigates efficient and secure approximate kNN query over a spatial data federation. We propose two secure algorithms with low communication cost and time complexity. We also prove that their approximation guarantees (*i.e.*, the accuracy of the query results) have non-trivial bounds. The experimental results on both synthetic and real datasets show that the query efficiency of our algorithms is significantly better than that of state-of-the-arts by a large margin. Results also show that our algorithms can often achieve a higher result accuracy.

Acknowledgements. We are grateful to anonymous reviewers for their constructive comments. This work is partially supported by the National Science Foundation of China (NSFC) under Grant No. U21A20516 and 62076017, the Beihang University Basic Research Funding No. YWF-22-L-531, the Funding No. 22-TQ23-14-ZD-01-001 and WeBank Scholars Program. Lei Chen's work is partially supported by National

Science Foundation of China (NSFC) under Grant No. U22B2060, the Hong Kong RGC GRF Project 16213620, RIF Project R6020-19, AOE Project AoE/E-603/18, Theme-based project TRS T41-603/20R, China NSFC No. 61729201, Guangdong Basic and Applied Basic Research Foundation 2019B151530001, Hong Kong ITC ITF grants MHX/078/21 and PRP/004/22FX, Microsoft Research Asia Collaborative Research Grant and HKUST-Webank joint research lab grants.

References

1. Amap: https://lbs.amap.com/. Accessed 30 Jan 2023
2. Bater, J., Elliott, G., Eggen, C., Goel, S., Kho, A.N., Rogers, J.: SMCQL: secure query processing for private data networks. PVLDB **10**(6), 673–684 (2017)
3. Bater, J., He, X., Ehrich, W., Machanavajjhala, A., Rogers, J.: Shrinkwrap: efficient SQL query processing in differentially private data federations. PVLDB **12**(3), 307–320 (2018)
4. Bater, J., Park, Y., He, X., Wang, X., Rogers, J.: SAQE: practical privacy-preserving approximate query processing for data federations. PVLDB **13**(11), 2691–2705 (2020)
5. Bonawitz, K., et al.: Practical secure aggregation for privacy-preserving machine learning. In: CCS, pp. 1175–1191 (2017)
6. Cai, D.: A revisit of hashing algorithms for approximate nearest neighbor search. IEEE Trans. Knowl. Data Eng. **33**(6), 2337–2348 (2021)
7. Choi, S., Ghinita, G., Lim, H., Bertino, E.: Secure knn query processing in untrusted cloud environments. IEEE Trans. Knowl. Data Eng. **26**(11), 2818–2831 (2014)
8. Evans, D., Kolesnikov, V., Rosulek, M.: A pragmatic introduction to secure multi-party computation. Found. Trends Priv. Secur. **2**(2–3), 70–246 (2018)
9. Gao, D., Tong, Y., She, J., Song, T., Chen, L., Xu, K.: Top-k team recommendation in spatial crowdsourcing. In: WAIM, pp. 191–204 (2016)
10. Jurczyk, P., Xiong, L.: Information sharing across private databases: secure union revisited. In: SocialCom/PASSAT, pp. 996–1003 (2011)
11. Keller, M.: MP-SPDZ: a versatile framework for multi-party computation. In: CCS, pp. 1575–1590 (2020)
12. Lei, X., Liu, A.X., Li, R.: Secure KNN queries over encrypted data: dimensionality is not always a curse. In: ICDE, pp. 231–234 (2017)
13. Li, W., Zhang, Y., Sun, Y., Wang, W., Li, M., Zhang, W., Lin, X.: Approximate nearest neighbor search on high dimensional data - experiments, analyses, and improvement. IEEE Trans. Knowl. Data Eng. **32**(8), 1475–1488 (2020)
14. Li, Y., Yuan, Y., Wang, Y., Lian, X., Ma, Y., Wang, G.: Distributed multimodal path queries. IEEE Trans. Knowl. Data Eng. **34**(7), 3196–3210 (2022)
15. Liu, C., Wang, X.S., Nayak, K., Huang, Y., Shi, E.: Oblivm: a programming framework for secure computation. In: S & P, pp. 359–376 (2015)
16. Mount, D.M., Arya, S.: Ann library. http://www.cs.umd.edu/mount/ANN/. Accessed 30 Jan 2023
17. Pan, X., et al.: Hu-fu: a data federation system for secure spatial queries. PVLDB **15**(12), 3582–3585 (2022)
18. Shi, Y., Tong, Y., Zeng, Y., Zhou, Z., Ding, B., Chen, L.: Efficient approximate range aggregation over large-scale spatial data federation. IEEE Trans. Knowl. Data Eng. **35**(1), 418–430 (2023)

19. Tao, Q., Zeng, Y., Zhou, Z., Tong, Y., Chen, L., Xu, K.: Multi-worker-aware task planning in real-time spatial crowdsourcing. In: DASFAA, pp. 301–317 (2018)

20. Tong, Y., et al.: Hu-fu: efficient and secure spatial queries over data federation. PVLDB **15**(6), 1159–1172 (2022)

21. Tong, Y., Zeng, Y., Zhou, Z., Chen, L., Xu, K.: Unified route planning for shared mobility: an insertion-based framework. ACM Trans. Database Syst. **47**(1), 2:1-2:48 (2022)

22. Tong, Y., Zhou, Z., Zeng, Y., Chen, L., Shahabi, C.: Spatial crowdsourcing: a survey. VLDB J. **29**(1), 217–250 (2020)

23. Vershynin, R.: High-Dimensional Probability: An Introduction with Applications in Data Science. Cambridge University Press, Cambridge (2018)

24. Volgushev, N., Schwarzkopf, M., Getchell, B., Varia, M., Lapets, A., Bestavros, A.: Conclave: secure multi-party computation on big data. In: EuroSys, pp. 3:1–3:18 (2019)

25. Wang, M., Xu, X., Yue, Q., Wang, Y.: A comprehensive survey and experimental comparison of graph-based approximate nearest neighbor search. PVLDB **14**(11), 1964–1978 (2021)

26. Wang, Y., et al.: Fed-LTD: towards cross-platform ride hailing via federated learning to dispatch. In: KDD, pp. 4079–4089 (2022)

27. Xie, D., Li, F., Yao, B., Li, G., Zhou, L., Guo, M.: Simba: efficient in-memory spatial analytics. In: SIGMOD, pp. 1071–1085 (2016)

28. Yuan, Y., Ma, D., Wen, Z., Zhang, Z., Wang, G.: Subgraph matching over graph federation. PVLDB **15**(3), 437–450 (2021)

TAGnn: Time Adjoint Graph Neural Network for Traffic Forecasting

Qi Zheng and Yaying Zhang[⊠]

The Key Laboratory of Embedded System and Service Computing,
Ministry of Education, Tongji University, Shanghai 200092, China
{zhengqi97,yaying.zhang}@tongji.edu.cn

Abstract. Spatial-temporal modeling considering the particularity of
traffic data is a crucial part of traffic forecasting. Many methods take
efforts into relatively independent time series modeling and spatial mining and then stack designed space-time blocks. However, the intricate
deep structures of these models lead to an inevitable increment in the
training cost and the interpretability difficulty. Moreover, most previous
methods generally ignore the meaningful time prior and the spatial correlation across time. To address these, we propose a novel Time Adjoint
Graph Neural Network (TAGnn) for traffic forecasting to model entangled spatial-temporal dependencies in a concise structure. Specifically,
we inject time identification (i.e., the time slice of the day, the day of the
week) which locates the evolution stage of traffic flow into node representation. Secondly, based on traffic propagation, we connect data across
time slices to generate the time-adjoint hidden feature and spatial correlation matrix, allowing the spatial-temporal semantics to be captured by
a simple graph convolution layer. And we introduce a time residual connection in generating predictions to capture the future traffic evolution.
Experiments on four traffic flow datasets demonstrate that our method
outperforms the state-of-the-art baselines efficiently.

Keywords: Spatial-temporal mining · Graph neural network · Traffic
forecasting

1 Introduction

Real-time and dynamic traffic forecasting is vital in the growing demand for
intelligent transportation services. Accurate traffic prediction greatly impacts
on urban spatial-temporal situation awareness for city management and travel
planning. As a widely concerned spatial-temporal data forecasting problem in
both academia and industry, traffic prediction has its uniqueness due to the
following temporal and spatial observations:

(1) **The time-prior information helps to locate the evolution of traffic dynamics in periodic changes.** Traffic flow naturally has its unique temporal characteristics. Flow series recorded in road sensors always show similar
periodic changes on daily and weekly scales. Intuitively, given a road section

with its historical traffic values and the time to predict (e.g. the peak hour on a weekday), a basic inference can be carried out easily. The time-prior information implies the innate traffic representation based on long-term observations. It could be beneficial in the traffic prediction tasks to directly use this intact and readily available time information (e.g. the time slice of the day, the day of the week) without much additional acquisition cost. Besides, based on the general temporal cycle pattern, modeling the change amount of traffic flow instead of the original value in a short future evolution may be easier for forecasting.

(2) **The influence range of spatial dependence (i.e. local to global) between road sensors varies according to the time span of traffic flow propagation.** As traffic state is transmitted along the road structure with time, traffic information of one location directly affects its spatially nearby neighbors in a short-term period (e.g. 5 min). However, when a traffic state (e.g. congestion) lasts a relatively long time, the scope of affected spatial locations will become wider. This means that the traffic status of one location can potentially influence that of distant locations over time as well as reflect the influence of historical states of locations far away from it. The closer two traffic locations are, the more often the spatial influence between them occurs within a short period. Conversely, spatial influence at more distant locations tends to persist over time due to traffic propagation [14]. How to extract global spatial connections related to time spans is nontrivial.

In recent years, deep learning-based models have been continuously proposed for capturing the temporal and spatial correlations in spatial-temporal series data. Many researchers utilize widely-used time-related models (e.g. Recurrent Neural Networks, Temporal Convolution Networks, and Transformer-liked models) to capture time dependencies. However, the physical time label describing the evolution stage of traffic flow series is usually neglected in these methods. Studies based on spatial mining made progress by considering varied adjacency relationships in short-term time slices but they generally ignore direct spatial influences that exist on different time spans. It is still a challenge to appropriately capture the entangled spatial and temporal dependencies in different space-time scales. Besides, the separate capture of temporal and spatial features hinders the exploration the space-time interaction in traffic propagation.

To address these issues, a Time Adjoint Graph neural network (TAGnn) for traffic forecasting is proposed in this work. The proposed model TAGnn can explicitly use the time-prior to increase the accuracy and reliability of prediction and dynamically mine the spatial-temporal dependencies from different space-time scales. The main contributions of this work are as follows:

(1) A new time encoding method is designed to make explicit use of time-prior information in the node representation, which provides considerable gains in accuracy. Besides, the time-residual connection is introduced in the generation of predicted values to capture the evolution of traffic flow.

(2) The spatial-temporal characteristics in each time slice are captured through novel yet simple spatial-temporal mining modules, which connect elements of input series from different time spans to extract spatial dependencies

across time. It is conducive to exploring spatial-temporal interaction on appropriate space-time scales.

(3) Extensive experiments on four public real-world traffic datasets show that the proposed method outperforms state-of-the-art baselines with less training time cost.

2 Related Work

Time analysis and spatial mining are two key parts of the traffic forecasting problem. Early methods [8,15] are computationally efficient but perform poorly in complex scenarios. RNN-based, CNN-based and Transformer-based [10] models [2,5,6,11,12] can extract short-term and long-term temporal correlations in time series. Some other methods [4,9,14] capture cross-time dependencies. However, these models devoted to time dependence extraction usually rely only on the implied time characteristics or location coding of input data itself, and rarely take into account the auxiliary gain of time prior information. As advanced research on spatial mining, an increasing number of methods are focusing on the non-local and dynamic spatial influence between road nodes [2,11,13]. However, it is still challenging to explore the global spatial-temporal interaction in an appropriate space and time scale. Two recent studies [7,14] have turned the spotlight on mining data connections across different time horizons. They put forward some new ideas about modeling spatial-temporal interactions with different mechanisms, but there may be redundancy in the construction of adjacency matrices and input representations.

3 Preliminary

A group of road sensors distributed in a road network can be formulated as a graph $\mathcal{G} = (\mathcal{V}, \mathcal{E})$. $\mathcal{V}$ is a set of nodes meaning the road sensors, $|\mathcal{V}| = N$. $\mathcal{E}$ is a set of edges. In the urban road network, the 24-h traffic data collected by sensors are generally aggregated at a certain frequency (e.g., once every 5 min), so that a day can be divided into multiple time slices (e.g., 288). The traffic data collected by one sensor $v_i \in \mathcal{V}$ is a time sequence composed of multiple discrete values $\{x_{v_i;t_1}, x_{v_i;t_2}, \ldots || x_{v_i;t_j} \in \mathbb{R}^C\}$, and C is the number of features. At a certain time slice t_i, traffic features of all N graph nodes form the feature matrix $X_{t_i} \in \mathbb{R}^{N \times C}$. The traffic forecasting problem on a road network $\mathcal{G}$ can be formulated as: Given historical observation $\mathbf{X} = [X_{t-p}, \ldots, X_{t-1}, X_t]$ in the past P time slices (here $p = P - 1$ for brevity), we aim to predict the traffic data $\mathbf{Y} = [X_{t+1}, \ldots, X_{t+2}, X_{t+Q}]$ in the next consecutive Q time slices. Here, time slice t denotes the most recent slice in the input series.

4 Methodology

The core idea of the proposed model TAGnn is to inject the time-prior information into node representation and connect data in different time spans to produce the time-adjoint hidden feature and the across-time global spatial correlation. Then the spatial-temporal feature can be captured by a simple graph convolution layer in each time slice. And a time-residual connection is introduced to generate predicted values for capturing the future evolution of traffic flow.

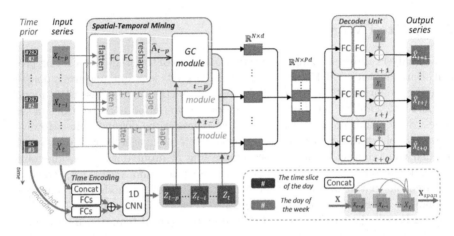

Fig. 1. The overall framework of the proposed TAGnn model.

The overall framework of the proposed TAGnn model is illustrated in Fig. 1. The input series $\mathbf{X} = [X_{t-p}, ..., X_{t-i}, ..., X_t] \in \mathbb{R}^{P \times N \times C}$ is reconstructed by data concatenation and combined with the corresponding time prior information $T_{slice} \in \mathbb{R}^{P \times C_1}$ (the time slice of the day, C_1 is the maximum number of time slices in one day, e.g. $C_1 = 288$ when each time slice is 5 min) and $T_{day} \in \mathbb{R}^{P \times 7}$ (the day of the week) in the *Time Encoding Module*, they are respectively transformed to hidden features and then aggregated. Then the time-adjoint hidden features $\mathbf{Z} = [Z_{t-p}, ..., Z_{t-i}, ..., Z_t] \in \mathbb{R}^{P \times N \times d}$ (d is the number of the hidden dimension) is generated through a convolution layer by time dimension. After that, each time slice owns a *Spatial-Temporal Mining (STM) Module*. It uses the input feature X_{t-i} in time slice $t - i$ and X_t in the most recent time slice t (X_t only when $i = 0$) to learn the spatial adjacency matrix $\hat{A}_{t-i}$ and then fed $\hat{A}_{t-i}$ and the time-adjoint hidden feature Z_{t-i} into the *Graph Convolution (GC) module*. Outputs of P STM modules are then merged and input to Q decoder units to generate the final prediction in the *Decoder Module*.

4.1   Time Encoding Module

The unique time properties shown in the periodic changes of traffic flow distinguish the traffic forecasting problem from other time series prediction problems. The adjoint temporal information (e.g. the time slice of the day, the day of the week) helps to locate the evolution stage of traffic flow in an intuitive way and readily available. Thus, we input two tensors T_{slice}, T_{day} indicating the temporal tags along with the traffic data into the deep model. All graph nodes in each time slice share the same temporal information, it can be normalized in the day and week scale. For instance, 00:05 am to 00:10 am on Thursday is the *second* time slice (5 min per slice) of the *fourth* day in the week. These two order numbers are encoded to vectors belonging to $\mathbb{R}^{C_1}$ and $\mathbb{R}^7$ by one-hot coding. For the whole input sequence, the adjoint time data are constructed as $T_{slice} \in \mathbb{R}^{P \times C_1}$ and $T_{day} \in \mathbb{R}^{P \times 7}$. Both time labels are transformed into hidden tensors through individual fully-connected layers respectively:

$$H_{slice} = \text{ReLU}(FC_{slice}(T_{slice})) = \text{ReLU}(T_{slice}W_1 + b_1) \in \mathbb{R}^{P \times d} \tag{1}$$

$$H_{day} = \text{ReLU}(T_{day}W_2 + b_2) \in \mathbb{R}^{P \times d} \tag{2}$$

where FC is the fully-connected layer, ReLU is the activation function, $W_1 \in \mathbb{R}^{C_1 \times d}$, $W_2 \in \mathbb{R}^{7 \times d}$ and $b_1, b_2 \in \mathbb{R}^d$ are the weights and biases of linear projections. H_{slice}, H_{day} are temporal embedding tensors.

For exploring the across-time influence of traffic propagation, we concatenate the items in each time slice $t-i, i \in \{0, 1, ..., p\}$ and the latest time slice t in input traffic data $\mathbf{X}$ to obtain the across-time data feature $\mathbf{X}_{span} \in \mathbb{R}^{P \times N \times 2C}$. Then this reconstructed data feature is transformed into a close dimension through two fully-connected layers:

$$(\mathbf{X}_{span})_i = (X_{t-i} || X_t) \in \mathbb{R}^{N \times 2C} \tag{3}$$

$$\mathbf{H}_{flow} = FC_{span}(\text{ReLU}(\mathbf{X}_{span}W_3 + b_3)) \in \mathbb{R}^{P \times N \times d} \tag{4}$$

where $||$ is the concatenation, $W_3 \in \mathbb{R}^{2C \times d}, b_3 \in \mathbb{R}^d$ are the parameters of linear projection, and $\mathbf{H}_{flow}$ is the transformed traffic feature with across-time representation. This traffic feature and time embedding tensors are added through a broadcasting mechanism since every traffic node shares the same time prior:

$$\mathbf{H}_{ST} = H_{slice} + H_{day} + \mathbf{H}_{flow} \in \mathbb{R}^{P \times N \times d} \tag{5}$$

here, H_{slice} and H_{day} are replicated along with the node dimension and then added with $\mathbf{H}_{flow}$. And $\mathbf{H}_{ST}$ is the deep representation with time-prior of all nodes. To better describe the local temporal pattern of traffic nodes, a convolution layer along with the time dimension is then carried out to obtain the time-adjoint hidden feature $\mathbf{Z}$ with local time trends:

$$\mathbf{Z} = \Theta_1 \star \mathbf{H}_{ST} + a \in \mathbb{R}^{P \times N \times d} \tag{6}$$

where $\star$ denotes the convolution operation, Θ_1 means the $1 \times k$ temporal kernel, k is the kernel size, and a means the bias. To maintain the original time length, we use the replication padding in the time dimension to operate equal-width convolution, which means the first and the last items in $\mathbf{H}_{ST}$ are replicated for the supplement to reach the target length. Here, $\mathbf{Z}$ can be regarded as a matrix sequence $[Z_{t-p}, ...Z_{t-i}, ..., Z_t]$, each item is one of the inputs to the following unit in each time slice.

4.2 Spatial Temporal Mining Modules

The spatial correlations among traffic nodes are time-varying. For a relatively long term, the spatial scope of the impact will become bigger. This motivates us to construct across-time combinations in input series for extracting the spatial-temporal impacts on future traffic evolution. We formulate the influence degree in each time slice as an adjacency matrix, which can be integrated with the time-hidden feature (described in Sect. 4.1) through a graph convolution layer.

Specifically, the raw input data X_{t-i} from each past time slice $t-i$ is connected with the most recent feature X_t and this combination is then flattened. Here features of every node are investigated individually through the flatten operation and their interaction can be mined by simple fully-connected layers. Thus two FC layers transform the flattened result into an adjacency matrix $\hat{A}_{t-i}$:

$$E_{t-i} = \begin{cases} FC_i^1(\text{flatten}(X_{t-i}||X_t)), & i \in \{1, ..., p\} \\ FC_i'(\text{flatten}(X_t)), & i = 0 \end{cases} \in \mathbb{R}^l \qquad (7)$$

$$A_{t-i}' = \tanh(FC_i^2(E_{t-i})) \in \mathbb{R}^{N^2} \qquad (8)$$

where $E_{t-i} \in \mathbb{R}^l$ is a hidden embedding with size l, and tanh is the activation function. A_{t-i}' is then reshaped into the learned adjacency matrix $\hat{A}_{t-i} \in \mathbb{R}^{N \times N}$ followed by a dropout layer with the rate ϕ. Each entry in $\hat{A}_{t-i}$ can be regarded as the spatial impact degree of a node in past time slice $t-i$ on the other node in the most recent time slice t.

After that, adjacency matrix $\hat{A}_{t-i}$ is fed into the Graph Convolution (GC) module along with the time-adjoint hidden feature Z_{t-i} in the corresponding time slice. In this work, a GC module consists of a spatial-domain graph convolution layer with the GLU activation and a residual connection from input:

$$GC(X, A) = (AX\theta_1 + \beta_1) \odot \sigma(AX\theta_2 + \beta_2) + X \qquad (9)$$

where $X \in \mathbb{R}^{N \times d}$ is the interested feature, $A \in \mathbb{R}^{N \times N}$ is the adjacency matrix, $\odot$ denotes element-wise product, σ means the sigmoid function, and $\theta_1, \theta_2 \in \mathbb{R}^{d \times d}, \beta_1, \beta_2 \in \mathbb{R}^d$ are the parameters of projections. The spatial-temporal feature in each time slice $t-i$ is extracted through individual GC module:

$$H_{t-i} = GC_i(Z_{t-i}, \hat{A}_{t-i}) \in \mathbb{R}^{N \times d} \qquad (10)$$

Finally, the outputs of STM modules in all P time slices are merged on feature dimension into $H_o \in \mathbb{R}^{N \times Pd}$.

4.3 Decoder Module

The Decoder Module converts the spatial-temporal features from different space-time scales into the final expected targets. To directly model the future evolution of traffic, we sum the input data X_t on the latest time slice t with the hidden features processed by two fully connected layers:

$$\hat{X}_{t+j} = FC_j^4(\text{ReLU}(FC_j^3(H_o))) + X_t \in \mathbb{R}^{N \times C} \qquad (11)$$

The setting of the time residual connection of X_t means the model aims to extract future changes based on the latest spatial-temporal situation.

Finally, the expected prediction can be obtained by concatenating the outputs of all Q target time slices:

$$\hat{\mathbf{Y}} = [\hat{X}_{t+1}, \hat{X}_{t+2}, ..., \hat{X}_{t+Q}] \in \mathbb{R}^{Q \times N \times C} \tag{12}$$

where $\hat{\mathbf{Y}}$ is the final predicted sequence.

5   Experiments

5.1   Datasets and Experiment Settings

To evaluate our model, we conduct experiments on four public datasets [2]: PEMS03, PEMS04, PEMS07, PEMS08. All datasets record highway traffic flow in four districts in California. The raw flow data are aggregated every 5 min, thus C_1 is 288. Consistent with previous studies [2,9], all data sets are divided into training, validation, and test sets in a ratio of 6:2:2. The implementation of the proposed model is under the PyTorch framework[1] on a Linux server with one Intel(R) Xeon(R) Gold 5220 CPU @ 2.20 GHz and one NVIDIA Tesla V100-SXM2 GPU card. We use one-hour historical data to predict the next hour's data for all datasets, which means length P and Q are both 12. The number of feature C is 1 for all datasets. We choose Mean Absolute Error (MAE) as the loss function and use Adam as the optimizer. The batch size is 32, the learning rate is 0.001, and the number of training epochs is 100. We use MAE, Mean Absolute Percentage Error (MAPE), and Root Mean Squared Error (RMSE) [1] to evaluate the prediction performance. The average results of 12 time slices are presented. We compare TAGnn with 10 classical and up-to-date methods: VAR [15], SVR [8], LSTM [3], DCRNN[5], STGCN [12], STFGNN [4], DSTAGCN [14], FCGAGA [7], Graph Wavenet (GWN) [11], and ASTGNN [2].

5.2   Experiment Results

The performance comparison of TAGnn and baseline methods is shown in Table 1, the best results are shown in **bold**. The proposed model TAGnn achieves better performance than baseline methods for most of cases. As Table 1 shown, compared with the most competitive method ASTGNN, our TAGnn achieves 9.28%, 5.92%, 6.69% improvements in terms of MAE, MAPE, RMSE on PEMS08. Likewise, TAGnn has the best performance of all three metrics on PEMS04 and PEMS07, the improvements are 2.25%, 2.19%, 2.55% and 3.24%, 5.16%, 2.86%, respectively. And TAGnn improves ASTGNN by 3.80% in terms of MAE on PEMS03. Instead of stacking many blocks, TAGnn uses a single and independent graph convolution layer in each time slice and captures the spatial-temporal interactive features in parallel through the recombination of input data.

---

[1] Our codes are available at https://github.com/zhuoshu/TAGnn.

Table 1. Performance comparison on PEMS datasets.

Models	PEMS03			PEMS04			PEMS07			PEMS08		
	MAE	MAPE (%)	RMSE	MAE	MAPE (%)	RMSE	MAE	MAPE (%)	RMSE	MAE	MAPE (%)	RMSE
VAR	23.65	24.51	38.26	23.75	18.09	36.66	75.63	32.22	115.24	23.46	15.42	36.33
SVR	21.97	21.51	35.29	28.70	19.20	44.56	32.49	14.26	50.22	23.25	14.64	36.16
LSTM	21.33	23.33	35.11	27.14	18.20	41.59	29.98	13.20	45.84	22.20	14.20	34.06
DCRNN	18.18	18.91	30.31	24.70	17.12	38.12	25.30	11.66	38.58	17.86	11.45	27.83
STGCN	17.49	17.15	30.12	22.70	14.59	35.55	25.38	11.08	38.78	18.02	11.40	27.83
STFGNN	16.77	16.30	28.34	19.83	13.02	31.88	22.07	9.21	35.80	16.64	10.60	26.22
DSTAGCN	15.31	14.91	25.30	19.48	12.93	30.98	21.62	9.10	34.87	15.83	10.03	24.70
FCGAGA	15.99	17.44	26.99	19.42	15.12	31.33	21.73	9.11	35.33	15.80	10.73	24.73
GWN	14.79	**14.32**	25.51	19.36	13.31	31.72	21.22	9.07	34.12	15.07	9.51	23.85
ASTGNN	14.78	14.79	**25.00**	18.60	12.36	30.91	20.62	8.86	34.00	15.00	9.50	24.70
TAGnn	**14.22**	14.76	25.04	**18.18**	**12.09**	**30.12**	**19.95**	**8.40**	**33.03**	**13.61**	**8.94**	**23.05**

It may help to avoid the difficulty of interpretation to some extent caused by exceedingly increasing the depth of the model. The explicit use of time-prior to node representation and the direct extraction of interactive spatial-temporal dependencies in different time span enables TAGnn to achieve better forecasting performance.

5.3   Ablation Studies and Efficiency Analysis

Our ablation studies are conducted to further validate the effectiveness of the time encoding module, the time residual connection, and the across-time graph convolution modules in the proposed TAGnn model. We design 3 variants: **(a) TAGnn w/o time prior**: It removes the time prior information of the input data. **(b) TAGnn w/o across-time connection**: It cancels the connection of data in each past time slice and the most recent time slice. **(c) TAGnn w/o time residual connection**: It removes the residual connection from the data in the latest time slice.

From the ablation results in Table 2, we can find: (1) Time prior provides considerable gains in traffic prediction tasks and the explicit use of time auxiliary information is helpful for traffic modeling. (2) It's beneficial to consider across-time influence in global spatial mining which is more in line with the characteristics of traffic propagation. (3) By using the time residual connection to learn the future evolution of traffic flow, the framework of TAGnn well describes the temporal evolution contexts and makes a better prediction.

To investigate the proposed model from the efficiency level, we compare the time consumption and convergence curves of TAGnn with ASTGNN and GraphWavenet due to their compelling performance on accuracy. Table 3 shows the training time of each epoch and the total inference time for the whole test data. Note that the training of ASTGNN has two stages (non-autoregressive one and autoregressive one). Figure 2 illustrates the convergence curves, which contain validation MAE in the training process of three models on one-hour prediction. The results demonstrate that TAGnn presents the lowest time consumption among the three methods and quickly achieves lower validation errors than both baseline methods.

Table 2. Performance comparison of the variants of TAGnn.

Models	PEMS04			PEMS08		
	MAE	MAPE(%)	RMSE	MAE	MAPE(%)	RMSE
GraphWavenet	19.36	13.31	31.72	15.07	9.51	23.85
ASTGNN	18.60	12.36	30.91	15.00	9.50	24.70
TAGnn w/o time prior	18.82	12.53	30.53	15.15	9.64	24.03
w/o across-time connection	18.42	12.26	30.30	13.81	9.12	23.30
w/o time residual connection	18.35	12.53	**29.98**	13.94	9.57	**23.05**
TAGnn	**18.18**	**12.09**	30.12	**13.61**	**8.94**	**23.05**

Table 3. Time consumption on PEMS04 dataset.

Models	PEMS04	
	Training (s/epoch)	Inference (s)
GraphWavenet	21.4	2.0
ASTGNN (stage 1)	84.6	42.5
ASTGNN (stage 2)	189.0	
TAGnn	19.2	1.4

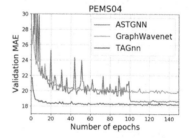

Fig. 2. Model convergence curves on PEMS04 dataset.

6   Conclusion

In this work, we propose a novel Time Adjoint Graph Neural Network (TAGnn) for traffic forecasting. To extract the spatial-temporal interactive features originated from traffic propagation, TAGnn provided a new perspective to process the time series which injects the time prior information into deep representation and connects data of the latest and each past time slices for the following node embedding and spatial modeling. By the data connection across time slices, TAGnn directly capture the long-term and short-term spatial-temporal propagation features. And the introduced time residual connection enables the model to capture the future evolution more easily. The empirical test results validate the accuracy and efficiency of the proposed model. Note the time prior used in this work is plain time identification information independent of the traffic data. When the traffic state changes differently than usual due to events or other factors, the time prior may not necessarily play an auxiliary role. Increasing the robustness of TAGnn against abrupt changes is our future work.

Acknowledgements. This work was supported in part by the National Key Research and Development Program of China under Grant No.2019YFB1704102 and the Fundamental Research Funds for the Central Universities of China under Grant No.2022-4-ZD-05.

References

1. Chicco, D., Warrens, M.J., Jurman, G.: The coefficient of determination r-squared is more informative than SMAPE, MAE, MAPE, MSE and RMSE in regression analysis evaluation. PeerJ Comput. Sci. **7**, 1–24 (2021)
2. Guo, S., Lin, Y., Wan, H., Li, X., Cong, G.: Learning dynamics and heterogeneity of spatial-temporal graph data for traffic forecasting. IEEE Trans. Knowl. Data Eng. **34**(11), 5415–5428 (2022)
3. Hochreiter, S., Schmidhuber, J.: Long short-term memory. Neural Comput. **9**(8), 1735–1780 (1997)
4. Li, M., Zhu, Z.: Spatial-temporal fusion graph neural networks for traffic flow forecasting. In: Proceedings of the AAAI Conference on Artificial Intelligence, vol. 35, no. 5, pp. 4189–4196 (2021)
5. Li, Y., Yu, R., Shahabi, C., Liu, Y.: Diffusion convolutional recurrent neural network: data-driven traffic forecasting. In: International Conference on Learning Representations (ICLR) (2018)
6. Liu, D., Wang, J., Shang, S., Han, P.: MSDR: multi-step dependency relation networks for spatial temporal forecasting. In: Proceedings of the 28th ACM SIGKDD Conference on Knowledge Discovery and Data Mining, pp. 1042–1050. KDD 2022, Association for Computing Machinery, New York, NY, USA (2022)
7. Oreshkin, B.N., Amini, A., Coyle, L., Coates, M.: FC-GAGA: fully connected gated graph architecture for spatio-temporal traffic forecasting. In: Proceedings of the AAAI Conference on Artificial Intelligence, vol. 35, pp. 9233–9241 (2021)
8. Ristanoski, G., Liu, W., Bailey, J.: Time series forecasting using distribution enhanced linear regression. In: Pei, J., Tseng, V.S., Cao, L., Motoda, H., Xu, G. (eds.) PAKDD 2013. LNCS (LNAI), vol. 7818, pp. 484–495. Springer, Heidelberg (2013). https://doi.org/10.1007/978-3-642-37453-1_40
9. Song, C., Lin, Y., Guo, S., Wan, H.: Spatial-temporal synchronous graph convolutional networks: a new framework for spatial-temporal network data forecasting. In: Proceedings of the AAAI Conference on Artificial Intelligence, vol. 34, pp. 914–921 (2020)
10. Vaswani, A., et al.: Attention is all you need. In: Advances in Neural Information Processing Systems, vol. 30. Curran Associates, Inc. (2017)
11. Wu, Z., Pan, S., Long, G., Jiang, J., Zhang, C.: Graph wavenet for deep spatial-temporal graph modeling. In: Proceedings of the Twenty-Eighth International Joint Conference on Artificial Intelligence, IJCAI-19, pp. 1907–1913 (2019)
12. Yu, B., Yin, H., Zhu, Z.: Spatio-temporal graph convolutional networks: a deep learning framework for traffic forecasting. In: Proceedings of the Twenty-Seventh International Joint Conference on Artificial Intelligence, IJCAI-18, pp. 3634–3640 (2018)
13. Zhang, Q., Chang, J., Meng, G., Xiang, S., Pan, C.: Spatio-temporal graph structure learning for traffic forecasting. In: Proceedings of the AAAI Conference on Artificial Intelligence, vol. 34, pp. 1177–1185 (2020)

14. Zheng, Q., Zhang, Y.: DSTAGCN: dynamic spatial-temporal adjacent graph convolutional network for traffic forecasting. IEEE Trans. Big Data **9**(1), 241–253 (2023)
15. Zivot, E., Wang, J.: Vector autoregressive models for multivariate time series. Model. Financ. Time Series with S-PLUS®, 385–429 (2006)

Adversarial Spatial-Temporal Graph Network for Traffic Speed Prediction with Missing Values

Pengfei Li, Junhua Fang[(✉)], Wei Chen, An Liu, and Pingfu Chao

Department of Computer Science and Technology, Soochow University,
Suzhou, China
20205227067@stu.suda.edu.cn, {jhfang,weichen,anliu,pfchao}@suda.edu.cn

Abstract. Traffic prediction plays a crucial role in constructing intelligent transportation systems. However, the acquired traffic data is often accompanied by missing values due to sensor faults in the data collection stage or communication failures in the data transmission stage. The lack of traffic data will bring great difficulties to the traffic prediction task. Existing traffic prediction models usually rely on the intactness of the data in terms of both spatial and temporal dimensions. In this paper, we propose an Adversarial Spatial-Temporal Graph Network model named ASTGnet, for traffic speed prediction. Our model acts like a multi-task learning framework to predict traffic speed and simultaneously impute missing values. This method effectively reduces the error accumulation between the imputation task and the prediction task. Moreover, we add the adversarial perturbation on the traffic data hidden state to enhance the robustness of the data representation embedding. We evaluate our model on real-world traffic datasets, and experimental results show that our framework has better prediction performance on datasets with missing values than baselines.

Keywords: Graph neural network · Spatial-Temporal data analysis · Time series forecast

1 Introduction

The advent of smart transportation systems in cities has greatly reduced the burden on urban traffic. Traffic prediction plays a vital role in intelligent transportation, allowing for the forecasting of future traffic conditions based on historical observation data. Accurate traffic prediction can aid in better urban planning and development, as well as provide valuable travel advice for individuals.

Benefit from advancements in sensor technology and big data, a large amount of traffic data is now available, providing ample opportunities for deep learning applications in traffic prediction. One of the most advanced methods for traffic prediction is Spatial-Temporal Graph learning. In Spatial-Temporal Graph learning, each road segment or sensor in a city is represented as a node, with the edges

X. Wang et al. (Eds.): DASFAA 2023, LNCS 13943, pp. 380–395, 2023.
https://doi.org/10.1007/978-3-031-30637-2_25

between nodes reflecting the spatial influence between road segments. Numerous methods for traffic prediction have been proposed by researchers in this field.

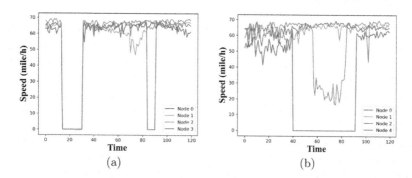

Fig. 1. Missing cases from **METR-LA** dataset. (a)Short-time data missing on the entire transportation network and partial network. (b) Long-term data missing on partial networks.

Spatial-Temporal Graph learning is a cutting-edge approach to traffic prediction that models the spatial-temporal correlations in the traffic network as a graph. Nodes in the graph represent road intersections or traffic sensors, and their spatial connections are represented as edges. By integrating GCN with RNN [14,25], CNN [9,22,24], or Attention [23,26], researchers have significantly improved the accuracy of traffic speed prediction compared to traditional statistical methods. These advances are thanks to the powerful graph model used in Spatial-Temporal graph learning.

However, real-world traffic data is often affected by sensor failures or communication errors, leading to a large number of missing values. The presence of missing values significantly impacts the prediction performance. Existing methods do not effectively handle missing values, with simple imputation strategies like padding with 0 or mean often leading to rapid degradation in cases with high missing rates. There are three main types of missing data in real-world traffic data, as shown in Fig. 1: (1) failures at the control center of the entire transportation network, (2) loss of partial sensor signals, and (3) partial device errors or power outages. A flexible and robust approach is therefore required to handle these various types of missing data in complex situations.

Another solution is to interpolate data and then make predictions. Existing data interpolation schemes includes matrix decomposition-based methods [20], tensor factorization-based methods [1], recurrent neural networks [3], self-attention-based methods [16] and *etc.* However, such a two-step solution may magnify the error and cause additional work. Therefore, a more efficient and effective solution is needed, which has a unified prediction framework that can model and impute missing values ata single step.

In an effort to tackle the issue of missing values in real-time traffic predictions, we present the Adversarial Spatial-Temporal Graph Network (ASTGnet) model. Our primary goal is to enhance the accuracy of traffic predictions when historical data contains missing values. ASTGnet utilizes an encoder-decoder architecture, combining GCN with spatial-temporal attention to discover complex spatial-temporal correlations in traffic data. We combine the imputation task with the prediction task for training. This training strategy can improve both imputation and prediction processes simultaneously, thereby reducing the problem of error accumulation caused by imputation errors to a certain extent. Inspired by adversarial learning, we perform adversarial perturbation on the hidden states of the traffic data, which leads to a more robust embedding representation of the data. The final prediction is obtained by combining the standard prediction decoder and the adversarial prediction module. We conducted experiments on two datasets and evaluated our approach in both imputation and prediction tasks. The results demonstrate that our approach outperforms baseline methods. The contribution of this study is summarized as follows:

- We propose a novel Adversarial Spatial-Temporal Graph Network (ASTGnet) for traffic prediction that deals with missing values in a unified manner. The proposed framework merges the imputation and prediction tasks into one network, thereby avoiding the accumulation of errors between the two processes.
- We design an adversarial training framework for traffic prediction. We make the hidden representation captured by the encoder process robust by adding perturbations to the traffic data embedding.
- Our experiments show that our framework has better prediction performance in the presence of missing data and also has a good imputation accuracy in the impute task.

The rest of this paper is organized as follows. Section 2 describes the related works. We then define the problem of traffic prediction in Section 3 and illustrate the proposed solution in Section 4. Section 5 shows the experiment results of this paper. Finally, we conclude the paper in Section 6.

2   Related Work

2.1   Traffic Predicition

Traffic prediction is an essential part of building intelligent transportation. Generally, traffic prediction methods can be classified into two categories including statistical methods and deep learning methods. Statistical methods, such as auto-regressive integrated moving average (ARIMA) [15], vector auto regressive model (VAR) [6] and support vector regression (SVR) [21], focus on forecasting individual time series, but they are limited by their reliance on stationary assumptions and are unable to handle the complex variability of traffic data.'

The advancement of artificial intelligence has facilitated the extensive utilization of deep learning methods in traffic prediction. Graph convolution [12], as exemplified in STGCN [24], has become a popular tool among researchers for modeling traffic by capturing spatial correlations. Meanwhile, temporal correlations are extracted through the application of one-dimensional convolution along the temporal axis. Another example is DCRNN [14], which views the dynamics of traffic flow as a diffusion process and replaces fully connected layers with diffusion convolution in its gated recursive unit. T-GCN [25] leverages graph convolution to capture spatial correlations and GRU to capture temporal correlations. Furthermore, the integration of the attention mechanism in GCN, as demonstrated in ASTGCN [9], enhances prediction accuracy, while GMAN [26] utilizes self-attention to dynamically understand spatial correlations.

However, when faced with real-world scenarios containing missing values, current methods tend to handle these values crudely, either by discarding them or using simple imputation techniques (e.g., padding with zeros or mean values). When the problem of missing values is severe, the performance of these models drops significantly.

2.2 Traffic Predicition with Missing Values

There are also many studies on missing value processing in time series forecasting, including KNN-based methods [17], matrix decomposition-based methods [20], tensor factorization-based methods [1], and clustering-based methods [13]. Many deep learning-based methods have also been proposed in recent years, such as RNN-based methods [2], denoising stacked autoencoders [5], GANs [18], and self-attention-based methods [16]. These imputation methods can preprocess incomplete data and then be used to predict traffic. However, these two-step solutions can amplify errors and create additional work.

In order to combine missing value repair and prediction tasks, some RNN-based methods are proposed based on GRU-D [3] to deal with multivariate time series of missing values. Although these RNN-based methods are able to capture the temporal correlation of time series, they ignore the spatial correlation of traffic data. Graph MarkovNet [4] treats traffic flow as a Markov process on the road network. It infers the missing traffic state through a step-by-step context. However, this process can only be autoregressive and cannot make long-term predictions. RIHGCN [27] proposes a unified framework based on GRU to combine missing value repair and traffic prediction. However, it still has two problems: (1) Recurrent structure, which leads to slow training speed and inference speed. (2) Although the imputation loss is calculated into the backpropagation, there is a lack of a certain regular term. When the imputed value has a considerable error, it will suffer from the error accumulation problem.

3   Preliminaries

In this section, we will introduce the definition and the problem statement.

Definition 1: Traffic Network. A traffic network is represented as a directed graph $G = (V, E, A)$, where $V = \{V_1, V_2, ..., V_N\}$ represents the set of sensors collecting traffic data, $N = |V|$ represents the number of sensors, and E refers to the set of edges, $A \in \mathbb{R}^{N \times N}$ is the spatial adjacency matrix representing the proximity or distance between nodes from V.

Definition 2: Traffic speed matrix. The traffic speed collected by the entire traffic G at time t is represented as $X_t = \{x_{t,V_1}, x_{t,V_2}, ..., x_{t,V_N}\} \in \mathbb{R}^{N \times C}$, where $x_{t,V_i} \in \mathbb{R}^C$ represents the speed of traffic collected by node V_i at time t, C is the number of traffic features (e.g., volume, speed).

Problem Statement: Traffic speed prediction. Given a series of historical traffic speed matrices $X_{1:P} = (X_1, ..., X_P) \in \mathbb{R}^{N \times P \times C}$ over the last P time slices, the traffic network graph G and an integer Q, the traffic speed prediction aims to generate another series of traffic speed matrices $X_{P+1:P+Q} = (X_{P+1}, ..., X_{P+Q}) \in \mathbb{R}^{N \times Q \times C}$ representing the traffic speed in the next Q time slices.

In addition, traffic data can have missing values due to a failed traffic sensor or the absence of trajectories passing the sensor. To represent the missing variables in X, the missing mask vector $M \in \mathbb{R}^{N \times P \times C}$ is introduced, where

$$M_{t,V} = \begin{cases} 1 & \text{if } X_{t,V} \text{ is observed.} \\ 0 & \text{if } X_{t,V} \text{ is missing.} \end{cases} \tag{1}$$

4   Methodology

In this section, we first present the framework of our proposed approach and introduce each component in detail.

4.1   Architecture Overview

The overall structure of our proposed model, ASTGnet, is depicted in Fig. 2. It consists of an encoder-decoder structure. We utilize a sophisticated encoder that extracts intricate spatial-temporal interdependencies from the incomplete traffic data, thereby yielding hidden states that are essential for the effective execution of downstream traffic-related tasks, such as data imputation and speed prediction. We could have conducted the prediction task in two separate steps: (1) missing data imputation and (2) using the imputed data for prediction. However, the continuous progression of mistakes from the imputation procedure to the prediction task detracts from the overall efficacy of the integrated framework. To mitigate this, we use the hidden state as an intermediate variable to extract features from the traffic data and use these features for both imputation and prediction. Our model can be seen as a unified framework that balances the trade-off between imputation and prediction tasks. Additionally, to make the traffic hidden state representation more robust, we perform an adversarial perturbation on the hidden states. Finally, we calculate the errors simultaneously for the imputation, the prediction, and the adversarial prediction results.

Traffic Data Imputation Module. We first train the imputation module in ASTGnet to learn the hidden representation of the original data with missing values, allowing it to handle the task of traffic data imputation. Let $X_{1:P}, M_{1:P}$ represent the inputs of the model. After they are inputted into the model, the

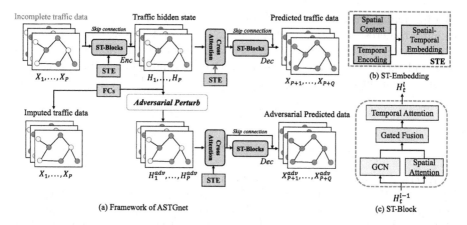

Fig. 2. Framework of propoesed ASTGnet. (a) The model consists of an encoder and decoders. The encoder and decoder consist of ST-Blocks and ST-Embeddings. (b) The ST-Embedding is composed of Spatial Context and Temporal Encoding. (c) The ST-block is made up of GCN and Spatial and Temporal Attention.

initial traffic data is transformed into a hidden state $H_{1:P} \in \mathbb{R}^{N \times P \times D}$ through two fully-connected layers, where D represents the hidden dimension. Next, H is transferred to the encoder for the deep hidden representation. The encoder primarily consists of ST-Blocks, which extract traffic spatial-temporal dependencies. The input data H is then propagated through the following rules, as detailed in Sect. 4.2.

$$H_l = \text{STB}(H_{l-1}) \tag{2}$$

where H_l is the output of l-th layer.

Here, we then obtain the important intermediate variable: the traffic hidden state $h \in \mathbb{R}^{N \times T \times D}$. We use h to estimate the traffic speed $\widehat{X} \in \mathbb{R}^{N \times T \times C}$ after two fully-connected layers.

Therefore, we fill the missing values in $X \in \mathbb{R}^{N \times T \times C}$ into the completed matrix $\tilde{X} \in \mathbb{R}^{N \times T \times C}$ as follows:

$$\tilde{X} = M \odot X + (1 - M) \odot \widehat{X} \tag{3}$$

where $\odot$ denotes element-wise multiplication.

To better obtain a hidden representation of the input data with missing values and to reconstruct the data, the model is trained by minimizing the mean absolute error (MAE) imputation loss:

$$\mathcal{L}_m = \frac{1}{P} \sum_{t=t_1}^{t=Q} \left| X(t) - \tilde{X}(t) \right| \tag{4}$$

The imputation error is added to the training process, which allows the model to better adjust the parameters to produce imputation results that are closer to the true values.

Traffic Data Prediction Module. The decoder of the prediction module includes the cross-attention part and the ST-Blocks. Cross attention is used to extract latent features of traffic data from hidden representation and transform past and future time through Spatial-Temporal Embedding (STE) to convert features of encoded vectors into future representations. It can be formulated as:

$$X^{(l)} = \text{Attention}(STE_{P+1:P+Q}, STE_{1:P}, X^{(l-1)}) \tag{5}$$

$$\text{Attention}(Q, K, V) = \text{softmax}\left(\frac{QK^T}{\sqrt{d_k}}\right)V \tag{6}$$

Afterward, similar to the imputation phase, the final prediction is obtained by passing the input through the ST-Block layers. In order to reduce the discrepancy between the predicted data $\tilde{X}P + 1 : P + Q$ and the actual data $XP + 1 : P + Q$, we adopt mean absolute error (MAE) as the optimization criterion for model training:

$$\mathcal{L}_p = \frac{1}{Q}\sum_{t=P+1}^{t=P+Q}\left|X(t) - \tilde{X}(t)\right| \tag{7}$$

Adversarial Training. Deep learning models always have the problem of overfitting. In order to improve the robustness of the traffic hidden state embedding, regularization methods are needed to solve the problem [10]. Here, we use adversarial training as an efficient method. Adversarial training regularizes the model by adding small perturbations to the embedding during training [7]. This adversarial perturbation reduces the confidence of the repaired embedding, making the model perform as poorly as possible.

After the encoder, we modify the embedding results in the direction of increasing the loss. Specifically, given our hidden representation embedding h, we denote the back-propagated gradient of the loss function L with respect to the embedding of the hidden representation as g, $g = \nabla_X (\mathcal{L}(f_\theta(X), Y))$ and the adversarial perturbation r_{adv} can be computed with the following formula:

$$r_{adv} = -\epsilon\frac{g}{\|g\|_2} \tag{8}$$

where ϵ is a hyperparameter that controls the scale of $l2$-norm perturbation. This adversarial perturbation r_{adv} can reduce the confidence of the hidden representation embedding to make it perform as badly as possible, so that the generalization of the traffic deep representation is stronger, and the input structure of the prediction model is more robust.

Before the adversarial prediction module, we add r_adv to the hidden representation embedding h, and then we use the perturbed embedding h_adv as the input of the prediction module for prediction, and finally we get the adversarial prediction result X_{adv}.

To make the whole model robust to such perturbations, we minimize the adversarial loss, defined as:

$$\mathcal{L}_{adv} = \frac{1}{Q} \sum_{t=P+1}^{t=P+Q} |X(t) - X_{adv}(t)| \tag{9}$$

When using adversarial training, the overall loss function can be defined as:

$$\mathcal{L} = \mathcal{L}_e + \lambda_1 \mathcal{L}_m + \lambda_2 \mathcal{L}_{adv} \tag{10}$$

where λ_1 and λ_2 are hyperparameters.

By using the imputation task as an auxiliary task and implementing adversarial perturbation on the hidden representation embeddings, we can obtain high-quality imputed data and expressive traffic data embeddings, which contributes to the accuracy of the traffic speed prediction task. This training strategy can improve the two processes of imputation and prediction simultaneously, thereby reducing the problem of error accumulation caused by imputation errors to a certain extent.

4.2   Spatial-Temporal Block

A Spatial-Temporal Block is composed of two parts: a spatial feature extraction part and a temporal feature extraction part. The spatial feature extraction part uses Graph Convolutional Network (GCN) and spatial attention mechanism to extract spatial features from the input data.

Graph Convolution. Graph Convolutional Networks broaden the purview of traditional convolution operations, incorporating graph structures and the capability to identify patterns that may not be immediately obvious in such structures. With T specified as either P or Q and $X^{(l)} \in \mathbb{R}^{N \times T \times D}$, the graph convolution operation in the l-th layer is explicitly outlined as follows:

$$GCN\left(X^{(l)}\right) = \sigma\left(AX^{(l-1)}W^l\right) \tag{11}$$

where $W^l \in \mathbb{R}^{D \times D}$, σ represent node representation weight matrix, and activation function, respectively. $A \in \mathbb{R}^{N \times N}$ usually represents the interactions between nodes and is defined as follows:

$$A = \widetilde{D}^{-\frac{1}{2}} \widetilde{A} \widetilde{D}^{-\frac{1}{2}} \tag{12}$$

where $\widetilde{A}$ is the graph adjacency matrix, and $\widetilde{D}_{ij} = \sum_j \widetilde{A}_{ij}$, i, j are nodes in the graph.

Spatial Attention Mechanism. The spatial attention mechanism complements the GCN by capturing the global spatial features. It has a larger field of view than GCN and is able to capture non-distance associations. The spatial attention module can be formulated as:

$$H_A = \text{Attention}(W_s^q X, W_s^k X, W_s^v X) \tag{13}$$

where $W_s^q, W_s^w, W_s^v \in \mathbb{R}^{D \times D}$ are learnable parameters.

Spatial Gated Fusion. To amplify the capability of the traffic network model in comprehending a broader spectrum of information, we bring together the outputs of the GCN module and the spatial attention (SA) module through a spatial gated fusion method. This process adaptively combines the information extracted by the GCN and SA modules. Let H_G and H_A be represented as the output of the GCN and SA modules, respectively, with dimensions of $\mathbb{R}^{N \times T \times D}$. These outputs are then harmoniously integrated through the following process:

$$
\begin{aligned}
H_F &= z \odot H_G + (1 - z) \odot H_A \\
z &= \sigma \left(H_G W_{z1} + H_A W_{z2} + b_z \right)
\end{aligned}
\tag{14}
$$

where H_F is the output of SGF, z is a gate that adaptively controls the information flow extracted by GCN and SA. $\odot$ is the element-wise product. σ is the activation function. $W_{z1}, W_{z2} \in \mathbb{R}^{N \times N}$ and $b_z \in \mathbb{R}^D$ are learnable parameters. The merged spatial features are then fed into the temporal feature extraction module.

Temporal Attention Mechanism. The correlation between traffic conditions and past observations necessitates the processing of data in the temporal dimension, in order to capture the intricacies of the temporal evolution of traffic. Utilizing temporal attention mechanisms, we are able to extract the most salient temporal features of the traffic data, through a methodology similar to that employed in the spatial attention mechanism. The output from the temporal module can be expressed as:

$$
H_T = \text{Attention}(W_t^q X, W_t^k X, W_t^v X)
\tag{15}
$$

where $W_t^q, W_t^w, W_t^v \in \mathbb{R}^{D \times D}$ are learnable parameters.

4.3   Spatial-Temporal Embedding

To obtain node information better, Spatial-Temporal embedding is added to our module as an auxiliary input. Spatial context matrix $\text{SE} \in \mathbb{R}^{N \times D_{SE}}$ is generated via node2vec [8], while temporal encoding matrix $\text{TE} \in \mathbb{R}^{(P+Q) \times D_{TE}}$ is created through one-hot encoding. Both matrices are then transformed into $\text{SE} \in \mathbb{R}^{N \times D}$ and $\text{TE} \in \mathbb{R}^{(P+Q) \times D}$ after two fully-connected layers. Finally, the spatial context and temporal encoding are fused to $\text{STE} \in \mathbb{R}^{(P+Q) \times N \times D}$.

5   Experiments

To evaluate the performance of our proposed model, we conducted extensive experiments on two real-world datasets. The following is a detailed introduction.

The following section provides a comprehensive overview of the datasets and experimental setup, including the introduction of the datasets, presentation of experimental baselines, and explanation of model details. Our experiments aim to answer the following research questions (RQs).

1) Does our model outperform other methods on traffic speed prediction tasks?
2) Does our method also work on the imputation tasks?
3) Is our training framework more advantageous than ordinary methods?

5.1 Datasets

To evaluate the performance of our proposed model, we carry out experiments using two real-world datasets: METR-LA and PEMS-BAY, published by DCRNN [14]. METR-LA consists of 207 sensors and PEMS-BAY has 325 sensors. The specifications of these datasets are outlined in Table 1.

In the preprocessing step, we apply Z-score normalization to both datasets. The data is split into three parts with 70% used for training, 10% for validation, and the remaining 20% for testing. The adjacent matrix is then calculated accordingly:

$$A_{i,j} = \begin{cases} \exp\left(-\frac{dist(V_i,V_j)^2}{\sigma^2}\right), & \text{if } \exp\left(-\frac{dist(V_i,V_j)^2}{\sigma^2}\right) \geq \epsilon \\ 0, & \text{otherwise} \end{cases} \quad (16)$$

where $dist(V_i, V_j)$ denotes the road network distance from sensor V_i to sensor V_j. σ is the standard deviation of distances and ϵ is the threshold to control the sparsity of the adjacency matrix.

Table 1. Dataset statistics.

Datasets	Nodes	Edges	Samples	Sample Rate
METR-LA	207	1515	34272	5min
PEMS-BAY	325	2369	52116	5min

5.2 Baselines

We compare ASGTnet with the state-of-the-art prediction models, including:

- HA: We calculate the average value for each timestamp, and use it as the prediction value for future timestamps.
- SVR: Support Vector Regression (SVR) is a machine learning method based on statistical theory, which can effectively solve the fitting problem of non-linear complex systems.
- LSTM: LSTM is an important method for time series forecasting with long and short-term memory.
- Diffusion Convolutional Recurrent Neural Network(DCRNN) [14]: DCRNN applies two-way diffusion graph convolution and gated recurrent units into an encoder-decoder model to capture the spatial-temporal dependencies.
- Graph WaveNet [22]: Graph WaveNet uses a learnable adjacency matrix and uses TCN instead of 1D convolution to capture complex time correlation.

- GMAN [26]: Graph multi-attention network, whose spatial attention dynamically assigns weights to nodes of each time slice.

These methods are based on the complete traffic data set and do not directly deal with the missing problem. We handle missing values by filling them with the mean of the observations, which is a common method used in existing traffic prediction models, and then utilize these baseline methods for prediction.

To evaluate the accuracy of our method in recovering missing traffic data, we compare the performance of our method with commonly used imputation methods, including mean filling, KNN, iterativeSVD [19], MICE [20]. Here, we refer to fancyimpute[1] for the imputation experiments.

For all baselines, we experiment with their default settings. Three metrics are adopted to evaluate performance: Mean Absolute Error (MAE), Root Mean Square Error (RMSE).

5.3   Experimental Setup

Our proposed model is implemented with PyTorch 1.7.0. All experiments are conducted on a single Nvidia RTX 3070Ti. The proposed model and all baselines use the historical traffic speed of the past $P = 12$ time steps (60 min) to predict that of $Q = 3, 6, 12$ time steps (15, 30, 60 min) ahead. The Adam Optimizer [11] is used for optimization with an initial learning rate of 0.001. The hidden dimension D in main blocks is 64. The hyperparameter ϵ which controls the scale of $l2$-norm perturbation in adversarial training is 2. In terms of the weight λ_2 of adversarial prediction loss, we choose an empirical parameter 0.3. The imputation loss will be discussed in a subsequent chapter.

5.4   Experimental Results

In this section, we demonstrate the effectiveness of ASTGnet with real-life traffic datasets. The experiments were designed to answer the following research questions(RQs).

Prediction Performance. We evaluated traffic prediction performance on two data sets with different missing rates. We compared the results for different missing rates of 20%, 40%, 60%, and 80%, representing the percentage of values that were randomly dropped in the historical data. We fill missing values with the mean. The results of the experiments are presented in Table 2. We can first find that the performance of all methods decreases as the missing rate increases. HA and SVR are always the worst because they fail to capture the complex nonlinearity of traffic. Compared with traditional machine learning algorithms, LSTM can better capture time series features, but it cannot capture the spatial correlation of traffic. Graph networks are state-of-the-art methods for traffic prediction. However, DCRNN, Graph WaveNet and GMAN perform not that

---

[1] https://github.com/iskandr/fancyimpute.

Table 2. Prediction accuracy comparison

Data	Method	20%		40%		60%		80%	
		MAE	RMSE	MAE	RMSE	MAE	RMSE	MAE	RMSE
METR-LA	HA	5.42	12.84	5.69	13.38	5.71	13.36	5.75	13.34
	SVR	5.12	9.05	5.69	9.78	6.43	11.33	6.91	12.62
	LSTM	4.01	7.94	4.06	7.97	4.26	8.38	4.92	9.54
	DCRNN	3.44	6.80	3.52	6.90	3.83	7.31	4.05	7.86
	Graph WaveNet	3.27	6.42	3.36	6.67	3.44	6.73	3.68	7.36
	GMAN	3.27	6.64	3.36	6.84	3.40	6.89	3.51	7.24
	ASTGnet	**3.09**	**6.31**	**3.14**	**6.46**	**3.17**	**6.55**	**3.29**	**6.80**
PEMS-BAY	HA	2.59	6.08	2.61	6.15	2.63	6.27	2.68	6.32
	SVR	3.63	6.30	3.79	6.60	3.84	6.74	3.86	6.77
	LSTM	2.16	4.89	2.17	4.95	2.32	5.27	2.75	6.23
	DCRNN	1.91	4.16	2.02	4.36	2.1	4.72	2.36	5.41
	Graph WaveNet	1.68	**3.53**	1.77	3.88	1.85	4.04	2.05	4.43
	GMAN	1.70	3.77	1.75	3.83	1.77	3.95	1.95	4.31
	ASTGnet	**1.65**	3.65	**1.70**	**3.75**	**1.72**	**3.89**	**1.88**	**4.20**

good on our datasets. This is because they were originally designed for full datasets and thus cannot handle missing data efficiently. Our ASTGnet achieves the best performance in almost all cases because our model can accurately impute missing values and capture hidden traffic patterns.

We also evaluated the performance of our method under different prediction lengths. We fixed the missing rate at 60% and compared our method to the state-of-the-art methods, Graph WaveNet and GMAN. The results, shown in Fig. 3, indicate that Graph WaveNet performs well in short-term prediction while GMAN performs better in long-term prediction due to its temporal attention mechanism. However, our method has stable and competitive performance in both long and short-term forecasts.

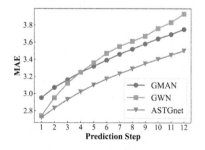

Fig. 3. MAE comparision on each step.

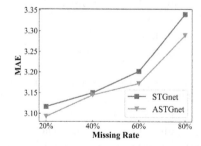

Fig. 4. MAE of adversarial strategy.

Table 3. Imputation accuracy comparison

Data	Method	20%		40%		60%		80%	
		MAE	RMSE	MAE	RMSE	MAE	RMSE	MAE	RMSE
METR-LA	Mean	1.36	4.81	2.79	6.81	4.10	8.35	5.49	9.69
	KNN	0.47	1.77	1.02	2.77	1.79	4.10	3.77	7.64
	IterativeSVD	0.60	2.14	1.25	3.17	2.14	4.41	3.89	7.01
	MICE	0.68	2.52	1.41	3.72	2.26	4.86	3.58	6.67
	ASTGnet	**0.46**	**1.76**	**0.94**	**2.66**	**1.42**	**3.40**	**2.37**	**5.10**
PEMS-BAY	Mean	1.01	3.93	2.02	5.56	3.03	8.35	4.06	7.92
	KNN	**0.17**	**0.69**	0.38	1.17	0.71	1.88	1.44	3.63
	IterativeSVD	0.29	1.17	0.61	1.71	0.99	2.29	1.96	4.15
	MICE	0.29	1.18	0.62	1.77	1.00	2.34	1.56	3.27
	ASTGnet	**0.17**	0.71	**0.36**	**1.11**	**0.64**	**1.67**	**1.08**	**2.62**

Imputation Performance Here, we randomly mask data with missing rates of 20%, 40%, 60%, and 80%, then use it as the ground truth to evaluate the imputation performance. Table 3 shows the difference between the imputed value and the real value. It is worth mentioning that although our task is a prediction task, we get the imputation results while obtaining the best prediction results.

It can be clearly seen that our method is better at inferring missing values compared to the currently widely used imputation methods including mean, KNN, IteraterSVD, and MICE. These methods are weak in capturing nonlinear relationships and complex Spatial-Temporal relationships. When the missing rate is low, ASTGnet performs similarly to KNN. As we increase the missing rate, the imputation effect of most methods decreases due to the reduction of observed data. However, ASTGnet still performs best due to its ability to capture spatial correlation. This means that our proposed ASTGnet also shows state-of-the-art performance on imputation tasks as an additional effect.

Impact of Our Framework. In this section, we investigate the effect of the adopted framework on the prediction task. Specifically, we use different training frameworks for training tasks separately. One set of results used the adversarial training trick, while the other did not. The effect of adversarial training can be seen in Fig. 4. The figure shows the performance comparison of two different frameworks with different missing rates. It is apparent that adversarial training leads to a noticeable improvement in most of the missing rate cases, with a more pronounced effect when the missing rate is substantial.

Additionally, we study the effect of the imputation task on the prediction task. We do performance studies on the value of λ_1, which controls the weight of imputation loss. According to Fig. 5, we can see that the imputation performance will continue to increase with the increasing value of λ_1. This shows that the model can better impute the missing values if we pay more attention to the imputation loss. Furthermore, we can see that when λ_1 is in the range of $(0, 0.7)$, the prediction performance will also improve, indicating that our imputation

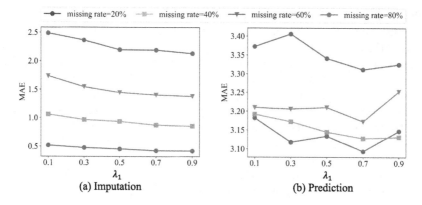

Fig. 5. Performance of (a) data imputation accuracy and (b) prediction accuracy on the weight of imputation loss on METR dataset.

task has a certain auxiliary role in the prediction task. When λ_1 is too high, our model will overfit the imputation task, which will have a negative effect on the main task of prediction. In short, if the weight of the imputation loss is set appropriately, the imputation task can have a positive effect on prediction.

Computational Efficiency. Finally, we also study the computational efficiency of ASTGnet. Table 4 shows the per epoch training time comparison between ASTGnet and the baseline models. ASTGnet outperforms DCRNN because of the recurrent structure of DCRNN. However, compared to other predictive models, the computational efficiency of ASTGnet is still weaker. This is due to two reasons: (1) The calculation of attention scores during the generation of traffic embeddings requires significant computational resources. (2) The adversarial training method involves freezing and copying parameters for certain networks, which also consumes computing resources. Despite these limitations, ASTGnet still boasts an acceptable level of computational efficiency.

Table 4. Model efficiency: training time per epoch (s)

Datasets	GWN	GMAN	LSTM	DCRNN	ASTGnet
METR-LA	86.1	93.5	253.3	318.6	213.4
PEMS-BAY	167.5	162.4	345.7	660.3	364.5

6   Conclusions

In this paper, we propose an adversarial Spatial-Temporal Graph network for traffic speed prediction with missing values. In the real world, the collected traffic data will inevitably have missing values. We propose an advanced Spatial-Temporal network that seamlessly integrates the data imputation process and traffic prediction into a unified framework, alleviating the problem of errors caused by imputation. Our proposed framework captures Spatial-Temporal correlations through a Spatial-Temporal attention network while simultaneously accomplishing data imputation and traffic prediction tasks. In addition, our designed adversarial training framework enhances the robustness of the data hidden representation, further reducing the impact of missing data. We evaluated the effectiveness of ASTGnet through comprehensive experiments on real-world traffic data with missing rates ranging from 20% to 80%. The experimental results show that our method outperforms existing methods in both imputation and prediction tasks. And it is effective to use the imputation task and the prediction task as complementary frameworks. Moreover, the adversarial perturbation on the traffic hidden representation results is feasible.

Acknowledgment. This work was supported by National Natural Science Foundation of China under grant (No.61802273, 62102277), Postdoctoral Science Foundation of China (No.2020M681529), Natural Science Foundation of Jiangsu Province (BK20210 703), China Science and Technology Plan Project of Suzhou (No. SYG202139), Postgraduate Research & Practice Innovation Program of Jiangsu Province (SJC X2_11342), Project Funded by the Priority Academic Program Development of Jiangsu Higher Education Institutions.

References

1. Baggag, A., et al.: Learning spatiotemporal latent factors of traffic via regularized tensor factorization: imputing missing values and forecasting. IEEE TKDE, pp. 2573–2587 (2019)
2. Cao, W., Wang, D., Li, J., Zhou, H., Li, L., Li, Y.: Brits: bidirectional recurrent imputation for time series. In: NeurIPS, vol. 31 (2018)
3. Che, Z., Purushotham, S., Cho, K., Sontag, D., Liu, Y.: Recurrent neural networks for multivariate time series with missing values. Sci. Rep. 8(1), 1–12 (2018)
4. Cui, Z., Lin, L., Pu, Z., Wang, Y.: Graph markov network for traffic forecasting with missing data. Transp. Res. Part C 117, 102671 (2020)
5. Duan, Y., Lv, Y., Liu, Y.L., Wang, F.Y.: An efficient realization of deep learning for traffic data imputation. Transp. Res. Part C 72, 168–181 (2016)
6. Eric Zivot, J.W.: Vector autoregressive models for multivariate time series, pp. 385–429 (2006)
7. Goodfellow, I.J., Shlens, J., Szegedy, C.: Explaining and harnessing adversarial examples. arXiv preprint arXiv:1412.6572 (2014)
8. Grover, A., Leskovec, J.: node2vec: scalable feature learning for networks. In: SIGKDD, pp. 855–864 (2016)
9. Guo, S., Lin, Y., Feng, N., Song, C., Wan, H.: Attention based spatial-temporal graph convolutional networks for traffic flow forecasting. In: AAAI, pp. 922–929 (2019)
10. Jin, D., Szolovits, P.: Advancing pico element detection in biomedical text via deep neural networks. Bioinformatics 36(12), 3856–3862 (2020)
11. Kingma, D.P., Ba, J.: Adam: a method for stochastic optimization. In: ICLR (2015)
12. Kipf, T.N., Welling, M.: Semi-supervised classification with graph convolutional networks. arXiv preprint arXiv:1609.02907 (2016)

13. Ku, W.C., Jagadeesh, G.R., Prakash, A., Srikanthan, T.: A clustering-based approach for data-driven imputation of missing traffic data. In: FISTS, pp. 1–6 (2016)
14. Li, Y., Yu, R., Shahabi, C., Liu, Y.: Diffusion convolutional recurrent neural network: data-driven traffic forecasting. In: ICLR (2018)
15. Lippi, M., Bertini, M., Frasconi, P.: Short-term traffic flow forecasting: an experimental comparison of time-series analysis and supervised learning. IEEE Trans. Intell. Transp. Syst. **14**(2), 871–882 (2013)
16. Ma, J., Shou, Z., Zareian, A., Mansour, H., Vetro, A., Chang, S.F.: Cdsa: cross-dimensional self-attention for multivariate, geo-tagged time series imputation. arXiv preprint arXiv:1905.09904 (2019)
17. Sun, B., Ma, L., Cheng, W., Wen, W., Goswami, P., Bai, G.: An improved k-nearest neighbours method for traffic time series imputation. In: Chinese Automation Congress (CAC), pp. 7346–7351 (2017)
18. Tang, X., Yao, H., Sun, Y., Aggarwal, C., Mitra, P., Wang, S.: Joint modeling of local and global temporal dynamics for multivariate time series forecasting with missing values. In: AAAI, pp. 5956–5963 (2020)
19. Troyanskaya, O., et al.: Missing value estimation methods for DNA microarrays. Bioinformatics **17**(6), 520–525 (2001)
20. Van Buuren, S., Groothuis-Oudshoorn, K.: mice: multivariate imputation by chained equations in R. J. Stat. Softw. **45**, 1–67 (2011)
21. Wu, C., Ho, J., Lee, D.: Travel-time prediction with support vector regression. IEEE Trans. Intell. Transp. Syst. **5**(4), 276–281 (2004)
22. Wu, Z., Pan, S., Long, G., Jiang, J., Zhang, C.: Graph wavenet for deep spatial-temporal graph modeling. In: IJCAI, pp. 1907–1913 (2019)
23. Xu, M., et al.: Spatial-temporal transformer networks for traffic flow forecasting. CoRR abs/2001.02908 (2020)
24. Yu, B., Yin, H., Zhu, Z.: Spatio-temporal graph convolutional networks: a deep learning framework for traffic forecasting. In: IJCAI, pp. 3634–3640 (2018)
25. Zhao, L., et al.: T-GCN: a temporal graph convolutional network for traffic prediction. IEEE Trans. Intell. Transp. Syst. **21**(9), 3848–3858 (2020)
26. Zheng, C., Fan, X., Wang, C., Qi, J.: GMAN: a graph multi-attention network for traffic prediction. In: AAAI, pp. 1234–1241 (2020)
27. Zhong, W., Suo, Q., Jia, X., Zhang, A., Su, L.: Heterogeneous spatio-temporal graph convolution network for traffic forecasting with missing values. In: ICDCS, pp. 707–717 (2021)

Trajectory Representation Learning Based on Road Network Partition for Similarity Computation

Jiajia Li[1], Mingshen Wang[1], Lei Li[2,4(✉)], Kexuan Xin[3], Wen Hua[3], and Xiaofang Zhou[2,4]

[1] Shenyang Aerospace University, Shenyang, China
lijiajia@sau.edu.cn
[2] The Hong Kong University of Science and Technology (Guangzhou), Guangzhou, China
thorli@ust.hk
[3] The University of Queensland, Brisbane, Australia
{uqkxin,w.hua}@uq.edu.au
[4] The Hong Kong University of Science and Technology, Hong Kong, China
zxf@cse.ust.hk

Abstract. In the tasks of location-based services and vehicle trajectory mining, trajectory similarity computation is the fundamental operation and affects both the efficiency and effectiveness of the downstream applications. Existing trajectory representation learning works either use grids to cluster trajectory points or require external information such as road network types, which is not good enough in terms of query accuracy and applicable scenarios. In this paper, we propose a novel partition-based representation learning framework *PT2vec* for similarity computation by exploiting the underlying road segments without extra information. To reduce the number of words and ensure that two spatially similar trajectories have embeddings closely located in the latent feature space, we partition the network into multiple sub-networks where each is represented by a word. Then we adopt the GRU-based seq2seq model for word embedding, and a loss function is designed based on spatial features and topological constraints to improve the accuracy of representation and speed up model training. Furthermore, a hierarchical tree index *PT-Gtree* is built to store trajectories for further improving query efficiency based on the proposed pruning strategy. Experiments show that our method is both more accurate and efficient than the state-of-the-art solutions.

Keywords: Similarity Query · Representation Learning · Seq2Seq Model

1 Introduction

With the rapid development of location acquisition in the *Internet of Vehicles* (IoV) technologies, large-scale trajectory data has been generated. The collected

X. Wang et al. (Eds.): DASFAA 2023, LNCS 13943, pp. 396–413, 2023.
https://doi.org/10.1007/978-3-031-30637-2_26

trajectory data is usually represented as a set of discrete GPS coordinate point sequences and is very useful for various trajectory mining tasks, such as trajectory similarity computation, k nearest neighbor (kNN) query, contact tracing [2], speed profile generation [7], and etc. Among them, the trajectory similarity calculation [14] is the basis of many trajectory applications, and several similarity measurements have been proposed, such as *Dynamic Time Wrapping* (DTW) [24], *Longest Common SubSequence* (LCSS) [16], and *Edit Distance Real sequence* (EDR) [4]. However, the traditional methods usually adopt dynamic programming to decide the best alignment for the pairwise point-matching of two trajectories, which leads to a quadratic computational complexity.

Representation learning has been extensively studied in various fields (text processing, CV et al.), and has recently attracted more and more attention in the context of trajectories [5,9,20,21,23], which aims to represent trajectories as vectors of a fixed dimension [12]. Then the similarity of the two trajectories can be computed based on the Euclidean distance of their vector representations, which only takes a constant time $O(|v|)$ ($|v|$ is the length of vector v).

To the best of our knowledge, *t2vec* [9] is the first deep learning based solution for similarity computation. By partitioning the space into cells, *t2vec* transforms a trajectory as a sequence of grid cells and then designs an RNN-based model to encode a transformed trajectory into a latent vector. In their experiments, *t2vec* showed to be more accurate and efficient than LCSS [16], EDR [4], and EDwP. However, the equal size of cell will lead to different granularity of representation because the distribution of trajectories is not uniform. Cells with many trajectory points will lack the expression of details of trajectory patterns, while cells with few trajectory points will reduce the ability to alleviate sparsity.

In order to reduce the storage cost and facilitate the trajectory compression, trajectories in road-network are often converted to road segments by map matching [3]. Thus, each trajectory is succinctly represented as a concatenation of road IDs. Using these roads as words and then learning word embedding can capture the topology information of the underlying road network. For instance, *Trembr* [5] maps the raw trajectories to roads first, and then exploits the road networks to learn the trajectory representation based on an encoder-decoder model *Traj2vec* to encode spatial and temporal properties and a neural network-based model *Road2Vec* to learn road segment embedding. However, external information such as road network type is difficult to obtain and limits its usage in practice.

In this paper, we propose a novel $\underline{P}$artition-based $\underline{T}$rajectory representation learning framework *PT2vec*, and it utilizes the underlying road segments like *Trembr* but adopts a different word definition method for word embedding that does not require external information. Firstly, since the road number is often large in real-life networks, the training time will be very long if the road ID is taken as the word directly. To reduce the number of words and ensure that two spatially similar trajectories have embeddings closely located in the latent feature space, we partition the network into multiple sub-networks where each is represented by a word. In this way, by adjusting the subnet size, a better balance can be obtained between training time and similarity accuracy.

Secondly, the loss function of the model should consider the trajectory's spatial characteristics and the network topology. To train the model, we design a loss function based on spatial and topological constraints to optimize the objective function. Specifically, outputs that are closer to the target road network are more accurate than those that are farther away. At the same time, to reduce the training cost, the topological structure is used to constrain the decoding process. In this way, an accurate representation of the trajectory can be generated.

Thirdly, to further efficiently support the similarity query, a tree index structure is built and a pruning strategy is proposed. Specifically, in the process of recursively partitioning the network, a hierarchical index structure *PT-Gtree* (like *G-tree* [27]) is constructed, and then the trajectories are stored in the smallest node that can enclose them. In this way, given a query trajectory, those trajectories that are far away can be pruned, which can reduce the computation of Euclidean distances. To sum up, the main contributions are: (1) We propose a road network constrained trajectory representation learning framework for similarity computation by exploiting the underlying roads without extra information. (2) We propose a novel seq2seq-based model and efficient similarity query algorithm based on the proposed *PT-Gtree* index. (3) We propose a loss function that considers trajectory spatial features and road network topology constraints to learn the representation of trajectories. (4) We conduct extensive experiments on real-world trajectory datasets and verify the superiority of our framework in terms of accuracy and efficiency compared with the baselines.

2   Related Work

The recent survey papers [12,13] comprehensively review the research work on trajectory data management, while we focus on trajectory similarity computation and representation learning.

2.1   Trajectory Similarity Computation

We divide the existing measurements into three main categories.

Pointwise Measures. It treats the trajectories as a geometric curve, and measures the similarity according to the curve using Euclidean distance, edit distance, and Hausdorff distance, such as DTW [24], LCSS [16], and EDR [4].

Segment-Based Similarity. It first converts trajectories to segments and then evaluates the similarity based on the segments. This approach has been shown to reduce sample mismatch effects as well as reduce complexity [12], such as DTW-based method [10], LORS [17], and EBD [18].

Vector Euclidean Distance based Similarity. Most of the above methods generally require dynamic programming to compute the optimal distances with ordering constraints, which leads to a quadratic computational complexity of

$O(n^2)$. As the trajectories can be represented as vectors of fixed dimensions, the similarity of two trajectories can be computed based on the Euclidean distance of the vectors and reduces the complexity to $O(n)$. *t2vec* is the first attempt at learning such representation, and *Trembr* exploits both moving behaviors and road networks. Both of them will be discussed in Sect. 2.2.

2.2 Trajectory Representation Learning

Most of the early trajectory representation methods are based on artificially designed features, so they are called *trajectory feature extraction*. For example, speed, road type, and length are extracted from the original trajectory and are combined together to represent the trajectory [26]. However, this method relies on expert knowledge and needs to re-select features for different scenarios, and the types of features are limited, which limits the practical use.

For trajectory data, it can be regarded as a kind of sequence data, so it is natural to consider the encoder-decoder models to learn the representations. *traj2vec* [22] uses a sliding window to group trajectory points into segments, extracts several artificially designed motion features, and learns the representations. *t2vec* [9] is the first known representation of trajectories learned using deep learning methods. It divides the space into grids and then converts trajectories into sequences of grid cells. A GRU-based code model is designed to encode cell sequences as latent vectors. However, *t2vec* cannot really capture the moving behavior of moving objects without considering the road network characteristics (topology) underlying the trajectory. *Trembr* [5] matches the trajectory to the road network to learn the representation. In the experiments, we compare with *t2vec* instead of *Trembr* for two reasons: i) the implementation part of *Trembr* requires all road types of the road network, which are not available in our dataset; ii) *Trembr* uses road segments as a word list, and the word space is too large and the training time is too long.

3 Preliminaries

3.1 Problem Definition

A road network is represented as $G(V, E)$, where V is a set of road vertices and $E \subseteq V \times V$ is a set of edges.

Definition 1 (Road Network Constrained Trajectory). *The trajectory is collected by the moving object constrained in the road network, and it consists of n coordinate points $T = \langle (lat_1, lon_1), ..., (lat_n, lon_n) \rangle$, where (lat_i, lon_i), $i \in [1, n]$, denotes a 2-dimensional latitude and longitude point, and n is the length of T.*

Definition 2 (Trajectory Similarity Query). *Given a trajectory set TS, and a query $Q = (T_q, f_{sim})$, where T_q is the query trajectory and f_{sim} is the similarity function (Hausdorff, Frechet, DTW, LCSS, EDR, ERP, vector Euclidean distances, etc.) of two trajectories, the trajectory similarity query returns the trajectory $T_i \in TS$, such that $f_{sim}(T_i, T_q) < f_{sim}(T_j, T_q)$, for all $T_j \in TS \setminus T_i$.*

Problem Statement. Given a set of road network constrained trajectory data, our goal is to learn a representation $v \in R^d$ (d is the dimension of Euclidean space) for each trajectory T such that the representation can reflect the spatial feature and topological information for trajectory similarity calculation, where the similarity function is based on the vector Euclidean distance, and can be calculated by the following Eq. 1.

$$dist(T_i, T_q) = \sqrt{\sum_{m=1}^{d}(T_i.v_m - T_q.v_m)^2} \qquad (1)$$

3.2   Sequence Encoder-Decoder

The RNN-based sequence-to-sequence model is used to deal with indefinite-length sequences, which mainly consist of an encoder and a decoder.

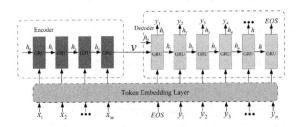

Fig. 1. GRU-based sequence encoder-decoder model

As shown in Fig. 1, the encoder encodes the input sequence $x = \langle x_1, ..., x_m \rangle$ into a fixed-dimensional vector v, which is in turn decoded back to $y = \langle y_1, ..., y_n \rangle$ by maximizing the conditional probability $P(y|x)$. As such, the hidden state h_t captures the sequential information in $\langle x, y_1, .., y_{t-1} \rangle$. To obtain the conditional probability $P(y|x) = P(y_1, .., y_n|x_1, .., x_m)$, we get the joint probability function $P(y_1, ..., y_n|x_1, ..., x_m) = P(y_1|x)\sum_{t=2}^{n} P(y_t|y_1, ..., y_{t-1}, v)$.

The decoder calculates $P(y_t|y_1, ..., y_{t-1}, v)$ at each position t accordingly. Specifically, at position t, the decoder transforms $y_1, ..., y_{t-1}$ and v into the hidden state h_t. This state preserves the sequence information of x and $y_1, ..., y_{t-1}$. And then y_t is predicted by h_t, where $h_t = f(y_{t-1}, h_{t-1})$ is calculated from the outputs of the previous positions h_{t-1} and y_{t-1}. Finally, $P(y_t = u|y_1, ..., y_{t-1}, v) = P(y_t = u|h_t) = \frac{exp(W_u \cdot h_t)}{\sum_{v \in V} exp(W_v \cdot h_t)}$, where W is the projection matrix that projects h_t from the hidden state space into the word list space, W_u denotes its u-th row, and V is the word list.

4   Partition-Based Trajectory to Vector, *PT2vec*

4.1   Framework of *PT2vec*

The representation of trajectory sequence units can use the word embedding method to learn, and then the representation of the entire trajectory can be obtained. This procedure is mainly divided into two steps: *word definition* and *word embedding*. Figure 2 illustrates the framework of our proposed *PT2vec* method.

Words Definition. *PT2vec* utilizes the underlying road segments to address the issues of low (non-uniform) sampling, and noisy sample points in raw trajectories. Since the number of edges in the road network is often large, if the edge ID is taken as the word directly, the training time will be very long. To reduce the number of words and ensure that two spatially similar trajectories should have embeddings closely located in the latent feature space, we partition the whole road network into multiple equal-sized sub-partitions spatially, and each sub-partition as well as each border edge is represented by a word (Step A).

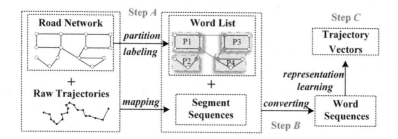

Fig. 2. Framework of PT2vec

Furthermore, to obtain the **trajectory (word) sequence**, each raw trajectory should be mapped onto the road network (Step B) as a sequence of road edges first, and any map-matching method can be applied here. Then according to the sub-partitions traversed by the trajectory and the defined words, the trajectory sequence can be prepared for word embedding.

Word Embedding. *PT2vec* adopts a GRU-based sequence encoder-decoder model to learn the representation of trajectory (Step C), and the loss function takes both the spatial features and topological information into account. Thus, the trajectories of different lengths can be represented as vectors of equal length.

4.2 Road Network Partition

We propose a word definition method based on the partition, where the whole network is recursively partitioned into multiple sub-partitions of the same size, and each sub-partition is represented by one word. In this way, edges in the same sub-partition are spatially close, and using the same word can guarantee that they are also close in vector space. The border edge that connects different sub-partitions is also regarded as a word, and thus can preserve the original topological information.

Definition 3 (Road Network Partition). *Given $G(V, E)$, a partition of G is a set of n^{k-1} subnets P_k at level k, $P_k{}^i = (V_i, E_i)(i \in [1, n^{k-1}])(k \in [1, h])$,* *where n is fan-out parameter, h is level, such that (1) $V = \cup_{i \in [1, n^{h-1}]} V_i$; (2) $\forall i \neq j, V_i \cap V_j = \phi$; (3) $\forall u, v \in V_i$, if $e_{u,v} \in E$, then $e_{u,v} \in E_i$.*

It is worth noting that if the two endpoints of an edge in the original road network are located in different sub-partitions, this edge does not belong to any sub-partition. Such an edge is called a border edge, which connects different sub-partitions. $BE(P_h{}^i)$ denotes the set of border edges of partition $P_h{}^i$.

Partition Method. A good partition should not only generate approximately equal-sized sub-networks, but also minimize the number of border edges. However, it has been proven that obtaining the optimal partitioning is NP-Hard. In this paper, we adopt the multilevel partitioning algorithm [6], which is a famous heuristics algorithm and used by many previous works [8,19,27]. It first reduces the network size by coarsening the vertices and edges, and then partitions on the coarsened graph using traditional network partitioning algorithms, e.g. Kernighan-Lin algorithm. Finally, it coarsens the sub-networks to generate the final partitions of the original network. Thus, the algorithm can guarantee that each sub-networks nearly has the same size.

Word List Generation. In the sequence encoder-decoder model, the input of the model should be a sequence of discrete labels. Hence, we number the sub-partitions as well as the border edges in order, and these labels are the corresponding words.

Figure 3(a) shows an example of a road network partition, $P_2 = \{P_2^1, P_2^2\}$ is one partition. Both sub-partitions P_2^1 and P_2^2 have 9 vertices and can be further recursively partitioned, hence, $P_3 = \{P_3^1, P_3^2, P_3^3, P_3^4\}$ is another partition with smaller sub-partitions, whose word labels are 1, 2, 3, and 4. Consider the complete partition P_3, it has five border edges, so we have $BE(P_3) = \{e_{v_2,v_6}, e_{v_5,v_{15}}, e_{v_4,v_{11}}, e_{v_{10},v_{16}}, e_{v_{12},v_{15}}\}$, and their labels are 5, 6, 7, 8, and 9.

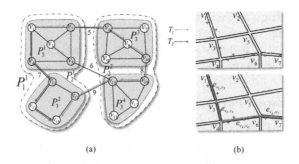

(a)   (b)

Fig. 3. Example of Road Network Partition and Road

4.3   From Raw Trajectory to Word Sequence

To obtain the word sequence, each raw trajectory should be mapped onto the road network to obtain a sequence of road edges first. Figure 3(b) shows an example of mapping the raw trajectory points onto the road network, and the matched trajectory is represented as the edges of the underlying road network.

Existing road network matching algorithms include HMMM, ST-Matching, IVMM, etc., and any of them can be applied. In this paper, we adopt the *ST-Matching* algorithm, who utilizes a road segment projection process to obtain candidate paths and construct a graph first and then selects the highest scoring path maximizing the global matching probability.

Take Fig. 3 as an example, for the partition P_3, the trajectory $T_2 = \{e_{v_1,v_4}, e_{v_4,v_{11}}, e_{v_{11},v_{13}}, e_{v_{13},v_{14}}\}$, traverses two sub-partitions P_3^1, P_3^2 and a border edge $e_{v_4,v_{11}}$. Their labels are 1, 2, 7 respectively, therefore, the word sequence of T_2 is $\{1, 7, 2\}$. Similarly, the word sequence of T_1 is $\{1, 5, 3, 8, 4, 9, 2\}$.

4.4   Trajectory Representation Learning

RNN-based encoder-decoder model is used to handle text sequence tasks, and it cannot capture the spatial features in the trajectory, nor consider the topology of the underlying road network of the trajectory. To address these issues, we propose a spatial and topology-based loss function to optimize the objective.

Existing Loss Function. To train the sequence encoder-decoder, a loss function is needed to optimize the objective, which is maximizing the joint probability of the output sequence. To achieve the maximum probability, the Maximum Likelihood Estimation (MLE) can be used, which is equivalent to minimizing the Negative Log-Likelihood (NLL). Hence, the loss function for processing text sequences is designed as: $maxP(y_1, ..., y_n | x_1, ..., x_m) = min(-logP(y_1, ..., y_n | x_1, ..., x_m))$ and $L = -log \prod_{t=1}^{n} P(y_t | y_1, ..., y_{t-1}, v)$.

The above loss function penalizes the output words with the same weight, however, words that are close to the target word or topologically connected to the target should have a higher output probability rather than equal. Take Fig. 4 as an example, assuming that the rectangles are the generated leaf sub-partitions and our decoding target word is y_1, it is obviously not a good strategy to penalize the output y_3 and y_9 with equal probability during decoding. And the words like y_2, y_3, and y_5 that are either adjacent to y_1 or closer to y_1 should be given a higher probability.

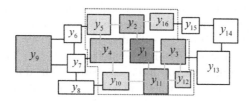

Fig. 4. Word Distance and Topological Relations

Designed Loss Function Based on Spatial and Topology. We propose
a loss function that only assigns probabilities to a small number of important
words around the target word, which not only considers distance and topological relations, but also reduces the amount of probability calculation, thereby
increasing the learning efficiency and effectiveness. In particular, according to
the generated sub-partitions connected by border edges, starting from the target word y_t, we heuristically select K words (denoted as $TK(y_t)$) that are close
to y_t in a breadth-first manner. As a result, these K words are mainly from
the surrounding sub-partitions and border edges. In addition, since the words
that are topologically directly connected to the target word (denoted as $T(y_t)$)
are more important, we assign a larger probability to them. Furthermore, the
probability of other words is directly assigned to 0, which also improves learning
efficiency. To sum up, our designed loss function is given in Eq. 2:

$$L = -\sum_{t=1}^{n} \sum_{u \in T(y_t)} w_{uy_t} log \frac{exp(W_u \cdot h_t)}{\sum_{v \in TK(y_t)} exp(W_v \cdot h_t)} \tag{2}$$

$$w_{uy_t} = \frac{exp(-D(u,y_t)/\lambda)}{\sum_{v \in T(y_t)} exp(-D(v,y_t)/\lambda)}$$

$D(u, y_t)$ represents the shortest network distance between words, and λ is a
distance scale parameter. The exponential kernel function is chosen as it decays
fast at the tail, which would encourage the model to learn to output a word near
the target word y_t.

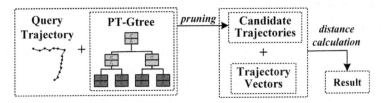

Fig. 5. Overview of the Query Processing

5   Trajectory Similarity Query

After all, trajectories are represented as equal-length vectors by *PT2vec*, the
similarity query only needs to compare the Euclidean distance between these
vectors and the query vector. Specifically, when a query trajectory is given,
the *PT2vec* model is first used to represent it as a vector. Then the Euclidean
distance between the query vector and each vector in the database is calculated,
and the closest trajectory is returned. The time complexity is $O(n + |v|)$, where
$|v|$ is the length of the vectors. In order to further speed up the query efficiency,
we use the pruning idea to filter out those trajectories that are far away from
the query trajectory in space. Particularly, we propose a hierarchical tree index
structure *PT-Gtree*, and store trajectories into tree nodes. Figure 5 presents the

overview of the similarity query processing based on *PT-Gtree*. In this section, we first introduce the structure and construction process of *PT-Gtree*, and then describe the query optimization algorithm *SQPru* based on *PT-Gtree*.

5.1   Partition Based Trajectory Index, *PT-Gtree*

Let's go back to the road network partition method. If we keep each level of the partition generated by the recursive process, we can build a hierarchical tree structure for the trajectory index in a top-down manner, like the *G-tree* [27].

Structure of *PT-Gtree*. The root node is the whole road network. The leaf nodes are sub-partitions used to generate words, and these sub-partitions will not have further sub-partitions. The non-leaf nodes are the intermediate results (sub-partitions) of the partition process, and the nodes on the same level are a complete partition of the original road network. The parent node in the tree is similar to the concept of the minimum bounding rectangle (MBR) in the *R-tree*, which can spatially contain the points and edges of all child nodes (sub-partitions). Intuitively, trajectories with high similarity are adjacent spatially in the road network. If a node can enclose all the segments of the specified trajectory T_q, those trajectories outside the node space may be less similar to T_q, so these can be filtered out safely. Therefore, we store each trajectory at the lowest level node that can enclose it for further pruning and refining.

Construction of *PT-Gtree*. The construction of the hierarchical index *PT-Gtree* needs two parameters: one is the fanout parameter n, which is to restrict the number of sub-partitions for each level of partitioning, and the other is the maximum number of vertices in a leaf node m, which is to limit the size of leaf partitions. As shown in Fig. 6, *PT-Gtree* is a highly balanced tree that: (1) Each tree node represents a sub-partition $P_k^i (i \in [1, n^{h-1}])(k \in [1, h])$; (2) Each internal node has n (≥ 2) child nodes (sub-partitions); (3) Each leaf node contains at most m (≥ 1) vertices, and all leaf nodes appear at the same level; (4) Each node maintains a trajectory list $P_k^i.TL$.

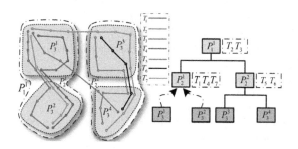

Fig. 6. The Constructed *PT-Gtree* of Fig. 3

Trajectory List. As described in Sect. 4.3, the raw trajectory will be mapped onto road segments and be represented by the segments sequence. One trajectory may pass through one or more leaf sub-partitions. For each trajectory T, we find the lowest level node that can enclose T in a bottom-up manner, that is, find the least common ancestor node (LCA) of the sub-partitions traversed by T. And then insert T into the trajectory list of the LCA.

As shown in Fig. 6, the leaf nodes of PT-$Gtree$ are $\{P_3^1, P_3^2, P_3^3, P_3^4\}$, and the internal nodes include P_1^1, P_2^1, P_2^2. For the trajectory T_1, it passes through P_3^1 and P_3^2. Their LCA is P_2^1, so we insert T_1 into $P_2^1.TL$. Similarly, other trajectories are all inserted into the corresponding LCAs as illustrated.

5.2 Similarity Query Algorithm, *SQPru*

Our proposed similarity query optimization algorithm *SQPru* adopts the filter-refine mechanism to perform the similarity query. Let TS be a trajectory set and T_q be a query trajectory. *SQPru* filters out those trajectories with a certain low similarity to T_q, and only accurately calculates (refines) the Euclidean distance from the candidate trajectories. The procedure is shown in Algorithm 1.

Candidate Trajectories, *CndT*. Let LCA_{T_q} be the LCA of T_q in PT-$Gtree$. All trajectories stored in LCA_{T_q} and its child nodes are regarded as candidate trajectories. Because their trajectory topology is similar to the query trajectory, they all have a certain possibility of becoming the query result(Lines 3–9 in Algorithm 1).

Refinement of *CndT*. For each trajectory T_i in *CndT*, its distance to T_q will be calculated based on the represented vectors. At last, one or k trajectories with the smallest distances will be returned by the similarity query or kNN query. (Lines 10–13 in Algorithm 1).

Algorithm 1: *SQPru*

Input: Trajectory dataset TS, Query trajectory $T_q = (v_1, v_2, ..., v_n)$, Built
      PT-$Gtree$ index G

Output: Trajectory T_{rst} with the most similarity with T_q

1 Initialize queue $PQ = \emptyset$, candidate set $CndT = \emptyset$;
2 Sub-partitions Set $spS = \text{leaf}(v_1) \cup \text{leaf}(v_2) \cup ..., \cup \text{leaf}(v_n)$;
3 $LCA_{T_q} = $ the LCA of nodes in spS in G; $PQ.\text{push}(LCA_{T_q})$;
4 **while** $PQ \neq \emptyset$ **do**
5     $n_i = PQ.\text{popfront}()$;
6     $CndT.\text{push}(n_i.TL)$;
7   **if** n_i *is a non-leaf node* **then**
8   **foreach** *child-node* n_c *of* n_i **do**
9             $PQ.\text{push}(n_c)$;

10 Trajectory embedding($CndT$) ;
11 **foreach** $T_i \in CndT$ **do**
12   Calculate $dist(T_i, T_q)$ according to Equation 1;
13 **return** Trajectory T_{rst} with the smallest $dist$;

For example, in Fig. 6, assuming that T_1 is the query trajectory, it can be seen that its LCA is P_2^1, whose trajectory list contains another two trajectories. Hence, the candidate trajectories set $CndT = \{T_4, T_7\}$. Their vector Euclidean distance between T_1 will be further calculated to obtain the best one.

6 Experimental Evaluation

6.1 Experimental Setup

Dataset: We use two real-world road networks and taxi datasets, as shown in Table 1. The trajectories of Porto [1] are collected over 19 months at a resolution of 15 s. Beijing is derived from T-drive [25]. The first 1 million trajectories in Porto and 150,000 in Beijing are selected for training, and the rest are used for testing. We chose Porto as the default dataset to test the effects of other parameters on the model. The parameters are list in Table 2 with the default settings labeled in bold. Here the default partition size m of Beijing is set to 16.

Table 1. Real World Maps and Trajectories

Map	# Edges	# Vertices	# Trajectories	#Edges/trajectory
Porto	**150,761**	**114,099**	**1,565,595**	**65**
Beijing	126,827	54,198	250,997	237

Table 2. Parameters

Param	Value	Param	Value
DB size	20k, 40k, 60k, 80k, **100k**	k	**20**, 30, 40, 50
m	16, 32, **64**, 128	r	**0**, 0.2, 0.4, 0.6
l	64, 128, **256**, 512		

Evaluation Criteria: We adopt the self-similarity comparison proposed in [11, 15] to evaluate the trajectory similarity. Specifically, for each trajectory T in a given trajectory dataset D, we take odd and even GPS sample points from T, create two interleaved trajectories T_a and T_b, and divide the dataset D into two data sets D_a and D_b. Next, we randomly select 1,000 trajectories from D_a and their corresponding trajectories in D_b to form test sets Q_a and Q_b. We use the mean rank as the accuracy measure for the most similar trajectory query. In addition, we also evaluate the average query response time for queries.

Methods: We implement the following methods for comparison: 1) *LCSS* [16], *EDR* [4], and *EBD* are three baselines. 2) For *traj2vec* and *t2vec*: we adjusted to the optimal parameter settings and the cell size of *t2vec* is 100m. 3) *PT2vec*: we choose 3-layer GRU and randomly select 10,000 trajectories as validation dataset. The training will be terminated if the loss does not decrease in 20,000 successive iterations. λ in Eq. 2 is fixed at 10. The parameter K in Sect. 4.4 is set to 100. We employ Adam stochastic gradient descent with an initial learning rate of 0.001. All results are the average from 1000 queries.

Query: We evaluate the performance using two similar queries. 1) Most similar trajectory query. For each query trajectory T_a, we use the Euclidean distance to find the most similar one from the target database. 2) k-NN query. We use *t2vec* and *PT2vec* to find the k-NN trajectories of each query trajectory from the target database as ground truth. Next, convert the query and target database by downsampling. Finally, for each transformed query, we use each method to find its k-NN from the target database, and then compare the result with the corresponding ground truth.

Implementation Details: *PT2vec* and *t2vec* are implemented in Julia and PyTorch, and trained using a Quadro RTX5000 GPU. *LCSS, EDR, EBD* and *traj2vec* are implemented in Python. The platform runs the Ubuntu 18.04 OS with an Intel Xeon W-2245@3.2GHz CPU.

6.2   Experimental Results

Effect of the Number of the Trajectories Data, DB Size. Table 3 shows the mean rank for the similarity queries over the Porto and Beijing (listed in parentheses) under different *DB size*. As expected, the mean rank of all methods increases with *DB size*, because more candidate trajectories will pose a challenge to pick the best one. Overall, the traditional methods (LCSS, EDR and EBD) perform worse than the representation learning-based methods, showing the advantages of deep learning techniques. Interestingly, the traditional methods have worse accuracy in Beijing, while the learning-based methods perform the opposite. It can be seen that the average length of the trajectories in Beijing is larger than in Porto. The longer trajectories can enable *traj2vec* to obtain more polymerization features and at the same time enable *t2vec* and *PT2vec* to learn more sufficiently after converting trajectories into word sequences and retaining more trajectory information. Therefore, better results can be obtained after the vector Euclidean distance calculation. On the contrary, long trajectories increase the probability of noise points, which makes alignment-based methods more difficult. In addition, our *PT2vec* is 2 to 7 times better than *t2vec*, and the performance is hardly affected by the map size and database size, verifying the effectiveness of the partition-based learning model.

Table 3. Mean rank v.s *DB size* over Porto and Beijing

DB size	LCSS	EDR	EBD	traj2vec	t2vec	PT2vec
20k	28.75 (44.5)	23.23 (24.53)	10.90 (10.83)	9.11(8.67)	2.46 (1.5)	**1.03 (1.04)**
40k	52.92 (85.9)	36.20 (48.78)	21.78 (21.10)	15.15(13.48)	4.03 (2.09)	**1.08 (1.09)**
60k	86.7 (123.6)	54.10 (71.18)	32.38 (31.02)	21.23(18.98)	5.55 (2.67)	**1.13 (1.14)**
80k	120.3 (163.9)	70.32 (93.24)	42.17 (40.46)	27.32(23.69)	6.69 (3.26)	**1.17 (1.17)**
100k	138.7 (190.3)	88.96 (115.28)	50.10 (49.48)	33.41(29.14)	8.45 (3.85)	**1.20 (1.20)**

Effect of the downsampling rate, r. As shown in Table 4, as r increases, the trajectory sample points become more and more sparse, causing the difference between T_a and T_a in the test trajectory data to become larger, so the average ranking of all methods increases. The mean ranks of *traj2vec*, *t2vec* and *PT2vec* vary more dramatically in Porto than in Beijing, due to the longer average length of trajectories in Beijing. Although r is increased, the trajectory can still maintain enough sample points. So the difference between T_a and T_a is small, and the model is less affected. Moreover, under all settings, *PT2vec* is 2 to 3 times more accurate than *t2vec*.

Table 4. Mean Rank v.s r over Porto and Beijing

	Porto (Beijing)			
r	**0**	**0.2**	**0.4**	**0.6**
LCSS	138.7 (190.3)	154.3 (215.57)	172.1 (236.8)	179.8 (245.2)
EDR	88.96 (115.3)	98.55 (129.5)	109.5 (150.4)	120.8 (166.3)
EBD	52.10 (49.48)	74.85 (71.56)	92.22 (86.23)	106.87 (100.58)
traj2vec	33.41 (29.14)	41.03 (35.99)	48.92 (42.84)	56.91 (48.98)
t2vec	8.45 (3.85)	9.39 (4.14)	11.21 (4.95)	15.92 (5.37)
PT2vec	**1.2 (1.2)**	**4.469 (1.31)**	**5.403 (1.38)**	**7.44 (2.21)**

Effect of the Leaf Partition Size, m. The structure of *PT-Gtree* determines the generation of words, which in turn affects the performance of *PT2vec*. We set the fan-out parameter n to 4, and changed the maximum allowable number of vertices m from 16 to 128 to test the performance of the model as shown in Table 5. Obviously, a larger m means fewer sub-partitions, and thus a smaller number of words and a shorter training time. Interestingly, the accuracy is best when $m = 64$, this is because too many sub-partitions will bring higher training complexity and make the model more prone to overfitting. Conversely, too few sub-partitions will result in a lower resolution, since many trajectories pass through the same sub-partition, making the distinction of trajectories more difficult.

Effect of the Dimensionality of Hidden Layers, l. The quality of the model *PT2vec* also depends on the dimensionality l of hidden layers in the encoder, which is also the length of the output vector $|v|$. Higher dimensionality is usually more expressive, but requires more training data to avoid overfitting. Table 5 shows the mean rank of *PT2vec* over Porto under different l and r. When l is increased from 64 to 256, the mean rank is improved, and when it is changed to 512, the mean rank increases, indicating that the model may be overfitting. Here overfitting happens because the limited amount of information contained in the training set is not enough to train all neurons in the hidden layer.

Table 5. Impact of m and l on PT2vec over Porto

	$m = 16$	$m = 32$	$m = 64$	$m = 128$	$l = 64$	$l = 128$	$l = 256$	$l = 512$
#words	32,502	18,219	11,813	6,856				
Mean Rank ($r = 0$)	1.33	1.63	**1.2**	4.78	1.26	1.25	**1.2**	1.28
Mean Rank ($r = 0.4$)	7.25	5.99	**5.4**	24.05	12.16	9.65	**5.4**	6.03
Training Time(s)	15,782	10,272	8,655	6,140				

Effect of the Pruning Strategy. Table 6 presents the mean rank and the average query response time of *PT2vec* and *PT2vec without pruning* (denoted as *Naive*) under different *DB size* to evaluate the effect of the proposed pruning strategy in Sect. 5.2. In the table, $|CndT|$ is the number of candidate trajectories obtained after pruning by *SQPru*. As shown, the pruning strategy can filter out about 40% of the trajectories in the database, thereby reducing the query response time by 40%. Meanwhile, the mean rank is hardly affected, reflecting the effectiveness and efficiency of the pruning-based query algorithm *SQPru*.

kNN Query. Figure 7 illustrates the accuracy of kNN query for *t2vec* and *PT2vec* under different k and sampling rate r over Porto and Beijing. It can be seen that when r is the same, as k increases, the accuracy of both models is slightly improved. This is because when k is larger, the searched vector space is larger, which enables the trajectories with larger distances due to the sampling rate to be searched, thereby improving the accuracy. Moreover, under all settings, *PT2vec* is 2% to 5% more accurate than *t2vec*, verifying the effectiveness and scalability w.r.t. k of our proposed methods.

Table 6. Performance of *SQPru* v.s *DB size* over Porto

	|CndT|	Response Time (ms)		Mean Rank	
DB size	PT2vec	Naive	PT2vec	Naive	PT2vec
20k	12k	4.68	2.09	1.04	1.03
40k	24k	8.5	4.16	1.09	1.08
60k	37k	13.32	6.04	1.13	1.13
80k	49k	18	8.32	1.17	1.17
100k	61k	22.46	10.29	1.22	1.2

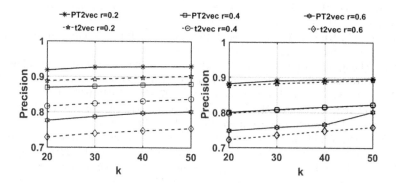

Fig. 7. Accuracy of *k*NN Query v.s *k* and *r*

7 Conclusion

In this paper, we study the representation learning problem for road network constrained trajectory data. We map the raw trajectory to the road network first reducing the storage cost of the trajectory, and use the spatial proximity between road segments to partition the road network in preparation for representation learning. We design a new GRU-based encoder-decoder model *PT2vec* to learn the representation of trajectories by exploiting the spatial features of trajectories and the topology of road networks. In addition, we design a new tree structure, *PT-Gtree*, to index trajectory data to improve the efficiency of trajectory query. Experiments show that our method achieves better performance in both accuracy and query efficiency compared with state-of-the-art solutions.

Acknowledgments. The research work was supported by Shenyang Young and Middle-aged Scientific and Technological Innovation Talent Program (grant# RC220504); Natural Science Foundation of Liaoning Education Department (grant# LJKZ0205); Hong Kong Research Grants Council (grant# 16202722); Natural Science Foundation of China (grant# 62072125, grant# 61902134); partially conducted in the JC STEM Lab of Data Science Foundations funded by The Hong Kong Jockey Club Charities Trust.

References

1. Challenge, T.S.T.P.: https://www.geolink.pt/ecmlpkdd2015-challenge/ (2015)
2. Chao, P., He, D., Li, L., Zhang, M., Zhou, X.: Efficient trajectory contact query processing. In: Jensen, C.S., et al. (eds.) DASFAA 2021. LNCS, vol. 12681, pp. 658–666. Springer, Cham (2021). https://doi.org/10.1007/978-3-030-73194-6_44
3. Chao, P., Xu, Y., Hua, W., Zhou, X.: A survey on map-matching algorithms. In: Borovica-Gajic, R., Qi, J., Wang, W. (eds.) ADC 2020. LNCS, vol. 12008, pp. 121–133. Springer, Cham (2020). https://doi.org/10.1007/978-3-030-39469-1_10
4. Chen, L., Özsu, M.T., Oria, V.: Robust and fast similarity search for moving object trajectories. In: Proceedings of the 2005 ACM SIGMOD, pp. 491–502 (2005)
5. Fu, T.Y., Lee, W.C.: Trembr: exploring road networks for trajectory representation learning. ACM TIST 11(1), 1–25 (2020)
6. Karypis, G., Kumar, V.: Analysis of multilevel graph partitioning. In: Conference on Supercomputing, pp. 29–29. IEEE (1995)
7. Li, L., Zheng, K., Wang, S., Hua, W., Zhou, X.: Go slow to go fast: minimal on-road time route scheduling with parking facilities using historical trajectory. VLDB J. 27(3), 321–345 (2018). https://doi.org/10.1007/s00778-018-0499-4
8. Li, F., Zhang, Q., Zhang, W.: Graph partitioning strategy for the topology design of industrial network. IET Commun. 1(6), 1104–1110 (2007)
9. Li, X., Zhao, K., et al., G.C.: Deep representation learning for trajectory similarity computation. In: Proceedings of the IEEE 34th ICDE, pp. 617–628 (2018)
10. Mao, Y., Zhong, H., Xiao, X., Li, X.: A segment-based trajectory similarity measure in the urban transportation systems. Sensors 17(3), 524 (2017)
11. Ranu, S., Deepak, P., Telang, A.D., et al.: Indexing and matching trajectories under inconsistent sampling rates. In: ICDE, pp. 999–1010. IEEE (2015)
12. Sheng, W., Zhifeng, B., Shane, C.J., Gao, C.: A survey on trajectory data management, analytics, and learning. CSUR 54(2), 1–36 (2021)
13. Sousa, R.S.D., Boukerche, A., Loureiro, A.A.: Vehicle trajectory similarity: models, methods, and applications. ACM Comput. Surv. (CSUR) 53(5), 1–32 (2020)
14. Su, H., Liu, S., Zheng, B., Zhou, X., Zheng, K.: A survey of trajectory distance measures and performance evaluation. VLDB J. 29(1), 3–32 (2020)
15. Su, H., Zheng, K., Wang, H., Huang, J., Zhou, X.: Calibrating trajectory data for similarity-based analysis. In: ACM SIGMOD, pp. 833–844 (2013)
16. Vlachos, M., Kollios, G., Gunopulos., D.: Discovering similar multidimensional trajectories. In: ICDE, pp. 673–684 (2002)
17. Wang, S., Bao, Z., Culpepper, J.S., Xie, Z., Liu, Q., Qin, X.: Torch: a search engine for trajectory data. In: The 41st ACM SIGIR, pp. 535–544 (2018)
18. Wang, S., Bao, Z., Culpepper, J., Sellis, T., Qin, X.: Fast large-scale trajectory clustering. Proc. VLDB Endowment 13(1), 29–42 (2019)
19. Wang, Y., Li, G., Tang, N.: Querying shortest paths on time dependent road networks. Proc. VLDB Endowment 12(11), 1249–1261 (2019)
20. Wang, Z., Long, C., Cong, G., Ju, C.: Effective and efficient sports play retrieval with deep representation learning. In: 25th ACM SIGKDD, p. 499–509 (2019)
21. Wang, Z., Long, C., Cong, G., Liu, Y.: Efficient and effective similar subtrajectory search with deep reinforcement learning. VLDB, p. 2312–2325 (2020)
22. Yao, D., Zhang, C., et al, Z.Z.: Trajectory clustering via deep representation learning. In: Proceedings of the IJCNN, pp. 3880–3887 (2017)
23. Yao, D. and Cong, G.e.a.: Computing trajectory similarity in linear time: a generic seed-guided neural metric learning approach. In: ICDE, p. 1358–1369 (2019)

24. Yi, B.K., Jagadish, H.V., Faloutsos, C.: Efficient retrieval of similar time sequences under time warping. In: ICDE, pp. 201–208 (1998)
25. Yuan, J., et al.: T-drive: driving directions based on taxi trajectories. In: SIGSPA-TIAL, pp. 99–108 (2010)
26. Zheng, Y., Liu, L., Wang, L., Xie, X.: Learning transportation mode from raw GPS data for geographic applications on the web. In: WWW, pp. 247–256 (2008)
27. Zhong, R., Li, G., Tan, K.L., Zhou, L., Gong, Z.: G-tree: an efficient and scalable index for spatial search on road networks. TKDE **27**(8), 2175–2189 (2015)

ISTNet: Inception Spatial Temporal Transformer for Traffic Prediction

Chu Wang, Jia Hu, Ran Tian$^{(\boxtimes)}$, Xin Gao, and Zhongyu Ma

College of Computer Science and Engineering, Northwest Normal University,
Lanzhou, China
{tianran,2019221875,mazybg}@nwnu.edu.cn

Abstract. As a typical problem in spatial-temporal data learning, traffic prediction is one of the most important application fields of machine learning. The task is challenging due to (1) Difficulty in synchronizing modeling long-short term temporal dependence in heterogeneous time series. (2) Only spatial connections are considered and a mass of semantic connections are ignored. (3) Using independent components to capture local and global relationships in temporal and spatial dimensions, resulting in information redundancy. To this end, we propose Inception Spatial Temporal Transformer (ISTNet). First, we design an Inception Temporal Module (ITM) to explicitly graft the advantages of convolution and max-pooling for capturing the local information and attention for capturing global information to Transformer. Second, we consider both spatially local and global semantic information through the Inception Spatial Module (ISM), and handling spatial dependence at different granular levels. Finally, the ITM and ISM brings greater efficiency through a channel splitting mechanism to separate the different components as the local or a global mixer. We evaluate ISTNet on multiple real-world traffic datasets and observe that our proposed method significantly outperforms the state-of-the-art method.

Keywords: Traffic prediction · Spatial-Temporal data prediction · attention mechanism

1 Introduction

Many countries have recently boosted the construction of smart cities to help realize the intelligent management of cities. Among them, traffic prediction is the most important part of smart city construction, and accurate traffic prediction can help relevant departments to guide vehicles reasonably [4], thereby avoiding traffic congestion and improving highway operation efficiency.

Traffic prediction is the classical spatial-temporal prediction problem, it is challenging due to the complex intra-dependencies (i.e., temporal correlations within one traffic series) and inter-dependencies (i.e., spatial correlations among

---

C. Wang and J. Hu—Equal contribution.

X. Wang et al. (Eds.): DASFAA 2023, LNCS 13943, pp. 414–430, 2023.
https://doi.org/10.1007/978-3-031-30637-2_27

multitudinous correlated traffic series) [1]. Traditional methods Vector Auto-Regressions (VARs) [12], Auto-Regressive Integrated Moving Average (ARIMA) [7] rely on the assumption of smoothness and fails to capture the complex spatial-temporal patterns in large-scale traffic data. With the rise of deep learning, researchers utilize convolutional neural networks (CNN) [24] and graph convolutional networks (GCN) [6] to capture spatial correlations based on grid structure and non-euclidean structure, respectively. For temporal correlations researchers model the local and global correlations of time series using CNN and recurrent neural networks (RNN) [3], respectively.

To joint spatial-temporal relationship, recent studies formulate the traffic prediction as a spatial temporal graph modeling problem, STGNN [20] and DGCRN [9] integrate GCN into RNN, STSGCN [18] constructs local spatial-temporal graphs by adjacent time steps, and STFGNN [10] constructs adaptive spatial-temporal fusion graphs using dynamic time regularization (DTW). While having shown the effectiveness of introducing the graph structure of data into a model, but there is still a lack of satisfactory progress in long-term traffic prediction, mainly due to the following two challenges. First, the current study is weak in modeling temporal dependence. Researchers typically capture local and global temporal dependence by using CNNs alone or the attention mechanism, with the former requiring stacking multiple layers to capture long sequences, and the latter's preference for global information jeopardizing its ability to capture local correlation. Some studies complement CNNs with attention by serial or parallel approaches, which can lead to information loss and information redundancy. It is a challenging problem to model local and global information synchronously and maintain high computational efficiency.

Second, there is a powerful correlation between traffic conditions on adjacent roads in a traffic network, and we can calculate the distance weights between nodes by a threshold Gaussian kernel function. However, traffic networks are spatially heterogeneous, for example, traffic conditions near schools in different areas are similar, but their distance weights may be zero, and this potential global spatial correlation is important. Some studies construct dynamic graph by feature matrix, but this relationship is not robust because the feature matrix changes with time. Simultaneous modeling of local and global spatial correlations is difficult with guaranteed computational efficiency.

To address the aforementioned challenges, we propose a Inception Spatial Temporal Transformer (ISTNet) to perform traffic prediction task. The core of ISTNet is the ST-Block, which contains Inception Temporal Module and Inception Spatial Module, Inception Temporal Module aims to augment the perception capability of ISTNet in the temporal dimension by capturing both global and local information in the data. To this end, the Inception Temporal Module first splits the input feature along the channel dimension, and then feeds the split components into local mixer and global mixer respectively. Here the local mixer consists of a max-pooling operation and a convolution operation, while the global mixer is implemented by pyramidal attention. Inception Spatial Module and Inception Temporal Module make the same segmentation in the channel dimension and feed into local mixer (local GCN) and global mixer (global GCN), respectively. In this way, ISTNet can effectively capture local and global information on the corresponding channel, thereby learning more comprehensive characterization. Further, ISTNet stacks a plurality of ST-Blocks with residual connections and generates multi-step prediction results at one time via an attention mechanism. We conducted extensive experiments on four real-world datasets and the experiments demonstrated that ISTNet achieves state-of-the-art performance. In summary, we summarize the contributions of this work as follows:

- We propose the Inception Temporal Module to model local and global temporal correlations, which grafts the merit of CNNs for capturing local information and attention for capturing global information to Transformer.
- We propose the Inception Spatial Module to model local and global spatial correlations, and construct a self-adaptive adjacency matrix that preserves hidden spatial dependencies. Inception Spatial Module can tackle spatial dependencies of nodes' information extracted by Inception Temporal Module at different granular levels.
- We propose the Inception Spatial Temporal Transformer (ISTNet) to perform the traffic prediction task, which stacks multiple ST-Blocks with residual connections and controls the ratio of local and global information by the frequency ramp structure.
- We evaluated ISTNet on four real datasets and the experimental results showed that ISTNet consistently outperformed all baselines.

2   Related Work

2.1   Traffic Prediction

Traffic prediction is a classical spatial-temporal prediction problem that has been extensively studied in the past decades [22,23]. Compared with statistical methods VAR [12] and ARIMA [7], deep learning methods Recurrent Neural Networks (RNNs) [3], Long-Short-Term-Memory networks (LSTM) [19] break away from the assumption of smoothness and are widely used with the advantage of modeling serial data. Temporal convolutional networks(TCN) [2,15] is also a representative work, it considers a large receptive field through dilated convolutions, and thus enable the model process very long sequence with fewer time. Lately, more and more researchers have started to focus on spatial-temporal modeling studies, ConvLSTM [16] using CNN and RNN to extract local spatial patterns among variables and long-term patterns of time series, respectively. ST-ResNet [24] designed a deep residual network based on CNN for urban pedestrian flow prediction, these works can effectively extract spatial-temporal features, but the limitation is that the input must be standard spatial-temporal grid data, which cannot be applied in non-Euclidean spatial structures.

2.2   Graph Neural Network

Traffic data is collected by sensors deployed in traffic networks, and recently researchers have used GNN to model the spatial correlation between different sensors. For instance, DCRNN [11] models traffic flow as a spatially diffusive process and combines diffusion convolution and GRU to model spatial and temporal correlations. STGCN [23] integrates GCN and gated temporal convolution into one module to learn spatial-temporal dependence. Graph WaveNet [22] proposed an adaptive adjacency matrix and spatially fine-grained modeling of the output of the temporal module via GCN, for simultaneously capturing spatial-temporal correlations. STJGCN [25] performs GCN operations between adjacent time steps to capture local spatial-temporal correlations, and further proposes an dilated causal GCN layer to extract information on multiple spatial-temporal scales. Further to enhance the ability of the model to capture spatial-temporal correlations, ASTGCN [4], ASTGNN [5] added a complex attention mechanism as a complement to GCN. In addition, MTGNN [21] and DGCRN [9] proposed graph generation algorithms that fully consider real-world uncertainties and enhance the generalization ability of the models. Although these methods improve the traffic prediction accuracy to a large extent, they do not explicitly define the local and global relationships of the traffic network, which may lead to prediction errors.

3   Preliminary

The target of traffic prediction is to predict the traffic flow in the future period based on the previous observations of N sensors in the traffic network. We construct a weighted graph $\mathcal{G} = (\mathcal{V}, \mathcal{E}, \mathcal{A})$ from the positions of the N sensors, where v is a set of $N = |\mathcal{V}|$ vertices, representing the sensors in the road network. $\mathcal{E}$ is the set of edges, $\mathcal{A} \in \mathbb{R}^{N \times N}$ is a weighted adjacency matrix representing the nodes proximity (e.g., road network distance of node pairs). We represent the traffic flow on the graph as a graphic signal $\mathcal{X} \in \mathbb{R}^{N \times C}$ (where C is the number of signals, the signal can be traffic volume, traffic speed, etc.). Furthermore, we express the traffic prediction problem as follows: given a sequence of observations from N sensors at P time steps, predicting the traffic data at Q future time steps by the function $\mathcal{F}$:

$$\{\mathcal{X}_{t-P+1}, \mathcal{X}_{t-P+2}, \cdots, \mathcal{X}_t\} = \mathcal{F}_\theta(\mathcal{X}_{t+1}, \mathcal{X}_{t+2}, \cdots, \mathcal{X}_{t+Q}; \mathcal{G}) \tag{1}$$

4   Methodology

Figure 1 shows our proposed ISTNet, which contains L ST-Blocks with residual connections and position encoding, and through a frequency ramp structure to control the ratio of local and global information of different blocks, lastly an attention mechanism generates multi-step prediction results at one time.

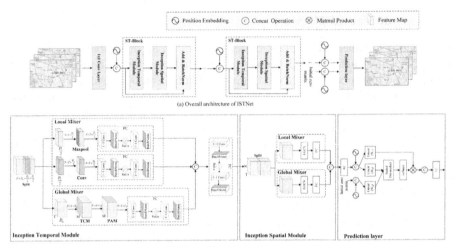

(a) Overall architecture of ISTNet

(b) Architecture of ITM, ISM and prediction layer

Fig. 1. The framework of Inception Spatial Temporal Trasnformer (ISTNet). (a) IST-Net consists of multiple ST-Blocks stacked on top of each other, each ST-Block is composed of inception temporal module and inception spatial module, and to synchronously capture local and global information in temporal or special dimensions. (b) Show details of ST-Block and prediction layer.

4.1   Inception Temporal Module

Inspired by the Inception module, we proposed an Inception Temporal Module (ITM) to model long-term and short-term patterns in the temporal dimension. As shown in Fig. 1, ITM first splits the input feature along the channel dimension, and then feeds the split components into local mixer and global mixer respectively. Here the local mixer consists of a max-pooling operation and a convolution operation, while the global mixer is implemented by a pyramidal attention.

Specifically, given the graph signal $\mathcal{X} \in \mathbb{R}^{N \times T \times D}$, it is factorized $\mathcal{X}$ into $\mathcal{X}^{local} \in \mathbb{R}^{N \times T \times D_L}$ and $\mathcal{X}^{global} \in \mathbb{R}^{N \times T \times D_G}$ along the channel dimension, where $D = D_L + D_G$. Then, $\mathcal{X}^{local}$ and $\mathcal{X}^{global}$ are assigned to local mixer and global mixer respectively.

Local Mixer in Temporal Dimension. Nodes normally present similar traffic conditions over a certain period of time, so it is necessary for us to focus on local contextual information. Considering the sharp sensitiveness of the maximum filter and the detail perception of convolution operation, we capture short-term temporal patterns in Local Mixer through a parallel structure. We divide the input $\mathcal{X}^{local} \in \mathbb{R}^{N \times T \times D_L}$ into $\mathcal{X}_M \in \mathbb{R}^{N \times T \times D_L/2}$ and $\mathcal{X}_C \in \mathbb{R}^{N \times T \times D_L/2}$ along the channel, then, $\mathcal{X}_M$ is embedded with a max-pooling and a fully-connected layer, and $\mathcal{X}_C$ is fed into a convolution layer and a fully-connected layer:

$$\mathcal{Y}_M^T = FC(MaxPool(\mathcal{X}_M)) \tag{2}$$

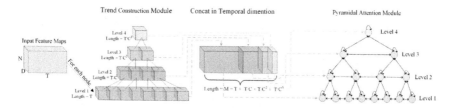

Fig. 2. Global Mixer in Temporal Dimension. We first obtain the trend representations of different levels by Trend Construction Module, then concatenate them and use them as initialized node representations of Pyramidal Attention Module.

$$\mathcal{Y}_C^T = FC(Conv(\mathcal{X}_C)) \tag{3}$$

where $\mathcal{Y}_M^T \in \mathbb{R}^{N \times T \times D_L/2}$ and $\mathcal{Y}_C^T \in \mathbb{R}^{N \times T \times D_L/2}$ are the output of local mixer.

Global Mixer in Temporal Dimension. There could be long-term effects of sudden traffic conditions at a node in a certain moment, and modeling long-term patterns in the temporal dimension is necessary. Inspired by [13], we propose a temporal attention mechanism with multiresolution structure, as shown in Fig. 2, consisting of two parts, Trend Construction Module (TCM) and Pyramidal Attention Module (PAM). TCM obtains local trend block and global trend block representations by successive convolution operations in the temporal dimension:

$$\begin{aligned}
\mathcal{X}_{TCM} = concat(&\mathcal{X}^{global}, \\
&\sigma(\Phi_1 * \mathcal{X}^{global}), \\
&\sigma(\Phi_2 * \sigma(\Phi_1 * \mathcal{X}^{global})), \\
&\sigma(\Phi_3 * \sigma(\Phi_2 * \sigma(\Phi_1 * \mathcal{X}^{global}))))
\end{aligned} \tag{4}$$

$\mathcal{X}^{global}$ is the component that is fed into the global mixer, where $*$ denotes a standard convolution operation, Φ is the parameters of the temporal dimension convolution kernel, and $\sigma(\cdot)$ is the activation function. $\mathcal{X}_G = \sigma(\Phi_3 * \sigma(\Phi_2 * \sigma(\Phi_1 * \mathcal{X}^{global})))$ is a global trend block representation. After obtaining several trend blocks with different levels (the higher the level, the larger the time range represented, e.g. hourly, daily, weekly), we concatenate them as the output of the TCM $\mathcal{X}_{TCM} \in \mathbb{R}^{B \times (T+T/C+T/C^2+T/C^3) \times N \times D_G}$, it has a time length of M and C is the convolution kernel size and step size.

Further, we construct a pyramidal structure that allows the time step to enjoy a long-term horizon by adding a connection between the time step and the trend block. In detail: (1) initializes the node representation of the Pyramidal structure; (2) add connections to the Pyramidal structure with the following strategy: at the same level, nodes are connected to adjacent nodes, and at different levels, connections are added between nodes with their parents nodes and children nodes, with other pairs of nodes that are not connected subjected to the mask operation; (3) information interaction via attention mechanism in Pyramidal. Specifically expressed as:

$$Q^h = mask(\frac{(reshape(\mathcal{X}_{TCM}^{i,h} Z_Q^h))(reshape(\mathcal{X}_{TCM}^{i,h} Z_K^h))}{\sqrt{d_k}}) \tag{5}$$

$$\mathcal{Y}^{i,h} = reshape(softmax(Q^h)(\mathcal{X}_{TCM}^{i,h} Z_V^h)) \tag{6}$$

$$\mathcal{Y}_G^T = \sum_{i=1}^{N} concat(\mathcal{Y}^{i,1}, \cdots, \mathcal{Y}^{i,h}) \tag{7}$$

$\mathcal{X}_{TCM}^{i,h} \in \mathbb{R}^{M \times D_G}$ is the vector representation of node v_i at the hth head, $Z_Q^h \in \mathbb{R}^{d_k \times d_k}$, $Z_K^h \in \mathbb{R}^{d_k \times d_k}$ and $Z_V^h \in \mathbb{R}^{d_k \times d_k}$ are learnable parameters of the hth header. $Q^h \in \mathbb{R}^{M \times M}$ is the mask matrix of the hth header, and $\mathcal{Y}^{i,h}$ is the vector representation of node v_i after the update at the hth head. Lastly, we merge the output of h headers and take the previous T time steps as the output of PAM, and the output graph signal is $\mathcal{Y}_G^T \in \mathbb{R}^{N \times T \times D_G}$.

Finally, the outputs of local mixer and global mixer are concatenated along the channel dimension, and the output of the ITM is obtained by a fully-connected:

$$\mathcal{Y}_T = FC(concat(\mathcal{Y}_M^T, \mathcal{Y}_C^T, \mathcal{Y}_G^T)) \tag{8}$$

$\mathcal{Y}_T \in \mathbb{R}^{N \times T \times D}$ is the graph signal output from the ITM.

4.2   Inception Spatial Module

As with ITM, we use local mixer and global mixer in Inception Spatial Module to model local and global correlations in spacial synchronously. In the same ST-Block, ISM and ITM have the same channel split ratio. $\mathcal{X}^{local} \in \mathbb{R}^{N \times T \times D_L}$ and $\mathcal{X}^{global} \in \mathbb{R}^{N \times T \times D_G}$ are assigned to local and global mixer respectively.

Local Mixer in Spatial Dimension. It is well known that there tends to be a strong correlation between road pairs that are located closer together and a weaker correlation at further distance. This is a local correlation at the spatial level, and to measure the interaction between different roads, we use the thresholded Gaussian kernel function distance to measure the proximity between different pairs of roads:

$$\mathcal{A}_{i,j} = exp(-\frac{dist(v_i, v_j)}{\mu^2}) \tag{9}$$

$$\mathcal{Y}_L^S = FC(\mathcal{A}\mathcal{X}^{local}W_1 + b_1) \tag{10}$$

If $dist(v_i, v_j) \leq \varepsilon$, $\mathcal{A}_{i,j} = 0$. Where $dist(v_i, v_j)$ represents the road network distance from sensor v_i to v_j, μ is the standard deviation, and ε is the threshold. $\mathcal{Y}_L^S \in \mathbb{R}^{N \times T \times D_G}$ is the output of local mixer.

Global Mixer in Spatial Dimension. In a traffic network, the traffic conditions of two nodes that are distant but have similar attributes (schools, apartments, etc.) are usually similar and can represent each other at a certain level. Distance-based adjacency matrix ignores this due to its local correlation limitation, therefore, we propose an adaptive adjacency matrix to learn this hidden relation. We start by initializing the spatial node embedding representation $E \in \mathbb{R}^{N \times N}$, and then construct the adjacency matrix by the dot product mechanism:

$$\mathcal{M} = softmax(EE^T/\sqrt{D}) \tag{11}$$

$$
\begin{aligned}
&for \quad i = 1, 2, \cdots, N \\
&\quad nodeId = argtopk(\mathcal{M}[i, :]) \\
&\quad \mathcal{M}[i, -nodeId] = 0
\end{aligned}
\tag{12}
$$

$$\mathcal{Y}_G^S = FC(\mathcal{M}\mathcal{X}^{global}W_2 + b_2) \tag{13}$$

$\mathcal{M} \in \mathbb{R}^{N \times N}$ is an adaptive adjacency matrix, which has a set of learnable parameters. We reserve top-k closest nodes as its neighbors for each node by $argtopk(\cdot)$ to reduce the computing cost of GCN. $\mathcal{Y}_G^S \in \mathbb{R}^{N \times T \times D_G}$ is the output of global mixer.

Finally, the outputs of local mixer and global mixer are concatenated along the channel dimension, and get the output of ISM by a fully-connected:

$$\mathcal{Y}_S = FC(concat(\mathcal{Y}_L^S, \mathcal{Y}_G^S)) \tag{14}$$

$\mathcal{Y}_S \in \mathbb{R}^{N \times T \times D}$ is the output of ISM.

4.3 Prediction Layer

We direct interaction by adding position embedding to future data and historical data. The position embedding contains two parts, one is the spatial position embedding, and we take a set of learnable spatial embedding representations which share parameters with E in Eq. (11). The other part is temporal embedding, by one-hot encoding of time-of-day, day-of-week and concatenation, and finally summing the two to get the spacial-temporal position embedding $U \in \mathbb{R}^{N \times T \times D}$. U_H and U_P denote the historical and future position embedding, respectively, and U_H is taken as part of the input in each ST-Block.

We utilize attention mechanism to generate multi-step prediction results at one time. It is worth noting that the historical feature F_H is obtained by concatenating the output of the ST-Block with U_H, and the future feature F_P is obtained by concatenating the future observations (preset to zero) with U_P. We use F_P as the query matrix and F_H as the key and value matrices to obtain the final prediction:

$$R^h = \frac{(reshape(F_P^{i,h}M_Q^h))(reshape(F_H^{i,h}M_K^h))}{\sqrt{d_k}} \tag{15}$$

$$\mathcal{Y}^{i,h} = reshape(softmax(R^h)(F_H^{i,h}M_V^h)) \tag{16}$$

$$\mathcal{Y} = \sum_{i=1}^{N} concat(\mathcal{Y}^{i,1}, \cdots, \mathcal{Y}^{i,h}) \qquad (17)$$

where h is the number of attention head, $M_Q^h \in \mathbb{R}^{d_k \times d_k}$, $M_K^h \in \mathbb{R}^{d_k \times d_k}$ and $M_V^h \in \mathbb{R}^{d_k \times d_k}$ are learnable parameters, $R^h \in \mathbb{R}^{M \times M}$ is the temporal attention matrix and $\mathcal{Y}^{i,h}$ is the updated vector representation of node v_i, and $\mathcal{Y}$ is the output of the model.

4.4 Frequency Ramp Structure

Previous investigations [14] proved that bottom layers prefer local information, while top layers play a more significant role in capturing global information. With deeper layers, lower layers can capture short-term and local patterns in spatial and temporal dimension, and also gradually gather local information to achieve a global understanding of the input. For this reason, we have set up a frequency ramp structure in ISTNet, from the first layer of the ST-Block to the lth layer, we progressively decrease the channel dimension of the local mixer and increase the channel dimension of the global mixer. More specifically, for each ST-Block, we define a channel ratio to better balance the local and global components. Hence, with the flexible frequency ramp structure, ISTNet enables effective modeling of local and global spatial-temporal correlation to make precise predictions.

Table 1. Dataset description.

Datasets	Time Range	Time Steps	Time Interval	Nodes
PEMS03	09/01/2018 - 11/30/2018	26202	5-min	358
PEMS04	01/01/2018 - 02/28/2018	16992	5-min	307
PEMS07	05/01/2017 - 08/31/2017	28224	5-min	883
PEMS08	07/01/2016 - 08/31/2016	17856	5-min	170

5   Experiments

5.1   DataSets

We validated the performance of ISTNet on four public traffic datasets. The PEMS03, PEMS04, PEMS07, and PEMS08 datasets published by [1, 10], which are from four districts, respectively in California. The detailed information is shown in Table 1. We used Z-score normalization to standardize the input of data.

On the four traffic flow datasets, we divide them into training set, validation set, and test set according to the ratio of 6:2:2, respectively. We use one hour's historical data (12 steps) to predict the next hour's data.

Our experimental environment is with a 24G memory Nvidia GeForce RTX 3090 GPU. We train the model by Adam optimize and set the initial learning rate to 0.01, batchsize to 128, the maximum epoch is 80, the window size of the max-pooling layer and convolution layer is 3. The default settings for the TCM convolution kernel and step size are [2, 2, 3], frequency ramp structure sets the channel ratios of the local mixer and global mixer for different layers to [2, 1, 1/2].

5.2   Baseline Methods

(1) VAR [12] is a traditional time series model capable of capturing the pairwise relationship of time series. (2) SVR [17] utilizes a linear support vector machine to perform regression. (3) FC-LSTM [19] is an encoder-decoder framework using long short-term memory (LSTM) with peephole for multi-step time-series prediction. (4) DCRNN [11] employs an encoder-decoder architecture and combines GRU with a diffusion graph convolutional network to predict traffic speed. (5) STGCN [23] models spatial and temporal correlations using GCN and CNN, respectively. (6) ASTGCN [4] models spatial and temporal correlations via GCN and CNN. (7) Graph WaveNet [22] combines adaptive graph convolution and dilated casual convolution to capture spatial-temporal correlation. (8) STSGCN [18] constructs a local spatio-temporal graph and captures local spatio-temporal correlations by spatio-temporal synchronous graph convolution. (9) GMAN [26] captures spatio-temporal correlations through an attention mechanism and designs a transformation layer to reduce error propagation. (10) DSTAGNN [8] proposed a dynamic spatial-temporal aware graph and a gated convolution capable of capturing multiple ranges.

5.3   Experiment Results

Prediction Performance Comparison. Table 2 shows the average prediction results of ISTNet and baseline in the next hour. From Table 2 we can observe that: (1) VAR, SVR and LSTM only consider temporal correlation and as a result perform poorly in spatial-temporal data prediction. (2) The graph-based model takes into account spatial information, so the performance is further improved. (3) DCRNN and STGCN are highly dependent on predefined graph structure, CNN-based Graph WaveNet and STSGCN have difficulties in capturing long-term dependence, and attention-based GMAN and DSTAGNN loses local information, and it leads to their poor overall performance. (4) ISTNet achieves state-of-the-art prediction performance on four traffic flow datasets, ISTNet has three advantages compared to Baseline, First, ISTNet captures both local and global pattern in a parallel manner in both spatial and temporal dimensions. Second, ISTNet introduces trend blocks to model long-term temporal dependence. Third, ISTNet achieves optimal performance by controlling the proportion of local and global information at different layers.

Table 2. Prediction performance of different models on traffic flow datasets.

Dataset	Metrics	VAR	SVR	FC-LSTM	DCRNN	STGCN	ASTGCN	Graph WaveNet	STSGCN	GMAN	DSTAGNN	ISTNet
PEMS03	MAE	19.72	19.77	19.56	17.62	19.76	18.67	15.67	17.51	15.52	15.57	**15.03±0.09**
	RMSE	32.38	32.78	33.38	29.86	33.87	30.71	26.42	29.05	26.53	27.21	**24.89±0.25**
	MAPE(%)	20.50	23.04	19.56	16.83	17.33	19.85	15.72	16.92	15.19	**14.68**	15.24±0.19
PEMS04	MAE	24.44	26.18	23.60	24.42	23.90	22.90	19.91	21.52	19.25	19.30	**18.51±0.03**
	RMSE	37.76	38.91	37.11	37.48	36.43	33.59	31.06	34.14	30.85	31.46	**30.36±0.10**
	MAPE(%)	17.27	22.84	16.17	16.86	13.67	16.75	13.62	14.50	13.00	12.70	**12.36±0.16**
PEMS07	MAE	27.96	28.45	34.05	24.45	26.22	28.13	20.83	23.99	20.68	21.42	**19.67±0.13**
	RMSE	41.31	42.67	55.70	37.61	39.18	43.67	33.62	39.32	33.56	34.51	**32.96±0.04**
	MAPE(%)	12.11	14.00	15.31	10.67	10.74	13.31	9.10	10.10	9.31	9.01	**8.57±0.20**
PEMS08	MAE	19.83	20.92	21.18	18.49	18.79	18.72	15.57	17.88	14.87	15.67	**14.08±0.05**
	RMSE	29.24	31.23	31.88	27.30	28.2	28.99	24.32	27.36	24.06	24.77	**23.27±0.12**
	MAPE(%)	13.08	14.24	13.72	11.69	10.55	12.53	10.32	11.71	9.77	9.94	**9.34±0.09**

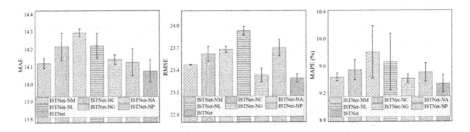

Fig. 3. The results of ablation study on the PEMS08 dataset.

Ablation Study. To further evaluate the effectiveness of each component in ISTNet, we conduct ablation studies on PEMS08 dataset. We named the variants of ISTNet as follows: (1) ISTNet-NM: We removed the maximum pooling layer from the ITM. (2) ISTNet-NC: We eliminate the convolutional layer from the ITM. (3) ISTNet-NA: We utilize a common temporal attention mechanism instead of PAM. (4) ISTNet-NL: We removed the local mixer from the ISM. (5) ISTNet-NG: We eliminate the global mixer from ISM. (6) ISTNet-NP: The prediction layer is replaced by a common multi-headed attention mechanism to investigate the impact of the prediction layer on model performance.

The experimental results are shown in Fig. 3. We observed that: (1) ISTNet-NM and ISTNet-NC show that removing the max-pooling layer and CNN layer in ITM degrades the performance of ISTNet due to their ability to efficiently extract local information from different perspectives. (2) ISTNet-NA demonstrates that PAM is critical to model performance because it can effectively model long-short-term temporal dependence and cope with temporal heterogeneity. (3) Distance-based graph and adaptive graph are highly effective, but for traffic prediction, the local information of the road is more significant than the global information. (4) The effectiveness of the prediction layer has been shown by the effect of ISTNet-NP, where we perform a direct interaction between future and historical observations, and this direct generation of multi-step prediction results facilitates the reduction of error propagation.

Long Term Prediction Performance. For evaluating the performance of ISTNet in long-term prediction, we predict the traffic data for the next 30, 60, 90, and 120 min on the PEMS04 dataset. The results are shown in Table 3, as the time step increases, ISTNet continues to outperform the most advanced baseline in the long-term, and the gap widens gradually, suggesting the effectiveness of ISTNet in long-term temporal modeling. Figure 4 shows the specific performance of the per prediction step.

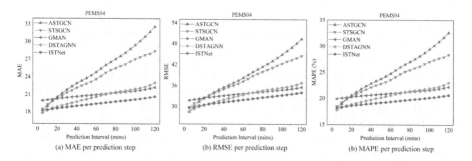

(a) MAE per prediction step   (b) RMSE per prediction step   (b) MAPE per prediction step

Fig. 4. Prediction performance comparison at each horizon on the PEMS04.

Table 3. Long-term prediction performance comparison of different models on PEMS04 dataset.

Model	Metrics	30 min	60 min	90 min	120 min	Average
ASTGCN	MAE	22.08±0.28	25.51±0.69	29.32±1.17	34.04±1.42	26.01±0.75
	RMSE	34.47±0.42	39.35±1.10	44.95±1.87	51.60±2.20	40.64±1.28
	MAPE (%)	14.70±0.10	16.84±0.19	19.28±0.28	22.49±0.31	17.22±0.19
STSGCN	MAE	21.66±0.36	24.04±0.41	26.70±0.52	29.07±0.64	24.35±0.47
	RMSE	34.56±0.75	37.98±0.72	41.91±0.75	45.45±0.90	38.46±0.79
	MAPE (%)	14.44±0.13	15.76±0.11	17.50±0.18	18.92±0.15	16.13±0.20
GMAN	MAE	20.50±0.01	21.02±0.04	21.55±0.08	22.29±0.05	21.08±0.05
	RMSE	33.21±0.42	34.18±0.48	35.09±0.56	36.13±0.54	34.24±0.49
	MAPE (%)	15.06±0.52	15.37±0.57	15.78±0.66	16.54±0.76	15.48±0.60
DSTAGNN	MAE	19.36±0.04	20.69±0.08	21.69±0.03	22.91±0.15	20.60±0.02
	RMSE	31.36±0.17	33.65±0.27	35.29±0.22	36.81±0.04	33.47±0.15
	MAPE (%)	12.88±0.02	13.54±0.03	14.22±0.01	15.04±0.05	13.58±0.02
ISTNet	MAE	**18.93±0.10**	**19.51±0.11**	**20.11±0.15**	**20.85±0.17**	**19.58±0.13**
	RMSE	**30.96±0.15**	**32.01±0.08**	**32.97±0.03**	**34.00±0.08**	**32.05±0.06**
	MAPE (%)	**12.85±0.18**	**13.18±0.13**	**3.57±0.10**	**14.25±0.12**	**13.26±0.14**

Computation Cost. We present the computation cost of ASTGCN, STSGCN, GMAN, DSTAGNN and ISTNet on the PEMS04 dataset in Table 4. We set the input and prediction step size to 24, batch-size to 16, and the rest of the parameters follow the best configuration in the text, and the five methods are tested uniformly in the Tesla V100 GPU environment.

Table 4. The computation cost on the PEMS04 dataset.

Method	Computation Time		Memory	
	Training (s/epoch)	Inference (s)	Parameter (M)	GPU Memory (G)
ASTGCN	72.92	19.83	0.48	3.62
STSGCN	302.09	60.86	5.98	8.58
GMAN	574.2	30.6	0.51	28.67
DSTAGNN	379.22	67.19	0.44	9.78
ISTNet	157.9	7.4	0.13	5.85

From Table 4 we observe that in terms of computation time, based on the spatial-temporal attention mechanism GMAN and DSTAGNN run the slowest in the training stage, ASTGCN with the removal of the cycle component is the fastest, and ISTNet has a moderate training speed but the fastest inference speed. In respect of memory consumption, ISTNet has the smallest number of parameters, which can avoid overfitting to a certain degree, and the maximum GPU usage of ISTNet is only higher than ASTGCN, by further considering the prediction accuracy (refer to Table 1 and Table 2), ISTNet shows superior ability in balancing predictive performances and time consumption as well as memory consumption.

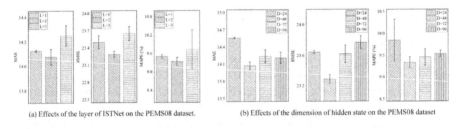

(a) Effects of the layer of ISTNet on the PEMS08 dataset.   (b) Effects of the dimension of hidden state on the PEMS08 dataset

Fig. 5. Parameter study.

Parameter Study. To further investigate the effect of hyperparameter settings on model performance, we conduct the study on the number of layers L, vector dimension D, and kernel size of TCM for ISTNet on the PEMS08 dataset. Except for the changed parameters, the other configurations remain the same.

We can observe from Fig. 5 that increasing the number of layers and the number of hidden units of ISTNet can improve the performance of the model, which is because increasing the number of layers expands the perceptual field of the nodes and the high-dimensional vectors can more fully express the hidden information. However, when the threshold is exceeded, the model performance gradually decreases, which means that an overfitting phenomenon occurs.

In addition, we have also investigated the convolutional kernel size of TCM. Here, we can choose to stack two or three CNN layers to build the global trend block. From Table 5 we observe that the model performs best when stacking three CNN layers and the kernel size is [2,2,3], which illustrates that a reasonable layer setting enables the representation range to be precise, and thus more conducive to modeling long-term dependence.

Table 5. Impact of kernel size in TCM on model performance.

Metric	TCM kernel size						
	[2, 6]	[3, 4]	[4, 3]	[6, 2]	[2, 3, 2]	[3, 2, 2]	[2, 2, 3]
MAE	14.13±0.05	14.13±0.06	14.18±0.05	14.15±0.13	14.12±0.07	14.08±0.08	**14.08±0.05**
RMSE	23.63±0.13	23.40±0.06	23.55±0.08	23.49±0.10	23.54±0.10	23.55±0.12	**23.37±0.12**
MAPE (%)	9.48±0.11	9.48±0.13	9.58±0.14	9.41±0.09	9.53±0.11	**9.33±0.01**	9.34±0.09

Visualization of PAM. We randomly visualized pyramidal attention scores of three nodes on the PEMS04 dataset. In Fig. 6 we can see that: node 10 pays more attention to the features of neighboring time steps, node 51 and node 89 give more attention to local trend blocks, which indicates that most nodes require more trend features, it can contribute to modeling long-term dependence.

Visualization of Prediction Results. We visualized the traffic flow of the node for the next day and compared it with the predicted value. As shown in Fig. 7, where the boxes are highlighted, we observe that ISTNet's prediction curves coincide with the true values, and the performance at the peaks demonstrates ISTNet's ability to make accurate predictions for challenging situations, which further illustrates the effectiveness of ISTNet in modeling traffic data.

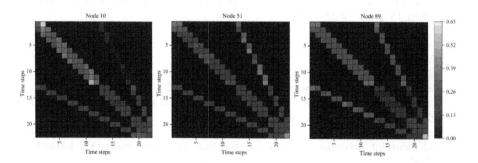

Fig. 6. Heatmap of pyramidal attention scores. The corresponding TCM kernel size is [2, 2, 3], and the total time step is 22.

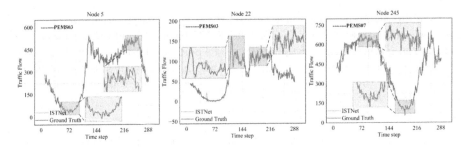

Fig. 7. Traffic prediction visualization on the PEMS03 and PEMS07 datasets.

6   Conclusion

In this paper, we propose a Inception Spatial Temporal Trasnformer (ISTNet) to perform the traffic prediction task. The core components of ISTNet include ITM and ISM, and ITM captures local and global correlations in temporal dimension with a parallel approach. In particular, we enhance the global sensing capability of the model by introducing trend blocks, which further improves the long-term prediction performance. ISM captures local and global correlations in traffic networks through distance-based graph and global dynamic graph. In addition, ISTNet controls the ratio of local and global information of ITM and ISM in different ST-Blocks by means of frequency ramp structure to achieve the best performance. Experiments on several traffic datasets show that ISTNet achieves the optimal performance, especially in long-term prediction, and ISTNet's performance is further improved. In the future, we will explore the application of ISTNet in the field of meteorological prediction and air quality prediction.

Acknowledgment. This work is supported in part by the National Natural Science Foundation of China (71961028), the Key Research and Development Program of Gansu (22YF7GA171), the Scientific Research Project of the Lanzhou Science and Technology Program (2018-01-58).

References

1. Bai, L., Yao, L., Kanhere, S., Wang, X., Sheng, Q., et al.: STG2Seq: spatial-temporal graph to sequence model for multi-step passenger demand forecasting. arXiv preprint arXiv:1905.10069 (2019)
2. Bai, S., Kolter, J.Z., Koltun, V.: An empirical evaluation of generic convolutional and recurrent networks for sequence modeling. arXiv preprint arXiv:1803.01271 (2018)
3. Connor, J.T., Martin, R.D., Atlas, L.E.: Recurrent neural networks and robust time series prediction. IEEE Trans. Neural Netw. **5**(2), 240–254 (1994)
4. Guo, S., Lin, Y., Feng, N., Song, C., Wan, H.: Attention based spatial-temporal graph convolutional networks for traffic flow forecasting. In: Proceedings of the AAAI Conference on Artificial Intelligence, vol. 33, pp. 922–929 (2019)

5. Guo, S., Lin, Y., Wan, H., Li, X., Cong, G.: Learning dynamics and heterogeneity of spatial-temporal graph data for traffic forecasting. IEEE Trans. Knowl. Data Eng. **34**(11), 5415–5428 (2021)
6. Kipf, T.N., Welling, M.: Semi-supervised classification with graph convolutional networks. arXiv preprint arXiv:1609.02907 (2016)
7. Kumar, S.V., Vanajakshi, L.: Short-term traffic flow prediction using seasonal ARIMA model with limited input data. Eur. Transp. Res. Rev. **7**(3), 1–9 (2015)
8. Lan, S., Ma, Y., Huang, W., Wang, W., Yang, H., Li, P.: DSTAGNN: dynamic spatial-temporal aware graph neural network for traffic flow forecasting. In: International Conference on Machine Learning, pp. 11906–11917. PMLR (2022)
9. Li, F., et al.: Dynamic graph convolutional recurrent network for traffic prediction: benchmark and solution. ACM Trans. Knowl. Discov. Data (TKDD) **17**, 1–21 (2021)
10. Li, M., Zhu, Z.: Spatial-temporal fusion graph neural networks for traffic flow forecasting. In: Proceedings of the AAAI Conference on Artificial Intelligence, vol. 35, pp. 4189–4196 (2021)
11. Li, Y., Yu, R., Shahabi, C., Liu, Y.: Diffusion convolutional recurrent neural network: data-driven traffic forecasting. arXiv preprint arXiv:1707.01926 (2017)
12. Lippi, M., Bertini, M., Frasconi, P.: Short-term traffic flow forecasting: an experimental comparison of time-series analysis and supervised learning. IEEE Trans. Intell. Transp. Syst. **14**(2), 871–882 (2013)
13. Liu, S., et al.: Pyraformer: low-complexity pyramidal attention for long-range time series modeling and forecasting. In: International Conference on Learning Representations (2021)
14. Raghu, M., Unterthiner, T., Kornblith, S., Zhang, C., Dosovitskiy, A.: Do vision transformers see like convolutional neural networks? In: Advances in Neural Information Processing Systems, vol. 34, pp. 12116–12128 (2021)
15. Sen, R., Yu, H.F., Dhillon, I.S.: Think globally, act locally: a deep neural network approach to high-dimensional time series forecasting. In: Advances in Neural Information Processing Systems, vol. 32 (2019)
16. Shi, X., Chen, Z., Wang, H., Yeung, D.Y., Wong, W.K., Woo, W.C.: Convolutional LSTM network: a machine learning approach for precipitation nowcasting. In: Advances in Neural Information Processing Systems, vol. 28 (2015)
17. Smola, A.J., Schölkopf, B.: A tutorial on support vector regression. Stat. Comput. **14**, 199–222 (2004)
18. Song, C., Lin, Y., Guo, S., Wan, H.: Spatial-temporal synchronous graph convolutional networks: a new framework for spatial-temporal network data forecasting. In: Proceedings of the AAAI Conference on Artificial Intelligence, vol. 34, pp. 914–921 (2020)
19. Sutskever, I., Vinyals, O., Le, Q.V.: Sequence to sequence learning with neural networks. In: Advances in Neural Information Processing Systems, vol. 27 (2014)
20. Wang, X., et al.: Traffic flow prediction via spatial temporal graph neural network. In: Proceedings of the Web Conference 2020, pp. 1082–1092 (2020)
21. Wu, Z., Pan, S., Long, G., Jiang, J., Chang, X., Zhang, C.: Connecting the dots: multivariate time series forecasting with graph neural networks. In: Proceedings of the 26th ACM SIGKDD International Conference on Knowledge Discovery & Data Mining, pp. 753–763 (2020)
22. Wu, Z., Pan, S., Long, G., Jiang, J., Zhang, C.: Graph wavenet for deep spatial-temporal graph modeling. arXiv preprint arXiv:1906.00121 (2019)
23. Yu, B., Yin, H., Zhu, Z.: Spatio-temporal graph convolutional networks: a deep learning framework for traffic forecasting. arXiv preprint arXiv:1709.04875 (2017)

24. Zhang, J., Zheng, Y., Qi, D.: Deep spatio-temporal residual networks for citywide crowd flows prediction. In: Thirty-First AAAI Conference on Artificial Intelligence (2017)
25. Zheng, C., Fan, X., Pan, S., Wu, Z., Wang, C., Yu, P.S.: Spatio-temporal joint graph convolutional networks for traffic forecasting. arXiv preprint arXiv:2111.13684 (2021)
26. Zheng, C., Fan, X., Wang, C., Qi, J.: GMAN: a graph multi-attention network for traffic prediction. In: Proceedings of the AAAI Conference on Artificial Intelligence, vol. 34, pp. 1234–1241 (2020)

SimiDTR: Deep Trajectory Recovery with Enhanced Trajectory Similarity

Yupu Zhang[2], Liwei Deng[2], Yan Zhao[4], Jin Chen[2], Jiandong Xie[5],
and Kai Zheng[1,2,3](✉)

[1] Yangtze Delta Region Institute (Quzhou), University of Electronic Science and
Technology of China, Chengdu, China
zhengkai@uestc.edu.cn

[2] School of Computer Science and Engineering, University of Electronic Science and
Technology of China, Chengdu, China
{zhangyupu,deng_liwei,chenjin}@std.uestc.edu.cn

[3] Shenzhen Institute for Advanced Study, University of Electronic Science and
Technology of China, Chengdu, China

[4] Department of Computer Science, Aalborg University, Aalborg, Denmark
yanz@cs.aau.dk

[5] Cloud Database Innovation Lab of Cloud BU, Huawei Technologies Co., Ltd.,
Chengdu, China
xiejiandong@huawei.com

Abstract. The pervasiveness of GPS-equipped smart devices and the accompanying deployment of sensing technologies generates increasingly massive amounts of trajectory data that cover applications such as personalized location recommendation and urban transportation planning. Trajectory recovery is of utmost importance for incomplete trajectories (resulted from the constraints of devices and environment) to enable their completeness and reliability. To achieve effective trajectory recovery, we propose a novel trajectory recovery framework, namely **D**eep **T**rajectory **R**ecovery with enhanced trajectory **Simi**larity (SimiDTR), which is capable of contending with the complex mobility regularity found in trajectories in continuous space. In particular, we design a rule-based information extractor to extract the spatial information related to an incomplete trajectory, which is then fed into a deep model based on attention mechanism to generate a tailored similar trajectory for the incomplete trajectory. Finally, we use a deep neural network model to recover the incomplete trajectory with the blessing of its similar trajectory. An extensive empirical study with real data offers evidence that the framework is able to advance the state of the art in terms of effectiveness for trajectory recovery, especially in scenes with sparse trajectory data.

Keywords: Trajectory Recovery · Trajectory Similarity · Sparse Data

X. Wang et al. (Eds.): DASFAA 2023, LNCS 13943, pp. 431–447, 2023.
https://doi.org/10.1007/978-3-031-30637-2_28

1   Introduction

The widespread use of mobile devices has resulted in a proliferation of trajectory data, which contain a wealth of mobility information that is critical to location-based services, e.g., route optimization [29] and travel time estimation [6].

Trajectory data use discrete spatial-temporal point pairs to describe the motion of objects in continuous time and space. Due to the limitations of equipment and environment such as equipment failure and signal missing, many trajectories are recorded at a low sampling rate or with missing locations, called *incomplete trajectories*. Too large sampling interval between two consecutive sampling points can lose detailed information and lead to high uncertainty [30], which affects downstream applications (e.g., indexing [12], clustering [20], and mining [11,24,25]) negatively. Therefore, it is important to recover missing spatial-temporal points for incomplete trajectories and reduce their uncertainty.

In general, previous studies on trajectory recovery can be divided into two directions. The first direction focuses primarily on modeling users' transition pattern among different locations to predict users' missing locations [4,7,8,10, 15,19,25–27]. The basic task is essentially a classification task, and the recovered trajectories are usually composed of locations or POIs. The second direction aims to recover the specific geographic coordinates of trajectories at the missing timestamps based on the incomplete trajectory data recorded [2,5,9,14,16–18,23,24,28]. The final rebuilt trajectories usually consist of precise (GPS or road network) coordinates. In this work, we focus on the second direction, i.e., recovering precise GPS coordinates for incomplete trajectories.

A straightforward approach for the second direction is to regard a single trajectory as two-dimensional time series directly and apply time series imputation methods to recover incomplete trajectories [2,5,9,16,17,28]. These methods exhaust all the precise information of a single incomplete trajectory when recovering it and work quite well when the proportion of the missing trajectory data is small. However, their effectiveness decreases significantly as the missing proportion increases, which means that they cannot deal with the sparse trajectory data. Another common solution for this problem is cell-based methods [14,18,23,24], which divide the space into discrete cells and then recover the missing trajectories described by cells. They further design different post-calibration algorithms to refine the results. These methods transform the trajectory recovery problem from infinitely continuous space into finite discrete space, which reduce the complexity of prediction to improve the capacity of modeling transition patterns. Although the cell-based methods can alleviate the problem of data sparsity to some extent, they only use the information contained in the incomplete trajectory instead of making full use of information coming from other trajectories. Besides, some extra noise and inaccurate information would inevitably be introduced since these methods use cells to represent trajectories. Furthermore, in the calibration stage, there is a lack of available information for getting accurate trajectory coordinates.

Exploiting the similarity among different trajectories to model the complex mobility regularity for incomplete trajectories, we propose a novel trajectory recovery framework, namely **D**eep **T**rajectory **R**ecovery with enhanced trajectory **Simi**larity (SimiDTR) to recover precise coordinates for trajectories. To address the problem of data sparsity, we carefully design a rule-based information extractor to tease out a raw similar trajectory, which has relevant spatial information about the given incomplete trajectory. This raw similar trajectory is the result of integrating information from several other related incomplete trajectories. Considering the properties (e.g., spatial bias, temporal bias, and temporal shifts) of trajectory data, we use an attention-based deep neural network model to sort out this raw similar trajectory and generate a similar trajectory tailored to the incomplete trajectory (i.e., the similar trajectory that does not actually exist but best fits the data of incomplete trajectory), which is used for recovering the incomplete trajectory. In order to make full use of the trajectory coordinate information, we perform the trajectory recovery in continuous space. Our contributions can be summarized as followed:

1) We propose a novel deep trajectory recovery framework with enhanced trajectory similarity, which directly recovers coordinates of trajectories in continuous space. To our knowledge, this is the first study that takes similar trajectory information into the deep learning model in the field of trajectory recovery.

2) To contend with the sparsity of incomplete trajectories, we design a rule-based information extractor that aims to generate a raw similar trajectory for the target incomplete trajectory and propose an attention-based deep neural network model to create a tailor-made similar trajectory based on the inherent characteristics of trajectory data for incomplete trajectory recovery.

3) We report on experiments using real trajectory data, showing that our proposal significantly outperforms state-of-the-art baselines in terms of effectiveness for trajectory recovery, especially in scenes with sparse trajectory data.

2 Related Work

Based on the objects to be recovered, the trajectory recovery can be classified into location recovery and coordinate recovery.

Location recovery aims to predict missing locations (e.g., POIs) for a trajectory [19,25,26]. Xi et al. [25] propose a framework named Bi-STDDP for location recovery, which takes into account bidirectional spatio-temporal dependency. AttnMove [26] uses attention mechanisms to inject aggregated historical trajectory information into recovery process. On the basis of AttnMove, PeriodicMove [19] takes into account the influence of trajectory shifting periodicity.

Taking the trajectory data as two-dimensional time series, the time series imputation methods [2,5,9,16,17,28] can be used to recover the missing coordinates (observations) in trajectories (time series). Some RNN-based [2,28], GAN-based [16,17] and VAE-based methods [9] exist, which can be used to recover coordinates. Generating the recovered trajectories represented by cells, following by a post-calibration algorithm to get the coordinates of trajectories, Cell-based methods [14,18,23,24] are proposed for this task. For example, Wei et al. [24]

build a top-k routes inference framework by aggregating trajectories. Ren et al. [18] propose a deep learning model, which utilizes a traditional seq2seq framework and attention mechanism. The former uses linear regression as the post-calibration module. The later inputs cell-level trajectories into a deep learning model and predicts the road segment ID and moving ratio directly.

In this work we will go further in the GPS coordinate recovery for trajectories by integrating spatial information of similar trajectories into the incomplete trajectory recovery, enabling more accurate coordinate recovery.

3   Problem Statement

Definition 1 (Location). *A location is denoted by $l = (lon, lat)$, where lon and lat represent its longitude and latitude, respectively.*

Definition 2 (Region). *We divide the geographical space into a set of discrete and disjoint square regions, denoted by R. Each region (also referred to as grid or cell), denoted by $r \in R$, is a small square with the length of Δr.*

Definition 3 (Trajectory Point). *A trajectory point is a sampled point from a moving object, denoted by $p = (lon, lat, t)$, where t denotes its timestamp.*

If the location of a trajectory point is known, it is *a recorded trajectory point* (*a recorded point* for short), and we can get the region r where the point falls. Otherwise, it is *a missing trajectory point* (*a missing point* for short), denoted by $\tilde{p}$, where the sampling timestamp is given but the location is unknown.

Definition 4 (Sampling Interval). *The sampling interval, denoted by ϵ, is the time difference between two consecutive trajectory points, which is generally set according to the sampling equipment.*

Ideally, the sampling interval of trajectory data is a fixed constant, but due to the inherent temporal bias of trajectory data, the sampling interval often changes near ϵ in most cases.

Definition 5 (Complete Trajectory). *A complete trajectory, denoted by $tr = p_1 \to \dots \to p_i \to \dots \to p_n$, is a sequence of recorded points sampled from a moving object, where p_i is the i-th trajectory point of tr.*

Definition 6 (Incomplete Trajectory). *An incomplete trajectory is composed of a sequence of recorded points and missing points, e.g., $\tilde{tr} = p_1 \to \tilde{p}_2 \to p_3 \to \tilde{p}_4 \to p_5 \to \cdots$, where $p_1, p_3, p_5 \cdots$.*

Problem Statement. Given a set of incomplete trajectories $\tilde{Tr}$ with sampling interval ϵ, our problem is to recover their missing coordinates.

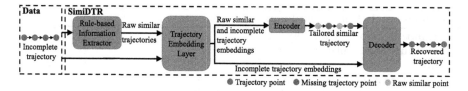

Fig. 1. The framework of SimiDTR model.

4 Methodology

4.1 Framework Overview

We propose a framework, namely **D**eep **T**rajectory **R**ecovery with enhanced trajectory **Simi**larity (SimiDTR), to recover coordinates for incomplete trajectories. It consists of four parts shown in Fig. 1.

Rule-Based Information Extractor. Given an incomplete trajectory and to inject more relevant spatial information into the subsequent components, we design a rule-based information extractor that fills in missing trajectory points of the incomplete trajectory in continuous space to generate a related trajectory (called raw similar trajectory), which is the result of integrating information from several other related incomplete trajectories.

Trajectory Embedding Layer. To encode the sequence of trajectory coordinates into high-dimensional embeddings and introduce sequential, temporal and spatial information simultaneously, we design a trajectory embedding layer.

Encoder. Consider that trajectory data exhibit inherent spatial bias, temporal bias, and temporal shifts even when they are collected under the same sampling interval [13], which means that the locations and timestamps of the corresponding trajectory points in two trajectories (even if capturing the same motion of an object) are usually not the same. To solve this issue, we feed the raw similar trajectory into an attention-based encoder to make it sorted out. The encoder takes the embeddings of the raw similar trajectory and incomplete trajectory as input and outputs a similar trajectory represented by high-dimensional vectors, which is tailored for the incomplete trajectory.

Decoder. In the recovery step, we send the high-dimensional vectors of the similar trajectory and the embedding of the incomplete trajectory to a decoder and finally use a linear layer to predict the missing coordinates.

4.2 Rule-Base Information Extractor

Figure 2(a) and Algorithm 1 shows the rule-based information extractor. We firstly divide the input trajectory into several segments. These trajectory segments have the same form that both sides are recorded points and the middle is missing points. Then the divided trajectory segments are sent to *Padding Module*

Algorithm 1: Rule-based Information Extractor

Input: Incomplete trajectory
Output: Raw similar trajectory

1 **do**
2 | Divide incomplete trajectory into segments;
3 | Fill in segments with padding module separately;
4 | Splice segments into trajectory;
5 **while** *Input and output of the loop are different*;
6 **if** *Unfilled points exist* **then**
7 | Fill in the missing points with linear interpolation;
8 **return** Filled trajectory (i.e. raw similar trajectory)

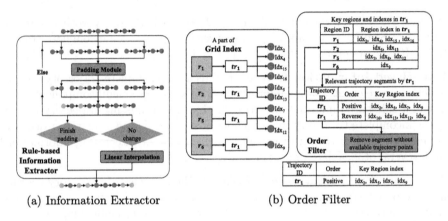

(a) Information Extractor   (b) Order Filter

Fig. 2. Information Extractor and Order Filter

separately. This module finds other relevant trajectory segments from dataset to fill in all or part of the missing points according to the information that the selected relevant trajectory segment contains. Next, we splice the outputs of padding module together and get the raw similar trajectory which is filled once. Since padding module only extracts the information from the most relevant trajectory segment in a single segment padding process, part of the missing points may not be filled because there is not enough information in the only one segment in the sparse scenario. So we regard the filled trajectory points as recorded points and continue to input the raw similar trajectory into padding module until all the missing points are filled or the outputs no longer changes. The output that no longer changes means that there is no relevant information available in dataset. In this case, we use the module *Linear Interpolation* to fill in the missing points.

Padding Module. The padding module takes a trajectory segment as input, and fill in all or part of its missing points at a time. We call this input trajectory

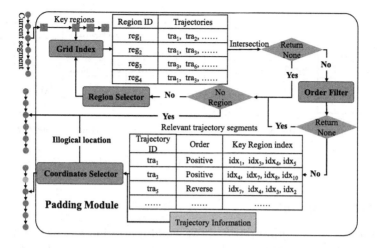

Fig. 3. The process of Padding Module.

segment current segment, and call its corresponding trajectory current trajectory. As shown in Fig. 3, we firstly change current segment into region-level. We call these recorded points key points and call their corresponding regions key regions. With the help of *Grid Index*, we can quickly get the IDs of trajectories (except current trajectory) through key regions and the indexes of these regions in these trajectories. Relying on *Order Filter*, we can remove trajectory segments with no available information and only retain the trajectory segment through key regions in the positive or reverse order. The trajectory segments output by order filter are called relevant trajectory segments. Taking relevant trajectory segments and the trajectory information as input, *Coordinates Selector* can get the proper longitude-latitude pairs to fill in the missing points. Here, the trajectory information is composed of the sequence of coordinates, regions, timestamps and indexes of all the incomplete trajectories. If there is no trajectory segment that we can utilize, we should reduce the number of key points and regions. *Region Selector* can help us remove one point/region, which allows us to continue above process. We will present the technical details of padding module in Sect. 4.3.

Linear Interpolation. If there is no relevant information available in the trajectory dataset to fill in the missing points, it's time to use linear interpolation to fill them. For missing points sandwiched in the middle by recorded points, we fill in with one-dimensional linear interpolation of longitude and latitude, respectively. If there are missing points at the beginning and end of the incomplete trajectory, we use the closest recorded points to fill in them.

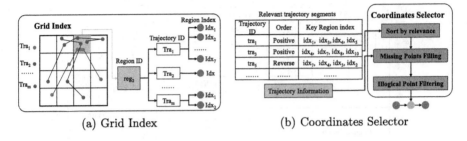

(a) Grid Index   (b) Coordinates Selector

Fig. 4. Grid Index and Coordinates Selector

4.3   Padding Module

Grid Index. To facilitate retrieval, we build a grid index shown in Fig. 4(a). We first divide the entire geographic space into disjoint regions. For each region we record the IDs of trajectories that whose recorded trajectory points fall inside it. For each trajectory, we record the indexes of the region, and sort them in the order that the region appears in the trajectory. Several indexes in a trajectory means the trajectory visits the region several times. Here, the index of regions is the same to the index of trajectory points.

Order Filter. The purpose of order filter is to only retain the trajectory segments through the key regions in the positive and reverse order, and to remove the trajectory segments that have no available information.

To achieve this purpose, we firstly use grid index to get the trajectories through key regions. For each such trajectory, we can get the indexes of key regions in this trajectory. Each region usually corresponds to several indexes, which means this trajectory passes through the region several times. For each region of these trajectories, we keep only one index and eventually form the relevant trajectory segments in the same or opposite order as the key regions. When searching for the positive relevant trajectory segment, we regard the smallest index of the first region as the index of the first point in relevant trajectory segment. In each subsequent step, we use the smallest but larger than the previous index of the region as the selected index. In the reverse case, we regard the largest index of the first region as the first index. In next steps, we select the largest but smaller than previous indexes. Lastly, we remove segments without available trajectory points.

Coordinates Selector. Coordinates selector takes relevant trajectory segments and trajectory information as input, output a raw similar trajectory segment.

Process. Here, the recorded points in relevant trajectory segment who fall in key regions are called relevant points. We firstly calculate the average distance between key points and relevant points. The smaller this value, the higher the correlation between them. Therefore, we sort relevant trajectory segments in order of this average distance from smallest to largest, as shown in Fig. 4. Then we select the GPS coordinate values from the most relevant trajectory segment

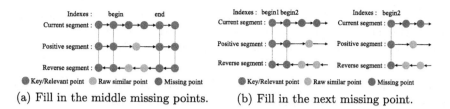

(a) Fill in the middle missing points.   (b) Fill in the next missing point.

Fig. 5. Scenario 1 and Scenario 2

to fill in the missing points. We design coordinate point filling strategies separately in the following three scenarios. **1) Scenario 1.** Both sides of the current segment are recorded points, with missing points in the middle. **2) Scenario 2.** The current segment is preceded by recorded points and followed by missing points. This occurs when the trajectory points at the end of the trajectory are missing. **3) Scenario 3.** The current segment is preceded by missing points and followed by recorded points. This occurs when the trajectory points at the beginning of the trajectory are missing. Finally, we judge whether these filled points are illogical. If the velocity of a moving object between any two trajectory points exceeds a threshold, we consider it illogical. Then we remove the filled missing points and use the left relevant trajectory segments to repeat the above process. We set threshold to $2\,km/min$ in scenario 1 and $1.5\,km/min$ in the others. If there is no logical points provided, we will not fill any of the missing points.

Missing Points Filling. We choose appropriate GPS coordinates for the missing points based on time ratio, denoted by $time\_ratio$. In Scenario 1, as shown in Fig. 5(a), we assume the indexes of the two key points which are adjacent to missing points are bc and ec. The indexes of their corresponding relevant points of positive and reverse segment are bp, ep and br, er separately. Then the time ratio of missing point is $time\_ratio_c^i = \frac{t_i - t_{bc}}{t_{ec} - t_{bc}}$, $bc < i < ec$. Here, i means the index of a missing point in incomplete trajectory. The time ratio of available points in positive and reverse segment are $time\_ratio_p^j = \frac{t_j - t_{bp}}{t_{ep} - t_{bp}}$, $bp < j < ep$ and $time\_ratio_r^k = \frac{t_{br} - t_k}{t_{br} - t_{er}}$, $er < k < br$. We select the closest $time\_ratio_p^j$ or $time\_ratio_r^k$ for each $time\_ratio_c^i$ and fill in $\tilde{p}_i$ with the recorded point corresponding to $time\_ratio_p^j$ or $time\_ratio_r^k$. In Scenario 2, as shown in Fig. 5, we only fill in the missing point which is adjacent to key points. If there are at least two key points, we assume the indexes of the two key points which are adjacent to missing points are $bc1$ and $bc2$. The indexes of their corresponding relevant points of positive and reverse segment are $bp1$, $bp2$ and $br2$, $br2$ separately. Then the time ratio of the first missing point is $time\_ratio_c^i = \frac{t_i - t_{bc2}}{t_{bc2} - t_{bc1}}$, $i = bc2 + 1$. The time ratio of available points in positive and reverse segment are $time\_ratio_p^j = \frac{t_j - t_{bp2}}{t_{bp2} - t_{bp1}}$, $bp2 < j$ and $time\_ratio_r^k = \frac{t_{br2} - t_k}{t_{br1} - t_{br2}}$, $k < br2$. Then we can find the closest $time\_ratio_p^j$ or $time\_ratio_r^k$ for $time\_ratio_c^i$. If there is only one key point, the three time ratios are $time\_ratio_c^i = t_i - t_{bc2}$, $i = bc2 + 1$, $time\_ratio_p^j = t_j - t_{bp2}$, $bp2 < j$ and

$time\_ratio_r^k = t_{br2} - t_k, \ k < br2$ separately. In scenario 3, we also only fill in the missing point which is adjacent to key points. Its calculation process is the opposite of Scenario 2, which are included in technical report.

Region Selector. We use region selector to remove a key point in the following cases, as there is no trajectory segment in dataset that is so similar to current segment. **Case 1.** With the exception of current segment, no other trajectory segment passes through all the key regions. **Case 2.** The Case 1 is satisfied, but none of these trajectory segments with available information pass through the key regions in positive or reverse order. **Case 3.** The Case 2 is satisfied, but there are some illogical trajectory points in the selected trajectory segments. We believe that the closer the recorded point to the missing points, the more critical it is. Everytime we remove the point furthest from the missing points.

4.4 Trajectory Embedding Layer

For the sake of narrative, we use $Concat(\cdot)$ to denote concatenating matrices horizontally, and use $\boldsymbol{W}$ and $\boldsymbol{b}$ to denote linear transformation matrix and bias. For an incomplete trajectory $tr = p_1 \to \cdots \to p_n, \ n \in N$, where n is the length of tr, N is length of the longest trajectory in dataset. The locations and timestamps of tr are $\boldsymbol{L} \in \mathbb{R}^{n \times 2}$ and $\boldsymbol{T} \in \mathbb{R}^{n \times 1}$.

Positional Embedding. Positional embedding is composed of sine and cosine functions following [21], which contains sequence information. We use $\boldsymbol{E}_{idx} \in \mathbb{R}^{n \times d}$ to represents the positional embedding of tr, and generate it as follows,

$$
\begin{cases}
\boldsymbol{E}_{idx}(i, 2j) = sin(\frac{i}{10000^{2j/d}}) \\
\boldsymbol{E}_{idx}(i, 2j+1) = cos(\frac{i}{10000^{2j/d}})
\end{cases}
\tag{1}
$$

where i denotes the i-th point of tr, j denotes the j-th dimension. Each positional embedding vector has d dimensions.

Timestamp Embedding. The timestamp embedding is mainly to inject time features of trajectories into the following module. For each timestamps t of $\boldsymbol{T}$, we can calculate that t is at the t_{min}-th minute of the current hour and the t_{sec}-th second of the current minute ($t_{min} \in \boldsymbol{T}_{min}, t_{sec} \in \boldsymbol{T}_{sec}$). Then we map them to the interval $[-0.5, 0.5]$ and follow a linear transformation, just as follows,

$$
\boldsymbol{E}_t = (Concat([\frac{\boldsymbol{T}_{min}}{59} - 0.5, \frac{\boldsymbol{T}_{sec}}{59} - 0.5]))\boldsymbol{W}_t^T + \boldsymbol{b}_t
\tag{2}
$$

where $\boldsymbol{W}_t \in \mathbb{R}^{d \times 2}$, $\boldsymbol{b}_t \in \mathbb{R}^{1 \times d}$ and $\boldsymbol{T}_{min}, \boldsymbol{T}_{sec} \in \mathbb{R}^{n \times 1}$.

Coordinate Embedding. In order to ensure that the missing points (padded by $\boldsymbol{0} \in \mathbb{R}^{1 \times 2}$) remain $\boldsymbol{0}$ ($\boldsymbol{0} \in \mathbb{R}^{1 \times d}$) when they are mapped to high-dimensional space, we use one-dimensional convolution to act as a mapping function as follows,

$$
\boldsymbol{E}_{GPS} = Conv1d(\boldsymbol{L})
\tag{3}
$$

where $Conv1d(\cdot)$ has no bias. The size of convolution kernel is 1.

Embedding Integration. Assume the raw similar trajectory of tr is denoted by tr_{rs}. Because it is simply a rigid combination of relevant spatial information, which means T doesn't fit tr_{rs} in the missing points, we don't introduce the timestamp embedding of tr_{rs} into the following module. In a word, the final embedding of tr_{rs} and tr are as following,

$$X_{rs} = E_{idx} + Conv1d(L_{rs}), \quad X = E_{idx} + E_t + E_{GPS} \tag{4}$$

where $L_{rs} \in \mathbb{R}^{n \times 2}$ is the locations of tr_{rs}, and $X_{rs}, X \in \mathbb{R}^{n \times d}$.

4.5   Encoder and Decoder

The framework of this module is shown in Fig. 6. Encoder takes the embedding of incomplete trajectory and its raw similar trajectory as input, outputs the tailored similar trajectory represented as a high-dimensional vectors, which is denoted by $X_s \in \mathbb{R}^{n \times d}$. Decoder takes X and X_s as input, and outputs the predicted trajectory. We use a linear transformation to get its GPS coordinates. Next, we firstly introduce the operation of each component, and then introduce the complete process of encoder and decoder.

Concat and Linear. This module is mainly to integrate the results of two multi-head attention in encoder. Suppose these two results are $O_{a1}, O_{a2} \in \mathbb{R}^{n \times d}$, the operations of this module is as following,

$$O_c = Concat([O_{a1}, O_{a2}])W_c + b_c \tag{5}$$

where $W_c \in \mathbb{R}^{2d \times d}$ and $b_c \in \mathbb{R}^{1 \times d}$. Above operations are noted by $O_c = CL(O_{a1}, O_{a2})$. Other underlying components are following [21], such as **Multi-Head Attention**, **Add and Norm** and **Feed Forward**, which are noted by $O_a = MHA(X_q, X_k, X_v)$, $O_A = LN(O_{Add})$ and $O_F = FFN(X_F)$, respectively. The detailed description are included in our technical report.

Encoder. In encoder, we are equivalent to refer the change in X, and use attention mechanism to generate a similar trajectory tailored to it. Encoder is composed of $B \in \mathbb{R}$ encoder blocks. For each block, its process is as following,

$$\begin{aligned} O_{a1} = MHA(X_{rs}, X, X), \quad O_{a2} = MHA(X_{rs}, X_{rs}, X_{rs}), O_c = CL(O_{a1}, O_{a2}) \\ O_A = LN(X_{rs} + Dropout(O_c)), X_{rs} = LN(O_A + Dropout(FFN(O_A))) \end{aligned} \tag{6}$$

We note the X_{rs} outputed by the last Encoder Block as X_s, which is the tailored similar trajectory of X.

Decoder. In decoder, We are equivalent to use attention mechanism to add additional relevant spatial information when recovering X. Decoder consists of $B \in \mathbb{R}$ decoder blocks. For each block, its process is as following,

$$O_{a1} = MHA(\boldsymbol{X}, \boldsymbol{X}, \boldsymbol{X}), \quad O_{A1} = LN(\boldsymbol{X} + Dropout(O_{a1}))$$
$$O_{a2} = MHA(O_{A1}, \boldsymbol{X_s}, \boldsymbol{X_s}), \quad O_{A2} = LN(O_{A1} + Dropout(O_{a2})) \quad (7)$$
$$\boldsymbol{X} = LN(O_{A2} + Dropout(FFN(O_{A2})))$$

Finally, we send the $\boldsymbol{X}$ outputed by the last decoder block into a linear layer to get the coordinates of the predicted trajectory:

$$\boldsymbol{L}_{pre} = \boldsymbol{X}\boldsymbol{W} + \boldsymbol{b} \quad (8)$$

where $\boldsymbol{W} \in \mathbb{R}^{d \times 2}, \boldsymbol{b} \in \mathbb{R}^{1 \times 2}$.

We simply use mean square error between the locations of predicted trajectory (denoted by $\boldsymbol{L}_{pre} \in R^{n \times 2}$) and the locations of true trajectory (denoted by $\boldsymbol{L}_{tar} \in \mathbb{R}^{n \times 2}$) as the loss function:

$$\mathcal{L} = \frac{1}{2n} \sum_{i=1}^{n} \sum_{j=1}^{2} (predict_{i,j} - target_{i,j})^2 \quad (9)$$

where $predict_{i,j}, target_{i,j}$ are elements of row i and column j of $\boldsymbol{L}_{pre}$ and $\boldsymbol{L}_{tar}$.

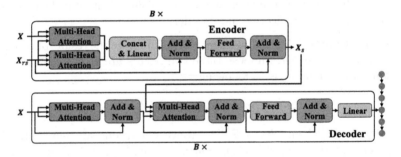

Fig. 6. The Schematic Diagram of Encoder and Decoder.

5   Experimental Evaluation

5.1   Experimental Setup

Datasets. To verify the effect of our proposed model, we use two real-world taxi trajectory datasets collected from Porto[1] and Shanghai[2], respectively.

For Porto dataset, we convert the interval from 15 s to 1 min. For Shanghai taxi dataset, we regard the stay points of taxis as boundaries to segment their entire day's trajectories. Next, we remove all trajectories that contain points being out of the latitude and the longitude ranges. If the number of trajectory points in the region is less than region point threshold, we remove such regions

---

[1] https://www.kaggle.com/c/pkdd-15-predict-taxi-service-trajectory-i.
[2] https://cse.hkust.edu.hk/scrg/.

together with points in them. Note that after preprocessing, the time interval of Porto is a constant, i.e., 60 s, while the time interval of Shanghai is variable. More detailed statistics is included in technical report. We divide each dataset into three parts with the splitting ratio of 7:1:2 as training, validation, and test set. Besides, we randomly retain $1 - ratio\%$ ($ratio = 30, 50, 70$) of points for each trajectory. We call $ratio\%$ coverage ratio in this paper.

Evaluation Methods. 1) Linear Interpolation (LI), 2) Raw Similar Trajectories (RST) that uses a rule-based information extractor, 3) SAITS [5] that is a time series imputation method, 4) Transformer [21], and 5) our SimiDTR method.

Metrics. In the evaluation phase, we copy the recorded points to the corresponding position in L_{pre} (i.e., the predicted trajectory by model) to get L_{eva} (i.e., the finally recovered trajectory). Then we calculate the four metrics, i.e., RMSE (Root Mean Squared Error), NDTW (Normalized Dynamic Time Warping) [1], I-LCSS (Improved Longest Common SubSequence) [22], and EDR (Edit Distance on Real sequence) [3] between L_{eva} and L_{tar} to evaluate the effectiveness of the above methods. These metrics are widely used to measure the similarity between trajectories in previous [14,23,24,30]. The smaller these four metrics are, the more accurate the method is.

5.2 Experimental Results

Overall. We report the performance of all the methods in Table 1. The best result in each column is bold, the second is underlined, and Trans, -S, -D, and -E mean Transformer, SimiDTR-S, SimiDTR-D and SimiDTR-E. We observe:

1) When time interval is fixed, LI and RST work better, even comparable to some deep learning models. However, when time interval is unfixed, the performance of them are generally lower than that of the deep learning models.
2) SAITS works well when the trajectory coverage ratio is small, but its performance decreases more significantly than the other deep learning models when the coverage ratio increases.
3) Although Transformer only uses information from incomplete trajectory itself, it achieves satisfying results regardless of coverage ratios.
4) Due to the combination of similar trajectory information and the powerful feature extraction capabilities of deep neural networks, the performance of SimiDTR is significantly improved compared with all other methods above.

Ablation Analysis. We remove one module at a time:

1) **SimiDTR-S**: Only encoder and decoder are used to recover trajectories. In this case, we replace the inputs of encoder and decoder with the embedding of incomplete trajectory.
2) **SimiDTR-D**: Only rule-based information extractor and encoder are used to recover trajectories. We send the embedding of the tailored similar trajectory output by encoder into a linear layer to get its trajectory coordinates.

Table 1. Performance Comparison on two Datasets

Metric	Datasets		LI	RST	SAITS	Trans	-S	-D	-E	SimiDTR
RMSE	Porto	30%	0.1486	0.1391	0.1385	0.1290	0.1084	0.1000	<u>0.0942</u>	**0.0822**
		50%	0.2838	0.2425	0.2597	0.1971	0.1895	0.1658	<u>0.1488</u>	**0.1343**
		70%	0.5306	0.3761	0.4749	0.3260	0.3262	0.2129	<u>0.1951</u>	**0.1821**
	Shanghai	30%	0.3547	0.4486	0.2651	0.2439	<u>0.1811</u>	0.2210	0.1813	**0.1750**
		50%	0.5221	0.6647	0.4575	0.3168	0.2525	0.3013	<u>0.2468</u>	**0.2210**
		70%	0.7468	0.9089	0.7445	0.4417	0.3643	0.3503	<u>0.3107</u>	**0.2769**
NDTW	Porto	30%	0.0649	0.0525	0.0606	0.0410	0.0351	<u>0.0278</u>	0.0304	**0.0256**
		50%	0.1570	0.1143	0.1420	0.0810	0.0775	<u>0.0590</u>	0.0613	**0.0536**
		70%	0.3349	0.2013	0.2911	0.1513	0.1519	0.0911	<u>0.0909</u>	**0.0836**
	Shanghai	30%	0.1452	0.1600	0.0901	0.0604	<u>0.0408</u>	0.0501	0.0427	**0.0395**
		50%	0.2663	0.2761	0.1989	0.0970	<u>0.0702</u>	0.0902	0.0733	**0.0625**
		70%	0.4368	0.4079	0.3919	0.1641	0.1189	0.1278	<u>0.1080</u>	**0.0906**
I-LCSS	Porto	30%	0.1173	0.0916	0.1117	0.0787	0.0666	<u>0.0475</u>	0.0591	**0.0459**
		50%	0.2533	0.1919	0.2423	0.1428	0.1367	<u>0.0987</u>	0.1182	**0.0983**
		70%	0.4232	0.3040	0.4048	0.2301	0.2337	<u>0.1574</u>	0.1666	**0.1533**
	Shanghai	30%	0.0773	0.1007	0.0897	0.0608	0.0612	**0.0387**	0.0565	<u>0.0514</u>
		50%	0.1734	0.2082	0.1880	0.1072	0.0940	<u>0.0851</u>	0.0999	**0.0793**
		70%	0.3223	0.3360	0.3318	0.1577	0.1342	<u>0.1232</u>	0.1312	**0.1132**
EDR	Porto	30%	0.1206	0.0955	0.1151	0.0811	0.0687	<u>0.0496</u>	0.0615	**0.0478**
		50%	0.2262	0.2071	0.2566	0.1518	0.1452	<u>0.1068</u>	0.1272	**0.1057**
		70%	0.4482	0.3365	0.4367	0.2502	0.2541	<u>0.1743</u>	0.1838	**0.1697**
	Shanghai	30%	0.0804	0.1061	0.0925	0.0638	0.0654	**0.0412**	0.0603	<u>0.0548</u>
		50%	0.1850	0.2281	0.2004	0.1152	0.1032	<u>0.0950</u>	0.1098	**0.0873**
		70%	0.3469	0.3820	0.3628	0.1717	0.1492	<u>0.1399</u>	0.1475	**0.1276**

3) **SimiDTR-E**: Only rule-based information extractor and decoder are used to recover trajectories. We directly send the raw similar trajectory obtained by the rule-based information extractor to the decoder without processing.

We report the results of these four models in Table 1 and observe that:

1) Our SimiDTR has the best comprehensive performance, which also shows that each module can improve the recovery;
2) Specifically, the three models with rule-based information extractor have a significant improvement compared with that without trajectory information when the coverage ratio is 70%, which shows the use of similar trajectory information can well alleviate the sparsity of trajectory data;

Visualization Comparison. In order to analyze the effect of the model more vividly, we select one sample trajectory and visualize the recovery results of each model as shown in Fig. 7 (More visualization is included in the technical report). Note that its coverage ratio are 70%, and we can observe that:

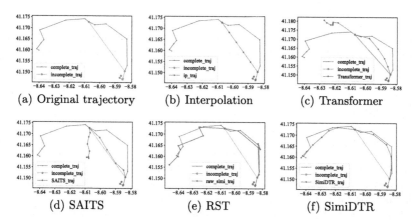

Fig. 7. Sample trajectories from Porto dataset

1) In the case of sparsity, because an incomplete trajectory contains little information, LI, Transformer and SAITS are difficult to recover trajectory segments with serious data loss. However, with the help of similar trajectory information, our model can recover the movement trend of the object.

2) The raw similar trajectory has a more similar change trend to the complete trajectory, but due to the spatial bias, temporal bias and temporal shifts between different trajectory data, it does not perform well in metrics. However, after it is sorted out by the neural network model, its effect will be greatly improved. The technical report[3] contains more details of algorithms and experiments.

6   Conclusion

We propose a novel deep trajectory recovery framework with enhanced trajectory similarity that enables the capture of complex mobility regularity for recovering incomplete trajectories.To counter data sparsity problems, the framework leverages the similarity between the target incomplete trajectory and other trajectories. In particular, it adopts a rule-based method that can extract a similar trajectory from the observed points of other incomplete trajectories for the given incomplete trajectory. Further, the framework exploits attention mechanism to generate a trajectory that is not actually existing but the most similar to the incomplete trajectory. The generated trajectory is used for recovering the incomplete trajectory. We conduct experiments on two datasets collected from real world. The experimental results show that this framework outperforms the state of the art in trajectory recovery performance, especially when data is sparse.

---

[3] https://github.com/111anonymity111/SimiDTR_DASFAA_technical_report/tree/
main.

Acknowledgements. This work is partially supported by NSFC (No. 61972069, 61836007 and 61832017), Shenzhen Municipal Science and Technology R&D Funding Basic Research Program (JCYJ20210324133607021), and Municipal Government of Quzhou under Grant No. 2022D037.

References

1. Agrawal, R., Faloutsos, C., Swami, A.: Efficient similarity search in sequence databases. In: Lomet, D.B. (ed.) FODO 1993. LNCS, vol. 730, pp. 69–84. Springer, Heidelberg (1993). https://doi.org/10.1007/3-540-57301-1_5
2. Cao, W., Wang, D., Li, J., Zhou, H., Li, L., Li, Y.: Brits: bidirectional recurrent imputation for time series. In: NIPS, vol. 31 (2018)
3. Chen, L., Özsu, M.T., Oria, V.: Robust and fast similarity search for moving object trajectories. In: SIGMOD, pp. 491–502 (2005)
4. Cheng, C., Yang, H., Lyu, M.R., King, I.: Where you like to go next: successive point-of-interest recommendation. In: IJCAI, pp. 2605–2611 (2013)
5. Du, W., Côté, D., Liu, Y.: SAITS: self-attention-based imputation for time series. arXiv preprint: arXiv:2202.08516 (2022)
6. Fang, X., Huang, J., Wang, F., Zeng, L., Liang, H., Wang, H.: ConSTGAT: contextual spatial-temporal graph attention network for travel time estimation at Baidu maps. In: SIGKDD, pp. 2697–2705 (2020)
7. Feng, J., Li, Y., Zhang, C., Sun, F., Meng, F., Guo, A., Jin, D.: DeepMove: predicting human mobility with attentional recurrent networks. In: WWW, pp. 1459–1468 (2018)
8. Feng, S., Li, X., Zeng, Y., Cong, G., Chee, Y.M., Yuan, Q.: Personalized ranking metric embedding for next new poi recommendation. In: IJCAI, pp. 2062–2068 (2015)
9. Fortuin, V., Baranchuk, D., Rätsch, G., Mandt, S.: GP-VAE: deep probabilistic time series imputation. In: AISTATS, pp. 1651–1661 (2020)
10. Gambs, S., Killijian, M.O., del Prado Cortez, M.N.: Next place prediction using mobility Markov chains. In: MPM, pp. 1–6 (2012)
11. Guo, S., et al.: Rod-revenue: seeking strategies analysis and revenue prediction in ride-on-demand service using multi-source urban data. TMC **19**(9), 2202–2220 (2019)
12. Hadjieleftheriou, M., Kollios, G., Tsotras, V.J., Gunopulos, D.: Efficient indexing of spatiotemporal objects. In: Jensen, S., et al. (eds.) EDBT 2002. LNCS, vol. 2287, pp. 251–268. Springer, Heidelberg (2002). https://doi.org/10.1007/3-540-45876-X_17
13. Hung, C.C., Peng, W.C., Lee, W.C.: Clustering and aggregating clues of trajectories for mining trajectory patterns and routes. PVLDB **24**(2), 169–192 (2015). https://doi.org/10.1007/s00778-011-0262-6
14. Liu, H., Wei, L.Y., Zheng, Y., Schneider, M., Peng, W.C.: Route discovery from mining uncertain trajectories. In: ICDMW, pp. 1239–1242 (2011)
15. Liu, Q., Wu, S., Wang, L., Tan, T.: Predicting the next location: a recurrent model with spatial and temporal contexts. In: AAAI, pp. 194–200 (2016)
16. Luo, Y., Cai, X., Zhang, Y., Xu, J., et al.: Multivariate time series imputation with generative adversarial networks. NIPS **31**, 1603–1614 (2018)
17. Luo, Y., Zhang, Y., Cai, X., Yuan, X.: E2GAN: end-to-end generative adversarial network for multivariate time series imputation. In: IJCAI, pp. 3094–3100 (2019)

18. Ren, H., Ruan, S., Li, Y., Bao, J., Meng, C., Li, R., Zheng, Y.: MTrajRec: map-constrained trajectory recovery via seq2seq multi-task learning. In: SIGKDD, pp. 1410–1419 (2021)
19. Sun, H., Yang, C., Deng, L., Zhou, F., Huang, F., Zheng, K.: PeriodicMove: shift-aware human mobility recovery with graph neural network. In: CIKM, pp. 1734–1743 (2021)
20. Ulm, M., Brandie, N.: Robust online trajectory clustering without computing trajectory distances. In: ICPR, pp. 2270–2273 (2012)
21. Vaswani, A., et al.: Attention is all you need. NIPS **30**, 5998–6008 (2017)
22. Vlachos, M., Kollios, G., Gunopulos, D.: Discovering similar multidimensional trajectories. In: ICDE, pp. 673–684 (2002)
23. Wang, J., Wu, N., Lu, X., Zhao, W.X., Feng, K.: Deep trajectory recovery with fine-grained calibration using Kalman filter. TKDE **33**(3), 921–934 (2019)
24. Wei, L.Y., Zheng, Y., Peng, W.C.: Constructing popular routes from uncertain trajectories. In: SIGKDD, pp. 195–203 (2012)
25. Xi, D., Zhuang, F., Liu, Y., Gu, J., Xiong, H., He, Q.: Modelling of bi-directional Spatio-temporal dependence and users' dynamic preferences for missing poi check-in identification. In: AAAI, no. 01, pp. 5458–5465 (2019)
26. Xia, T., et al.: AttnMove: history enhanced trajectory recovery via attentional network. arXiv preprint: arXiv:2101.00646 (2021)
27. Xie, M., Yin, H., Wang, H., Xu, F., Chen, W., Wang, S.: Learning graph-based poi embedding for location-based recommendation. In: CIKM, pp. 15–24 (2016)
28. Yoon, J., Zame, W.R., van der Schaar, M.: Estimating missing data in temporal data streams using multi-directional recurrent neural networks. TBME **66**(5), 1477–1490 (2018)
29. Yuan, J., Zheng, Y., Xie, X., Sun, G.: T-drive: enhancing driving directions with taxi drivers' intelligence. TKDE **25**(1), 220–232 (2011)
30. Zheng, Y.: Trajectory data mining: an overview. TIST **6**(3), 1–41 (2015)

Fusing Local and Global Mobility Patterns for Trajectory Recovery

Liwei Deng[2], Yan Zhao[4], Hao Sun[5], Changjie Yang[6], Jiandong Xie[6], and Kai Zheng[1,2,3(✉)]

[1] Yangtze Delta Region Institute (Quzhou), University of Electronic Science and Technology of China, Chengdu, China
zhengkai@uestc.edu.cn
[2] School of Computer Science and Engineering, University of Electronic Science and Technology of China, Chengdu, China
deng_liwei@std.uestc.edu.cn
[3] Shenzhen Institute for Advanced Study, University of Electronic Science and Technology of China, Chengdu, China
[4] Aalborg University, Aalborg, Denmark
yanz@cs.aau.dk
[5] Peking University, Beijing, China
sunhao@stu.pku.edu.cn
[6] Cloud Database Innovation Lab of Cloud BU, Huawei Technologies Co., Ltd., Chengdu, China
xiejiandong@huawei.com

Abstract. The prevalence of various mobile devices translates into a proliferation of trajectory data that enables a broad range of applications such as POI recommendation, tour recommendation, urban transportation planning, and the next location prediction. However, trajectory data in real-world is often sparse, noisy and incomplete. To make the subsequent analysis more reliable, trajectory recovery is introduced as a preprocessing step and attracts increasing attention recently. In this paper, we propose a neural attention model based on graph convolutional networks, called TRILL, to enhance the accuracy of trajectory recovery. In particular, to capture global mobility patterns that reflect inherent spatiotemporal regularity in human mobilities, we construct a directed global location transition graph and model the mobility patterns of all the trajectories at point level using graph convolutional networks. Then, a self-attention layer and a window-based cross-attention layer are sequentially adopted to refine the representations of missing locations by considering intra-trajectory and inter-trajectory information, respectively. Meanwhile, an information aggregation layer is designed to leverage all the historical information. We conduct extensive experiments using real trajectory data, which verifies the superior performance of the proposed model.

1 Introduction

The prevalence of GPS-enabled devices and wireless communication technologies generates massive trajectories containing a wealth of human mobility

© The Author(s), under exclusive license to Springer Nature Switzerland AG 2023
X. Wang et al. (Eds.): DASFAA 2023, LNCS 13943, pp. 448–463, 2023.
https://doi.org/10.1007/978-3-031-30637-2_29

information, which is the cornerstone of a wide range of downstream applications, such as points of interest (POI) recommendation [24,30], tour recommendation [22,36], urban transportation planning [38], and the next location prediction [4]. However, due to privacy concerns or other factors, such as device failure, the GPS-enabled devices cannot record locations of users continuously. As a result, the trajectory data inevitably have lots of missing locations, which affects the performance of these downstream applications. For instance, it is difficult for a recommender system to generate satisfying POIs for a specific user, when the preferences of the user are not well captured from the incomplete trajectories [31]. As for urban transportation planning, most of the existing methods are proposed based on complete and high-quality trajectories [33], which means that they require full observations (i.e., locations) of trajectories, thus they are unable to effectively utilize incomplete trajectories [15]. In other words, when there are missing locations in trajectories, the performace of these methods will deteriorate sharply [15]. Therefore, recovering the missing locations in trajectories is of utmost importance to deliver reliable analysis results of human mobilities.

Existing studies for trajectory recovery can be roughly devided into two directions. The first direction treats this problem as time series imputation, where the latitude and longitude values at each timestamp are treated as two-dimensional time series [2,18]. Although these studies can get acceptable performance when the percentage of missing locations in trajectories is low, they have poor performance with the increase of missing locations since they do not take the human mobility patterns into consideration. The other direction is to model users' transition among different locations at different time slots and to generate the missing locations according to the highest transition probability from the observed locations [17,31]. These studies consider the intrinsic and natural characteristics such as periodicity and repeatability [5,10,21,31]. For example, AttnMove [31] leverages a multi-stage attention mechanism to recover the missing locations based on the historical trajectories. Sun et al. [23] proposes PeriodicMove, in which day-level graphs are constructed for modeling complex transition patterns among locations. Our work falls in the second direction.

To achieve effective trajectory recovery, we still face two main challenges.

- *Challenge I: how to combine individual historical trajectories and global mobility patterns when recovering trajectories?* Despite the inspiring results obtained by the above models [23,31] in the second direction, they only leverage individual historical trajectories to recover the missing locations for the current trajectory, while the global mobility patterns among locations are ignored. Prior studies find human mobilities have global mobility patterns with inherent high-level spatial and temporal regularity [10] (i.e., people have high probability to repeat similar travel patterns) and highly skewed travel distribution [37] (i.e., different people often take similar routes when traveling between certain locations). Modeling these patterns can effectively relieve the data sparsity problem and improve the accuracy of trajectory recovery.
- *Challenge II: how to fully explore the relationship between the current trajectory and historical ones?* The current studies fail to explore the relationship

between the current trajectory and historical ones [10], or just explore the relationship by simple operations among historical trajectories [31]. For example, AttnMove [31] fuses multiple historical trajectories into one by extracting the locations with the highest visiting frequency in the corresponding time slot, in which the naive fusion operation will break the transition pattern in historical trajectories and hinder the modeling of current and historical trajectories.

To address the above challenges, we propose a neural attention model based on graph convolutional networks for Trajectory Recovery by fusIng Local and gLobal mobility patterns, called TRILL. More specifically, we first construct a transition graph that records the transition relationships among locations using all the trajectories in the training data at a global level. Then, a vanilla graph convolutional network is adopted to enrich the location information by aggregating the representations of neighbors (*Challenge I*). For *Challenge II*, we leverage a self-attention layer to get appropriate representations of locations using intra-trajectory information, and adopt a window-based cross-attention and information aggregation layer to fully explore the relationship between current and historical trajectories.

Overall, our contributions can be summarized as follows:

- We propose a novel neural attention model with graph convolutional network, called TRILL, to recover missing locations of trajectories. In TRILL, global mobility patterns are captured by a constructed global transition graph and graph convolutional networks. To the best of our knowledge, TRILL is the first model that considers global mobility patterns for trajectory recovery.
- We propose a multi-level attention and information aggregation layer to fully explore the relationship between the current and historical trajectories, where a novel window-based cross-attention mechanisim is designed to capture the local features (i.e., the informaion of neighbour locations) of the missing locations.
- We conduct extensive experiments on two real-world datasets, which verifies the effectiveness of the proposed model. TRILL outperforms the state-of-the-art models by 12.4% and 3.2% in terms of recovery accuracy (i.e., Recall@1) on Geolife and Foursquare datasets, respectively.

2   Related Works

In this section, we present the related works, i.e., human mobility recovery and neural networks on graphs, and the necessary preliminaries followed by the formal definition of the problem to be solved.

2.1   Human Mobility Recovery

The solutions for human mobility recovery can be divided into two lines as stated above. Our proposed model can fall into the second lines, i.e., generating missing locations by modeling transition among different locations at different time slot. The core idea in these methods is to learn appropriate representations of missing locations in the current trajectory, which can be used to accurately calculate the probability of the observed locations to be the missing locations. DeepMove [8] predicts the missing location using the previous locations, which ignores the future locations in current trajectory. AttnMove [31] considers the abundent informations from historical trajectories, in which multiple locations in a specific time slot from historical trajectories are fused into one with the highest visiting frequency and then attention mechanisms are adopted to obtain the representations of missing locations. PeriodicMove [23] considers the multilevel periodicity and shifting periodicity of human mobility, and achieves the state-of-the-art performance. However, the missing locations existing in historical trajectories cannot be dealt well by PeriodicMove. Despite the inspiring results they got, there still exists some drawbacks shown in the former section. Taking these factors into account, we propose a novel attention neural model with graph convolutional network, i.e., TRILL, for human mobility recovery.

2.2 Attention Mechanisms and Graph Convolutional Networks

To recover trajectories, vanilla sequential models, such as LSTM and GRU, are hardly adopted due to the restrictions of these model, e.g., the input should be complete. Different from these models, attention-based sequential modeling methods have the ability to deal with missing values by simply setting the attention weights of missing steps to the minimum. Besides, previous studies [26,27,29,31] have shown the priority of attention mechanisms in terms of efficiency and effectivenss in many fields, such as ViT [7] in computer vision and BERT [6] in natural language processing. PeriodicMove [23], currently the best performing trajectory recovery method in the same setting with ours, is purely based on attention module. Thus in this paper, we adopt multi-level attention layer to fully explore the relationship between current and historical trajectories. Except the modeling in trajectory level, the transition relation among locations is also helpful in trajectory recovery [23]. For example, PeriodicMove construct user-specific graph to model the complex trasition in user's behaviour. However, the global mobility patterns, extracting from all the non-missing consecutive records, is ignored. To deal with this, a global transition graph is constructed in TRILL, which is needed to be encoded into vectors to be used by the later modules. There are plenty of graph embedding methods, such as DeepWalk [19], LINE [25], and CARE [12], which are task-agnostic and thus usually sub-optimal. Recently, graph convolutional networks [13] get a lot of attention due to the power to deal with graph, which can be easily trained in end-to-end fashion [1,14]. In addition, the ability of GCN for modeling relation among nodes has been shown in previous studies [23,29]. Therefore, we adopt the vanilla graph convolutional networks [13] to encode the nodes into embeddings, due to its stable performance on many tasks.

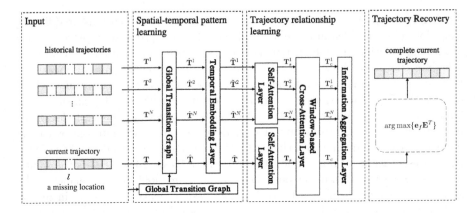

Fig. 1. Framework Overview

2.3   Preliminaries

Definition 1 (Trajectory). *A trajectory, denoted as T, is a sequence of times-tamped locations with the form of $((l_1, t_1), (l_2, t_2), ..., (l_n, t_n))$, where l_i ($1 \leq i \leq n$) stands for the location of a POI at time stamp t_i.*

It is worthy noting that the locations of POIs may not be recorded over fixed intervals. For example, a user goes to a store (the location of which is denoted by l_1) at 8:00 am, a company (the location of which is denoted by l_2) at 8:30 am, and a restaurant (the location of which is denoted by l_3) at 11:30 am. As a result, the trajectory can be represented by $((l_1, 8{:}00), (l_2, 8{:}30), (l_3, 11{:}30))$.

Definition 2 (Incomplete Trajectory). *A trajectory T is incomplete when missing locations exist.*

For the above example, if we do not know the location of the POI where the user stays at 9:00 am, the trajectory is incomplete, which is represented as $((l_1, 8{:}00), (l_2, 8{:}30), (l, 9{:}00), (l_3, 11{:}30))$, where l denotes the missing location.

Definition 3 (Trajectory Recovery). *Given an incomplete trajectory T and a set of historical trajectories $\{T_1, T_2, ..., T_N\}$, we aim to recover the missing locations for T.*

3   Framework and Methodology

We propose a framework, namely TRILL, to recover trajectories. We first give an overview of the framework and then provide specifics on each component in the framework.

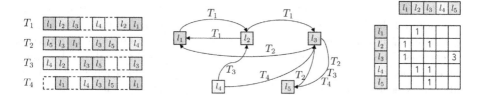

Fig. 2. An exmaple of graph construction

3.1 Model Overview

Taking a set of historical trajectories and a trajectory to be recovered as input, the framework mainly consists of three components: spatio-temporal pattern learning, trajectory relationship learning, and trajectory recovery, as shown in Fig. 1.

In the spatio-temporal pattern learning component, we first construct a global transition graph based on the input trajectory data and then use Graph Convolutional Networks (GCNs) to capture the spatial mobility patterns among all trajectories. Specifically, the original trajectory embeddings, $\{T^i\}$ (the embeddings of historical trajectories) and T (the embedding of the trajectory to be recovered), are converted into $\hat{T}^i$ and $\hat{T}$, respectively, by GCN, where $1 \leq i \leq N$ and N denotes the number of historical trajectories. After that, we design a temporal embedding layer to learn the individual temporal pattern, which takes the periodicity and repeatability of human mobility into account. The trajectory relationship learning component includes two self-attention layers, a window-based cross-attention layer, and an information aggregation layer, which are to fully explore the relationship between intra- and inter-trajectories. Finally, in the trajectory recovery component, we adopt a dot-product operation between the embedding of the missing location e_f and all the candidate locations E^T to predict the missing location.

3.2 Spatio-Temporal Pattern Learning

In this section, we devise a Spatio-temporal Pattern Learning module, where a global transition graph and GCNs are adopted to capture the spatial mobility patterns at a global level, and a temporal embedding layer is developed to learn the individual temporal pattern.

Global Transition Graph. We first construct a global transition graph that contains the spatial mobility patterns among all the trajectories. Specifically, an empty graph G is firstly constructed. Then, we loop all the trajectories once. In particular, for each trajectory, we sequentially retrieve all its locations, where an edge between the current location and the next location will be added to G unless one of these locations is marked as missing. Accordingly, an adjacency matrix of the constructed graph is obtained, where each entry is the weight of the two particular locations. An example of global transition graph construction is illustrated in Fig. 2. The top of this figure presents four incomplete trajectories,

$\{T_1, T_2, T_3, T_4\}$ in the training set, in which each grid denotes a location in a time slot and the grid with dotted line represents a missing location. The middle part in Fig. 2 is the directed graph constructed based on the four incomplete trajectories. The number of trajectories along an edges is the weight of the edge. For example, the weight of the edge from location l_1 to location l_2 is 1 since only one trajectory (i.e., T_1) has the transition from l_1 to l_2. The weight of the edge from l_3 to l_5 is 3 since trajectories T_2, T_3, and T_5 have the transition from l_3 to l_5. A larger weight denotes a stronger relationship between the connected locations. An adjacency matrix of the constructed graph is shown at the bottom of this figure, in which each entry is the weight of the two particular locations and the values of empty grids equal to zero.

Graph Convolutional Networks. After the global transition graph is constructed, vanilla GCNs [13] are used to encode the spatial mobility patterns into embeddings of locations. Specifically, supposing that the adjacency matrix of G is denoted as $\boldsymbol{A} \in \mathbb{R}^{N \times N}$, we firstly apply symmetric normalization, which converts A to $\hat{\boldsymbol{A}} = \boldsymbol{D}^{-\frac{1}{2}} \tilde{\boldsymbol{A}} \boldsymbol{D}^{-\frac{1}{2}}$, where $\tilde{\boldsymbol{A}} = \boldsymbol{A} + \boldsymbol{I}_N$, $\boldsymbol{I}_N$ is the identity matrix, and $D_{ii} = \sum_j \tilde{A}_{jj}$. After that, a matrix form of graph convolution operation can be formulated as follows:

$$E_g = ReLU(\hat{\boldsymbol{A}} \boldsymbol{E} \boldsymbol{W}) \tag{1}$$

where $\boldsymbol{E}$ denotes the embeddings of all the locations, $\boldsymbol{W} \in \mathbb{R}^{d \times d}$ is a learnable weight matrix, and $ReLU$ is a non-linear activation function. Then, we concatenate $\boldsymbol{E}$ and $\boldsymbol{E}_g$ and denote the embeddings of locations after concatenating as $\hat{\boldsymbol{E}}$, i.e., $\hat{\boldsymbol{E}} = Cat(\boldsymbol{E}_g, \boldsymbol{E})$.

Temporal Embedding Layer. Due to the fact that proposed TRILL does not contain any recurrent or convolutional neural networks, we borrow the idea of Transformer [26] to integrate spatial-temporal dependency. Specifically, for each time slot t, its embedding is generated as follows:

$$\begin{aligned} e_t(2k) &= sin(t/10000^{2k/2d}) \\ e_t(2k+1) &= cos(t/10000^{2k/2d}) \end{aligned} \tag{2}$$

where k presents the k-th dimension and $\boldsymbol{e}_t \in \mathbb{R}^{2d}$ is the embedding of time slot t.

To fuse spatial and temporal information, we adopt a simple strategy that calculates the sum of time and location embeddings. Taking the j-th historical trajectory as example, we suppose that $T^j = (p_0^j, p_1^j, \cdots, p_n^j)$ and $T_t^j = (t_0^j, t_1^j, \cdots, t_n^j)$ are the j-th historical location sequence and the corresponding time slot sequence, respectively, where t_j is the time slot of p_j. By looking up $\hat{\boldsymbol{E}}$, this trajectory can be converted into vectors, $\hat{\boldsymbol{T}}^j = (\hat{\boldsymbol{p}}_0^j, \hat{\boldsymbol{p}}_1^j, \cdots, \hat{\boldsymbol{p}}_n^j) \in \mathbb{R}^{n \times 2d}$. Based on that, the fusion operation can be expressed.

$$\tilde{\boldsymbol{p}}_i^j = \hat{\boldsymbol{p}}_i^j + \boldsymbol{e}_i^j \tag{3}$$

where $\tilde{\boldsymbol{p}}_i^j$ is the fusion representation of location p_i^j and time slot t_i^j. We denote the fusion embedding of j-th historical trajectory as $\tilde{\boldsymbol{T}}^j$.

3.3   Trajectory Relationship Learning

Masked Self-Attention Layer. Previous studies such as [23] only focus on the missing locations in current trajectory and ignore the fact that there are missing locations in historical trajectories, which will affect the performance of trajectory recovery. Taking this into consideration, we learn the appropriate representations of each location including the missing locations in trajectories using a Masked Multi-head Self-Attention (MMSA) mechanism by exploring the intra-trajectory information, which can be expressed as follows:

$$MSA^h(\tilde{\boldsymbol{T}}^j) = softmax(\frac{\boldsymbol{M}^j \otimes ((\boldsymbol{W}_q^h \tilde{\boldsymbol{T}}^j{}^T)(\boldsymbol{W}_k^h \tilde{\boldsymbol{T}}^j))}{\sqrt{d}})(\boldsymbol{W}_v^h \tilde{\boldsymbol{T}}^j)$$

$$MMSA(\tilde{\boldsymbol{T}}^j) = \boldsymbol{W}_s Cat([MSA^1(\tilde{\boldsymbol{T}}^j), ..., MSA^m(\tilde{\boldsymbol{T}}^j)]) \tag{4}$$

where $\boldsymbol{W}_q^h \in \mathbb{R}^{2d \times 2d}$, $\boldsymbol{W}_k^h \in \mathbb{R}^{2d \times 2d}$, $\boldsymbol{W}_v^h \in \mathbb{R}^{2d \times 2d}$, and $\boldsymbol{W}_s \in \mathbb{R}^{2d \times 2md}$ are the trasformation matrices of head h, m is the number of heads, $softmax$ is used to normalize the values of each row, $\otimes$ is the element-wise product and $\boldsymbol{M}^j \in \mathbb{R}^{n \times n}$ is an binary indicator matrix used to presents the non-missing locations, i.e., $\forall_i \boldsymbol{M}^j[i,:] = \boldsymbol{m}^j$, where $\boldsymbol{m}^j[k]$ is 1 iff k-th location of $\tilde{\boldsymbol{T}}^j$ is non-missing. We denote the embedding of the j-th historical trajectory after MMSA as $\boldsymbol{T}_s^j$.

Similar to historical trajectories, we apply the same operations for the current trajectory, i.e., adding temporal embedding and applyling self-attention that does not share weights with self-attention of historical trajectories, to update embeddings by capturing the sequential context information in intra-trajectory. We denote the embedding of current trajectory after MMSA as $\boldsymbol{T}_s$.

Window-Based Cross-Attention Layer. A previous study [23] shows that there are shifting periodicity in human mobility, which means that a person may do the same thing, e.g., going to company, in different time slots of two days. [23] solves this problem by considering all the information from historical trajectories. However, intuitively, shifting periodicity will only happen in the near time slots. For example, a user usually arrives at company at 9:00 a.m. But for some reasons, such as traffic jams or getting up late, the user arrives at company at 9:40 a.m. today. If the user does not go to work at weekend, the behavior pattern of the user should change accordingly, in which modeling this pattern should not be expected unless it exists in training dataset. Based on this observation, we believe that aggregation information from all the time slots in historical trajectories will lead to lots of noise.

To solve this problem, a Window-based Cross-Attention (WCA) mechanism is proposed. Specifically, we constrain the information from historical trajectories to current trajectory in a window. Taking the j-th historical trajectory and current trajectory as example, WCA can be formally expressed by the following equations.

$$\boldsymbol{W}_{att}^c = mask(\frac{(\boldsymbol{W}_{qc}\boldsymbol{T}_s{}^T)(\boldsymbol{W}_{kc}\boldsymbol{T}_s^j)}{\sqrt{d}})$$

$$WCA^h(\boldsymbol{T}_s, \boldsymbol{T}_s^j) = softmax(\boldsymbol{W}_{att}^c)(\boldsymbol{W}_{vc}\boldsymbol{T}_s^j) \tag{5}$$

where $\boldsymbol{W}_{qc} \in \mathbb{R}^{2d \times 2d}$, $\boldsymbol{W}_{kc} \in \mathbb{R}^{2d \times 2d}$, $\boldsymbol{W}_{vc} \in \mathbb{R}^{2d \times 2d}$ are the transformation matrices, and $mask(\boldsymbol{M})$ is used to set the values, where the distance to diagonal is larger than the window size w, to a minimum, i.e., $-1e10$. Therefore, the weights of attention that are not contained in the window, will be set to zero by $softmax$. Similar to self-attention above, we adopt multi-head to stabilize the training process and improve the performance for trajectory recovery.

Moreover, as analyzed above, AttnMove [31] fails to sufficiently explore the relationship between current and historical trajectories due to their naive fusion operation. So in TRILL, to fully explore the information, historical trajectories are not fused before inputting into the model. We leverage this module to explore the information among the current and historical trajectories. We denote $\boldsymbol{T}_c^j$ as the output of window-based cross-attention layer by inputting the j-th historical trajectory and current trajectory. For all historical trajectories, we can get a set of representations, i.e., $\{\boldsymbol{T}_c^1, \boldsymbol{T}_c^2, ..., \boldsymbol{T}_c^N\}$, in which $\forall_j \boldsymbol{T}_c^j$ contains the information of the current trajectory and the j-th historical trajectory.

Information Aggregation Layer. To leverage all the historical information at the same time, an information aggregation layer is proposed, in which the representation of the current trajectory, i.e., $\boldsymbol{T}_s$, is treated as an anchor to sum all the representations of histories based on the similarity between the current trajectory and historical trajectories. This layer can be denoted as follows:

$$\alpha_i^j = \boldsymbol{q}^T \sigma(\boldsymbol{W}_c \boldsymbol{t}_{si} + \boldsymbol{W}_h \boldsymbol{t}_{si}^j)$$

$$\boldsymbol{e}_{f_i} = \boldsymbol{W}_f Cat([\boldsymbol{t}_{si}, \sum_{i=1}^{N} \alpha_i^j \boldsymbol{t}_{si}^j]) \qquad (6)$$

where $\boldsymbol{W}_c \in \mathbb{R}^{2d \times 2d}$, $\boldsymbol{W}_h \in \mathbb{R}^{2d \times 2d}$, $\boldsymbol{q}^T \in \mathbb{R}^{1 \times 2d}$, $\boldsymbol{W}_f \in \mathbb{R}^{d \times 2d}$ are trainable parameters, and $\boldsymbol{e}_{f_i}$ is the final representation of i-th point in the current trajectory, which is generated by fully exploring the information from current and historical trajectories.

Similar to a previous study [31], cross-entropy loss is used to train our model, which can be denoted as follows:

$$\boldsymbol{p} = softmax(\boldsymbol{e}_{f_i} \boldsymbol{E}^T)$$

$$l = \sum_{k=1}^{D} \sum_{i \in S} \sum_{l=1}^{M} y_l \log(p_l) \qquad (7)$$

where D is the size of dataset, i.e., the number of current trajectory and historical trajectories pairs, S represents a set of missing locations in the current trajectory, M is the number of candidate locations, y_l equals to 1 iff the l-th location is the missing point, and p_l is the l-th value of $\boldsymbol{p}$, which represents the probability of the l-th location to be the missing point.

3.4   Trajectory Recovery

To recover the missing locations in trajectories, we follow the prior studies [31] to define the recovery function. Suppose that the representation of missing location

Table 1. Statistics of Datasets

Dataset	City	Duration	#Users	#Loc	#Traj Pair
Geolife	Beijing	5 years	83	1124	3912
Foursquare	Tokyo	11 months	841	1411	2286

l_i in current trajectory is $e_{f_i} \in \mathbb{R}^d$ and $E \in \mathbb{R}^{M \times d}$ denotes the embeddings of candidate locations M. The missing location can be identified as follows.

$$l_i = \arg\max\{e_i E^T\} \tag{8}$$

where $\arg\max$ is a function to find the index of the maximum value in the vector, which means that we aim to find the location whose embedding is the most similar to e_{f_i} based on Euclidean distance.

4   Experiments

In this section, we present the datasets and competitors, followed by the experimental results and the corresponding analysis.

4.1   Datasets

Geolife. [39] This dataset is collected in the Geolife project by Microsoft Research Asia. It contains trajectories generated by 182 users from April 2007 to August 2012.

Foursquare. [32] This dataset is collected from Foursquare API from April 2012 to February 2013. Following previous work [23], we normalize the timestamp into one week while keeping the original order of trajectories.

Pre-processing. To represent the locations (including longitude and latitude) from GPS data, we adopt a simple strategy that is used commonly in spatial data analytics [11,16,34]. Specifically, we partition the space into cells of equal size, i.e., $0.25\,\mathrm{km}^2$, for both two datasets, and treat each cell as a location. All GPS points falling into the same cell are then mapped to the same location. However, our proposed model is not dependent on this grid-splitting way and can handle other splitting methods, e.g., semantic splitting method [31]. Following the previous studies [3,31], we set the interval of time slot as 30 min. Moreover, we filter out the locations appearing less than 5 times and trajectories with less than 36 locations. The statistics of two datasets after pre-processing are summarized in Table 1.

4.2   Baselines

We compare the proposed model with eight representative baselines.

TOP [17]. In TOP, the most popular locations in the training set are used as the recovery for each user.

Markov [9]. Markov is a widely-used method for human mobility prediction, which regards all the visited locations as states and builds a transition matrix capturing the first order transition probabilities between them.

PMF [20]. PMF is a conventional collaborative filtering method based on a user location matrix and probabilistic matrix factorization.

LSTM [17]. LSTM is a deep-learning-based model, in which historical locations are fed into LSTM to predict the missing locations in the next time slot.

BiLSTM [35]. BiLSTM extends LSTM by introducing a bidirectional LSTM and considers the spatial-temporal constraints among locations.

DeepMove [8]. DeepMove jointly models user preferences and spatial-temporal dependency for the next location prediction. We use the prediction results for recovery.

AttnMove [31]. AttnMove considers abundant information from historical trajectories, in which various attention mechanisms are adopted to model the patterns of human mobility.

PeriodicMove [23]. PeriodicMove is the latest state-of-the-art trajectory recovery model, which takes many factors, such as complex transition patterns among locations and periodicity of human mobility, into consideration.

4.3   Experimental Settings

To evaluate performance of the above methods, we randomly mask some non-missing locations, i.e., 10 locations in the current trajectory, as ground-truth to recover followed by PeriodicMove [23]. We sort each user's trajectories by time and take the first 60% as the training set, the following 20% as the validation set, and the rest as the testing set. We employ three widely-used metrics, Recall@K, Dis@K [31], and Mean Average Precision (MAP) [28] to evaluate performance, where Dis@K is the smallest geographical distance between the centers of locations in the top-K ranked list and the ground truth location. The smaller the value of Dis@K is, the better the performance will be.

For the hyper-parameters of the proposed model, we set hidden dimension d, number of heads m, window size, learning rate, maximal running epochs, and batch size to 128, 4, 6, 1e–3, 200, and 50, respectively. To update training parameters of the proposed model, Adam optimizer with cosine annealing learning rate scheduler is adopted. To prevent model from overfitting, we apply a dropout mechanism after the information aggregation layer with drop rate 0.3, and use an early stopping strategy with patience 15, i.e., the training procedure stops when the Recall@1 on the validation set does not improve for 15 consecutive epochs. For deep-learning-based baselines, we set the same hidden dimension and batch size to ours, and keep other hyper-parameters in line with the original papers.

Table 2. Overall Performance Comparison in Terms of Recall@K, Dis@K, and MAP

		Recall@1	Recall@5	Recall@10	Dis@1	Dis@5	Dis@10	MAP
Geolife	Top	0.1148	0.2451	0.3166	7863	6259	5176	0.1812
	Markov	0.1417	0.3263	0.3974	6909	4974	4259	0.2304
	PMF	0.1941	0.3436	0.4059	6506	4389	3555	0.2752
	LSTM	0.2086	0.3917	0.4720	6318	3928	3068	0.2965
	BiLSTM	0.2285	0.4538	0.5773	6209	3620	2255	0.3298
	DeepMove	0.3045	0.5380	0.6371	5370	2052	1358	0.4131
	AttnMove	0.3920	0.6696	0.7213	5342	2007	975	0.5046
	PeriodicMove	0.4199	0.6893	0.7681	4209	1443	863	0.5385
	TRILL	**0.4721**	**0.7563**	**0.8364**	**3484**	**1112**	**603**	**0.5985**
	%Improv.	**12.4%**	**9.7%**	**8.8%**	**17.2%**	**22.9%**	**30.1%**	**11.1%**
Foursquare	Top	0.0865	0.1673	0.2268	8427	4919	3483	0.1347
	Markov	0.1090	0.2010	0.2575	8345	4402	3125	0.1792
	PMF	0.1215	0.2468	0.2887	8116	3971	3229	0.2358
	LSTM	0.1393	0.2540	0.3143	7913	3804	2801	0.2519
	BiLSTM	0.2323	0.3968	0.4703	6206	2745	1849	0.3154
	DeepMove	0.2612	0.4631	0.5337	5189	2648	1649	0.3789
	AttnMove	0.2975	0.5172	0.5746	4942	2396	1482	0.4078
	PeriodicMove	0.3125	0.5534	0.6264	4704	1758	1197	0.4245
	TRILL	**0.3227**	**0.5636**	**0.6372**	**4639**	**1650**	**1074**	**0.4341**
	%Improv.	**3.2 %**	**1.8%**	**1.7%**	**1.3%**	**6.1%**	**10.2 %**	**2.2%**

4.4 Overall Performance

We report the overall performance in Table 2, through which we have the following observations.

1) The methods that only model the first order information among locations, i.e., TOP, Markove, and PMF, get the worst performance compared with others on both datasets.

2) The LSTM-based models, i.e., LSTM, BiLSTM, and DeepMove, perform worse than AttnMove and PeriodicMove. Among the LSTM-based models, DeepMove equipped with an attention mechanism is slightly better than others.

3) AttnMove and PeriodicMove, which are purely based on an attention module, beat other baselines in terms of all the metrics on two datasets. This finding further verifies that attention modules have a stronger ability to deal with missing locations compared with vanilla LSTM.

4) Taking a series of factors (e.g., shifting periodicity and sparsity of trajectories) into account, PeriodicMove achieves better performance over AttnMove, which proves that considering the intrinsic characteristics of human mobility is helpful for missing location recovery.

5) The proposed model, TRILL, outperforms all the baselines for all evaluation metrics on both datasets with a large margin especially in Geolife dataset. Specifically, recall of TRILL outperforms the best baseline (i.e., PeriodicMove) by 8.8%~12.4% on Geolife and 1.7%~3.2% on Foursquare, respectively. In terms of Dis@K, TRILL outperforms PeriodicMove by 17.2%~30.1% and 1.3%~10.2% on the two datasets. Moreover, TRILL gets the best performance in terms of MAP, which achieves a large margin compared with PeriodicMove on both datasets. These great improvements indicate the effectiveness of TRILL in trajectory recovery.

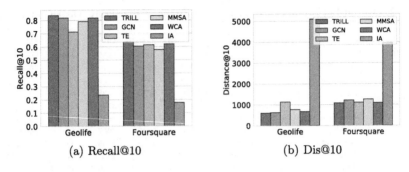

(a) Recall@10   (b) Dis@10

Fig. 3. Ablation study

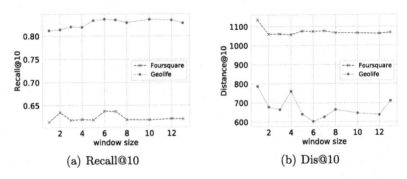

(a) Recall@10   (b) Dis@10

Fig. 4. The effects of window size on both datasets

4.5 Ablation Study

We analyze the effects of each component of the proposed model and conduct an ablation study on two datasets by removing each component from TRILL. Specifically, we evaluate five TRILL variants: 1) **GCN** that removes the graph convolutional networks, where a linear layer is added to enlarge the dimension of trajectory representation to $2d$; 2) **TE** that removes the temporal embedding layer; 3) **MMSA** that removes the self-attention layer; 4) **WCA** that removes the window-based cross attention layer; 5) **IA** that replaces the information aggregation layer with arithmetic mean. The results are shown in Fig. 3. We can see that TRILL outperforms all the variants, which indicates the importance of each component. Moreover, the performance degrades after removing the GCNs, which demonstrates that modeling the mobility patterns among trajectories is useful for trajectory recovery. Besides, MSSA and WCA have poor performance, which shows the effectiveness of sufficiently exploring the relationship among intra- and inter-trajectories. The performance of IA drops significantly, which empirically illustrates that various noises exist in historical trajectories, and treating them equally is unwise.

4.6 Sensitivity Analysis

We analyze the sensitivity of window sizes in cross attention. Figure 4 shows the results of Recall@10 and Dis@10 by increasing the window sizes on both datasets. We can see that TRILL achieves the best Recall@10 when the window size is set to 6, which indicates that too conservative (small) and too aggressive (large) strategies are not suitable. This is because when the window size is too small, there is no sufficient information to recover the missing locations. When the window size is too large, the irrelevant locations will introduce a great deal of noises. Both of these situations will decrease the performance of human mobility recovery.

4.7 Robustness Analysis

To demonstrate the robustness of the proposed model, we compare TRILL with two best-performing baselines, i.e., AttnMove and PeriodicMove, by varying dropping ratios on Geolife. Specifically, we increase the percentage of missing locations in historical trajectories from 20% to 80%. The results are demonstrated in Table 3.

Table 3. Performance w.r.t. Dropping Ratios on Geolife

Missing Rate		20%	40%	60%	80%
AttnMove	Recall@10	0.7117	0.6985	0.6785	0.6160
	Dist@10	987	1037	1174	1371
	MAP	0.4815	0.4657	0.4226	0.4112
PeriodicMove	Recall@10	0.7451	0.7392	0.7186	0.6857
	Dist@10	884	954	1059	1176
	MAP	0.5175	0.4750	0.4413	0.4076
TRILL	Recall@10	0.8216	0.8038	0.7627	0.7436
	Dist@10	682	720	915	1089
	MAP	0.5760	0.5534	0.5111	0.5044

From the results, we can see that with the increase of missing ratio, the performances of all models drop. However, TRILL consistently outperforms the two baselines since TRILL not only captures the mobility pattern among trajectories with the transition graph and GCNs, but also fully explores the information between intra- and inter-trajectories through a self-attention layer and a window-based cross-attention layer. The improvements in robustness experiment demonstrate the effectiveness of the proposed model again.

5 Conclusion

We propose and offer a neural attention model based on graph convolutional networks, called TRILL, which aims to recover the missing locations for trajectories. In order to model spatio-temporal patterns among trajectories, we use GCNs to capture the spatial mobility patterns among all trajectories and design a temporal embedding layer to learn the individual temporal pattern for each trajectory. To sufficiently explore the relationship between the current trajectory and historical ones, a self-attention layer, a window-based cross-attention layer and an information aggregation layer are proposed to learn the relationship between intra- and inter-trajectories. An extensive empirical study with real datasets demonstrates the superiority of TRILL compared with the state-of-the-art methods and the effectiveness of each component in TRILL.

Acknowledgements. This work is partially supported by NSFC (No. 61972069, 61836007 and 61832017), Shenzhen Municipal Science and Technology R&D Funding Basic Research Program (JCYJ20210324133607021), and Municipal Government of Quzhou under Grant No. 2022D037.

References

1. Abu-El-Haija, S., Kapoor, A., Perozzi, B., Lee, J.: N-GCN: multi-scale graph convolution for semi-supervised node classification. In: UAI (2019)
2. Alwan, L.C., Roberts, H.V.: Time-series modeling for statistical process control. J. Bus. Econ. Stat. **6**(1), 87–95 (1988)
3. Chen, G., Viana, A.C., Fiore, M., Sarraute, C.: Complete trajectory reconstruction from sparse mobile phone data. EPJ Data Sci. **8**(1), 1–24 (2019). https://doi.org/10.1140/epjds/s13688-019-0206-8
4. Chen, M., Zuo, Y., Jia, X., Liu, Y., Yu, X., Zheng, K.: CEM: a convolutional embedding model for predicting next locations. TITS **22**, 3349–3358 (2021)
5. Cho, E., Myers, S.A., Leskovec, J.: Friendship and mobility: user movement in location-based social networks. In: KDD (2011)
6. Devlin, J., Chang, M.W., Lee, K., Toutanova, K.: BERT: pre-training of deep bidirectional transformers for language understanding. arXiv preprint arXiv:1810.04805 (2019)
7. Dosovitskiy, A., et al.: An image is worth 16x16 words: transformers for image recognition at scale. arXiv preprint arXiv:2010.11929 (2021)
8. Feng, J., et al.: Deepmove: predicting human mobility with attentional recurrent networks. In: WWW (2018)
9. Gambs, S., Killijian, M.O., del Prado Cortez, M.N.: Next place prediction using mobility Markov chains. In: MPM 2012 (2012)
10. González, M.C., Hidalgo, C.A., Barabasi, A.L.: Understanding individual human mobility patterns. Nature **453**(7196), 779–782 (2008)
11. Güting, R.H., Schneider, M.: Realm-based spatial data types: the ROSE algebra. VLDB J. **4**, 243–286 (2005)
12. Keikha, M.M., Rahgozar, M., Asadpour, M.: Community aware random walk for network embedding. arXiv preprint arXiv:1710.05199 (2018)
13. Kipf, T., Welling, M.: Semi-supervised classification with graph convolutional networks. arXiv preprint arXiv:1609.02907 (2017)
14. Li, G., Xiong, C., Thabet, A.K., Ghanem, B.: DeeperGCN: all you need to train deeper GCNs. arXiv preprint arXiv:2006.07739 (2020)
15. Li, L., Li, Y., Li, Z.: Efficient missing data imputing for traffic flow by considering temporal and spatial dependence. Transp. Res. Part C Emerg. Technol. **34**, 108–120 (2013)
16. Li, X., Zhao, K., Cong, G., Jensen, C.S., Wei, W.: Deep representation learning for trajectory similarity computation. In: ICDE, pp. 617–628 (2018)

17. Liu, Q., Wu, S., Wang, L., Tan, T.: Predicting the next location: a recurrent model with spatial and temporal contexts. In: AAAI (2016)
18. Moritz, S., Bartz-Beielstein, T.: imputeTS: time series missing value imputation in R. R J. **9**, 207 (2017)
19. Perozzi, B., Al-Rfou, R., Skiena, S.: Deepwalk: online learning of social representations. In: KDD (2014)
20. Salakhutdinov, R., Mnih, A.: Probabilistic matrix factorization. In: NIPS (2007)
21. Schneider, C., Belik, V., Couronné, T., Smoreda, Z., González, M.C.: Unravelling daily human mobility motifs. J. R. Soc. Interface **10** (2013)
22. Su, H., Cong, G., Chen, W., Zheng, B., Zheng, K.: Personalized route description based on historical trajectories. In: CIKM (2019)
23. Sun, H., Yang, C., Deng, L., Zhou, F., Huang, F., Zheng, K.: Periodicmove: shift-aware human mobility recovery with graph neural network. In: CIKM (2021)
24. Sun, K., Qian, T., Chen, T., Liang, Y., Nguyen, Q.V.H., Yin, H.: Where to go next: modeling long- and short-term user preferences for point-of-interest recommendation. In: AAAI (2020)
25. Tang, J., Qu, M., Wang, M., Zhang, M., Yan, J., Mei, Q.: Line: Large-scale information network embedding. In: WWW (2015)
26. Vaswani, A., et al.: Attention is all you need. arXiv preprint arXiv:1706.03762 (2017)
27. Velickovic, P., Cucurull, G., Casanova, A., Romero, A., Lio', P., Bengio, Y.: Graph attention networks. arXiv preprint arXiv:1710.10903 (2018)
28. Wang, J., Wu, N., Lu, X., Zhao, W.X., Feng, K.: Deep trajectory recovery with fine-grained calibration using Kalman filter. TKDE **33**, 921–934 (2021)
29. Wu, S., Tang, Y., Zhu, Y., Wang, L., Xie, X., Tan, T.: Session-based recommendation with graph neural networks. arXiv preprint arXiv:1811.00855 (2019)
30. Xi, D., Zhuang, F., Liu, Y., Gu, J., Xiong, H., He, Q.: Modelling of bi-directional spatio-temporal dependence and users' dynamic preferences for missing poi check-in identification. In: AAAI (2019)
31. Xia, T., et al.: Attnmove: history enhanced trajectory recovery via attentional network. In: AAAI (2021)
32. Yang, D., Zhang, D., Zheng, V.W., Yu, Z.: Modeling user activity preference by leveraging user spatial temporal characteristics in LBSNs. IEEE Trans. Syst. Man Cybern.: Syst. **45**, 129–142 (2015)
33. Zhan, X., Zheng, Y., Yi, X., Ukkusuri, S.V.: Citywide traffic volume estimation using trajectory data. TKDE **29**(2), 272–285 (2016)
34. Zhang, Y., Liu, A., Liu, G., Li, Z., Li, Q.: Deep representation learning of activity trajectory similarity computation. In: ICWS, pp. 312–319 (2019)
35. Zhao, J., Xu, J., Zhou, R., Zhao, P., Liu, C., Zhu, F.: On prediction of user destination by sub-trajectory understanding: a deep learning based approach. In: CIKM (2018)
36. Zhao, P., Xu, X., Liu, Y., Sheng, V., Zheng, K., Xiong, H.: Photo2Trip: exploiting visual contents in geo-tagged photos for personalized tour recommendation. In: MM (2017)
37. Zhao, Y., Shang, S., Wang, Y., Zheng, B., Nguyen, Q.V.H., Zheng, K.: Rest: a reference-based framework for spatio-temporal trajectory compression. In: SIGKDD, pp. 2797–2806 (2018)
38. Zheng, Y., Capra, L., Wolfson, O., Yang, H.: Urban computing: concepts, methodologies, and applications. TIST **5**, 38:1–38:55 (2014)
39. Zheng, Y., Xie, X., Ma, W.Y.: GeoLife: a collaborative social networking service among user, location and trajectory. IEEE Data Eng. Bull. **33**, 32–39 (2010)

Explicit Assignment and Dynamic Pricing of Macro Online Tasks in Spatial Crowdsourcing

Lin Sun[1], Yeqiao Hou[1], and Zongpeng Li[1,2(✉)]

[1] School of Computer Science, Wuhan University, Wuhan, China
{lin.sun,houyeqiao}@whu.edu.cn
[2] Insititue for Network Sciences and Cyberspace, Tsinghua University, Beijing, China
zongpeng@tsinghua.edu.cn

Abstract. The past decade has witnessed rapid development of wireless communication and ubiquitous availability of mobile devices. Spatial crowdsourcing (SC), which requires workers to physically move to specific locations to perform tasks, has gained unprecedented popularity. A typical class of SC tasks is macro tasks, which demand the cooperation of multiple workers with specific skills. Existing studies on macro task assignment often generate an assignment strategy where worker skill set can fully cover the required skills, leading to ambiguous division of labor among workers and unreasonable static distribution of rewards. This work studies the explicit assignment and dynamic pricing (EADP) problem of macro tasks, and considers both online explicit task assignment (skill used and realtime workload) and dynamic reward distribution. Aiming to maximize social welfare of the SC system, we develop a pricing mechanism based on primal-dual technique, and propose an online batch-based framework coupled with an effective dual subroutine to assign macro SC tasks. Through theoretical analysis, we prove that the pricing mechanism is truthful and individually rational, and our online framework guarantees a good parameterized competitive ratio, as well as polynomial time complexity. Extensive experiments demonstrate our algorithms' effectiveness and efficiency.

Keywords: Spatial Crowdsourcing · Explicit Task Assignment · Dynamic Pricing · Primal-Dual

1 Introduction

With the proliferation of mobile devices and the development of sharing economy, spatial crowdsourcing (SC) has drawn significant attention from academia and industry. A common SC task processing flow is based on server-assigned tasks (SAT) mode [6]. Commercial SC platforms, *e.g.*, Gigwalk[1] and gMission[2], collect

---

[1] https://www.gigwalk.com
[2] https://gmission.github.io/

X. Wang et al. (Eds.): DASFAA 2023, LNCS 13943, pp. 464–480, 2023.
https://doi.org/10.1007/978-3-031-30637-2_30

SC tasks from task requesters, assign workers to move to specific locations to perform tasks, and distribute rewards to them for doing the labor.

Certain spatial tasks, such as traffic monitoring (Waze[3]) and flash delivery (Uber Eats[4]), are quite straightforward and rely solely on vehicles and sensing devices (denoted as "micro tasks"). Most existing studies on task assignment focus on micro tasks [4,13,15,16,19]. In contrast, some tasks (referred to as "macro tasks") are more complex and demand multiple professional workers with specialized skills to accomplish, as exemplified by cleaning after house decoration, holding a ceremony, making a move, and yard maintenance. For example, if a task requester publishes a task of holding a ceremony on gMission, drivers, porters, stage designers, and light and sound technicians may be needed. If the task is admitted, the platform will assign workers with specific skills to perform the task, charge from the requester and distribute rewards to workers reasonably. Thus, how to appropriately handle complex macro tasks, including worker allocation and reward distribution, is a natural yet challenging problem.

Firstly, in practice, both tasks and workers appear dynamically and stochastically on the SC platform, demanding immediate response, which implies that the task assignment mechanism needs to be globally online [13]. Secondly, a rational task assignment should ensure the interests of the SC platform and the assigned workers, which requires a well-designed pricing mechanism. In the context of SC, such mechanism should further consider the spatio-temporal characteristics [14]. The reward of workers should dynamically reflect the supply and demand of workers in the SC market, calling for dynamic pricing. Thirdly, a variety of fundamental constraints, including skill, budget, time, location, *etc.*, should be taken into account while making the allocation decision. Fourthly, the assignment strategies for macro tasks need to be explicit, *i.e.*, the skill used by the worker and the corresponding workload should be determined, which ensures well-organized task execution and high-quality task completion. This is also the basis for the pricing mechanism, since skills provided by workers and the workload are central indicators for computing worker rewards.

To this end, we consider a basic problem in the SC system, namely Explicit Assignment and Dynamic Pricing (EADP) for macro online tasks, which considers both explicit task assignment and dynamic pricing mechanism design under the constraints of time, skill, and budget. We further illustrate the EADP problem through a motivating example.

Example 1. Assume that the SC platform includes two tasks $\{j_1, j_2\}$ and four available workers $\{i_1, i_2, i_3, i_4\}$, as shown in Fig. 1. Note that workload is measured in number of time units. For instance, task j_1's first subtask (skill s_1) takes six time slots to accomplish.

Take the assignment of task j_1 for example. For simplicity, we make the following assumptions: (i) task j_1 and all workers emerge simultaneously; (ii) all workers can arrive at the location of j_1 prior to any task's start time; and (iii) all

---

[3] https://www.waze.com

[4] https://www.ubereats.com

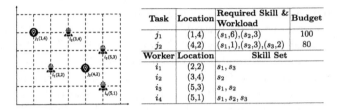

Task	Location	Required Skill & Workload	Budget
j_1	(1,4)	$(s_1,6),(s_2,3)$	100
j_2	(4,2)	$(s_1,1),(s_2,3),(s_3,2)$	80
Worker	**Location**	**Skill Set**	
i_1	(2,2)	s_1,s_3	
i_2	(3,4)	s_2	
i_3	(5,3)	s_1,s_2	
i_4	(5,1)	s_1,s_2,s_3	

Fig. 1. An example of the EADP problem.

workers' expiration time is later than j_1's completion deadline, which naturally guarantees that assigned workers are workable for any task duration time.

Concerning skill constraints, the first subtask of j_1 requires skill s_1, and workers $\{i_1,i_3,i_4\}$ are available for selection. The candidate worker set for the second subtask (skill s_2) is $\{i_2,i_3,i_4\}$. For time constraints, all subtasks must be done before j_1's deadline. Suppose j_1 starts to be performed at the third time slot and is required to finish by the fifth time slot. As a result, the first subtask of j_1 can only be assigned to all the available workers $\{i_1,i_3,i_4\}$ to ensure it is completed before the fifth time slot. Furthermore, the SC platform and workers need to be profitable. Specifically, j_1's budget should cover the total rewards of its workers, and each worker's reward ought to cover the transportation cost to j_1's location.

Motivated by the example above, we formalize the EADP problem, which makes explicit online decisions on macro task assignments and dynamically decides worker rewards respecting time, skill, and budget constraints, to maximize social welfare of the SC systems.

In this work, we first formulate the EADP problem into a combinatorial optimization problem, and translate it into an integer linear program (ILP). Then, we design an online pricing mechanism to reward workers, which is budget-balanced and guarantees both truthfulness and individual rationality. Next, we present a general batch-based framework and a dual subroutine to tackle the EADP problem with a provable parameterized competitive ratio. In conclusion, our main contributions are as follows:

- We formulate the problem of explicit assignment and dynamic pricing for online macro tasks into a social welfare maximization problem.
- We reformulate the welfare maximization problem into an ILP, and develop a dynamic pricing mechanism reflecting the supply and demand of workers based on the primal-dual framework.
- We propose a batch-based framework and a dual subroutine to tackle the EADP problem, which can achieve a good parameterized competitive ratio.
- We conduct extensive experiments on both real and synthetic datasets and demonstrate the effectiveness and efficiency of our primal-dual algorithm.

2   Related Work

In recent years, with the proliferation of mobile devices and wireless networks, spatial crowdsourcing has gained widespread popularity. Most existing works focus on the SC task assignment problem in micro scenarios, where tasks can be completed by a single skill worker. Tong et al. [13] propose four algorithms with constant competitive ratios under global online scenarios. Some studies [4,19] consider the issue of fairness, aiming to minimize the payoff difference among workers while maximizing the average worker payoff. Recent literature [15] proposed a mechanism considering both privacy protection and single-worker-multi-task assignment. Nevertheless, these problems are all different from the EADP problem and are not applicable to macro tasks. Besides, most existing studies on the pricing mechanism design target micro tasks, such as ride sharing [2,3], citizen sensing [17,18], and thus cannot apply to our problem.

The more closely related studies are [5,7,9,11,12], which also focus on multi-skilled SC task assignments. We summarize the main differences between them and this work as follows:

1. Competitive ratio is primary metric used to assess the effectiveness of online algorithms. Our algorithms guarantee a provable parameterized competitive ratio, while other online approaches [7,9,11] lack theoretical analysis and performance guarantee in this regard.
2. The online pricing mechanism of the EADP problem is truthful, individually rational, and budget-balanced, guaranteeing the interests of the SC platform and workers. However, [5,7], and [9] do not consider the distribution of worker rewards, and [11,12] decide worker rewards statically.
3. In contrast to [5,7,9,11,12], the division of labor for each worker, including skill used and workload, is clear and unambiguous in our solution.

3   Problem Statement

In this section, we first present the formal definition of the EADP problem for macro online tasks, then formulate EADP into a mathematical optimization problem. Important notations are summarized in Table 1.

3.1   Problem Definition

Spatial Crowdsourcing System Overview. The SC system contains three types of entities: the SC platform, multi-skilled workers, and macro tasks. The SC platform hosts a group of multi-skilled online workers. Upon the arrival of a new task, the platform first decides whether to accept it and if yes, assigns workers to the task. Once allocation decision is made, the assigned workers will be offline temporarily and travel to the task's location to accomplish it. When the task is completed, the platform collects payment from the task publisher, and distributes rewards to the designated workers. When a worker completes the task assigned, he/she will log back to the platform, waiting for the next task.

Table 1. Notations

Inputs	Descriptions
W, J, S	worker set, task set and skill set
T	system time span
i, j, k, t	indices of worker, task, skill, and time slot
$l_i(l_j)$	location of worker i (task j)
$r_i(r_j)$	release time of worker i (task j)
$d_i(d_j)$	expiration time (completion deadline) of worker i (task j)
B_j	budget of task j
D_j^k	skill-k subtask's workload of task j
R_{ij}	reward of worker i for performing task j
$z_{ij}(t)$	whether worker i is performing task j at time slot t
$dist(l_i, l_j)$	distance between worker i and task j
v	global traveling velocity of workers
T_j^k	execution time of task j's subtask k
γ	average traveling cost per time slot
$c^k(t)$	num of online workers occupying skill k at time slot t
$\eta^k(t)$	num of workers who are performing tasks using skill k at time slot t
p^k	initial unit price of skill k
Decisions	Descriptions
x_j	whether task j is accepted
a_j	start execution time of accepted task j
u_{ij}	whether worker i is allocated to task j
y_{ij}^k	whether worker i works for skill-k subtask of task j

Definition 1. (Multi-Skilled Workers): The worker set, denoted as W, involves plenty of multi-skilled workers. A worker i is defined as $< r_i, d_i, l_i, S_i >$, which means worker i appears or returns to the platform at time slot r_i and plans to leave at time slot d_i. l_i indicates i's current location. Besides, worker i owns a set of skills S_i.

Definition 2. (Complex Macro Tasks): The task set, denoted as J, includes many complex macro tasks. A task j can be expressed as $< r_j, d_j, l_j, S_j, D_j, B_j >$, which implies task j with location l_j appears at time slot r_j and is expected to be completed before d_j. Besides, j has a group of subtasks, where each subtask k requires a certain skill in S_j to execute and needs to run D_j^k time slots. Moreover, each task j is associated with a budget B_j to pay the SC platform for successful execution.

Definition 3. (Social Welfare): Under the context of EADP, social welfare SW, i.e., the total utility of the SC system, is the aggregated worker payoff plus the SC platform's profit.

We define R_{ij} as the reward of worker i to perform task j. Let a binary x_j indicate whether task j is accepted by the platform ($x_j = 1$) or not ($x_j = 0$) and binary u_{ij} equals 1 if worker i is assigned to task j, and 0 otherwise. Thus the total payoff for all workers can be formulated as $\sum_{i \in [W]} \sum_{j \in [J]} x_j \cdot (u_{ij} \cdot R_{ij})$ and the SC platform's profit can be formulated as $\sum_{j \in [J]} x_j \cdot (B_j - \sum_{i \in [W]} u_{ij} \cdot R_{ij})$. SW is the sum of these two types of utility, shown as Eq. (1).

$$SW = \sum_{j \in [J]} x_j \cdot B_j \tag{1}$$

Definition 4. (EADP Problem): Over the system time span T, workers and tasks appear on the SC platform randomly. At each time slot t, the platform collects a group of online workers W_t and a set of tasks J_t waiting for assignment. Then, it schedules each task in J_t and makes the following decisions: (i) Whether task j can be admitted, denoted by x_j. (ii) When to perform task j, denoted by an integer variable a_j. (iii) Whether worker i in W_t to be allocated to task j, indicated by u_{ij}. (iv) Which subtask to perform for each assigned worker, indicated by a binary variable y_{ij}^k: if worker i is assigned to perform skill-k subtask, $y_{ij}^k = 1$; otherwise, $y_{ij}^k = 0$.

Besides, the EADP problem aims to assign multi-skilled workers to macro tasks while meeting time, skill and budget constraints to maximize social welfare (Discussed in Sect. 3.2).

3.2 Problem Formulation

Based on the above definitions, the social welfare maximization problem can be formulated as follows.

$$\max \sum_{j \in [J]} x_j \cdot B_j \tag{2}$$

subject to:

$$if \ u_{ij} \cdot y_{ij}^k = 1, then \ k \in [S_j \cap S_i], \forall j \in [J], \forall i \in [W] \tag{3}$$

$$u_{ij} \cdot \left\{ t + \frac{dist(l_i, l_j)}{v} \right\} \leq min(a_j, d_i), \forall i \in [W], \forall j \in [J] \tag{4}$$

$$u_{ij} \cdot y_{ij}^k \cdot (a_j + T_j^k) \leq min(d_i, d_j), \forall j \in [J], \forall k \in [S_j], \forall i \in [W] \tag{5}$$

$$\sum_{i \in [W]} u_{ij} \cdot y_{ij}^k \geq 1, \forall j \in [J], \forall k \in [S_j] \tag{6}$$

$$\sum_{j \in [J]} \sum_{i \in [W]} z_{ij}(t) \cdot y_{ij}^k \leq c^k(t), \forall k \in [S_j], \forall t \in [T] \tag{7}$$

$$\sum_{i \in [W]} u_{ij} \cdot R_{ij} \leq B_j, \forall j \in [J] \tag{8}$$

$$u_{ij} \cdot \left[R_{ij} - \gamma \cdot \frac{dist(l_i, l_j)}{v} \right] \geq 0, \forall i \in [W], \forall j \in [J] \tag{9}$$

$$x_j, u_{ij}, y_{ij}^k, z_{ij}(t) \in \{0, 1\}, a_j \in [T], \forall i \in [W], j \in [J], k \in [S] \tag{10}$$

Constraint (3) shows worker professionalism. Each assigned worker should use the skill that he/she possesses and the task requires. (4) ensures that workers arrival time is earlier than the task's start time and cannot exceed their own expiration time. For simplicity, we adopt Euclidean distance to measure the distance between worker i and task j, denoted as $dist(l_i, l_j)$. Global variable v represents worker velocity, and integer variable t refers to the current time slot. (5) guarantees the accepted task j can be completed before its deadline and each subtask's completion time is no later than the assigned worker's expiration time. Note that T_j^k refers to the execution time of task j's subtask k, which is the ratio of k's workload to the number of assigned workers. (6) requires each subtask of an accepted task j should be assigned at least one worker. (7) requires the number of busy workers using skill k can not exceed the number of online workers mastering skill-k (skill-k workers), denoted as $c^k(t)$.

Constraints (8) and (9) restricts the profitable of the SC platform and SC workers. (8) shows the total compensation that the SC platform needs to pay task j's workers can not exceed j's budget. Besides, for any worker i assigned to perform task j, i's reward R_{ij} needs to be sufficient to pay for the traveling cost of i to j's location, as depicted in constraint (9), where the parameter γ represents the traveling cost per time slot.

Theorem 1. *(Hardness of EADP)*. *The EADP problem is NP-hard.*

Proof. Please see our supplementary material [1].

4   Dynamic Pricing Mechanism

To efficiently solve the social welfare maximization problem (2), we need a judicious pricing mechanism that safeguards the interests of both the SC platform and the workers. In this section, we first reformulate problem (2) into an ILP, relax the ILP into LP, and then formulate its dual LP. Finally, a well-designed price function is proposed to determine worker rewards appropriately.

4.1   Problem Reformulation

The social welfare maximization problem (2) involves integer variables and non-conventional constraints (3)–(5) that are not of a packing or covering nature. To tackle these challenges, we simplify problem (2) into a packing structure. More specifically, we use I_j to denote the set of feasible schedules for task j, each of which represents the set of decisions $(x_i, a_j, u_{ij}, y_{ij}^k, \forall i \in [W], j \in [J], k \in [S_j])$ satisfying constraints (3)–(6)(8)–(10) simultaneously. Then, problem (2) can be reformulated as the following ILP (11).

$$\max \sum_{j \in [J]} \sum_{\iota \in I_j} x_{j\iota} \cdot B_j \tag{11}$$

subject to:

$$\sum_{j\in[J]}\sum_{\iota\in[I_j]} x_{j\iota} \cdot z_{ij}^{\iota}(t) \cdot y_{ij}^{k\iota} \le c^k(t), \forall k \in [S_j], i \in [W], t \in [T], \tag{12}$$

$$\sum_{\iota\in I_j} x_{j\iota} \le 1, \forall j \in [J], \tag{13}$$

$$x_{j\iota} \in \{0,1\}, \forall j \in [J], \iota \in I_j. \tag{14}$$

In the above ILP, binary variable $x_{j\iota}$ indicates whether task j is admitted and assigned according to schedule ι or not, $\forall j \in [J], \iota \in I_j$. $z_{ij}^{\iota}(t)$ represents whether worker i is performing task j at time slot t or not in schedule ι. Likewise, $y_{ij}^{k\iota}$ denotes whether worker i is assigned to the skill-k subtask of task j according to ι. Obviously, constraint (12) is equivalent to constraint (7). Constraints (13) and (14) cover other constraints in problem (2). Thus, problem (2) and (11) are equivalent, so we can obtain a feasible solution to (2) by addressing (11).

To solve ILP (11), we next apply the primal-dual technique on (11) by relaxing $x_{j\iota} \in \{0,1\}$ to $x_{j\iota} \ge 0$ and introducing dual variables $\alpha^k(t)$ and β_j to constraints (12) and (13), respectively. The dual LP of the relaxed (11) is:

$$\min \sum_{j\in[J]} \beta_j + \sum_{t\in[T]}\sum_{k\in[S]} \alpha^k(t) \cdot c^k(t) \tag{15}$$

subject to:

$$\beta_j \ge B_j - \sum_{t\in[T]}\sum_{i\in[W]}\sum_{k\in[S_j]} \alpha^k(t) \cdot z_{ij}^{\iota}(t) \cdot y_{ij}^{k\iota}, \forall j \in [J], \iota \in I_j \tag{16}$$

$$\alpha^k(t) \ge 0, \forall t \in [T], k \in [S] \tag{17}$$

$$\beta_j \ge 0, \forall j \in [J]. \tag{18}$$

If we interpret $\alpha^k(t)$ as the unit reward of a skill-k worker at time slot t, then $\sum_{t\in[T]}\sum_{i\in[W]}\sum_{k\in[S_j]} \alpha^k(t) \cdot z_{ij}^{\iota}(t) \cdot y_{ij}^{k\iota}$ is the total rewards of workers assigned to task j in schedule ι. The right hand side (RHS) of (16) is task j's budget minus the total expenses for handling task j, i.e., the profit of the SC platform. According to complementary slackness, if task j is accepted, i.e., $x_{j\iota} = 1$, the associated dual constraint (16) should be tight. Combed with constraint (18), the following should hold to minimize the dual objective: $\beta_j = \max\{0, \max_{\iota\in I_j} \text{RHS of (16)}\}$.

Hence, β_j can be finely interpreted as the revenue of the SC platform for launching task j according to the best schedule ι^*:

$$\iota^* = \arg\max_{\iota\in I_j} \text{RHS of (16)}. \tag{19}$$

4.2   Price Function Design

To obtain the optimal task assignment strategy, i.e., compute equation (19), we need to determine the unit reward of workers. Here, a thoughtful price function is presented to determine worker rewards dynamically as follows:

$$\alpha^k(t) = p^k \cdot \lambda^{\frac{\eta^k(t)}{c^k(t)}}.\tag{20}$$

We define a new variable $\eta^k(t)$ to denote the number of skill-k workers who are performing tasks at time slot t. Recall that $c^k(t)$ denotes the number of online skill-k workers at time slot t. Consequently, the exponential part of λ reflects the supply and demand relationship of the SC system to skill k. Let p^k indicate the initial unit price of skill k. In practical application, it can be set by the SC platform according to the prior knowledge of the SC market. Therefore, the marginal price of skill k starts at p^k and increases exponentially as the ratio of $\eta^k(t)$ to $c^k(t)$ increases but does not exceed $\lambda \cdot p^k$. When most online skill-k workers are busy, the marginal price of k is relatively high. When there are plenty of free skill-k workers, the marginal price is naturally low, which is more rational than static pricing. It is worth emphasizing that if the SC platform does not yet have a worker with skill k ($c^k(t) = 0$), the associated task will be rejected directly, and thus the above price function won't be employed.

5 Online Algorithm Design

In this section, we first design a basic batch-based framework to address the social welfare maximization problem in Sect. 5.1. Section 5.2 presents a dual subroutine to calculate the best assignment strategy for macro SC tasks, *i.e.*, tackle problem (19). Finally, we conduct theoretical analysis in Sect. 5.3.

5.1 Online Batch-Based Framework

Inspired by the observations of the previous section, we design a standard batch-based online framework in A_{BAF} to generate task scheduling schemes at each time slot.

In A_{BAF} (Fig. 2), the SC platform first collects tasks waiting for assignment, the online multi-skilled workers, and computes the number of online skill-k workers at time slot t (line 3). Then the marginal price of each skill is updated (line 4). During the decision process of each task, the optimal allocation is given by function Find (line 6). If task utility β_j is positive, the task will be accepted, and the decision variables are updated as well as $\eta^k(t)$ and $\alpha^k(t)$ (lines 7–12). Otherwise, if β_j is not positive, which implies the task can not be scheduled at the current time slot, it is put into the next iteration if its deadline has not expired (line 13), and is rejected otherwise (line 14).

Function Find iterates over every possible task start time bt (line 17) until it finds the best assignment. It first obtains candidate workers for task j respecting to constraints (3) and (4) if it starts to be performed at time slot bt (line 18). If the candidate worker set W_c is empty, or its skill set cannot cover task j's required skills, jump directly to the next possible start time (line 19). Then, dual subroutine (A_{dual}) computes a schedule ι and the corresponding worker rewards $cost$ (line 20). Next, the best schedule ι^* and the utility β_j will be updated if the schedule returned by A_{dual} is feasible (lines 21–24).

5.2   Subroutine for Finding Best Assignment

Due to constraint imposed by B_j, the welfare maximization problem in (19) for task j is equivalent to the following cost minimization problem:

$$\min \sum_{i\in[W]} \sum_{t\in[T]} \sum_{k\in[S_j]} \alpha^k(t) \cdot z_{ij}(t) \cdot y_{ij}^k \tag{21}$$

subject to:

$$\eta^k(t) + \sum_{i\in[W]} z_{ij}(t) \cdot y_{ij}^k \le c^k(t), \forall k \in [S_j], t \in [T] \tag{22}$$

Constraints (3)–(6)(8)–(10), where $x_j = 1$.

To address the above cost minimization problem, we develop a greedy dual algorithm (Fig. 3) to generate assignment strategies for macro tasks.

The basic idea of A_{dual} (Fig. 3) is that we schedule the subtask with the fewest candidate workers each time until task j's subtasks are all assigned. The outer while loop is executed until all subtasks are scheduled (line 3). For each subtask k, we first initialize the SC platform cost $cost^*$ and the final assigned worker set W^{k*}, and then calculate the minimum number of workers min_worker (line 5). Next, we iterate over all possible numbers of workers for subtask k in the inner for loop (line 6). Under each number of workers, we first calculate the corresponding duration time dt of subtask k (line 7) and then filter out workers that don't meet constraint (5) from W_c^k (line 8). The final selected workers for subtask k and the associated cost for rewarding workers are returned by function Choose (line 10). Next, we update W^{k*} and $cost^*$ if Choose returns a lower cost schedule (lines 11). After assigning subtask k, the assignment decisions are updated in line 12. Then, we remove subtask k from S_j, filter out k's workers from the candidate worker set (line 13), and schedule the next subtask.

Function Choose returns the final assigned workers W_k and their total rewards $cost^k$ for subtask k. It first calculates the reward of each worker in W_c^k (line 16) and updates W_c^k by filtering out workers whose traveling cost exceeds the reward (lines 17–18). Next, workers with a smaller skill set are given higher priority to be selected (line 19), for we want to reserve more versatile workers for future complex tasks. Finally, $cost^k$ is calculated and returned along with W_k (lines 20–21).

5.3   Theoretical Analysis

Due to space limitation, the proof of Theorem 3, Theorem 4 and Lemma 1 are shown in our supplementary material [1].

Theorem 2. (Efficiency of the Pricing Mechanism.) *The pricing mechanism is truthful, individually rational, and budget-balanced.*

Algorithm 1: Batch-based Allocation Framework A_{BAF}

Input: T

Output: $x_j, a_j, u_{ij}, y_{ij}^k, \forall j \in [J], \forall i \in [W], \forall k \in [S]$

1 **Initialize:** $\eta^k(t) = 0, \alpha^k(t) = p^k, \forall k \in [S], t \in [T]$;

2 **foreach** time slot $t \in [T]$ **do**

3   Update $J_t, W_t, c^k(t), \forall k \in [S]$;

4   Update $\alpha^k(t)$ according to (20);

5   **foreach** task $j \in [J_t]$ **do**

6   Best schedule ι^*, utility $\beta_j = \text{Find}(t, j, W_t, \alpha^k)$;

7   **if** $\beta_j > 0$ **then**

8   Set $x_j = 1$;

9   Set a_j, u_{ij}, y_{ij}^k according to schedule $\iota^*, \forall i \in \iota^*, k \in [S_j]$;

10   Update $\eta^k(t) = \eta^k(t) + \sum_{i \in \iota^*} y_{ij}^k, \forall k \in [S_j], t \in \iota^*$;

11   Update $\alpha^k(t)$ according to (20), $\forall t \in \iota^*$;

12   Launch task j according to ι^*;

13   **else if** $\beta_j < 0$ and $t + 1 < d_j$ **then** Merge task j into J_{t+1};

14   **else** Set $x_j = 0$ and reject task j;

15 **Function** $\text{Find}(t, j, W_t, \alpha^k)$:

16   **Initialize:** $\iota^* = \emptyset, \beta_j = -\infty$;

17   **for** $b_\iota = t + 1$ to d_j **do**

18   Filter out the eligible worker set W_e from W_t according to (3),(4);

19   **If** $W_e = \emptyset$ or $S_j \notin S_{W_e}$ **then** continue;

20        $\iota, cost = A_{dual}(bt, \alpha^k, \text{task } j, W_e)$;

21   **if** every subtask is assigned in ι **then**

22           $\iota^* = \iota$;

23           $\beta_j = B_j - cost$;

24   break;

25   **return** ι^*, β_j;

Algorithm 2: Dual Subroutine A_{dual}

Input: task j, bt, α^k, W_e

Output: best schedule ι and cost

1 **Initialize:** $\iota = \emptyset, cost = 0$;

2 Merge available skill-k worker into W_e^k for all $k \in [S_j]$;

3 **while** $S_j \neq \emptyset$ **do**

4   Select subtask k with the fewest candidate workers;

5     $cost^{k*} = +\infty, W^{k*} = \emptyset, min\_worker = D_j^k/(d_j - bt)$;

6   **for** $num\_worker = min\_worker$ to $|W_e^k|$ **do**

7        $dt = D_j^k/num\_worker$;

8   Filter out worker i from W_e^k if $bt + dt > d_i$;

9   **if** $|W_e^k| < num\_worker$ **then** continue;

10        $cost^k, W^k = $

         Choose($bt, dt, \alpha^k, j, k, W_e^k, num\_worker$);

11   **if** $cost^{k*} > cost^k$ **then** $cost^{k*} = cost^k, W^{k*} = W^k$;

12     $cost = cost + cost^{k*}, \iota = \iota \cup W^{k*}$;

13   Filter out k from S_j and update $W_e^{k'}, \forall k' \in [S_j]$;

14 **return** $\iota, cost$;

15 **Function** Choose($bt, dt, \alpha^k, j, k, W_e^k, num\_worker$):

16   Compute $R_{ij} = \sum_{t \in [bt, dt]} \alpha^k(t)$ for $i \in [W_e]$;

17   **foreach** $i \in W_e^k$ **do**

18   **if** $R_{ij} < \gamma \cdot \frac{dist(l_i, l_j)}{v}$ **then** $W_e^k \setminus i$;

19   Merge num_worker workers with the smallest skill set into W^k;

20     $cost^k = \sum_{i \in W^k} R_{ij}$;

21   **return** $cost^k, W^k$;

Fig. 2. Online Framework A_{BAF} **Fig. 3.** Dual Subroutine A_{dual}

Proof. Each skill-k worker's reward at time slot t is computed according to the price function (20) in line 16 of A_{dual}, which is decided by the initial unit price of k, the current number of busy skill-k workers and the number of online skill-k workers. Therefore, in practical application, even if a worker is allowed to submit his/her expected unit price to the SC platform, he/she cannot obtain more payoff by submitting false information. This guarantees our proposed pricing mechanism is truthful.

Constraint (9) ensures that workers' reward for performing task j covers travel cost to j, and function Choose ensures the selected workers satisfy this constraint (line 18 in A_{dual}). Thus, $R_{ij} > 0$ always exists as long as $u_{ij} > 0$, which means our pricing mechanism satisfies individual rationality.

A_{BAF} examines all the returned schedules, and if the task utility β_j is not positive, it will not be employed. Hence, the pricing mechanism is budget-balanced and always ensures the SC platform and workers are profitable.

Theorem 3. *(Correctness of Online Algorithms.)* A_{BAF} *together with* A_{dual} *produces a feasible solution to problem (2), (11), (15).*

Theorem 4. *(Polynomial Running Time)* A_{BAF} *together with* A_{dual} *runs in polynomial time for deciding admission and making assignment strategy upon arrival of each task j with time complexity* $O(Tsm(m\log m + s))$.

Theorem 5. *(Parameterized Competitive Ratio.)* *The competition ratio is the upper bound on the maximum social welfare to that achieved by our online algorithms in the EADP problem. For problem (2),* A_{BAF} *is* $\log\lambda$-*competitive.*

Proof. Let OPT be the optimal value of (2) and (11). P_j and D_j denote the objective values of (11) and (15) returned by A_{BAF} after handling task j. P_0 and D_0 indicate the initial values of (11) and (15). Note that $P_0 = 0$ and $D_0 = 0$. After handling all tasks, the final objective values of (11) and (15) returned by A_{BAF} are denoted as P_J and D_J. Then we have the following lemma:

Lemma 1. *If there is $D_j - D_{j-1} \leq log\lambda(P_j - P_{j-1}), \forall j \in [J]$, and if $P_0 = D_0 = 0$, then A_{BAF} is $log\lambda$-competitive.*

We next prove that the price function (20) satisfies $D_j - D_{j-1} \leq log\lambda(P_j - P_{j-1}), \forall j \in [J]$. For any task $j \in [J]$, if task j is rejected, the inequality $D_j - D_{j-1} \leq log\lambda(P_j - P_{j-1})$ holds naturally. If task j is successfully assigned with schedule ι, i.e., $x_{j\iota} = 1, P_j - P_{j-1} = B_j$. We define $\alpha_j^k(t)$ as the unit reward of skill-k worker after task j is assigned and $\eta_j^k(t)$ denotes the number of busy skill-k workers after handling task j. Thus we have the following equation for all $t \in \iota, k \in [S]$.

$$\alpha_j^k(t) - \alpha_{j-1}^k(t) = p^k \cdot \lambda^{\frac{\eta_j^k(t)}{c^k(t)}} - p^k \cdot \lambda^{\frac{\eta_{j-1}^k(t)}{c^k(t)}} = p^k \cdot \lambda^{\frac{\eta_{j-1}^k(t)+z_{ij}^\iota(t)\cdot y_{ij}^{k\iota}}{c^k(t)}} - p^k \cdot \lambda^{\frac{\eta_{j-1}^k(t)}{c^k(t)}}$$

For any $x \in [0,1]$, the inequality $2^x - 1 \leq x$ holds, then we have

$$\alpha_j^k(t) - \alpha_{j-1}^k(t) = \alpha_{j-1}^k(t)log\lambda^{\frac{z_{ij}^\iota(t)\cdot y_{ij}^{k\iota}}{c^k(t)}}$$

Since $x_{j\iota} = 1$, constraint (16) is tight. Then we can derive

$$D_j - D_{j-1} = \beta_j + \sum_{t\in[T]} \sum_{k\in[S_j]} (\alpha_j^k(t) - \alpha_{j-1}^k(t))c^k(t)$$

$$\leq B_j - \sum_{t\in[T]} \sum_{i\in[W]} \sum_{k\in[S_j]} \alpha_{j-1}^k(t)z_{ij}^\iota(t)y_{ij}^{k\iota} + \sum_{t\in[T]} \sum_{i\in[W]} \sum_{k\in[S_j]} (\alpha_{j-1}^k(t)log\lambda^{\frac{z_{ij}^\iota(t)\cdot y_{ij}^{k\iota}}{c^k(t)}})c^k(t)$$

$$= B_j + (log\lambda - 1) \sum_{t\in[T]} \sum_{i\in[W]} \sum_{k\in[S_j]} \alpha_{j-1}^k(t)z_{ij}^\iota(t)y_{ij}^{k\iota}$$

Combining (16) and (18), we have $B_j \geq \sum_{i\in[W]} \sum_{k\in[S_j]} \alpha_{j-1}^k(t)z_{ij}^\iota(t)y_{ij}^{k\iota}$.

Therefore, $D_j - D_{j-1} \leq log\lambda B_j$ holds, i.e., $D_j - D_{j-1} \leq log\lambda(P_j - P_{j-1})$. According to Lemma 1, we can conclude the A_{BAF} is $log\lambda$-competitive.

6 Performance Evaluation

6.1 Experimental Setup

Datasets. We evaluate our proposed approach on two datasets using both real and synthetic data. For real dataset (RDS), we use Meetup from [8], which was crawled from meetup.com. Specifically, we extract data from the Hong Kong region (with latitude from $22.209°$ to $22.609°$ and longitude from $113.843°$ to

Table 2. Experiment Parameters

Parameters	RDS	SDS
the number of workers $|W|$	**3256**	1K, 3K, **5K**, 7K, 9K
the number of tasks $|T|$	**1248**	1K, 2K, **3K**, 4K, 5K
mean of task budgets	200, 300, **400**, 500, 600	**300**
parameter λ in price function	5, 8, **10**, 12, 15	**10**
workers' velocity v	**0.035**	0.1, **0.2**, 0.3, 0.4
unit traveling cost γ	**2.25**	**2.25**, 4.75, 6.25, 8.25

114.283°). The original dataset's users, events, and tags are regarded as workers, tasks, and skills. Besides, we take the intersection of tasks' skill set and workers' skill set as the global skill set S and remove irrelevant skills of workers and tasks. For synthetic dataset (SDS), the global skill set contains 50 skills randomly selected from RDS and the locations of workers and tasks are initialized in a 2D data space $[0,1]^2$, following a uniform distribution. For RDS and SDS, workers and tasks appear or leave randomly over the system time span fixed at 50. Each task's subtasks' workloads are randomly selected from 5 to 20. The initial price of each skill follows a uniform distribution with a range $[1, 3]$. We also generate task budgets using a Gaussian distribution with a standard deviation of 100. In addition, the number of skills that a worker possesses, or a task requires is randomly selected within $[1,10]$. The evaluated parameters are shown in Table 2 and the default values are in bold font.

Benchmarks and Procedures. No existing methods on macro task assignment can be directly applied to the general EADP problem. We propose a baseline algorithm and extend two related approaches to conduct experiments. Note that the three comparison approaches decide worker rewards statically, and thus we uniformly generate each worker's skill k's price ranging from p^k to $\lambda \cdot p^k$. Specifically, we evaluate the effectiveness (social welfare) and efficiency (running time) of the following algorithms:

1. Baseline: The general idea of this method is to assign as many workers as possible to each task to accelerate task completion.
2. GTA: The main point of Greedy Task Assignment algorithm (GTA) in literature [16] is to give higher priority to tasks with higher utility per unit of workload and workers with lower traveling costs.
3. DASC_Game: DASC_Game [10] treats task assignment as a game process, and the state of reaching Nash equilibrium is regarded as the final result.
4. BAF: Our proposed framework based on primal-dual technique.

We test the effect of one parameter in each experiment, and the remaining parameters are set to default values. All the experiments were run on AMD R5 CPU @2.30 GHz with 16GB RAM in Python. Due to space limitation, please refer to the supplementary material [1] for a detailed description of the experimental parameter settings and approaches.

6.2   Experiments on SDS

To evaluate the effects of the number of workers and tasks, global unit transportation cost γ, and workers' velocity v on BAF, and three comparison algorithms, we conduct experiments on the synthetic dataset.

Effect of the Number of Workers $|W|$. As shown in Fig. 4 (a), all methods' welfare increases when the number of workers gets larger. It is reasonable because each task has more candidate workers to select from and is more probable to be scheduled. BAF outperforms the other tested approaches for it iterates all possible numbers of workers for each subtask and dynamically adjusts worker rewards. In Fig. 4 (b), the running time stably rises with the increase of $|W|$ as more workers are to be checked when scheduling tasks. Furthermore, the difference in running time of the four algorithms is not obvious. Specifically, the time cost of GTA and Baseline are relatively higher than DASC_Game and BAF. The potential reason is that some tasks are scheduled multiple times due to their low effectiveness.

Effect of the Number of Tasks $|T|$. In Fig. 5 (a), the social welfare increases when the number of tasks grows, and BAF still outperforms the other methods. The running time of the four approaches in Fig. 5 (b) still shows an upper trend when $|T|$ increases, similar to the above results of the effect of $|W|$. What's more, Baseline and GTA are more time-consuming than DASC_Game and BAF.

Effect of the Global Unit Transportation Cost γ. It can be observed from Fig. 6 (a) that the welfare achieved by Baseline reduces with the increase of γ. When γ gets larger, the traveling cost of workers rises, and thus there are fewer candidate workers who can satisfy constraint (9), and due to the low effectiveness of Baseline, the welfare is lower. While GTA always selects the minimum number of workers, and DASC_Game and BAF are more effective, these three approaches are less impacted by the increasing γ. Similar to the previous results, BAF continues to perform best. In Fig. 6 (b), GTA costs the most time in all scenarios except when γ is 8.25. The running time of Baseline increases with the increase of γ, and when γ is 8.25, it costs the most time. The time-consuming difference between BAF and DASC_Game is not obvious.

Effect of the Global Velocity v. As illustrated in Fig. 7 (a), when v changes from 0.1 to 0.4, the welfare obtained by GTA and BAF remains stable, and Baseline's welfare increases at first and then keeps stable as well. The reason is that although there are fewer candidate workers when v is low, this effect can be compensated by delaying the task start execution time. Additionally, DASC_Game is more susceptible to the increasing v. This is because when the search space of candidate workers is larger, it can reach a higher quality of Nash equilibrium resulting in better performance, but still worse than BAF. In Fig. 7 (b), the running time of BAF increases when v gets larger. The reason is that

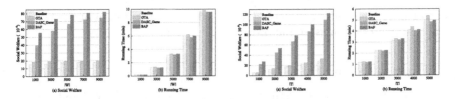

Fig. 4. Effect of the Number of Workers

Fig. 5. Effect of the Number of Tasks

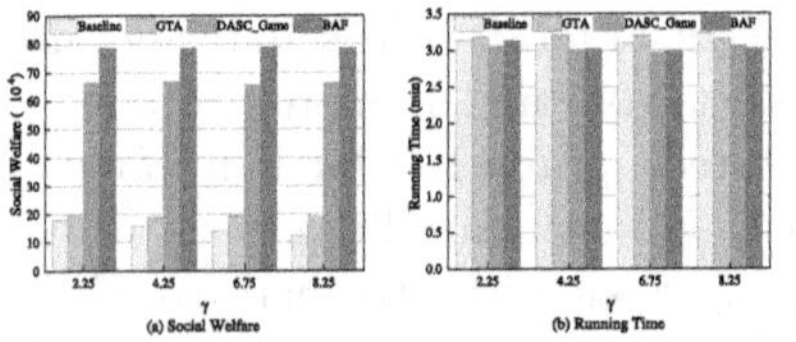

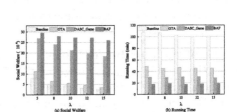

Fig. 6. Effect of the Unit Traveling Cost γ

Fig. 7. Effect of the Global Velocity v

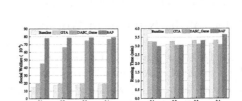

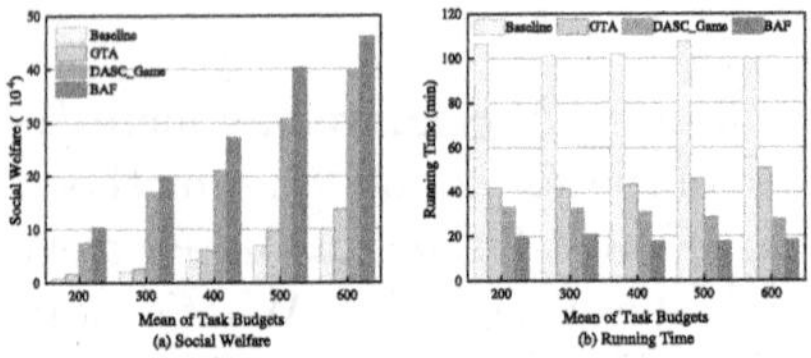

Fig. 8. Effect of the Mean Task Budget

Fig. 9. Effect of λ in Price Function

there are more candidate workers to be checked, and it takes more time to decide how many workers to assign to each subtask.

6.3   Experiments on RDS

We further conduct experiments to evaluate the effects of the mean task budget and the parameter λ in the price function on the real dataset.

Effect of the Mean Task Budget. In Fig. 8 (a), the social welfare increases with the budget mean increase in all approaches. The reason behind it is self-evident, i.e., a task with a higher budget has a higher probability of being admitted by the SC platform, which leads to higher social welfare. Besides, BAF still obtains the highest welfare. As shown in Fig. 8 (b), Baseline is the most time-consuming, and BAF costs the least time, which demonstrates BAF's superior performance under a large global skill set.

Effect of the Parameter λ. From Fig. 9 (a), we can see that the social welfare shows a downward trend when λ rises in all approaches. In particular, GTA,

DASC_Game, and Baseline are more sensitive to the increase of λ. The reason is that the increase of λ results in a higher reward for each worker and fewer strategies satisfying budget constraint (8). Therefore, some tasks are rejected due to the insufficient budget. Moreover, BAF is not as sensitive to the increase of λ as the other three methods, for it decides worker rewards in real-time according to the number of available workers and total online workers, which can better cope with the changing budget than static pricing. In Fig. 9 (b), Baseline is the most time-consuming, and BAF continues to be the most efficient approach, further demonstrating BAF's high efficiency when the global skill set is large.

Finally, we summarize the experimental findings: (i) BAF strikes a balance between effectiveness and efficiency when global skill set S is small. When S is large, BAF shows the best performance of four practical approaches (highest welfare and lowest time consumption). (ii) The overall performance of the three algorithms based on static pricing (Baseline, GTA, DASC_Game) is inferior to that of dynamic pricing (BAF).

7 Conclusion

This work studied the EADP problem in spatial crowdsourcing, which focuses on macro task assignment and dynamic pricing mechanism design. We first define the EADP problem and model EADP into a combinatorial optimization problem. Next, we formulate it into an ILP and construct a functional pricing mechanism based on the primal-dual framework. Moreover, an online batch-based framework are presented, aiming to maximize the social welfare of the SC system. Finally, we conduct extensive experiments on real dataset and synthetic dataset. As future work, we plan to generalize the problem model to a broader set of crowdsourcing problems, and hope the primal-dual framework demonstrated in this work can shed light on other related problems in the field.

Acknowledgement. This work has been supported in part by National Key R&D Program of China (2022YFB2901300), China Telecom, and NSFOCUS (2022671026).

References

1. Supplementary material. https://1drv.ms/b/s!Al10zQV2U0Nlgl1AyDhZmkGyT t2K
2. Banerjee, S., Johari, R., Riquelme, C.: Pricing in ride-sharing platforms: a queueing-theoretic approach. In: EC, pp. 639. ACM (2015)
3. Chen, M., Shen, W., Tang, P., Zuo, S.: Optimal vehicle dispatching for ride-sharing platforms via dynamic pricing. In: Comp. Proc. of the The Web Conf. 2018, pp. 51–52 (2018)
4. Chen, Z., Cheng, P., Chen, L., Lin, X., Shahabi, C.: Fair task assignment in spatial crowdsourcing. Proc. VLDB Endow. **13**(12), 2479–2492 (2020)
5. Cheng, P., Lian, X., Chen, L., Han, J., Zhao, J.: Task assignment on multi-skill oriented spatial crowdsourcing. TKDE **28**(8), 2201–2215 (2016)

6. Kazemi, L., Shahabi, C.: Geocrowd: enabling query answering with spatial crowd-sourcing. In: 20th Int. Conf. Adv. Geographic Inform. Syst., pp. 189–198. ACM (2012)
7. Liang, Y., Wu, W., Wang, K., Hu, C.: Online multi-skilled task assignment on road networks. IEEE Access **7**, 57371–57382 (2019)
8. Liu, X., He, Q., Tian, Y., Lee, W.C., McPherson, J., Han, J.: Event-based social networks: linking the online and offline social worlds. In: SIGKDD, pp. 1032–1040. ACM (2012)
9. Liu, Z., Li, K., Zhou, X., Zhu, N., Gao, Y., Li, K.: Multi-stage complex task assignment in spatial crowdsourcing. Inf. Sci. **586**, 119–139 (2022)
10. Ni, W., Cheng, P., Chen, L., Lin, X.: Task allocation in dependency-aware spatial crowdsourcing. In: ICDE, pp. 985–996. IEEE (2020)
11. Song, T., Xu, K., Li, J., Li, Y., Tong, Y.: Multi-skill aware task assignment in real-time spatial crowdsourcing. GeoInformatica **24**(1), 153–173 (2020)
12. Song, T., Zhu, F., Xu, K.: Specialty-aware task assignment in spatial crowdsourc-ing. In: Fleuriot, J., Wang, D., Calmet, J. (eds.) AISC 2018. LNCS (LNAI), vol. 11110, pp. 243–254. Springer, Cham (2018). https://doi.org/10.1007/978-3-319-99957-9_19
13. Tong, Y., Zeng, Y., Ding, B., Wang, L., Chen, L.: Two-sided online micro-task assignment in spatial crowdsourcing. TKDE **33**(5), 2295–2309 (2021)
14. Tong, Y., Zhou, Z., Zeng, Y., Chen, L., Shahabi, C.: Spatial crowdsourcing: a survey. VLDB J. **29**(1), 217–250 (2020)
15. Wang, H., Wang, E., Yang, Y., Wu, J., Dressler, F.: Privacy-preserving online task assignment in spatial crowdsourcing: a graph-based approach. In: INFOCOM, pp. 570–579. IEEE (2022)
16. Xia, J., Zhao, Y., Liu, G., Xu, J., Zhang, M., Zheng, K.: Profit-driven task assign-ment in spatial crowdsourcing. In: IJCAI, pp. 1914–1920. IJCAI (2019)
17. Xu, Y., Xiao, M., Wu, J., Zhang, S., Gao, G.: Incentive mechanism for spatial crowdsourcing with unknown social-aware workers: a three-stage stackelberg game approach. In: TMC, p. 1 (2022)
18. Zhao, D., Li, X.Y., Ma, H.: How to crowdsource tasks truthfully without sacrificing utility: online incentive mechanisms with budget constraint. In: INFOCOM, pp. 1213–1221. IEEE (2014)
19. Zhao, Y., Zheng, K., Guo, J., Yang, B., Pedersen, T.B., Jensen, C.S.: Fairness-aware task assignment in spatial crowdsourcing: game-theoretic approaches. In: ICDE, pp. 265–276. IEEE (2021)

Systems and Optimization

InstantChain: Enhancing Order-Execute Blockchain Systems for Latency-Sensitive Applications

Jianfeng Shi[1,2], Heng Wu[2], Diaohan Luo[1,2], Heran Gao[1,2], and Wenbo Zhang[2,3(✉)]

[1] University of Chinese Academy of Sciences, Beijing, China
[2] Institute of Software, Chinese Academy of Sciences, Beijing, China
{shijianfeng20,wuheng,luodiaohan21,
gaoheran19,zhangwenbo}@otcaix.iscas.ac.cn
[3] State Key Laboratory of Computer Sciences, Institute of Software, Chinese Academy of Sciences, Beijing, China

Abstract. As an emerging replicated transactional processing system, blockchain is being adopted by more and more industries. Some blockchain-based applications require low latency and high throughput. However, the ordering phase is being a bottleneck of the *order-execute* blockchain. Poor ordering will result in low throughput due to transaction conflicts, and poor quality of service for latency-sensitive applications because their transactions are not processed within their expected times. Therefore, this paper enhances *order-execute* blockchain systems to minimize latency and maximize parallelism by considering transaction latency requirements, conflicts and varied execution times. First, we propose a parallelism-oriented and latency-sensitive transaction ordering method. Then, we propose a deep reinforcement learning-based method to determine optimal block parameters. Finally, we propose an adaptive packing method based on space occupancy that enables packed transactions to fully exploit parallelism. Experimental results show that our implemented system (**InstantChain**) outperforms 3 related systems under both real and simulated datasets.

Keywords: Blockchain · Instant transactions · Parallel execution

1 Introduction

More and more industries are attracted to develop blockchain-based decentralized applications (DApps) to improve collaboration efficiency. Unfortunately, the inherent nature of blockchains, namely the replicated transaction processing model and the involvement of consensus protocols, makes them difficult to compare with traditional transaction systems in terms of throughput and latency.

As shown in Fig. 1, some DApps' transactions (Txs) are insensitive to latency, such as storing a certificate to the blockchain. On the contrary, some DApps' Txs have very high expectations for low latency. For decentralized exchanges, high latency will make some Txs of buying or selling financial assets meaningless,

X. Wang et al. (Eds.): DASFAA 2023, LNCS 13943, pp. 483–498, 2023.
https://doi.org/10.1007/978-3-031-30637-2_31

because when a Tx is executed, the price of the asset it wants to buy or sell has changed significantly. For decentralized online games, high latency will greatly impair the players' experience. Therefore, it is valuable to reduce the latency of these latency-sensitive Txs.

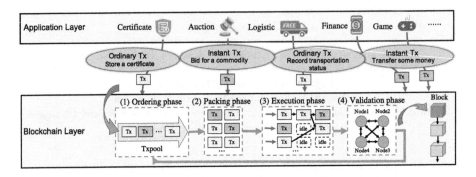

Fig. 1. Transaction processing model of *order-execute* blockchain systems.

The *order-execute* architecture is widely adopted by blockchains, such as FISCO BCOS [9], one of the most popular permissioned blockchains in China. As shown in Fig. 1, the blockchain node packs a batch of Txs from the transaction pool (Txpool) and executes them. Obviously, the ordering phase and the packing phase will significantly affect the completion time of Txs in the execution phase.

Most *order-execute* blockchains use the first-come-first-serve (FCFS) strategy to order and pack Txs [8]. We observe the following 3 challenges that limit these blockchains to achieve low-latency and high-throughput transaction processing.

Challenge 1: Instant Txs are not processed within their expected times.

Those transactions that are sensitive to latency and need to be processed immediately are called *instant Txs* (or *real-time Txs*). On the contrary, those transactions that are not sensitive to latency are called *ordinary Txs*. In order to know how the ordering phase and the packing phase affect the blockchain's throughput and latency, we send 20,000 Txs (10% are instant Txs) to FISCO BCOS. As shown in Fig. 2(a), even if there is no conflict between Txs, there are still 1292 instant Txs whose latency requirements have not been met.

Challenge 2: Conflicts and varied execution times between Txs make it difficult for each batch of packed Txs to fully exploit parallelism during the execution phase.

Figure 2(a) shows that conflicts between Txs will result in a significant drop in the throughput and an increase in the number of Txs that do not meet their latency requirements. This is because, as shown in the execution phase in Fig. 1, some Txs must wait for their conflicting Txs to complete execution before they can begin execution. Some execution threads are idle, making the parallelism provided by the multi-core processor not fully utilized.

(a) Varing the conflict rate under block size = 1000 txs

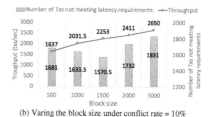
(b) Varing the block size under conflict rate = 10%

Fig. 2. Impact of conflict rate and block size on throughput and latency.

Challenge 3: The default block size fails to achieve the optimal balance between throughput and latency of a blockchain system.

The block size is an important parameter for a blockchain system, which limits the maximum number of Txs that each block is allowed to contain. If the block size is too small, more blocks will be generated, and if the block size is too large, the execution time and the transmission time of each block will be longer. As shown in Fig. 2(b), both cases will increase the confirmation latency of Txs, leading to an increase of Txs that do not meet their latency requirements.

The above observations motivate us to enhance *order-execute* blockchain systems from three aspects. (1) How to order Txs in the Txpool so that higher priority is assigned to those Txs that are more urgent and more conducive to increasing parallelism. (2) How to pack Txs from the Txpool so that the parallelism provided by the multi-core processor can be fully utilized in the execution phase and the confirmation latency of instant Txs can be minimized. (3) How to dynamically adjust the block size so that the blockchain system can increase throughput while maintaining low latency.

Contributions. We make the following new contributions.

- An efficient transaction ordering method based on DAG (directed acyclic graph) and ordered queue: It can give higher priority to the Txs that have higher latency requirements and are more conducive to parallelism.
- A dynamic block size adjustment method based on deep reinforcement learning (DRL): It can achieve the optimal balance between the overall throughput and the confirmation latency of instant Txs.
- An adaptive packing method based on space occupancy: It enables packed Txs to fully utilize parallelism while packing as many instant Txs as possible.
- We implement a blockchain system **InstantChain**. Using a real dataset on Ethereum and a simulated dataset with different conflict rates and different percentage of instant Txs, InstantChain achieves higher throughput and lower confirmation latency for instant Txs compared to 3 other systems.

The rest of this paper is organized as follows. Section 2 formalizes the problem of maximizing throughput and minimizing instant Txs' latency. Section 3 describes our proposed methods. Section 4 evaluates the implementation. Related work is reviewed in Sect. 5. Section 6 concludes the paper.

2   Problem Statement

2.1   Transaction Processing Model of the *Order-Execute* Blockchain

Transaction: The basic unit of changing the states on the blockchain ledger [11]. For a transaction Tx_i, its arrival time is at_i, its processing time is pt_i that can be inferred from the types and number of its operational instructions, or from its previous processing experience, and its expected latency is el_i. If $el_i \neq 0$, Tx_i is an instant Tx, otherwise Tx_i is an ordinary Tx.

Txpool: A place to temporarily store Txs submitted by clients of DApps. All Txs in the Txpool are denoted by the set $Txpool = \{Tx_1, Tx_2, ..., Tx_n\}$.

Block: The basic unit of state synchronization among blockchain nodes, containing a batch of Txs and their execution results. The process of processing Txs is actually dividing the Txs in the Txpool into multiple blocks $B = \{B_1, B_2, ..., B_m\}$ and processing them in sequence. The time taken to complete processing all Txs within B_k is bt_k. The time taken to complete consensus (i.e. propagation and validation) on B_k among blockchain nodes is ct_k.

Conflict: Two Txs conflict if the read set of one Tx and the write set of another Tx intersect, or if their write sets intersect. In FISCO BCOS [9], developers define mutually exclusive parameters (i.e. shared variables) before deploying smart contracts, so we can derive the read-write set of a Tx based on this prior knowledge. In order to ensure correctness, conflicting Txs cannot be executed in parallel, but can only be executed serially.

DAG (Directed Acyclic Graph): A way to manage conflicting Txs. Each vertex in the DAG represents a Tx. If two Txs are conflicting, there is an edge or a path between them, which means they should be executed serially. A Tx with in-degree equal to 0 is called a *ready Tx*, and this Tx will be executed before all Txs that conflict with it. A Tx with out-degree equal to 0 is called an *exit Tx*.

Throughput: The average number of Txs successfully processed and recorded to the blockchain ledger per second.

$$Throughput = \frac{n}{\sum_{k=1}^{m}(bt_k + ct_k)} \qquad (1)$$

Latency: The transaction confirm latency (tcl_i) [22] of Tx_i contains 3 parts. (1) Transaction queuing delay (tqd_i): the waiting time of Tx_i in the Txpool. (2) Block generation delay (bgd_i): mainly includes the total time spent processing these Txs within the block B_k ($Tx_i \in B_k$). (3) Block consensus delay (bcd_i): the time it takes for the block B_k to complete consensus among blockchain nodes.

$$tcl_i = tqd_i + bgd_i + bcd_i \qquad (2)$$

2.2   Offline-Version Problem Formulation

The objective of our paper consists of two parts: (1) Maximize the throughput, i.e. minimize the sum of processing times and consensus times of all blocks.

(2) Minimize the sum of transaction confirmation latencies of all instant Txs. Therefore, the offline-version of the problem we study can be expressed as follows.

$$\mathbf{InstantChain}: \quad min \quad U = \sum_{k=1}^{m}(bt_k + ct_k) + \alpha \sum_{Tx_p \in insTxs} tcl_p \tag{3}$$

$$\text{s.t.} \quad \sum_{k=1}^{m} Y_{jk} = 1 \qquad j = 1, \ldots n \tag{4}$$

$$0 < \sum_{j=1}^{n} Y_{jk} \leq blockSize \qquad k = 1, \ldots m \tag{5}$$

In objective function (3), $insTxs$ is the set of instant Txs in the Txpool. Const. (4) specifies that each Tx can only be assigned to one block. $Y_{jk} \in \{0/1\}$ indicates whether Tx_j is packed into B_k. Const. (5) claims that the number of Txs in each block is greater than 0 and does not exceed the block size $blockSize$.

It can be seen that this problem is NP-hard, namely it is difficult to find a polynomial-time algorithm to solve this problem. In addition, the problem we study is actually online and real-time. If the time it takes to make scheduling decisions is too high, it will significantly affect the overall latency. Solving this problem is challenging and meaningful. Therefore, we propose InstantChain.

3   InstantChain

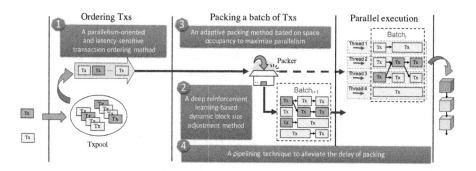

Fig. 3. System structure of InstantChain.

As shown in Fig. 3, InstantChain enhances *order-execute* blockchain systems through four methods. (1) A parallelism-oriented and latency-sensitive transaction ordering method: Those Txs that are beneficial to improve parallelism and reduce latency will be able to be given a higher priority. (2) A DRL-based dynamic block size adjustment method: By sensing the transaction status in the Txpool and the environment of the blockchain system, the packer will be able to dynamically determine the optimal block parameters. (3) An adaptive packing method based on space occupancy to maximize parallelism: On the premise of packing as many instant Txs as possible, the packer will have the ability to pack a batch of Tx that can fully exploit parallelis (4) A pipelining technique: The delay caused by the packing phase will be effectively alleviated.

3.1 A Parallelism-Oriented and Latency-Sensitive Transaction Ordering Method

To efficiently order Txs in the Txpool, as shown in Fig. 4, we use a DAG to manage all ordinary Txs, and use an ordered queue to manage all instant Txs. If a new ordinary Tx is received, it will be added to the DAG and the edges will be created between it and the exit Txs that conflict with it. If a new instant Tx is received, it will be added to the ordered queue according to its urgency. We define the priority of each Tx based on the following insights.

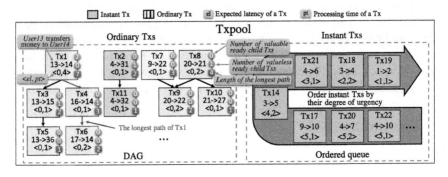

Fig. 4. Efficiently order Txs using a DAG and an ordered queue.

(1) Give higher priority to the Tx with higher latency requirement. Because the earlier it is packed, the smaller its transaction confirmation latency.

Definition 1. (Degree of urgency). We define the degree of urgency (UD) of a transaction Tx_i by its expected latency el_i.

$$UD_i = \begin{cases} \frac{1}{el_i} & if \ el_i \neq 0 \\ 0 & if \ el_i = 0 \end{cases} \tag{6}$$

In Fig. 4, $UD_{19} = 1$ and $UD_{18} = 0.5$, so Tx_{19} should be packed before Tx_{18}.

(2) Give higher priority to the Tx that can generate more ready Txs after being packed. The packing phase always packs ready Txs (i.e. Txs with *in-degree = 0*), so the more Txs are ready, the easier the packing phase is to pack a batch of Txs that can fully exploit parallelism.

Definition 2. (Degree of contribution to parallelism). In the DAG, after Tx_i is packed, it will release 3 kinds of child Txs. (I) $VARC_i$ is the number of valuable ready Txs (*in-degree = 0* and *out-degree>0*), they can release more ready Txs after being packed. (II) $VLRC_i$ is the number of valueless ready Txs

($in\text{-}degree = 0$ and $out\text{-}degree = 0$). (III) NRC_i is the number of non-ready Txs ($in\text{-}degree \neq 0$). We define the ability of Tx_i to produce ready Txs after being packed as the degree of contribution to parallelism (CPD_i).

$$CPD_i = VARC_i + \mu VLRC_i + \nu NRC_i \tag{7}$$

For example, in Fig. 4, $VARC_7 = 0$, $VLRC_7 = 0$, $VARC_8 = 0$, $VLRC_8 = 1$, so Tx_8 should be packed before Tx_7.

(3) Give higher priority to the Txs on the longest path (i.e. critical path) in the DAG. Because the Txs on each edge in the DAG are conflicting and can only be executed serially. If the ready Tx on the longest path is not selected for execution, this may increase the total processing time.

Definition 3. (Length of the longest path). In the DAG, a path from Tx_i to an exit Tx is called a path of Tx_i. The length of a path is the sum of processing times of all Txs on the path. LP_i is the length of Tx_i's longest path.

$$LP_i = pt_i + \max_{Tx_j \in Children(Tx_i)} \{LP_j\} \tag{8}$$

In Fig. 4, $\{Tx_1, Tx_4, Tx_6\}$ is the longest path of Tx_1, so $LP_1 = 7$.

In summary, the priority of each Tx can be defined as follows.

$$priority_i = UD_i + \beta CPD_i + \gamma LP_i \tag{9}$$

3.2 A DRL-Based Dynamic Block Size Adjustment Method

The block sizes set by most blockchains are one-dimensional, e.g. 1000 Txs in FISCO BCOS v2.7.2 [9]. They do not take into account the characteristics of parallel execution of Txs. So we redefine the block size from the perspective of parallelism. As shown in Fig. 5, the proposed block size consists of 3 parts.

- The optimal number of parallel threads (Considering the different processing capabilities of blockchain nodes).
- The optimal block data size (Considering the network bandwidth).
- The optimal block execution time (Considering the status of instant Txs and ordinary Txs in the Txpool).

Manually adjusting parameters places high demands on blockchain developers, requiring them to understand the characteristics of Txs and the impact of different parameter combinations. In addition, we observe that the process of packing a batch of Txs for execution and consensus in the blockchain is a typical Markov decision process. Therefore, we convert the optimal blockchain parameter configuration problem into the problem of obtaining the maximum return in reinforcement learning. As shown in Fig. 5, the agent tells the environment (i.e. blockchain system) how to change the blockchain parameters. Then the blockchain nodes pack, execute, and validate Txs. The agent observes the state of the blockchain system and computes the reward. The agent repeats the above steps, eventually enabling the blockchain system to find the optimal block size based on the characteristics of its own environment.

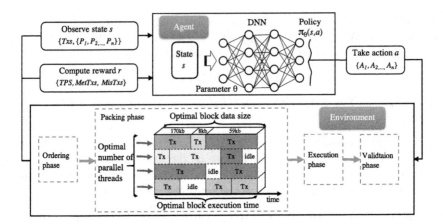

Fig. 5. Adaptively adjusting the block size based on DRL.

(1) State space S. The state s contains the status of Txs in Txpool and the blockchain parameters that you want to adjust, namely $s = \{Txs, \{p_1, p_2, ..., p_n\}\}$, where p_i represents the value of the ith parameter.

(2) Action space A. The action $a = \{a_1, a_2, ..., a_n\}$ contains the adjustment of each blockchain parameter. For example, $a = \{+10, -100, ...\}$ means that p_1 needs to be increased by 10, p_2 needs to be decreased by 100, etc.

(3) Reward R. We design a dynamic reward function to compute the immediate reward r obtained by the agent taking action a under state s. For the ith packing, $metTxs_i$ is the instant Txs meeting their latency requirements, $unmetTxs_i$ is the instant Txs not meeting their latency requirements, TPS_i is the system throughput, and t_i is the moment when blockchain nodes complete consensus on the block that contains this batch of packed Txs.

$$r_i = TPS_i + \varrho * \frac{|metTxs_i|}{|misTxs_i| + |metTxs_i|} - \zeta * \sum_{Tx_j \in misTxs_i} \{t_i - at_j - el_j\} \quad (10)$$

3.3   An Adaptive Packing Method Based on Space Occupancy

The process of packing Txs can be abstracted as a multi-layer bin packing problem. Determining how to pack a batch of Txs that can maximize parallelism is equivalent to determining how to minimize the area of vacancies in the bin.

As shown in Fig. 6(a), Each block is considered as a bin with the number of layers equal to the optimal number of parallel threads and the length of each layer equal to the optimal block execution time suggested in Sect. 3.2. Each Tx is treated as a rectangular piece with length equal to its processing time. In this bin, Txs cannot overlap, and conflicting Txs cannot be placed in parallel. Therefore, Txs within the same layer are executed serially in order, and Txs between different layers can be executed in parallel without conflict.

A continuous space not occupied by Txs in a layer will be defined as a *space* in the format $threadNum : [startTime, endTime]$. For example, in Fig. 6(c), $totalSpaces = \{1 : [1, 2], 1 : [4, 5], 2 : [3, 5], 3 : [1, 5], 4 : [0, 5]\}$.

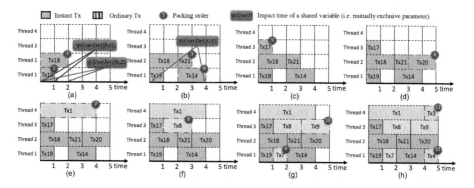

Fig. 6. Pack a batch of Txs that can maximize parallelism. (a–d) represent the first stage to pack instant Txs. (e–f) represent the second stage to pack ordinary Txs.

For each shared variable v_j contained in these packed Txs, its impact time is recorded in the form of $\psi(v_j) = \{[startTime1, endTime1], ...\}$. For example, in Fig. 6(a), the shared variables of Tx_{18} is $\{User3, User4\}$, so $\psi(User3) = \{[0,2]\}$. In Fig. 6(b), Tx_{14} is packed into the bin. The shared variables of Tx_{14} is $\{User3, User5\}$, so $\psi(User3) = \{[0,4]\}$.

For a transaction Tx_i to be packed into the bin, we need to find a space $targetSpa$ that has a subspace $subSpa$ whose length is equal to the processing time of Tx_i, and those Txs parallel to $subSpa$ do not conflict with Tx_i.

$$\exists targetSpa \in totalSpaces, \exists subSpa \in targetSpa,$$
$$subSpa.length = Tx_i.pt, \forall v_j \in Tx_i.vars, \ subSpa \cap \psi(v_j) = \emptyset \quad (11)$$

(1) If $targetSpa$ is not found, this means that Tx_i cannot be executed in parallel with these Txs in the bin at this moment, so it is necessary to temporarily add Tx_i to the set $excludedTxs$ in order to exclude it from this packing.

$$excludedTxs \cup = \{Tx_i\} \quad (12)$$

(2) If $targetSpa$ is found, place Tx_i into $targetSpa$. This may result in two smaller remaining spaces $leftSpace$ and $rightSpace$, which we add to $totalSpaces$.

$$totalSpaces \cup = \{leftSpace, rightSpace\} \quad (13)$$

Then, update the impact time of each shared variable v_j contained in Tx_i.

$$\forall v_j \in Tx_i.vars, \ \psi(v_j) \cup = [subSpa.startTime, subSpa.endTime] \quad (14)$$

We divide the packing process into 2 stages. Figure 6 depicts the process of packing a batch of Txs from the Txpool shown in Fig. 4.

(1) *Stage 1* is responsible for packing as many instant Txs as possible, which will ensure the quality of service (QoS) for instant Txs. In Fig. 6(a), Tx_{19} with

the highest priority will be packed from the ordered queue first, followed by Tx_{18}. Because there is no conflict between them, they can be placed in parallel. In Fig. 6(b), Tx_{21} can only be placed after Tx_{18} because there is a conflict between them (i.e. the shared variable $User4$). In Fig. 6(d), until there are no suitable instant Tx in the ordered queue that can be packed into the bin, the first stage ends.

(2) *Stage 2* tries its best to fill the vacancies caused by *Stage 1* by packing ordinary Txs. In Fig. 6(e), Tx_1 with the highest priority is packed from the DAG and placed into the bin. Repeating the above step, we will get a batch of Txs as shown in Fig. 6(h), and these Txs will be able to fully utilize the parallel capabilities provided by the multi-core processor.

3.4 A Pipelining Technique to Alleviate the Delay of Packing

Our proposed packing method is more complex than the original FCFS-based method. The delay of packing cannot be negligible. As shown in Fig. 7, we use the pipelining technique to parallelize the packing phase and the execution phase. While the current batch of Txs is about to be processed by the execution phase, the packer starts to pack the next batch of Txs.

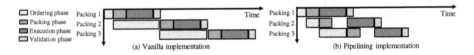

Fig. 7. Alleviate the delay of the packing phase through pipeline technique.

4   Experimental Evaluation

4.1   Experimental Setup

Baseline Blockchains. InstantChain is implemented on top of FISCO BCOS v2.7.2, and we use the following 3 blockchains to evaluate its performance.

- *FCFS:* A blockchain that orders and packs Txs based on their arrival times, i.e. FISCO BCOS v2.7.2.
- *Latency-first:* A blockchain that orders Txs in descending order based on their expected latencies, then packs the Txs with the highest priority.
- *Serial-mode:* Unlike the other three blockchains that support parallel execution, this blockchain executes Txs serially to emulate Ethereum.

Datasets. We use the test tool [9] provided by FISCO BCOS for testing.

◇ *Simulated dataset:* We generate multiple loads by combining different conflict rates and different percentage of instant Txs. We regard the processing time of a *transfer* Tx with 2 read operations and 2 write operations as 1. Each load contains 50,000 Txs, and the distribution of execution times of these Txs is $[20\% \rightarrow 1, 20\% \rightarrow 2, 20\% \rightarrow 3, 20\% \rightarrow 4, 20\% \rightarrow 10]$. The expected latency of an instant Tx is taken from [1.5s, 2s, 2.5s, 3s]. Txs with different execution times, conflicts, and different expected latencies follow a uniform distribution in each load.

⋄ *Real dataset:* Between 10/4/2022 and 11/4/2022, **Tether USDT Stablecoin** (contract address: 0xdac17f958d2ee523a2206206994597c13d831ec7) was the most popular smart contract on Ethereum. We extract the most recent 100,000 transaction logs of this contract from 3/11/2022 to 4/11/2022 as our real dataset.

4.2 Overall Performance

We vary the conflict rate (CR), the percentage of instant Txs (PIT), the transaction arrival rate (TAR), and the consensus latency (CL), respectively, to see how the throughput, the number of Txs not meeting latency requirements and the average transaction confirmation latency change accordingly.

Impact of conflict rate: As shown in Fig. 8(a), with an increase of conflict rate from 0% to 30%, all blockchains exhibit a decrease in throughput, but InstantChain always achieves higher throughput than other blockchains. The reason is that InstantChain can reasonably distribute conflicting Txs into different blocks, enabling each block to fully exploit parallelism. When the conflict rate is 10%, compared to FCFS, InstantChain can increase the throughput by 31.6%, reduce the number of Txs not meeting latency requirements by 97.4%, and reduce the average transaction confirmation latency by 22.5%.

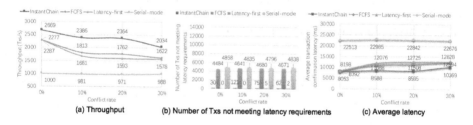

Fig. 8. Varying the conflict rate under PIT = 10%, TAR = 8000 txs/s, CL = 200 ms.

Impact of percentage of instant Txs: As shown in Fig. 9(b), the increase in the percentage of instant Txs results in a dramatic increase in the number of Txs not meeting latency requirements in FCFS. However, InstantChain can not only pack the most urgent instant Txs first, but also distribute conflicting instant Txs to different blocks, which enables InstantChain to achieve latency satisfaction rate close to that of the latency-first blockchain.

Impact of transaction arrival rate: Figure 10 shows an increase in transaction arrival rate will lead to an increase in average transaction confirmation latency, but InstantChain is able to keep the number of Txs not meeting latency requirements very low.

Impact of consensus latency: Multiple scenarios can lead to increased consensus latency in blockchains, such as changes in network bandwidth, or an

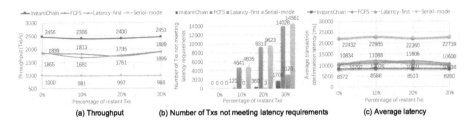

Fig. 9. Varying the percentage of instant Txs under $CR = 10\%$, $TAR = 8000\,\mathrm{txs/s}$, $CL = 200\,\mathrm{ms}$.

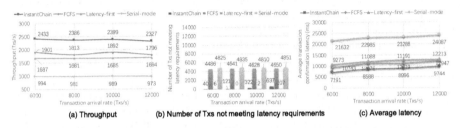

Fig. 10. Varying the transaction arrival rate under $CR = 10\%$, $PIT = 10\%$, $CL = 200\,\mathrm{ms}$.

increase in the number of blockchain nodes participating in the PBFT consensus. As shown in Fig. 11, varying the consensus latency from 200 ms to 350 ms, InstantChain still outperforms other blockchains.

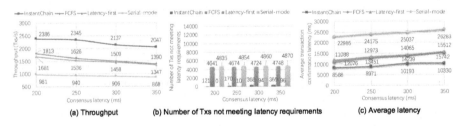

Fig. 11. Varying the consensus latency under $CR = 10\%$, $PIT = 10\%$, $TAR = 8000$ Txs/s.

Performance under real dataset: Since Txs on Ethereum do not have the property of expected latency, we randomly set 5% of these Txs in the dataset to be instant Txs. We conduct experiments by varying the transaction arrival rate under $CL = 200\,\mathrm{ms}$. As shown in Fig. 12, InstantChain outperforms other blockchains in terms of throughput and average transaction confirmation latency. When the transaction arrival rate is 8000 Txs/s, compared to FCFS, InstantChain can increase the throughput by 30.9%, reduce the number of Txs not meeting latency requirements by 74.2%, and reduce the average transaction confirmation latency by 26.2%.

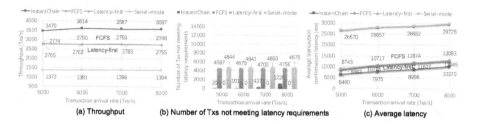

Fig. 12. Performance under a real dataset containing 100,000 Txs on Ethereum.

4.3 Bottleneck Analysis

We conduct a comprehensive analysis of the experiment under $CR = 10\%$, $PIT = 10\%$, $TAR = 8000\,Txs/s$, $CL = 200\,ms$. As shown in Fig. 13(a), in InstantChain, instant Txs are processed first, which guarantees QoS for instant Txs. The size of each block is dynamic. For most blocks, the Txs within them can fully utilize the parallelism provided by the multi-core processor when they are executed. Figure 13(b) shows that InstantChain can reduce the execution time but increase the importing time (i.e. ordering time). Since the consensus time occupies a very large proportion, if the consensus time is too large or the execution time is too small, the optimization effect of InstantChain will be weakened.

5   Related Work

Efficient utilization of parallelism. Conflicts between Txs is the major factor that prevent blockchains from fully utilizing parallelism to process Txs.

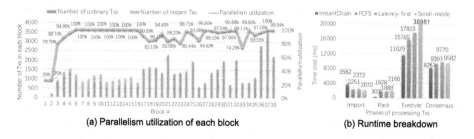

Fig. 13. Bottleneck Analysis.

- *Execute-order blockchains.* Hyperledger Fabric checks for read-write conflicts in the validation phase, which might result in transaction aborts. Fabric++ [18] introduces a transaction reordering mechanism, which can reduce the number of serialization conflicts between Txs within a block, and abort conflicting Txs in the execution phase and in the ordering phase. FabricSharp [17] efficiently reduces the transaction abort rate by applying transactional analysis of OCC databases. FabricCRDT [14] can automatically merge conflicting Txs by using CRDT techniques without losing updates.
- *Order-execute blockchains.* ParBlockchain [1] generates a dependency graph for the Txs within a block during the ordering phase, so that Txs can be executed in parallel during the execute phase without rolling back. SChain [7] proposes a scalable order-execute-finalize paradigm to exploit intra-block concurrency and inter-block concurrency. Nathan et al. [15] implement a blockchain relational database with two different transaction flows (i.e. order-execute and execute-order-in-parallel). Anjana [2] introduces DiPETrans and OptSmart. DiPETrans groups the Txs within a block into independent shards and processes these shards in parallel. OptSmart saves the non-conflicting Txs in a concurrent bin and conflicting Txs in a block graph to make validators deterministically re-execute the batch of Txs in parallel.

Supporting instant Txs. Many recent studies change the transaction processing architecture or the consensus protocol of the blockchain to support instant Txs. However, our approach does not break the blockchain architecture or sacrifice its original properties. Rahasak [3] introduces a validate-execute-group architecture and adopts the Apache Kafka-based consensus to support real-time transaction. Min et al. [13,23] propose a collaborative mechanism to improve the throughput and support instant Txs. Goel et al. [8] propose a weighted fair queueing strategy for ordering Txs that can support differentiated QoS for multiple transaction types. Sokolik et al. [21] propose an age-aware fair approach to ordering Txs. CATP-Fabric [25] introduces a conflicting transaction mitigating solution for Fabric, considering latency requirements of IoT applications.

Block size optimization. Many recent studies optimize block size from the perspective of balancing throughput, latency, and revenue. However, none of them optimize block size from a parallelism perspective. Singh et al. [19,20] propose meta-heuristic algorithms to find an optimal block size with optimal transmission time and optimal block composition time. Chen et al. [6] and Jiang et al. [10] find the optimal block size in a PoW-based blockchain by weighing the reward with the successful mining probability. Wilhelmi et al. [22] propose an end-to-end latency model and derive an estimation of optimum block size. Chacko et al. [5] analyzes the impact of block size on transaction failures in Hyperledger Fabric. Liu et al. [12] and Xu et al. [24] use DRL to adjust the block size. RainBlock [16] safely increases the block size by addressing the I/O bottleneck in transaction processing. Meta-Regulation [4] can adaptively change the block size and the creation interval for Bitcoin.

6 Conclusion

In order to enable *order-execute* blockchain systems to improve the QoS for instant Txs, we propose *InstantChain*. A transaction ordering method based on DAG and ordered queue is proposed, which gives higher priority to Txs that have higher latency requirements and are more conducive to parallelism. Then, a DRL-based dynamic block size adjustment method is introduced to balance maximizing throughput and minimizing latency. Finally, we propose an adaptive packing method based on space occupancy to enable each batch of packed Txs to fully exploit parallelism. Experimental results show that InstantChain can achieve lower latency and higher throughput compared to other blockchains.

Acknowledgements. This work is supported by the Provincial Key Research and Development Program of Shandong, China (No. 2021CXGC010101), the National Natural Science Foundation of China (No. 61872344 and No. 61972386).

References

1. Amiri, M.J., Agrawal, D., El Abbadi, A.: ParBlockchain: leveraging transaction parallelism in permissioned blockchain systems. In: IEEE 39th International Conference on Distributed Computing Systems (ICDCS), pp. 1337–1347 (2019)
2. Anjana, P.S.: Efficient parallel execution of block transactions in blockchain. In: Proceedings of the 22nd International Middleware Conference: Doctoral Symposium, pp. 8–11 (2021)
3. Bandara, E., Liang, X., Foytik, P., Shetty, S., Ranasinghe, N., De Zoysa, K.: Rahasak—scalable blockchain architecture for enterprise applications. J. Syst. Architect. **116**, 102061 (2021)
4. Cao, M., Wang, H., Yuan, T., Xu, K., Lei, K., Wang, J.: Meta-regulation: adaptive adjustment to block size and creation interval for blockchain systems. IEEE J. Sel. Areas Commun. 1 (2022)
5. Chacko, J.A., Mayer, R., Jacobsen, H.A.: Why do my blockchain transactions fail? A study of hyperledger fabric. In: Proceedings of the 2021 International Conference on Management of Data, pp. 221–234 (2021)
6. Chen, J., Cheng, Y., Xu, Z., Cao, Y.: Evolutionary equilibrium analysis for decision on block size in blockchain systems. In: Du, D.-Z., Du, D., Wu, C., Xu, D. (eds.) COCOA 2021. LNCS, vol. 13135, pp. 180–194. Springer, Cham (2021). https://doi.org/10.1007/978-3-030-92681-6_15
7. Chen, Z., et al.: SChain: a scalable consortium blockchain exploiting intra-and inter-block concurrency. Proc. VLDB Endow. **14**(12), 2799–2802 (2021)
8. Goel, S., Singh, A., Garg, R., Verma, M., Jayachandran, P.: Resource fairness and prioritization of transactions in permissioned blockchain systems (industry track). In: Proceedings of the 19th International Middleware Conference Industry (2018)
9. Huizhong, L., Chenxi, L., Haoxuan, L., Xingqiang, B., Xiang, S.: An overview on practice of FISCO BCOS technology and application. Inf. Commun. Technol. Policy **46**(1), 52 (2020)
10. Jiang, S., Wu, J.: Bitcoin mining with transaction fees: a game on the block size. In: 2019 IEEE International Conference on Blockchain, pp. 107–115 (2019)

11. Liu, J., Wan, S., He, X.: Alias-Chain: improving blockchain scalability via exploring content locality among transactions. In: 2022 IEEE International Parallel and Distributed Processing Symposium (IPDPS), pp. 1228–1238 (2022)

12. Liu, M., Yu, F.R., Teng, Y., Leung, V.C., Song, M.: Performance optimization for blockchain-enabled industrial internet of things (IIoT) systems: a deep reinforcement learning approach. IEEE Trans. Industr. Inform. (2019)

13. Min, X., Li, Q., Liu, L., Cui, L.: A permissioned blockchain framework for supporting instant transaction and dynamic block size. In: 2016 IEEE Trustcom/BigDataSE/ISPA, pp. 90–96 (2016)

14. Nasirifard, P., Mayer, R., Jacobsen, H.A.: FabricCRDT: a conflict-free replicated datatypes approach to permissioned blockchains. In: Proceedings of the 20th International Middleware Conference, pp. 110–122 (2019)

15. Nathan, S., Govindarajan, C., Saraf, A., Sethi, M., Jayachandran, P.: Blockchain meets database: design and implementation of a blockchain relational database. In: Proceedings of the VLDB Endowment, vol. 12, no. 11 (2019)

16. Ponnapalli, S., et al.: RainBlock: faster transaction processing in public blockchains. In: 2021 USENIX Annual Technical Conference (USENIX ATC 2021), pp. 333–347 (2021)

17. Ruan, P., Loghin, D., Ta, Q.T., Zhang, M., Chen, G., Ooi, B.C.: A transactional perspective on execute-order-validate blockchains. In: Proceedings of the 2020 ACM SIGMOD International Conference on Management of Data (2020)

18. Sharma, A., Schuhknecht, F.M., Agrawal, D., Dittrich, J.: Blurring the lines between blockchains and database systems: the case of hyperledger fabric. In: Proceedings of the 2019 International Conference on Management of Data (2019)

19. Singh, N., Vardhan, M.: Computing optimal block size for blockchain based applications with contradictory objectives. Procedia Comput. Sci. **171** (2020)

20. Singh, N., Vardhan, M.: Multi-objective optimization of block size based on CPU power and network bandwidth for blockchain applications. In: Nath, V., Mandal, J.K. (eds.) Proceedings of the Fourth International Conference on Microelectronics, Computing and Communication Systems. LNEE, vol. 673, pp. 69–78. Springer, Singapore (2021). https://doi.org/10.1007/978-981-15-5546-6_6

21. Sokolik, Y., Rottenstreich, O.: Age-aware fairness in blockchain transaction ordering. In: IEEE/ACM 28th International Symposium on Quality of Service (2020)

22. Wilhelmi, F., Barrachina-Muñoz, S., Dini, P.: End-to-end latency analysis and optimal block size of proof-of-work blockchain applications (2022)

23. Xie, W., et al.: ETTF: a trusted trading framework using blockchain in e-commerce. In: 2018 IEEE 22nd International Conference on Computer Supported Cooperative Work in Design (2018)

24. Xu, S., et al.: Deep reinforcement learning assisted edge-terminal collaborative offloading algorithm of blockchain computing tasks for energy internet. Int. J. Electr. Power Energy Syst. **131**, 107022 (2021)

25. Xu, X., et al.: Mitigating conflicting transactions in hyperledger fabric-permissioned blockchain for delay-sensitive IoT applications. IEEE Internet Things J. **8**(13), 10596–10607 (2021)

DALedger: Towards High-Performance Transaction Processing for Collaborative Decentralized Applications

Junkai Wang[1], Zhiwei Zhang[1]([⊠]), Shuai Zhao[1], Jiang Xiao[2], Ye Yuan[1], and Guoren Wang[1]

[1] Beijing Institute of Technology, Beijing, China
{wangjunkai,zwzhang,shuaizhao,yuan-ye}@bit.edu.cn
[2] Huazhong University of Science and Technology, Wuhan, China
jiangxiao@hust.edu.cn

Abstract. Despite recent intensive research, existing blockchain systems still have limitations in supporting decentralized applications. In particular, although existing blockchain systems execute internal transactions of different applications concurrently, it is difficult to process the concurrent control considering both the cross-application transactions and internal transactions. The reason is that each application can only access the transactions related to it, and the conflicts between all transactions cannot be detected by the applications. To improve the concurrency of blockchain systems for decentralized applications, we propose a novel blockchain named as DLedger, which is designed based on the directed acyclic graph ledge structure. DALedger supports the concurrent execution of not only internal transactions in the same application but also internal transactions and cross-DApp transactions. We prove that cycles in the dependency graph must have a special dangerous structure, which can be detected in a single application's partial dependency graph. Based on that, we propose a novel concurrency control mechanism to resolve concurrency conflicts, while ensuring serializability. We conduct more extensive experiments compared with state-of-the-art blockchain ledgers for decentralized applications. Experimental results show that our method outperforms existing works significantly.

Keywords: Blockchain · Data Confidentiality · Concurrency Control.

1 Introduction

As a decentralized technology, blockchain has been developed rapidly with the rise of Bitcoin [9]. More and more enterprises design their blockchain-based applications, and use it as an infrastructure for various tasks including data sharing [8], authentication [6], etc. While these applications are running independently, they need to interact with each other. For example, in order to share data between applications, they need to construct the ledger for the shared data, to ensure that only the required application can access the data. Within each decentralized application(DApp) [1,12], they reach a consensus on the transaction ordering and data among all the participants. On the other hand, to deal with the collaboration of DApps, such as the task of data sharing, they use smart contracts to represent the agreement. As there is no centralized controller, the collaborations are also conducted as transactions in the blockchain.

However, under the condition of data privacy, throughput and scalability issues, especially the low performance in collaboration situations, hinder the development of DApps. Specifically, the permissioned blockchain system has the characteristics of decentralized, tamper-evident, traceable, and can be used in each DApp. However, it is difficult to apply such approaches to the collaboration situation, with cross-DApp transactions. The reason is that, for each DApp, it should only be able to access the data of its own, and other DApps shared with it. However, in permissioned blockchain, the ledger, transactions and smart contracts should be agreed by all nodes, leads to data confidentiality problems.

To deal with these challenges, many works have been conducted. Caper [1] ensures the data confidentiality by dividing data into internal data and public data, and structuring blockchain ledger as a directed acyclic graph (DAG). Caper reaches a consensus on the order of transactions before execution. Although transactions of different DApps can be processed concurrently, the execution and consensus of both internal and cross-DApp transactions in the same DApp is serial, resulting in all internal transactions will be blocked while a cross-DApp transaction arrived. Hyperledger Fabric [2] improves performance by executing the transactions in parallel and using private data to prevent the leaks of state data. However, the complete ledger in Fabric is still stored on each node, which does not satisfy the data confidentiality of DApps collaboration.

Considering these, in this paper, we propose our decentralized applications collaboration ledger DALedger. As in collaboration among multiple DApps, the guarantee of data confidentiality is the first requirement. We use the ledger structure of DAG, in which the share data among different DApps will appear in the corresponding ledger of each DApps. To improve the performance by supporting concurrent execution and ordering, we propose the dependency graph construction based on all the transactions among multiple DApps. Notice that each DApp can only detect a partial dependency graph constructed by its own transaction, and the cross-DApp transaction related to it. To make the cross-DApp transaction concurrent control, we propose a subgraph-based concurrent control protocol to ensure serializability by running independently on each node. The contributions of this paper are as follows:

- We propose the construction of dependency graph, and prove that the cycles in the complete dependency graph must have a special dangerous structure which can be detected within a single application.
- we propose an optimistic concurrency control framework for DApps and implement a decentralized applications collaboration ledger DALedger to support highly concurrent in execution and consensus.
- We implemented our approaches and experimental results reveal that our method outperforms the existing work significantly.

The rest of the paper is organized as follows. Section 2 reviews the background and related works, and Sect. 3 gives a system overview of DALedger and analyzes the concurrency control algorithm. The experimental results are reported in Sect. 4. Finally, Sect. 5 concludes this paper.

2   Background and Related Work

Compared with the order-execute architecture of the blockchain [13], Fabric [2] first proposes an execute-order-validate (EOV) architecture to improve the concurrency of the blockchain system, consists of three phases: execution, ordering, and validation. The execution organizations simulate the transaction requests to generate the read-set and the write-set for the transaction. Then the ordering organizations reach a consensus on the order of all transactions. Finally, the execution organizations validate transactions based on the total order [10,14].

Caper [1] first proposed a blockchain to supports DApp collaboration. The blockchain ledger of Caper consists of the following two types of transactions:

Internal transactions are executed and agreed within the application, follow the logic of the application's internal smart contract. Due to confidentiality requirements, internal transactions are not visible to other applications, so only internal data can be written, and public data cannot be modified.

Cross-DApp transactions involve multiple applications and are visible to all applications, ordered by all DApps. According to the smart contract between DApps, cross-DApp transactions can achieve DApp collaboration by reading and writing public data, but it cannot access the internal data of applications.

3   Decentralized Application Ledger

In this section, we introduce the system overview of DALedger. In this scenario, it is necessary to support concurrent transaction processing while ensuring the confidentiality of the internal data of each application and the consistency of the blockchain ledger. DALedger supports concurrent execution and concurrent consensus of internal transactions and cross-DApp transactions in DApps, consists of three types of nodes including *client, orderer* and *executor*.

3.1   DALedger Workflow

Transaction Processing Workflow. As shown in Fig. 1, in DALedger, firstly, the client sends transactions to the executor of its corresponding application. The executor process transactions concurrently according to the latest snapshot in the ledger, and generates the execution results containing the read sets and write sets of the transactions. Afterward, the execution results of internal transactions are sent to the orderers. The orderers apply the consensus protocol to internal transactions and cross-DApps concurrently. Orderers construct an index to ensure the serializability of all the transactions. Notice that, orderers

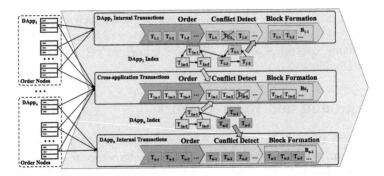

Fig. 1. The workflow of order phase.

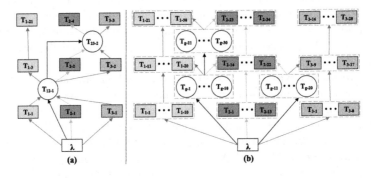

Fig. 2. Differences in transaction execution between (a) Caper and (b) DALedger.

of different applications can only see their own internal transactions and related cross-DApp transactions. At last, a block is constructed, and sent to executors of the corresponding application for validation.

As shown in Fig. 2, DALedger supports the parallel execution of internal transactions (T_{1-1} to T_{1-10}), parallel execution of cross-DApp transactions (T_{g-1} to T_{g-10}), and parallel consensus on internal transactions and cross-DApp transactions (T_{1-1} and T_{g-10}). DALedger ensures that all transactions are serializable by detecting special structures. Then, we introduce the concurrent control protocol of DALedger. We use cross-Tx to denote the cross-DApp transaction, and internal-Tx to denote the internal transaction.

3.2 Serializability Analysis

Definition 1 (Blockchain snapshot). *A blockchain snapshot is the data state after a committed block denoted as a sequence number $[CSS, ISS]$ to represent the block height of cross-Tx and internal-Tx.*

We define the start timestamp of T_i as the blockchain snapshot it read (StartTs(T_i)), and the end timestamp of it as the block contains T_i and its order in the block (EndTs(T_i)). Given two transactions T_1 and T_2, they are concurrent if their lifecycle are overlap. We define the dependencies described in [5]:

- $n - ww$: if T_1 updates a key X_1, a non-concurrent transaction T_2 updates the immediate successor version X_2, then $T_1 \xrightarrow{n-ww} T_2$.
- $c - ww$: if T_1 updates a key X_1, a concurrent transaction T_2 updates the immediate successor version X_2, then $T_1 \xrightarrow{c-ww} T_2$.
- $n - wr$: if T_1 updates a key X_1, T_2 reads the version X_1, then $T_1 \xrightarrow{n-wr} T_2$.
- $n - rw$: if T_1 updates a key X_1, a non-concurrent T_2 reads the previous version X_0, then $T_1 \xrightarrow{n-rw} T_2$.
- $c - rw$: if T_1 updates a key, a concurrent T_2 reads the previous version X_0, then $T_1 \xrightarrow{c-rw} T_2$.

Remark. For all containing c-ww dependency, according to the "First-Committer-Wins" (FCW) rule [3], we abort the latter committed one in c-ww. Thus, if there is a dependency $T_1 \longrightarrow T_2$ between two concurrent transactions, we can conclude the dependency is c-rw. Based on the definition of internal-Tx and cross-Tx introduced in Sect. 2, we summarize the following lemma:

Lemma 1. *Only* n-wr *dependency exists in the dependencies from cross-Txs to internal-Txs and only* n-rw *and* c-rw *exist in the dependencies from internal-Txs to cross-Txs, and no dependency between internal-Tx of different application.*

Based on [10], we define **strong serializability** as the committed results of transactions are consistent with the serial execution in the proposed order, while **serializability** means that the committed results are consistent with the serial execution in some order. We have the following theorem:

Theorem 1. *To ensure the consistency of cross-DApp transactions in different applications, cross-DApp transactions must satisfy strong serializability.*

Proof. Obviously, if the cross-Txs satisfy strong serializability, then the cross-Txs in different application ledgers are consistent. If the cross-Txs only satisfy serializability, the cross-Tx may not be consistent in the ledgers of different DApps. Because each application can only detect a subgraph of the complete dependency graph. Removing cycles in the subgraph cannot guarantee the complete dependency graph is acyclic. Thus, the cross-Txs must be strong serializable.

According to the above discussion, the **complete dependency graph (CDG)** is composed of all internal-Txs and cross-Txs, where the vertexes represent the transactions and the edges represent dependencies. In particular, non-serializable transactions form cycles in the CDG.

Partial Dependency Graph (PDG). Since internal-Txs are only visible in the corresponding DApp, each DApp only maintains one partial dependency graph instead of the CDG, composed of its internal-Txs, all cross-Txs and edges between them. Figure 3 shows the CDG of two applications, in which $r-X/w-X$ means read or write the key X and PDG is represented by a dashed box.

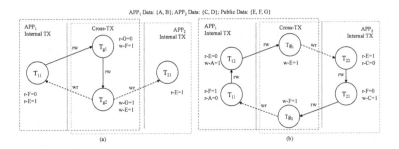

Fig. 3. Complete Dependency Graph.

3.3   Dealing with Non-serializable Transactions

Based on the discussion above, we find that a special structure can be extracted to eliminate cycles in the dependency graph, at the condition that each application can only detect a subgraph of the CDG. We have the following theorem:

Theorem 2. *Every cycle implies non-serializable in CDG contains at least one dangerous structure consists of three transactions: $T_1 \xrightarrow{c-rw} T_2 \xrightarrow{c-rw} T_3$ (T_1 and T_3 can be the same transaction). In addition, T_3 is the first committed transaction in the cycle and this structure can be detected in at least one application's PDG.*

Proof. For any arbitrary cycle in the CDG, choose T_3 committed first and assume the predecessor of T_3 is T_2, and the predecessor of T_2 is T_1. (1) If T_3 is a cross-Tx. First, we suppose T_2 is not concurrent with T_3: if T_2 commits before T_3 starts, contradicts T_3 committed first, or T_2 starts after T_3 commits, contradicts T_2 is the predecessor of T_3. Thus, T_2 must be concurrent with T_3, the dependency is *c-rw* by Remark. If T_2 is a cross-Tx, it will be aborted according to Theorem 1, so we can conclude T_2 is an internal-Tx. We suppose T_1 is not concurrent with T_2: if T_2 commits before T_3 starts, then $End(T_1) < Start(T_1) < End(T_3)$, contradicts T_3 committed first, or T_1 starts after T_2 commits, contradicts T_1 is the predecessor of T_2. Thus, T_1 must be concurrent with T_2, the dependency is *c-rw* and we can conclude T_1 is an internal-Tx by Lemma 1. (2) If T_3 is an internal-Tx, similarly, T_2 is concurrent with T_3, and T_1 is concurrent with T_2. (3) Based on the above proof, T_1 and T_2 must be internal-Txs. $T_1 \xrightarrow{c-rw} T_2 \xrightarrow{c-rw} T_3$ must be in the *PDG* of the same DApp by Lemma 1.

Based on Theorem 2, we design the DApps collaborative index construction (DAppIndex) to detect conflicts. It consists of $M_{w\to s}$, $M_{r\to s}$, and M_{flag}, where $M_{w\to s}$ and $M_{r\to s}$ denote the map between the key to an transactions list updating or reading it, and M_{flag} denote the map between the key to a list including the transactions updating this key while having an originating rw-dependency.

(1) **Cross-Tx processing (strong serializable).** Algorithm 1 shows the concurrency control algorithm for cross-Txs. For transactions to be detected, first, for each key in the read set, We detect whether it is modified by other

---

Algorithm 1: Concurrency Control of Cross-DApp Transactions

Input: read set $\{r\}_g$, write set $\{w\}_g$, the sequence number of global snapshot CSS which corresponds to H_g

1 **for** $r \in \{r\}_g$ **do**
2 | **if** $\exists\, i$ s.t. $i \in M_{w \to s}[r] \wedge i \geq CSS$ **then** return failed;
3 **for** $w \in \{w\}_g$ **do**
4 | **if** $\exists\, i$ s.t. $i \in M_{w \to s}[w] \wedge i \geq CSS$ **then** return failed;
5 **for** $r \in \{r\}_g$ **do** $M_{r \to s}[r] \leftarrow CSS + 1$;
6 **for** $w \in \{w\}_g$ **do** $M_{w \to s}[w] \leftarrow CSS + 1$;

Algorithm 2: Concurrency Control of Internal Transactions

Input: read set $\{r\}_{l,g}$, write set $\{w\}_l$, the sequence number CSS which corresponds to H_g, the sequence number ISS which corresponds to H_l

1 $flag1 \leftarrow false, flag2 \leftarrow false$
2 **for** $w \in \{w\}_l$ **do**
3 | **if** $\exists\, i$ s.t. $i \in M_{w \to s}[w] \wedge i \geq ISS$ **then** return $failed$;
4 | **if** $\exists\, i$ s.t. $i \in M_{r \to s}[w] \wedge i \geq ISS$ **then** $flag1 \leftarrow true$;
5 **for** $r \in \{r\}_{l,g}$ **do**
6 | **if** $r \in Internal\ state \wedge (\exists\, i$ s.t. $i \in M_{w \to s}[r] \wedge i \geq CSS)$ **then**
7 | | $flag2 \leftarrow true$
8 | | **if** $\exists\, i$ s.t. $i \in M_{flag}[r] \wedge i \geq ISS$ **then** $flag1 \leftarrow true$;
9 | **if** $r \in Public\ state$ **then** $flag2 \leftarrow true$;
10 **if** $flag1 == true \wedge flag2 == true$ **then** return $failed$;
11 **if** $flag2 == true$ **then**
12 | **for** $w \in \{w\}_l$ **do** $M_{flag}[w] \leftarrow ISS + 1$;
13 **for** $r \in \{r\}_l$ **do** $M_{r \to s}[r] \leftarrow ISS + 1$;
14 **for** $w \in \{w\}_l$ **do** $M_{w \to s}[w] \leftarrow ISS + 1$;

---

committed concurrent transactions. if true then returns failed (Lines 1–2). Then for each key in the write set, we detect whether it will overwrite the key of other concurrent transactions (Lines 3–4). Then update all indexes (Lines 5–6).

(2) Internal-Tx processing (serializable). Then we introduce our cycle elimination algorithm. According to Theorem 2, serializability can be achieved by ensuring that there is no dangerous structure in any PDG. Algorithm 2 shows the concurrency control algorithm for internal-Txs. We first abort those overwrite the key in write set of concurrent transactions (Lines 3). Then we check whether it modifies the key in read set of concurrent transactions. We detect whether the transaction is in a dangerous structure based on two flags: (a) $Flag_1$: If the transaction writes a key in the read set of concurrent transactions, or updates a key in the read set of the transaction has rw-dependency on other transactions, set $Flag_1$ to true. (b) $Flag_2$: If the transaction reads a key in the write set of a concurrent transaction, or reads public data(to avoid divergence caused by asynchrony), set $Flag_2$ to true (Line 4–9). If both flags are true, abort this transaction (Lines 10),or update all indexes and return success (Lines 11–14).

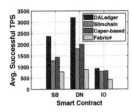 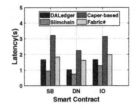 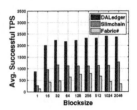

Fig. 4. Throughput.   **Fig. 5.** Latency.   **Fig. 6.** Blocksize.

4   Performance Evaluation

We implemented the prototype of DALedger and deployed it on four servers with
Intel(R) Core (TM) i7 CPU and 32G RAM with SGX. We use the method in [14]
to support the off-chain execution, and support the processing of different kinds
of data, including relational data, graph data, text data and key-value data.

- Caper-based [1]: We implemented a caper-model with the global consensus
  using a set of orderers.
- Fabric# [10]: Fabric# is an EOV architecture permissioned blockchain system
  based on Fabric supporting multiple applications.
- Slimchain [14]: Another EOV architecture blockchain using stateless design
  to achieve off-chain storage.

4.1   Impact of Smart Contract

We used three kinds of smart contracts in Blockbench [4] contains 10% cross-
DApp transactions as the workload, equals to the typical setting of distributed
database [11]. Figure 4 shows the throughput when using different contracts.
When using Smallbank (SB), DALedger reaches the highest throughput of 2363
tps, Caper reaches 1436 tps, and Slimchain and Fabric# reach 1281 tps and
590 tps respectively. When using Donothing (DN), the result is similar to that
in SB. When using Ioheavy (IO), the throughput of each system is not very
different, proves the benefits of parallelism decline when the hardware becomes a
bottleneck. Figure 5 shows the latency under the full load of each smart contract.
Figure 6 shows the throughput of each system under different blocksize. In other
expriments, the blocksize of each system is set to the best.

4.2   Impact of Hot Conflict Account

We use a smart contract read two accounts and write another two accounts, and
set 1% accounts as hot accounts to evaluate the impact of conflict. By default,
Read and write have a 10% probability of accessing hot accounts. Figure 7
shows the trend of the throughput of each system under different write hot

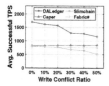
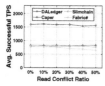
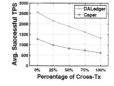
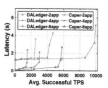

Fig. 7. Write Hot Conflict Account.　　**Fig. 8.** Read Hot Conflict Account.　　**Fig. 9.** Cross-Tx Ratio.　　**Fig. 10.** Scalability Evaluation.

ratio. The performance of each EOV system gradually decrease while Caper has not changed significantly. DALedger always maintains the highest throughput. Figure 8 shows the trend of the throughput of each system under different read hot ratio. All throughput has not decreased significantly. Figure 9 shows the trend of throughput as the proportion of cross-DApp transactions changes. DALedger has the highest throughput of 2555 tps when all transactions are internal and decline to 1323 tps. Caper reaches a peak of 1280 without any cross-DApp transactions and decline to 622 tps when all transactions are cross-DApp transactions.

4.3 Scalability Evaluation

Figure 10 shows the latency of DALedger and Caper with different number of DApps under different throughput. Result shows both DALedger and Caper have good scalability, the latency of Caper is benefited from streaming transaction processing [7], and DALedger achieves higher peak throughput, which benefits from the concurrent execution and consensus.

5　Conclusion

In this paper, we propose DALedger, a permissioned blockchain which supports both internal and cross-DApp transactions of collaborating DApps. We found the transactions can be processed concurrently, even each DApp does not know the all conflicts. We design the corresponding concurrency control algorithm to ensure serializability and data consistency. Our experiments show that DALedger has better performance and scalability, compared with the existing works.

Acknowledgements. Zhiwei Zhang is supported by National Key Research and Development Program of China (Grant No. 2020YFB1707900, No. 2021YFB2700700), National Natural Science Foundation of China (Grant No. 62072035), Open Research Projects of Zhejiang Lab (Grant No. 2020KE0AB04) and CCF-Huawei Database System Innovation Research Plan (Grant No. CCF-HuaweiDBIR2021007B). Jiang Xiao is supported by National Key Research and Development Program of China under Grant (No. 2021YFB2700700), Key Research and Development Program of Hubei Province (No. 2021BEA164), National Natural Science Foundation of China

(Grant No. 62072197). Ye Yuan is supported by the NSFC (Grant Nos. 61932004, 62225203, U21A20516). Guoren Wang is supported by the NSFC (Grant Nos. 61732003, U2001211).

References

1. Amiri, M.J., Agrawal, D., Abbadi, A.E.: Caper: a cross-application permissioned blockchain. Proc. VLDB Endowment **12**(11), 1385–1398 (2019)
2. Androulaki, E., et al.: Hyperledger fabric: a distributed operating system for permissioned blockchains. In: Proceedings of the Thirteenth EuroSys Conference, pp. 1–15 (2018)
3. Cahill, M.J., Röhm, U., Fekete, A.D.: Serializable isolation for snapshot databases. ACM Trans. Database Syst. (TODS) **34**(4), 1–42 (2009)
4. Dinh, T.T.A., Wang, J., Chen, G., Liu, R., Ooi, B.C., Tan, K.L.: Blockbench: a framework for analyzing private blockchains. In: Proceedings of the 2017 ACM International Conference on Management of Data, pp. 1085–1100 (2017)
5. Fekete, A., Liarokapis, D., O'Neil, E., O'Neil, P., Shasha, D.: Making snapshot isolation serializable. ACM Trans. Database Syst. (TODS) **30**(2), 492–528 (2005)
6. Hammi, M.T., Hammi, B., Bellot, P., Serhrouchni, A.: Bubbles of trust: a decentralized blockchain-based authentication system for IoT. Comput. Secur. **78**, 126–142 (2018)
7. István, Z., Sorniotti, A., Vukolić, M.: Streamchain: do blockchains need blocks? In: Proceedings of the 2nd Workshop on Scalable and Resilient Infrastructures for Distributed Ledgers, pp. 1–6 (2018)
8. Kokoris-Kogias, E., Alp, E.C., Gasser, L., Jovanovic, P., Syta, E., Ford, B.: Calypso: private data management for decentralized ledgers. Cryptology ePrint Archive (2018)
9. Nakamoto, S.: Bitcoin: A peer-to-peer electronic cash system. Decentralized Bus. Rev. 21260 (2008)
10. Ruan, P., Loghin, D., Ta, Q.T., Zhang, M., Chen, G., Ooi, B.C.: A transactional perspective on execute-order-validate blockchains. In: Proceedings of the 2020 ACM SIGMOD International Conference on Management of Data, pp. 543–557 (2020)
11. Thomson, A., Diamond, T., Weng, S.C., Ren, K., Shao, P., Abadi, D.J.: Calvin: fast distributed transactions for partitioned database systems. In: Proceedings of the 2012 ACM SIGMOD International Conference on Management of Data, pp. 1–12 (2012)
12. Weber, I., Xu, X., Riveret, R., Governatori, G., Ponomarev, A., Mendling, J.: Untrusted business process monitoring and execution using blockchain. In: La Rosa, M., Loos, P., Pastor, O. (eds.) BPM 2016. LNCS, vol. 9850, pp. 329–347. Springer, Cham (2016). https://doi.org/10.1007/978-3-319-45348-4_19
13. Wood, G., et al.: Ethereum: a secure decentralised generalised transaction ledger. Ethereum Proj. Yellow Pap. **151**(2014), 1–32 (2014)
14. Xu, C., Zhang, C., Xu, J., Pei, J.: Slimchain: scaling blockchain transactions through off-chain storage and parallel processing. Proc. VLDB Endowment **14**(11), 2314–2326 (2021)

Efficient Execution of Blockchain Transactions Through Deterministic Concurrency Control

Huahui Xia[1], Jinchuan Chen[1,2(✉)], Nabo Ma[1], Jia Huang[3], and Xiaoyong Du[1,2]

[1] School of Information, Renmin University of China, Beijing 100872, China
{hhxia,jcchen,naboma,duyong}@ruc.edu.cn
[2] Key Laboratory of Data Engineering and Knowledge Engineering (MOE), Beijing, China
[3] Institute of High Performance Computing (IHPC), Agency for Science, Technology and Research (A*STAR), 1 Fusionopolis Way, #16-16 Connexis, Singapore 138632, Republic of Singapore
huang_jia@ihpc.a-star.edu.sg

Abstract. Concurrently executing Blockchain transactions can make good use of modern hardware and improve system performance. Recent works mainly utilize a dependency graph to represent the partial order among conflicting transactions. All participants can then execute the transactions concurrently according to an identical order and keep consistent with each other. However generating dependency graphs is quite time-consuming, and during this process some transactions have to be re-executed multiple times. In this paper, we adopt deterministic concurrency control to quickly compute the partial order. Instead of one large graph, we partition the transactions into a sequence of batches. In this way, there is no need to deal with the inter-batch conflicts in finding the partial order. We also propose a two stage approach (DVC) to find a partial order with high degree of parallelism. DVC does not need to re-execute transactions, and can find an approximate optimal partial order by solving an equivalent MinVWC problem. We integrate the proposed techniques into an open-source system and compare them with several advanced solutions. As shown by the experimental results, our approaches can significantly reduce the costs of computing the partial order and obtain a schedule of high parallelism degree.

Keywords: Blockchain · Concurrency control · Deterministic

1 Introduction

The term "blockchain" is created by Nakamoto as a fundamental data structure to maintain an identical database among multiple distributed and trustless servers in the Bitcoin system. All the transactions to update the database are recorded in a series of *blocks*. Each block contains the hashed value of its predecessor, and is logically linked one by one to form a *chain*.

Both Bitcoin and Ethereum utilize a *proposing and attestation* paradigm to process transactions inside a block, which is also adopted by many permissioned

© The Author(s), under exclusive license to Springer Nature Switzerland AG 2023
X. Wang et al. (Eds.): DASFAA 2023, LNCS 13943, pp. 509–518, 2023.
https://doi.org/10.1007/978-3-031-30637-2_33

blockchains [1–4]. Firstly a specific node is chosen to be the *proposer* by a computation contest or a predefined order. The proposer needs to select a set of transactions to compose a block and broadcast the block to all the other nodes, which are called *attestors*.In order to ensure the correctness, each attestor should execute the transactions and reject a block if it contains any invalid transactions.

In conventional blockchains, both the proposer and attestors need to sequentially execute the transactions to ensure consistency among all nodes. Classical DBMS concurrency control methods, like optimistic concurrency control (OCC), cannot be directly applied, because they are based on competition for computation resources. Different nodes may have different schedules even though they all adopt an identical concurrency control strategy.

Sequential execution cannot benefit from modern hardware like multi-core processors and this has greatly limited the performance of blockchains. In recent years, the issue of concurrently executing blockchain transactions attracts many research interests [1–9]. Most works require the proposer to prescribe a partial order among conflicting transactions, in the form of a *dependency graph*. Attestors can then execute these transactions concurrently according to the order.

For this purpose, the proposer needs to execute the transactions and find the conflicts among them. Hence this kind of approaches is more suitable for permissioned blockchains, where the proposer is required to undertake extra workload of finding an appropriate partial order to help the attestors.

Unfortunately, finding optimal dependency graphs has been proven to be a NP-hard problem [4]. To avoid the expensive dependency graphs, we adopt the idea of *deterministic database* [10], where transactions are executed concurrently and deterministically in several replicas. Instead of generating one large dependency graph, we try to partition the transactions into *a sequence of batches*. For any two transactions from different batches, the one with smaller batch number always precedes the other. Therefore, it is no longer necessary to identify inter-batch conflicts and the workload of proposing will be significantly reduced. Moreover, storing batch boundaries is much cheaper than storing dependency graphs.

To make good tradeoff between the time costs of proposing and attestation, we propose a two stage approach, called *DVC*. In the first stage, we execute all candidate transactions in parallel by deterministic concurrency control and most of them will be put into some batches. The time complexity of the first stage is $O(n)$. In the second stage, we try to partition the remaining (usually highly correlated) transactions. We find that the problem of finding an optimal partition can be transformed to the MinVWC problem (minimum vertex weighted coloring) [11]. Hence we adapt a state-of-the-art algorithm [11] of solving MinVWC to our scenario.

We further investigate the problem of repeatedly executing transactions. In previous works [4], the proposer has to re-execute an aborted transaction since a transaction may access different data items when running on different database vesions. Whereas in our work, we only need to deal with intra-batch conflicts. With deterministic concurrency control, the attestors can cocurrently execute a block containing intra-bach conflicts and still keep consistency with each other. Therefore, the proposer does not need to re-execute aborted transactions.

We will discuss related works in Sect. 2 and then define the problem and introduce the Baseline approach in Sect. 3. Next we will propose the two stage approach in Sect. 4, and report the experimental results in Sect. 5. Finally we will conclude this paper in Sect. 6.

2   Related Works

Dependency Graph Based Approaches. Dickerson et al. propose a dependency graph based concurrent execution model [1], which is also adopted by [2–5, 7–9]. Instead of just packaging transactions into a block, a proposer needs to also specify precedence relations between conflicting transactions in the form of a *dependency graph*, which is a directed and acyclic graph (DAG).

Two transactions may have conflicts if they access an identical data item, and a proposer needs to obtain the write and read sets of each transaction to identify conflicts. In [1], a proposer executes transactions speculatively in parallel with Software Transactional Memory (STM), which can output the read/write sets and a serializable concurrent schedule. Some other works also adopt this idea but with different kinds of transaction memories, such as object-based TM [2] and hardware TM [5]. Another way to find the read/write sets is static analysis [8]. It would be much faster than executing transactions but may face problems if read/write sets change with different database states.

Jin et al. try to carefully choose precedence relations and generating a graph which can be scheduled with high degree of parallelism [4]. The proposer needs to further divide the graph into several sub-graphs and broadcast intermediate read/write sets to alleviate lock contests. But this method requires expensive computations in the proposing phase.

Determinsitic Database. This paper borrows some ideas from deterministic database, i.e. partitioning transactions into batches [10]. But there are two major differences. Firstly, there are no Byzantine nodes in determinsitic database management systems. Secondly, the schedulers will not execute transactions, what they do is just to presecribe a partial order. As aforementioned, the partial order will impact the degree of parallelism. In our solution the proposer needs to find an appropriate partial order by executing transactions.

Recently, Peng et al. propose to improve the performance of Blockchain by deterministic transaction processing [12]. An *Epoch Server* is introduced to compose a block and the transactions are executed by all participants concurrently in a deterministic order of transaction IDs. Similar to [10], they do not consider the issue of finding optimal partial order, either.

3   System Overview and Baseline Approach

In this section, we will give a brief overview of the network structure and transaction processing workflow. Then we introduce our baseline approach.

3.1   System Overview

Network Structure. Our blockchain network is composed of one *proposer* and many *attestors*. The post of proposer periodically rotates among all nodes. A BFT protocol like *PBFT* is installed in the network to ensure consistency. Each node has equal chances to take a leadership role, and is equipped with similar computation and communication resource.

Transaction Processing Workflow. In the beginning, the proposer on duty reaps a set of transactions, and partitions them into a sequence of batches, $B_1, \cdots, B_m$. Next, it packages and broadcasts a block consisting of these batches. Subsequently, all attestors will execute and verify the transactions.

3.2   Baseline Approach

Our Baseline approach consists of two major phases. In the proposing phase, the proposer will execute some transactions concurrently and obtain a series of batches. Then in the attestation phase, all attestors are able to execute each batch of transactions in parallel and keep consistency with each other.

Generating Batches by Deterministic OCC (proposer). As illustrated in Algorithm 1, the proposer needs to repeatedly fetch some transactions from the input transaction set T (Step 4). Step 5 invokes a procedure DOCC (Deterministic OCC, Algorithm 2) to find a set $B \subseteq S$ which can be executed simultaneously, i.e. a *batch*. Next, we append B to the result list and remove it from T. The loop will continue till T is empty.

Algorithm 1: Generate Batches (proposer)

Input: T, a set of transactions.
Output: $\mathbb{B}$, a sequence of batches.

1 $\mathcal{D} \leftarrow$ local state DB
2 $\mathbb{B} \leftarrow \emptyset$
3 **while** T *Not Empty* **do**
4       $S \leftarrow RandomFetch(T, k)$ //k is the maximum value of batch size
5       $B, \mathcal{D} \leftarrow \text{DOCC}(S, \mathcal{D})$
6       $\mathbb{B}.append(\{B\})$
7       $T \leftarrow T \backslash B$
8 **return** $\mathbb{B}, \mathcal{D}$

Algorithm 2 illustrates a modified optimistic concurrency control method adapted from [10]. Steps 2 to 6 execute all transactions in parallel. The set writes[x] stores the ID of the transaction with the first priority to update x. In DOCC the committing priority is decoded in the function *isPrecedent*. The Baseline approach determines the priority by comparing the IDs, i.e. the one with smaller ID always precedes the other. We will revisit this issue in the next section. Steps 9–12 try to find all transactions which can be put into the current batch.

The function *hasDependency* is to examine whether a transaction is dependent on some transactions. Basically, for any input transaction $T_j \in \boldsymbol{S}$, *hasDependency* returns true iff there exist another transaction $T_i \in \boldsymbol{S}$ and $T_i \prec T_j$, where $\prec$ is the *precedence relation* between two transactions as defined in the below.

Definition 1. *Precedence Relation. For any two transactions T_i and T_j,*

- *If T_i and T_j have no conflicts, i.e. $(R[i] \cap W[j]) \cup (W[i] \cap R[j]) \cup (W[i] \cap W[j]) = \emptyset$, there are no precedence relations between T_i and T_j, denoted by $T_i \perp T_j$.*
- *Otherwise, $T_i \prec T_j$ iff $BNo(T_i) < BNo(T_j)$, or $i < j \wedge BNo(T_i) = BNo(T_j)$, where $BNo(T_i)$ is the index number of the batch containing T_i.*

Here $R[i]$ and $W[i]$ are the the read set and write set of a transaction T_i respectively. Note that since $T_i \prec T_j$, T_j is required to be committed *after* T_i, which is a critical setting to ensure the determinism among different nodes. We also want to mention that here we ignore the write-after-read dependency because a transaction can safely update a data item which has been read by some earlier transaction [10].

Algorithm 2: Deterministic OCC (proposer)

Input: S, a set of transactions. $\mathcal{D}$, original state DB.
Output: $\boldsymbol{B} \subseteq \boldsymbol{S}$ a batch. $\mathcal{D}$, updated state DB.

```
 1  Do in parallel
 2  │   Execute each transaction T ∈ S with D.
 3  │   for each x written by T do
 4  │   │   put x into W[T.ID]
 5  │   │   if isPrecedent(T,writes[x]) then
 6  │   │   └   writes[x] ← T.ID
 7  B ← ∅
 8  Do in parallel
 9  │   Check each tansaction T ∈ S
10  │   if ! hasDependency(R[T.ID], W[T.ID], writes) then
11  │   │   Put T into B
12  │   └   Commit T to D.
13  return B, D
```

Verify a Block (attestor). On receiving a block, each attestor should execute all the transactions concurrently in order to judge its validity. Next, if any batch contains conflicts, this block will be rejected. Hence the proposer cannot inject any incorrect blocks into the blockchain, and our approach still satify the BFT requirement.

Discussions. The most distinctive feature of the Baseline approach is the adoption of deterministic concurrency control. According to the definition of precedence relation(Definition 1), there is no need to check the conflicts between two transactions from different batches. Therefore, the workload of proposers will be much lower compared with the works in [1,4].

Despite these metrits, the Baseline approach has, nevertheless, several problems. The first problem is the repetition of executing the same transaction. Note that if a transaction cannot be committed (Step 10 of Algorithm 2), it should be re-executed each time it is selected (Step 5 of Algorithm 1) because its read/write sets may change. The problem of repeatedly executing transactions comes from the "no conflicts in batch" requirement. However, this requirement is too strict.

Another problem is that the Baseline approach pays no attention to the time costs of executing transactions when generating batches, which may lead to poor use of multi-core processors. According to our experimental results, both transaction independency and the distribution of running costs will affect the utilization rate of CPU. We will discuss this issue in the next section.

4   DVC: A Two-Stage Appoach Based on DOCC and Vetex-Coloring

In this section, we introduce a two-stage approach to deal with the problems listed in Sect. 3.2. Generally speaking, the process of generating batches can be divided into two stages as listed in Algorithm3.

In the first stage (Algorithm 3), we try to generate some batches by DOCC (Steps 5 to 9). In fact this part is quite similar to the Baseline approach. The only difference is that, the proposer does not need to re-execute transactions (Sect. 4.1).

We also modify the DOCC procedure by introducing a new equation to determine precedence. The original metric of determining precedence (Definition 1) only compares transaction IDs. From our observation, transactions with larger running costs should be put together to improve the usage of multi-core processors. We thus propose a new metric to reorder transactions (Sect. 4.2).

The transactions returned by DOCC will be put into some batch (Step 7), and the remaining ones are inserted into a weighted transaction conflict graph $\mathcal{G}$ (Step 9). Each node of $\mathcal{G}$ is attached with a positive weight as the time cost of executing the corresponding transaction. After that, in the second stage (Step 10), we try to partition the remaining transactions into a set of batches, by finding an approximate optimal partition of the aborted transactions in $\mathcal{G}$. The details will be explained in Sect. 4.3.

4.1   Tolerating Intra-batch Conflicts

During the attestation phase, the attestors need to execute all transactions in parallel and then detect possible conflicts. When meeting a batch containing conflicts, the attestors can re-execute this batch instead of simply rejecting it.

Firstly, an attestor will execute all transactions inside a batch in parallel and record their read/write sets. Next, these transactions will be committed one by one by the order of their IDs. If a transaction T is found to have conflicts with any transactions preceding it, T will be re-executed and its read/write sets will be updated.

Algorithm 3: Generate Batches by DVC(Proposer)

Input: T, a set of transactions.
Output: $\mathbb{B}$, a sequence of batches.

1 Init $\mathcal{G}$ as an empty graph.
2 $\mathbb{B} \leftarrow \emptyset$
3 $\mathcal{D} \leftarrow$ local state DB
4 **while** $\mathbb{T}$ *Not Empty* **do**
5     $S \leftarrow RandomFetch(T, k)$
6     $B, \mathcal{D} \leftarrow \mathrm{DOCC}(S, \mathcal{D})$
7     $\mathbb{B}.append(\{B\})$
8     $T \leftarrow T \backslash S$
9   insertVetecies$(\mathcal{G}, S \setminus B)$
10 $\mathbb{B}.append(\mathrm{LocalSearch}(\mathcal{G}))$
11 **return** $\mathbb{B}$

4.2 Reordering Transactions to Improve Parallelism

According to our observations, it is valuable to avoid putting a transaction with high cost inside a small batch, because it may bring down the usage of a multicore processor. For this purpose, we adopt a simple but effective heuristic to reorder the conflicted transactions. If two transactions have conflicts, the one with higher cost will have priority. To implement this heuristic, we only need to rewise the second clause of Definition 1 as the following.

– Otherwise, $T_i \prec T_j$ iff $BNo(T_i) < BNo(T_j)$ or $(c[i] > c[j]\|\|i < j \wedge c[i] = c[j]) \wedge BNo(T_i) = BNo(T_j)$, where $c[i]$ is the time cost of executing T_i.

4.3 Finding an Optimal Partition

At the end of the first stage, all the aborted transactions are put into a weighted TCG, i.e. $\mathcal{G}$. Now we explain how to partition this set of correlated transactions.

    If we want to partition a set of transactions into several batches, the total cost of executing these batches are at least the sum of the most time-consuming transaction inside each batch. Formally speaking, given a set of transactions T, we need to partition T into $B_1, \cdots B_m$, such that the transactions inside each B_i are pairwise independent, and the value of $\sum_i max_{T \in B_i} c(T)$ should be minimized, where $c(T)$ is the time cost of executing T.

    The above problem is exactly the Minimum Vetex Weighted Coloring problem (MinVWC) [11]. A *feasible coloring* is a partition $\{V_1, \cdots, V_k\}$ of the node set $\mathbf{V} = \cup_{i=1}^{k} V_i$, such that for any edge $< u, v >$ inside the graph, u and v are not inside the same partition. The target of MinVWC is find a feasible coloring which minimizes $\sum_{i=1}^{k} max_{v \in V_i} c(v)$.

    It is not hard to see that with the undirected graph $\mathcal{G}$, the above problem is exactly mapped to an MinVWC. Therefore we could find an optimal partition by solving MinVWC.

Unfortunately, MinVWC is a known NP-hard problem [11], we thus adopt an advanced approximate method, *RedLS* [11]. However, our scenario is somehow different from classical MinVWC. The *RedLS* algorithm is designed for large graphs with millions of nodes. *RedLS* usually requires several minutes to find a good coloring for this kind of huge graphs, while we can only spend at most dozens of milliseconds. The graphs in our scenario are much smaller and it is not necessary to compute the maximum cliques as suggested by the original *RedLS*.

5   Performance Evaluation

In this section, we report the experimental results on the performance of our proposed approaches.

5.1   Experiment Setting

Implementation. We implement our methods and two competing methods [4, 10] based on an open source platform, Tendermint ®[1]. We also adopt an open source implementation of the RedLS algorithm [11]. All the codes are written in Go1.16.13, and are running on CentOS® 7.

Hardware Configuration. We deploy the nodes on the cloud platform in the National Supercomputer Center in Guangzhou[2]. Each virtual machine is equipped with a 4-core processor of 2.2 MHz and 4 GB main memory.

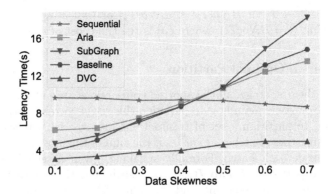

Fig. 1. Effect of Data Skewness

Compared Methods. We compare our proposed methods with several recent works, Aria [10] is a deterministic database system. We implement and rewise it as a simple method to concurrently execute Blockchain transactions.

---

[1]   https://tendermint.com/
[2]   http://www.nscc-gz.cn/

Datasets. We utilize a synthetic dataset SmallBank, which is also adopted in [4,12] to evaluate the performance of blockchains. In the begining of our simulation, the dataset has one million accounts. The pattern of accessing data items follows the Zipfian distribution with varying data skewness.

We randomly collect ten thousands of transactions from Ethereum®, and adapt the time costs of SmallBank transactions with the distribution of the GAS requirements in Ethereum transactions. Each Ethereum transaction needs to specify its maximum amount of GAS, which can be regarded as a reasonable estimation of its real time cost.

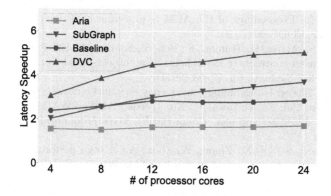

Fig. 2. Effect of # of Cores

5.2 Experimental Results

Now we are ready to report the results of our simulations. Each point of the following figures is an average of 50 runs.

Effect of Data Skewness. As illustrated in Fig. 1, DVC is the fastest approach, whose average time cost is about a third of the competitors. Aria, Baseline and SubGraph have similar performance. Their time costs increase quite fast when data skews.

Effect of # of Cores. In this experiment, we increase the number of cores from 4 to 24, in order to evalute the degree of parallelism. Figure 2 shows the speedup of the parallel approaches over Sequential. As shown in Fig. 2, the latency of DVC decreases with more cores. The highest speedup is about 5X when each machine is equipped with 20+ cores. Both Baseline and SubGraph performs better than Aria, whose speedup is about 1.8X.

6   Conclusion

In this paper, we propose a two stage approach to efficiently execute Blockchain transactions in parallel. We try to reduce the workload of proposers by partitioning the transactions into a sequence of batches. Also we adopt an advanced approximate algorithm of solving MinVWC to find optimal partition. We conduct a series of experiments, and illustrate that the proposed approach achieved better performance than the competitors in different scenarios.

Acknowledgments. This work is partially supported by National Key R&D Program of China (2022YFB2702100), National Science Foundation of China (U1911203).

References

1. Dickerson, T., Gazzillo, P., Herlihy, M., Koskinen, E.: Adding concurrency to smart contracts. In: Proceedings of the ACM Symposium on Principles of Distributed Computing. PODC 2017 (2017)
2. Anjana, P.S., Attiya, H., Kumari, S., Peri, S., Somani, A.: Efficient concurrent execution of smart contracts in blockchains using object-based transactional memory. In: Networked Systems, pp. 77–93 (2021)
3. Zhang, A., Zhang, K.: Enabling concurrency on smart contracts using multiversion ordering. In: Cai, Y., Ishikawa, Y., Xu, J. (eds.) APWeb-WAIM 2018. LNCS, vol. 10988, pp. 425–439. Springer, Cham (2018). https://doi.org/10.1007/978-3-319-96893-3_32
4. Jin, C., Pang, S., Qi, X., Zhang, Z., Zhou, A.: A high performance concurrency protocol for smart contracts of permissioned blockchain. IEEE Trans. Knowl. Data Eng. **34**, 1–1 (2021)
5. Li, Y., et al.: Fastblock: Accelerating blockchains via hardware transactional memory. In: 2021 IEEE 41st International Conference on Distributed Computing Systems (ICDCS) (2021)
6. Bartoletti, M., Galletta, L., Murgia, M.: A true concurrent model of smart contracts executions. In: Bliudze, S., Bocchi, L. (eds.) COORDINATION 2020. LNCS, vol. 12134, pp. 243–260. Springer, Cham (2020). https://doi.org/10.1007/978-3-030-50029-0_16
7. Baheti, S., Anjana, P.S., Peri, S., Simmhan, Y.: DiPETrans: a framework for distributed parallel execution of transactions of blocks in blockchains. Concurrency Comput. Pract. Experience **34**(10), e6804 (2022)
8. Amiri, M.J., Agrawal, D., El Abbadi, A.: Parblockchain: Leveraging transaction parallelism in permissioned blockchain systems. In: 2019 IEEE 39th International Conference on Distributed Computing Systems (ICDCS), pp. 1337–1347 (2019)
9. Fang, M., Zhang, Z., Jin, C., Zhou, A.: High-performance smart contracts concurrent execution for permissioned blockchain using SGX. In: 2021 IEEE 37th International Conference on Data Engineering (ICDE), pp. 1907–1912 (2021)
10. Lu, Y., Yu, X., Cao, L., Madden, S.: Aria: a fast and practical deterministic OLTP database. Proc. VLDB Endow. **13**(12), 2047–2060 (2020)
11. Wang, Y., Cai, S., Pan, S., Li, X., Yin, M.: Reduction and local search for weighted graph coloring problem. In: Proceedings of the Thirty-Fourth AAAI Conference on Artificial Intelligence, pp. 2433–2441. AAAI 2020 (2020)
12. Peng, Z., et al.: Neuchain: a fast permissioned blockchain system with deterministic ordering. Proc. VLDB Endow. **15**(11), 2585–2598 (2022)

Anole: A Lightweight and Verifiable Learned-Based Index for Time Range Query on Blockchain Systems

Jian Chang, Binhong Li, Jiang Xiao[(✉)], Licheng Lin, and Hai Jin

National Engineering Research Center for Big Data Technology and System, Services Computing Technology and System Lab, Cluster and Grid Computing Lab, School of Computer Science and Technology, Huazhong University of Science and Technology, Wuhan 430074, China
{j_cahng,binhong,jiangxiao,lichenglin,hjin}@hust.edu.cn

Abstract. Time range query is essential to facilitate a wide range of blockchain applications such as data provenance in the supply chain. Existing blockchain systems adopt the storage-consuming tree-based index structure for better query performance, however, fail to efficiently work for most blockchain nodes with limited resources. In this paper, we propose *Anole*, a lightweight and verifiable time range query mechanism, to present the feasibility of building up a learned-based index to achieve high performance and low storage costs on blockchain systems. The key idea of *Anole* is to exploit the temporal characteristics of blockchain data distribution and design a tailored lightweight index to reduce storage costs. Moreover, it uses a digital signature to guarantee the correctness and completeness of query results by considering the learned index's error bounds, and applies batch verification to further improve verification performance. Experimental results demonstrate that *Anole* improves the query performance by up to 10× and reduces the storage overhead by 99.4% compared with the state-of-the-art vChain+.

Keywords: Blockchain · Time range query · Learned index · Lightweight

1 Introduction

Blockchain has become a promising distributed ledger technology for multiple parties to engage and share a decentralized tamper-proof database [13]. Blockchain systems record the transactions between these parties in timestamped and chronologically linked data blocks [11]. Time range query is the most fundamental query type on blockchain systems that retrieves data blocks within a given time interval. It has played a pivotal role in supporting trustworthy data provenance and traceability for many crucial applications, such as supply chain [8], smart manufacturing [9], healthcare [12]. For example, blockchain time range query can support efficient and trustworthy contact tracing, vaccine logistics, and donations during the COVID-19 outbreak. Donors hope to query

© The Author(s), under exclusive license to Springer Nature Switzerland AG 2023
X. Wang et al. (Eds.): DASFAA 2023, LNCS 13943, pp. 519–534, 2023.
https://doi.org/10.1007/978-3-031-30637-2_34

about their donations (Bitcoin or Ethereum, etc.) within a given time interval, so as to know the whereabouts of the donation funds. The client sends Q = ([Addr:2AC03E7F], [2022.06.18, 2022.09.17], [in/out]) to get the transactions for address 2AC03E7F from June 18, 2022, to September 17, 2022.

A lightweight and verifiable time range query mechanism is desirable and important for blockchain systems. While blockchain full nodes (e.g., resource-rich servers) store the ever-growing immutable data blocks, they can retrieve reliable results upon the full history. In practice, however, most users on blockchain are light nodes (e.g., mobile users with constrained resources), who can only rely on full nodes by proceeding with remote queries. Since the full nodes can be malicious in the trustless blockchain, light nodes must further verify the query results to ensure the correctness and completeness. Moreover, the indexing for accelerating query processing will bring in additional and even unacceptable storage overhead [24].

Prior studies [14,18,20] have exploited the tree-based index structure for facilitating verifiable blockchain Boolean range query. The authors built a *verification object* (VO) set and reconstructed the Merkle tree root to verify the query results. Since the tree-based indexing query mechanism trades off storing the large-size VO for query performance, it is not affordable for lightweight nodes with limited storage space. Furthermore, it incurs an excessively time-consuming traverse process (i.e., each tree element must be equally traversed), leading to significant performance degradation.

In this paper, we explore an alternative approach towards lightweight and verifiable blockchain time range query. Our key insight is that the data blocks of a blockchain exhibit a temporal distribution by nature, as the time interval between block generation is determined by the underlying consensus protocol. For example, the Bitcoin system generates a block about every ten minutes with the *Proof-of-Work* (PoW) consensus. It allows us to build a regression function between the timestamp (i.e., key) and block height (i.e., the value of the position). Inspired by [19], our design takes the advantage of learned index that supports efficient time range query tailored to the blockchain data.

However, it is challenging to build up an appropriate blockchain learned index that can simultaneously achieve high performance and low storage costs. The first *challenge 1* is how to build a novel index structure based on the temporal distribution of blockchain. The malicious operations in a trustless blockchain system can prohibit the procedure of block generation, such as the selfish mining attacks [2,10] that delay publication of blocks (i.e., 5% of blocks were generated for more than 30 min). The second *challenge 2* is even after we successfully construct a tailored learned index, how to design a lightweight and efficient method to verify the query results? Because of diverse underlying storage models, it is not possible to directly apply the tree-based VO on learned index. The learned index can locate the position of the query data in a list, but it is difficult to map to the position in the *merkle hash tree* (MHT) to get VO. Hence, the conventional MHT-based verification approaches are not suitable for learned index structures.

To address the above challenges, we present *Anole*, a lightweight and verifiable time range query mechanism that explores the use of learned index structures on blockchain systems. It includes three major components: (1) To automatically learn the relationship between the timestamp and block height, *Anole* leverages piecewise linear functions to build a novel layered learned index structure, which can reduce the storage cost by orders of magnitude by storing function argument. (2) *Anole* proposes the aggregate signature that greatly reduces the VO size and alleviates the burden of data transmission in the blockchain network compared to the classical tree-based approaches. (3) We further develop a lightweight batch verification for ensuring the correctness and completeness, while retaining high performance. We have implemented a fully functional, open-sourced query prototype of *Anole*.

In summary, we make the following contributions:

- We propose *Anole*, a novel learned index-based lightweight and efficient mechanism for blockchain time range query. To the best of our knowledge, this is the first layered learned index structure that can automatically bound the query results by temporal distribution patterns. Moreover, it can reduce the storage overhead of building indexes, and improve query performance.
- We also develop two optimizations - aggregation signature and batch verification - which significantly reduce the verification overhead. It not only keeps the VO size lightweight and guarantees the verifiability of query results.
- Experimental results demonstrate that *Anole* significantly outperforms the state-of-the-art vChain+ [18] in terms of storage overhead and query performance.

The rest of this paper is organized as follows. We outline the related work in Sect. 2, prior to introducing the system overview of Anole in Sect. 3. Section 4 and Sect. 5 present the detailed design of layered learned index and lightweight verification process of Anole. Comprehensive evaluation is shown in Sect. 6. Finally, we conclude this work in Sect. 7.

2   Related Work

In this section, we review the most relevant range query techniques over traditional database and discuss state-of-the-art blockchain verifiable query mechanisms in Table 1.

Database Range Queries. Considering the impact of data partitioning on large-scale data processing, Yue et al. [22] proposed a time-based partitioning technology to provide range query operations on large-scale trajectory data. To further improve the query efficiency, learned-based techniques for data query are used in database [1,17]. The structure of the FITing-Tree [7] is very similar to the traditional B+ tree, and the difference is that its leaf nodes store the start key and slope of each segment. PGM-index [6] optimized FITing-Tree from data segmentation, insertion, and deletion, which can achieve better query and update time efficiency. ALEX [5] proposed another scheme to support insertion operation in which a newly arrived key keeps the array in order at the predicted position gap. These learned indexes can efficiently support range query operations in traditional databases but fail to offer verifiable queries in a trustless blockchain environment with malicious nodes.

Table 1. The comparison of Anole with existing query mechanisms

Category	Approach	Time range	Lightweight		Verification	
			Index size	VO size	Correctness	Completeness
Distributed database	FITing-Tree [7]	✔	✔	✖	✔	✖
	PGM-index [6]	✔	✔	✖	✔	✖
	ALEX [5]	✔	✔	✖	✔	✖
Blockchain	LineageChain [15]	◉	✖	◉	✔	✔
	GEM²-Tree [23]	✖	◉	✖	✔	✔
	P²B-Trace [14]	◉	✖	◉	✔	◉
	LVQ [4]	✖	◉	◉	✔	✔
	vChain+ [18]	✔	◉	◉	✔	✔
	Anole	✔	✔	✔	✔	✔

NOTE: ✔ : support   ◉ : poor support   ✖ : not support

Blockchain Verifiable Queries. Efficient query and processing of a large number of time series data have attracted unprecedented attention from both industry and academia [21]. Shao et al. [16] presented an authentication range query scheme based on *Trusted Execution Environment* (TEE). However, due to the limited secure memory space, existing TEEs cannot easily handle large-scale applications. Besides, GEM²-tree [23] designed a two-level index structure, which is gas-efficient and effective in supporting authenticated queries. LVQ [4] presented a new Bloom filter based on sorted Merkle Tree to achieve lightweight verifiable but not support time range query. Unfortunately, the maintenance cost of the tree-based *authenticated data structure* (ADS) is relatively heavy for light nodes to support the verification procedure. Moreover, LineageChain [15] provided a new skip list index to support efficient provenance. To overcome the practical problem of public key management in vChain, vChain+ [18] proposed a new *sliding window accumulator* (SWA) to reduce the public key storage overhead of accumulators. However, their design takes up a large amount of storage space for VO and computing overhead, which requires high node configuration.

3   System Overview

Figure 1 shows the overview of *Anole* system consisting of three actors: miners, full nodes, and clients. The miners, responsible for packaging data into blocks and appending new blocks to the blockchain, are considered trusted third parties because of the rewarding scheme of the blockchain system. The full nodes store both block headers and data, responsible for processing the client's query requests. When the clients send a query request, the full nodes take advantage of the learned index to find the location of request data, and then return the query result and corresponding digital signature as VO for the clients to verify. The clients, who store only block headers, use the public key and VO to check the completeness and correctness of the query result.

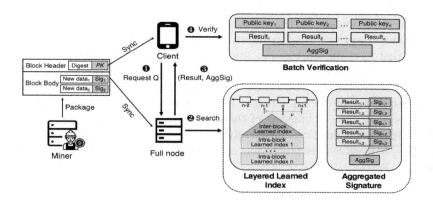

Fig. 1. The system overview of *Anole*

Suppose a client C submits a request Q to a full node for retrieving the transactions during last two weeks on blockchain (Step ❶ in Fig. 1). To ensure query efficiency, the full node utilizes learned index for retrieval, that is, the block height range of the element is quickly located through the inter-block learned index, and the query results that meet the conditions are searched through the intra-block learned index (step ❷). Multiple digital signatures are combined into one aggregated signature to reduce the overall VO size. After query execution, the full node assembles a tuple, including the result and the *aggregated signature* (AggSig), and sends it to the client (Step ❸). Upon receiving all the results and aggregated signature from the full node, the client obtains the corresponding public keys by synchronizing the block header to verify the returned results and VO in batch (Step ❹).

4   Learned Index-Based Time Range Queries

In this section, we propose a layered learned index to meet *challenge 1*, which captures mapping relationships between timestamps and block height. Furthermore, we discuss efficient query execution and theoretical analysis of error bound.

4.1   Layered Learned Index

Layer 1: Inter-block Learned Index. The main idea of our method is to extract the mapping relationship between the data through a piecewise linear function. We propose a dynamic piecewise linear regression algorithm based on the distribution of timestamp and block height, which is linear in run time. *Anole* introduces error bound to ensure that all the data of the learned index can be retrieved, including the occurrence of outliers that deviate from the normal distribution. More specifically, we define the error bounds for dynamic piecewise linear regression by three parallel lines: the start function that a line formed by two starting points, the upper boundary function that the regression function plus the error bound, and the lower boundary function that the regression function minus the error bound as shown in Fig. 2. Points 1 and 2 are the two basic points that form the starting function. Point 3 is inside the two boundaries, and the current regression does not need to be updated. As point 4 is outside the updated regression boundary (i.e., the actual value of point 4 is under the lower boundary based on the error bound μ), a new piecewise regression function will be generated. The combination of these two boundary functions gives the edges of the regression function. Intuitively, the two boundary functions represent a sequence of feasible linear regressors around the beginning of the regression function.

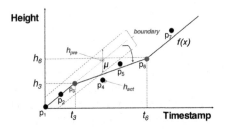

Fig. 2. The dynamic regression of inter-block learned index

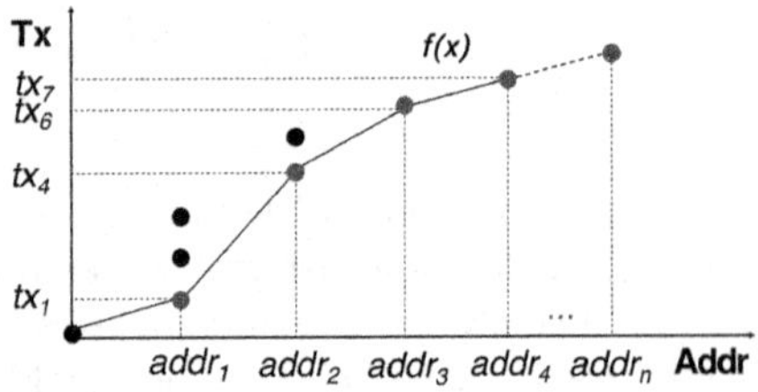

Fig. 3. The mapping of intra-block learned index

Layer 2: Intra-block Learned Index. One block usually contains numerous transactions (e.g., bitcoin averages over 1,000 transactions per block). For an active address, in addition to having multiple transactions between different blocks, there may also be multiple transactions within a block. Based on this observation, we construct an intra-block learned index to optimize the query time in block. The intra-block index is constructed by blockchain miners in an aggregated manner based on identical address transactions. Figure 3 shows the block with an intra-block index. It sorts and aggregates the transactions corresponding to each address, and the first transaction of the identical address is used as the aggregation point (e.g., tx_1 for $addr_1$, and tx_4 for $addr_2$). Then, the intra-block learned index is constructed based on the sorted transactions and different aggregation points. In particular, since the amount of data in a single block is small and does not need to be updated, the error bound of intra-block learned index can be set to 0 to achieve precise positioning.

4.2 Efficient Query Execution

In this section, we start by considering a particular timestamp and focusing on
the point query with learned index for ease of illustration. Then we extend it
to the time range query condition to show how to process time range query
requests efficiently. Anole maintains the parameters of each segment, including the starting point, the slope, and the intercept. The timestamp value is an
increment property, hence we use a variant of the binary search algorithm to
quickly search the segment where the key value is located. The time complexity is $O(log_2(n))$ of searching for the specific segment, where n is the length of
segment list. Once the segment is found with the corresponding timestamp, we
can locate the block height of the given timestamp by calculating the function
in this segment. Recall that when creating a piecewise linear function, the actual
position of the key is kept in a range (i.e., error bound μ) from the position calculated by the regression function. According to the slope and intercept parameter
of segment s, we can obtain the predicted position of the given timestamp t by
calculating the following equation:

$$pred_h(t) = t * s.slope + s.intercept \tag{1}$$

The true location of the elements can be restricted to the error bound once the
learned index is constructed by dynamic piecewise linear function. Consequently,
after the predicted position is obtained by calculation, sequential traversal is
used to perform a local retrieval for the blocks within the error bound. The
true position of the given timestamp t and error bound μ is calculated by the
following equation:

$$true_h(t) \in [pred_h(t) - \mu, pred_h(t) + \mu] \tag{2}$$

Time range query is a special case of range query that has the subsidiary
conditions. It requires to check whether each item is within a particular given
time range as shown in Algorithm 1. Hence, unlike point query, for a time range
query, the condition has a great impact on the total running time and the result
size. The main difference between the two query types is that the time range
query requires finding two endpoints of the given time range. Since the segment
satisfies the given error bound, the overhead of finding the key within the segment
can be restricted. To be more precise, the time complexity of searching keys
within the bound is $O(1 + 2 * \mu)$. The linear function either points to the store
key consecutively in the same segment or exists in adjacent segments, where the
segmented points are sorted by value. Therefore, Anole can first simply start the
scan at the start point position of the time range within the error bound, and
then traverse the adjacent segments to the end point of the range.

Algorithm 1: Time Range Query

---

  input : $Q = (addr, <t_1, t_2>)$
  output: $R = (TxSet, VO)$
1   $(seg_1, seg_2) \longleftarrow VBS(t_1, t_2)$;
2   $pre\_h_i \longleftarrow seg_i.slope * t_i + seg_i.intercept \quad (i = 1, 2)$;
3   **for** h **in** $(pred\_h_1 - error)..(pred\_h_2 + error)$ **do**
4     $Tx.id \longleftarrow intra\_fn(addr)$;
5   **if** Tx $exist$ **then**
6       $TxSet = TxSet.append(Tx)$;
7       $VO = aggr(Tx.sign)$;
8   **else**
9       $TxSet = prior(Tx.id) + next(Tx.id)$;
10      $VO = prior(Tx.id).sign + next(Tx.id).sign$
11   **end**
12 **end**
13 **return** $R = (TxSet, VO)$

4.3   Theoretical Analysis

The error bound has an impact on the query efficiency and the size of the index. Inevitably, the following question naturally arises: how to choose the error bound? To deal with this trade-off, we define an assessment model to choose a "suitable" error bound during the learned index construction. There are two essential indicators that can be optimized regarding the error bound: the query performance impact on the system (i.e., query latency) and the storage overhead of the system (i.e., index size).

 The value of error bound affects the segment's generation and the searching precision (i.e., the large error bound has fewer segments produced and lower precision). We let N_μ denote the segment's amount. The query latency for computation estimated by the error bound μ can be calculated by the following equation, where α is the number of function parameters, β is the intra-block transaction data, and c is the system delay of the instruction on given hardware (e.g., 10 ns).

$$Latency(\mu) = c * (log_2(N_\mu) + \alpha + 2\mu * \beta) \tag{3}$$

 For a given error bound μ, we can evaluate the storage cost of the learned index (in Byte) using the following equation. The first part of the formula is the size of the index parameter (i.e., the slope and intercept parameter, each 2 bytes), and the second part is the space occupied by the digital signature within the error bound (64 bytes/block), which is discussed in the verification section.

$$Storage(\mu) = N_\mu * \alpha * 2B + 2\mu * 64B \tag{4}$$

Based on these two cost estimation formulas, it can be obtained that the minimum storage for the learned index that meets a specific latency demand $L(ms)$ or the minimum error bound for the learned index that satisfies a given storage budget $S(bytes)$. Therefore, the most suitable μ is given by the following expression, where E denotes a set of possible error bounds (e.g., $E = \{1, 2, 5, 10\}$).

$$\mu = argmin \begin{cases} Storage(\mu) & | & Latency(\mu) \leq L \\ Latency(\mu) & | & Storage(\mu) \leq S \end{cases}, \quad \mu \in E \qquad (5)$$

5   Lightweight Verification

In this section, we propose a lightweight verification scheme based on the learned index and digital signatures to meet *challenge 2*, which omits the step of re-traversing in MHT and enables dynamic half-aggregation to reduce the VO size. Furthermore, we optimize the digital signatures to achieve efficient batch verification on the basis of ensuring security to remedy the defect of long verification time.

5.1   Lightweight VO Design

Aggregated Signature. We divide the aggregated signatures into two parts: inter-block signature aggregation and intra-block signature aggregation. Intra-block signature aggregation is done by miners when packing blocks. As in the example shown in Fig. 4, miners bundle transactions with the same address, e.g., join all transactions with an address of $452daksm$, $O = o_5||o_6||o_7||o_8||o_9$. Then miners sign the object O to obtain the signature (R, s). In this way, it greatly reduces the size of the verification object in a block, as each block returns only one signature (R, s) for a certain address. To further reduce the size of VO, Anole proposes a dynamic inter-block half-aggregation technique. As the block height will change with the time range, Anole assigns the inter-block signature aggregation to the full node. With this method, the full node can dynamically aggregate signatures according to the query request and not return redundant data for verification. The implementation is shown in Fig. 5, where λ is the security parameter, g is the generator of a cyclic group G, (Pk, sk) is the keypair where Pk is the public key and sk is the private key, σ is the signature which can be decomposed into (R, s), and m is the message. It is worth noting that the coefficient L can effectively prevent a malicious full node from tampering with query data or signatures, as it commits to each signature, message, and public key.

Batch Verification. Compared with the MHT-based authentication method, the digital signature eliminates unnecessary data and greatly reduces the size of the verification object, but it pays a price in terms of verification time. To solve this problem, we design a batch verification method based on the Straus

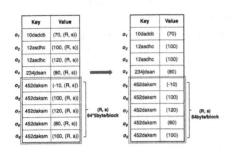

Fig. 4. Example of intra-block aggregated signatures

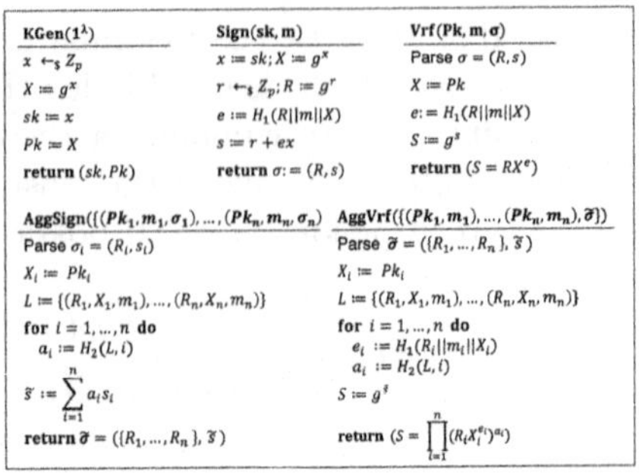

Fig. 5. The scheme of inter-block aggregated schnorr signature

algorithm [3] to speed up the verification. The basic authentication method is to verify the signature of query results according to the aggregated signature by the following equation:

$$g^{\tilde{s}} = \prod_{i=0}^{n} (R_i X_i^{e_i})^{a_i} \tag{6}$$

We analyze the computation overhead of each step in verification and observe that exponentiation in a cyclic group is time-consuming. To deal with this problem, we combine verification with the Straus algorithm to reduce the time overhead of exponentiation. To describe more formally, we transform the right side of the verification equation into the expression:

$$\prod_{i=0}^{n} (R_i X_i^{e_i})^{a_i} = \prod_{i=0}^{2n} M_i^{t_i} \tag{7}$$

The verification algorithm performs the following. First, we choose a radix 2^c for the algorithm to compute. Generally, $c = 5$ is appropriate for 256-bit scalars. Second, we precompute the quantity $M_i, 2M_i, ..., (2^c - 1)M_i$ to reduce the cost of subsequent calculations. Third, we recursively compute $\lfloor t_i/2^c \rfloor$ until $t_i = 0$ and record the remainder $N_j = \prod_{i=0}^{n} M_i^{t_i \bmod 2^c}$, where j is the number of divisions performed. Finally, we can recursively compute the result R by $R_{j-1} = R_j^{2^c} N_j$ until $j = 1$. In this way, the total cost of identifying the forgeries among n signatures at a 2^b security level is compressed from roughly $2^b n$ multiplications to roughly $(2 + 8/lgb)nb$ multiplications.

5.2 Verifiable Query Processing

In this section, we focus on how to generate the verification object for clients to verify the completeness and correctness of query results. We use concrete examples to illustrate the security of the validation process. The query type of Anole is the time range query for a specific address, which is in the form of $Q = <[t_1, t_2], key>$.

Completeness. The completeness proof generated by full nodes consists of two aspects: timestamp and the key of data. For timestamp, the full node returns the left and right boundaries to prove that the query results contain all blocks satisfied the requirement for timestamp. To process the query request, the full node uses the inter-block learned index to locate the block heights (h_1, h_2) corresponding to t_1 and t_2 respectively. Then, check the timestamp of the block height in the interval $[h_1 - \mu - 1, h_1 + \mu + 1]$ and $[h_2 - \mu - 1, h_2 + \mu + 1]$, where μ is the error bound of the intra-learned index. In this way, the full node can easily find the left and right boundaries (h_l, h_r) and add them to VO. Note that h_l and h_r are the maximum block height with the timestamp smaller than t_1 and the minimum block height with the timestamp larger than t_2. Figure 6 depicts an example of completeness proof for timestamp when the time range query $Q = <[20210513, 20210514], addr>$. According to the calculation results in the interval $2([h_1 - 2, h_1 + 2], \mu = 1)$, we find the maximum block height 114503 as left boundary h_l, of which the timestamp is smaller than 20210513. We can find h_r in the same way and the completeness proof for the timestamp is $[h_l = 114503, h_r = 114508]$.

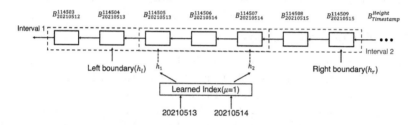

Fig. 6. The completeness proof of timestamp

For the part of key (i.e., address), we divide the completeness proof into two aspects: existence proof and inexistence proof. Existence proof should be able to prove that all the query results are exactly existed inside the block and no key-related transactions are omitted. Since miners are considered as trusted third parties, the aggregated signature (R_i, s_i) that miners generate can be used as the existence proof for the transactions in block i.

For the inexistence proof, we also use the boundary determination principle to prove the key is inexistent in a block. To generate the inexistence proof, we require miners to add an additional attribute p to each transaction when they sort the block data and package the block. The attribute p represents the position of the transaction in the block. When the client queries for a particular transaction, the full node uses the learned index to find the position p_1, where it would be if it exists. As the transactions are sorted lexicographically during block packaging, the full node only returns the corresponding signatures of transactions with positions $p_1 - 1$ and p_1 as boundaries (i.e., the value of key_{p_1} is greater than the value of key_{query}). To keep the VO lightweight, we only return a single signature and transaction for each boundary as inexistence proof.

Correctness. Correctness requires that all the query results are satisfied and correct. Consider the VO is $< h_1, h_2, s_{sum}, (R_{addr})_{height} * n) >$. To verify the correctness of query results, the client checks the timestamps of the block in $[h_1, h_2]$ and the key of query results to ensure query results are satisfied. Then, the client can achieve batch verification by calculating the Eq. 6. Due to the uncontrollable discrete logarithm problem and collision resistance of hash functions, it is hard for full nodes to forge signatures, especially when the private key of signature is not known. It means the query results are correct if the verification is passed.

6 Evaluation

The public Bitcoin data set is used in the experiments, which is extracted from the Bitcoin system from June 18, 2022, to September 17, 2022. It contains 13361 blocks, and the transaction is reorganized as <block height, address, in/out, amount, timestamp>. The miners sign the data and store the public key in the block header (32 bytes). We perform experiments on a server Intel Xeon (Ice Lake) Platinum 8369B with 3.5 GHz CPUs, running CentOS 7.6 with 16 cores and 32 GB memory, and the Anole system is programmed in Rust language.

6.1 Index Construction Cost

Figure 7 reveals the index construction overhead and block generation for the miner with the different number of blocks, involving the CPU running time and the storage cost of index. In Anole, the number of blocks is set from 256 to 13361 with the error bound as 2. For vChain+, we set the same block size, and the parameters of fanout and time window for SWA-B+-tree are set to 4 and 2 respectively. The error bound of learned index is set to 2 according to the impact analysis in the following section. From Fig. 7(a), we can see that the running time of index construction in Anole is shorter than vChain+ and less than 4 times during the number of blocks is under 13361. Beyond that, Anole yields a smaller index structure compared with vChain+, as shown in Fig. 7(b). This is not unexpected because the accumulator employed in vChain+ takes a lot of storage space compared with the function parameters used in Anole to realize data location.

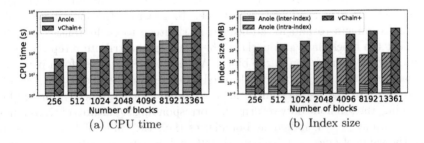

(a) CPU time   (b) Index size

Fig. 7. The comparison of index construction cost

6.2   Impact of Error Bounds

In the following, we assess the effects of various error bounds on the CPU processing time, index size, and latency performance of point queries. We conduct the scalability experiments and utilize the data set from 2048 to 13361 blocks. Figure 8(a) and Fig. 8(b) show the performance of CPU time and index size with error bounds varied from 1 to 10. It shows that the CPU time remains almost constant as the error bound increases, and the index size degrades from fast to slow with the increase of error bound. Next, we evaluate the impact of error bounds on latency. Figure 8(c) reveals the query latency of different block heights and the error bound varies from 1 to 10 with the maximum of blocks at 13361. We can observe that the query latency oscillates slightly when the error bound is less than 3, and then increases with the error bound because more block traversals are needed within a large error bound.

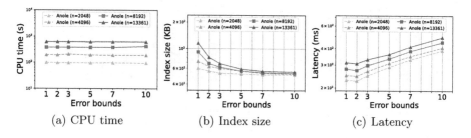

| (a) CPU time | (b) Index size | (c) Latency |

Fig. 8. The impact of error bounds (with the different number of blocks)

6.3   Query Performance

Figure 9 and Fig. 10 show the point and time range query performance of Anole during the block height from 1024 to 13361 and the time range from 4 to 24 h. Three types of query metrics, including query latency, verification time, and VO size are compared. Anole keeps the point query latency within 0.1 s as the block height increases. It has 1/100 latency compared with Anole w/o inter-block learned index and 1/10 lower than vChain+ in Fig. 9(a), respectively. When processing time range queries, we observe that Anole and Anole with inter-block index greatly exceed vChain+ when the time range is under 8 h in Fig. 10(a). The reason is that the inter-block learned index can quickly locate the time range, and combine the intra-block index to find elements. As the verification process for Anole in Fig. 9(b) and Fig. 9(c), the aggregated signature is returned when the query result exists, which reduces the verification time and VO size. In vChain+, the verification object is generated by ACC.Prove, which includes lots of set operations. The reason we can summarize is that the size of digital signature in Anole is smaller than the tree-based ADS generated by vChain+. As shown in Fig. 10(b) and Fig. 10(c), when the time range is within 24, the verification time and VO size of Anole are under 0.05 s and 5 KB, but these performance indicators of vChain+ are 0.6 s and 17 MB, respectively. We also experiment the throughput of Anole w/o aggregation, which is 70% of Anole as the large VO size with single signature leads to heavy network overhead.

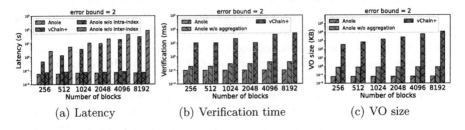

(a) Latency   (b) Verification time   (c) VO size

Fig. 9. The comparison of point query performance

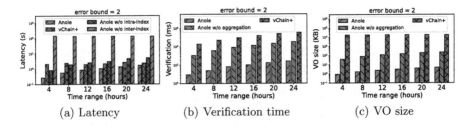

(a) Latency   (b) Verification time   (c) VO size

Fig. 10. The comparison of range query performance

7   Conclusion

We present *Anole*, the first lightweight and verifiable learned index-based blockchain time range query mechanism. *Anole* incorporates three key designs: (1) the layered learned index that captures the dynamic temporal distribution of data blocks in the untrusted blockchain environment; (2) the aggregated digital signature technique to reduce the size of the returned verifiable objects; and (3) the design of batch verification to speed up verification while guaranteeing the integrity and correctness of query results. Evaluation of our *Anole* prototype demonstrates that it achieves 10× average speedup and significantly reduces the storage overhead by 99.4%, in comparison with the state-of-the-art vChain+. We hope that our first step in introducing the learned index structures in blockchain query will seed the ground for further exploration on this topic.

Acknowledgement. This work was supported by National Key Research and Development Program of China under Grant No. 2021YFB2700700, Key Research and Development Program of Hubei Province No. 2021BEA164, National Natural Science Foundation of China (Grant No. 62072197), Key-Area Research and Development Program of Guangdong Province No. 2020B0101090005, Knowledge Innovation Program of Wuhan-Shuguang.

References

1. Bi, W., Zhang, H., Jing, Y., He, Z., Zhang, K., Wang, X.: Learning-based optimization for online approximate query processing. In: Bhattacharya, A., et al. (eds.) Database Systems for Advanced Applications. (DASFAA 2022). LNCS, vol. 13245, pp. 96–103 (2022). https://doi.org/10.1007/978-3-031-00123-9_7

2. Bissias, G., Levine, B.: Bobtail: improved blockchain security with low-variance mining. In: Proceedings of the 2020 Network and Distributed System Security (NDSS) Symposium, pp. 1–16 (2020)

3. Chen, C., Chen, X., Fang, Z.: Addition chains of vectors (problem 5125). Am. Math. Monthly **70**(1), 806–808 (1964)

4. Dai, X., et al.: LVQ: a lightweight verifiable query approach for transaction history in Bitcoin. In: Proceedings of the 40th International Conference on Distributed Computing Systems (ICDCS), pp. 1020–1030 (2020)

5. Ding, J., et al.: ALEX: an updatable adaptive learned index. In: Proceedings of the 2020 International Conference on Management of Data (SIGMOD), pp. 969–984 (2020)

6. Ferragina, P., Vinciguerra, G.: The PGM-index: a fully-dynamic compressed learned index with provable worst-case bounds. In: Proceedings of the 2020 International Conference on Very Large Data Bases (VLDB), pp. 1162–1175 (2020)

7. Galakatos, A., Markovitch, M., Binnig, C., Fonseca, R., Kraska, T.: FITing-Tree: a data-aware index structure. In: Proceedings of the 2019 International Conference on Management of Data (SIGMOD), pp. 1189–1206 (2019)

8. Han, R., et al.: Vassago: efficient and authenticated provenance query on multiple blockchains. In: Proceedings of the 40th International Symposium on Reliable Distributed Systems (SRDS), pp. 132–142 (2021)

9. Hewa, T., Braeken, A., Liyanage, M., Ylianttila, M.: Fog computing and blockchain-based security service architecture for 5G industrial IoT-enabled cloud manufacturing. IEEE Trans. Industr. Inform. **18**(10), 7174–7185 (2022)

10. Hou, C., et al.: SquirRL: automating attack analysis on blockchain incentive mechanisms with deep reinforcement learning. In: Proceedings of the 2021 Network and Distributed System Security (NDSS) Symposium, pp. 1–18 (2021)

11. Jin, H., Xiao, J.: Towards trustworthy blockchain systems in the era of 'internet of value': development, challenges, and future trends. Sci. China Inf. Sci. **65**(153101), 1–11 (2022)

12. Liu, L., Li, X., Au, M.H., Fan, Z., Meng, X.: Metadata privacy preservation for blockchain-based healthcare systems. In: Bhattacharya, A., et al. (eds.) Database Systems for Advanced Applications (DASFAA 2022). LNCS, vol. 13245, pp. 404–412. Springer, Cham (2022). https://doi.org/10.1007/978-3-031-00123-9_33

13. Nakamoto, S.: Bitcoin: a peer-to-peer electronic cash system (2008). https://bitcoin.org/bitcoin.pdf

14. Peng, Z., Xu, C., Wang, H., Huang, J., Xu, J., Chu, X.: P^2b-trace: privacy-preserving blockchain-based contact tracing to combat pandemics. In: Proceedings of the 2021 International Conference on Management of Data (SIGMOD), pp. 2389–2391 (2021)

15. Ruan, P.C., Chen, G., Dinh, T.T.A., Lin, Q., Ooi, B.C., Zhang, M.H.: Fine-grained, secure and efficient data provenance on blockchain systems. In: Proceedings of the 2019 International Conference on Very Large Data Bases (VLDB), pp. 975–988 (2019)

16. Shao, Q., Pang, S., Zhang, Z., Jing, C.: Authenticated range query using SGX for blockchain light clients. In: Nah, Y., Cui, B., Lee, S.-W., Yu, J.X., Moon, Y.-S., Whang, S.E. (eds.) DASFAA 2020. LNCS, vol. 12114, pp. 306–321. Springer, Cham (2020). https://doi.org/10.1007/978-3-030-59419-0_19

17. Vaidya, K., Chatterjee, S., Knorr, E., Mitzenmacher, M., Idreos, S., Kraska, T.: SNARF: a learning-enhanced range filter. In: Proceedings of the 2022 International Conference on Very Large Data Bases (VLDB), pp. 1632–1644 (2022)

18. Wang, H., Xu, C., Zhang, C., Xu, J.L., Peng, Z., Pei, J.: vChain+: optimizing verifiable blockchain Boolean range queries (technical report). In: Proceedings of the 2021 International Conference on Management of Data (SIGMOD), pp. 1–14 (2021)

19. Wu, N., Xie, Y.: A survey of machine learning for computer architecture and systems. ACM Comput. Surv. **55**(3), 1–39 (2022)

20. Xu, C., Zhang, C., Xu, J.L.: vChain: enabling verifiable Boolean range queries over blockchain databases. In: Proceedings of the 2019 International Conference on Management of Data (SIGMOD), pp. 141–158 (2019)

21. Yagoubi, D., Akbarinia, R., Masseglia, F., Palpanas, T.: Massively distributed time series indexing and querying. IEEE Trans. Knowl. Data Eng. **32**(1), 108–120 (2018)

22. Yue, Z., Zhang, J., Zhang, H., Yang, Q.: Time-based trajectory data partitioning for efficient range query. In: Liu, C., Zou, L., Li, J. (eds.) DASFAA 2018. LNCS, vol. 10829, pp. 24–35. Springer, Cham (2018). https://doi.org/10.1007/978-3-319-91455-8_3

23. Zhang, C., Xu, C., Xu, J., Tang, Y., Choi, B.: GEM^2-tree: a gas-efficient structure for authenticated range queries in blockchain. In: Proceedings of the 2019 IEEE 35th International Conference on Data Engineering (ICDE), pp. 842–853 (2019)

24. Zhang, H., Andersen, D., Pavlo, A., Kaminsky, M., Ma, L., Shen, R.: Reducing the storage overhead of main-memory OLTP databases with hybrid indexes. In: Proceedings of the 2016 International Conference on Management of Data (SIGMOD), pp. 1567–1581 (2016)

Xenos: Dataflow-Centric Optimization to Accelerate Model Inference on Edge Devices

Runhua Zhang[1], Hongxu Jiang[1(✉)], Fangzheng Tian[1], Jinkun Geng[2], Xiaobin Li[1],
Yuhang Ma[1], Chenhui Zhu[1], Dong Dong[1], Xin Li[1], and Haojie Wang[3]

[1] Beihang University, Beijing, China
jianghx@buaa.edu.cn
[2] Stanford University, Stanford, United States
[3] Tsinghua University, Beijing, China

Abstract. In this paper, we propose Xenos, a high-performance edge platform for model inference. Unlike the prior works which mainly focus one *operator-centric* optimization, Xenos can automatically conduct *dataflow-centric* optimization on the computation graph and accelerate inference in two dimensions. Vertically, Xenos develops *operator linking* technique to improve data locality by restructuring the inter-operator dataflow. Horizontally, Xenos develops *DSP-aware operator split* technique to enable higher parallelism across multiple DSP units. Our evaluation proves the effectiveness of Xenos' vertical and horizontal dataflow optimization, which reduce the inference time by 15.0%–84.9% and 17.9%–89.9%, respectively.

Keywords: edge-based inference · dataflow-centric · DSP-aware · data locality

1 Introduction

Edge devices are widely applied nowadays and becoming a prevalent scenario for deep learning applications [10,14,16]. These edge devices have heterogeneous configurations of hardware resources (e.g. memory, computation, etc.) and usually require real-time responsiveness, i.e. the duration of the inference cannot take too long. Existing solutions (e.g. TVM [4]) execute model inference inefficiently on these platforms, and fail to satisfy the responsiveness requirement.

With a deep dive into numerous typical model inference workflows, we have identified two main reasons for the inference inefficiency.

(1) **Inefficient dataflow scheduling.** The dataflow scheduling of model inference can seriously spoil memory locality. Take the typical CNN inference as an example, after completing the inference of each layer, the computation operators output the feature maps to the shared memory region. These feature map elements will serve as the input for the next layer, and be fed to multiple DSP units for the following inference computation. However, there is a mismatch between the data layout (output from the prior layer) and the data access sequence (required by the next layer). In other words, DSP units are not reading in the sequential order as what was written previously. Therefore, while reading the feature maps, DSP units suffer from bad data locality and require *unnecessarily* much read operation, which leads to non-trivial overheads and prolongs the inference time.

X. Wang et al. (Eds.): DASFAA 2023, LNCS 13943, pp. 535–545, 2023.
https://doi.org/10.1007/978-3-031-30637-2_35

(2) **Hardware-Oblivious parallelism.** Different edge devices are usually equipped with heterogeneous computation resources and memory hierarchies. A fixed model partition scheme simply ignores the resource conditions and fails to fit the memory hierarchy and/or cannot fully utilize the computation resource. During the model inference process, parameters are frequently swapped in and out. Only a few digital signal processing (DSP) computing units (DSP cores/slices[1]) are active and undertaking the computation tasks, whereas the majority remains idle, waiting for the dependent data. Such partition schemes can waste much computation power and yield no satisfying performance.

Motivated by the drawbacks of existing frameworks, we propose Xenos in this paper, which (1) restructures the inference dataflow between adjacent operators to yield better locality; (2) takes hardware information (DSP units and memory hierarchy) into account and develop *DSP-aware operator split* technique for higher parallelism and more efficient data access. We summarize our contributions as below.

- **Framework.** Xenos is a complete end-to-end framework which focuses on dataflow-centric optimization instead of operator-centric optimization. Thus, it can effectively accelerate the edge inference without introducing new operators (i.e., adding extra programming effort).
- **Vertical dataflow optimization.** Xenos restructures the inference dataflow so that each DSP units can read the input sequentially, thus preserving the spatial data locality during the inference.
- **Horizontal dataflow optimization.** Xenos conducts DSP-aware operator split, so that the feature map and operator parameters can be partitioned across multiple DSP units and fit into their memory, thus improving the computation parallelism and load balance.
- **Evaluation.** We conduct comparative experiments on different platforms showing that Xenos can reduce the inference time by 17.9%–84.9% and 15.9%–89.9%, respectively.

2 Background and Motivation

2.1 Data Locality in Inference Computation

While executing the inference computation, the DSP units usually read the data in a non-sequential order due to the data layout mismatch between different operators. We use an example of a *depthwise separable convolution*, which consists of a depthwise convolution followed by a pointwise convolution, to illustrate this.

In Fig. 1, feature map Fm, which has 2 channels $C1$ and $C2$, is the output of a depthwise convolution and the input of a pointwise convolution. The depthwise convolution write Fm in a width-first order, while the pointwise convolution read Fm in a channel-first order, leading to data layout mismatch between these two operators. Thus, simply partitioning the model will cause the inference process to suffer from poor data locality, because each DSP unit requires different data blocks when writing and reading the feature maps.

---

[1] Mutli-Core DSP uses the term "DSP core" whereas FPGA uses "DSP slice". We use "DSP unit" as the general term in the following description.

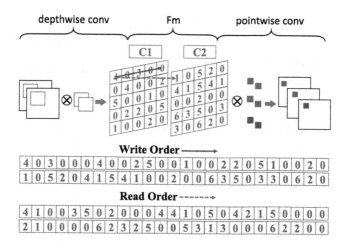

Fig. 1. Inefficient dataflow scheduling with bad locality

2.2 Hardware Resource Heterogeneity

When executing a model inference task on edge devices, the most efficient way is to place the whole model in high-level memory (e.g., L2 memory). However, such strategies tend to be non-trivial in practice due to the limited memory on edge devices. Although prior works [9, 11, 13] have proposed different pruning approaches, the model size is still too large to fit into the memory hierarchy on edge devices, leading to serious performance degrade for inference tasks. For example, as a lightweight inference model designed for edge hardware, MobileNet [11] still has many layers whose sizes of feature maps and kernels are larger than the size of L2 memory (e.g., 512 KB on TMS320C6678) or the size of shared memory (e.g., 4 MB on TMS320C6678). Simply executing this model without partition will lead to significant inefficiency.

As edge devices can be various and possess heterogeneous configurations of resources running models with different sizes, there is no "one-size-fits-all" partition scheme to fit all scenarios. Manually-tuned scheme highly relies on human expertise, which can be very inefficient and causes a long deployment period. Existing automated solutions, such as TASO [12] and PET [15], enumerate every possible partition scheme in a large search space, which prolongs the deployment time. Moreover, these frameworks are limited by graph size due to the complexity of their algorithm, which make them miss many optimization opportunities. Therefore, we are motivated to design vertical dataflow optimization, which takes the hardware resource information into account, and automatically generates a desirable partition scheme to both fully utilize the computation power and well fit into the memory hierarchy.

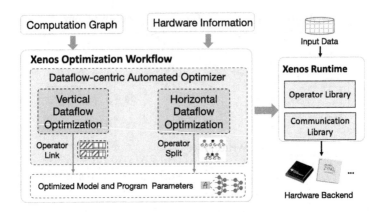

Fig. 2. The architecture of Xenos

2.3 Existing Drawbacks and Our Motivation

Reviewing the existing frameworks, we find that they mainly adopt operator-centric optimization, i.e., they optimize the computation graphs by replacing the operators with some fused/split ones to improve the computation inefficiency. We argue such optimization strategies are not globally sufficient and cannot save the performance overheads caused by the spoiled data locality and hardware heterogeneity. Therefore, we turn to develop a new framework, Xenos, and implement its optimization from a dataflow perspective. Compared with the existing works, Xenos enjoys the following advantages.

(1) Xenos possesses a rich operator library and each operator supports multiple dataflow patterns. Given an under-optimized computation graph, Xenos can automatically choose the dataflow pattern for each operator to improve the inference performance. Compared with the operator-centric optimization adopted by prior works, Xenos's dataflow-centric approach conducts more in-depth optimization and facilitates the continuous maintenance.

(2) Xenos preserves the data locality with the operator linking technique. During the inference computation, Xenos can automatically derive the optimal dataflow pattern for the operators, so as to make a match between (a) the writing order of intermediate parameters output from the operators and (b) the reading order of the subsequent operators. Thus, Xenos can avoid the costly cache misses throughout the whole inference process.

(3) Xenos fully leverages the computation/memory resource with the DSP-aware operator split technique. Xenos partitions the input tensors across multiple DSP units and further splits them to fit into the private memory of every DSP unit. Thus, it improves the parallelism and reduces the overheads of data fetch.

3   Xenos Design

3.1   Architecture Overview

Figure 2 illustrates the architecture of Xenos. Given an under-optimized model, Xenos conducts an automatic optimization workflow based on the hardware information of the edge device. Afterwards, Xenos outputs an optimized model equivalent to the original model. During runtime, both the input data and the optimized model are fed into Xenos Runtime. Xenos Runtime employs multiple DSP units to run the inference task atop its high-performance operator library and communication library.

As in typical frameworks (e.g., TASO and PET), Xenos' optimization workflow conducts operator fusion during the preprocessing stage to reduce some inefficient computation (not shown in Fig. 2). However, such basic optimization is insufficient for high-performance inference. Therefore, Xenos provides two key techniques, namely, operator linking and DSP-aware operator split to implement vertical and horizontal dataflow optimization.

3.2 Vertical Optimization: Operator Linking

During the inference, Xenos runtime, after finishing the computation task of one operator, should output the feature map to the shared memory in the desirable layout, so that these data can be read sequentially with good locality during the inference computation with subsequent operators. Otherwise, the data access would suffer from serious overheads due to cache misses. Xenos incorporates the operator linking technique to preserve the data locality. We exemplify the technique with a linked operator (Conv1x1 + AvgPooling2x2) in Fig. 3.

In Fig. 3, the input feature map is generated by the previous operator which has 4 output channels, so the input feature map consists of four matrices (marked with different colors). Without dataflow optimization, the four matrices are placed into the memory one by one in a row-based manner, with the red one placed first, and the blue one placed last. However, during the inference computation, the fused operator needs to read 1×1 feature map from each matrix every time, and then computes the average on every 2×2 square after the convolution. As magnified on the right side of Fig. 3, the dataflow goes through the four matrices every time (Conv1x1 computation). On each matrix, it follows the zigzag pattern (AvgPooling computation), leading to the restructured dataflow. From Fig. 3 we can see that the unoptimized dataflow suffers from compulsory cache misses for each data access. By contrast, the optimized dataflow completely matches the write/read order during the inference, so it maximizes the data locality and enables more efficient data fetch.

To maintain desirable data locality, the previous operator should be aware of the data access pattern of the subsequent operator when it outputs the feature map. With such awareness, Xenos can modify the data layout following the access pattern, instead of simply outputting the matrices one by one. Xenos uses the operator linking technique to attain this, and we describe the details below.

Before running the inference model, Xenos scans the computation graph and generates the metadata to describe the dataflows in the computation graph. Then, Xenos analyzes the metadata and identify the specific patterns (i.e., a sequence of adjacent operators) that can spoil data locality. After finding such inefficient patterns, Xenos will modify the metadata to change the dataflow between these adjacent operators. The metadata is fed into the inference engine. During runtime, the inference engine can know from the metadata the data access pattern of the subsequent operator. Therefore, it can write the feature map according to the optimized dataflow with data locality preserved (see Fig. 3).

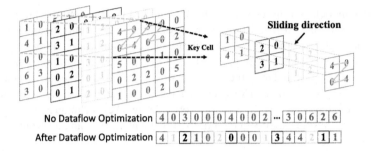

No Dataflow Optimization   4 0 3 0 0 0 4 0 0 2 ... 3 0 6 2 6

After Dataflow Optimization   4   2 1 0   0 0 0   3 4 4   1 1

Fig. 3. Operator linking to optimize dataflow

Notably, the operator linking technique can also incur data redundancy when conducting the computation of standard convolution, because it replicates some parameters of the feature map to avoid the subsequent operator from looking back. However, the memory sacrifice proves to be worthwhile, because the performance benefit brought by the restructuring outweighs the additional memory cost and the inference workflow is effectively accelerated (shown in Sect. 4.2).

3.3   Horizontal Optimization: DSP-Aware Operator Split

Xenos incorporates DSP-Aware operator split technique to partition the inference workload for higher parallelism. Xenos' split focuses on two aspects. First, it needs to partition the feature map across multiple DSP units, so that the multiple DSP units can share the inference workload. Second, it needs to split the operator parameters into the memory hierarchy of the edge device, so that the parameter fetch can be more efficient.

Partition Feature Map. Xenos partitions the feature map[2] in three dimensions, namely, the input feature map height (inH), the input feature map width (inW), and the output feature map channel ($outC$). We dismiss the input-channel-based (inC) partition since the inC-based partition performs extra reduction and introduces more computation overheads. Considering that the feature maps are stored in the shared memory with size of 4 MB, which is far beyond the input channel size in typical models, we usually do not need to partition along input channel.

Xenos prioritizes $outC$-based partition due to its less complexity: Xenos simply distributes the kernel parameters to different DSP units, and these kernel parameters will be placed into the L2 memory of DSP units. All DSP units can access the feature map located in the shared memory. On the other hand, inH-based scheme and inW-based scheme partition the remaining feature map after $outC$-based partition, and they usually require special handling of the boundary rows/columns. Only if the kernels cannot be evenly distributed across DSP units, DOS will seek further partition by inH/inW.

---

[2] As a special case, when the feature map is too large to be held by the shared memory, Xenos will first slice the feature map as a preprocessing step, and then the split procedure continues to partition the sliced feature map.

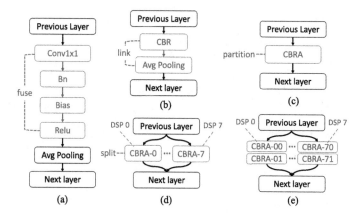

Fig. 4. Illustration of Xenos's optimization

If imbalance still exists after the triple partition, DOS will randomly assign the remaining feature map (workload) to different DSP units.

While the operator partition procedure distributes the inference workload across multiple DSP units, the single DSP unit may still fail to conduct the inference work efficiently, because the operator parameters assigned to it is too large to fit into the L2 memory (e.g. the CNN model has very large-sized kernels). To address that, Xenos needs to do further split of the operator parameters according to the memory resource of the DSP unit. We explain the split of operator parameters next.

Split Operator Parameters. Xenos splits the large-sized operator parameters into smaller chunks so that they can be placed into the private L2 memory of the DSP unit. Thus, the parameter fetch can become more efficient and the inference time can be reduced.

Xenos follows a certain priority for each dimension when performing parameter splitting, to guarantee that minimum computation overhead is introduced after splitting. Taking the popular CNN model as an example, which have four dimensions for parameters, i.e., output channel (K), input channel (C), kernel height (R), and kernel width (S). Splitting at K dimension will not introduce any extra computation, while splitting at the other three dimensions requires additional reduction operation to aggregate the results on these dimensions, which introduces extra computation overhead. Thus, Xenos will first try to split the parameters at K dimension, and then C, R and S if only splitting K dimension is not enough to fit the parameter in L2 memory.

Equation 1 gives an example of output-channel-based split. Since the large-sized parameters (W and B) can not be put into the L2 memory, Xenos performs fine-grained split of the operator: W is split into W_1 and W_2, and B is split into B_1 and B_2. After that, parameters can be distributed into the L2 memory of two different DSP units. Equation y_1 and y_2 can be jointly executed on two DSP units in parallel, or on one DSP unit one by one. The output $y_1(x_i)$ and $y_2(x_i)$ are automatically joined together afterwards, without performing any data layout transformation operators. Other types of splitting work in a similar way.

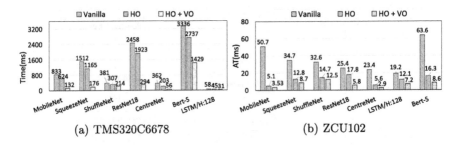

Fig. 5. Inference time comparison

$$y(x_i) = Wx_i + B \implies \begin{cases} y_1(x_i) = W_1x_i + B_1 \\ y_2(x_i) = W_2x_i + B_2 \end{cases} \tag{1}$$

Exemplar Optimization. Figure 4 gives an example of how Xenos optimizes one part of the computation graph of MobileNet [11]. Xenos first performs typical operator fusion on operator Conv1x1 (convolution), Bn (batch normalization), Bias, and Relu, and generates a fused operator CBR (Fig. 5(a)), then links CBR and AvgPooling together to form the operator CBRA (Fig. 5(b)). Next, Xenos uses DOS technique to partition the feature map across DSP units (Fig. 5(d)). After that, the operator parameters are still too large to fit each DSP unit's L2 memory, so Xenos continues to split the operator parameters into smaller chunks (Fig. 5(e)) and finalizes the optimization.

4 Experimental Evaluation

4.1 Experiment Setting

Testbeds: We employ two testbeds for evaluation: (1) a Multi-Core DSP device, which is equipped with 2 TMS320C6678-type nodes and a high-speed image collector, which are directly connected via SRIO; (2) the ZCU102-type FPGA device, with code generated by High-Level Synthesis (HLS) Compiler. We use TMS320C6678 and ZCU102 to refer to them for simplicity.

Benchmarks: We choose 7 typical models as the benchmarks, i.e. MobileNet, SqueezeNet, ShuffleNet, ResNet18, CentreNet, LSTM and Bert.

Baselines: We compare the inference time between the complete Xenos solution and two baselines. As shown in Fig. 5, one baseline is named Vanilla since it involves no horizontal optimization (HO) or vertical optimization (VO). The other baseline is named HO since it only adopts horizontal optimization.

4.2 Inference Time Comparison

TMS320C6678. In Fig. 5(a), compared with Vanilla, HO reduces the inference time by 17.9%–43.9%, which demonstrates the acceleration brought by higher computation parallelism. We further evaluate the performance benefit of VO by comparing the HO baseline and the full Xenos solution. We can see that the VO further reduces the inference time by 30.3%–84.9%, which demonstrates the performance benefit brought by the data locality improvement.

ZCU102. Figure 5(b) also demonstrate the performance benefit of HO and VO on ZCU102. Similarly, compared with the Vanilla baseline, HO can reduce the inference time by 37.0%–89.9%. Compared to the HO baseline, VO further reduces the inference time by 15.0%–67.4%.

Comparing Fig. 5(a) and Fig. 5(b), we can easily notice that HO contributes more inference time reduction on TMS320C6678, whereas VO contributes more on ZCU102. The reasons are explained in two main aspects.

(1) VO is more effective on TMS320C6678 than ZCU102. This is because a large number of LUT resources are used on ZCU102 to implement data mapping, therefore, the memory access efficiency has already been very high even without VO. By contrast, TMS320C6678 is not equipped with such a utility, so the memory access efficiency can be seriously damaged when the inference workload breaks the data locality. VO, however, helps to preserve the data locality with dataflow restructuring and becomes the main contributor.

(2) HO is more effective on ZCU102 than TMS320C6678. This is because ZCU102 has much more DSP units than TMS320C6678. While TMS320C6678 only has 8 DSP units, ZCU102 can allocate thousands of DSP units to participate in the computation. Therefore, the management of model partition and parallel execution becomes more essential to the efficiency of ZCU102. Since the Vanilla baseline is not equipped with a proper partition scheme, it fails to exploit the abundant computation resource. In contrast, HO works with the optimized operators and achieves high utilization of DSP computation resource, which can significantly reduce the inference time.

5 Related Work

Graph-Level Optimization. TensorFlow [3] with XLA [2] and TensorRT [1] optimize the computation graph by transformation rules designed by domain experts. TASO [12] and PET [15] can automatically optimize computation graphs but enlarges the search space. When there is a lack of hardware knowledge, simply applying the above optimization cannot achieve promising performance. Thus, Xenos uses architecture-aware approaches (DSP-aware operator split and operator linking) to perform more in-depth optimization. Other works (e.g., Fela [7], ElasticPipe [6]) adopt flexible parallelism for computation/communication saving in distribute setting. Such mechanisms can be incorporated in the distributed version of Xenos as our future work.

Inference on Edge Hardware. AOFL parallelization [17] accelerates edge inference by dynamically selecting the parallelism based on network conditions. Xenos, by contrast, considers more about the resource (memory and computation) conditions on the edge device (HO), and conducts deeper optimization to the inter-operator dataflow (VO). Mema [5] enhances the scheduling policy to run multiple inference jobs without additional edge resources. While Xenos currently focuses on accelerating single inference job, we believe its strategies can become compatible to work in multiple-job scenario with some adaption. [8] employs both cloud resource and edge devices to jointly undertake one big inference task. It would be an interesting direction for Xenos to leverage cloud resource to accelerate the edge-based inference and we leave it as our future work.

6   Conclusion and Future Work

We present Xenos, which incorporates dataflow-centric optimization strategies to accelerate edge-based inference. We conduct comprehensive experiments with 7 benchmarks on two typical platforms, Multi-Core DSP(TMS320C6678) and FPGA(ZCU102). Evaluation results demonstrate the effectiveness of both Xenos' vertical and horizontal dataflow optimization. Our future works include (1) the continuous development of distributed Xenos and (2) the incorporation of TASO/PET's optimization techniques to further boost Xenos' performance.

Acknowledgement. This work is supported by the National Key Research and Development Program of China (No. 2021ZD0110202). Haojie Wang is supported by the Shuimu Tsinghua Scholar Program.

References

1. Nvidia tensorrt. https://developer.nvidia.com/tensorrt
2. XLA: Optimizing compiler for ML. https://www.tensorflow.org/xla
3. Abadi, M., et al.: TensorFlow: a system for large-scale machine learning. In: OSDI (2016)
4. Chen, T., et al.: {TVM}: an automated {End-to-End} optimizing compiler for deep learning. In: OSDI (2018)
5. Galjaard, J., et al.: MemA: fast inference of multiple deep models. In: PerCom Workshops (2021)
6. Geng, J., et al.: ElasticPipe: an efficient and dynamic model-parallel solution to DNN training. In: ScienceCloud (2019)
7. Geng, J., et al.: Fela: incorporating flexible parallelism and elastic tuning to accelerate large-scale DML. In: ICDE (2020)
8. Grulich, P.M., et al.: Collaborative edge and cloud neural networks for real-time video processing. In: VLDB (2018)
9. He, Y., et al.: AMC: AutoML for model compression and acceleration on mobile devices. In: ECCV (2018)
10. Hochstetler, J., et al: Embedded deep learning for vehicular edge computing. In: SEC (2018)
11. Howard, A.G., et al.: MobileNets: efficient convolutional neural networks for mobile vision applications. arXiv preprint arXiv:1704.04861 (2017)
12. Jia, Z., et al.: TASO: optimizing deep learning computation with automatic generation of graph substitutions. In: SOSP (2019)

13. Polino, A., et al.: Model compression via distillation and quantization. arXiv preprint arXiv:1802.05668 (2018)
14. Suo, K., et al.: Keep clear of the edges : an empirical study of artificial intelligence workload performance and resource footprint on edge devices. In: IPCCC (2022)
15. Wang, H., et al.: PET: optimizing tensor programs with partially equivalent transformations and automated corrections. In: OSDI (2021)
16. Wang, H., et al.: User preference based energy-aware mobile AR system with edge computing. In: INFOCOM (2020)
17. Zhou, L., et al.: Adaptive parallel execution of deep neural networks on heterogeneous edge devices. In: SEC (2019)

Accelerating Recommendation Inference via GPU Streams

Yuean Niu[1,2], Zhizhen Xu[1,2], Chen Xu[1,2(✉)], and Jiaqiang Wang[3]

[1] East China Normal University, Shanghai, China
{yaniu,zhizhxu}@stu.ecnu.edu.cn, cxu@dase.ecnu.edu.cn
[2] Shanghai Engineering Research Center of Big Data Management, Shanghai, China
[3] Tencent, Shenzhen, China
jiaqiangwang@tencent.com

Abstract. Deep Learning based recommendation is common in various recommendation services and widely used in the industry. To predict user preferences accurately, state-of-the-art recommendation models contain an increasing number of features and various methods of feature interaction, which both lengthen inference time. We observe that the embedding lookup and feature interaction of different features in a recommendation model is independent of each other. However, current deep learning frameworks (e.g., TensorFlow, PyTorch) are oblivious to this independence, and schedule the operators to execute sequentially in a single computational stream. In this work, we exploit multiple CUDA streams to parallelize the execution of embedding lookup and feature interaction. To further overlap the processing of different sparse features and minimize synchronization overhead, we propose a topology-aware operator assignment algorithm to schedule operators to computational streams. We implement a prototype, namely *StreamRec*, based on TensorFlow XLA. Our experiments show that StreamRec is able to reduce latency by up to 27.8% and increase throughput by up to 52% in comparison to the original TensorFlow XLA.

Keywords: CUDA Stream · Operator Assignment · Parallelization · Inference Service · Recommendation model

1 Introduction

Recommendation plays an important role in modern web applications, including advertisements, search engines, e-commerces, etc. Deep learning based recommendation models, which are extensively used in the industry, integrate neuron network architecture to improve the accuracy of recommendation services.

As shown in Fig. 1, deep learning-based recommendation models are often composed of four parts, including sparse feature embedding lookup, dense feature processing, feature interaction, and prediction layer. In recent years, industrial DL-based recommendation models increase the amount of features and adopt various feature interaction methods to improve the accuracy of inference results. The feature interaction parts often include transformer, attention and

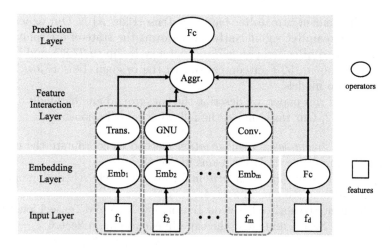

Fig. 1. General Recommendation Model Architecture

other computationally intensive computation operations to extract more useful information from features. These intensive computations make CPU no longer meet the demand of inference service, leading to the popularity of GPU inference, e.g., DeepRecSys [7] and Hercules [10]. Besides, we have an observation that the embedding lookup and feature interaction of different features is often independent of each other in recommendation models. Each group looks up the embedding table independently and applies individual feature interaction methods, which means they could execute in parallel.

However, this parallel opportunity is not fully exploited by existing solutions. Current deep learning frameworks including TensorFlow, PyTorch and so on sequentially schedule the operators to a single computation stream in GPU. Recent work [5,6,9] has proposed multi-process service (MPS in short) that serves multiple models simultaneously on a single GPU. Using MPS, different models execute on one GPU using different computation streams, leading to the improvement of throughput. However, MPS faces the side effect of increasing both GPU memory and latency. In particular, MPS has to maintain multiple copies of a model, causing the near linear growth of the memory footprint with parallelism. In addition, it still does not take advantage of the parallelism opportunities of independence between sparse feature processing, and fails to reduce the inference latency.

To bridge the gap between the potential parallelism of DL-based recommendation models and existing parallel methods, we propose a stream-based parallel approach. It assigns the processing of individual features to different GPU streams for parallel execution. Moreover, we observe that stream-based parallelism incurs synchronization overhead due to inefficient operator assignments, leading to performance loss compared to the original case. To alleviate this issue, we design a topology-aware operator assignment algorithm to assign operators to proper computational streams and eliminate synchronization overhead. In particular, we implement StreamRec based on TensorFlow XLA. Our experimental results show StreamRec significantly outperforms the state-of-art solutions. As an example, StreamRec reduces 99% tail latency by up to 27.8% and increases throughput by up to 52% in comparison to the original TensorFlow XLA on recommendation models.

In the rest of this paper, we describe the background and highlight our motivations in Sect. 2. Our work makes the following contributions.

- We propose *stream-based parallelization* method to coordinate the inference operator on multiple CUDA streams in Sect. 3.
- We propose *topology-aware algorithm* for appropriately scheduling operators to computational streams in Sect. 4.
- We implement StreamRec based on TensorFlow XLA and TensorFlow Serving illustrated in Sect. 5 and show the experiment results in Sect. 6.

In addition, we introduce related work in Sect. 7 and summarize our work in Sect. 8.

2   Background and Motivation

In this section, we show the general characteristics of recommendation models and figure out new optimization opportunities based on these characteristics.

2.1   Recommendation Inference

To improve the accuracy of inference results and the user experiences of recommendations, state-of-the-art recommendation models adopt DL-based solutions widely. Figure 1 depicts a generalized architecture of DL-based recommendation models with dense and sparse features as inputs. The structure consists of four components, sparse feature embedding lookup layers, dense feature processing layers, feature interaction layers, and prediction layers.

Specifically, sparse feature embedding layers are responsible to transform high-dimensional features into low-dimensional embedding vectors. Dense feature processing layers usually use multiple fully connected layers to extract features, which is not the focus of this article. The feature interaction layer would first organize the embedding features into multiple groups. Each group applies an individual feature interaction method, such as Attention [17] and Transformer [3], to extract useful information from intra-group feature embeddings. The outputs of the constituent feature interaction modules are then concatenated to form a final output of the feature interaction layer. Prediction layers usually apply fully-connected layers to provide final outputs.

Industrial DL-based recommendation models use an increasing number of features and various methods of feature interaction [8,16], which makes the model more computationally intensive. For instance, recently proposed models, e.g., SIM [15], CAN [2], contain more than 20 feature interaction modules and more than 300 feature fields. Besides, for transformer, attention, and other computationally intensive computation parts, CPU no longer meets the demand for inference service and produces a deteriorating service latency. DeepRecSys [7] and Hercules [10] show that GPU inference has much lower latency than CPU with proper scheduling.

2.2 Motivation

We explore typical recommendation models and popular deep-learning frameworks, and have the following observations.

❶ **The embedding lookup and feature interaction of different sparse features in the recommendation model is independent of each other.** As shown in Fig. 1 by the dotted line box part, each group of the feature looks up embedding tables independently and applies individual feature interaction methods. Therefore, there is no sequential relationship among their executions, which means they could execute in parallel.

Besides, the embedding lookup part and feature interaction part occupy an increasing proportion with the rise of embedding table number and complex architecture [10,16]. Therefore, accelerating the execution of these parts can significantly reduce the overall latency.

❷ **Operators in the model are executed serially on the GPU, constrained by the current GPU scheduling policy of deep learning frameworks.** Taking the computation graph in Fig. 2a as an example, we consider v_1, v_3, v_5 as the embedding table lookup operators, v_2, v_4, v_6 as the feature interaction operators and v_7 as the prediction operator. Current deep learning frameworks sequentially schedule the operators to the same stream as shown in Fig. 2b, even if these operators could be executed simultaneously. Clearly, there is a gap between the potential parallelism of the recommendation models and existing methods. Hence, we need to find a new method to parallelize the embedding table lookup and feature interaction operators.

3 Parallel Inference

To take advantage of the independent processing between different features of the recommendation model, we try a coarse-grained parallel approach, i.e., MPS, and further explore the possibility of fine-grained parallelism. In Sect. 3.1, we describe the limitation of coarse-grained parallelization. In Sect. 3.2, we propose a stream-based fine-grained parallelization method.

3.1 Limitation of Coarse-Grained Parallelization

One intuitive solution to increase parallelism is to run multiple models in parallel. Abacus [5], GSLICE [6] and Olympian [9] have proposed some efficient ways for the parallel running of multiple models by CUDA stream or MPS techniques. As shown in Fig. 3a, two models are loaded into the GPU and process inference

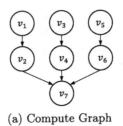

(a) Compute Graph

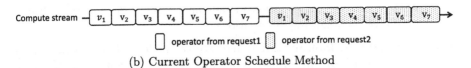

(b) Current Operator Schedule Method

Fig. 2. Example of a Compute Graph and a Schedule

requests respectively. CUDA stream is a sequence of operations executed on the device in the order. While operations within a stream are guaranteed to execute in the determined order, operations in different streams can be interleaved and even run concurrently. Moreover, the MPS is an alternative implementation of the CUDA API, which transparently enables cooperative multi-process CUDA applications to utilize Hyper-Q capabilities on the GPUs. It allows CUDA kernels to be processed concurrently on the same GPU.

Although MPS allows multiple models to run simultaneously and increases the parallelism, it suffers from several drawbacks. First, the embedding lookup and feature interaction of different sparse features are still serial in their respective compute streams, as shown in Fig. 3a. Moreover, due to the parallel execution of multiple models, the execution time of the operators in each request is longer than when they are executed separately. In other words, since the MPS approach improves throughput by increasing parallelism, it may also impact tail latency negatively. In addition, since MPS has to maintain multiple copies of a model, it intensifies the consumption of GPU memory, both from the model parameters and the intermediate results of the inference. Therefore, further optimization should be achieved by more fine-grained parallelism on each model.

3.2   Stream-Based Fine-Grained Parallelization

Benefiting from the increasing performance, Nvidia GPUs support multiple CUDA streams for simultaneous computation. Existing work [6] has run multiple models simultaneously using different CUDA streams. Operators of the same model can execute in parallel on different streams. Therefore, we can assign the processing of individual sparse features rather than different models to different GPU streams for parallel execution.

For example, the embedding table lookup operators (i.e., v_1, v_3 and v_5) and feature interaction operators (i.e., v_2, v_4 and v_6) in Fig. 2a can execute in parallel.

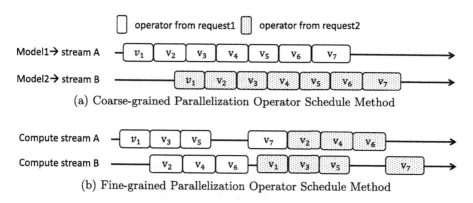

Fig. 3. Comparison of Different Parallel Inference

This two kinds of operators are assigned to different streams for simultaneous execution, as shown in Fig. 3b. Hence, operators v_2, v_4, and operator v_3, v_5 belonging to the same request can execute in parallel. Further, in practice, except for assigning the processing of individual sparse features into different streams, other parts could also exist parallelism opportunities to execute in parallel and the same approach can apply to the entire computational graph.

Since it is unnecessary to maintain multiple copies of a model at the same time, the stream-based fine-grained parallel approach reduces GPU memory footprint compared to MPS. Besides, as shown in Fig. 2, fine-grained parallelism significantly reduces the latency due to the parallel execution of a single request. However, the improvement in throughput compared to coarse-grained parallelism is minor.

It is worthy noticed that there are gaps between operators due to synchronization caused by operator dependencies as shown in Fig. 3b. Besides, coarse-grained and fine-grained parallelism are not contradictory with each other. Alternatively, it is possible to use the two parallel methods together. We will make a further study in Sect. 6.3.

4 Operator Assignment

In this section, we first describe the high synchronization overhead caused by the naive round-robin operator assignment algorithm and its impact on inference performance. Then, we propose a topology-aware operator assignment algorithm based on the greedy search to mitigate the performance loss due to inefficient operator assignment.

4.1 Round-Robin Operator Assignment

Round-robin is a simple but effective algorithm that assigns operators to each stream in equal portions and circular order. Since it provides reliable and fair

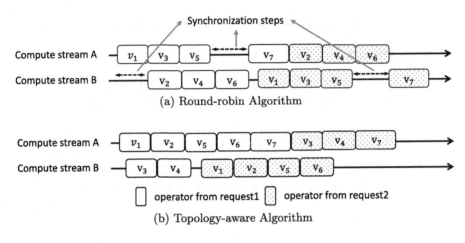

(a) Round-robin Algorithm

(b) Topology-aware Algorithm

Fig. 4. Operator Assignment

assignments, MPS uses it to assign operators to different streams. While the round-robin operator assignment algorithm shows good performance in MPS, it may incur synchronization overhead when handling stream-based parallelism operators and thus impact both latency and throughput.

Figure 4a depicts the operator assignment result using the round-robin algorithm and indicates the synchronization overhead due to ineffective operator assignment. The computation graph of the request is the same as the Fig. 2a, which contains seven operators, i.e., $v_1, v_1, \ldots, v_7$. For simplicity, we denote the i-th CUDA stream as s_i in the following. According to the round-robin operator assignment algorithm, these operators need to be assigned to two streams, s_1 and s_2. Operator v_1 is firstly assigned to s_1. Then, following the round-robin algorithm, operators v_2 are assigned to s_2 sequentially. Once assigned, v_1 could immediately start to execute, whereas v_2 has to wait for the end of all its dependencies (v_1 in this case), leading to a synchronization step before v_2 (denoted as the dashed line in Fig. 4a). This synchronization step postpones the execution of v_2 and prevents other operators from executing in s_2. There are similar synchronization steps before v_7 from both requests. Synchronization steps incur overhead, which eventually leads to a decrease in parallelism and a reduction of inference performance.

4.2   Topology-Aware Operator Assignment

The synchronization steps in round-robin operator assignment is incurred by the dependency of the topology of compute graph. However, the round-robin operator assignment is oblivious to the topology. To alleviate the synchronization, topology-aware operator assignment should avoid assigning an operator to stream until all its precedent operators have finished. To achieve that, we propose a greedy-based algorithm in this section.

Algorithm 1: Topology-Aware Operator Assign Algorithms

Data: computational graph G, max number of streams s, current stream index i,

1 $operator\_queue \leftarrow G.operators()$;
2 $seen\_ops \leftarrow []$;
3 **foreach** $operator$ in $operator\_queue$ **do**
4      $assigned\_streams \leftarrow []$;
5   **foreach** $seen\_op$ in $seen\_ops$ **do**
6          $stream\_num \leftarrow seen\_op.stream()$;
7   **if** $stream\_num$ not in $assigned\_streams$ and not $G.IsConnected(operator, seen\_op)$ **then**
8              $assigned\_streams.append(stream\_num)$;
9   **end**
10   **end**
11   **for** $i{=}1$ to s **do**
12   **if** i not in $assigned\_streams$ **then**
13             $operator.AssignStream(i)$;
14   **end**
15   **end**
16   **if** not $operator.IsAssign()$ **then**
17         $operator.RoundRobinAssignStream()$;
18   **end**
19     $seen\_ops.append(operator)$;
20 **end**

Figure 4b demonstrates the assignment process using the topology-aware algorithm, which indicates how it could eliminate synchronization steps. Firstly, the topology-aware algorithm identifies independent operators v_1, and v_3 in the computation graph and assigns them to different streams sequentially. Then, operators v_2 to v_6 are assigned to the according stream where their predecessors are. Finally, since v_7 depends on all other operators, it is assigned to the first available stream.

Following the topology-aware algorithm, v_2 would no longer occupy s_2 because it has to wait for the end of v_1. Consequently, operators v_3 could execute in s_2, rendering higher parallelism using CUDA streams. The full exploitation of stream-based parallelism could increase GPU utilization and improve overall inference performance.

As illustrated in Algorithm 1, the topology-aware operator assignment algorithm first scans all operators in the topological graph. It then determines whether each operator depends on an already traversed operator based on the topology of the computation graph. If so, these operators are only allowed to be assigned to streams which their precedent operators are assigned to (Line 5–10). For those operators that do not depend on any other operators that have been traversed previously, the topology-aware algorithm assigns them to the first available stream using the round-robin method (Line 11–18).

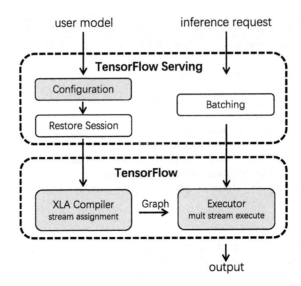

Fig. 5. System Architecture of StreamRec

5 Implementation

We implement StreamRec based on TensorFlow 2.9 and use TensorFlow Serving 2.9 as the service layer for inference. We add some new parameter configuration settings, including whether to enable multiple compute streams and the maximum number of streams. As shown in Fig. 5, given a user model and maximum stream number, TensorFlow Serving loads the configuration and serves the model by TensorFlow. When a request arrives, it is batched in TensorFlow Serving, and then the request is handed off to TensorFlow runtime for computation.

We modified the XLA (Accelerated Linear Algebra) module of TensorFlow to implement stream-based parallelization and topology-aware operator assignment algorithm. TensorFlow XLA is a domain-specific compiler for linear algebra that can accelerate TensorFlow models. Since the inference performance with XLA is significantly better than native TensorFlow in both latency and throughput, we modify the XLA module rather than original TensorFlow runtimes. We added the stream assignment algorithm to the XLA compilation module (Sect. 4) and modified the XLA execution module to execute models in multiple streams (Sect. 3) according to the results obtained from the compilation stage. The TensorFlow XLA module compiles the model when the model is loaded in TensorFlow Serving, including stream assignment and other optimization. After the TensorFlow Serving receives an inference request, the TensorFlow XLA execution module is responsible for performing the calculation through multiple streams.

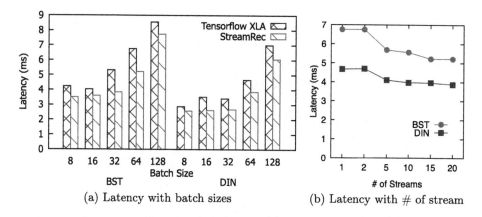

(a) Latency with batch sizes

(b) Latency with # of stream

Fig. 6. Overall Latency Comparison

6   Evaluation

In this section, we first introduce our experimental setup in Sect. 6.1. Then, we show the overall improvement of the stream-based fine-grained parallelization in different models and parameters and discuss how to set each parameter reasonably in Sect. 6.2. Next, we show the comparison of the stream-based fine-grained parallelization and coarse-grained parallelization MPS in Sect. 6.3. Finally, we present the performance improvement from the topology-aware operator assignment and explain the reason in Sect. 6.4.

6.1   Experimental Setup

Server Architecture. We deploy StreamRec on the server with two Intel(R) Xeon(R) Gold 6240R processors with 24 CPU cores and 256 GB DRAM in total, a 4TB hard disk, and a single V100 GPU with 32 GB DRAM. We use CUDA version 11.4.

Models and Datasets. Two models with different model structures are used in our experiments, namely, DIN [17] and BST [3]. Among them, DIN also contains the Attention structure as feature interaction layers, and BST contains the Transformer structure as feature interaction layers. We use the open source datasets Taobao[1] as experimental workloads. Note that we quadrupled the size of the data set in order to simulate larger scale and practical data.

Metrics. We use 99% tail latency, which is widely used in inference systems, and throughput under 10 ms SLA constraints as our performance measures. By default, when we mention throughput later, we mean the throughput under the 10 ms SLA limit.

---

[1] https://tianchi.aliyun.com/dataset/649.

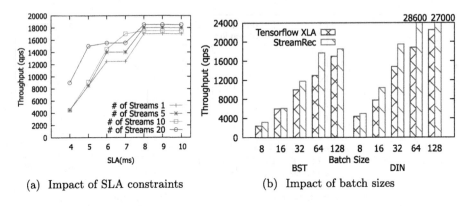

(a) Impact of SLA constraints   (b) Impact of batch sizes

Fig. 7. Latency and throughput with different numbers of streams

6.2 Overall Performance

In this section, we evaluate the overall performance improvement of the Stream-Rec. First, we show the performance gains with various batch sizes and models. Then, we evaluate the performance improvements regarding different number of streams.

Figure 6a provides the 99% tail latency of the original TensorFlow XLA and StreamRec in various batch sizes and models. We can see that for both models, StreamRec makes a performance improvement compared to the original Tensor-Flow XLA, regarding different number of streams. For BST, StreamRec reduces tail latency by up to 27.8% while also reducing up to 25.6% tail latency for DIN.

We further investigate how stream-based parallelization contributes to overall performance improvement by evaluating inference performance regarding different number of stream when batch size is 64. Figure 6b indicates that, as the number of stream rises, the tail latency of both two models gradually decreases.

Besides 99% tail latency, we also evaluate SLA constraint throughput of StreamRec. Figure 7a demonstrates how throughput changes with latency constraint. It clearly indicates that the increasing number of streams contributes to throughput regardless of latency constraints. In addition, as throughput of each numbers of streams stops increasing when latency constraint is 10 ms, we choose 10 ms as the SLA constraint in latter experiments. Under the SLA constraint, we demonstrate the throughput improvement in Fig. 7b, which indicates that StreamRec could outperforms TensorFlow XLA in various models and batch sizes. Since StreamRec further exploits parallelism, it improves the throughput under SLA constraint compared to the original TensorFlow SLA. The results show that StreamRec increases throughput by up to 52%.

In general, the performance is best when the number of streams is taken to the maximum parallelism of the model, both in terms of tail latency and throughput.

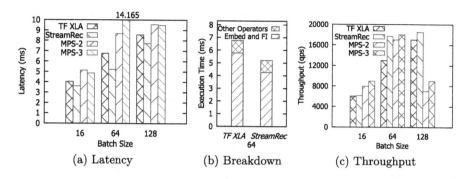

Fig. 8. Performance of the StreamRec and MPS

6.3   Efficiency of Stream-Based Parallelization

To show the performance improvement of stream-based parallelism by exploiting the independence between different feature processing, we compare our stream-based fine-grained parallel approach with the original TensorFlow XLA and coarse-grained parallel approach, i.e., MPS. For simplicity, we employ MPS-# to represent the model execution in parallel using by MPS technique, where # is the number of models. When we experiment with MPS, the GPU resources are equally allocated to each model.

Latency and Throughput. We evaluate the latency and throughput of StreamRec, MPS and origin TensorFlow XLA. As shown in Fig. 8a, compared to MPS and the original TensorFlow XLA, SBP always has the lowest tail latency at all batch sizes. In particular, the stream-based parallel approach has a 22.9%–66.4% performance improvement in latency compared to MPS-2 and 23.0%–171.7% improvement in latency compared with MPS-3. Coarse-grained parallelism using MPS makes tail latency extremely unstable due to competition for GPU resources.

To explore whether our performance improvement comes from parallel execution of sparse feature processing, we further measure the execution time of each part of a model and compared it with the original TensorFlow XLA as shown in Fig. 8b. Compared with the original TensorFlow XLA, 99% of the performance improvement from stream-based parallelism comes from embedding lookup and feature interaction.

Besides, stream-based parallelism does not always perform better than MPS in terms of throughput as shown in Fig. 8c. When batch size is equal to 16, MPS-2 and MPS-3 outperforms StreamRec by 31.1% and 47.5% respectively in throughput. However, their throughputs are similar when the batch size is equal to 64. And when the batch size is equal to 128, the throughput of StreamRec exceeds the throughput of MPS. This is due to the fact that as the batch size and the computation volume increases, MPS causes severe resource competition, resulting in an increasing number of cases where the latency exceeds 10 ms, further leading to a drop in throughput.

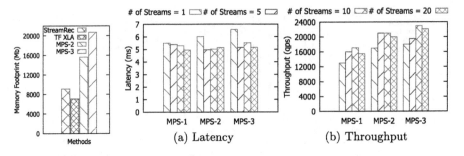

Fig. 9. Memory   **Fig. 10.** Performance of Co-using StreamRec and MPS

GPU Memory Footprint. In addition to latency, we also compare the GPU memory footprint with the original TensorFlow XLA and MPS as shown in Fig. 9. StreamRec increases the GPU memory footprint by 29.3% compared to the original TensorFlow XLA, while MPS-2 and MPS-3 increase by 121.9% and 194.1% respectively. Since the MPS method needs to store the parameters of multiple models and more intermediate results in GPU memory, there is a larger increase in memory usage than StreamRec.

Discussion. Since stream-based parallelism does not always improve throughput compared to MPS, we try to use both coarse-grained and fine-grained parallelism. As shown in in Fig. 10, the stream-based method mitigates tail latency boost from MPS, while increasing throughput over 23.5% with proper stream number. The latency decreases up to 21.0% in MPS-2 and 27.7% in MPS-3. Therefore, regardless of the number of model execution using by MPS, stream-based parallelism method reduces the tail latency and further increase the throughput.

6.4   Efficiency of Topology-Aware Operator Assignment

In this section, we first demonstrate the performance drop incurred by round-robin operator assignment algorithm, and how topology-aware operator assignment alleviates overhead and speeds up inference. We study tail latency of round-robin algorithm and topology-aware algorithm in Fig. 11a.

For round-robin algorithm, as the number of streams increases, the latency gradually rises due to increasing synchronization overhead. When the number of stream is greater than ten, benefiting from parallel execution, the latency reduces slightly. In a nutshell, the latency shows a trend of deterioration with the growth of the number of streams. In particular, the latency of round-robin operator assignment will never outperform the latency of original TensorFlow XLA. Since topology-aware algorithm eliminates the synchronization overhead, the latency decreases when the number of streams is greater than five. Consequently, it reaches an improvement up to 27.0% compared to the round-robin algorithm. Figure 11b illustrates the throughput, which performs a similar trend as latency Due to synchronization overhead of round-robin algorithm, through-put decreases as stream-based parallelism increases. Topology-aware algorithm hides the overhead and shows a better performance.

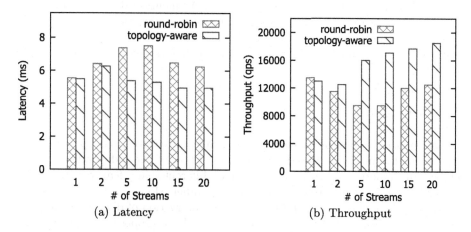

(a) Latency   (b) Throughput

Fig. 11. Latency and throughput with different operator assignment

7   Related Work

This section describes related work on DL-based recommendation services and inference parallelism methods.

Recommendation Model Service. Many online serving systems have been proposed for deep learning recommendation models in recent years. They are optimized for the characteristics of recommendation models. DeepRecSys [7] and Hercules [10] focus on how to use GPUs to accelerate the inference of rec-ommendation models by dynamically deciding the placement strategy of models under heterogeneous clusters based on real-time workload. Observing the access pattern of the embedding table, JiZhi [12], HET [13] and FAE [1] optimize the access strategy of the embedding table. They use GPU memory to cache fre-quently accessed entries, and other cold entries are placed into main memory. All of the above works are orthogonal to our work, and none of them take advan-tage of the feature that the processing of sparse features is independent of each other for optimization.

Inference Parallelism. The methods to improve the parallelism of inference could be divided into three main areas. First, the batch processing method is widely used in various inference systems, such as TensorFlow Serving [14], and Clipper [4]. Some dynamic tuning batch size methods also have been proposed. This kind of parallelism work is orthogonal to our work. Second, model-level parallelism is proposed and optimized by some work such as Abacus [5], GSLICE [6], and Olympian [9]. They suggested some efficient ways for the parallel running of multiple models by CUDA stream or MPS techniques. They optimize the memory footprint of various models and reduce the interaction of tail latency when different models are executed together. As mentioned above, these work still has some problems and can coexist with our optimization. Finally, stream-level parallelism is also used by Nimble [11]. However, it only focuses on the optimization for computer vision models rather than the recommendation model in our work.

8   Conclusion

In this paper, we propose a new solution, StreamRec, to speed up recommendation model inference. Based on the observation of the independence of the embedding lookup and feature interaction of different sparse features, we employ multiple CUDA streams to execute them in parallel. Moreover, for maximize the overlap of the embedding lookup and feature interaction of different sparse features and minimize the synchronization overhead, we propose a topology-aware stream assignment algorithm for scheduling operators to computational streams. StreamRec reduces tail latency by up to 27.8% and improves up to 40% throughput in comparison to the original TensorFlow XLA.

Acknowledgements. This work was supported by the National Natural Science Foundation of China (No. 62272168), and CCF-Tencent Open Fund (RAGR20210110).

References

1. Adnan, M., Maboud, Y.E., Mahajan, D., Nair, P.J.: Accelerating recommendation system training by leveraging popular choices. Proc. VLDB Endow. (PVLDB) **15**(1), 127–140 (2021)
2. Bian, W., et al.: Can: feature co-action network for click-through rate prediction. In: Proceedings of the Fifteenth ACM International Conference on Web Search and Data Mining (WSDM), pp. 57–65 (2022)
3. Chen, Q., Zhao, H., Li, W., Huang, P., Ou, W.: Behavior sequence transformer for e-commerce recommendation in alibaba. In: Proceedings of the 1st International Workshop on Deep Learning Practice for High-Dimensional Sparse Data (DLP-KDD), pp. 1–4 (2019)
4. Crankshaw, D., Wang, X., Zhou, G., Franklin, M.J., Gonzalez, J.E., Stoica, I.: Clipper: a low-latency online prediction serving system. In: Proceedings of the 14th USENIX Symposium on Networked Systems Design and Implementation (NSDI), pp. 613–627 (2017)
5. Cui, W., Zhao, H., Chen, Q., Zheng, N., et al.: Enable simultaneous DNN services based on deterministic operator overlap and precise latency prediction. In: Proceedings of the International Conference for High Performance Computing, Networking, Storage and Analysis (SC), pp. 15:1–15:15 (2021)

6. Dhakal, A., Kulkarni, S.G., Ramakrishnan, K.K.: GSLICE: controlled spatial sharing of GPUs for a scalable inference platform. In: Proceedings of the ACM Symposium on Cloud Computing (SoCC), pp. 492–506 (2020)
7. Gupta, U., et al.: DeepRecSys: a system for optimizing end-to-end at-scale neural recommendation inference. In: Proceedings of the 47th ACM/IEEE Annual International Symposium on Computer Architecture (ISCA), pp. 982–995 (2020)
8. Gupta, U., Wu, C., Wang, X., Naumov, M., et al.: The architectural implications of Facebook's DNN-based personalized recommendation. In: Proceedings of the IEEE International Symposium on High Performance Computer Architecture (HPCA), pp. 488–501 (2020)
9. Hu, Y., Rallapalli, S., Ko, B., Govindan, R.: Olympian: scheduling GPU usage in a deep neural network model serving system. In: Proceedings of the 19th International Middleware Conference (Middleware), pp. 53–65 (2018)
10. Ke, L., Gupta, U., Hempstead, M., Wu, C., Lee, H.S., Zhang, X.: Hercules: heterogeneity-aware inference serving for at-scale personalized recommendation. In: Proceedings of the IEEE International Symposium on High-Performance Computer Architecture (HPCA), pp. 141–144 (2022)
11. Kwon, W., Yu, G., Jeong, E., Chun, B.: Nimble: lightweight and parallel GPU task scheduling for deep learning. In: Advances in Neural Information Processing Systems 33: Annual Conference on Neural Information Processing Systems (NeurIPS) (2020)
12. Liu, H., et al.: JIZHI: a fast and cost-effective model-as-a-service system for web-scale online inference at Baidu. In: Proceedings of the 27th ACM Conference on Knowledge Discovery and Data Mining (SIGKDD), pp. 3289–3298 (2021)
13. Miao, X., et al.: HET: scaling out huge embedding model training via cache-enabled distributed framework. Proc. VLDB Endow. (PVLDB) 15(2), 312–320 (2021)
14. Olston, C., et al.: Tensorflow-serving: flexible, high-performance ML serving. CoRR abs/1712.06139 (2017)
15. Pi, Q., et al.: Search-based user interest modeling with lifelong sequential behavior data for click-through rate prediction. In: Proceedings of the 29th ACM International Conference on Information & Knowledge Management (CIKM), pp. 2685–2692 (2020)
16. Zhang, Y., Chen, L., Yang, S., Yuan, M., et al.: PICASSO: unleashing the potential of GPU-centric training for wide-and-deep recommender systems. In: Proceedings of the 38th IEEE International Conference on Data Engineering (ICDE), pp. 3453–3466 (2022)
17. Zhou, G., et al.: Deep interest evolution network for click-through rate prediction. In: Proceedings of the Thirty-Third AAAI Conference on Artificial Intelligence (AAAI), pp. 5941–5948 (2019)

Mitigating Data Stalls in Deep Learning with Multi-times Data Loading Rule

Derong Chen[1], Shuang Liang[1(✉)], Gang Hu[2], Han Xu[2], Xianqiang Luo[2], Hao Li[1], and Jie Shao[1,3]

[1] University of Electronic Science and Technology of China, Chengdu, China
{chenderong,shuangliang,hao_li}@std.uestc.edu.cn, shaojie@uestc.edu.cn
[2] Huawei Data Storage, Huawei Technologies Co., Ltd., Chengdu, China
{hugang27,xuhan17,luoxianqiang}@huawei.com
[3] Shenzhen Institute for Advanced Study, UESTC, Shenzhen, China

Abstract. With the growth of AI data scale, most deep learning jobs separate data storage and computation tasks. Therefore, I/O optimization has gradually become an important issue for training optimization of deep learning. Recent studies focus on I/O optimization for those deep learning methods with one-time data loading rule, where each data is used only once per epoch. However, these methods cannot deal with some jobs with multi-times data loading rule (e.g., meta-learning). By analyzing the characteristic of multi-times data loading, we design a simple, intuitive and effective cache replacement strategy called steady cache strategy. This strategy utilizes a cache to mitigate data stalls and converts the data placement problem to a 0/1 knapsack problem. To our best knowledge, we are the first to mitigate data stalls in AI jobs with multi-times data loading rule and our method is suitable for multi-job scenario. Our experiments demonstrate that the steady cache strategy achieves great improvement over the LRU strategy.

Keywords: Data stalls · Steady cache strategy · Multi-times data loading rule

1 Introduction

Training deep learning models requires a lot of computation and I/O resources. Many prior investigations focus on how to reduce the computation time of the models [1,4,7,10]. Besides, some researchers study how to reduce I/O latency for AI jobs [3,8,12,20], which is usually called data stalls. With the growth of AI data scale, the training procedure of huge deep learning models is heterogeneous, which separates data storage and computation tasks. Therefore, mitigating data stalls is critical to speed up the overall training process.

To better explain data stalls, we first introduce the data input pipeline of heterogeneous deep learning as shown in Fig. 1. In many cases, a large dataset is stored in remote storage device and fetched (F in Fig. 1) to computation node during the training process. In computation node, training data are preprocessed (P in Fig. 1) by CPU, and then sent to GPU. Most prior works focus on

X. Wang et al. (Eds.): DASFAA 2023, LNCS 13943, pp. 562–577, 2023.
https://doi.org/10.1007/978-3-031-30637-2_37

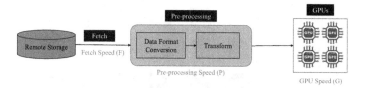

Fig. 1. Data input pipeline of deep learning. Training data are stored in remote storage device. When deep learning training starts, training data need to be fetched from remote storage device and pre-processed before sent to GPUs.

reducing GPU computation time, which makes GPU speed (G in Fig. 1) very fast. However, as GPU speed (G) speeds up, the overall training time is dominated by I/O time: time spent obtaining data from the remote storage device to the GPU. In Fig. 1, when $G \gg F + P$, GPUs spend much time waiting for data, rather than computation, which causes a serious waste of GPU resources. This phenomenon is called data stalls, indicating time is spent waiting for data to be fetched and pre-processed. In the presence of data stalls, increasing the GPU speed (G) cannot reduce the overall training time. In this case, we should pay more attention to reducing I/O time.

In order to mitigate data stalls, we should first examine the data loading rules of deep learning. The data loading rules are divided into two categories: one-time data loading rule and multi-times data loading rule. One-time data loading rule is a very simple and common rule used in most traditional deep learning training scenarios [5,17]. This rule divides training process into many epochs, and when an epoch is done, the indexes of the overall training data need to be shuffled to obtain a different index order in next epoch. In each epoch, every training data is used only once. Therefore, each data in the dataset has the same number of accesses during the whole training process. Although one-time data loading rule is such a widely used rule, it cannot suit all deep learning methods (e.g., meta-learning [6]).

Figure 2 shows the multi-times data loading rule based on few-shot learning [19] (which belongs to meta-learning) with a toy training dataset. Few-shot learning adopts episodic training strategy [9], which has multiple individual episodes in the training process. In each episode, training data are randomly sampled from the overall dataset. Figure 2 shows a toy few-shot learning task with two different episodes. The classes and images in blue box form the sub-dataset of episode 1, and the classes and images in orange box form the sub-dataset of episode 2. Both sub-datasets are formed by a randomly sampling strategy. Therefore, unlike one-time data loading rule, where every image has the same number of accesses, multi-times data loading rule makes different images have different numbers of accesses. For example, the images in both blue and orange boxes in Fig. 2 have two accesses compared with other sampled images. From the perspective of storage, the frequently accessed images are hot or warm data, while other images are cold data. Figure 3 shows the amounts of data corresponding to different numbers of accesses based on a few-shot learning task with

Fig. 2. The illustration of multi-times data loading rule. This toy training dataset contains 5 classes, and each class contains 7 images. Images in blue box form a sub-dataset of one episode, and images in orange box form a sub-dataset of another episode. While some images are selected only once, there are two images selected twice, which are in both blue and orange boxes. (Color figure online)

18000 episodes. This task adopts an image dataset with 64 classes and 38400 images by using multi-times data loading rule. We can see that some images are accessed more than 70 times, while some images are accessed less than 30 times.

This observation inspires us to design a unique cache replacement strategy called steady cache strategy for deep learning with multi-times data loading rule. Some researchers have explored how to mitigate data stalls in deep learning with one-time data loading rule [8,12,14]. However, mitigating data stalls in the scenario of multi-times data loading rule has not been investigated, and this paper aims to tackle this new problem.

The proposed steady cache strategy has four characteristics: 1. Count the frequency of data usage in advance. Before starting deep learning training process, we collect the data usages of all training data for the deep learning job. This information is needed for the next data loading task; 2. Convert data loading task into the 0/1 knapsack problem. 0/1 knapsack aims at selecting items to fit into a fixed size knapsack so as to maximize the total value of the items to be loaded. While loading data from a remote storage device to the cache, we can follow the 0/1 knapsack algorithm, by which we select the most valuable data based on their frequencies of data usages and their sizes, and then load them to the fixed size cache; 3. Train AI models with steady cache strategy. Our steady cache strategy has a non-elimination characteristic. After the data loading, data in the cache would not be replaced by other data during the training process, which avoids significant replacement overhead; 4. Steady cache strategy is suitable for multi-job scenario. In many cases, multiple jobs are running in system. For example, deep learning experts often try many different combinations of hyper-parameters and run some deep learning jobs simultaneously to look for the best hyper-parameter [13,15]. Among these different jobs, some jobs are highly data-hungry, while others are not. Prior works ignored this phenomenon and equally treated every deep learning job. In contrast, we design a

Fig. 3. The amounts of data corresponding to different numbers of accesses. Multi-times data loading rule is applied on 38400 images, and we collect statistics on 18000 episodes.

reasonable resource allocation method in the steady cache strategy based on the data-hungry level of each job.

This paper makes the following key contributions:

- We analyze the characteristics of data flow of deep learning and we divide the data loading rules into two categories: one-time data loading rule and multi-times data loading rule. We conduct an experiment to show that multi-times data loading rule results in cold, warm and hot data.
- To our best knowledge, we are the first to mitigate data stalls for deep learning with multi-times data loading rule, while many prior works focus on mitigating data stalls for deep learning with one-time data loading rule.
- We design a cache replacement strategy called steady cache strategy to mitigate data stalls in deep learning with multi-times data loading rule. Our cache is designed for both single-job and multi-job scenarios. We conduct a series of experiments to verify the effectiveness of our method. Our steady cache strategy achieves great improvement in both single-job and multi-job scenarios compared with traditional methods.

2 Background

2.1 Optimizing I/O Performance in Deep Learning

Recently, many works aim to optimize I/O performance in deep learning [3,8, 12,20]. They point out that the I/O speed greatly limits the training speed of deep learning. Yang and Cong [20] show that data loading can dominate the overall training time of deep learning and propose a locality-aware data loading method to improve the I/O performance. Demir and Sayar [3] improve small image file processing performance of Hadoop Distributed File System (HDFS). They use a case study of face detection and propose two approaches: converting the images into single large-size file by merging and combining many images for a single task without merging. This idea of merging files helps a lot in HDFS system, because HDFS is designed for storing and processing large-size files but not good at storing and processing mass small files. Zhu et al. [21] also focus on improving I/O performance of deep learning in HDFS file system. They mention that the huge number of small files makes HDFS suffer a severe I/O performance penalty. Therefore, they propose pile-HDFS based on the idea of file aggregating approach, where pile is the I/O unit in pile-HDFS. Both works above utilize the idea of merging in HDFS. In contrast to these works, our method is file system agnostic. Kumar and Sivathanu [8] summarize four characteristics of deep learning training: shareability, random access, substitutability and predictability. They design an informed cache named Quiver for deep learning training jobs in a cluster of GPUs, which significantly improves throughput of deep learning workloads. Mohan et al. [12] categorize and analyze data stalls in deep learning, and provide a comprehensive analysis of data input pipeline of deep learning training. They point out that page cache strategy in operating system is not effective for the data flow of deep learning, and propose a new MinIO cache strategy to mitigate data stalls. However, both of them only consider the deep learning with one-time data loading rule but ignore those with multi-times data loading rule. Our work focuses on this unsolved problem.

2.2 Data Loading Rule in Episodic Training Strategy

Some deep learning training processes (e.g., meta-learning) do not use one-time data loading rule. These deep learning methods leverage episodic training strategy. In each episode, they obtain training data by randomly sampling data from a large dataset. Few-shot learning is one of them which uses episodic training strategy. This paper aims to do a case study on how to mitigate data stalls in the few-shot image classification scenario. Compared with traditional deep learning, the data loading rule of few-shot learning is more complicated. In few-shot learning, we have a basic dataset D_{train} with classes C_{train} used for training. In each episode, we randomly sample data from D_{train} to form a sub-dataset. Specifically, we randomly select N classes from C_{train}, and then randomly select K images from each selected class. Therefore, for one episode, we have a sub-dataset with $N \times K$ images. Generally, a training process contains thousands of episodes. Some images are repeatedly selected many times, while others are selected a few times. Therefore, unlike one-time data loading rule, multi-times data loading rule leads to cold, warm and hot data.

2.3 LRU Cache Strategy

As we all know, Least Recently Used (LRU) [2] is the most common cache replacement strategy for many I/O cache replacement systems. One of its advantage is that it can achieve a sub-optimal and acceptable effect on many different jobs. The simplicity and efficiency of its implementation is also an outstanding advantage. The core idea of LRU is that if the data have been accessed recently, the chance of being accessed in the future is higher. This access rule works well for most jobs, but not for AI jobs. For those AI jobs with one-time data loading

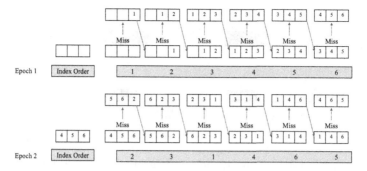

Fig. 4. LRU strategy for AI training with one-time data loading rule. In epoch 1, the cache is filled with data [4, 5, 6]. In epoch 2, because of the new index order, every data cached in epoch 1 is not used in the beginning, thus causing cache missing.

rule, every data is accessed only once in an epoch. If the scale of data accessed in one epoch is much larger than the scale of data that cache can contain, every access causes cache missing and the data fetched from disk to the cache is not used again in the current epoch. Although the same data will be accessed in next epoch, the indexes of the training data are shuffled, so the index order is randomly different from previous epochs. At the beginning of the next epoch, there is little chance of accessing the data in the cache. The worst situation is that all the data cached are not used in the beginning of next epoch, thus causing cache missing on every access. Figure 4 shows the worst situation. Fortunately, this situation does not occur in every epoch, so there are still some cache hits in the LRU strategy. As for AI tasks with multi-times data loading rule, LRU is still not the optimal strategy. Because the data used for training in an episode is completely random, the core idea of LRU is not suitable for this scenario either.

3   Method

In this section, we introduce our method in detail. Due to the unique cache replacement strategy of our method, we call it steady cache. We discuss single-job scenario and multi-job scenario in Sect. 3.1 and Sect. 3.2, respectively.

3.1   Single-Job Scenario

AI training relies heavily on GPU hardware. Most AI researchers do not have GPU clusters to run multiple AI training jobs. Usually, they only have one GPU in their PC or server to run single AI training job. Therefore, we intend to first investigate this simpler but more general scenario relative to multi-job scenario. Our method uses cache resource and includes three steps: 1. Count the frequency of data usage in advance; 2. Load the most valuable data into steady cache; 3. Perform AI training with steady cache strategy, which does not replace the data in the cache.

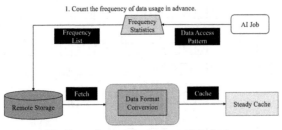

Fig. 5. The first two steps of steady cache strategy. For the specified AI job, we collect its index order and count the frequency of data usage, then form a frequency list and send it to the remote storage device. We select the most valuable data to load into the cache according to the 0/1 knapsack algorithm. Before loading the data into cache, we convert the data format.

Figure 5 shows the first two steps. In step 1, we collect the index order in advance by running the multi-times data loading rule, but not actually loading the data to the computation node. We calculate the usage frequency of data based on the index order and form a frequency list, and then send the frequency list to the remote storage device. In traditional deep learning models, index order is generated during the training process. When the model requires a batch of data, a group of indexes is randomly generated and data are obtained based on the indexes. After the obtained batch of data are used up, the preceding operation is repeated to ensure that the model continuously obtains new training data. The fetch of one batch data relies on the index generated randomly. Since the indexes of all batches required for this complete training are generated at once, the index order of the data can be determined before the training starts.

As mentioned in Sect. 2.2, multi-times data loading rule results in cold, warm, and hot data. Before loading the most valuable data into our steady cache, we need to judge the hot and cold data based on the indexes of all the batches collected. Obviously, the frequency of an index indicates how hot and cold the corresponding data is. Therefore, we calculate the frequency of all data and form a frequency list. Indexes would not be generated again in the subsequent training process. Instead, indexes are collected before training and used as the index order during the training process, which ensures that hot and cold characteristics of data do not change.

In step 2, we select the most valuable data in the frequency list to load into the cache. This problem is very similar to the 0/1 knapsack problem, which solves the problem of how to put a batch of items with different weights and values into a fixed capacity knapsack to maximize the total value of the items. In our data loading problem, many similar concepts can be mapped to the 0/1 knapsack problem. For example, the sizes of data in a dataset are different, which can be mapped to the weights of the items in the 0/1 knapsack problem. The usage frequencies of different data collected in step 1 are also different, and cache can maximize the I/O efficiency by caching more hot data. Therefore, the data frequently used is more valuable. The usage frequencies of the data can be mapped to the values of the items in the 0/1 knapsack problem.

While cache capacity is fixed, the data loading problem can be considered as finding a placement strategy that maximizes the value of the data in the cache. We take the size of the i-th data as the weight W_i and take the frequency of the i-th data as the value V_i in this special 0/1 knapsack problem. The total cache capacity is considered as the capacity C of the knapsack. Therefore, we need to solve the following optimization problem:

$$\max \sum_{i=1}^{n} X_i V_i, \quad \text{s.t.} \quad \sum_{i=1}^{n} X_i W_i \leq C, \quad X_i \in \{0, 1\}. \tag{1}$$

n indicates the total number of data (e.g., number of training images). $X_i = 1$ indicates loading the i-th data and $X_i = 0$ indicates not loading the i-th data into the cache. The greedy algorithm cannot get the optimal solution, so the dynamic programming algorithm can be used to solve this problem. We define sub-problem as $P(i, E)$, which means to select items whose total weight does not exceed E from the first i items. Each item is either selected or not selected to maximize the total value $M(i, E)$. Considering the i-th item, there are only two possibilities: to select or not to select. If it is selected, the problem $P(i, E)$ turns into $P(i - 1, E - W_i)$. If it is not selected, the problem $P(i, E)$ turns into $P(i - 1, E)$. The optimal selection is made by comparing these two ways and we choose the way that maximizes the total value:

$$M(i, E) = \max \{M(i - 1, E), M(i - 1, E - W_i) + V_i\}. \tag{2}$$

According to the above analysis, we can get the following recursion formula:

$$M(i, E) = \begin{cases} 0 & \text{if } i = 0 \text{ or } E = 0, \\ M(i - 1, E) & \text{if } W_i > E, \\ \max \{M(i - 1, E), V_i + M(i - 1, E - W_i)\} & \text{otherwise.} \end{cases}$$
$$\tag{3}$$

The table filling method is a classical method to solve dynamic programming problem. Considering M as a two-dimensional table with size $n \times C$, $M(i, E)$ indicates row i, column E in the table M. Algorithm 1 shows the dynamic programming algorithm of 0/1 knapsack based on the table filling method. The input includes the number of data n, the sizes of different data $W_1, ..., W_n$, the values of different data $V_1, ..., V_n$, and the cache capacity C. The output is a filled table M. We first use 0 to initialize row 0 (in lines 1–3) and column 0 (in lines 4–6) of the table M, and then traverse the table again (in lines 7–15). As mentioned above, $M(i, E)$ corresponds to the sub-problem $P(i, E)$. If the weight W_i is greater than E, we set $X_i = 0$ and turn the sub-problem $P(i, E)$ into $P(i-1, E)$, so $M(i, E) = M(i - 1, E)$ (in lines 9–10). If the weight W_i is smaller than or equal to E, we compare the value of $M(i - 1, E)$ and $M(i - 1, E - W_i) + V_i$ and select the larger one to fill in $M(i, E)$ (in lines 11–12). If $M(i - 1, E)$ is selected, we set $X_i = 0$, otherwise set $X_i = 1$. During the execution of the algorithm, we

Algorithm 1. Dynamic programming algorithm of 0/1 knapsack.

Input: $n, W_1, ..., W_n, V_1, ..., V_n, C$
Output: M
1: **for** $E = 0$; $E <= C$; $E + +$ **do**
2:     $M(0, E) = 0$
3: **end for**
4: **for** $i = 1$; $i <= n$; $i + +$ **do**
5:     $M(i, 0) = 0$
6: **end for**
7: **for** $i = 1$; $i <= n$; $i + +$ **do**
8:     **for** $E = 1$; $E <= C$; $E + +$ **do**
9:         **if** $W_i > E$ **then**
10:             $M(i, E) = M(i - 1, E)$
11:         **else**
12:             $M(i, E) = \max\{M(i - 1, E), M(i - 1, E - W_i) + V_i\}$
13:         **end if**
14:     **end for**
15: **end for**
16: **return** M

---

can easily get the values of $X_1, ..., X_n$, and then we can load the i-th data where $X_i = 1$ to the steady cache.

As shown in Fig. 5, data format conversion in step 2 is a very important operation. We do not load the data in original format, because it is not the format needed by AI training. We convert the data into the tensor format which is exactly the data format needed by AI models. In training process, the data from cache do not need to be transformed to tensor again, but only need some random transformation, which accelerates the I/O speed from CPU to GPU (the data are pre-processed in CPU). Converting data format also has a disadvantage: large cache size is required because the size of tensor is larger than the original format.

Figure 6 shows the last step of our steady cache strategy. After the first two steps, the AI job is ready to start training. In the real training process, the AI job accesses the data according to the index order collected earlier. If the data is stored in the cache, it can be quickly transformed and fetched to the GPU. If the data is not stored in the cache, it can be fetched from the remote storage device and converted to tensor first, which costs more time. Different from the LRU strategy, the cached data would not be replaced when using the steady cache strategy. This steady cache strategy may be bad for common jobs, but is helpful for AI jobs, because it avoids useless cache replacement overhead.

Overall, our method includes three steps. The frequency lists of all episodes are collected ahead and the 0/1 knapsack algorithm is utilized to decide which data should be loaded into the steady cache. While the LRU strategy may cause replacement overhead, the status of steady cache remains unchanged, thus achieving steady improvement compared with LRU.

3.2 Multi-job Scenario

Our method is also suitable for multi-job scenario. AI training researchers may run multiple AI jobs with different hyper-parameters at the same time to search

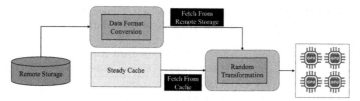

3. AI training with non-elimination cache strategy.

Fig. 6. The third and last step of steady cache strategy. In the training process, data stored in the cache can be directly sent to GPU, while data stored in remote storage device is transmitted over the network and transformed to tensor before being sent to GPU. The non-elimination characteristic of our steady cache strategy avoids useless cache replacement overhead.

the best hyper-parameter. Different jobs have different index orders and different model sizes. As mentioned above, some AI jobs are data-hungry, while others are not. For those AI jobs with larger model sizes, the computation time is longer than those with smaller model sizes. Longer computation time reduces the I/O latency, because if the I/O latency remains the same, the more computation time means the less waiting time for GPU. We use model size to evaluate the data-hungry level of AI job.

$$\beta_i = \frac{S_i}{\sum_{i=1}^{h} S_i}, \tag{4}$$

where β_i denotes the data-hungry coefficient of the i-th AI job, and S_i denotes the model size of the i-th AI job. h is the number of jobs. The value of β_i is between 0 and 1. As mentioned in Sect. 3.1, our method generates frequency list for each AI job. We use $Flist_i$ to denote the frequency list of the i-th AI job. We should consider all lists together to get a global optimal list $Flist^*$, which guides the data loading order from remote storage device to the cache.

$$Flist^* = \sum_{i=1}^{h} \beta_i \times Flist_i. \tag{5}$$

Different AI jobs contribute differently to $Flist^*$ based on their data-hungry coefficient. The higher data-hungry level means the larger data-hungry coefficient β_i and the greater contribution to the global optimal list $Flist^*$. In simple words, the cache is more likely to store data for the more data-hungry AI jobs. After getting $Flist^*$, the selected data stored in cache can be obtained by Algorithm 1 mentioned in Sect. 3.1. Then, we can assume the cache consumption of multiple AI jobs as a large single AI job, and follow the last two steps of single AI job

Table 1. The I/O time (seconds) in single-job scenario using different strategies. Cache proportion is the ratio of cache size to the scale of total data.

Cache proportion	MiniImageNet			CUB-200-2011		
	zero cache	LRU	steady cache	zero cache	LRU	steady cache
20%	1225	1230	990	218	221	176
40%	1225	1117	736	218	207	132
60%	1225	994	510	218	181	91
80%	1225	690	273	218	126	49
90%	1225	455	158	218	80	28
100%	1225	15	15	218	3	3

to achieve the I/O optimization of multiple AI jobs. In training process, these multiple AI jobs are running at the same time and share the memory of the same cache.

4   Implementation and Evaluation

4.1   Experimental Setup

Our study aims to mitigate data stalls in deep learning with multi-time data loading rule. One of the scenarios using this strategy is few-shot learning. Therefore, we perform a case study of few-shot image classification in this paper. The detail of the data sampling strategy of this scenario is mentioned in Sect. 2.2. The number of classes N is set to 5 and the number of images per class K is set to 20, so 100 images are randomly sampled in each episode. Note that, our method is not limited to this scenario but general for all scenarios of this type. We use two servers to implement the scenario of the heterogeneous storage computing architecture. Server A is regarded as the remote storage device, which has 48 cores of Intel Xeon Silver 4214R CPU @ 2.40 GHz. Server B is regarded as the computation node, which has 48 cores of Intel Xeon Silver 4210 CPU @ 2.20 GHz and 2 GPUs of NVIDIA GeForce RTX 3090. We take Secure File Transfer Protocol (SFTP) as the file transfer protocol from A to B. We use ResNet [5] as the deep learning model to achieve I/O-intensive scenario based on PyTorch deep learning framework.

4.2   Datasets

We use two few-shot learning datasets to evaluate our method: MiniImagenet [18] and CUB-200-2011 [16]. MiniImagenet consists of 60,000 images with 100 classes and each class contains 600 images. CUB-200-2011 consists of 11,778 images from 200 different bird classes. Both of them are stored in remote storage device (server A) and fetched to computation node (server B).

4.3   Evaluation of Single-Job Scenario

We evaluate three different strategies: zero cache, LRU, and steady (our method). Besides LRU, zero cache is another baseline of our method, which indicates that

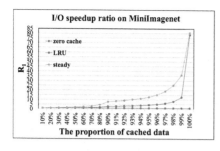
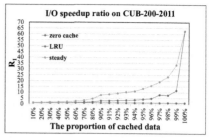

Fig. 7. The relationship between I/O speedup ratio (R_1) and the proportion of cached data. The left side illustrates results on MiniImagenet, and the right side illustrates results on CUB-200-2011. As the proportion of cached data increases, the gap between LRU and steady becomes larger in almost all proportions except 100%.

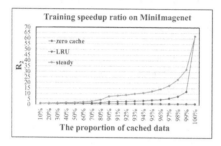
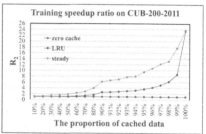

Fig. 8. The relationship between training speedup ratio (R_2) and the proportion of cached data. The left side illustrates results on MiniImagenet, and the right side illustrates results on CUB-200-2011. As the proportion of cached data increases, the gap between LRU and steady becomes larger in almost all proportions except 100%.

no cache resources are available, so all data need to be fetched from the remote storage device during the training process. We compare the I/O speedup ratio R_1 and the training speedup ratio R_2 with the change of the proportion of cached data under different strategies.

$$R_1^{lru} = \frac{T_1^{zero}}{T_1^{lru}}, \quad R_1^{steady} = \frac{T_1^{zero}}{T_1^{steady}}. \tag{6}$$

$$R_2^{lru} = \frac{T_2^{zero}}{T_2^{lru}}, \quad R_2^{steady} = \frac{T_2^{zero}}{T_2^{steady}}. \tag{7}$$

R_1 and R_2 denote the I/O and training speedup ratios. T_1 and T_2 denote the I/O and training time (training time includes I/O and computation time) of a complete training process. The superscript denotes which strategy we choose.

Table 1 shows the I/O time (seconds) in single-job scenario using different strategies. We collect training information for 150 episodes and count the total I/O time. We present experimental results for several representative groups of the proportion of cached data. Figure 7 and Fig. 8 show the I/O speedup ratio and training speedup ratio on MiniImagenet and CUB-200-2011 by three different

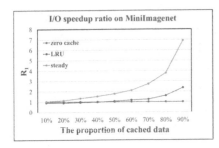
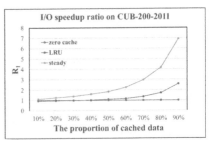

Fig. 9. The relationship between I/O speedup ratio (R_1) and the proportion of cached data. This figure shows the results without data format conversion.

Table 2. The average I/O time (seconds) of four jobs on CUB-200-2011 using different strategies. Cache proportion is the ratio of cache size to the scale of total data.

Cache proportion	zero cache	LRU	steady cache
20%	565.75	707.5	486
40%	565.75	557.75	376
60%	565.75	400	260.5
80%	565.75	254.75	162.75
90%	565.75	173.75	121.5
100%	565.75	86	86

cache replacement strategies. LRU strategy performs better than zero cache strategy in the single AI training job, which demonstrates the effectiveness of cache. However, the LRU strategy is only sub-optimal, while our steady strategy shows great improvement over LRU. Because our AI job is I/O-intensive, I/O time ≫ computation time, and thus Fig. 7 and Fig. 8 are very similar. In Fig. 7 and Fig. 8, the data are converted into tensor format ahead before loading into the cache. We conduct extensive experiment to observe the performance of LRU and steady without data format conversion. Figure 9 shows the experimental results without data format conversion on MiniImagenet and CUB-200-2011. The results show that LRU is still inferior to our method. With a small proportion of the cached data, the LRU strategy causes frequent replacement operations, resulting in a catastrophic I/O performance. In extreme cases, as shown in Fig. 9, the LRU strategy performs even worse than fetching data directly from a remote storage device (zero cache strategy). Only when the proportion of cached data is large, LRU is better than zero cache. In contrast to this, our steady cache strategy is robustly better than the LRU strategy. Note that, when the proportion of cached data is 100%, LRU no longer incurs replacement overhead, which leads to the same speedup ratio with the steady cache strategy.

4.4 Evaluation of Multi-job Scenario

Some AI jobs may be running at the same time. We should reduce the I/O time by considering them together. We use CUB-200-2011 to evaluate 4 AI jobs

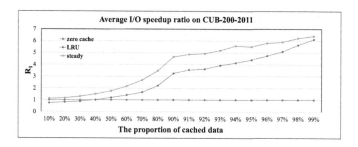

Fig. 10. The relationship between I/O speedup ratio (R_1) and the proportion of cached data. This figure shows the average results in multi-job scenario.

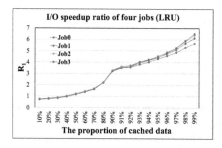
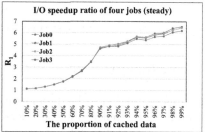

Fig. 11. The relationship between I/O speedup ratio (R_1) and the proportion of cached data. This figure shows the results of each job in multi-job scenario.

with different model sizes: ResNet18, ResNet50, ResNet101 and ResNet152. We show the average I/O speedup ratio of these AI jobs by using three different cache strategies: zero cache, LRU and steady. Furthermore, We also show the I/O speedup ratio for each AI job separately.

Table 2 shows the average I/O time (seconds) of four jobs under different strategies. Figure 10 shows the average I/O speedup ratio of four AI jobs. The results show that our steady cache strategy is still better than the LRU strategy in multi-job scenario. While LRU suffers from the replacement cost when the proportion of cached data is small, our strategy achieves steady improvement. Figure 11 shows the results of four AI jobs with different AI models. The left side shows the results using the LRU strategy, and the right side shows the results using the steady cache strategy. There are steady improvements for each AI job.

4.5   More Baselines for Comparison

In this section, we select two additional cache replacement algorithms for comparison. LRU-K extends the judgment standard of "recently used 1 time" to "recently used K times" for LRU. That is to say, data that has not reached K times of access will not be cached. In our experiment, the value of K is set to 2. Adaptive Replacement Cache (ARC) improves LRU by balancing access time priority and access frequency priority [11]. Figure 12 shows the results on

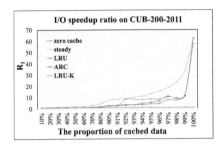
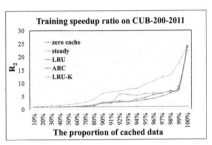

Fig. 12. Comparison of steady cache and multiple baselines.

CUB-200-2011. It can be seen that LRU-K and ARC perform better than LRU, but are still worse than our steady cache.

5   Conclusion

In this paper, we divide the data loading rules into two categories: one-time data loading rule and multi-times data loading rule. While existing studies has tries to mitigate data stalls of classical deep learning models with one-time data loading rule, we propose a new steady cache strategy to mitigate data stalls with multi-times data loading rule. Our steady cache strategy is a three-step approach, including pre-counting data usage, cache placement using the 0/1 knapsack algorithm, and AI training with non-elimination characteristic. Additionally, we extend our approach from a single-job scenario to a multi-job scenario. We conduct a series of experiments on two datasets to verify the effectiveness of our steady cache strategy. The results show that the proposed steady cache strategy achieves great improvements in both single-job and multi-job scenarios.

Acknowledgements. This work is supported by the National Natural Science Foundation of China (No. 62276047).

References

1. Chen, T., Xu, B., Zhang, C., Guestrin, C.: Training deep nets with sublinear memory cost. CoRR abs/1604.06174 (2016)
2. Chrobak, M., Noga, J.: LRU is better than FIFO. Algorithmica **23**(2), 180–185 (1999)
3. Demir, I., Sayar, A.: Hadoop optimization for massive image processing: case study face detection. Int. J. Comput. Commun. Control **9**(6), 664–671 (2014)
4. Hashemi, S.H., Jyothi, S.A., Campbell, R.H.: TicTac: accelerating distributed deep learning with communication scheduling. In: Proceedings of Machine Learning and Systems 2019, MLSys 2019 (2019)
5. He, K., Zhang, X., Ren, S., Sun, J.: Deep residual learning for image recognition. In: 2016 IEEE Conference on Computer Vision and Pattern Recognition, CVPR 2016, pp. 770–778 (2016)

6. Hospedales, T.M., Antoniou, A., Micaelli, P., Storkey, A.J.: Meta-learning in neural networks: a survey. IEEE Trans. Pattern Anal. Mach. Intell. **44**(9), 5149–5169 (2022)

7. Jayarajan, A., Wei, J., Gibson, G., Fedorova, A., Pekhimenko, G.: Priority-based parameter propagation for distributed DNN training. In: Proceedings of Machine Learning and Systems 2019, MLSys 2019 (2019)

8. Kumar, A.V., Sivathanu, M.: Quiver: an informed storage cache for deep learning. In: 18th USENIX Conference on File and Storage Technologies, FAST 2020, pp. 283–296 (2020)

9. Li, D., Zhang, J., Yang, Y., Liu, C., Song, Y., Hospedales, T.M.: Episodic training for domain generalization. In: 2019 IEEE/CVF International Conference on Computer Vision, ICCV 2019, pp. 1446–1455 (2019)

10. Lin, Y., Han, S., Mao, H., Wang, Y., Dally, B.: Deep gradient compression: reducing the communication bandwidth for distributed training. In: 6th International Conference on Learning Representations, ICLR 2018 (2018)

11. Megiddo, N., Modha, D.S.: ARC: a self-tuning, low overhead replacement cache. In: Proceedings of the FAST 2003 Conference on File and Storage Technologies (2003)

12. Mohan, J., Phanishayee, A., Raniwala, A., Chidambaram, V.: Analyzing and mitigating data stalls in DNN training. Proc. VLDB Endow. **14**(5), 771–784 (2021)

13. Prabu, S., Thiyaneswaran, B., Sujatha, M., Nalini, C., Rajkumar, S.: Grid search for predicting coronary heart disease by tuning hyper-parameters. Comput. Syst. Sci. Eng. **43**(2), 737–749 (2022)

14. Pumma, S., Si, M., Feng, W., Balaji, P.: Scalable deep learning via I/O analysis and optimization. ACM Trans. Parallel Comput. **6**(2), 6:1–6:34 (2019)

15. Quijano, A.J., Nguyen, S., Ordonez, J.: Grid search hyperparameter benchmarking of BERT, ALBERT, and LongFormer on DuoRC. CoRR abs/2101.06326 (2021)

16. Ren, M., et al.: Meta-learning for semi-supervised few-shot classification. In: 6th International Conference on Learning Representations, ICLR 2018 (2018)

17. Vaswani, A., et al.: Attention is all you need. In: Advances in Neural Information Processing Systems 30: Annual Conference on Neural Information Processing Systems 2017 (2017)

18. Vinyals, O., Blundell, C., Lillicrap, T., Kavukcuoglu, K., Wierstra, D.: Matching networks for one shot learning. In: Advances in Neural Information Processing Systems 29: Annual Conference on Neural Information Processing Systems 2016, pp. 3630–3638 (2016)

19. Wang, Y., Yao, Q., Kwok, J.T., Ni, L.M.: Generalizing from a few examples: a survey on few-shot learning. ACM Comput. Surv. **53**(3), 63:1–63:34 (2020)

20. Yang, C., Cong, G.: Accelerating data loading in deep neural network training. In: 26th IEEE International Conference on High Performance Computing, Data, and Analytics, HiPC 2019, pp. 235–245 (2019)

21. Zhu, Z., Tan, L., Li, Y., Ji, C.: PHDFS: optimizing I/O performance of HDFS in deep learning cloud computing platform. J. Syst. Archit. **109**, 101810 (2020)

Internet Public Safety Event Grading and Hybrid Storage Based on Multi-feature Fusion for Social Media Texts

Die Hu[1], Yulai Xie[2(✉)], Dan Feng[1,3(✉)], Shixun Zhao[3], and Pengyu Fu[3]

[1] Wuhan National Laboratory for Optoelectronics, Huazhong University of Science and Technology, Wuhan, China
{hudie,dfeng}@hust.edu.cn
[2] School of Cyber Science and Engineering, Hubei Engineering Research Center on Big Data Security, Huazhong University of Science and Technology, Wuhan, China
ylxie@hust.edu.cn
[3] School of Computer Science and Technology, Huazhong University of Science and Technology, Wuhan, China
sxzhao@hust.edu.cn

Abstract. With the rapid development of mobile Internet technology, various public events appear on social media platforms and attract a lot of attention. Since netizens can express their opinions freely on the Internet, some events that cause negative public opinion seriously threaten public security, which are called Internet Public Safety events (IPSe). Existing solutions use a single metric to realize event detection, which has a high false detection rate for Internet Public Safety events. In addition, they lack countermeasures after the outbreak of public opinion, resulting in inefficient information management. This paper proposes a novel Internet Public Safety event grading method based on multi-feature, which measures events from the three perspectives of heat, emotion, and sensitivity. In order to improve the retrieval efficiency of events information, we implement a smart dynamic hot/cold data migration mechanism using a hybrid storage system containing solid-state drives and hard-disk drives, which realizes real-time data adjustment in the storage layer and ensures efficient events management. In the experiments, we verify our method through several real internet events, and the results show that our method achieves the state-of-the-art accuracy in the detection of Internet Public Safety events and realizes efficient retrieval with a low query overhead.

Keywords: Internal Public Safety · Social Text · Event Grading · Hybrid Storage

1 Introduction

Sina Weibo is the most widely used social platform in China. It supports the dissemination of social events and allows people to post their opinions on various online events. Since social platforms are not strict about uploading content, there

X. Wang et al. (Eds.): DASFAA 2023, LNCS 13943, pp. 578–587, 2023.
https://doi.org/10.1007/978-3-031-30637-2_38

are some events that would cause negative public opinion seriously [1–3], called Internet public safety events (IPSe).

Existing methods determine the occurrence of new events by monitoring new words and word frequencies in the social text stream. Most studies [4–9] that extract and analyze social media data only consider emotional factors or event heat for ranking public events. For instance, EC method [4] introduces emotional factors for judgment. RWDM method [8] observes the heat changes in the data constantly and uses past data to eliminate some interference in the analysis of the current data.

However, some other factors, such as event sensitivity, may directly indicate the nature of an event and thus are also very critical for Internet public safety events analysis. It is necessary to take into account all the important characteristics (e.g., heat, emotion and sensitivity) to identify Internet public safety events. In addition, the huge amount of social network data has posed a huge challenge to information storage. Existing storage management methods [10,11] mainly focus on calculating the social relationship and providing reliability, security and high throughput. These methods don't take into account how these social network data can be organized to be accessed fast, which is very important for public safety event analysis and pre-alarm [12].

Our motivation is to develop a system that integrates social event extraction, accurate event grading and high efficient storage management for Internet public safety events. In this paper, we propose a novel public safety-oriented system that provides efficient hot event extraction, accurate event grading and high efficient hybrid storage management.

The contributions of this paper are as follows:

- We design a novel multi-feature fusion-based event grading scheme that can accurately identify Internet public safety events on social networks. It takes into account the heat, emotion and sensitivity of social events, and has a more efficient performance in identifying social events compared with existing methods.
- We implement a hybrid storage framework that can take advantage of the fast characteristics of SSD devices to store social events. The dynamic hot/cold data migration mechanism can adjust the storage layout in real-time according to the status of social events, which realizes the efficient query of social events.
- Experiments on a large number of real-world Weibo data show that our method performs better performance than the existing method on social event grading, and the hybrid storage can greatly improve the event query efficiency and reduce the time consumption.

2   Related Work

The social text sentiment analysis is mainly used to mine users' opinions on Internet public safety events. There are a lot of studies on the emotional calculation of network texts [6,7,9]. Cavdur et al. [13] created a second-order stochastic model to address the allocation of relief facilities in sudden disasters and validated them in earthquake cases. Rawluk et al. [14] proposed a conceptual framework for defining, identifying and analyzing public values in sudden natural disasters and conducting a value analysis of the 2009 Victorian forest fire. Existing event discovery research focuses on probability-based topic models and text-based clustering algorithms. Mi et al. [15] used the probabilistic model to extract popular content from Weibo social media, and then clustered data using Kmeans to extract the topic. Kolajo et al. [16] improves event detection by facilitating the understanding of implicit semantics embedded in social media streams.

These methods has not taken into account the combination of event heat, emotion analysis and event sensitivity analysis, all of which are important to identify a public safety event. In addition, how to store the data of Internet public safety events to improve query performance has not been considered in the current methods.

3   Design and Implementation

Figure 1 illustrates the overall design architecture of our system. The system is composed of three modules: event extraction, events grading and hybrid storage. Event extraction module can extract social events from massive social text data.

Event grading module realizes the determination of Internet public safety events in social events through multi-feature fusion of heat, emotion and sensitivity. Finally, the hybrid storage module achieves efficient storage based on the scores of social events. It realizes the adaptive adjustment of social events in the storage layer through two kinds of storage media with different characteristics.

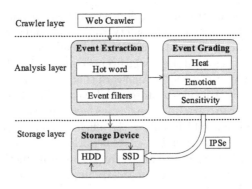

Fig. 1. Architecture of our system.

3.1   Social Event Extraction

Social event is an abstract concept, which is composed of one or a group of interrelated social texts. In this paper, we extract the social events by analyzing the hot words in the social texts.

Hot Word Extraction. When calculating the frequency of a word, we need to consider its change in frequency in the current time window compared to the previous one. This incremental information represents the current heat of a word and is expressed as the ratio of the frequency in the current time window to the frequency in the previous consecutive time windows.

$$H(w) = \frac{f(w, T_j)}{f(w, T_{j-k}) + f(w, T_{j-k+1}) + \cdots + f(w, T_{j-1})} \tag{1}$$

where w represents a word, T_j represents the current time window, and $F(w, T_j)$ represents the appearing frequency of w in the previous window T_j.

In order to improve the processing efficiency of text data, we use the co-occurrence model to identify synonyms in texts.

$$p(w_1|w_2) = \frac{f(w_1, w_2)}{f(w_1)} + \frac{f(w_1, w_2)}{f(w_2)} \tag{2}$$

where $f(w_1)$ and $f(w_2)$ represent the frequency of w_1 and w_2, respectively. $f(w_1, w_2)$ represents the frequency of w_1 and w_2 in the same record.

Event Extraction. We use the mutual information between hot words to generate a series of word sets that can represent the complete social events and use them as the representation of the events. To measure the ability of a word set to represent a social event, we calculate the potential power of the candidate word set $W = \{w_1, w_2, ...w_n\}$.

$$Power(W) = \sum_{i=1}^{N} \frac{f(w_j, w_i)}{f(w_i)} \tag{3}$$

where N represents the length of the candidate word set W.

3.2 Internet Public Safety Events Grading

To recognize the Internet public safety events from the massive social data, we design an efficient event grading strategy based on multi-feature fusion. It calculates the risk score of a social event from three aspects: heat, emotion and sensitivity.

Heat Score. For an event in social network, its relevant microblogs count is a continuous derivable function $f = f(t)$, which shows the total number of microblogs at time t. The maximum number of microblogs for this event is assumed to be K, the amount of change in the cumulative volume of blogs is proportional to the cumulative quantity f and the network free space $(1 - \frac{f}{K})$:

$$f(t) = \frac{K}{1 + (\frac{K}{f} - 1)e^{-rt}} \tag{4}$$

where r is the maximum growth rate.

We count the number of related microblogs posts per hour, then calculate the number of related microblogs posts per unit time, and define it as the heat score for a period. The formula is as follows:

$$HS = \frac{\Delta g}{\Delta t} \tag{5}$$

where HS is the heat score for a period of time, Δg is the number of related microblogs posts during a period and Δt is the space of time.

Emotion Score. For each microblog, we use a vector to represent the emotional orientation of the record in the three dimensions of positive, neutral and negative respectively. Then, we add up the emotional values of the corresponding microblogs of the same event, and normalize them uniformly to obtain the emotional scores of each event. Assuming that the microblogs of an event is $\{b_1, b_2, ...b_m\}$ and the emotional value of each microblog is $\{\vec{b_1}, \vec{b_2}, ...\vec{b_m}\}$, the emotion score of the event is:

$$ES = \sum_{i=1}^{m} [-1, 0, 1] \cdot \vec{b_i} \tag{6}$$

If the emotion of an event is positive, it means that the event has a positive impact on the public, which indicates that the event will not develop into an Internet public safety event that harms social mood.

Sensitivity Score. There are various types of events that happen on social networks, such as "entertainment news", "sports competitions", "political events" and so on. Among them, "entertainment news" and "sports competitions" will not evolve into Internet public safety events that harm social mood, although they are widely discussed and have negative emotions. It's worth paying attention to events with sensitive words, such as the names of national leaders or natural disasters. In this part, we collect the sensitive words in the dataset as a dictionary and use them to calculate the sensitivity score of the events. In the dictionary of sensitive words, each word has a preset value. Assuming that the sensitive words of an event are $\{w_1, w_2, ...w_n\}$ and the corresponding values are $\{v_1, v_2, ...v_n\}$, then the sensitive score of an event is:

$$SS = \sum_{i=1}^{n} w_i.count * v_i \tag{7}$$

where, $w_i.count$ refers to the number of occurrences of w_i in the event.

Summary. After the three factors of heat, emotion and sensitivity are obtained, we need to normalize the three scores respectively and convert each score into a value between 0 and 1. The final score for an Internet public safety event is expressed as:

$$S = a * HS + b * ES + c * SS \tag{8}$$

where, HS, ES and SS represent the heat score, emotion score and sensitivity score of the event. a, b and c respectively represent the weights of these three indicators in the evaluation of event grade.

3.3 Hybrid Storage for Social Events

Efficient retrieval of Internet public safety events play an important role when there is an incident that causes public opinion to erupt. However, with the spread of social events, there will be a huge amount of social text data generated on the Internet due to the opinions expressed by netizens, which causes a very low retrieval efficiency. The horizontal hybrid storage architecture is shown in Fig. 2.

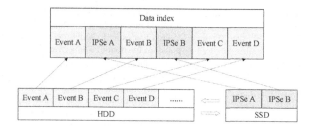

Fig. 2. Horizontal hybrid storage architecture

Considering that Internet public safety events have a high demand for real-time retrieval, we store them in the efficient storage device, SSD. The non-Internet public safety events will be stored in lower-priced HDD to ensure storage capacity for massive social data. In a horizontal hybrid storage structure, we only need to traverse SSD when a query request for Internet public safety events arrives, which improves query efficiency throughout the system. We implement the following the migration strategy of hot and cold data. When a new Internet public safety event arrives, we calculate the storage space available in the SSD. If the SSD has sufficient space to store the new public safety event, we directly store the event in the SSD. If the SSD space is insufficient, the data of the least recently accessed public safety event or the lowest level of the public safety event is migrated to HDD until the SSD has enough space. However, when the cold public safety data in the HDDs have been accessed multiple times, the cold data will become hot data and be migrated to SSDs.

4   Evaluation

Dataset. We crawled some microblog data from Sina Weibo[1] in the two periods which are from October 20 to December 2018 and in April 2019 respectively and got 2,336,453 records of microblogs. The average length of records is 24 Chinese words. In Weibo dataset, the maximum event has 18623 records and the minimum event has 114 records. There are 69 social events, and 31 of them are Internet public safety events. Table 1 shows the six representative social events extracted from Weibo dataset.

Table 1. Representative social events description.

ID	Event	Records	Event description
E1	DG insults Chinese	18,623	In November 2018, DG released a video of insulting China, which caused widespread discussion on Weibo.
E2	Jianjinfu is Domestic Violence	18,661	In November 2018, Chinese star Jiang Jinfu beat his Japanese girlfriend Zhong Puyouhua, and the netizens discussed it extensively.
E3	Meng Wanzhou was arrested	16,299	In December 2018, Huawei's CFO Meng Wanzhou was arrested in Canada. The Chinese Embassy asked the United States and Canada to immediately restore the personal freedom of Meng Wanzhou.
E4	Zhang Shoucheng is death	14,921	In December 2018, the Chinese-American physicist Zhang Shoucheng passed away. The venture capital company founded by Zhang Shoucheng was accused by the US government to help China acquire the cutting-edge technology and related intellectual property rights of the United States.
E5	Mercedes-Benz oil spill	3,708	In April 2019, Xi'an female car owners bought a series of rights-defending events triggered by Mercedes-Benz and Mercedes-Benz oil spills.
E6	Sichuan Liangshan Forest Fire	2,750	In April 2019, a sudden fire broke out in the forests of Liangshan, Sichuan, and many people rescued the fire

Baselines. We compare our method with existing baselines: Emotion Centroid (EC) algorithm and Remarkable Word Detecting Method (RWDM). EC [4] method analyzes mood swings in tweets and divides them into six categories: happiness, anger, sadness, fear, evil and surprise. Each emotion type is represented as a one-dimensional spatial model, with one emotion word for each tweet. If the match is successful, the dimension is 1, otherwise it is 0. RWDM [8] can be used to find hotspot events. It focuses on two issues: 1) How to quantify the importance of a word in Twitter. 2) How to evaluate this quantification to detect a significant word.

---

[1] https://weibo.com/.

4.1   Comparisons with Existing Works

Figure 3 illustrates the overall performances of social events grading approaches on Weibo dataset. For multi-feature fusion-based method, the parameters a, b, and c are set to $1, 1, 2$ respectively. Our multi-feature fusion-based method performs better than the EC method and the RWDM method in terms of both accuracy and $F1$ value. There are many hot events happening on the social network all the time. Some entertainment news-related events will trigger fierce discussion among netizens and generate extremely high heat, but these events will not cause bad public opinion and threaten social sentiment. However, the EC method and RWDM method identify entertainment news-related events as Internet public safety events. The multi-feature fusion-based event grading method can more accurately identify which are Internet public safety events, and which hot events don't constitute Internet public safety events. Although the recall rate of multi-feature fusion-based method is lower than that of the EC method, the accuracy rate is twice that of the EC method, and the $F1$ value increases by about 45% compared to EC method and RWDM method.

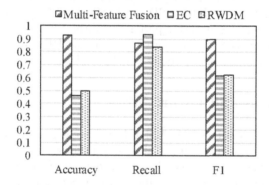

Fig. 3. The performance of social events extraction with different methods.

4.2   Performance of Hybrid Storage

Since the retrieval speed of Internet public safety events is of great significance in practical application, we need to build an efficient storage architecture for social text data. In the above cases, the data are stored in the MySQL database. Figure 4 records the average time it takes to retrieve the six events in diverse storage modes. For three Internet public safety events (E1, E3 and E6), compared to the HDD mode, the query time in the "HDD+SSD" mode is shortened by 85.2%–85.4%. This is because all the Internet public safety event data are gotten from SSD in the "HDD+SSD" mode. And when all data of those events in SSD is stored together, "HDD+SSD+Event" mode, the retrieval time can be shortened by 94.1%–96.1%.

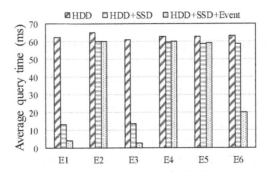

Fig. 4. Event retrieval time under different storage modes. The "HDD" mode queries the event from HDD. The "HDD+SSD" mode gets all the Internet public safety events data from SSD. The "HDD+SSD+Event" mode stored the records of Internet public safety events together on the SSD.

For other non-Internet public safety events, as in our storage policy, none of them has been stored in SSD. So the query performance is basically the same in two query cases "HDD+SSD" and "HDD+SSD+Event". The retrieval in "HDD+SSD" achieves a better performance than in HDDs. This is because only the non-public safety events are stored in the HDD in the "HDD+SSD" case, thus the retrieval scope becomes smaller and query performance is better.

5   Conclusions and Future Work

This paper presents multi-feature fusion based event grading strategy that can accurately identify Internet public safety events on social network. It extracts hotspot social events from the massive social text data through hot word extraction and event filtering. Then it combines sentiment analysis with event trend analysis to pick out potential Internet public safety events from the extracted social events by calculating the event score. In addition, it uses an efficient hybrid storage management strategy on HDDs and SSDs. The hot/cold data migration can greatly improve the event query efficiency and reduce the time consumption. The experimental results demonstrate the effectiveness of our system.

In future work, we would like to explore how different weight ratios of heat, emotion and sensitivity affect the calculation of public safety event score. We will also analyze the access patterns for different events and make a more intelligent storage layout for Internet public safety events.

Acknowledgements. This work was supported in part by the National Science Foundation of China under Grant No. 61972449, U1705261, and 61821003, and the Fundamental Research Funds for the Central Universities HUST under Grant No. 2021JYCXJJ049.

References

1. Chen, J., et al.: Inductive document representation learning for short text clustering. In: Hutter, F., Kersting, K., Lijffijt, J., Valera, I. (eds.) ECML PKDD 2020. LNCS (LNAI), vol. 12459, pp. 600–616. Springer, Cham (2021). https://doi.org/10.1007/978-3-030-67664-3_36
2. Hu, D., Feng, D., Xie, Y.: EGC: a novel event-oriented graph clustering framework for social media text. Inf. Process. Manage. **59**(6), 103059 (2022)
3. Zhao, L., Liu, Y., Zhang, M., Guo, T., Chen, L.: Modeling label-wise syntax for fine-grained sentiment analysis of reviews via memory-based neural model. Inf. Process. Manag. **58**(5), 102641 (2021)
4. Akcora, C.G., Bayir, M.A., Demirbas, M., Ferhatosmanoglu, H.: Identifying breakpoints in public opinion. In: Proceedings of the First Workshop on Social Media Analytics, pp. 62–66. ACM (2010)
5. Amato, F., Cozzolino, G., Mazzeo, A., Romano, S.: Detecting anomalies in twitter stream for public security issues. In: 2016 IEEE 2nd International Forum on Research and Technologies for Society and Industry Leveraging a better tomorrow (RTSI), pp. 1–4. IEEE (2016)
6. Jiang, D., Luo, X., Xuan, J., Xu, Z.: Sentiment computing for the news event based on the social media big data. IEEE Access **5**, 2373–2382 (2016)
7. Mo, H., Meng, X., Li, J., Zhao, S.: Terrorist event prediction based on revealing data. In: 2017 IEEE 2nd International Conference on Big Data Analysis (ICBDA), pp. 239–244. IEEE (2017)
8. Sato, K., Wang, J., Cheng, Z.: Detecting real-time events using tweets. In: Computational Intelligence (2017)
9. Xu, Z., et al.: Social sensors based online attention computing of public safety events. IEEE Trans. Emerg. Top. Comput. **5**(3), 403–411 (2017)
10. Kokoulin, A., Dadenkov, S.: Distributed storage system for imagery data in online social networks. In: Proceedings of the Application of Information and Communication Technologies (AICT) (2015)
11. Paul, T., Lochschmidt, N., Salah, H., Datta, A., Strufe, T.: Lilliput: a storage service for lightweight peer-to-peer online social networks. In: Proceedings of the Computer Communication and Networks (ICCCN) (2017)
12. Hu, D., Feng, D., Xie, Y., Xu, G., Gu, X., Long, D.: Efficient provenance management via clustering and hybrid storage in big data environments. IEEE Trans. Big Data **6**(4), 792–803 (2020)
13. Cavdur, F., Kose-Kucuk, M., Sebatli, A.: Allocation of temporary disaster response facilities under demand uncertainty: an earthquake case study. Int. J. Disaster Risk Reduction **19**, 159–166 (2016)
14. Rawluk, A., Ford, R.M., Neolaka, F.L., Williams, K.J.: Public values for integration in natural disaster management and planning: a case study from Victoria, Australia. J. Environ. Manage. **185**, 11–20 (2017)
15. Wen-Li, M.I., Sun, Y.X.: Microblog hot topics discovery method based on probabilistic topic model. Comput. Syst. Appl. **8**, 163–167 (2014)
16. Kolajo, T., Daramola, O.J., Adebiyi, A.A.: Real-time event detection in social media streams through semantic analysis of noisy terms. J. Big Data **9**(1), 90 (2022)

A Self-decoupled Interpretable Prediction Framework for Highly-Variable Cloud Workloads

Bingchao Wang[1,2,3], Xiaoyu Shi[1,2(✉)], and Mingsheng Shang[1,2]

[1] Chongqing Key Laboratory of Big Data and Intelligent Computing, Chongqing Institute of Green and Intelligent Technology, Chinese Academy of Sciences, Chongqing 400714, China
xiaoyushi@cigit.ac.cn
[2] Chongqing School, University of Chinese Academy of Sciences, Chongqing 400714, China
[3] Chongqing Key Laboratory of Computational Intelligence, Chongqing University of Posts and Telecommunications, Chongqing 400065, China

Abstract. Cloud workloads prediction plays a crucial role in the various tasks of cloud computing, such as resource scheduling, performance optimization, cost management, etc. However, current time series prediction methods suffer instability and inefficiency issues when addressing cloud workloads, due to the high variability of workload patterns and the high fluctuation within a workload. To address these issues, we propose DeIP4CW, a Self-Decoupled Interpretable Prediction framework for highly-variable Cloud Workloads. It can accurately forecast future job arrival rates in a cloud environment. The core idea of DeIP4CW is to first introduce the periodic and residual states as hidden variables to decouple complicated dependencies in cloud workload signals. Then it adopts a deep expansion learning framework with the block structure to perform workload prediction layer by layer. Each block consists of some *periodic* modules and some *compensation* modules. The *periodic* module with a self-attention mechanism can effectively capture the global trend of cloud workload, while the *compensation* module is employed to compensate for the local volatility information. Moreover, our two customized modules also have interpretable abilities, such as attributing the predictions to either global trends or local compensation. We conduct extensive experiments on the real-world cloud workload traces to evaluate the effectiveness of the proposed DeIP4CW. The experimental results demonstrate the DeIP4CW achieves significant improvements over the best baseline in most cases, and the error reduction can even reach up to 20.66%.

Keywords: Cloud computing · Workload prediction · Time series forecasting · Deep learning

1 Introduction

Cloud computing has become one of the most prevailing computing paradigms in IT society [1]. With the unique characteristic of virtualization technologies,

X. Wang et al. (Eds.): DASFAA 2023, LNCS 13943, pp. 588–603, 2023.
https://doi.org/10.1007/978-3-031-30637-2_39

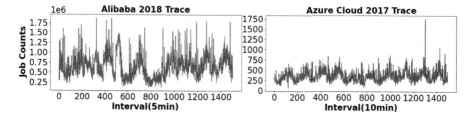

Fig. 1. Cloud workload traces. (a) Alibaba-Cluster-trace-v2018, (b) Azure cloud 2A017 trace.

cloud computing promises an on-demand provisioning mode to satisfy the variety of service level agreements (SLAs) between cloud service providers (CSPs) and end users. In the cloud environment, cloud workloads are usually unstable and not fixed in terms of submitted job rate and arriving time. The variations of cloud workload lead to the over/under-provisioning of resources, which causes unnecessary resource wasting or poor SLAs [2,3]. Therefore, it requires the CSPs to rapidly adjust the resource provisioning solutions for meeting SLAs while saving usage costs.

To achieve these objectives, accurately predicting future job arrival rate is crucial for the performance and cost optimization task in cloud computing. Generally, CSPs can estimate the amount of resources to be allocated in advance through the workload prediction method, so as to prevent waste and poor SLAs caused by over/under-provisioning resources. However, the high variability and fluctuation of cloud workloads make it difficult to accurately predict future job arrival rates:

- **High variability of workload patterns.** Workload patterns in cloud environments usually vary on different cloud-based applications. Figure 1 shows the workload traces collected from Alibaba and Microsoft Azure cloud data centers. Obviously, Alibaba's workload pattern exhibits random changes in terms of job arrival time and rate. Compared to Ali's cluster data, cloud workloads from Azure show a mixture of characteristics with high fluctuation and seasonality [4].
- **High fluctuation within a cloud workload.** Cloud workloads usually show a high dynamic and unexpected burst regarding the submitted job rates and arriving time. As shown in Fig. 1, it demonstrates that the workloads of the cluster fluctuate significantly at different times of the day. Especially cloud workloads will increase dramatically in a short time when an important live show or emergency event happened.

To address the aforementioned challenges, the complicated dependencies hidden in cloud workloads signals should be effectively captured. Based on the learned dependencies between adjacent signals, several methods have been proposed to design cloud workload predictors. Most of the existing methods focus on probabilistic or statistical theories to build their workload predictor, which only considers the linear dependence on historical observations [5,6]. With their strong ability in sequence data analysis, recurrent neural networks (RNN) have been employed to capture the complicated dependencies hidden in cloud workloads [7–9]. However, these RNN-based methods show poor performance on workloads with unexpected bursts, because of the strong assumption on the dependence between adjacent observations. Thus they cannot make an accurate prediction on highly-variable cloud workloads. In this paper, we propose DeIP4CW, a self-decoupled interpretable prediction framework for highly-variable cloud workloads. We argue that the complicated dependencies of cloud workload signals consist of the inherent periodicity and the correlation between adjacent historical observations. Therefore, the core idea of DeIP4CW is to decouple the complicated dependencies into two dedicated parts, by introducing the periodic and residual states as hidden variables. Inspired by residual learning, we design a deep expansion learning framework with a block structure to achieve this goal, which can perform the workload prediction layer by layer. Each block includes two customized modules, named *periodic* module and *compensation* module. The *periodic* module adopts the self-attention network to capture the global trend and periodic signal hidden in workloads. To cope with the residual signal, the *compensation* module then leverage a couple of one-dimensional convolutional neural networks to capture the local features. By adjusting the number of two modules, the complicated dependencies of cloud workloads can be easily handled.

We divide the three datasets from Alibaba and Azure into nine datasets with different time granularities to test the performance of our DeIP4CW. The results show that DeIP4CW reduces the error rate by an average of 9.00% compared to the primary baseline, and reduces the error rate by up to 20.66% on the dataset azure2019_60min. Besides, we also conduct ablation experiments and hyperparameter-sensitive experiments to verify the effectiveness of our proposed method.

We summarize our major contributions as follows.

- We introduce a self-decoupled prediction framework for highly-variable cloud workloads, which is important but neglected by existing approaches. We decompose the complicated dependencies of cloud workloads into a combination of periodic and residual signals.
- We propose DeIP4CW, a deep expansion learning framework with block structure, to perform cloud workload prediction layer by layer. Each block is composed of two customized modules to capture the global trend and local fluctuation signals. Moreover, these customized modules also have interpretable abilities.
- We conduct extensive experiments on real-world cloud workload trace to validate the effectiveness and adaptively ability of DeIP4CW. The results demonstrate the DeIP4CW outperforms SOTA methods when addressing the highly-variable cloud workloads, and we also visualize the prediction results to interpret model behaviors.

The rest of this paper is organized as follows. The related works are presented in Sect. 2, and the problem formulation is given in Sect. 3. We present the DeIP4CW in Sect. 4 and evaluate its performance in Sect. 5. The conclusion is given in Sect. 6.

2   Related Work

Workload prediction has received extensive attention in cloud computing. At an early stage, researchers focus on employing probabilistic and statistical methods to conduct workload predictors, including auto-regressive model (AR) [10], auto-regressive moving average model (ARMA) [11], exponentially weighted moving averages [12], and others [13,14]. These methods can only capture the linear dependence on historical observations and do not fit well with fluctuations in time series data. As a result, most of them show poor SLAs when addressing high-volatility cloud workloads.

With the development of artificial neural networks, researchers try to learn the complicated dependencies of cloud workloads by building complex neural network structures. Gao et al. [5] introduced a clustering based workload prediction method, which first clusters all the tasks into several categories and then train a neural network-based prediction model for each category respectively. Jitendra et al. [15] offered a workload prediction model using a neural network with a self-adaptive differential evolution algorithm. However, this method is mainly designed for the workload of the HTTP server. The workload pattern of the HTTP server has a strong seasonality, while the workload pattern of the cloud server usually is random. Thus, it is difficult to accurately predict the cloud workload. To deal with the could workloads, Shivani et al. [4] presented a novel time-series forecasting model called *WGAN-gp*, which adopts a Transformer network as a generator and a multi-layer perceptron as a critic. However, the computational complexity will increase for long sequence data. Chen et al. [16] proposed *L-PAW* that leverages the gated recurrent unit (GRU) to achieve accurate prediction for cloud workloads. Meanwhile, Jayakumar et al. [7] proposed *LoadDynamics* that employs the LSTM model to build their self-optimization generic prediction framework for cloud workloads, which can automatically optimize its internal parameters for different workload patterns. However, these RNN-based methods often perform poorly on highly-variable cloud workloads, because of the strong assumption on the dependence between adjacent observations.

3   Problem Formulations

Generally, cloud workload is a count of job arrivals sampled by a physical server or virtual machine (VM) at a constant sampling frequency. Therefore, cloud workload prediction is essentially a time series forecasting problem. Let x_i denote the job arrival counts (JAC) on the cloud servers at the i th moment. The classic autoregressive prediction formula is to project the historical observation

$x_{t-L:t} = [x_{t-L}, \ldots, x_{t-1}]$ into the JAC at the next moment x_t [17]. In [4,18], the Markov property is directly followed to predict the JAC at the next moment, and we also follow this assumption. That is, the JAC at the next moment is only dependent on $x_{t-L:t}$ and has nothing to do with $x_{0:t-L}$. Thus, the workload prediction can be defined as:

$$x_t = f_\theta(x_{t-L:t}) + \epsilon_t, \tag{1}$$

where L is the length of the historical observation sequence, $f_\theta : \mathbb{R}^L \to \mathbb{R}^1$ is a mapping function parameterized by θ, and ϵ_t denotes a value of independent and identically distributed Gaussian noise. Existing methods for period time series (PTS) adopt a few different instantiations of f_θ. For example, Jayakumar et al. [7] employs an LSTM that can automatically optimize its internal parameters. Feng et al. [19] adopted a deep learning prediction model designed with a deep belief network (DBN) composed of multiple-layered restricted Boltzmann machines (RBMs) and a regression layer to fit future workload to historical series. However, Calheiros et al. [14] employs the traditional mathematical and statistical method ARIMA.

4   DeIP4CW

In this section, we introduce the proposed DeIP4CW framework in detail. First, we start with a decoupled formulation of Eq. (1) in Sect. 4.1. Then, we illustrate the details of the deep expansion structure in Sect. 4.2. Last, the composition and neural structure of the *periodic* module and the *compensation* module are explained in Sect. 4.3 and Sect. 4.4.

4.1   The Decoupled Formulation

In response to two challenges of predicting cloud workloads, i.e., high variability of workload patterns and high fluctuation within a workload, we introduce a new function to explore the diversified local fluctuation compositions, besides the main prediction function f_θ in Eq. (1). As a result, we decouple the cloud workload prediction into two independent parts (i.e., f_θ and g_φ), by redefining (1). f_θ is used to capture the variant periodic patterns of cloud workloads, and g_φ is designed to compensate for diversified local fluctuation compositions. Thus, the cloud workload prediction can be formatted as:

$$x_t = f_\theta(x_{t-L:t}) + g_\varphi(x_{t-L:t} - f_\theta(x_{t-L:t})) + \epsilon_t, \tag{2}$$

where $g_\varphi : \mathbb{R}^L \to \mathbb{R}^1$ is a mapping function parameterized by φ that produces a prediction from the difference between the prediction of f_θ and the true input.

4.2   The Deep Expansion Structure

To capture the complicated dependencies (i.e., variant periodic patterns and diversified local fluctuation compositions) of cloud workloads, the key is to trade

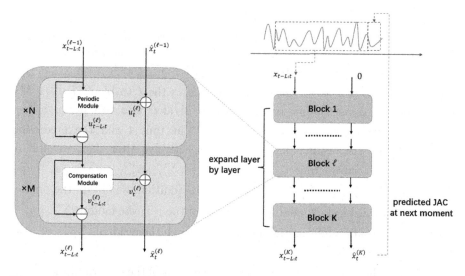

Fig. 2. The whole figure describes the deep expansion structure, and the left part describes the detailed data flow.

off between model capacity and generalization. To achieve this target, deep expansion learning has attracted lots of attention for time-series prediction. N-Beats [20] first proposed a deep expansion structure in time series forecasting, which is a deep neural architecture based on bi-directional residual links. It has demonstrated superior performances on time series forecasting than the traditional methods. Based on N-Beats, DEPTS [17] further introduced the periodic state to decouple the complicated dependencies of time series workloads for a more efficient and accurate prediction. Inspired by these successful examples and combined with the characteristics of cloud workload, we develop a novel framework DeIP4CW based on deep expansion learning for cloud workload prediction. DeIP4CW is also a deep neural network architecture with a block structure. Based on the redefined decoupled formulation (2), each block contains two kinds of customized and parameterized modules: *periodic* module f_θ and *compensation* module g_φ. Figure 2 depicts the architecture of DeIP4CW. Among them, the number of two kinds of modules in each block is controlled by the corresponding hyperparameters N and M. Each block has two residual branches, where $x_{t-L:t}^{(\ell)}$ is the residual after the decomposition of the previous ℓ layers, and $\hat{x}_t^{(\ell)}$ is the accumulation of the prediction results of the previous ℓ layers, $\ell \in [1, K]$.

First, we introduce the update equation of $x_{t-L:t}^{(\ell)}$, whose purpose is to exclude the information that the previous ℓ layers are used for prediction. To be more concrete, $f_\theta^{(\ell j)}$ takes $x_{t-L:t}^{(\ell-1)} - \sum_{q=0}^{j-1} u_{t-L:t}^{(\ell q)}$ as input, where $f_\theta^{(\ell j)}$ denotes the j th *periodic* module in the ℓ th block. The output of *periodic* module can be divided into two parts: $u_{t-L:t}^{(\ell j)}$ and $u_t^{(\ell j)}$, $u_t^{(\ell j)}$ is the prediction based on the current input information, and the other part $u_{t-L:t}^{(\ell j)}$ is called the backcast, which represents the information used to generate the current prediction $u_t^{(\ell j)}$. $u_{t-L:t}^{(\ell q)} = 0$ when $j = 1$, $j \in [1, N]$. For the *compensation* module, its purpose is to continue to extract information from the residual after N *periodic* modules processing. The input of $g_\varphi^{(\ell k)}$ is $x_{t-L:t}^{(\ell-1)} - \sum_{j=0}^{N} u_{t-L:t}^{(\ell j)} - \sum_{p=0}^{k-1} v_{t-L:t}^{(\ell p)}$, where $g_\varphi^{(\ell k)}$ denotes k th *compensation* module in the ℓ th block, $k \in [1, M]$. The output of *compensation* module is also divided into two parts: $v_{t-L:t}^{(\ell k)}$, $v_t^{(\ell k)}$, where represents prediction result $v_t^{(\ell k)}$ at time t and the prediction result $v_{t-L:t}^{(\ell k)}$ by the residual information, respectively. After that, we update $x_{t-L:t}^{(\ell)}$ by further subtracting $v_{t-L:t}^{(\ell k)}$ from $x_{t-L:t}^{(\ell-1)} - \sum_{j=0}^{N} u_{t-L:t}^{(\ell j)} - \sum_{p=0}^{k-1} v_{t-L:t}^{(\ell p)}$ as $x_{t-L:t}^{(\ell)} = x_{t-L:t}^{(\ell-1)} - \sum_{j=0}^{N} u_{t-L:t}^{(\ell j)} - \sum_{k=0}^{M} v_{t-L:t}^{(\ell k)}$. In addition, we obtain the prediction results of this block by adding the output of the *periodic* module $\sum_{j=0}^{N} u_t^{(\ell j)}$ and the compensation information proposed through the residuals $\sum_{k=0}^{M} v_t^{(\ell k)}$ as $\hat{x}_t^{(\ell)} = \sum_{j=0}^{N} u_t^{(\ell j)} + \sum_{k=0}^{M} v_t^{(\ell k)}$.

Note that the input of the ℓ th block is the output of the $\ell-1$ th block $x_{t-L:t}^{(\ell-1)}$, where the input of the first block is the real historical observation $x_{t-L:t}$, and the output of the last block is the final prediction result $\hat{x}_t$. According to the above description, the structure in Fig. 2 can be summarized as the Eq. (3):

$$x_{t-L:t} = x_{t-L:t}^{(0)} = x_{t-L:t}^{(K)} + \sum_{i=1}^{K}(\sum_{j=1}^{N} u_{t-L:t}^{(ij)} + \sum_{k=1}^{M} v_{t-L:t}^{(ik)}),$$

$$\hat{x}_t = \hat{x}_t^{(K)} = \sum_{i=1}^{K}(\sum_{j=1}^{N} u_t^{(ij)} + \sum_{k=1}^{M} v_t^{(ik)}),$$

(3)

where $x_{t-L:t}^{(K)}$ is considered irrelevant information for prediction and will be discarded after the last layer.

Connections and Differences to N-Beats. N-Beats directly uses the historical sequence $x_{t-L:t}$ to predict the value x_t at the next moment. Unlike them, we decouple the historical sequence into two states, period and residual to predict respectively. This brings about changes in the network structure. Compared with N-Beats, which only contains a fully connected network in each block, we introduce a self-attention mechanism and a one-dimensional convolutional neural network to model the periodicity and residual respectively.

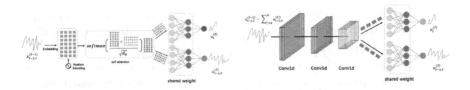

Fig. 3. The *periodic* module   **Fig. 4.** The *compensation* module.

4.3   The Periodic Module

To model the complicated dependences of workload sequence on periods, we model variant periodic patterns in the sequence through a parameterized function f_θ. With the success of Transformer in many fields, the self-attention mechanism has demonstrated a strong ability in capturing the long-term dependencies [21,22]. Thus, we employ the self-attention mechanism to construct our periodic module.

The structure of the *periodic* module is shown in Fig. 3. Taking ℓ th block, for example, adopts the learned workload feature $x_{t-L:t}^{(\ell-1)}$ after previous $\ell-1$ blocks as input, and output two parts $u_t^{(\ell)}$ and $u_{t-L:t}^{(\ell)}$. In detail, it first encode $x_{t-L:t}^{(\ell-1)}$ with the position information as $x_{t-L:t}^{(\ell-1)'}$. Then $x_{t-L:t}^{(\ell-1)'}$ is fed into the self-attention network to calculate the correlation between each value of the whole sequence. After that, it adopts two standard fully connected networks with ReLU activation to calculate the final prediction result $u_t^{(\ell)}$ and $u_{t-L:t}^{(\ell)}$ respectively. Two neural network shares the same parameters and uses linear projection functions to fit periodic functions and prevailing trends contained in series.

4.4   The Compensation Module

To solve the problem of inaccurate prediction caused by mutation of highly-variable workload, we propose another parameterized function g_φ to extract the remaining information after *periodic* module prediction to compensate for local prediction results. In other methods, the result of the *periodic* module is used directly as the final result [4,10]. However, we found that the data used to predict the final outcome was not fully extracted, especially the highly-variable workload. Therefore, we decouple it into periodic and residual states and propose a *compensation* module to compensate the local prediction result for the global prediction result.

So we introduce a 1D convolutional neural network (Conv1d), which is responsible for extracting local features to compensate for information not noticed during global prediction. Due to the highly-variable of cloud workload, the load values at each moment are not highly correlated, and Conv1d is very effective in this regard.

As with the input of the *periodic* module, we only discuss the case of $k = 1$, so the input of the *compensation* module is $x_{t-L:t}^{(\ell-1)} - \sum_{j=0}^{N} u_{t-L:t}^{(\ell j)}$. The specific demonstration is shown in Fig. 4. First, using a three layers Conv1d, we can obtain local features of residual information that cannot be extracted by multiple *periodic* modules. Then, two fully connected networks with ReLU activation with shared parameters and two linear mapping functions are used to obtain forecast and backcast respectively:

$$feature^1_{feature\_size^1} = AvgPool1d(ReLU(Conv1d^1(x^{\ell-1}_{t-L:t} - \sum_{j=0}^{N} u^{(\ell j)}_{t-L:t}))),$$

$$feature^2_{feature\_size^2} = AvgPool1d(ReLU(Conv1d^2(feature^1_{feature\_size^1}))),$$

$$feature^3_{feature\_size^3} = AvgPool1d(ReLU(Conv1d^3(feature^2_{feature\_size^2}))),$$

$$feature^4_{feature\_size^4} = FC(feature^3_{feature\_size^3}),$$

$$v^{(\ell)}_{t-L:t} = LINEAR(feature^4_{feature\_size^4}),$$

$$v^{(\ell)}_t = LINEAR(feature^4_{feature\_size^4}),$$

$$(4)$$

5 Experiments

Our experiments aim to address two questions: 1). How much benefit does DeIP4CW gain from predicting on highly-variable workload compared to the optimal algorithm? 2). What interpretability can be brought through our two custom modules? To answer the first question, we conduct extensive experiments on real datasets in Sect. 5.2. Then we experiment with the second question in Sect. 5.3 and analyze it for specific cases.

5.1 Experiments Setup

Workload Datasets. We collected three cloud workloads with different fluctuation datasets from two CSPs to evaluate the performance of the algorithm. As shown in Fig. 5, the workload value of alibaba 2018[1] is extremely large, the periodicity is not obvious, and there will be huge fluctuations at a certain moment. Cloud VM workloads from Azure[2] show a mixture of characteristics having high fluctuations and seasonality [4].

The different granularity of cloud workloads will also have a greater impact on workload characteristics. Therefore, we divided the selected three datasets into different time granularities, resulting in 9 different workloads. Among them, we use 1, 5, and 10 min intervals for alibaba 2018, use 10, 30, and 60 min for azure 2017, and 1, 30, and 60 min for azure 2019.

Hyperparameters. We use *Pytorch* to implement our algorithm. When training the algorithm, we use Adam optimizer on NVIDIA GeForce RTX 1080 machine, learning rate = 0.01, loss = L1Loss, epochs = 8000. We use grid search to find the optimal hyperparameters for training DeIP4CW, and their search ranges are shown in Table 1. For other baseline algorithms, we expand the search range based on the default parameters to achieve the best prediction

---

[1] https://github.com/alibaba/clusterdata.
[2] https://github.com/Azure/AzurePublicDataset.

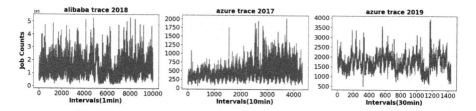

Fig. 5. Cloud Workload Traces. (a) Alibaba Cluster Workload Traces (b) Azure 2017 VM Traces (c) Azure 2019 VM Traces

Table 1. Hyperparameter search space of DeIP4CW algorithm.

Workload	Prediction module size(N)	Residual-aware module size(M)	Block size(K)	Lookback size	Batch size	Kernel size
alibaba 2018	[1, 2]	[1, 2]	[2-30]	[10-48]	[16-1024]	[2-32]
azure 2017				[16-72]	[32, 64]	[2-18]
azure 2019					[32, 64, 128]	

effect. For our network structure, there are several key hyperparameters that have a great influence on the final prediction results, and we conduct a series of experiments on them in Sect. 5.2.

Evaluation Metric. We use Mean Absolute Percentage Error (MAPE) and Root Mean Square Error (RMSE) to compare the prediction accuracy of different algorithms on all datasets. These two metrics are usually used to measure the similarity between the predicted results and the actual values [4,17].

$$MAPE = 100 \times \frac{1}{n} \sum_{i=1}^{n} |\frac{\widetilde{y}_i - y_i}{y_i}|, RMSE = \sqrt{\frac{1}{n} \sum_{i=1}^{n} (\widetilde{y}_i - y_i)^2}, \qquad (5)$$

where n is the total number of data points, $\widetilde{y}_i$ represents predicted JAC at time step i, and y_i represents actual JAC at time step i.

Baselines. We adopt the SOTA deep learning time series forecasting architecture N-Beats [20] as the primary baseline. Furthermore, we use WGAN-gp Transformer [4] for highly-variable cloud workload prediction, which is one of the SOTA cloud workload prediction methods. Of course, we also take into account traditional algorithms such as RNN [23], LSTM [8], GRU [24], ARIMA [6], etc., and fully compare their positions in the field of cloud workload prediction.

Besides, to understand the critical designs of our proposed method, we conduct ablation experiments in Sect. 5.2. We employ three variants of DeIP4CW:

- **DeIP4CW-OP:** The *compensation* module is removed from the entire structure so that the algorithm only retains the *periodic* module.
- **DeIP4CW-OR:** Similarly, the *periodic* module is removed from the entire structure, so that the algorithm only retains the *compensation* module.

Table 2. MAPE and RMSE of DeIP4CW and other baseline algorithms on different datasets. "%Improve." represents the improvement of DeIP4CW over the best-performing of all baseline algorithms. We highlight the best results in bold and sub-optimal results with underlining.

Workload Interval	alibaba 2018						azure 2017						azure 2019					
	1min		5min		10min		10min		30min		60min		1min		30min		60min	
	MAPE	RMSE	MAPE	RMSE	MAPE	RMSE	MAPE	RMSE	MAPE	RMSE	MAPE	RMSE	MAPE	RMSE	MAPE	RMSE	MAPE	RMSE
DeIP4CW	**26.3**	**471.77**	**18.68**	**201.08**	**18.2**	**334.66**	**33.58**	**240.07**	**23.07**	**438.69**	**14.32**	**596.71**	**17.86**	**72.08**	**14.57**	**310.08**	**11.71**	**505.70**
N-Beats	27.18	509.76	20.03	228.15	18.64	377.86	<u>36.13</u>	254.76	23.59	<u>448.06</u>	17.26	698.58	18.58	76.71	17.72	435.29	14.76	723.33
WGAN	36.9	569.6	31.84	221.05	31.18	444.81	36.72	302.48	28.08	517.08	17.66	591.55	33.36	120.04	19.30	337.39	15.21	629.09
LSTM	<u>27.01</u>	482.68	20.31	221.86	<u>18.31</u>	357.09	36.53	278.12	24.62	470.75	15.25	598.44	18.40	76.63	15.67	323.24	12.55	<u>519.95</u>
RNN	27.61	514.86	20.75	240.46	18.83	<u>355.42</u>	37.27	311.68	24.92	489.09	<u>14.55</u>	<u>570.83</u>	19.32	78.23	16.68	338.92	12.9	560.28
GRU	26.5	<u>490.48</u>	<u>20.01</u>	<u>204.88</u>	18.78	385.10	36.96	278.68	<u>23.08</u>	459.39	15.59	<u>573.05</u>	<u>18.34</u>	79.74	<u>15.27</u>	<u>315.17</u>	<u>12.18</u>	536.25
ARIMA	30.3	493.03	25.2	226.63	24.43	417.91	47.23	<u>254.30</u>	26.68	459.50	20.14	718.70	20.7	<u>75.56</u>	21.45	463.24	18.19	792.65
%Improve	↓0.75%	↓3.81%	↓6.64%	↓1.85%	↓0.60%	↓5.90%	↓7.05%	↓5.59%	–	↓2.09%	↓1.58%	↑4.53%	↓2.61%	↓4.60%	↓4.58%	↓1.64%	↓3.85%	↓2.74%

Table 3. Performance comparisons of DeIP4CW, DeIP4CW-OP, DeIP4CW-OR, and DeIP4CW-SW. We highlight the best results in bold.

Workload	DeIP4CW-OP		DeIP4CW-OR		DeIP4CW-SW		**DeIP4CW**	
	MAPE	RMSE	MAPE	RMSE	MAPE	RMSE	**MAPE**	**RMSE**
alibaba_1min	27.68	547.24	26.98	538.86	27.28	558.51	**26.3**	**471.77**
alibaba_5min	26.79	259.32	19.84	203.92	25.13	232.15	**18.68**	**201.08**
alibaba_10min	22.10	426.94	18.43	342.95	22.19	408.96	**18.20**	**334.66**
azure2017_10min	39.37	265.42	35.85	262.64	37.50	278.14	**33.58**	**240.07**
azure2017_30min	28.69	559.85	23.56	447.20	24.85	508.34	**23.07**	**438.69**
azure2017_60min	22.90	972.31	**14.07**	619.80	20.31	891.40	14.32	**596.71**
azure2019_1min	20.34	88.98	18.25	72.40	19.76	90.49	**17.86**	**72.08**
azure2019_30min	20.75	438.01	15.54	330.78	17.9	371.21	**14.57**	**310.08**
azure2019_60min	18.45	756.72	12.25	523.82	17.75	704.87	**11.71**	**505.70**

- **DeIP4CW-SW:** Swap the positions of the *periodic* module and the *compensation* module, so that the data is processed locally before global prediction.

5.2 Evaluation Results

Comparing DeIP4CW with Competitors. Table 2 shows the prediction errors of DeIP4CW and other baseline algorithms on different datasets. We can find that our algorithm outperforms other algorithms in almost all cases. In particular, on the azure2019_60min dataset, DeIP4CW achieves a 20.66% reduction in prediction error compared to the primary baseline. In order to show the final prediction effect more intuitively, Fig. 6 draws the predicted data and real data of alibaba_1min and azure2017_60min dataset. For a highly-variable dataset like alibaba 2018, the prediction errors of all algorithms are large, but our algorithm can still achieve the best results, due to the cooperation between the *periodic* module and the *compensation* module. We can observe that for the azure2019 dataset, which exhibits very good periodicity, DeIP4CW can also achieve an accurate result.

We note that while our algorithm's MAPE outperforms other baselines on most datasets, RMSE is a bit weaker on individual datasets. Because the RMSE

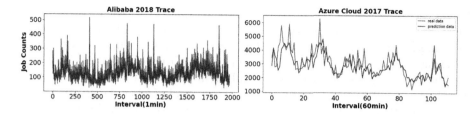

Fig. 6. The predicted data and real data of alibaba_1min and azure2017_60min.

is more sensitive to larger values, and when there are values in the prediction that are much larger than the true value, the RMSE will be larger. Because N-Beats only uses a fully connected network, it cannot capture complex workload pattern changes, but it still achieves good results compared to traditional mathematical statistics methods. RNN and LSTM have storage units that can store more complex historical sequence information, the prediction effect is better than N-Beats, and even the RMSE of individual datasets exceeds our method. WGAN-gp Transformer is one of the most advanced cloud workload prediction algorithms. During the training process, we found that in many cases, its final prediction result is a straight line close to the median line of data fluctuations. Overall, unlike GRU and LSTM, which only perform well on some datasets, our method can handle data with various workload patterns.

Ablation Tests. Then, let us focus on Table 3. We can see that on the highly-variable datasets of azure2017 and alibaba, DeIP4CW-OP cannot predict the data well from a global perspective by using the *periodic* module alone. However, on the dataset with a strong periodicity of azure2019, the effect can be better. Nonetheless, it still has a large error rate compared to our method, which indicates that only predicting the periodicity from the global loses a lot of local information, proving the importance of our proposed *compensation* module. The effect of DeIP4CW-OR using the *compensation* module alone is the closest to the effect of DeIP4CW, and exceeds the performance of DeIP4CW on azure2017_60min. However, its prediction accuracy still falls short of ours. It shows that for different workload patterns, only using local information to predict the results is inaccurate, which proves the importance of our proposed *periodic* module. Although the performance of DeIP4CW-SW in which the positions of the *periodic* module and the *compensation* module are exchanged is better than that of DeIP4CW-OP, it still has a very wrong, which explains why we put the *periodic* module responsible for global prediction at the beginning and then pass the local features to extract useful information from the residuals. In all of our designs, including the *periodic* module, the *compensation* module and the placement of both play a critical role in ultimately accurately predicting future workloads.

Hyperparameter Analysis. In order to verify the sensitivity of our method to hyperparameters, we selected several datasets and conducted hyperparameters analysis experiments. The final experimental result is shown in Fig. 7. From Fig. 7(a), we can know that when the size of deep expansion layers is too small, the information related to the prediction result cannot be extracted fully. On the contrary, if the size is too large, there will be content unrelated to the prediction in the extracted information, which will eventually affect the accuracy of the prediction result. It should be noted that the optimal number of expansion layers corresponding to the workload with different characteristics is different. It can be seen from Fig. 7(b) that when the ratio of the number of *periodic* modules to the number of *compensation* modules is 1:1, a very low prediction error rate can be obtained. When the number of two modules is unbalanced, the whole structure does not work well. Figure 7(c) shows the relationship between the number of different CNN layers and the final result. We can clearly see that when the number of CNN layers is too small or too many, the local features it extracts either have insufficient information, or the information contains unrelated noises, so we end up using a 3-layer Conv1d in our neural network.

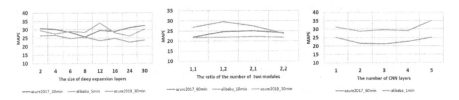

Fig. 7. (a) The effect of different depth expansion layers on different datasets. (b) The prediction results of different combinations of *periodic* modules and *compensation* modules numbers. (c) The prediction performance of different convolutional layers.

5.3   Interpretability

Classical time series forecasting divides the sequence components into seasonal, trend and Gaussian noise [20]. Benefiting from our proposed *periodic* module and *compensation* module, it can bring some interpretability. From the Eq. (3), we attribute the final prediction result to two parts global trends $\sum_{j=1}^{N} u_t^{(\ell j)}$ and local compensation $\sum_{k=1}^{M} v_t^{(\ell k)}$. According to their share of the final result, we can infer the fluctuation characteristics of the current cloud workload. In more detail, if the data has a large fluctuation and the periodicity is not obvious, then the effect of the *periodic* module has a high probability of having a poor effect. And Sect. 5.2 proves that the *compensation* module using Conv1d can extract local features to compensate for the loss that the *prediction* module does not work well.

Figure 8 shows the results we verified in our experiments. For highly-variable aibaba_1min workload, the prediction module works poorly, but thanks to our proposed *compensation* module, they can extract the information missed by the *periodic* module. For azure2017_60min workload with obvious periodicity, the *periodic* module can extract most of the data, and the useful information will account for a small proportion of the remaining residuals. This also fully proves that our network structure can reflect the fluctuation characteristics of the data in the final prediction results.

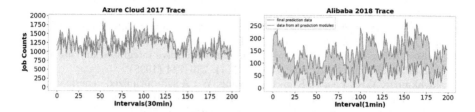

Fig. 8. The orange line in the figure represents the final prediction result, and the blue line represents the sum of the results produced by all prediction modules. The orange shading represents the sum of the information extracted by all *compensation* modules. (Color figure online)

6   Conclusion

In this paper, we propose a self-decoupled interpretable method for predicting highly-variable cloud workloads. Our core contribution is to introduce periodic and residual states to address the incomplete extraction of effective information caused by highly-variable workload predictions and propose a *compensation* module to extract it specifically. Interestingly, the two modules we propose also bring some interpretability. Finally, extensive experiments are conducted on cloud workloads to demonstrate the effectiveness of our method in handling highly-variable data.

Acknowledgments. This work is supported in part by the National Natural Science Foundation of China under Grants 62072429, in part by the Chinese Academy of Sciences "Light of West China" Program, and in part by the Key Cooperation Project of Chongqing Municipal Education Commission (HZ2021008, HZ2021017), and the "Fertilizer Robot" project of Chongqing Committee on Agriculture and Rural Affairs.

References

1. Luo, C., et al.: Correlation-aware heuristic search for intelligent virtual machine provisioning in cloud systems. In: Proceedings of the AAAI Conference on Artificial Intelligence, vol. 35, pp. 12363–12372 (2021)
2. Ran, L., Shi, X., Shang, M.: SLAs-aware online task scheduling based on deep reinforcement learning method in cloud environment. In: 2019 IEEE 21st International Conference on High Performance Computing and Communications; IEEE 17th International Conference on Smart City; IEEE 5th International Conference on Data Science and Systems (HPCC/SmartCity/DSS), pp. 1518–1525. IEEE (2019)

3. Zhao, Z., Shi, X., Shang, M.: Performance and cost-aware task scheduling via deep reinforcement learning in cloud environment. In: Troya, J., Medjahed, B., Piattini, M., Yao, L., Fernández, P., Ruiz-Cortés, A. (eds.) ICSOC 2022. LNCS, vol. 13740, pp. 600–615. Springer, Cham (2022). https://doi.org/10.1007/978-3-031-20984-0_43

4. Arbat, S., Jayakumar, V.K., Lee, J., Wang, W., Kim, I.K.: Wasserstein adversarial transformer for cloud workload prediction. arXiv preprint arXiv:2203.06501 (2022)

5. Gao, J., Wang, H., Shen, H.: Machine learning based workload prediction in cloud computing. In: 2020 29th International Conference on Computer Communications and Networks (ICCCN), pp. 1–9. IEEE (2020)

6. Nelson, B.K.: Time series analysis using autoregressive integrated moving average (ARIMA) models. Acad. Emerg. Med. **5**(7), 739–744 (1998)

7. Jayakumar, V.K., Lee, J., Kim, I.K., Wang, W.: A self-optimized generic workload prediction framework for cloud computing. In: 2020 IEEE International Parallel and Distributed Processing Symposium (IPDPS), pp. 779–788. IEEE (2020)

8. Graves, A.: Long short-term memory. In: Graves, A. (ed.) Supervised Sequence Labelling with Recurrent Neural Networks. SCI, vol. 385, pp. 37–45. Springer, Heidelberg (2012). https://doi.org/10.1007/978-3-642-24797-2_4

9. Kumar, J., Goomer, R., Singh, A.K.: Long short term memory recurrent neural network (LSTM-RNN) based workload forecasting model for cloud datacenters. Procedia Comput. Sci. **125**, 676–682 (2018)

10. Dinda, P.A., O'Hallaron, D.R.: Host load prediction using linear models. Cluster Comput. **3**(4), 265–280 (2000)

11. Vecchia, A.V.: Maximum likelihood estimation for periodic autoregressive moving average models. Technometrics **27**(4), 375–384 (1985)

12. Winters, P.R.: Forecasting sales by exponentially weighted moving averages. Manag. Sci. **6**(3), 324–342 (1960)

13. Roy, N., Dubey, A., Gokhale, A.: Efficient autoscaling in the cloud using predictive models for workload forecasting. In: 2011 IEEE 4th International Conference on Cloud Computing, pp. 500–507. IEEE (2011)

14. Calheiros, R.N., Masoumi, E., Ranjan, R., Buyya, R.: Workload prediction using ARIMA model and its impact on cloud applications' QoS. IEEE Trans. Cloud Comput. **3**(4), 449–458 (2014)

15. Kumar, J., Singh, A.K.: Workload prediction in cloud using artificial neural network and adaptive differential evolution. Futur. Gener. Comput. Syst. **81**, 41–52 (2018)

16. Chen, Z., Hu, J., Min, G., Zomaya, A.Y., El-Ghazawi, T.: Towards accurate prediction for high-dimensional and highly-variable cloud workloads with deep learning. IEEE Trans. Parallel Distrib. Syst. **31**(4), 923–934 (2019)

17. Fan, W., et al.: DEPTS: deep expansion learning for periodic time series forecasting. In: International Conference on Learning Representations (2021)

18. Salinas, D., Flunkert, V., Gasthaus, J., Januschowski, T.: DeepAR: probabilistic forecasting with autoregressive recurrent networks. Int. J. Forecast. **36**(3), 1181–1191 (2020)

19. Qiu, F., Zhang, B., Guo, J.: A deep learning approach for VM workload prediction in the cloud. In: 2016 17th IEEE/ACIS International Conference on Software Engineering, Artificial Intelligence, Networking and Parallel/Distributed Computing (SNPD), pp. 319–324. IEEE (2016)

20. Oreshkin, B.N., Carpov, D., Chapados, N., Bengio, Y.: N-beats: neural basis expansion analysis for interpretable time series forecasting. In: International Conference on Learning Representations (2019)

21. Wu, H., Xu, J., Wang, J., Long, M.: AutoFormer: decomposition transformers with auto-correlation for long-term series forecasting. In: Advances in Neural Information Processing Systems, vol. 34, pp. 22419–22430 (2021)
22. Zhou, H., et al.: Informer: beyond efficient transformer for long sequence time-series forecasting. In: Proceedings of the AAAI Conference on Artificial Intelligence, vol. 35, pp. 11106–11115 (2021)
23. Medsker, L.R., Jain, L.C.: Recurrent neural networks. Des. Appl. **5**, 64–67 (2001)
24. Cho, K., et al.: Learning phrase representations using RNN encoder-decoder for statistical machine translation. In: EMNLP (2014)

TieComm: Learning a Hierarchical Communication Topology Based on Tie Theory

Ming Yang[1], Renzhi Dong[1], Yiming Wang[1], Furui Liu[4], Yali Du[5],
Mingliang Zhou[6], and Leong Hou U[1,2,3(✉)]

[1] SKL of Internet of Things for Smart City, University of Macau, Zhuhai, China
ryanlhu@um.edu.mo
[2] Department of Computer Information Science, University of Macau, Zhuhai, China
[3] Centre for Data Science, University of Macau, Zhuhai, China
[4] Zhejiang Lab, Hangzhou, China
[5] King's College London, London, UK
[6] College of Computer Science, Chongqing University, Chongqing, China

Abstract. Communication plays an important role in Internet of Things that assists cooperation between devices for better resource management. This work considers the problem of learning cooperative policies using communications in Multi-Agent Reinforcement Learning (MARL), which plays an important role to stabilize agent training and improve the policy learned by enabling the agent to capture more information in partially observable environments. Existing studies either adopt a prior topology by experts or learn a communication topology through a costly process. In this work, we optimize the communication mechanism by exploiting both local agent communications and distant agent communications. Our solution is motivated by tie theory in social networks, where strong ties (close friends) communicate differently with weak ties (distant friends). The proposed novel multi-agent reinforcement learning framework named TieComm, learns a dynamic communication topology consisting of inter- and intra-group communication for efficient policy learning. We factorize the joint multi-agent policy into a centralized tie reasoning policy and decentralized conditional action policies of agents, based on which we propose an alternative updating schema to achieve efficient optimization. Experimental results on Level-Based Foraging and Blind-particle Spread demonstrate the effectiveness of our tie theory based RL framework.

Keywords: Reinforcement Learning · Multi-agent System · Communication Topology · Cooperation · Social Welfare

1 Introduction

Multi-agent reinforcement learning (MARL) is widely used in industrial scenarios where agents (e.g., IoT devices) are required to make intelligent decision-making for higher long-term returns. Learning to communicate effectively (under partial observation settings) among agents has shown crucial to strengthen inter-agent

X. Wang et al. (Eds.): DASFAA 2023, LNCS 13943, pp. 604–613, 2023.
https://doi.org/10.1007/978-3-031-30637-2_40

collaboration and consequently improve the quality of policies learned by MARL. Since the quality of communication structure directly affects the effectiveness of the communication [4,16], previous work either directly adopts a prior communication topology or learns a new communication structure from scratch. However, existing studies are limited in the learning of multi-agent systems. In the studies of prior topology, including fully connected [6], star topology [18], tree topology [11], and static graph [2,5,8,22], the information sharing is restricted by the structure of prior knowledge. For example, [10] only adopts a one-layer graph neural network, where an agent only communicates with its one-hop neighbors. In the studies of learning novel communicate topology [3,4,12,17], the space for topology exploration grows exponentially with the number of agents. Suppose there are n agents in the environment, the number of possible typologies is up to $2^{n(n-1)}$. Such a huge space restricts the quality of the learned topology.

In this work, we attempt to learn a topology based on the logic that agents should communicate differently with each other in terms of their distances. Our idea is inspired by tie theory [7], an influential theory in social network analysis and mathematical sociology [1]. Particularly, the tie theory studies information-carrying connections, called **ties**, between people. Interpersonal ties, generally, come in three varieties: strong, weak, or absent. People in the same group are usually connected by strong ties and share common knowledge. On the other hand, weak ties link different groups to carry additional information.

To construct a good communication graph, we first build a base graph g using the neighborhood relationship and then adopt a reasoning policy φ to generate a communication topology $\hat{g}$ that removes edges based on ties strength. We treat the isolated graphs in $\hat{g}$ as strong ties and learn the message passing by a graph neural network. In addition, we model the message passing between two strong ties (communities) as a weak tie communication, which is learned by an attention mechanism. Finally, we propose an iterative learning framework to optimize the reasoning policy φ together with the action-choosing policy. We denote this novel MARL communication framework for cooperative problems as **TieComm**. To the best of our knowledge, we are the first work to build a communication mechanism based on the tie theory, which not only provides reasoning of the communication but also increases the effectiveness of the learning process.

The main contributions are three-fold. First, we propose TieComm, an end-to-end framework that updates the communication-based policies and communication policy simultaneously by gradient descent. Second, we propose a novel communication approach, consisting of the overlay message-passing mechanism and the corresponding topology learning motivated by the tie theory. Third, experimental results on Level-Based Foraging and Blind-particle Spread environments show that our proposed method outperforms the state-of-the-art methods, which verifies the effectiveness of our proposed method.

2   Background

In this paper, we consider a cooperative problem in multi-agent reinforcement learning, which can be formed as a Decentralised Partially Observable Markov Decision Process (Dec-POMDP) [13]. A Dec-POMDP can be denoted by a tuple $\langle \mathcal{N}, \boldsymbol{S}, \boldsymbol{A}, \boldsymbol{O}, P, \boldsymbol{R}, \gamma \rangle$, where $\mathcal{N}$ is the set of the agents indexed from 1 to n. $\boldsymbol{S}$ is the set of states where the agents can stay at each time step. $\boldsymbol{A} = \left\{ A^1, A^2, \cdots, A^n \right\}$ are the action space of agents. $\gamma \in [0, 1)$ is a discount factor.

In Dec-POMDP, the agents only have partial observations and make decisions relying on aggregating messages $\hat{o}_t^i$ from communication function $\Psi^i : \boldsymbol{S} \to \boldsymbol{O}$. At each time step t, each agent chooses an action $a_t^i \in A^i$ following its own policy $\pi^i \left(a_t^i \mid \hat{o}_t^i \right) : S \times A^i \to [0, 1]$, forming the joint action $\boldsymbol{a}_t = \left\{ a_t^i \right\}_{i=1}^n$. Then, the system randomly transits to next state s_{t+1} following the state transition probability distribution $P \left(s_{t+1} \mid s_t, \boldsymbol{a}_t \right) : S \times A \times S \to \mathbb{R}^+$. Each agent receives its private reward r^i following reward function $r \left(s_t, a_t \right) : S \times A \to \mathbb{R}$. The joint policy $\boldsymbol{\pi}$ is denoted as $\boldsymbol{\pi} = \left\{ \pi^i \right\}_{i=1}^n$. Each agent i is decentralized execution, attempting to maximize its cumulative discounted reward:$\eta^i \left(\pi^i \right) = \mathbb{E} \left[R_t^i \right]$, where $R_t^i = \sum_{t=l}^\infty \gamma^l r^i \left(s_{t+l}, a_{t+l} \right)$. Correspondingly, the state-action Q-function and the value function of each agent are defined by $Q_\pi^i(s, a^i) = \mathbb{E}_\pi \left[R_t^i \mid s_t = s, a_t = a^i \right]$ and $V_\pi^i(s) = \mathbb{E}_\pi \left[R_t^i \mid s_t = s \right]$. respectively. The global Q-function is

$$Q_\pi(s, a) = \mathbb{E}_\pi \left[\sum_i^n R_t^i \mid s_t = s, a_t = a \right] \tag{1}$$

3   Methodology

3.1   Communication Mechanism of Agents

Our first objective is to find a communication mechanism, i.e., a topology, for multi-agent cooperation. Finding a good graph topology is difficult as the search space (e.g., the number of possible topologies) grows exponentially to the number of agents. A possible solution is to build a base communication topology g by manual rules and then refine g by optimization techniques (Fig. 1).

Tie theory, a sociological theory on the social network, also indicates the importance of the topology structure in the information diffusion process [14]. To address the distant diffusion in the tie theory, there are two kinds of communication connections, (1) strong ties that exchange information in a dense community and (2) weak ties that pass critical information in between communities. Motivated by the tie theory, we attempt to refine g into a hierarchical communication structure that consists of strong ties and weak ties. In the learning process, the final observation $\hat{o}^i$ of each agent i should consider not only its local observation o^i but also strong tie messages m_s^i and weak tie messages m_w^i to pick an action a^i. The corresponding communication function $\Psi(\hat{o}^i \mid \boldsymbol{o}, \hat{g})$ of each agent i therefore is defined by:

$$\Psi(\hat{o}^i \mid \boldsymbol{o}, \hat{g}) := o^i \cup m_s^i(\boldsymbol{o}, \hat{g}) \cup m_w^i(\boldsymbol{o}, \hat{g}) \tag{2}$$

where $\hat{g}$ is a hierarchical communication topology consisting of strong ties and weak ties.

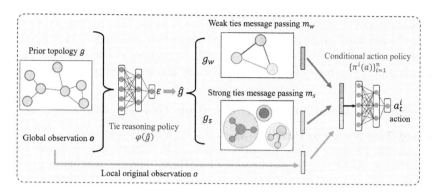

Fig. 1. Framework of TieComm: Tie reasoning policy $\varphi(\hat{g})$ partitions the prior topology g into $\hat{g}$, consisting of g_w for weak tie message passing and g_s for strong tie message passing by selecting the parameter ε. The conditional action policy $\{\pi^i(a)\}_{i=1}^{n}$ then chooses the appropriate action for each agent after the communication process.

3.2   Constructing Ties Graph $\hat{g}$ by Reasoning Policy φ

To construct the hierarchical topology $\hat{g}$, we introduce a function φ to optimize the prior topology g. In reinforcement learning, the communication topology should change dynamically as the environment state changes for better cooperation. Hence, the function φ should consider both global observation o and the prior topology g at each step time t:

$$\hat{g}_t \sim \varphi(\varepsilon \mid g_t, o_t) \tag{3}$$

where ε is the action for tie reasoning policy φ (to be discussed shortly). Note that we omit t for clarity in the subsequent discussion.

Definition 1. *Tie strength H_{ij} is defined as a measure of the closeness of the relationship between two agents:* $H_{ij} := \sum_{1}^{\xi} \frac{|N_i^{\xi} \cap N_j^{\xi}|\xi^{-1}}{\sqrt{|N_i^{\xi}|\xi^{-1}} \times \sqrt{|N_j^{\xi}|\xi^{-1}}}$ *where N_i^{ξ} and N_j^{ξ} are the sets of ξ-th hop neighbors of vertices i and j, respectively.*

In the [21], the strength of a tie is measured by the closeness between two connected vertices i and j by a set similarity function, which does not consider the normalization of common friends and intimate differences between friends. In this work, the closeness is calculated by Ochiai coefficient [15], a variant version of cosine similarity coefficient for sets (see Definition 1). In addition, we introduce the inverted number of hops ξ^{-1} to reflect the effect of neighbor distances.

To classify the type of ties, we introduce a hyperparameter ε, which serves as the threshold to distinct strong ties or weak ties:

$$\mathcal{E}_{i,j} := \begin{cases} \mathcal{E}_s, & \text{if } H_{ij} \geq \varepsilon; \\ \mathcal{E}_w, & \text{if } H_{ij} < \varepsilon; \end{cases} \tag{4}$$

where $\mathcal{E}_s$ denotes strong ties while $\mathcal{E}_w$ denotes weak ties. As Eq.(4) shows, ε plays the key role in generating the communication topology $\hat{g}$ and should be adaptive to the environment state. We will introduce how to optimize the tie reasoning policy φ based on the reinforcement learning feedback shortly.

3.3   Message Communications Based on Ties Graph $\hat{g}$

Strong Ties Message Passing on g_s: For each isolated graph g_s in $\{\mathcal{V}, \mathcal{E}_s\} \subseteq \hat{g}$, g_s should be denser than the prior topology g. Thus, a classic graph neural network, e.g., GAT [20], is good enough for strong message passing in g_s:

$$m_s^i := \sigma \left(\frac{1}{K} \sum_{k=1}^{K} \sum_{j \in \mathcal{N}_i} \alpha_{ij}^k \mathbf{W}^k o^j \right) \tag{5}$$

where k is attention head, σ is activation function, α is attention score, $\mathcal{N}$ is the connected neighbor vertices and $\mathbf{W}$ is weight matrix. Note that the number of graph neural network layers can be small (e.g. 1 layer in this work) since the strong ties graph is a dense graph.

Weak Ties Messages Passing on g_w: Following the tie theory, weak ties carry critical information among communities, which may provide additional information for agent decision-making. In this work, we pick a representative agent for each strong ties graph and form an overlay graph g_w in between the representative agents, which aims to encourage communication across all strong ties. To pick a representative agent r, we first introduce a concept called *agent strength*:

Definition 2. *Agent strength Δ_i^ξ is defined as* $\Delta_i^\xi := \sum_\xi \sum_{j \in N_i^\xi} \xi^{-1} H_{ij}$.

The agent strength Δ_i^ξ describes the relative importance of an agent in the community. The larger the value, the wider the range of information obtained by the agent. The agent of higher strength value should carry more information in the corresponding strong ties. Accordingly, the representative agent r is the one who has the maximum agent strength in the community.

Definition 3. *Representative agent r_c of community c is define as* : $r_c := \arg\max_{i \in c} \Delta_i^\xi$

We then form an overlay graph g_w that consists of all representative agents and the edges between them. As g_w is a fully connected graph, we decide to apply attention mechanism [19] for the message communication.

$$m_w^i := \sigma \left(\frac{1}{K} \sum_{k=1}^{K} \sum_{j \in g_w} \alpha_{ij}^k \mathbf{W}^k m_s^{r_j} \right) \tag{6}$$

In summary, the level of communication in between strong ties graphs is decided by the attention weights, which are optimized by each agent action policy π_i. The details on optimization will be introduced in the following.

3.4   Optimizing Reasoning Policy φ and Action Policy π^i

We attempt to simultaneously learn a tie reasoning policy φ for choosing ε on a prior topology g, forming a hierarchical topology $\hat{g}$, and the conditional action policy $\{\pi^i\}_{i=1}^{n}$, which chooses action after communication. By including the two kinds of policies, the joint policy π of the multi-agent system is factorized into conditional action policy $\pi^i\left(a^i \mid \hat{o}^i; \theta\right)$ of each agent and a tie reasoning policy $\varphi(\varepsilon \mid o, g; \omega)$:

$$\pi(a, \varepsilon \mid o, g; \theta, \omega) = \prod_{i=1}^{n} \pi^i\left(a^i \mid \phi^i(o, \hat{g}); \theta\right) \varphi(\varepsilon \mid o, g; \omega) \tag{7}$$

Besides, the objective of the cooperative MARL problem is to maximize the global Q value, which is expressed as:

$$\pi(a, \varepsilon \mid o, g; \theta, \omega) = \operatorname{argmax}_{\pi} \mathbb{E}_\pi\left[Q(o, a)\right] \tag{8}$$

Note that we can use global observation o to replace global state s in practice.

Since the joint policy π is factorized by two policies represented by two sets of parameters. To optimize the joint policy, we can iteratively optimize two policies.

Conditional Action Policy Gradient: To update the parameter θ in conditional action policy, we fix the ω in tie reasoning policy. In addition, we use a fully decentralized multi-agent Actor-Critic framework [9], in which the critic network ϕ takes local observations after communication instead of the global observation as input. Hence the derivative of the objective function reads as:

$$\nabla_\theta J(\pi^i) = \mathbb{E}[Q^i(\hat{o}^i, a^i)\nabla_\theta \ln \pi_\theta^i(a^i \mid \hat{o}^i))] \tag{9}$$

The loss function is correspondingly expressed as:

$$\mathcal{L}_Q(\phi) = \mathbb{E}_{(o_t^i, a_t^i) \sim \mathcal{D}}\left[\left(y_t^i - Q^i\left(\hat{o}_t^i, a_t^i; \phi\right)\right)^2\right] \tag{10}$$

with $y_t^i = r^i + Q^i\left(\hat{o}_{t+1}^i, a_{t+1}^i; \bar{\phi}\right)$ where $\bar{\phi}$ is frozen critic network used for stable training and $\mathcal{D}$ is the replay buffer for sampling.

Tie Reasoning Policy Gradient: We fix the parameter θ in each agent's conditional action policy π^i and update the parameter ω in graph reasoning policy φ. Hence, we can conclude the following proposition:

Proposition 1. *Given Eq. (8) and Eq. (7), the update rule for the tie reasoning gradient can be devised as follows:*

$$\nabla_\omega J(\varphi) = \mathbb{E}\left[Q(o, a)\nabla_\omega \ln \varphi_\omega\left(\varepsilon \mid g, o\right)\right] \tag{11}$$

Proposition 1 indicates that the tie reasoning policy should consider all the possible actions of each agent who attempts to react and improve its policy toward the global optimal return, which can encourage cooperation among agents. In practice, for graph reasoning policy, we use a centralized critic ψ and

Table 1. Brief description of environment settings

Environment	Agents	Types	Distribution	Targets	Map
BPS(1)	9	2	$[4, 5]$	2	–
BPS(2)	12	3	$[3, 4, 5]$	3	–
BPS(3)	15	3	$[5, 5, 5]$	3	–
LBF(1)	6	1	$[3, 3]$	4	10×10
LBF(2)	9	2	$[4, 5]$	6	12×12

take global observation o and global actions a as input to estimate the global Q. The corresponding loss function is expressed as:

$$\mathcal{L}_Q(\psi) = \mathbb{E}_{(o_t, a_t) \sim \mathcal{D}} \left[(y_t - Q(o_t, a_t; \psi))^2 \right] \tag{12}$$

with $y_t = \sum_i^n r^i + Q(o_{t+1}, a_{t+1}; \bar{\psi})$ where $\bar{\psi}$ is the frozen centralized critic network used for stable training and $\mathcal{D}$ is the replay buffer for sampling.

4   Numerical Experiments

We attempt to investigate the following research questions: **RQ1:** Can TieComm obtain competitive performance compared with a variety of state-of-the-art communication methods? **RQ2:** What is the effect of weak ties messages and strong ties messages?

4.1   Experiment Settings

Evaluation Environment: We modify two popular multi-agent environments, Blind-particle Spread (BPS) and Level-based Foraging (LBF), to secure the partially observable requirement for cooperative learning. The agents in both environments need to keep communicating with nearby and distant agents to avoid collisions and navigate to the targets. All methods are evaluated on them with three different random seeds. Table 1 shows a brief description of the environment's settings.

Baselines: We consider a variety of state-of-the-art communication methods as benchmarks including: **CommNet** [18], **IC3Net** [17], **TarMAC** [3], **MAGIC** [12]. We also create three variant versions of our approach for ablation study, including: **TieComm-n/p**, which directly utilizes the prior topology, blocking the proposed tie reason policy. **TieComm-n/w**, which blocks the weak ties messages. **TieComm-n/s**, which blocks the strong ties messages.

4.2   Results and Discussion

RQ1 Performance: Figure 2 depicts the converge speed of all models. Table 2 show the performance in all environments. In almost all settings, our TieComm

Table 2. Summary of the performance of trained MARL policies. Best values in bold.

	BPS(1)	BPS(2)	BPS(3)	LBF(1)	LBF(2)
CommNet	-451.6 ± 69.2	-601.8 ± 43.6	-865.4 ± 79.4	17.6 ± 0.6	27.4 ± 1.4
IC3Net	-473.8 ± 62.0	-653.9 ± 60.1	-832.4 ± 89.5	17.4 ± 0.6	27.3 ± 0.9
TarMAC	-431.4 ± 63.1	-666.7 ± 55.3	-880.8 ± 46.7	17.8 ± 0.5	27.5 ± 1.0
MAGIC	-431.2 ± 39.5	-652.7 ± 55.6	-790.6 ± 53.9	17.7 ± 0.6	26.7 ± 1.3
TieComm-n/w	-374.4 ± 38.1	-451.9 ± 51.1	-697.1 ± 58.8	18.7 ± 0.5	$\mathbf{28.1 \pm 0.6}$
TieComm-n/s	466.1 ± 55.1	-624.3 ± 34.3	-842.6 ± 66.5	18.5 ± 0.6	27.8 ± 0.8
TieComm-n/p	-446.4 ± 45.6	-488.0 ± 30.3	699.4 ± 58.3	15.6 ± 0.3	23.6 ± 0.4
TieComm	$\mathbf{-306.3 \pm 28.9}$	$\mathbf{-388.7 \pm 48.3}$	$\mathbf{-554.7 \pm 64.8}$	$\mathbf{18.7 \pm 0.5}$	27.9 ± 1.0

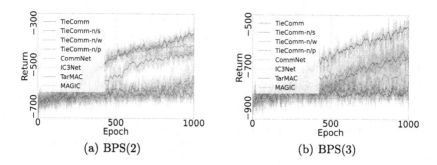

(a) BPS(2)   (b) BPS(3)

Fig. 2. Learning curves on BPS. TieComm (red) is best. (Color figure online)

Table 3. Average number of collisions during an episode on LBF.

Environment	TieComm-n/w	TieComm-n/s	TieComm-n/p	TieComm
LBF(1)	15.3 ± 1.9	16.5 ± 0.8	7.9 ± 0.4	14.8 ± 1.6
LBF(2)	24.7 ± 1.9	25.7 ± 1.6	16.7 ± 0.8	22.0 ± 1.7

achieves better performance than the baselines. As the number of agents increases and settings become more challenging, the performance gap between the baselines and ours becomes larger. Obviously, the topology learned by our TieComm is more effective and plays a more critical role in the learning process if more agents participate in the communication.

RQ2 Effect of Weak Ties and Strong Ties: We investigate the power of weak ties and strong ties in TieComm by evaluating our variant versions (called TieComm-n/w and TieComm-n/s), which block weak ties messages and strong ties messages by noise respectively. We find that in the LBF environment, some methods can achieve almost the same performance but the learned policies are different. As the Table 3 shows, TieComm-n/s has more collisions while Tiecomm has the least under similar return performance. It is because our method has strong ties messages, which force agents to pay more attention to nearby agents. In addition, as Fig. 2 shows, even if there exists no weak tie, the performance of the Tiecomm-n/m is still worthy of recognition, even better than some methods. In TieComm-n/m, the problem is that the agents cannot access the messages from distant agents, leading to late in finding landmarks or foods far away. Thus, TieComm-n/w is dominated by our complete version method.

5   Conclusion

In this work, we propose a novel approach TieComm, which learns an overlay communication topology for multi-agent cooperative reinforcement learning inspired by tie theory. We exploit the topology into strong ties (nearby agents) and weak ties (distant agents) by our reasoning policy. The optimization of the proposed system is achieved by decomposing the learning into the joint policy of the conditional action policies of agents and the tie reasoning policy of topology, which is optimized jointly by an iterative process. The experimental results verify the effectiveness of our proposed method, which compares our method with a variety of state-of-the-art methods on two representative multi-agent reinforcement learning environments, Level-Based Foraging, and Blind-particle Spread.

Acknowledgement. Ming Yang thanks to the support by the Science and Technology Development Fund Macau SAR 0015/2019/AKP, 0031/2022/A, SKL-IOTSC-2021–2023, the Research Grant of University of Macau MYRG2022-00252-FST, Wuyi University Hong Kong and Macau joint Research Fund 2021WGALH14, the National Natural Science Foundation of China under grant 62176027, the General Program of the National Natural Science Foundation of Chongqing under grant cstc2020jcyj and msxmX0790, and the Human Resources and Social Security Bureau Project of Chongqing under grant cx2020073. Experiments were conducted at SICC supported by SKL-IOTSC, University of Macau.

References

1. Arney, C.: Linked: how everything is connected to everything else and what it means for business, science, and everyday life. Math. Comput. Educ. **43**(3), 271 (2009)
2. Chu, T., Chinchali, S., Katti, S.: Multi-agent reinforcement learning for networked system control. In: International Conference on Learning Representations (2019)
3. Das, A., et al.: TarMAC: targeted multi-agent communication. In: International Conference on Machine Learning, pp. 1538–1546. PMLR (2019)
4. Du, Y., et al.: Learning correlated communication topology in multi-agent reinforcement learning. In: Proceedings of the 20th International Conference on Autonomous Agents and MultiAgent Systems, pp. 456–464 (2021)
5. Du, Y., et al.: Scalable model-based policy optimization for decentralized networked systems. In: 2022 IEEE/RSJ International Conference on Intelligent Robots and Systems (IROS), pp. 9019–9026. IEEE (2022)
6. Foerster, J.N., Assael, Y.M., de Freitas, N., Whiteson, S.: Learning to communicate with deep multi-agent reinforcement learning. In: Proceedings of the International Conference on Neural Information Processing Systems, pp. 2145–2153 (2016)

7. Granovetter, M.S.: The strength of weak ties. Am. J. Sociol. **78**(6), 1360–1380 (1973)
8. Gupta, S., Hazra, R., Dukkipati, A.: Networked multi-agent reinforcement learning with emergent communication. In: Proceedings of the International Joint Conference on Autonomous Agents and Multiagent Systems, AAMAS, vol. 2020, pp. 1858–1860 (2020)
9. Iqbal, S., Sha, F.: Actor-attention-critic for multi-agent reinforcement learning. In: International Conference on Machine Learning, pp. 2961–2970. PMLR (2019)
10. Jiang, J., Dun, C., Huang, T., Lu, Z.: Graph convolutional reinforcement learning. In: International Conference on Learning Representations (2020)
11. Jiang, J., Lu, Z.: Learning attentional communication for multi-agent cooperation. In: Advances in Neural Information Processing Systems, vol. 31, pp. 7254–7264 (2018)
12. Niu, Y., Paleja, R., Gombolay, M.: Multi-agent graph-attention communication and teaming. In: Proceedings of the 20th International Conference on Autonomous Agents and MultiAgent Systems, pp. 964–973 (2021)
13. Oliehoek, F.A.: Decentralized POMDPs. In: Wiering, M., van Otterlo, M. (eds.) Reinforcement Learning. Adaptation, Learning, and Optimization, vol. 12, pp. 471–503. Springer, Heidelberg (2012). https://doi.org/10.1007/978-3-642-27645-3_15
14. Onnela, J.P., et al.: Structure and tie strengths in mobile communication networks. Proc. Natl. Acad. Sci. **104**(18), 7332–7336 (2007)
15. Romesburg, H.C.: Cluster Analysis for Researchers. Wadsworth Inc, Belmont, CA (1984)
16. Ruan, J., et al.: GCS: graph-based coordination strategy for multi-agent reinforcement learning. In: Proceedings of the 21st International Conference on Autonomous Agents and Multiagent Systems (AAMAS), pp. 1128–1136 (2022)
17. Singh, A., Jain, T., Sukhbaatar, S.: Learning when to communicate at scale in multiagent cooperative and competitive tasks. In: International Conference on Learning Representations (2019)
18. Sukhbaatar, S., Fergus, R., et al.: Learning multiagent communication with backpropagation. In: Advances in Neural Information Processing Systems, vol. 29, pp. 2244–2252 (2016)
19. Vaswani, A., et al.: Attention is all you need. In: Advances in Neural Information Processing Systems, vol. 30 (2017)
20. Veličković, P., Cucurull, G., Casanova, A., Romero, A., Liò, P., Bengio, Y.: Graph attention networks. In: International Conference on Learning Representations (2018)
21. Weng, L., Karsai, M., Perra, N., Menczer, F., Flammini, A.: Attention on weak ties in social and communication networks. In: Lehmann, S., Ahn, Y.-Y. (eds.) Complex Spreading Phenomena in Social Systems. CSS, pp. 213–228. Springer, Cham (2018). https://doi.org/10.1007/978-3-319-77332-2_12
22. Zhang, K., Yang, Z., Başar, T.: Decentralized multi-agent reinforcement learning with networked agents: Recent advances. Front. Inf. Technol. Electron. Eng. **22**(6), 802–814 (2021). https://doi.org/10.1631/FITEE.1900661

Privacy Computing

Rewriting-Stego: Generating Natural and Controllable Steganographic Text with Pre-trained Language Model

Fanxiao Li[1], Sixing Wu[2(✉)], Jiong Yu[1], Shuoxin Wang[1], BingBing Song[3],
Renyang Liu[3], Haoseng Lai[1], and Wei Zhou[2]

[1] Engineering Research Center of Cyberspace, Yunnan University, Yunnan, China
lifanxiao@mail.ynu.edu.cn
[2] National Pilot School of Software, Yunnan University, Yunnan, China
{wusixing,zwei}@ynu.edu.cn
[3] School of Information Science and Engineering, Yunnan University, Yunnan, China

Abstract. Data transmission security and privacy play a crucial role in the era of information technology. Although the widely-used data encryption technique can ensure security, it can be easily detected and blocked by the observation system because the encrypted data format is quite different from the normal data. This work focuses on linguistic steganography, hiding a secret text in another normal stego text to ensure security and decrease the risk of being detected simultaneously. Rather than following the existing edit-based or generation-based paradigm, we propose a novel rewriting-based *Rewriting-Stego*, which tries to hide a secret text in the stego text by rewriting the given cover text. This paradigm integrates the advantages of both the edit-based paradigm and the generation-based paradigm, bringing higher information capacity without losing naturalness and controllability. Extensive experimental results on three public datasets have demonstrated the effectiveness of our *Rewriting-Stego* in terms of multiple metrics.

Keywords: steganography · linguistic steganography

1 Introduction

The Internet's rapid development arouses concerns about security and privacy because unauthorized attackers can easily intercept the transmitted data in non-dedicated networks. As shown in Fig. 1, data encryption is the most widely used security technique. The *sender* first uses a key to encrypt the data; then, the ciphertext can be transmitted via the Internet, and only the *receiver* who has another key can correctly decrypt the ciphertext. Nonetheless, the data format of ciphertext is quite different from the normal data, which may cause the vigilance of the observation system [1], and the data transmission may be blocked. Unlike data encryption, data steganography hides the secret message in a stego message and keeps a normal data format. Thus, data steganography

X. Wang et al. (Eds.): DASFAA 2023, LNCS 13943, pp. 617–626, 2023.
https://doi.org/10.1007/978-3-031-30637-2_41

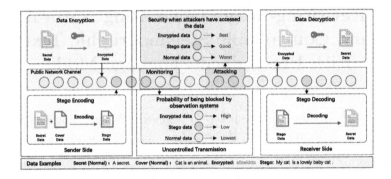

Fig. 1. The comparison between data encryption and data steganography

can reduce the vigilance from the observation system and has received much attention in security communication [2], watermarking [5,16], etc.

This paper studies linguistic steganography [4,13,17], which hides a secret text in the stego (i.e., *steganographic*) text given a cover text. Roughly, prior works are either the *edit-based* [14] or the *generation-based* [4,15]. Edit-based methods design a special encoding strategy to hide the secret text in some selected positions of the cover text via editing. For example, given a synonym dictionary, replacing a word with the 3rd ranked synonym can hide 2-bit[1] information 11. However, to ensure the naturalness of the stego text, the information capacity of the carried secret message is always limited. The average BPT (bit per token) is always less than 1.0. With the development of language models (LM) [11,12], generation-based methods have become mainstream. Generation-based methods first use a cover text to initialize the state of the backbone LM; subsequently, the LM generates a sequel text as the stego text based on the given secret bit stream and the decoding strategy. For example, at each generation step, the LM first outputs a probability distribution of the next token; then, rather than selecting the most possible token or randomly sampling a token, the strategy assigns bit encoding codes to candidate tokens according to the rank of the corresponding probability and selects the token whose bit code equals to the current secret bit code. This paradigm can achieve a higher information capacity (>1 BPT). However, the generated stego text 1) always lacks naturalness and 2) is hard to control the content because the supervision is limited during the generation process, increasing the chance to raise the vigilance of the observation system.

With such challenges in mind, we propose a novel *rewriting-based* method *Rewriting-Stego*, which rewrites the cover text and lets the outputted paraphrased text as the stego text. We regard the rewriting as a denoising sequence-to-sequence task; namely, given an input text, the model should denoise the unwanted information and then generate an output text with the same semantics but different word usage. In our context, the cover text is the input, the

---

[1] Generally, the stego text is represented as a bit stream.

stego text is the output, and the secret text serves as a restriction to denoise. Consequently, we choose a pre-trained sequence-to-sequence BART [8] as our backbone language model. We propose a Plug-and-Play Group-Wise Masked Decoding Strategy to hide the secret message without affecting the structure of the backbone BART. We group the vocabulary of the BART into 2^n groups; each group has a unique n-bit code, and each token only belongs to a group. Thus, the backbone BART first encodes the given cover text; then, in the decoding stage, we can hide n-bit secret information in each generated token by masking tokens whose group bit-id is not equal to the current secret information. Then, we propose to use a text-based Condition Codes to explicitly hint the desired length of the stego text in the encoding stage and a beam-search-based *Beam-then-Rank* to select higher-quality stego text in the decoding stage. Finally, we propose Adaptive Fine-Tuning to help the backbone adjust to the rewriting-based linguistic steganography task. Intuitively, the proposed rewriting-based paradigm combines the advantages of the previous two paradigms. On the one hand, similar to the edit-based methods [14], the generated stego text is highly similar to the cover text, bringing higher naturalness and controllability. On the other hand, similar to the generation-based methods [4,13,15], the rewriting-based method can reach higher information capacity because all tokens in the generated stego text can hide secret information. Our code is available at https://github.com/cheslee15/Rewriting-Stego.

2 Methodology

2.1 Problem Definition

Linguistic steganography task can be formulated as the *sender* hides a secret message S into a cover text Y (a text about other normal topics) and obtains a stego text Y' via the pre-defined invertible strategy. The stego text Y' is very similar to the original cover text Y because they describe a similar topic and use a similar format on the surface text. Thus, the *observation system* can hardly detect the anomaly that exists in the stego text. Unlike the observation system, the *reciever* can restore the secret message S from the stego text Y' via the pre-defined invertible strategy.

2.2 Rewriting Paradigm

Rather than following the generation-based or edit-based paradigm, this paper proposes a novel **rewriting-based** paradigm. Similar to the generation-based paradigm, the proposed rewriting-based paradigm employs a language model (LM) to generate the stego text. However, rather than generating a sequel text to hide S, our rewriting-based paradigm rewrites the cover text and regards the obtained *paraphrased text* as the stego text. By definition, a paraphrased text Y' has the same semantics as the original cover text Y but different word usage.

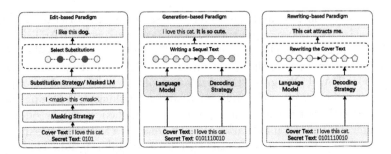

Fig. 2. The comparison among three different linguistic steganography paradigms.

Thus, similar to the edit-based paradigm, the generation process of the para-phrased text is highly supervised by the original cover text, bringing higher nat-uralness and controllability. In addition, compared to the edit-based paradigm, our rewriting-based paradigm can edit all tokens in the cover text, bringing higher information capacity without losing naturalness and controllability. Con-sequently, the essentials of a rewriting-based method are 1) a rewriting language model and 2) a strategy to hide the secret text S.

2.3 Methodology

Backbone Model. We employ BART [8], a denoising auto-encoder for Seq2Seq tasks, as our backbone language model. In the pre-training stage, given an orig-inal text $X_{original}$, a corrupted $X_{corrupt}$ is subsequently synthesized by adding the manually defined noises to the original text $X_{original}$. Then, the objective of BART is to restore $X_{original}$ given the corrupted $X_{corrupt}$. Consequently, BART is very suitable for the rewriting-based paradigm because it has the ability to denoise the input text and generate a higher-quality paraphrased text.

Encoding with the Conditional Code. In linguistic steganography, the length of the stego text Y' depends on the length of the secret message S, rather than the length of the inputted cover text Y. If the generated stego text is incomplete or strange, it may increase the risk of being blocked. To allevi-ate this issue, we propose to encode the cover text with a *Conditional Code*, which involves an explicit length signal from the secret text S. Given a cover text $Y = (y_1, y_2, \cdots, y_n)$, *Rewriting-Stego* first employs the BART encoder to encode and obtain $\mathbf{H} = Encoder([Y; ConditionalCode])$ where the input is the concatenation of the cover text Y and a conditional code $ConditionalCode$. Pre-vious studies [7] have shown the potential of promoting text in the pre-trained language model. Inspired by this, our $ConditionalCode$ uses a promoting text *'Generate a sentence of length L by paraphrasing the content on the left.'* to explicitly indicate the length of Y' should be L.

Plug-and-Play Group-Wise Masked Decoding Strategy. Subsequently, the decoder of BART continues to generate the stego text Y'. To hide the secret message S into the generated stego text Y', we propose a *Plug-and-Play Group-Wise Masked Decoding Strategy*. This strategy neither modify the network structure nor restrict the selection of the decoding algorithm.

Similar to previous works, the secret message S should be encoded as a bit stream (i.e., a sequence of bits), and the total bit length of the secret message S is denoted as l. Then, we assume each token $\in Y'$ hides n bits secret message. Subsequently, in each generation time step t, we generate a stego token y'_t to hide the current secret message bits $S_{(t-1)*n:t*n}$:

$$y'_t = DecodingStrategy(S_{(t-1)*n:t*n}, P(y'_t|Y'_{1:t}; Y; ConditionalCode))$$

$$P(y'_t|Y'_{1:t}; Y; ConditionalCode) = Softmax(MLP(Decoder(Y'_{1:t}, \mathbf{H})))$$

$$(1)$$

where $P(y'_t|Y'_{1:t}; Y; ConditionalCode)$ is the current token prediction probability distribution over the vocabulary, which is outputted by the BART decoder; the vocab predictor MLP is a feed-forward neural network. The generation process is restricted by the current secret bits $S_{(t-1)*n:t*n}$ and the *Decoding Strategy*. As illustrated in Table 1, *Rewriting-Stego* divides the vocabulary into 2^n groups and assigns an n-bits group bit id. If the vocabulary has $|V|$ tokens in total, then each group has an n-bit id and $\frac{|V|}{2^n}$ tokens. Thus, each token in the vocabulary corresponds to a deterministic n-bit code.

Table 1. Vocabulary-based grouping strategy (Modulo Operation).

Group Bit ID	Tokens	Group Bit ID	Tokens
00	{"be":0, "it":4...}	01	{"this":1, "who":5...}
10	{"he":2, "an":6...}	11	{"from":3, "much":7...}

Subsequently, the current secret message bits $S_{(t-1)*n:t*n}$ can be uniquely aligned to one vocabulary group G_t. Then, *Decoding Strategy* will mask a token probability to zero if this token is excluded by the aligned group G_t. For example, if the current secret message bits are 01, we mask the token probabilities in the other groups (00,10,11) to 0. After masking the invalid tokens, the following generation process is the same as the original BART.

Beam-then-Rank. *Rewriting-Stego* can freely select greedy search, beam search, or any other common algorithm to select the final prediction of Y'_t. Thus, to improve the generation quality, we design a *Beam-then-Rank*: 1) we first use beam search to generate K stego candidates; 2) then, we use an external GPT2 model to estimate the PPL and then select the best candidate.

Restoring. The receiver who knows the vocabulary grouping can restore the secret message S from the stego text Y' by checking the group bit id.

Fine-Tuning. We believe conducting fine-tuning can deliver better performance. Thus, we synthesize a fine-tuning dataset via the data augmentation technique. We first sample one million high-quality instances F whose perplexity (PPL) is greater than 20 and less than 200. Then, we employ the augmentation tool [9] to synthesize a parallel perturbed dataset, which includes 8 word-level perturbation operations: 1) random insertion, 2) random substitution, 3) synonym substitution, 4) antonym substitution, 5) word decomposition, 6) deletion, 7) transposition, and 8) random combination of the preceding methods. Finally, we mix the augmented data with the original data and randomly select one million data as input and the original text of these data as labels to form the fine-tuning dataset. In the fine-tuning process, we randomly masked about 75% of the input. We adopt AdamW as the optimizer; the batch size is set to 512; the learning rate is set to 3.5e−5, and the warm-up strategy is used with a warm-up number of 8000 steps. Finally, to avoid over-fitting, we only fine-tune 1 epoch.

3   Experiments

3.1   Settings

We evaluate models on three public datasets, namely, *Large Movie Review Dataset (**Movie**)* [10], *All the News (**News**), Sentiment140 (**Tweet**)* [6]. For all datasets, raw texts are first converted to lowercase, and HTML tags and most punctuations are removed. All texts are tokenized by the NLTK tools, and then sentences whose length is below 5 or above 200 are filtered. Next, several methods are selected as baselines, where **Bins** [4], **Huffman** [15], and **Saac** [13]are generation-based methods, **Masked-Stega** [14] is an edit-based method. The first three generation-based baseline methods are implemented through the source code released by Saac [13]. Masked-Stega is implemented through the official source code, and we set p to 0.01. For Bins, we set b to be 4, and the corresponding number of bins is 16. For Huffman, we build the Huffman tree with the top 128 likely tokens. For Saac, we chose the imperceptibility gap δ to be 0.01. For our method, we use *bart-base*[2] as the backbone, and the beam width is set to be 50. For all models, we have sampled 5000 texts as the cover texts and another 5000 texts as the secret messages, and use the following metrics: **1) Bits per Token (BPT)** measures the information capacity of stego text, it reports the average number of hided secret bits per token(word) in the generated stego text. A larger BPT indicates that the method can carry more secret information in the same-length stego text; **2) Perplexity (PPL)** is a language modeling metric that measures the quality of the given text from the perspective of probability. A smaller PPL means that the generated sentences are more natural.

---

[2] BART, 140M, https://huggingface.co/facebook/bart-base.

Here, we use a pre-trained *gpt2-medium* as the backbone language model; **3) Mean and Variance** measures the imperceptibility (anti-steganalysis ability). For a stego text, we mask each token to [MASK] in turn and use BERT [3] to get the sorted word predictions at the masked position. Then we collect the position of each original token in the sorted predictions. Finally, we calculate the mean and the variance of positions of the original words. A smaller mean and variance indicate higher imperceptibility, and the generated stego text is easier to avoid the detection of the observation system; **4) Detection Accuracy (ACC)** also measures the imperceptibility. We fine-tuned a BERT as a classifier to detect whether the text is stego text. We sampled 30,000 texts from the three mentioned datasets as the normal texts and generated 30,000 stego texts using the Arithmetic Coding [17].

3.2   General Results

Table 2. Evaluation Results. BPT_{C+S} considers the transmission of the cover text if required, but BPT_S does not. *: the secret message can not be entirely encoded.

Dataset	Method	BPT_S	BPT_{C+S}	PPL	Mean	Variance	Acc
Movie	Vanilla	0.0	0.0	114.9	162.4	2.7e04	1.6%
	Bins	4.0	4.0	332.2	170.4	1.2e05	67.6%
	Huffman	4.8	0.79	284.6	224.6	3.4e05	71.5%
	Saac	**5.3**	0.98	635.4	319.3	2.9e05	72.9%
	Masked-Stega*	0.16	0.16	126.4	148.5	2.5e04	**7.1%**
	Rewriting-Stego	1.0	1.0	105.2	121.4	**2.1e04**	13.7%
	Rewriting-Stego	2.0	2.0	**72.5**	**71.8**	2.7e04	20.3%
	Rewriting-Stego	4.0	**4.0**	130.1	87.1	6.6e04	23.4%
News	Vanilla	0.0	0.0	92.7	175.5	2.8e04	2.1%
	Bins	4.0	4.0	424.5	237.2	2.4e05	59.6%
	Huffman	4.7	0.76	346.7	269.9	3.8e05	75.5%
	Saac	**5.1**	0.93	586.9	327.5	2.9e05	78.1%
	Masked-Stega*	0.25	0.25	119.8	159.2	**2.5e04**	**2.2%**
	Rewriting-Stego	1.0	1.0	114.2	152.9	2.6e04	8.9%
	Rewriting-Stego	2.0	2.0	**101.6**	**119.3**	4.2e04	18.9%
	Rewriting-Stego	4.0	**4.0**	158.1	124.7	1.4e05	20.6%
Tweet	Vanilla	0.0	0.0	183.5	180.0	5.1e04	13.9%
	Bins	4.0	4.0	908.8	278.2	3.3e05	46.4%
	Huffman	4.7	0.87	1223.9	366.7	7.6e05	40.6%
	Saac	**5.4**	1.16	1924.3	369.9	6.0e05	51.8%
	Masked-Stega*	0.16	0.16	196.4	158.3	4.5e04	31.1%
	Rewriting-Stego	1.0	1.0	69.7	70.1	**2.1e04**	**18.4%**
	Rewriting-Stego	2.0	2.0	**62.1**	**44.4**	2.2e04	19.2%
	Rewriting-Stego	4.0	**4.0**	137.3	109.3	1.4e05	42.1%

Table 2 reports the result. It must be noted that Masked-Stega can not entirely hide the secret message in most cases because it strictly requires the length of the cover text is linearly related to the length of the secret message (about 3X-5X longer than the secret message).

Comparison with Generation-based: When BPT_S is set to 4, *Rewriting-Stego* can outperform generation-based baselines with similar information capacity. *Rewriting-Stego* have lower scores in Mean, Variance, and Acc, indicating the generated stego text will raise less attention from the observation system. *Rewriting-Stego* also has a significantly lower PPL. It shows *Rewriting-Stego* can generate more natural stego text. Finally, generation-based models need the cover text to initialize the backbone language model when restoring the secret message; thus, we have to consider the transmission of the cover text at the same time. *Rewriting-Stego* does not have this issue, bringing higher real-world information capacity in terms of BPT_{C+S}. **Comparison with Edit-based:** The edit-based Masked-Stega has significantly lower information capacity than others. If we set the BPT_S to 1.0, *Rewriting-Stego* has at least 3–4 times higher information capacity, and the overall performance is still better than Masked-Stega in most comparisons. The major advantage of Masked-Stega is the lower detection accuracy (Acc), which may be better than *Rewriting-Stego* in some specific anti-steganalysis systems. However, *Rewriting-Stego* has better performance in almost other metrics. **Comparison with the Vanilla:** We also evaluate the human-generated cover text in the same way. Besides the detection accuracy, our *Rewriting-Stego* has better performance. Such results are not strange because 1) We find such human-generated cover texts have various noises; 2) The backbone model of *Rewriting-Stego* is BART, which is also a denoising model. Thus, our *Rewriting-Stego* can denoise the input along with generating the stego text.

3.3 More Studies

Ablation Study. We conduct experiments to analyze the impact of the following terms: 1) Fine-Tune, 2) Conditional Code, and 3) Beam-then-Rank. The first *base* model uses the pre-trained BART with no advanced technique. Afterward, we gradually add the Fine-Tine (*base+FT*), Conditional Code (*base+FT+CC*), and Beam-then-Rank (*base+FT+BR*). As reported in Fig. 3: 1) The naive *base* still has a competitive performance compared to baselines (see Table 2), showing the notable advantage of rewriting-paradigm; 2) After the fine-tuning, the scores have notable improvements in all metrics. It shows this procedure will help *Rewriting-Stego* to generate higher-quality stego text; 3) The Conditional Code mainly helps *Rewriting-Stego* reduce the Mean and the Variance; 4) Beam-than-Rank has also notably improved the performance. This is why we design a plug-and-play decoding strategy to be compatible with most decoding algorithms.

Case Study. We have sampled two examples in Table 3. It can be observed that: 1) When $BPT_S=1$, the stego text generated by *Rewriting-Stego* is highly similar to the cover text, which verified the superior naturalness and controllability of our approach; 2) With the increasing of BPT_S, the generated stego texts become

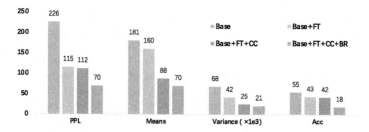

Fig. 3. Ablation Study on the Tweet Dataset. BPT_S is set to 1.0 in the experiments.

Table 3. Examples of stego texts. Limited by the space, we omitted the padding part.

24-bit Secret Message : (1,0,0,0,1,0,0,1,0,1,1,0,1,0,1,0,1,1,1,1,1,0,1,1)		
Cover Text	Method	Stego Text
sometimes we bring the story to you, sometimes you have to go to the story.	Rewriting-Stego $BPT_S=1$	sometimes we bring it to your mind but sometimes you have to go back to it to find the story you want to tell...
	Rewriting-Stego $BPT_S=2$	but we have a different way of thinking about what the story...
	Rewriting-Stego $BPT_S=4$	some of all the story lines...
	Bins $BPT_S=4$	somewhere and we have lots different...
	Masked-Stega $(BPT_S=0.16)$(Encrypted message :10001)	sometimes we read the story to you , sometimes you have to stick to the story...

shorter, and the similarity to the cover text is weaker; 3) The generation-based Bins and our *Rewriting-Stego* can entirely hide the secret message but need the padding operation to make the stego text complete; the edit-based Masked-Stega can generate a complete stego text but can not entirely hide the secret message.

4   Conclusion

This paper proposes *Rewriting-Stego*, a novel rewriting-based linguistic steganographic approach. *Rewriting-Stego* aims to improve the information capacity of the stego text without losing the naturalness and controllability. We use a pre-trained BART as the backbone l model and propose a *Plug-and-Play Group-Wise Msked Decoding Strategy* to rewrite the given cover text and hide the secret message in the obtained paraphrased text (stego text). Besides, *Rewriting-Stego* uses a text-based *Condition Code* and *Beam-then-Rank* strategy to deliver better performance. Experimental results show that *Rewriting-Stego* outperforms the baselines in most metrics.

Acknowledgment. This work was supported in part by the National Natural Science Foundation of China under Grant 62162067 and 62101480, in part by the Yunnan Province Science Foundation under Grant No. 202005AC160007, No. 202001BB050076, and Research and Application of Object detection based on Artificial Intelligence, in part by the Applied Basic Research Foundation of Yunnan Province under Grant 202201AT070156, in part by the Fund project of Yunnan Province Education Department "Generating mnatural and controllable steganographic text based on language model".

References

1. Bernaille, L., Teixeira, R.: Early recognition of encrypted applications. In: Uhlig, S., Papagiannaki, K., Bonaventure, O. (eds.) PAM 2007. LNCS, vol. 4427, pp. 165–175. Springer, Heidelberg (2007). https://doi.org/10.1007/978-3-540-71617-4_17

2. Bi, X., Yang, X., Wang, C., Liu, J.: High-capacity image steganography algorithm based on image style transfer. Secur. Commun. Netw. **2021** (2021)

3. Devlin, J., Chang, M., Lee, K., Toutanova, K.: BERT: pre-training of deep bidirectional transformers for language understanding. In: NAACL (2019)

4. Fang, T., Jaggi, M., Argyraki, K.J.: Generating steganographic text with LSTMs. In: ACL (2017)

5. Garg, M., Gupta, S., Khatri, P.: Fingerprint watermarking and steganography for ATM transaction using LSB-RSA and 3-DWT algorithm. In: ICCN, pp. 246–251 (2015)

6. Go, A., Bhayani, R., Huang, L.: Twitter sentiment classification using distant supervision. CS224N project report, Stanford **1**(12) (2009)

7. Lester, B., Al-Rfou, R., Constant, N.: The power of scale for parameter-efficient prompt tuning. arXiv preprint arXiv:2104.08691 (2021)

8. Lewis, M., et al.: BART: denoising sequence-to-sequence pre-training for natural language generation, translation, and comprehension. In: ACL (2020)

9. Ma, E.: Nlp augmentation (2019)

10. Maas, A.L., Daly, R.E., Pham, P.T., Huang, D., Ng, A.Y., Potts, C.: Learning word vectors for sentiment analysis. In: ACL (2011)

11. Mikolov, T., Karafiát, M., Burget, L., Cernockỳ, J., Khudanpur, S.: Recurrent neural network based language model. In: Interspeech, vol. 2 (2010)

12. Qiu, X.P., Sun, T.X., Xu, Y.G., Shao, Y.F., Dai, N., Huang, X.J.: Pre-trained models for natural language processing: a survey. Sci. China Technol. Sci. **63**(10), 1872–1897 (2020). https://doi.org/10.1007/s11431-020-1647-3

13. Shen, J., Ji, H., Han, J.: Near-imperceptible neural linguistic steganography via self-adjusting arithmetic coding. In: EMNLP (2020)

14. Ueoka, H., Murawaki, Y., Kurohashi, S.: Frustratingly easy edit-based linguistic steganography with a masked language model. In: NAACL (2021)

15. Yang, Z., Guo, X., Chen, Z., Huang, Y., Zhang, Y.: RNN-stega: linguistic steganography based on recurrent neural networks. IEEE Trans. Inf. Forensics Secur. **14**(5) (2019)

16. Zhang, C., Benz, P., Karjauv, A., Sun, G., Kweon, I.S.: UDH: universal deep hiding for steganography, watermarking, and light field messaging. In: NeurIPS (2020)

17. Ziegler, Z.M., Deng, Y., Rush, A.M.: Neural linguistic steganography. In: EMNLP-IJCNLP (2019)

Towards Defending Against Byzantine LDP Amplified Gain Attacks

Yukun Yan[1], Qingqing Ye[2], Haibo Hu[2], Rui Chen[1(✉)], Qilong Han[1], and Leixia Wang[3]

[1] Harbin Engineering University, Harbin, China
{yanyukun,ruichen,hanqilong}@hrbeu.edu.cn
[2] Hong Kong Polytechnic University, Kowloon, Hong Kong SAR, China
{qqing.ye,haibo.hu}@polyu.edu.hk
[3] Renmin University of China, Beijing, China
leixiawang@ruc.edu.cn

Abstract. Local differential privacy (LDP) has been widely used to collect sensitive data from distributed users while preserving individual privacy. However, very recent studies show that LDP is vulnerable to manipulation and poisoning attacks. Maximal gain attack (MGA) is one of the most fundamental examples. In this paper, we take one step further to introduce a novel type of attacks called *Byzantine LDP amplified gain attacks* (BLAGA) that is precisely derived from the randomness of an LDP protocol, unveiling LDP's inherent conflict between privacy and security. We show that MGA is a special case of BLAGA. Subsequently, we propose a defense framework that makes use of a data-driven approach to automatically identify the target items via multi-round data collection. It differs from existing solutions in that it does not require any prior knowledge, which is normally difficult to acquire in practical settings. Finally, we perform extensive experiments on various datasets to show that our defense framework can well preserve the utility of heavy hitter identification with effective security protection.

Keywords: Local differential privacy · Byzantine users · Manipulation attack

1 Introduction

With the increasing awareness of data privacy in the era of big data, local differential privacy (LDP) has been widely studied as a promising mechanism to collect sensitive data from distributed users while preserving individual privacy. It has also shown its practical value in a few large-scale real-world systems (e.g., Chrome [7], iOS [15]). Most of the existing efforts focus on improving the data utility of LDP protocols for different data analysis tasks [3,16,19,20]. Only until very recently, some pioneering works [4,5,18] have started to explore the security aspect of LDP, showing that LDP is vulnerable to manipulation attacks [5] and poisoning attacks [4,18].

These attacks can substantially undermine some of the building blocks of data analysis, for example, *heavy hitter identification* [2,3,7,17] whose goal is to

X. Wang et al. (Eds.): DASFAA 2023, LNCS 13943, pp. 627–643, 2023.
https://doi.org/10.1007/978-3-031-30637-2_42

identify the top-k items that are the most frequent among users. Cao *et al.* [4] show that a maximal gain attack is able to promote arbitrary attacker-chosen items (called target items) to appear in the top-k heavy hitter list by introducing a small number of fake users. With such an attack, an attacker can make a phishing web page as a popular default homepage of Chrome or manipulate top-k recommendation results in his/her favor [8]. In practice, maximal gain attacks are relatively easy to detect and defend because all fake users adopt the same attack pattern. In this paper, we take one step further to introduce a new type of more generalized attacks based on the existence of *Byzantine users*, which is more difficult to detect but still easy to perform. Unlike the fake users in a maximal gain attack, Byzantine users aim to not only bias the resulting heavy hitters, but also prevent themselves from being detected. We provide an analysis of input-manipulation and output-manipulation attacks [10,11] and pinpoint a niche space where the randomness of an LDP protocol can be used to amplify attack effects. Based on this niche space, we design *Byzantine LDP amplified gain attacks* (BLAGA). We show that the maximal gain attack is actually a special case of BLAGA.

Defending against BLAGA is technically challenging for a few reasons. First, Byzantine users normally exhibit different attack patterns, making them difficult to be detected. Second, due to the amplification effect, it becomes harder to achieve a reasonable balance between privacy and security (e.g., a more strict privacy requirement will lead to a more severe attack). Third, in practice, it is infeasible to obtain much prior knowledge to guide the defense. In addressing these challenges, we propose a defense framework that makes use of a data-driven approach to automatically identify the target items via multi-round data collection. In particular, we propose a parameter expectation maximization (PEM) method to separate target items and non-target items by modeling their distributions in an unsupervised manner. To improve accuracy in the absence of prior knowledge, we propose a two-stage parameter estimation method, which stacks two PEM modules. In addition, we design a post-processing method based on the difference of the data distributions of target and non-target items to further improve the accuracy of detected target items.

We summarize our key contributions as follows:

- We formulate a novel type of BLAGA attacks against LDP, which is a more generalized version of maximal gain attacks. It is of particular theoretical interest because it is precisely derived from the randomness of an LDP protocol, unveiling LDP's inherent conflict between privacy and security.
- We propose a novel defense framework that accurately learns the target items in an unsupervised manner. Since MGA is a special case of BLAGA, our defense framework also provides a more realistic solution to MGA without requiring prior knowledge.
- We perform extensive experiments on three benchmark datasets and different instantiations of BLAGA to demonstrate that our defense framework can well preserve data utility while reasonably balancing privacy and security under LDP.

2   Preliminaries

2.1   Local Differential Privacy

In recent years, local differential privacy (LDP) has been the de-facto status quo to protect individual privacy when user data is collected by an untrusted data collector. Intuitively, LDP allows a user to specify the extent to which his/her data can be distinguished by the data collector via a user-chosen privacy parameter ε. We give a formal definition below.

Definition 1 (Local Differential Privacy). *A randomized algorithm $\mathcal{R}$ satisfies $\varepsilon-$local differential privacy ($\varepsilon-$LDP) if for any two inputs $x, x' \in \mathcal{D}$ and for any output $y \subseteq Range(\hat{\mathcal{D}})$, it holds that*

$$e^{-\varepsilon} \leq \frac{\Pr(\mathcal{R}(x) \in y)}{\Pr(\mathcal{R}(x') \in y)} \leq e^{\varepsilon}. \tag{1}$$

The smaller ε is, the stronger privacy protection is provided.

2.2   Unary Encoding Protocols

In the literature, a series of protocols have been proposed to achieve LDP, such as generalized randomized response (GRR) [9], unary encoding (UE) [16], and optimized local hashing (OLH) [16]. Among different proposals, the family of UE protocols strikes a reasonable balance between efficiency and data utility in practical scenarios [6]. The general idea of UE protocols is to encode an item i to a d-bit binary vector $\mathbf{e}_i$ whose bits are all 0 except the i-th bit being 1 and perturb each bit independently. Specifically, if a bit is of value 1, it remains 1 with probability p; otherwise, it is flipped from 0 to 1 with probability q. There are two popular versions of UE protocols, namely optimized unary encoding (OUE) and symmetric unary encoding (SUE) [16]. In OUE, $p = \frac{1}{2}$ and $q = \frac{1}{e^{\varepsilon}+1}$; in SUE, $p = \frac{e^{\varepsilon/2}}{e^{\varepsilon/2}+1}$ and $q = \frac{1}{e^{\varepsilon/2}+1}$. For heavy hitter identification, after receiving the perturbed frequency f'_i of item i, the data collector derives the estimated frequency by $\hat{f}_i = \frac{f'_i - q}{p - q}$. The top-$k$ heavy hitters are then generated by ranking the estimated frequencies of all items. Without loss of generality, in this paper we assume that users utilize UE protocols to protect their sensitive data.

3   Byzantine LDP Amplified Gain Attack

We assume that Byzantine attackers have the same capability as genuine users. However, they can collude together, send arbitrary data in the encoded space to the data collector, and influence the estimated statistics in their favor [13]. Let N be the number of genuine users, M the number of Byzantine attackers, and $\beta = \frac{M}{N+M}$ the proportion of Byzantine attackers. In many real-world applications with a large number of users, it is reasonable to assume that β will be a small number (e.g., less than 10%). For example, consider the case of iOS, which has over 1 billion users worldwide. It is highly unlikely that 100 million users are Byzantine attackers.

The goal of a gain attack for heavy hitter identification is to promote target items into the top-k list by increasing their estimated frequencies. We denote the entire item set by $I = \{i_1, i_2, \cdots, i_d\}$, its original frequency set by $F = \{f_1, f_2, \cdots, f_d\}$ and the set of target items by $T = \{t_1, t_2, \cdots, t_r\} \subseteq I$. An attack's effect is measured by the frequency gain of target items via $G = \sum_{t \in T} \mathbb{E}[\Delta \tilde{f}_t]$, where $\mathbb{E}[\Delta \tilde{f}_t]$ represents the expected increase of the estimated frequency of item t after the attack.

3.1   Maximal Gain Attack

We first review maximal gain attack (MGA) [4], whose goal is to maximize the frequency gain of target items. Under the UE protocols, MGA initializes a vector $\mathbf{y}_u$ of fake user u as a binary vector with $\mathbf{y}_{u,t} = 1$ for $t \in T$ and all other bits being 0. In this section, we slightly abuse the notation to let M denote the number of fake users. To make the data sent by fake users statistically identical to that of genuine users, MGA randomly samples $l = \lfloor p + (d - 1)q - r \rfloor$ non-target bits of the vector $\mathbf{y}_u$, where r is the number of target items, and sets them to 1. The target item t's gain is

$$G_t = \frac{\sum_{u=N+1}^{N+M} \mathbb{1}_{\mathbf{y}_u}(t)}{(N+M)(p-q)} - \frac{M(f_t(p-q) + q)}{(N+M)(p-q)}, \tag{2}$$

where $\mathbb{1}_{\mathbf{y}_u}(t)$ is the probability of the t-th position of $\mathbf{y}_u$ being 1. Since in MGA $\mathbb{1}_{\mathbf{y}_u}(t) = 1$ if user u is a fake user, we have the maximal gain $G_t = \frac{M(1 - f_t(p-q) - q)}{(N+M)(p-q)}$.

While MGA is effective, it is just an extreme attack method to boost a target item into the top-k heavy hitter list. There are also many other attack methods that can accomplish this goal. In addition, since all fake users perform the same attack pattern on target items, MGA is relatively easy to detect. In the presence of Byzantine attackers, it is possible to launch successful attacks while better preventing them from being detected.

3.2   Byzantine LDP Amplified Gain Attack

To better understand the unique effect of LDP in gain attacks, we analyze the relationship between gain attacks and the following two classes of attacks on LDP protocols. **Input-manipulation attacks** [10,11] assume that attackers can launch attacks during the *input phase* of LDP protocols. Attackers can control the local system environment, but the LDP protocol is a black box for them. **Output-manipulation attacks** [10,11] assume that attackers can launch attacks during the *output phase* of LDP protocols. Attackers need not only to control the local system environment but also to fully understand the underlying LDP protocol. For example, MGA belongs to output-manipulation attacks

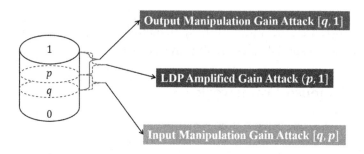

Fig. 1. An illustration of $\mathbb{1}_{\mathbf{y}_u}(t)$ of different types of gain attacks.

because it requires attackers to know the specific values of the parameters p and q if UE protocols are used.

According to Eq. (2), if we want to achieve positive gain, we can quantify $\mathbb{1}_{\mathbf{y}_u}(t)$ of input-manipulation attacks and output-manipulation attacks under UE protocols. Since $f_t \in [0,1]$, we have $\mathbb{1}_{\mathbf{y}_u}(t) \in [q,p]$ for input-manipulation attacks and $\mathbb{1}_{\mathbf{y}_u}(t) \in [q,1]$ for output-manipulation attacks. It can be seen that both input-manipulation attacks and output-manipulation attacks can make the probability of the t-th position of $\mathbf{y}_u$ being 1 fall into the interval of $[q,p]$. Since in an input-manipulation attack an attacker can input *any* falsified value in the input domain, it is interesting to ask *what allows an output-manipulation attack to output the unique interval* $(p,1]$. Our answer is the inherent randomness of an LDP protocol that amplifies attack effects.

In this paper, we follow the setting of gain attacks [4] that aim to promote unpopular target items into the top-k list, and propose a novel type of gain attacks called *Byzantine LDP amplified gain attacks* (BLAGA), which allows attackers to craft their perturbed data to make $\mathbb{1}_{\mathbf{y}_u}(t)$ fall into the interval of $(p,1]$. We illustrate the relationships of different types of gain attacks in terms of $\mathbb{1}_{\mathbf{y}_u}(t)$ in Fig. 1. BLAGA is of particular research value because it precisely unveils LDP protocols' inherent conflict between privacy and security. Next, we explain how to perform BLAGA. BLAGA sets p'_t, the probability of setting the bit representing t to 1, to be larger than p. All Byzantine users set p'_t together via collusion. Here we require $p'_t > p$ because only in this way it amplifies the attack effects of an LDP protocol. Then BLAGA randomly samples l' non-target bits and set them to 1 in a UE protocol. Here $l' = \lfloor p + (d-1)q - r \cdot \bar{p}' \rfloor$, where $\bar{p}' = \frac{1}{r} \sum_{t=1}^{r} p'_t$. It is worth noting that when $\forall t \in T, p'_t = 1$, BLAGA becomes MGA.

BLAGA can also be naturally extended for the scenarios of multi-round data collection. Let n be the number of rounds of data collection, and $\frac{\varepsilon}{n}$ be the privacy budget of a single round. Each round of data collection is independent of each other. In this case, we can use the average estimated frequency from multiple rounds $\frac{1}{n} \sum_{m=1}^{n} \hat{f}_{tm}$ as the final estimated frequency of item i. To obtain positive gain for target item t, BLAGA only requires $\frac{1}{n} \sum_{m=1}^{n} p'_{tm} > p$, instead of each $p'_{tm} > p$.

As such, BLAGA provides attackers with a wide range of attacks with varying trade-off between success rate and the probability of being detected. In Sect. 5.3, we consider three special cases of BLAGA and study their security properties.

4   Our Defense Framework

Before discussing our defense framework, we identify the technical challenges to defend against BLAGA in real-world scenarios. In practice, one does not have prior knowledge about Byzantine attacks, such as which items are attacked, which users are Byzantine attackers, or even the proportion of Byzantine attackers. In addition, one does not have background knowledge about the raw data either (e.g., the original data distribution). Therefore, our key idea is to introduce multi-round data collection to learn more information in order to compensate for the absence of prior knowledge, and make use of a data-driven approach to learn essential information via data collection. We also make use of two properties of BLAGA: (1) the target items are prechosen and should not alter in different rounds; (2) since the attackers' goal is to increase the data collector's estimated frequency of target items so that they enter the heavy hitter list, they aim to obtain positive gain for each target item. We leave more complicated scenarios in future work.

Based on the above observations, we put forward our defense framework that consists of three major steps. It first performs multiple rounds of data collection using a UE protocol, then uses the collected data to mine target items in an unsupervised manner, and finally introduces a post-processing mechanism to further improve the accuracy of identified target items. Once we find the target items, they are not considered in generating the top-k heavy hitter list, and thus attacks can be avoided. Our defense framework is illustrated in Fig. 2. In what follows, we focus on explaining the second and third steps.

4.1   Model Target Items and Non-target Items

With the data obtained from the multi-round data collection process, our goal is to discover the statistical differences between target and non-target items. Let n be the number of rounds. As shown in Fig. 2, we sum up every user's data from each round (either 0 or 1) and obtain a new value X_{ui} for each (user u, item i) pair. Let L_i be the list of all users' X_{ui}. It can be observed that L_i is composed of three parts. The first two come from genuine users and the third from Byzantine attackers. Since each round of data collection uses an LDP protocol with privacy budget ε/n, we can think of the data from genuine users as two Bernoulli distributions, meaning that the result of genuine users whose raw data is 1 (resp. 0) will form a Bernoulli distribution with parameters of n and p (resp. q). These two parts can be denoted by $(1 - \beta) \cdot f_i \cdot B(n, p)$ and $(1 - \beta) \cdot (1 - f_i) \cdot B(n, q)$ for item i, respectively.

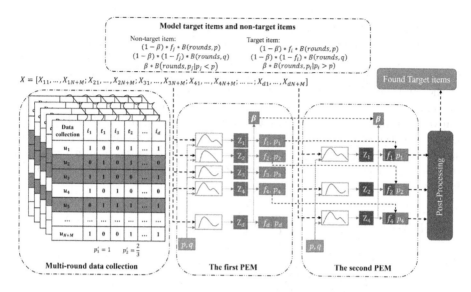

Fig. 2. The overall defense framework.

As for the third part from Byzantine attackers, since attackers can adopt different p'_t for different target item t in different rounds of data collection, the Byzantine attackers' data will form a Poisson Binomial distribution. Since the Poisson Binomial distribution with parameters $\{p'_{t1}, p'_{t2}, \cdots, p'_{tn}\}$ can be approximated as the Bernoulli distribution with parameter $\frac{1}{n}\sum_{m=1}^{n} p'_{tm}$ [14] and $\bar{p}'_t = \frac{1}{n}\sum_{m=1}^{n} p'_{tm} > p$, we can denote this part of data by $\beta \cdot B(n, \bar{p}'_t | \bar{p}'_t > p)$. Next, we consider the data of non-target items from attackers. Recall that, in m-th round of data collection, BLAGA samples $l'_m = \lfloor p + (d-1)q - r \cdot \bar{p}'_{-m} \rfloor$ non-target bits, where $\bar{p}'_{-m} = \frac{1}{r}\sum_{t=1}^{r} p'_{tm}$, and sets them to 1. Then the average probability of observing 1 on non-target items is $\bar{p}'_{nt} = \frac{1}{n}\sum_{m=1}^{n}\frac{l'_m}{d-r}$, and thus we can deduce that $\bar{p}'_{nt} < p$. We also use Bernoulli approximation to denote this part of data as $\beta \cdot B(n, \bar{p}'_{nt} | \bar{p}'_{nt} < p)$.

Therefore, we can observe that whether the parameter p_i of the Bernoulli distribution from attackers' data is larger than p in LDP protocols can determine whether item i is a target item.

4.2 Two-Stage Parameter Estimation Method

Since there is no ground truth to indicate target items, we use an unsupervised method to capture the difference between target and non-target items. The Expectation Maximization (EM) algorithm [12] provides a simple yet effective solution. Its key idea is to effectively find the maximum likelihood estimation by performing expectation (E) steps and maximization (M) steps iteratively when there are latent variables.

Based on the EM algorithm, we propose a novel parameter expectation maximization (PEM) method, which can simultaneously optimize all items' parameters. We set each item's parameter as $\{p_i, f_i, \beta\}$, where p_i and f_i are specific to every item while β is common among all items. We give the pseudo-code of the PEM method in Algorithm 1. The method takes $P^0 = [p_1^0, p_2^0, \cdots, p_d^0]$, $F^0 = [f_1^0, f_2^0, \cdots, f_d^0]$, β^0 and dataset X as input, where P^0, F^0 and β^0 are initial values that will be explained later. We denote the complete data by (X, Z), where X is the observed data and Z is the latent data. For each item, we construct $\alpha_{i1} = (1 - \beta) \cdot f_i$, $\alpha_{i2} = (1 - \beta) \cdot (1 - f_i)$, $\alpha_{i3} = \beta$, $\tilde{p}_{i1} = p$, $\tilde{p}_{i2} = q$, and $\tilde{p}_{i3} = p_i$. Then we have the maximum likelihood estimation and Q function as follows:

$$\log p(X, Z|P, F, \beta) = \log \prod_{i=1}^{d} \prod_{k=1}^{3} \prod_{j=1}^{N+M} \left[\alpha_{ik} C_n^{x_{ij}} (\tilde{p}_{ik})^{x_{ij}} (1 - \tilde{p}_{ik})^{n-x_{ij}} \right]^{z_{ijk}} \quad (3)$$

$$Q(P, F, \beta; P^t, F^t, \beta^t) = \sum_{i=1}^{d} \sum_{k=1}^{3}$$

$$\left\{ \sum_{j=1}^{N+M} \hat{z}_{ijk}^t \log \alpha_{ik}^t + \sum_{j=1}^{N+M} \hat{z}_{ijk}^t [\log C_n^{x_{ij}} + x_{ij} \log \tilde{p}_{ik}^t + (n - x_{ij}) \log(1 - \tilde{p}_{ik}^t)] \right\} \quad (4)$$

The E-step computes latent data $\hat{z}_{ijk}^t$ by the observed data X and the current P^t, F^t and β^t (Lines 2–3). The M-step uses Eq. (4) to calculate P^{t+1}, F^{t+1} and β^{t+1} (Lines 4–7) that maximize the expected likelihood as the inputs to the next round E-step. When meeting the convergence condition, the algorithm returns the estimated parameters $\hat{P}$, $\hat{F}$, and $\hat{\beta}$ (lines 9).

Since PEM can depict the distribution of each item, we can leverage it to distinguish target items from non-target items. However, due to the lack of prior knowledge, we can only empirically initialize the input values of PEM. As a result, using $\hat{P}$ from a single PEM as the criterion will make some non-target items mistakenly identified as target items. When attackers have strong attack capabilities, some target items are more likely to exhibit the same data distribution as those of some non-target items in the true top-k heavy hitter list. Specifically, we have the following theorem.

Theorem 1. *If a target item with frequency f_t and a non-target item with frequency f_{nt} exhibit the same data distribution on the collected data, $f_{nt} > f_t + \beta$.*

Proof. Assume that b_t (resp. b_{nt}) is the original number of 1 from genuine users for target item t (resp. non-target item nt) with $\bar{p}_t'$ (resp. $\bar{p}_{nt}'$). Then, we have:

$$\frac{b_t \cdot p + (N - b_t)q + M \cdot \bar{p}_t'}{N + M} = \frac{b_{nt} \cdot p + (N - b_{nt})q + M \cdot \bar{p}_{nt}'}{N + M} \quad (5)$$

$$\Rightarrow \frac{b_{nt}}{N} = \frac{b_t + \frac{M(\bar{p}_t' - \bar{p}_{nt}')}{p-q}}{N} > \frac{b_t}{N} + \frac{M}{N + M} \cdot \frac{(\bar{p}_t' - \bar{p}_{nt}')}{p - q} \quad (6)$$

Algorithm 1. Parameter Expectation Maximization (PEM)

Input: $X = [x_{11}, \cdots, x_{1N+M}; \cdots; x_{d1}, \cdots, x_{dN+M}]$, $P^0 = [p_1^0, p_2^0, \cdots, p_d^0]$, $F^0 = [f_1^0, f_2^0, \cdots, f_d^0]$, β^0

Output: $\hat{P} = [\hat{p}_1, \hat{p}_2, \cdots, \hat{p}_d]$, $\hat{F} = [\hat{f}_1, \hat{f}_2, \cdots, \hat{f}_d]$, $\hat{\beta}$

1: **while** not converge **do**

2:    **E step:**

3:    $$\hat{z}_{ijk}^t = \frac{\alpha_{ik}^t \cdot C_n^{x_{ij}} (\bar{p}_{ik}^t)^{x_{ij}} (1 - \bar{p}_{ik}^t)^{n - x_{ij}}}{\sum_{k=1}^{3} \alpha_{ik}^t \cdot C_n^{x_{ij}} (\bar{p}_{ik}^t)^{x_{ij}} (1 - \bar{p}_{ik}^t)^{n - x_{ij}}}$$

4:    **M step:**

5:    $$f_i^{t+1} = \frac{\sum_{j=1}^{N+M} \hat{z}_{ij1}^t}{\sum_{j=1}^{N+M} \hat{z}_{ij1}^t + \sum_{j=1}^{N+M} \hat{z}_{ij2}^t}$$

6:    $$p_i^{t+1} = \frac{\sum_{j=1}^{N+M} \hat{z}_{ij3}^t \cdot x_{ij}}{\sum_{j=1}^{N+M} \hat{z}_{ij3}^t \cdot n}$$

7:    $$\beta^{t+1} = \frac{\sum_{i=1}^{d} \sum_{j=1}^{N+M} \hat{z}_{ij3}^t}{\sum_{i=1}^{d} (\sum_{j=1}^{N+M} \hat{z}_{ij1}^t + \sum_{j=1}^{N+M} \hat{z}_{ij2}^t + \sum_{j=1}^{N+M} \hat{z}_{ij3}^t)}$$

8:    $\hat{P} = P^{t+1}, \hat{F} = F^{t+1}, \hat{\beta} = \beta^{t+1}$

9: **return** $\hat{P}, \hat{F}, \hat{\beta}$

Since $\bar{p}_t' > p$, $\bar{p}_{nt}' = \frac{1}{n} \sum_{m=1}^{n} \frac{\lfloor p + (d-1)q - r \cdot \bar{p}_{-m}' \rfloor}{d - r}$, and $r \geq 1$, we have $f_{nt} > f_t + \beta$.

Therefore, to overcome the above problem, we propose to stack two PEM modules and design a two-stage parameter estimation method, which is given in Algorithm 2. We initialize each p_i in P as p (the parameter of the UE protocol), each f_i in F as $\frac{1}{d}$, and β as β_0 (Line 1). In our experiments, we set $\beta_0 = 0.05$, which is half of the maximal value discussed in Sect. 3. Next, we feed X, P, F, β into the PEM algorithm and get estimated parameters $\hat{P}$, $\hat{F}$ and $\hat{\beta}$ (Line 2). The main goal of the first PEM module is to identify as many target items as possible (i.e., a high recall).

Next, we build the suspect datasets I_s, X_s and P_s from I, X, and $\hat{P}$ (Lines 3–7), which retain the knowledge captured from the first PEM module. In general, the more accurate initialized parameters of an EM algorithm are, the more accurate the results will be. Therefore, by Theorem 1 and the fact that normally $f_t < \beta$ in practice, we craft each f in F_s as $2\hat{\beta}$ and apply the second PEM module (Lines 8–9). Then, we utilize $\hat{P}_s$ to generate the candidate target item set I_{ca}, along with its frequency set $\hat{F}_{ca}$ (Lines 10–13). Finally, we use a post-processing method to get the final target item set $\hat{T}$ (Lines 14–15), which will be explained in the next section.

4.3   Post-processing

This post-processing mechanism aims to further eliminate the false positive rate of the candidate target item set I_{ca}. We first convert the L_i of each

---

Algorithm 2. Two-Stage Parameter Estimation Method

Input: $X = [x_{11}, \cdots, x_{1N+M}; \cdots; x_{d1}, \cdots, x_{dN+M}]$, the entire item set $I = \{i_1, i_2, \cdots, i_d\}$, UE protocol parameter: p, number of rounds n, threshold η, β_0

Output: Target item set $\hat{T}$

  1: Initialization: $P = [p_1, p_2, \cdots, p_d | p_i = p]$, $F = [f_1, f_2, \cdots, f_d | f_i = \frac{1}{d}]$, $\beta = \beta_0$.

  2: $\hat{P}, \hat{F}, \hat{\beta} = \text{PEM}(X, P, F, \beta)$

  3: **for** $j = 1 \to d$ **do**

  4:      **if** $\hat{p}_j > p$ **then**

  5:          $I_s \Leftarrow I_s \cup \{i_j\}$

  6:          $X_s \Leftarrow X_s \cup \{X_{j1}, \cdots, X_{jN+M}\}$

  7:          $P_s \Leftarrow P_s \cup \{\hat{p}_j\}$

  8: $F_s = [f_1, f_2, \cdots, f_{len(I_s)} | f = 2\hat{\beta}]$

  9: $\hat{P}_s, \hat{F}_s, \hat{\beta}_s = \text{PEM}(X_s, P_s, F_s, \hat{\beta})$

10: **for** $j = 1 \to len(I_s)$ **do**

11:      **if** $\hat{p}_{sj} > p$ **then**

12:          $I_{ca} \Leftarrow I_{ca} \cup \{i_{sj}\}$

13:          $\hat{F}_{ca} \Leftarrow \hat{F}_{ca} \cup \{\hat{f}_{sj}\}$

14: $\hat{T} = PostProcess(X, I_{ca}, \hat{F}_{ca}, \eta, p, n)$

15: **return** $\hat{T}$

---

item i to the ratio of counting form $RC = \{rc_1, rc_2, \cdots, rc_d\}$, where $rc_i = \frac{1}{N+M} [count(L_i, value)]^n_{value=0}$ and $count(L_i, value)$ gives the number of occurrences of $value$ in L_i. As an illustration, we consider three special cases of BLAGA (i.e., BGA, UGA and MGA) on the Zipf dataset (see Sect. 5.1 for more details) and draw the corresponding RC in Fig. 3, where each curve represents an item. We can have the following observation: the curves of target items and non-target items exhibit different patterns, especially when attackers launch MGA.

Our idea is to leverage such difference to further reduce the false positive rate. For each $value$ in the range of $[\lceil np \rceil, n]$, we sort all items according to $\frac{1}{N+M} \{count(L_i, value)\}^d_{i=1}$ at each $value$ in ascending order, and obtain the number of $n - \lceil np \rceil$ rankings for each item. As $value$ increases from $\lceil np \rceil$ to n, if an item's ranking does not go up, it is considered a non-target item. In addition, we add one more restriction that requires the frequency of item j to meet $\hat{F}_{ca}[j] > \eta$, where η is a pre-selected threshold, to prevent true target items from being removed. Items meeting the above requirements will be removed from I_{ca}. Finally, we denote the rest of I_{ca} as the final target item set $\hat{T}$.

5   Experimental Results

5.1   Experimental Setup

Datasets: We utilize the following three datasets in our experiments.

- **Zipf:** We synthesize a dataset with 100 items and 100,000 users following the Zipf's distribution with same parameters used in [16].

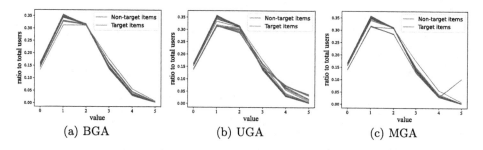

Fig. 3. *RC* on Zipf under BGA, UGA and MGA with $n = 5$.

- **IPUMS[1]:** We select the data of IPUMS 2021 and treat the city attribute as the item each user holds, which results in 102 items and 318,808 users.
- **FIRE[2]:** The FIRE dataset was collected by the San Francisco Fire Department in 2021, recording information about calls for service. We use the same filtering method as [4], resulting in 266 items and 645,312 users.

Attack Methods: In addition to MGA, we consider another two special cases of BLAGA.

- **Boundary gain attack (BGA):** Attackers set p'_t for each target item to $p + \xi$ in each round of data collection, where ξ is a small constant. We set ξ to 0.02 in the experiments.
- **Uniform gain attack (UGA):** Attackers sample p'_t for each target item from the uniform distribution in interval $(p, 1]$ in every round of data collection.

Among BGA, UGA and MGA, BGA uses the least amount of randomness from LDP and is the weakest attack in BLAGA. In contrast, MGA uses the maximal amount of randomness introduced by LDP and form the strongest attack in BLAGA. UGA provides a medium level of attacks in the BLAGA family.

Default Setting: For heavy hitter identification, we set k (the number of heavy hitters) to 10. Attackers adopt BGA as the default attack method. We set $\beta = 0.05$, and randomly sample r items from 20% of the entire item set with the lowest original frequency as the target item set T. Generally, we set r equal to k. All experimental results that do not rely on our defense framework are based on a single round of data collection, where the entire privacy budget is used. Due to the space limitation, we only report the experimental results based on the OUE protocol. Similar results can be observed on SUE. All experimental results are the average of five runs.

Metrics: We use the *success rate* of an attack as the metric to measure security, which is calculated as a fraction of target items appearing in the estimated top-k item list. To measure the utility of heavy hitter identification, we adopt *F1*,

---
[1] https://doi.org/10.18128/D010.V11.0.
[2] http://bit.ly/336sddL.

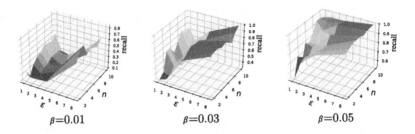

Fig. 4. The recall rate with different ε values and different n values on Zipf.

which measures whether the ground truth is ranked among the top-k items, and *NCR* [16], which is a position-aware ranking metric and assigns higher scores to higher positions.

5.2 Determination of Important Parameters

Q1: How to Balance Privacy and Security? The first PEM function in Algorithm 2 aims to find as many target items as possible. This step is the foundation for everything that follows, and thus it needs to achieve excellent performance. We denote the recall rate between T and I_s as an evaluation metric. The relevant parameters are the total privacy budget ε and the number of rounds n. As shown in the Fig. 4, we experiment on Zipf under $\beta = \{0.01, 0.03, 0.05\}$, respectively. The experimental results show that with the increase of ε initially the recall rates decrease and then start to increase. The reason for this phenomenon is that when $\frac{\varepsilon}{n}$ is a small value in single-round data collection, the parameters p and q will be similar. It makes each item's distribution impossible to explicitly model f_i. Reducing a random variable can benefit parameter estimation and improve the accuracy of identified target items. However, we do not recommend this because when ε becomes smaller, it will bring substantial utility degradation, making our defense framework less practical. Among various ε values, we can observe that $\varepsilon = 8$ can accurately identify target items while providing meaningful privacy protection. Therefore, we choose $\varepsilon = 8$ as the default value for all subsequent experiments. In addition, considering the factor of communication cost, we select $n = 5$ as the default setting.

Q2: How to Set η in Post-processing? Since the post-processing method aims to eliminate non-target items from I_{ca}, we choose the F1 score between T and $\hat{T}$ as the evaluation metric. As shown in Fig. 5, the post-processing method can improve performance in most cases. While different η values lead to different extent of performance improvement, almost all η values can outperform the baseline. Based on the results in Fig. 5, we choose $\eta = \frac{1}{d}$ by default.

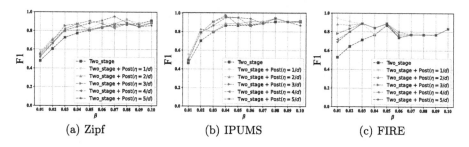

Fig. 5. The F1 score under OUE with $\eta = \{1, 2, 3, 4, 5\} * \frac{1}{d}$ on Zipf.

5.3 Security of Heavy Hitter Identification

In this set of experiments, we perform empirical analysis of the security aspects for the heavy hitter identification task under different β values or different r values. We will consider the theoretical analysis in our future work.

We first compare the three attack methods in Fig. 6. Since the three attack methods bring different levels of gain, an attacker can choose the most suitable attacks based on his/her expected success rate. For example, as shown in Fig. 6 (a), (b), (c), if $\beta = 0.05$ and attackers' goal is to promote 6 out of the 10 target items into the top-k list, all three attack methods can achieve the goal. In this case, we consider BGA as the best choice for attackers because they are less likely to expose themselves. If attackers' goal is to promote 8 or even 9 out of the 10 target items into the top-k list, UGA or MGA should be chosen. In Fig. 6 (d), (e), (f), if r ranges from 1 to 7 with $\beta = 0.05$, we can also observe that three attack methods result in 100% success rate. But when r goes beyond 7, attackers can choose an appropriate attack method based on their demand. Therefore, we believe BLAGA, as a general type of attacks, can provide the flexibility needed by attackers for different real-world scenarios.

Next, we evaluate the effectiveness of our defense framework against the three attack methods. As shown in Fig. 6, after adopting our defense framework, we can reduce the success rate less than 20% in most cases. When attackers adopt UGA or MGA, we can restrict the success rate to 0. When attackers adopt BGA, the curves exhibit irregular shapes as β or r increases. We deem that is due to the effects of the post-processing method, which improves F1 scores at the cost of recall rates. In Fig. 6(a) and (b), when $\beta = 0.01$, there is a little bit of negative optimization. The reason for this phenomenon is that when attacks have feeble effects, the success rate is already very small even without defense. As we divide the total privacy budget equally into n pieces, it can promote the attack and slightly improve success rate. However, this phenomenon is acceptable as the success rate is still low.

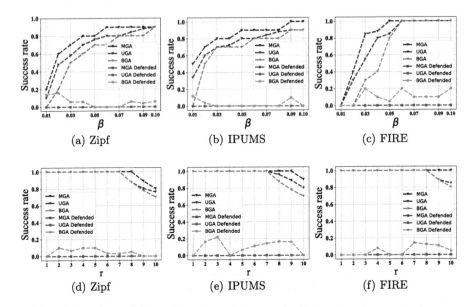

Fig. 6. Impact of β (with $r = 10$) and r (with $\beta = 0.05$) on the three attack methods for heavy hitter identification.

5.4   Utility of Heavy Hitter Identification

Similarly, due to the space limitation, we only report experimental results on FIRE. The conclusions drawn from FIRE are also valid on other datasets. We

Table 1. Impact of β or r on FIRE for the utility of heavy hitter identification.

Dataset	FIRE									
Params	β					r				
	0.01	0.03	0.05	0.08	0.1	1	3	5	7	9
F1(raw)	1.0	1.0	1.0	1.0	1.0	1.0	1.0	1.0	1.0	1.0
F1(MGA)	**1.0**	0.15	0.0	0.0	0.0	0.9	0.7	0.5	0.3	0.1
Defense	0.9	**0.9**	1.0	1.0	1.0	1.0	1.0	1.0	0.94	1.0
F1(UGA)	**1.0**	0.45	0.05	0.0	0.0	0.9	0.7	0.5	0.3	0.2
Defense	0.83	**0.82**	0.9	0.98	1.0	1.0	1.0	0.92	0.9	0.9
F1(BGA)	**1.0**	0.7	0.22	0.0	0.0	0.9	0.7	0.5	0.3	0.2
Defense	0.9	**0.8**	0.8	0.88	0.8	1.0	1.0	0.92	0.94	0.84
NCR(raw)	1.0	1.0	1.0	1.0	1.0	1.0	1.0	1.0	1.0	1.0
NCR(MGA)	**1.0**	0.182	0.0	0.0	0.0	0.982	0.891	0.709	0.409	0.164
Defense	0.936	**0.9**	1.0	1.0	1.0	1.0	1.0	1.0	0.907	1.0
NCR(UGA)	**1.0**	0.673	0.091	0.0	0.0	0.982	0.891	0.709	0.491	0.345
Defense	0.882	**0.827**	0.836	0.997	1.0	1.0	1.0	0.869	0.836	0.836
NCR(BGA)	**1.0**	0.891	0.365	0.0	0.0	0.982	0.891	0.727	0.491	0.345
Defense	0.982	**0.909**	0.822	0.983	0.945	1.0	1.0	0.927	0.985	0.8436

present the utility of heavy hitter identification under different β values or different r values in Table 1. F1(raw) and NCR(raw) represent the utility scores without attacks. We measure F1 score and NCR under the three attack methods before or after using the defense framework to defend against the corresponding attacks.

With β increasing, data utility under different attack methods declines at different rates. After applying the defense framework, the utility scores first decrease and then gradually increase or flatten out. The reason is that when target items suffer from more severe attacks, they will be easier to be distinguished from the entire item set. While there is slightly negative optimization when $\beta = 0.01$, the defense framework can still relatively well preserve the utility of heavy hitter identification. With r increasing, we can observe similar trends under attacks. This is because we use $\beta = 0.05$ in all attack methods, which can boost target items into top_k list no matter which attack method is used. After leveraging the defense framework, all utility scores are higher than before, which also proves that our defense framework can preserve the utility of heavy hitter identification.

6   Related Work

LDP has gained increasing attention in recent years. The mainstream research direction is to study how to improve data utility under LDP. As a result, a series of LDP protocols have been proposed [1,2,7,9,16]. Some typical examples include GRR [9], UE [16], OLH [16] and HM [2]. Cormode et al. [6] provide a comprehensive comparison study of different LDP protocols. Very recently, some pioneering studies have unveiled LDP's security vulnerability. Cheu et al. [5] report that LDP protocols are subject to manipulation attacks and a smaller privacy budget and a larger input domain will bring more severe damage. Cao et al. [4] formulate data poisoning attacks as an optimization problem and design three countermeasures. Wu et al. [18] further demonstrate a novel data poisoning attack on key-value data, which simultaneously attacks the frequency and mean values, and design two corresponding countermeasures with strong prior knowledge. Kato et al. [10] propose a novel verifiable LDP protocol based on multi-party computation techniques, which can verify the completeness of executing an agreed randomization mechanism for every data provider. Different from the existing works, we focus on studying the amplification effects of LDP protocols in the presence of Byzantine users and introduce new countermeasures without requiring any prior knowledge.

7   Conclusion

Recent studies have shown that existing LDP protocols are, unfortunately, vulnerable to manipulation attacks and poisoning attacks. In this paper, we studied LDP's inherent conflict between privacy and security, and introduced a novel type of BLAGA attacks that leverages the randomness of an LDP protocol to amplify attack effects in the presence of Byzantine users. The well-known MGA attack is a special case of BLAGA. Consequently, we proposed a novel defense framework that makes use of a data-driven approach to learn the target items using parameter expectation maximization. Unlike existing solutions, our defense framework does not require any prior knowledge. Extensive experimental results on three datasets show that our defense framework can provide strong security guarantee while maintaining high data utility.

Acknowledgments. This work was supported by the National Key R&D Program of China under Grant No. 2020YFB1710200, the National Natural Science Foundation of China (Grant No. 62072136, 62072390, 62102334 and 92270123), and the Research Grants Council, Hong Kong SAR, China (Grant No. 15222118, 15218919, 15203120, 15226221, 15225921, 15209922 and C2004-21GF).

References

1. Acharya, J., Sun, Z., Zhang, H.: Hadamard response: estimating distributions privately, efficiently, and with little communication. In: Proceedings of the 22nd International Conference on Artificial Intelligence and Statistics (2019)
2. Bassily, R., Nissim, K., Stemmer, U., Guha Thakurta, A.: Practical locally private heavy hitters. In: Advances in Neural Information Processing Systems, vol. 30 (2017)
3. Bassily, R., Smith, A.: Local, private, efficient protocols for succinct histograms. In: Proceedings of the 47th Annual ACM Symposium on Theory of Computing (2015)
4. Cao, X., Jia, J., Gong, N.Z.: Data poisoning attacks to local differential privacy protocols. In: Proceedings of the 30th USENIX Security Symposium (2021)
5. Cheu, A., Smith, A., Ullman, J.: Manipulation attacks in local differential privacy. In: Proceedings of the 42nd IEEE Symposium on Security and Privacy (2021)
6. Cormode, G., Maddock, S., Maple, C.: Frequency estimation under local differential privacy. Proc. VLDB Endow. **14**(11), 2046–2058 (2021)
7. Erlingsson, Ú., Pihur, V., Korolova, A.: RAPPOR: randomized aggregatable privacy-preserving ordinal response. In: Proceedings of the 21st ACM Conference on Computer and Communications Security (2014)
8. Gunes, I., Kaleli, C., Bilge, A., Polat, H.: Shilling attacks against recommender systems: a comprehensive survey. Artif. Intell. Rev. **42**(4), 767–799 (2014)
9. Kairouz, P., Oh, S., Viswanath, P.: Extremal mechanisms for local differential privacy. In: Advances in Neural Information Processing Systems, vol. 27 (2014)
10. Kato, F., Cao, Y., Yoshikawa, M.: Preventing manipulation attack in local differential privacy using verifiable randomization mechanism. In: Barker, K., Ghazinour, K. (eds.) DBSec 2021. LNCS, vol. 12840, pp. 43–60. Springer, Cham (2021). https://doi.org/10.1007/978-3-030-81242-3_3
11. Li, X., Gong, N.Z., Li, N., Sun, W., Li, H.: Fine-grained poisoning attacks to local differential privacy protocols for mean and variance estimation. arXiv preprint arXiv:2205.11782 (2022)
12. Moon, T.K.: The expectation-maximization algorithm. IEEE Signal Process. Mag. **13**(6), 47–60 (1996)
13. Prakash, S., Avestimehr, A.S.: Mitigating byzantine attacks in federated learning. arXiv preprint arXiv:2010.07541 (2020)

14. Tang, W., Tang, F.: The Poisson binomial distribution - old & new. Stat. Sci. **1**(1), 1–12 (2022)
15. ADP Team: Learning with privacy at scale. Apple Mach. **J1**(8), 1–25 (2017)
16. Wang, T., Blocki, J., Li, N., Jha, S.: Locally differentially private protocols for frequency estimation. In: Proceedings of the 26th USENIX Security Symposium (2017)
17. Wang, T., Li, N., Jha, S.: Locally differentially private heavy hitter identification. IEEE Trans. Dependable Secure Comput. **18**(2), 982–993 (2019)
18. Wu, Y., Cao, X., Jia, J., Gong, N.Z.: Poisoning attacks to local differential privacy protocols for key-value data. In: Proceedings of the 31st USENIX Security Symposium (2022)
19. Yang, J., Cheng, X., Su, S., Chen, R., Ren, Q., Liu, Y.: Collecting preference rankings under local differential privacy. In: Proceedings of the 35th IEEE International Conference on Data Engineering (2019)
20. Ye, Q., Hu, H., Meng, X., Zheng, H.: PrivKV: key-value data collection with local differential privacy. In: Proceedings of the 40th IEEE Symposium on Security and Privacy (2019)

Authenticated Ranked Keyword Search over Encrypted Data with Strong Privacy Guarantee

Ningning Cui[1(✉)], Zheli Deng[1], Man Li[2], Yuliang Ma[3], Jie Cui[1], and Hong Zhong[1]

[1] Anhui University, Hefei 230601, China
{20210,cuijie,zhongh}@ahu.edu.cn, e21301268@stu.ahu.edu.cn
[2] Deakin University, Geelong, VIC 3220, Australia
liama@deakin.edu.au
[3] Northeastern University, Shenyang 110819, China
mayuliang@mail.neu.edu.cn

Abstract. In the past decades, with the development of cloud computing, ranked keyword search, which devotes to find the most relevant results, has been extensively studied in outsourcing domain. However, due to a mass of sensitive information containing in the outsourced data, the issue of privacy has become the main brunt. Existing works primarily resort to searchable encryption to protect the privacy but do not consider the access pattern and search pattern, which can be used to infer the privacy information. Moreover, since the cloud server may be malicious, result integrity also needs to be considered. Therefore, in this paper, we study the problem of secure and authenticated ranked keyword search, called SARKS. Specifically, we first propose a framework that integrates d-differential privacy, erasure coding, and oblivious traverse to achieve access pattern and search pattern protection and meanwhile propose a scheme based on merkle hash tree to realize the correctness and completeness of the query results. To accelerate the performance, we further propose an improved scheme by adopting clustering method. Finally, the formal security analysis is conducted and the empirical evaluation over the real-world dataset has demonstrated the feasibility and practicability of our proposed schemes.

Keywords: Ranked keyword search · Privacy preserving · Result verification

1 Introduction

With the fast development of cloud computing, more and more cloud services are applied to outsourcing scenarios duo to its capacity of storage and computation, such as *Amazon* and *Google* cloud platform. Attracted by these appealing advantages, individuals and enterprises are inspired to outsource their data to the cloud. However, since massive sensitive information (e.g., electronic medical records) is involved in the outsourced data, data security has become the sharp brunt [3,4]. In addition, the cloud server as a third-party delegate may be malicious and return incorrect or forged results to the query users who are unaware of these undesirable behaviors. As such, it is imperative to safeguard data privacy while guaranteeing the integrity of query results.

© The Author(s), under exclusive license to Springer Nature Switzerland AG 2023
X. Wang et al. (Eds.): DASFAA 2023, LNCS 13943, pp. 644–660, 2023.
https://doi.org/10.1007/978-3-031-30637-2_43

Table 1. Summary of Existing Works

Method	Data privacy	Query privacy	Access pattern	Search pattern	Correctness	Completeness
Li [9]	√	√	×	×	×	×
Liu [10]	√	√	×	×	×	×
Dai [5]	√	√	×	×	×	×
Wan [13]	√	√	×	×	√	×
Li [8]	√	√	×	×	√	×
Liu [11]	√	√	×	×	√	×
Our scheme	√	√	√	√	√	√

'√' means it satisfies the condition;   '×' means it fails to satisfy the condition.

Motivated by these, in this paper, we study the problem of secure and authenticated ranked keyword search (**SARKS**), which is a ubiquitous application in information retrieval. Our designed scheme has three salient features: (1) Privacy. The scheme should ensure the data privacy and query privacy, especially access pattern and search pattern. (2) Verifiability. The scheme should ensure the correctness and completeness of the query results. (3) Efficiency. The scheme should achieve as efficient as sub-linear search time complexity.

Recently, there exist lots of researchers put their efforts to address security concerns. On one hand, to prevent potential data leakage, some inspiring Searchable Symmetric Encryption (SSE) based designs for ranked keyword search schemes in [1,5,6,9,10, 12,16] are used to data utilization and confidentiality. However, none of these schemes provide the access pattern and search pattern guarantee, which has been demonstrated that with some prior knowledge about the dataset, the attackers can do the inference attack and even recover the content of the query with the access pattern information [2]. On the other hand, to guarantee the result integrity, [7] has proposed a secure and verifiable multiple keywords search scheme, but it cannot support ranked search. [8,11, 13] have studied the verifiable ranked keyword search over encrypted data. But all of them only support correctness verification and do not support completeness verification. For ease of exhibition, the summary of the main works is exhibited in Table 1. Intuitively, there is no existing work to ensure the complete indispensable conditions. Therefore, it is imperative to design an efficient scheme for dealing with the issue of **SARKS**.

In this paper, we first formally formulate the problem of secure and authenticated ranked keyword search (**SARKS**). To cope with this issue, there still exist two key challenging problems: (1) *How to achieve efficient ranked keyword search while guaranteeing the access pattern and search pattern?* To do this, we first design a strategy that integrates d-differential privacy, erasure coding, and oblivious traverse to achieve access pattern and search pattern protection. To achieve sub-linear retrieval, we combine this strategy with a tree-based index while leveraging improved ASPE encryption to compute the ranked results in the form of ciphertext. Moreover, to further accelerate the performance, we propose a hierarchical clustering scheme tailored for pruning efficiently. (2) *How to design an efficient verification strategy along with the query processing?* To achieve this, based on merkle hash tree (MHT), we propose a verification scheme that can be applied to the above tree-based secure index while guaranteeing the correctness and completeness of the query results.

To sum up, our main contributions are listed as follows.

- To the best of our knowledge, we are the first one to define and study the problem of secure and authenticated ranked keyword search with access pattern and search pattern guarantee.
- We propose a novel framework that not only achieves access pattern and search pattern protection and meanwhile realizes the result verification of the correctness and completeness.
- We propose an optimized scheme based on a hierarchical clustering method to improve the performance of the query and verification.
- We theoretically analyze the security and conduct extensive empirical evaluations over real-world dataset to demonstrate the efficiency and feasibility of our schemes.

2   Related Work

There exist a lot of works to deal with privacy-preserving keywords search. [2,7,14] proposed an inverted index-based scheme. But these schemes can only deal with single keyword search and cannot support the ranked functionality and verification for multiple keywords search. Whereupon, [1,5,6,9,10,12,16] studied the multi-keyword ranked search. Specifically, Cao et al. [1] first proposed privacy-preserving multi-keyword ranked search scheme based on ASPE [15] and vector space model. After this, [12,16] proposed ranked multi-keyword search schemes with ASPE. However, ASPE is proven to be insecure that cannot guarantee known-plaintext attack. On the basis of their work, to enhance security, Li et al. [9] combined TF-IDF rule with improved ASPE. However, the search complexity is linear and cannot preserve the access pattern. In addition, Ding et al. [6] proposed a random traversal algorithm, which enables to generate different access paths over the index with access pattern guarantee but reduces the precision of the query results. More importantly, all of the above schemes cannot guarantee the correctness and completeness of the returned results.

To prevent the malicious cloud server, Wan et al. [13] designed a verifiable and privacy-preserving multi-keyword scheme based on homomorphic encryption to verify the correctness of the results but the efficiency is low. To support effective authentication, Ge et al. [7] proposed a new accumulative authentication tag (AAT) based on symmetric key encryption, which generates an identity authentication tag for each keyword and facilitates the verification of query results. In addition, Li et al. [8] and Liu et al. [11] proposed a verifiable multi-keyword search scheme based on RSA accumulator but incurs a huge computational overhead and [8] cannot support ranked functionality. Unfortunately, most of them just support the correctness of the results and still suffer from the threat of completeness.

3   Problem Formulation and Preliminaries

In this section, we first introduce our system model, then formally formulate the problem definition and security model, and finally present some preliminaries.

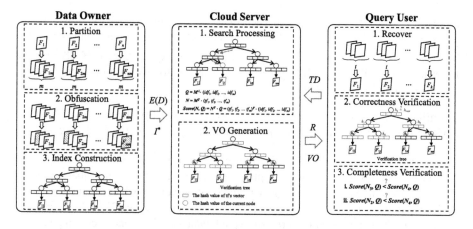

Fig. 1. The architecture of system model.

3.1 System Model

As shown in Fig. 1, our system model consists of three entities: Data Owner (DO), Cloud Server (CS), and Query User (QU). Specifically, the responsibility of each entity is presented as follows:

- **Data Owner.** DO is first responsible for generating and distributing the secret key. Then, DO builds the secure index $\mathcal{I}^*$, encrypts the dataset $E(\mathcal{D})$ and sends them to CS.
- **Cloud Server.** CS is responsible for performing the query processing after receiving the trapdoor $\mathcal{TD}$ from QU. Meanwhile, CS generates the verification object ($\mathcal{VO}$) during the query processing. After that, CS returns the result set $\mathcal{R}$ and verification object ($\mathcal{VO}$) to QU.
- **Query User.** QU requests a query, encrypts it to form the trapdoor and sends it to CS. Upon receiving the result set $\mathcal{R}$ and verification object ($\mathcal{VO}$) from CS, QU verifies the result integrity.

3.2 Problem Definition

In this paper, we study the problem of secure and authenticated ranked keyword search, called **SARKS**. For a given database $\mathcal{D} = \{\mathcal{F}_1, \mathcal{F}_2, ..., \mathcal{F}_n\}$, each file is modeled as a two-tuple $\mathcal{F} = \langle id, \{w\}_{\mathcal{F}} \rangle$, where id represents the identifier of the file $\mathcal{F}$ and $\{w\}_{\mathcal{F}}$ represents a set of keywords contained in the file $\mathcal{F}$. The set of keywords of database $\mathcal{D}$ is denoted as $\mathcal{W} = \cup_{i=1}^{n}\{w\}_{\mathcal{F}_i}$. Here, to realize ranked keyword search, we adopt the vector space model (VSM) along with TF×IDF rule, which is commonly used by related works [9,12]. In this model, TF is the term frequency that represents the frequency of keyword w_j occurring in file $\mathcal{F}_i$ and can be calculated as $tf_{i,j} = \frac{1+ln|w_j^i|}{|\mathcal{F}_i|}$, where $|w_j^i|$ denotes the number of keywords w_j occurring in file $\mathcal{F}_i$ and $|\mathcal{F}_i|$ denotes the size of file $\mathcal{F}_i$. IDF is inverse document frequency and can be computed as $idf_j = ln(1 + \frac{n}{n_j})$, where n_j denotes the number of files containing the keyword w_j. In addition, each file $\mathcal{F}_i$ can be regarded as a vector, i.e.,

$\mathcal{F}_i = \{tf_{i,1}, tf_{i,2}, ..., tf_{i,|\mathcal{W}|}\}$, and each query $\mathcal{Q}$ can also be regarded as a vector, i.e., $\mathcal{Q} = \{idf_1, idf_2, ..., idf_{|\mathcal{W}|}\}$. Thus, the relevance evaluation function is defined as $score(\mathcal{F}_i, \mathcal{Q}) = \mathcal{F}_i \cdot \mathcal{Q} = \sum_{j=1}^{|\mathcal{W}|} tf_{i,j} \cdot idf_j$. Here, given a query $\mathcal{Q}$, our SARKS intends to retrieve top-k files with the highest scores. Specifically, we formulate our SARKS scheme in the following definition.

Definition 1 (SECURE AND AUTHENTICATED RANKED KEYWORD SEARCH, SARKS). *A SARKS scheme Π is constituted of five polynomial time algorithms:*

- **Setup**$(1^\lambda) \rightarrow (sk_1, sk_2, SK, K)$. It accepts security parameter λ as input and returns secret keys sk_1 for encrypting index, sk_2 for encrypting query, SK for encrypting database, and K for generating cryptographic hash as outputs.
- **IndexGen**$(\mathcal{D}, sk_1, SK, K) \rightarrow (E(\mathcal{D}), \mathcal{I}^*)$. It accepts the database $\mathcal{D}$, secret keys sk_1, SK, K as inputs and returns encrypted database $E(\mathcal{D})$ and secure index $\mathcal{I}^*$ as outputs.
- **TrapGen**$(\mathcal{Q}, sk_2) \rightarrow \mathcal{TD}$. It accepts the query $\mathcal{Q}$ and secret key sk_2 as inputs and outputs the trapdoor $\mathcal{TD}$.
- **Search**$(\mathcal{TD}, \mathcal{I}^*, E(\mathcal{D})) \rightarrow (\mathcal{R}, \mathcal{VO})$. It accepts the trapdoor $\mathcal{TD}$, secure index $\mathcal{I}^*$, and encrypted database $E(\mathcal{D})$ as inputs and outputs result set $\mathcal{R}$ and verification object $\mathcal{VO}$.
- **Verify**$(\mathcal{R}, \mathcal{VO}, h_{root}, \mathcal{TD}) \rightarrow (true/false)$. It accepts result set $\mathcal{R}$, verification object $\mathcal{VO}$, hash value of root node in $\mathcal{I}^*$, and trapdoor $\mathcal{TD}$ as inputs and returns true or false.

3.3 Security Model

In our security model, we mainly focus on two aspects of threats from CS. On one hand, CS is curious and untrusted, who intends to capture private information during the query processing; on the other hand, CS is malicious, who attempts to forge the query results or returns incorrect data.

For the former case, we adopt the widely used simulation-based model [17], which implements a real game and an ideal game between probabilistic polynomial-time (PPT) adversary $\mathcal{A}$, denoted as $\mathsf{Real}_\mathcal{A}^\Pi$ and $\mathsf{Ideal}_\mathcal{A}^\Pi$, respectively. The scheme SARKS Π is secure if the two games are indistinguishable with non-negligible probability, which is defined as follows.

Definition 2 (SECURITY OF SARKS). *A SARKS scheme Π is secure under simulation-based model, if for any probabilistic polynomial time adversary $\mathcal{A}$, there exists a simulator $\mathcal{S}$ such that:*

$$|Pr[\mathsf{Real}_\mathcal{A}^\Pi(1^\lambda) = 1] - Pr[\mathsf{Ideal}_{\mathcal{A},\mathcal{S}}^\Pi(1^\lambda) = 1]| \leq negl(\lambda)$$

In addition, the detailed privacy requirements are as follows.

- **Data Privacy.** The plaintext information from the database and index should be protected from the cloud server.
- **Query Privacy.** The plaintext information from user queries should be protected from the cloud server.

- **Access Pattern.** The indexes which encrypted files match the query should be protected from the cloud server.
- **Search Pattern.** The information whether the current query has been searched before should be protected from the cloud server.

For the latter case, to resist malicious adversary, two requirements need to be satisfied to guarantee result integrity.

- **Correctness.** All returned results or files are not forged and originate from the original database $\mathcal{D}$.
- **Completeness.** All returned results or files are real top-k results with the most highest relevance scores.

Note that, in our setting, DO and QU are both trusted and QU is restricted from compromising with CS. Moreover, there may be other passive or active attacks such as denial of service. Since our work mainly focuses on privacy and efficiency in ranked keyword search, those attacks are beyond the scope of this paper.

3.4 Preliminaries

Erasure Coding. Erasure coding (EC) is one of coding fault tolerance technologies, which is first used to data recovery. The basic idea of EC is to obtain $m - t$ block check elements from t block original data elements through certain coding calculation. For the m block elements, when any $m - t$ block elements are missing, the original data can be recovered from the remaining t block elements.

In this paper, erasure coding is applied to partition a file into t blocks equally, and then $m - t$ check blocks with the same size are generated such that the original file can be reconstructed from any t of these m blocks.

Improved ASPE Scheme. The original ASPE technique was proposed to securely calculate the scalar product between two vectors, i.e., (i) the vector V_N in the database; and (ii) the query vector V_Q. A typical feature of ASPE is that its public key and secret key are a pair of an invertible matrix $(\mathcal{M}^T, \mathcal{M}^{-1})$, which are applied to encrypt the V_N and V_Q, respectively. In addition, to protect more privacy, a random asymmetric splitting process and adding artificial dimensions are adopted further, that is, the invertible matrix is further divided into two parts, respectively: $(\mathcal{M}_1^T, \mathcal{M}_2^T)$ and $(\mathcal{M}_1^{-1}, \mathcal{M}_2^{-1})$. And correspondingly, the two vectors V_N and V_Q are also divided into two parts. However, [9] and [12] point out that the original scheme of ASPE fails to resist the known-plaintext attack (KPA) due to its lack of randomness in the inner product calculation.

In the following, to resist the KPA, we adopt the improved ASPE scheme [17], which complements multiplicative and additive noises to the inner product.

Cryptographic Hash Function. A cryptographic hash function $h(\cdot)$ takes an arbitrary-length string as its input and outputs a fixed-length bit string, i.e., $h : \{0,1\}^i \rightarrow \{0,1\}^j$. Here, we adopt the one-way hash function $h(\cdot)$ with secret key K to generate hash value of the nodes in secure index. For ease of presentation, hereafter, we use $h(\cdot)$ represents $h(\cdot, K)$. More importantly, cryptographic hash function is collision-resistant, that is, it is difficult to find two different messages, m_1 and m_2, such that $h(m_1) = h(m_2)$.

4   Our Proposed Scheme

4.1   Scheme Overview

The goal of our SARKS is to find the top-k files with the highest scores while supporting privacy-preserving and result verification. Specifically, as shown in Fig. 1, to safeguard the access pattern, the key idea is to obfuscate the real pattern with false positives and false negatives. To this end, we first leverage erasure coding to partition each file into t blocks with $m - t$ check blocks. For each block, we utilize an obfuscation mechanism to obfuscate the keywords in blocks such that the access pattern obfuscation scheme achieves d-differential privacy. After these, to support efficient search, we leverage a tree-based index structure, in which each node is allocated with a TF vector encrypted by the improved ASPE scheme (i.e., $\mathcal{M}_1^T$, $\mathcal{M}_2^T$). Corresponding, the query is modeled as an IDF vector encrypted by the improved ASPE scheme (i.e., $\mathcal{M}_1^{-1}$, $\mathcal{M}_2^{-1}$), which makes the calculation of the relevance scores secure. Meanwhile, during the query processing, we use a verification mechanism to generate the verification object and verify the results following the way of the merkle hash tree.

4.2   Pattern Obfuscation

Access Pattern. As discussed above, the essence of the obfuscation is to introduce false positives and false negatives. Given a database with n files, we can use a $m \times n$-bits vector (i.e., after erasure encoding) to represent the access pattern. For example, when $m = 2$ and $n = 2$, the access pattern θ can be modeled as '0101', where '0' denotes the file block is not accessed and '1' denotes the file block is accessed and returned. Here, to obfuscate the access pattern, we introduce a flipping operation following [2], that is, flipping some bits of θ from 1 to 0 (i.e., false negatives) and flipping some bits of θ from 0 to 1 (i.e., false positives). For example, given an access pattern $\theta =$ '0101', the obfuscated access pattern ω can be '1101'. To better present the difference between two access patterns, Hamming distance is a promising option, which represents the number of different characters in the identical position for two strings with the same length. For example, given two access patterns $\theta =$ '0101' and $\theta' =$ '0110', the hamming distance $d_h(\theta, \theta')$ is 2. The goal of our access pattern obfuscation mechanism is to realize the indistinguishability of θ and θ' from ω. Therefore, based on d-differential privacy [2], we have the following definition.

Definition 3. *An access-pattern obfuscation mechanism $\mathcal{AP} : \Theta \rightarrow \Omega$ achieves d-differential privacy, if for any $\theta, \theta' \in \Theta$ and $\mathcal{O} \subseteq \Omega$, such that:*

$$Pr[\mathcal{AP}(\theta) \in \mathcal{O}] \leq e^{\epsilon d_h(\theta, \theta')} \cdot Pr[\mathcal{AP}(\theta') \in \mathcal{O}]$$

where ϵ represents the privacy budget that is a positive number. Intuitively, the smaller ϵ is, the more privacy is guaranteed but the overhead is larger. In the following, we customize an access pattern obfuscation mechanism $\mathcal{AP}$ for ranked keyword search with d-differential privacy. To be specific, during the search processing, when occurring bit flipping, it means the ranking score has been changed. Let p denotes the probability flipping bit 1 to 1 and q denotes the probability flipping bit 0 to 1. Whereupon, we can conclude that,

$$\begin{cases} Pr[\omega_i = 1|\theta_i = 1] = p \\ Pr[\omega_i = 0|\theta_i = 1] = 1 - p \end{cases} \Leftrightarrow \begin{cases} Pr[N(w) \neq 0|N(w) \neq 0] = p \\ Pr[T_{min}|N(w) \neq 0] = 1 - p \end{cases}$$

$$\begin{cases} Pr[\omega_i = 1|\theta_i = 0] = q \\ Pr[\omega_i = 0|\theta_i = 0] = 1 - q \end{cases} \Leftrightarrow \begin{cases} Pr[T_{max}|N(w) = 0] = q \\ Pr[N(w) = 0|N(w) = 0] = 1 - q \end{cases}$$

where $N(w)$ represents the number of keyword w in a file block, T_{min} is a threshold that makes the ranking score of a file block be the minimum, and T_{max} is a threshold that makes the ranking score of a file block be the maximum. In addition, based on the observation, we have $Pr[\omega_i = 1|\theta_i = 0] < Pr[\omega_i = 1|\theta_i = 1]$ and $Pr[\omega_i = 1|\theta_i = 0] < Pr[\omega_i = 0|\theta_i = 1]$, it is equivalent to $q < p$ and $q < 1 - p$. Therefore, we have the following theorem.

Theorem 1. *Our access pattern obfuscation mechanism $\mathcal{AP}$ satisfies ϵd_h-differential privacy when $\epsilon = ln\frac{p}{q}$.*

Proof. It is equivalent to demonstrate $Pr[\mathcal{AP}(\theta) = \omega] \leq e^{\epsilon d_h(\theta,\theta')} \cdot Pr[\mathcal{AP}(\theta') = \omega]$ for any $\theta, \theta' \in \Theta$ and $\omega \in \Omega$. Hence, we can get that,

$$\frac{Pr[\mathcal{AP}(\theta) = \omega]}{Pr[\mathcal{AP}(\theta') = \omega]} = \frac{\Pi_{i=1}^{mn} Pr[w_i|\theta_i]}{\Pi_{i=1}^{mn} Pr[w_i|\theta'_i]} = \Pi_{\theta \neq \theta'} \frac{Pr[w_i|\theta_i]}{Pr[w_i|\theta'_i]}$$

$$\leq \Pi_{\theta \neq \theta'} max\{\frac{p}{q}, \frac{q}{p}, \frac{1-p}{1-q}, \frac{1-q}{1-p}\}$$

$$= (\frac{p}{q})^{d_h(\theta,\theta')} = e^{ln\frac{p}{q} \cdot d_h(\theta,\theta')}$$

Therefore, the Theorem 1 has been proven.   □

Search Pattern. To preserve the search pattern, the key step is to obfuscate the search path such that the adversary cannot recognize two queries by the traverse path. For example, in a tree-based index, for the same query, the traverse path is different. To this end, integrated with the erasure coding, we propose a random traverse scheme using one-hot vector. To be specific, given a file $\mathcal{F}_i$, after erasure coding, the file is partitioned into m file blocks. Here, a file is initialized as a m-bits vector and each file block is regarded as one bit in a one-hot vector. If the bit is '1', it means the file block is selected; otherwise, the file block is unselected. For example, given a file $\mathcal{F}_i$ with $m = 4$ and $t = 2$, its file blocks are $\mathcal{F}_{i1}, \mathcal{F}_{i2}, \mathcal{F}_{i3}, \mathcal{F}_{i4}$ and the corresponding one-hot vectors are '1000', '0100', '0010', '0001'. Moreover, to facilitate the search, the returned results should contain enough file blocks to recover the original file. Therefore, for the user's query, it is also viewed as a m-bits vector. For the 't' selected file blocks, the corresponding bits are set to '0' and for the '$m - t$' unselected file blocks, the corresponding bits are set to a random negative value that makes the file block unselected. In this way, for each query, there exist at least t file blocks returned to recover the original file. Meanwhile, for the same query, the traverse path is diverse, which is incurred by the user's random selection to safeguard the search pattern.

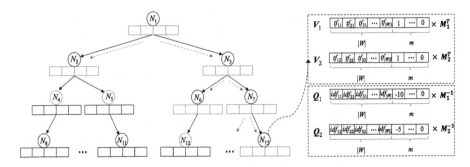

Fig. 2. Example of the secure index (red color represents the traverse path, green circle represents the node hash contained in $\mathcal{VO}$, and green rectangle represents the bit vector contained in $\mathcal{VO}$). (Color figure online)

4.3 Index Construction

Intuitively, the search can be implemented over the encrypted file blocks one by one, but the search complexity is $O(n)$ and is time-consuming. To accelerate the performance, we propose a tree-based secure index, which supports privacy-preserving and result verification simultaneously.

As shown in Fig. 2, the structure of the secure index is a binary tree. The main difference is that: (1) for the leaf node LN, it can be represented as a two-tuple $<V_{LN}, h_{LN}>$, where V_{LN} is a bit vector consisting of two parts: i) $|\mathcal{W}|$-bits vector indicating the tf_i value of each keyword w_j in the file block; ii) m-bits vector indicating the position of the file block in the one-hot vector, and h_{LN} is the hash value of the leaf node and can be computed as $h_{LN} = h(h(\mathcal{F}_{ij})|h(V_{LN}))$, where '|' is a string concatenator. (2) for the non-leaf node N, it can be represented as a quadruple $<V_N, h_N, pt_l, pt_r>$, where V_N is also a bit vector consisting of two parts: i) $|\mathcal{W}|$-bits vector indicating the maximum of the tf_i value of each keyword w_j in the current subtree; ii) m-bits vector, in which each bit is calculated by the logic *and* operation of corresponding bits from its children's node. h_N is the hash value of the non-leaf node and can be computed as $h_N = h(h(N_l)|h(N_r)|h(V_N))$, where $h(N_l)$ and $h(N_r)$ are the hash values of its left child and right node, respectively. In addition, pt_l and pt_r are the pointer to its left child and right node, respectively.

After the index construction, it is encrypted by the improved ASPE. Specifically, DO first partitions the bit vector V into two parts, i.e., V_1 and V_2, based the partition vector $v \in \{0,1\}^{|V|}$ in the improved ASPE. If the $v[i] = 0$, the partitioned vectors are identical with the original vector, i.e., $V_1[i] = V_2[i] = V[i]$; if the $v[i] = 1$, the sum of the partitioned vectors is identical with the original vector, i.e., $V_1[i] + V_2[i] = V[i]$. After this, each bit vector is encrypted by two invertible matrixes in the improved ASPE, i.e., $\mathcal{M}_1^T, \mathcal{M}_2^T, \widetilde{V} = \{\mathcal{M}_1^T V_1, \mathcal{M}_2^T V_2\}$. Note that, correspondingly, the bit vector V in the hash value is also in the form of two parts.

4.4 Trapdoor Generation

For a user query $\mathcal{Q} = \{\mathcal{W}_Q, k\}$, where $\mathcal{W}_Q = \{w_1, w_2, ..., w_q\}$, before submitting it to CS, QU first generates the trapdoor based on the improved ASPE. Specifically, the

---

Algorithm 1: QUERY PROCESSING AND VERIFICATION GENERATION

Input: Secure index $\mathcal{I}^*$, Trapdoor $\mathcal{TD}$;
Output: Result set $\mathcal{R}$, Verification object $\mathcal{VO}$;
CS:

1 Initiate a priority queue q ;
2 q.enqueue($root$) ;
3 **while** *the first k entries of q are not leaf nodes* **do**
     Node $N \leftarrow$ dequeue the non-leaf node with the highest score ;
     q.enqueue(N's children) ;
     $\mathcal{VO} \leftarrow V_N$;

4 $\mathcal{R} \leftarrow$ the fist k entries of q ;
5 $\mathcal{VO} \leftarrow$ the bit vector V of the fist k entries ;
6 **while** *q is not empty* **do**
     Node $N \leftarrow q.dequeue()$;
     if *N is non-leaf node* **then**
          $\mathcal{VO} \leftarrow N$.children's hash values ;
          $\mathcal{VO} \leftarrow V_N$;
     else
          $\mathcal{VO} \leftarrow E(\mathcal{F})$;
          $\mathcal{VO} \leftarrow V_N$;

7 Return $\mathcal{R}$ and $\mathcal{VO}$;

---

query $\mathcal{Q}$ is also viewed as a bit vector $\mathcal{Q}$ consisting of two parts: i) $|\mathcal{W}|$-bits vector indicating the inverse document frequency idf_i value of each keyword w_j; ii) m-bits vector, in which each bit is determined by the user's random selection that $m - t$ of m bits are set to a random negative value that makes the file block unselected while t selected bits are set to 0. Based on the partitioned vector v, the query bit vector $\mathcal{Q}$ is also divided into two parts, $\mathcal{Q}_1$ and $\mathcal{Q}_2$. If the $v[i] = 0$, the sum of the partitioned vectors is identical with the original vector, i.e., $\mathcal{Q}_1[i] + \mathcal{Q}_2[i] = \mathcal{Q}[i]$; If the $v[i] = 1$, the partitioned vectors are identical with the original vector, i.e., $\mathcal{Q}_1[i] = \mathcal{Q}_2[i] = \mathcal{Q}[i]$. After this, the query bit vector $\mathcal{Q}$ is encrypted by two invertible matrixes in the improved ASPE, i.e., $\mathcal{M}_1^{-1}, \mathcal{M}_2^{-1}, \widetilde{\mathcal{Q}} = \{\mathcal{M}_1^{-1}\mathcal{Q}_1, \mathcal{M}_2^{-1}\mathcal{Q}_2\}$. At last, QU obtains the trapdoor $\mathcal{TD} = \{\widetilde{\mathcal{Q}}, k\}$.

4.5 Query Processing and Verification Generation

On the cloud side, upon receiving the trapdoor $\mathcal{TD}$ from QU, it traverses the secure index $\mathcal{I}^*$ using trapdoor $\mathcal{TD}$ in a depth-first way. As shown in Algorithm 1, the verification generation is along with the query processing. Specifically, the algorithm initials a priority queue q to record all accessed nodes to generate the verification information while indicating the query results (Line 1). Then, the root node is enqueued into q (Line 2). Next, the algorithm checks whether the first k entries of q are leaf nodes or not. If they are, it means the first k entries are the final results with the highest relevance scores. Note that the relevance score can be computed $score(\widetilde{V}, \widetilde{\mathcal{Q}}) = \widetilde{V} \cdot \widetilde{\mathcal{Q}} = \mathcal{M}_1^T V_1 \cdot \mathcal{M}_1^{-1} \mathcal{Q}_1 + \mathcal{M}_2^T V_2 \cdot \mathcal{M}_2^{-1} \mathcal{Q}_2 = V_1^T \cdot \mathcal{Q}_1 + V_2^T \cdot \mathcal{Q}_2$. Otherwise, the algorithm continues to traverses the index, dequeues the non-leaf node with the highest score, adds its children into q for the further check and adds its bit vector v_N into verification object $\mathcal{VO}$ (Line 3). After this, the algorithm inserts the results into $\mathcal{R}$ and inserts their bit vectors into $\mathcal{VO}$ (Lines 4–5). In addition, we also need to insert the rest of nodes of q into verification object. If the node is non-leaf node, the algorithm adds the hash values of its children's nodes and its bit vector into verification $\mathcal{VO}$. Otherwise, the algorithm adds its encrypted file block and its bit vector into verification $\mathcal{VO}$ (Line 6). Finally, the algorithm will not terminate until q is empty.

4.6 Verification Processing

Upon receiving the result set $\mathcal{R}$ and verification object $\mathcal{VO}$, QU needs to verify the correctness and completeness of the query results due to the malicious CS.

Correctness. Recall that correctness refers to all returned file blocks that are not forged and originate from the original database. To this end, we utilize the one-way crypto-graphic hash similar to the merkle hash tree (MHT) to achieve this goal and the key step is to recompute the root hash of our index. Specifically, in $\mathcal{VO}$, for the accessed non-leaf nodes N, if its children are not reached, $\mathcal{VO}$ consists of its bit vector V_N and its left and right children's hash values h_l, h_r; otherwise, $\mathcal{VO}$ consists of its bit vector V_N. For the accessed leaf node LN, if it is not the result, $\mathcal{VO}$ consists of its bit vector V_{LN} and its encrypted file block $E(\mathcal{F})$; otherwise, $\mathcal{VO}$ consists of its bit vector V_{LN}. Based on $\mathcal{VO}$ and $\mathcal{R}$, QU first decrypts the encrypted file blocks, computes the file hash value $h(\mathcal{F})$, and further computes the node hash h_{LN}. In a bottom-up way, QU can compute the root hash $\widehat{h_{root}}$. Finally, if the calculated hash $\widehat{h_{root}}$ matches the h_{root} from DO, it means the correctness is verified; vice versa.

Completeness. Recall that completeness refers to all returned files being real top-k results with the highest relevance scores. Since all bit vectors of accessed nodes are contained in $\mathcal{VO}$, QU can recompute the scores and compare them with the k-th score. If all recomputed scores are lower than the k-th score, we say that the completeness is verified.

4.7 Clustering Based Optimization Scheme

To further accelerate the performance, we propose an optimized scheme based on hier-archical k-means clustering method. The main idea is that we first regard all file blocks as a cluster and based on clustering rule, we partition the file blocks into two clusters as its left child node and right child node. Following this way, for each level, we parti-tion the cluster into two clusters until each cluster contains one file block. The detailed process is as follows.

Step 1. First, DO chooses two initial cluster vectors $\{c_1, c_2\}$, then allocates each bit vector V to a cluster c_i such that the predefined distance $d(c_i, V)$ is minimum. Here, we adopt the cosine distance as the clustering rule and can be computed as $d(c_i, V) =$
$$\frac{\sum_{j=1}^{|V|} c_i[j] V[j]}{\sqrt{\sum_{j=1}^{|V|} c_i[j]^2} \sqrt{\sum_{j=1}^{|V|} V[j]^2}}.$$

Step 2. Then, for the next level, the clusters $\{c_1, c_2\}$ is further divided into two clusters, i.e., $\{c_1, c_2, c_3, c_4\}$. Iteratively, the partition will not stop until each cluster contains one file block. Note that, during the clustering process, the number of file blocks in each cluster is unequal, which will incur the unbalance of the binary tree. Therefore, it is necessary to redistribute the file blocks. Here, assuming $|c_1| > |c_2|$, we sort the file blocks as ascending order based on the cosine distance and allocate the last $\frac{|c_1| - |c_2|}{2}$ to c_2.

As such, combined with clustering scheme, compared with the basic method, the file blocks with similar relevance score will be clustered together, which enables an efficient pruning.

5   Security Analysis

In this section, to demonstrate the security, we adopt the widely used simulation-based model in the real and ideal experiments. Here, we first define the obfuscated leakage function $\mathcal{L} = \{\mathcal{L}_1, \mathcal{L}_2\}$ as follows.

- $\mathcal{L}_1 = \{n, |V|, \Delta(\mathcal{I}^*)\}$. n is the number of file blocks, $|V|$ is the size of the bit vector, $\Delta(\mathcal{I}^*)$ is the structure of index $\mathcal{I}^*$.
- $\mathcal{L}_2 = \{ap(\mathcal{TD}), sp(\mathcal{TD})\}$. $ap(\mathcal{TD})$ is the access pattern with d-differential obfuscation and $sp(\mathcal{TD})$ is the search pattern with $\frac{t}{m}$ obfuscation.

We first define the real view $\mathsf{Real}_{\Pi}^{\mathcal{A}}(1^\lambda)$ and simulated view $\mathsf{Sim}_{\Pi}^{\mathcal{A}}(1^\lambda)$.

$\mathsf{Real}_{\Pi}^{\mathcal{A}}(1^\lambda)$: Simulator $\mathcal{S}$ initializes the security parameter λ and invokes **Setup** function. Then, adversary $\mathcal{A}$ chooses a dataset $\mathcal{D}$ and sends it to simulator $\mathcal{S}$. After that, simulator $\mathcal{S}$ runs the **IndexGen** function to construct the index. In addition, adversary $\mathcal{A}$ randomly conducts a polynomial number of queries and obtains the trapdoor from simulator $\mathcal{S}$. Finally, the adversary $\mathcal{A}$ implements the experiment and returns a bit b.

$\mathsf{Sim}_{\Pi}^{\mathcal{A}}(1^\lambda)$: Adversary $\mathcal{A}$ outputs a dataset $\mathcal{D}$. Given $\mathcal{L}_1$, simulator $\mathcal{S}$ simulates the encrypted database and secure index and sends them to $\mathcal{A}$. Then, for each query, $\mathcal{S}$ simulates the trapdoor from $\mathcal{L}_2$. Finally, the adversary $\mathcal{A}$ implements the experiment and returns a bit b'.

Theorem 2. *Our SARKS is simulation-secure with obfuscated leakage $\mathcal{L}$ if the improved ASPE resists known-plaintext attack (KPA).*

Proof. The sketchy proof is illustrated as follows. Given $\mathcal{L}_1$, simulator $\mathcal{S}$ can simulate an indistinguishable index that has the same structure while bit vector and node hash are random strings owing the same size with the real one. Moreover, based on $\mathcal{L}_2$, $\mathcal{S}$ can simulate the trapdoor and ciphertexts correspondingly. Since the outputs are identical, based on the simulator $\mathcal{S}$, no probability polynomial-time (PPT) adversary $\mathcal{A}$ can distinguish simulated view $\mathsf{Sim}_{\Pi}^{\mathcal{A}}(1^\lambda)$ from real view $\mathsf{Real}_{\Pi}^{\mathcal{A}}(1^\lambda)$. Therefore, Theorem 2 has been proven.   □

Theorem 3. *Our proposed SARKS scheme satisfies verifiability.*

Proof. For the correctness verification, assuming that there exist a file block $\mathcal{F}_{ij}$ contained in the returned results is tampered by CS. When QU rebuilds the root hash, QU first computes the hash $h(\mathcal{F}_{ij})$ of the file block. Due to collision-resistance of cryptographic hash, the root hash h_{root} cannot be rebuilt based on the tampered $\mathcal{F}_{ij}$. For the completeness verification, assuming there exists a file block $\mathcal{F}_{ij}$ that is a result but is not

Table 2. Parameter Settings (Bold values are default values)

Name	Setting
# of files n	**2,000** 4,000 6,000 8,000 10,000
# of query keywords qk	**2** 4 6 8 10
# of keyword dictionary $|\mathcal{W}|$	**1,000** 1,200 1,400 1,600 1,800 2,000
query k	**5** 10 15 20 25
privacy budget ϵ	**28** 34 40
# of file blocks (m, t)	**(6, 2)** (8, 3) (10, 4)

returned. Due to the bit vector of its parent node verified by root hash, if its sibling node is a result, the bit vector of this node must be included in $\mathcal{VO}$. Then, QU can compute the score and check whether this node is a result or not. If its sibling node is not a result, the parent node's bit vector of this node must be included in $\mathcal{VO}$. QU can also compute the score and check whether the score of the parent node is larger than the k-th result. By this way, the missed result can be found. Therefore, Theorem 3 has been proven. □

6 Experimental Evaluation

6.1 Experimental Setup

Algorithms. To evaluate the performance of our proposed schemes, we compare with two state-of-the-art privacy-preserving ranked keyword search schemes, i.e., EDMRS [16] and MRSF [9]. In addition, we denote our basic solution as SARKS and denote our optimized solution as SARKS+.

Parameters. In the following, we evaluate the performance over varying parameters: (1) the number of files n from 2,000 to 10,000; (2) the number of query keywords qk from 2 to 10; (3) the size of the keyword dictionary $|\mathcal{W}|$ from 1,000 to 2,000; (4) the query k from 5 to 25; (5) the privacy ϵ budget; and (6) the number of file blocks (m, t). The detailed setting is shown in Table 2.

Setup. In experiments, our algorithms are executed in java language and are conducted on a Windows 10 professional machine with Intel(R) Core(TM) i7-11700K@2.50 GHz CPU and 48 GB RAM. Our evaluation is implemented on a real-world dataset: Enron Email Dataset[1], in which we randomly choose a certain number of documents.

6.2 Evaluation of Search Time

In this subsection, we evaluate the search time with varying the number of files n, the number of query keywords qk, and the size of keyword dictionary $|\mathcal{W}|$ and the results are shown in Fig. 3, Fig. 4, and Fig. 5, respectively. We can observe that with the number of files n, the number of query keywords qk, and the size of keyword dictionary $|\mathcal{W}|$ increasing, the search time is growing linearly and the running time of all schemes are similar but our schemes generate verification information during the query processing and SARKS+ has the best performance than others.

---

[1] https://www.cs.cmu.edu/enron/.

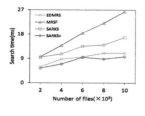

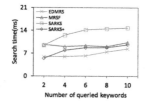

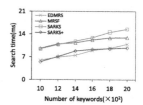

Fig. 3. n **vs.** Search time.   **Fig. 4.** qk **vs.** Search time.   **Fig. 5.** $|\mathcal{W}|$ **vs.** Search time.

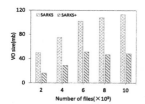

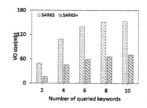

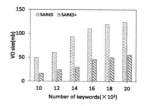

Fig. 6. n **vs.** $\mathcal{VO}$ size.   **Fig. 7.** qk **vs.** $\mathcal{VO}$ size.   **Fig. 8.** $|\mathcal{W}|$ **vs.** $\mathcal{VO}$ size.

6.3 Evaluation of $\mathcal{VO}$ Size

Due to **EDMRS** and **MRSF** not involving in the verification, in Fig. 6, Fig. 7, and Fig. 8, we evaluate the $\mathcal{VO}$ size with varying the number of files n, the number of query keywords qk, and the size of keyword dictionary $|\mathcal{W}|$, respectively, for our schemes. Specifically, we can observe that the $\mathcal{VO}$ size is growing with the increased n, qk, and $|\mathcal{W}|$ and **SARKS+** has a smaller $\mathcal{VO}$ size than **SARKS**. This is because based on the clustering method, **SARKS+** can prune more efficiently, which makes the number of accessed nodes smaller.

6.4 Evaluation of Verification Time

In experiment, we also evaluate the verification time with varying the number of files n, the number of query keywords qk, and the size of keyword dictionary $|\mathcal{W}|$. Since the $\mathcal{VO}$ size is larger, the verification time is larger correspondingly. Hence, we can observe that there exist a similar trends between $\mathcal{VO}$ size and verification time. In addition, our optimized scheme **SARKS+** still has a better performance than the basic scheme **SARKS** (Figs. 9, 10 and 11).

6.5 Evaluation of (m, t)

Figure 12, Fig. 13, and Fig. 14 show that different (m, t) in erasure coding has an obvious effect on the performance. In detail, with (m, t) becoming larger, the search time, $\mathcal{VO}$ size and verification time is increasing. For example, when (m, t) is from $(6, 2)$ to $(10, 4)$, the increasing rates of the search time, $\mathcal{VO}$ size, and verification time of

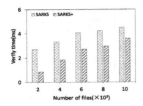

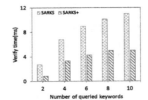

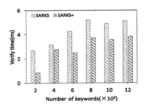

Fig. 9. n **vs.** Verification time.

Fig. 10. qk **vs.** Verification time.

Fig. 11. $|\mathcal{W}|$ **vs.** Verification time.

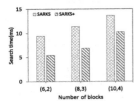

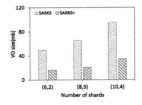

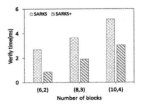

Fig. 12. (m, t) **vs.** Search time.

Fig. 13. (m, t) **vs.** $\mathcal{VO}$ size.

Fig. 14. (m, t) **vs.** Verification time

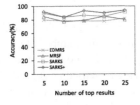

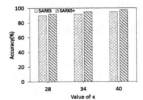

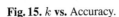

Fig. 15. k **vs.** Accuracy.

Fig. 16. ϵ **vs.** Accuracy.

SARKS and SARKS+ are 45% and 88%, 91% and 112%, 91% and 200%, respectively. This is because more file blocks need to be searched and more information needs to insert into the verification object.

6.6   Evaluation of Accuracy

In this subsection, we evaluate the search precision affected by the query k and privacy budget ϵ. In Fig. 15, when k is growing, compared with other approaches, our approaches SARKS+ and SARKS have the higher accuracy. For example, when k is 25, the accuracy of EDMRS, MRSF, SARKS+, and SARKS is 82%, 80%, 92%, and 94%, respectively. In addition, for our schemes, the accuracy is affected by the privacy budget. In Fig. 16, we can observe that the privacy budget ϵ is larger, the added noise is smaller and hence the accuracy is higher.

7   Conclusion

In this paper, we investigated the problem of secure and authenticated ranked keyword search. In specific, we proposed a novel framework that supports secure ranked keyword search by leveraging erasure coding and d-differential privacy to achieve access pattern and search pattern guarantee. Meanwhile, we put forward a verification mechanism to ensure the result integrity. Further, we devised an optimized scheme by introducing clustering method to accelerate the performance. Finally, we theoretically demonstrated that SARKS is secure under the simulation-based model. Moreover, massive experimental evaluations have been conducted on real-world dataset to present the superiority of SARKS.

Acknowledgement. This work was supported by National Natural Science Foundation of China (62011530-046, U1936220, 62002054), Industry-university-research Innovation Fund for Chinese Universities (2020ITA03009), Ten Thousand Talent Program (ZX20200035), and Excellent Youth Foundation of Anhui Scientific Committee (2108085J31).

References

1. Cao, N., Wang, C., Li, M., Ren, K., Lou, W.: Privacy-preserving multi-keyword ranked search over encrypted cloud data. IEEE Trans. Parallel Distrib. Syst. **25**(1), 222–233 (2014)
2. Chen, G., Lai, T., Reiter, M.K., Zhang, Y.: Differentially private access patterns for searchable symmetric encryption, pp. 810–818. IEEE (2018)
3. Cui, N., Li, J., Yang, X., Wang, B., Reynolds, M., Xiang, Y.: When geo-text meets security: privacy-preserving Boolean spatial keyword queries, pp. 1046–1057. IEEE (2019)
4. Cui, N., Yang, X., Chen, Y., Li, J., Wang, B., Min, G.: Secure Boolean spatial keyword query with lightweight access control in cloud environments. IEEE Internet Things J. **9**(12), 9503–9514 (2022)
5. Dai, H., Yang, M., Yang, T.G., Xiang, Y., Hu, Z., Wang, H.: A keyword-grouping inverted index based multi-keyword ranked search scheme over encrypted cloud data. IEEE Trans. Sustain. Comput. **7**(3), 561–578 (2022)
6. Ding, X., Liu, P., Jin, H.: Privacy-preserving multi-keyword top-$k similarity search over encrypted data. IEEE Trans. Dependable Secur. Comput. **16**(2), 344–357 (2019)
7. Ge, X., et al.: Towards achieving keyword search over dynamic encrypted cloud data with symmetric-key based verification. IEEE Trans. Dependable Secur. Comput. **18**(1), 490–504 (2021)
8. Li, F., Ma, J., Miao, Y., Jiang, Q., Liu, X., Choo, K.K.R.: Verifiable and dynamic multi-keyword search over encrypted cloud data using bitmap. IEEE Trans. Cloud Comput. 1 (2021). https://doi.org/10.1109/TCC.2021.3093304
9. Li, J., Ma, J., Miao, Y., Yang, R., Liu, X., Choo, K.R.: Practical multi-keyword ranked search with access control over encrypted cloud data. IEEE Trans. Cloud Comput. **10**(3), 2005–2019 (2022)
10. Liu, G., Yang, G., Bai, S., Wang, H., Xiang, Y.: FASE: a fast and accurate privacy-preserving multi-keyword top-k retrieval scheme over encrypted cloud data. IEEE Trans. Serv. Comput. **15**(4), 1855–1867 (2022)
11. Liu, Q., Tian, Y., Wu, J., Peng, T., Wang, G.: Enabling verifiable and dynamic ranked search over outsourced data. IEEE Trans. Serv. Comput. **15**(1), 69–82 (2022)

12. Miao, Y., Zheng, W., Jia, X., Liu, X., Choo, K.K.R., Deng, R.: Ranked keyword search over encrypted cloud data through machine learning method. IEEE Trans. Serv. Comput. 1 (2022). https://doi.org/10.1109/TSC.2021.3140098

13. Wan, Z., Deng, R.H.: VPSearch: achieving verifiability for privacy-preserving multi-keyword search over encrypted cloud data. IEEE Trans. Dependable Secur. Comput. **15**(6), 1083–1095 (2018)

14. Wang, C., Cao, N., Ren, K., Lou, W.: Enabling secure and efficient ranked keyword search over outsourced cloud data. IEEE Trans. Parallel Distrib. Syst. **23**(8), 1467–1479 (2012)

15. Wong, W.K., Cheung, D.W., Kao, B., Mamoulis, N.: Secure KNN computation on encrypted databases, pp. 139–152. ACM (2009)

16. Xia, Z., Wang, X., Sun, X., Wang, Q.: A secure and dynamic multi-keyword ranked search scheme over encrypted cloud data. IEEE Trans. Parallel Distrib. Syst. **27**(2), 340–352 (2016)

17. Zheng, Y., Lu, R., Guan, Y., Shao, J., Zhu, H.: Achieving efficient and privacy-preserving set containment search over encrypted data. IEEE Trans. Serv. Comput. **15**(5), 2604–2618 (2022)

Combining Autoencoder with Adaptive Differential Privacy for Federated Collaborative Filtering

Xuanang Ding[1], Guohui Li[2], Ling Yuan[1(✉)], Lu Zhang[3], and Qian Rong[1]

[1] School of Computer Science and Technology, Huazhong University of Science and Technology, Wuhan, China
{dingxuanang,cherryyuanling,qianrong}@hust.edu.cn
[2] School of Software Engineering, Huazhong University of Science and Technology, Wuhan, China
guohuili@hust.edu.cn
[3] School of Cyber Science and Engineering, Huazhong University of Science and Technology, Wuhan, China
luzhang_cs@hust.edu.cn

Abstract. Recommender systems provide users personalized services by collecting and analyzing interaction data, undermining user privacy to a certain extent. In federated recommender systems, users can train models on local devices without uploading raw data. Nevertheless, model updates transmitted between the user and the server are still vulnerable to privacy inference attacks. Several studies adopt differential privacy to obfuscate transmitted updates, but they ignore the privacy sensitivity of recommender model components. The problem is that components closer to the original data are more susceptible to privacy leakage. To address this point, we propose a novel adaptive privacy-preserving method combining autoencoder for federated collaborative filtering, which guarantees privacy meanwhile maintaining high model performance. First, we extend the variational autoencoder (VAE) to federated settings for privacy-preserving recommendations. Additionally, we analyze the privacy risks of the variational autoencoder model in federated collaborative filtering. Subsequently, we propose an adaptive differential privacy method to enhance user privacy further. The key is to allocate less privacy budget for sensitive layers. We apply a metric based on model weights to determine the privacy sensitivity of each layer in the autoencoder. Then we adaptively allocate the privacy budget to the corresponding model layer. Extensive experiments and analysis demonstrate that our method can achieve competitive performance to non-private recommender models meanwhile providing fine-grained privacy protection.

Keywords: Federated collaborative filtering · Autoencoder model · Privacy sensitivity · Adaptive differential privacy

Supported by the National Natural Science Foundation of China under Grant No. 62272180.

X. Wang et al. (Eds.): DASFAA 2023, LNCS 13943, pp. 661–676, 2023.
https://doi.org/10.1007/978-3-031-30637-2_44

1   Introduction

With the explosion of information, recommender systems (RS) can alleviate information overload by helping users find content that satisfies individualized preferences [3]. Collaborative filtering (CF) [10,11,30] provides personalized recommendations by modeling user data. Traditional recommendation models need to collect and centrally store user data, which may lead to user privacy issues. With the enactment of privacy laws such as GDPR[1], enterprises are restricted from collecting user-sensitive data. The trade-off between utility and privacy raises new challenges for the application of recommender systems. As a distributed machine learning paradigm, Federated Learning (FL) [21] can overcome the privacy dilemma of centralized recommender systems to some extent.

Combining federated learning and recommender systems [12,35], users can train recommender model without uploading sensitive data to the server. Federated recommender systems usually treat each user device as a client. Each client trains the recommender model locally and only uploads the model updates to the server. Then, the server collects all model updates and aggregates a new global model for delivery to the clients. Only model parameters are transmitted between the server and clients, and sensitive interaction data never leaves the user's device. Motivated by federated learning, FCF [2] is the first federated collaborative filtering framework based on matrix factorization, which updates the user feature vector on the client and shares the item feature matrix with the server. Following FCF, a series of matrix factorization-based works [4,9,18,34] are proposed to achieve privacy-preserving personalized collaborative filtering.

Compared with matrix factorization, deep neural networks have achieved remarkable achievements in recommender systems due to their strong fitting ability. Some recent works [24,26,33] have initially explored the application of deep neural networks in federated recommender systems. Nevertheless, these methods require the client to keep some personalized model components. To address this limitation, we extend the state-of-the-art variational autoencoder (VAE) [16] to the federated settings to provide personalized collaborative filtering. In the federated variational autoencoder, the client trains the encoder and decoder individually, and the server simultaneously collects model updates from all components to train a general federated variational model. The client does not need to train the personalized model component additionally.

The emergence of gradient leakage attacks [4,36] raises new privacy concerns for federated recommender systems. Attackers can infer sensitive user information from the uploaded gradient without hacking the user's device. Differential Privacy (DP) [7,25] can obfuscate the output distributions obtained by two neighboring inputs, making it challenging to distinguish slight data changes. The core idea is to add subtle perturbations to the data, which can effectively resist inference attacks. Most existing methods [20,22,33,34] apply differential privacy to federated recommender systems with a fixed privacy budget. However, they ignore the privacy risks posed by model architectures. Model components

---

[1] https://gdpr-info.eu.

closer to the original user input involved a higher privacy leakage risk [5,8,29]. Hence, it is necessary to provide different levels of privacy protection inside the model.

In this paper, we propose an adaptive private federated collaborative filtering method, named ADPFedVAE, which integrates the *Adaptive Differential Privacy* mechanism into *Federated Variational AutoEncoder*. ADPFedVAE employs variational autoencoder as the underlying recommender model and utilizes a federated training framework to ensure that user interactions do not leave the local device. To further enhance user privacy, ADPFedVAE adopts a novel adaptive differential privacy mechanism to perturb local model updates. With ADPFedVAE, two limitations can be tackled: 1) clients do not have to keep personalization components, and 2) the protection of more important model components is enhanced. The main contributions are summarized as follows:

- We propose a privacy-preserving federated collaborative filtering method ADPFedVAE, which combines the variational autoencoder model with federated learning to learn decentralized user data.
- We analyze the privacy risks of the variational autoencoder model and propose a novel adaptive differential privacy mechanism, AdaptiveDP. AdaptiveDP utilizes weight magnitude to identify the privacy sensitivity of model layers and allocate adaptive privacy budgets.
- We conduct extensive experiments and analyses on three real-world datasets. The results show that our method can achieve competitive performance meanwhile providing fine-grained privacy guarantees.

2 Preliminaries

Differential Privacy is first proposed by [6,7], which can prevent malicious attackers from inferring user privacy information through queries. The detailed definition of ϵ-Differential Privacy (ϵ-DP) is given as follows:

Definition 1 (ϵ-Differential Privacy). *A randomized algorithm $M : D \rightarrow R$ with input domain D and range R satisfies ϵ-differential privacy, if for two arbitrary neighboring private inputs $d, d' \in D$ and output range $S \subset R$, the following equation holds:*

$$Pr\left[M(d) \in S\right] \leq e^{\epsilon} \cdot Pr\left[M(d') \in S\right], \tag{1}$$

where $Pr[\cdot]$ denotes the probability, ϵ is the privacy budget of differential privacy and $\epsilon > 0$.

Equation 1 shows that the privacy budget ϵ controls the level of privacy protection, and the smaller value of ϵ provides a stricter privacy guarantee. In federated recommender systems, the client utilizes local data to train the model and uploads the updated model or gradient to the server for aggregation. Thus, each client needs a local randomizer to protect the transmitted model. For the local version of differential privacy in federated clients, we define ϵ-Local Differential Privacy (ϵ-LDP) as follows:

Definition 2 (ϵ-Local Differential Privacy). *A randomized algorithm $M : D \rightarrow R$ satisfies ϵ-local differential privacy, if for two arbitrary neighboring inputs x and x' in D, the following equation holds:*

$$Pr\left[M(x) = Y\right] \le e^\epsilon \cdot Pr\left[M(x') = Y\right], \qquad (2)$$

where Y denotes the output of randomized algorithm $M(\cdot)$.

Local differential privacy provides privacy protection by injecting controlled random noise into the shared model, ensuring that an attacker cannot accurately estimate sensitive user data. The Laplace mechanism [6] is widely used to achieve ϵ-differential privacy by adding Laplace noise. We define the sensitivity and Laplace Mechanism as follows:

Definition 3 (Sensitivity). *For any pair of neighboring inputs $d, d' \in D$, the sensitivity Δf of query function $f(\cdot)$ is defined as follows:*

$$\Delta f = \max_{d,d'} \| f(d) - f(d') \|_1, \qquad (3)$$

where the sensitivity Δf denotes the maximum change range of function $f(\cdot)$.

Definition 4 (Laplace Mechanism). *For any query function $f(\cdot)$ and input d, the Laplace mechanism $M(\cdot)$ satisfying ϵ-differential privacy is defined as:*

$$M(d) = f(d) + Lap\left(0, \frac{\Delta f}{\epsilon}\right), \qquad (4)$$

where $Lap\left(0, \frac{\Delta f}{\epsilon}\right)$ is a random noise vector following the Laplace distribution.

Several federated recommender systems [23, 24, 26, 33] employ deep neural networks as the base recommender system. The training of deep neural networks can be viewed as a multi-step sequential computational process while federated clients parallelly train and upload models in each round of global aggregation. The composition property of differential privacy [1, 14] provides guarantees for the privacy of deep federated recommenders. We introduce the sequential composition theorem and parallel composition theorem [14] as follows:

Theorem 1 (Sequential Composition Theorem). *Given a sequence of randomized algorithm $\langle M_1, M_2, \ldots, M_n \rangle$, each M_i provides ϵ_i-differential privacy. Then the sequence of $\langle M_1(d), M_2(d), \ldots, M_n(d) \rangle$ satisfies $(\sum_{i=0}^{n} \epsilon_i)$-differential privacy.*

Theorem 2 (Parallel Composition Theorem). *Given a sequence of randomized algorithm $\langle M_1, M_2, \ldots, M_n \rangle$, each M_i provides ϵ_i-differential privacy, and $\langle d_1, d_2, \ldots, d_n \rangle$ are arbitrary disjoint subsets of the input domain D. Then the sequence of $\langle M_1(d_1), M_2(d_2), \ldots, M_n(d_n) \rangle$ satisfies $(\max\{\epsilon_i\}_{i=0}^{n})$-differential privacy.*

3   Methodology

In this section, we first introduce the variational autoencoder, the underlying recommendation module in our approach. Next, we describe the overall training procedure of our ADPFedVAE method. Finally, we provide the details of the adaptive differential privacy mechanism.

3.1   Variational Autoencoder for Local Recommendation

As the most popular method, collaborative filtering provides promising recommendations by modeling the user-item interaction history. The variational autoencoder(VAE) [16] is a state-of-out-art work for CF method based on implicit feedback. The model architecture of VAE is shown in Fig. 1.

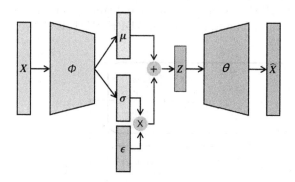

Fig. 1. Variational autoencoder model for collaborative filtering.

Given a recommender system with users $u \in \{1, 2, \ldots, U\}$ and items $i \in \{1, 2, \ldots, I\}$, VAE first samples a K-dimensional user latent representation z_u from a standard Gaussian prior for each user u. Then, the model utilizes a non-linear function $f_\theta(\cdot)$ to transform z_u to a probability distribution over all items. The VAE assumes that observed interactions x_u are sampled from a multinomial distribution:

$$z_u \sim \mathcal{N}(0, I_K), \qquad \pi(z_u) \propto \exp\{f_\theta(z_u)\},$$
$$x_u \sim \text{Mult}(N_u, \pi(z_u)), \tag{5}$$

where $\pi(z_u)$ is the probability vector produced by the softmax over the transformed latent representation z_u, N_u denotes the total interaction number of user u. The log-likelihood for user u conditioned on the latent representation z_u is $\log p_\theta(x_u|z_u) = \sum_i x_{ui} \log \pi_i(z_u)$. VAE introduces variational inference [13] to learn the model in the Eq. 5, and the posterior $p_\theta(x_u|z_u)$ is approximated by the variational distribution $q_\phi(z_u|x_u) = \mathcal{N}(\mu_\phi(x_u), \text{diag}\{\sigma_\phi^2(x_u)\})$, where both $\mu_\phi(x_u)$ and $\sigma_\phi^2(x_u)$ are K-dimensional vectors parametrized by ϕ.

Ultimately, $p_\theta(x_u|z_u)$ and $q_\phi(z_u|x_u)$ are combined to compose a variational autoencoder, and the evidence lower bound (ELBO) of VAE is defined as:

$$\mathcal{L}_\beta(x_u;\theta,\phi) = \mathbb{E}_{q_\phi(z_u|x_u)}[\log p_\theta(x_u|z_u)] - \beta \cdot \mathrm{KL}(q_\phi(z_u|x_u)\|p_\theta(z_u)), \quad (6)$$

where $\mathrm{KL}(\cdot)$ is the Kullback-Leibler divergence, β is the parameter controlling the strength of regularization. In particular, VAE introduces dropout [31] at the input layer to reduce the risk of overfitting. To eliminate the stochasticity caused by sampling, VAE adopts the reparametrization trick [15], and then the model can be optimized using gradient descent.

3.2 ADPFedVAE

In this work, we propose an adaptive private federated collaborative filtering method, ADPFedVAE, which extends the variational autoencoder to federated mode. In ADPFedVAE, clients collaborate to train a global model without undermining their privacy. The framework of ADPFedVAE is shown in Fig. 2.

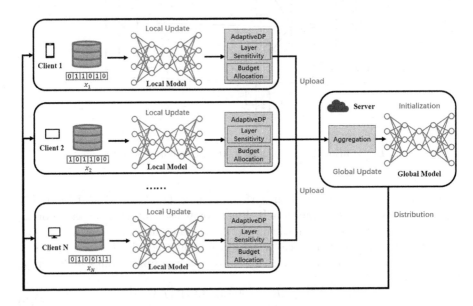

Fig. 2. The Framework of ADPFedVAE. The server coordinates the clients to train locally and aggregates the global model. The client adds adaptive noise to the model update and uploads it to the server. The interactions never leave the local device.

The framework of ADPFedVAE adopts the client-server architecture: clients use their private data to update the local model and upload the updates to the server; the server aggregates all client updates to generate a new global model and distributes it to clients for training. We assume that all clients are available at any time, and the training procedure of ADPFedVAE is listed as follows.

(1) **Initialization.** The server initializes the global VAE model $w_0^g = (\theta_0, \phi_0)$, and configures relevant hyper-parameters.

(2) **Distribution.** At the beginning of each epoch t, the server randomly selects K clients to participate in training and then distributes the latest global model w_t^g to each selected client.

(3) **Local Update.** After receiving the global model w_t^g, each client uses local interactions x_u to train the model individually via optimizing the loss in Eq. 6. The update of local model $w = (\theta, \phi)$ is as follows:

$$w \leftarrow w - \gamma \nabla_w \mathcal{L}_\beta(x_u; w), \tag{7}$$

where γ denotes the local learning rate. Each client completes at least one epoch of local training to get the model update $\Delta_{w_t^u} \leftarrow w - w_t^g$.

(4) **Upload.** To enhance data privacy, the client adds random noise to model updates using an adaptive differential privacy mechanism (AdaptiveDP). Then, the client uploads the perturbed update $\tilde{\Delta}_{w_t^u}$ to the server.

(5) **Global Update.** At the end of each epoch t, the server collects all model updates and performs aggregation to build a new global model w_{t+1}^g. Specifically, ADPFedVAE adopts the FedAdam [28] to update the global model:

$$\Delta_{w_t^g} \leftarrow \beta_1 \Delta_{w_{t-1}^g} + (1 - \beta_1)(\frac{1}{|S|} \sum_{u \in S} \tilde{\Delta}_{w_t^u}),$$

$$v_t \leftarrow \beta_2 v_{t-1} + (1 - \beta_2)\Delta_{w_t^g}^2, \tag{8}$$

$$w_{t+1}^g \leftarrow w_t^g + \eta \frac{\Delta_{w_t^g}}{\sqrt{v_t} + \tau},$$

where S denotes the selected client set, $0 < \beta_1, \beta_2 < 1$, $0 < \eta < 1$ denotes the global learning rate, and $0 < \tau < 1$ e.g. 10^{-8} to avoid divide by 0. Then, the server distributed the model w_{t+1}^g for next training epoch.

The above process is repeated until the recommendation model converges or the training terminates. We summarize the overall training process of ADPFed-VAE in Algorithm 1. Next, we introduce the adaptive differential privacy mechanism (AdaptiveDP) in detail.

3.3 Adaptive Differential Privacy Mechanism

In conventional federated recommender systems, clients upload model updates to the server, and the private data do not leave the device. Although our underlying model VAE adopts the architecture of deep neural networks and adds dropout to the input layer, attackers can still infer user-sensitive information from the uploaded gradients [5,36]. The main reason is that the encoder and decoder of the VAE model consist of multi-layer perceptrons. Moreover, the fully connected layers cannot resist gradient leakage attacks [5]. Given a linear layer, input $x \in \mathbb{R}^m$ and output $y \in \mathbb{R}^n$, the forward and backpropagation are as follows:

$$y = xW + b,$$
$$\nabla_W \mathcal{L} = x \cdot \nabla_y \mathcal{L}, \tag{9}$$

Algorithm 1. ADPFedVAE

Input: Global training epochs T, local training epochs E, client sampling number K, global learning rate η, local learning rate γ, privacy budget ϵ, clipping bound δ

Output: Trained global model $w^g = (\theta, \phi)$

1: Server initializes global model $w_0^g = (\theta_0, \phi_0)$
2: **for** each epoch $t = 0, 1, ..., T$ **do**
3:     $S \leftarrow$ randomly select K clients
4:     **for** each client $u \in S$ **in parallel do**
5:         $w \leftarrow w_t^g$
6:         **for** each local epoch **do**
7:             $w \leftarrow w - \gamma \nabla_w \mathcal{L}_\beta(x_u; w)$
8:         **end for**
9:         $\Delta_{w_t^u} \leftarrow w - w_t^g$
10:        $\tilde{\Delta}_{w_t^u} \leftarrow \text{ADAPTIVEDP}(\Delta_{w_t^u}, \epsilon, \delta)$
11:        send perturbed $\tilde{\Delta}_{w_t^u}$ to server
12:    **end for**
13:    $\Delta_{w_t^g} \leftarrow \beta_1 \Delta_{w_{t-1}^g} + (1 - \beta_1)(\frac{1}{|S|} \sum_{u \in S} \tilde{\Delta}_{w_t^u})$
14:    $v_t \leftarrow \beta_2 v_{t-1} + (1 - \beta_2) \Delta_{w_t^g}^2$
15:    $w_{t+1}^g \leftarrow w_t^g + \eta \frac{\Delta_{w_t^g}}{\sqrt{v_t + \tau}}$
16: **end for**
17: **return** $w^g = (\theta, \phi)$

---

where $W \in \mathbb{R}^{m \times n}$ and $b \in \mathbb{R}^n$ are weights and biases, $\nabla_W \mathcal{L}$ and $\nabla_y \mathcal{L}$ are layer gradients. For any factor $x^i \in x$, we have:

$$\left[\frac{\partial \mathcal{L}}{\partial W^{i1}}, \frac{\partial \mathcal{L}}{\partial W^{i2}}, \cdots, \frac{\partial \mathcal{L}}{\partial W^{in}} \right] = x^i \left[\frac{\partial \mathcal{L}}{\partial y^1}, \frac{\partial \mathcal{L}}{\partial y^2}, \cdots, \frac{\partial \mathcal{L}}{\partial y^n} \right],$$

$$x^i = \frac{\partial \mathcal{L}}{\partial W^{ik}} \left(\frac{\partial \mathcal{L}}{\partial y^k} \right)^{-1}, \quad 1 \le k \le n, \tag{10}$$

Hence, the input x can be uniquely determined by the layer gradients. Although the dropout added to the input layer approximates differential privacy [26], the output of the VAE model is the same as the original input, which brings the privacy hazard. Many federated recommender systems employ LDP techniques to perturb model updates by assigning a fixed privacy budget and adding noise to the model. When learning the encoder-decoder structure of VAE, the layers close to the input and output encode the sensitive attributes of the client [29]. Consequently, layers at different positions in VAE have distinct privacy risks.

The motivation for Adaptive Differential Privacy mechanism (AdaptiveDP) is to discriminate the privacy sensitivity of different layers and provide more protection to more sensitive layers. Within a fixed privacy budget ϵ, AdaptiveDP allocates distinct privacy budgets to each layer according to Theorem 1 to provide fine-grained protection. The AdaptiveDP mechanism mainly consists of the following two steps:

Algorithm 2. Adaptive Differential Privacy

Input: Updated gradients g, privacy budget ϵ, clipping bound δ
Output: Perturbed gradients $\tilde{g}$

1: **function** ADAPTIVEDP(g, ϵ, δ)
2:     **for** each layer $i = 0, 1, \ldots, L$ **do**
3:         $s[i] \leftarrow \frac{\|g_i\|_2}{\sum_{i \in L} \|g_i\|_2}$
4:     **end for**
5:     $\hat{g} \leftarrow clip(g, \delta)$
6:     **for** each layer $i = 0, 1, \ldots, L$ **do**
7:         $\epsilon_i \leftarrow s[i] \cdot \epsilon$
8:         $\tilde{g}_i \leftarrow \hat{g}_i + Lap\left(0, \frac{2\delta}{\epsilon_i}\right)$
9:     **end for**
10:     **return** $\tilde{g}$
11: **end function**

Layer Sensitivity Identification. As mentioned above, the model structure of VAE has serious privacy risks. Attackers can infer private data from layer updates uploaded by clients. The key is to identify the privacy sensitivity of each layer. Inspired by [19], the model layer close to the original input has larger weight magnitudes and contains more private user attributes. AdaptiveDP adopts a weight-based metric to identify the privacy sensitivity of each layer in VAE. Specifically, the client completes local update and calculates the 2-norm of the local model layers, then identifies the privacy sensitivity of each layer i:

$$s[i] \leftarrow \frac{\|g_i\|_2}{\sum_{i \in L} \|g_i\|_2}, \tag{11}$$

where $s[i]$ denotes the privacy sensitivity of layer i, g_i denotes the gradient of layer i, L denotes the number of model layers. After identifying layer privacy sensitivity, we can allocate privacy budgets and perturb local model updates.

Adaptive Budget Allocation. Given the total privacy budget ϵ, the privacy budget ϵ_i for each layer is allocated as $s[i] \cdot \epsilon$. Following [20,27,34], we clip each layer update g_i with a threshold δ and adaptively add Laplace noise to satisfy ϵ_i−local differential privacy, the process is as follows:

$$\hat{g}_i \leftarrow clip(g_i, \delta),$$
$$\tilde{g}_i \leftarrow \hat{g}_i + Lap\left(0, \frac{2\delta}{s[i] \cdot \epsilon}\right). \tag{12}$$

where $clip(\cdot)$ denotes the clipping function. The overall procedure of AdaptiveDP mechanism is shown in Algorithm 2.

Subsequently, the client uploads the perturbed updates to the server for aggregation. After the clipping and perturbation of AdaptiveDP, the uploaded updates contain less sensitive information, and the protection of critical layers is strengthened. Each layer i in the local update satisfies ϵ_i−local differential privacy for each client. Thus, the client's local update g satisfies ϵ−local differential privacy according to Theorem 1. In the federated recommender system, the data is decentralized stored on user device. For the global epoch, each epoch of model aggregation also satisfies $\epsilon-$local differential privacy according to Theorem 2. Finally, our proposed ADPFedVAE satisfies $\epsilon-$local differential privacy.

4   Experiments

4.1   Experimental Settings

Datasets. To evaluate our proposed method, we conduct experiments on three real-world datasets: MovieLens-1M[2] (ML-1M for short), Lastfm-2K[3] (LastFm for short), Steam[4]. The statistics of the three datasets are summarized in Table 1. From ML-1M to Steam, the number of users/items and sparsity keep increasing. Following [16], we transform all data to implicit feedback in the form of 0/1.

Table 1. Statistics of the evaluation datasets.

Dataset	#User	#Item	#Interaction	Sparsity
ML-1M	6,040	3,706	1,000,209	0.9553
LastFm	1,892	17,632	92,834	0.9972
Steam	12,393	5,155	129,511	0.9980

Evaluation. To evaluate the performance of ADPFedVAE, we first group the interaction data by user ID to meet the settings of the federated recommender system. For each dataset, we randomly divide 80% of each user's interactions into the training set, and the remaining interactions are used to test/validate the models. To measure the performance, We use Recall@N and the truncated normalized discounted cumulative gain (NDCG@N) to compare the predicted rank with the true rank. After that, our method generates a ranked top-N list over all items to evaluate the metrics mentioned above, where $N \in \{20, 100\}$.

Parameter Settings. In ADPFedVAE, we set the architecture of global VAE model to $[I \rightarrow 600 \rightarrow 200 \rightarrow 600 \rightarrow I]$, where I is the total number of items and 200 is the size of user latent representation. For the global update, the number of global training epochs T is set to 400, the number of sampling clients K is set to 250, and we set the FedAdam parameters $\eta = 0.001, \beta_1 = 0.9, \beta_2 = 0.999$. For the local update, the local learning rate γ is set to 0.01. For the adaptive differential privacy mechanism, the clipping bound δ is set to 0.5, and the privacy budget is $\epsilon \in \{1, 5, 10, 15, 20, 25, 30\}$.

---

[2] http://grouplens.org/datasets/movielens/1m/.
[3] https://grouplens.org/datasets/hetrec-2011/.
[4] https://www.kaggle.com/tamber/steam-video-games/.

Table 2. Performance comparison on the evaluation datasets.

Dataset	Method	Centralized	Federated		
		VAE	FCF	FedVAE	ADPFedVAE
ML-1M	NCDG@100	$0.4242_{\pm 0.0016}$	$0.2467_{\pm 0.0005}$	$0.4242_{\pm 0.0012}$	$0.4120_{\pm 0.0007}$
	Recall@100	$0.5615_{\pm 0.0008}$	$0.4018_{\pm 0.0004}$	$0.5657_{\pm 0.0008}$	$0.5427_{\pm 0.0011}$
	NCDG@20	$0.3457_{\pm 0.0024}$	$0.1475_{\pm 0.0006}$	$0.3411_{\pm 0.0021}$	$0.3492_{\pm 0.0017}$
	Recall@20	$0.3456_{\pm 0.0018}$	$0.1684_{\pm 0.0008}$	$0.3522_{\pm 0.0016}$	$0.3433_{\pm 0.0012}$
LastFm	NCDG@100	$0.3366_{\pm 0.0013}$	$0.2379_{\pm 0.0006}$	$0.3268_{\pm 0.0012}$	$0.3106_{\pm 0.0028}$
	Recall@100	$0.4789_{\pm 0.0011}$	$0.4206_{\pm 0.0016}$	$0.4784_{\pm 0.0009}$	$0.4473_{\pm 0.0023}$
	NCDG@20	$0.2483_{\pm 0.0014}$	$0.1400_{\pm 0.0003}$	$0.2353_{\pm 0.0014}$	$0.2260_{\pm 0.0027}$
	Recall@20	$0.2562_{\pm 0.0010}$	$0.1711_{\pm 0.0007}$	$0.2473_{\pm 0.0006}$	$0.2324_{\pm 0.0025}$
Steam	NCDG@100	$0.2410_{\pm 0.0004}$	$0.1021_{\pm 0.0004}$	$0.2413_{\pm 0.0006}$	$0.2251_{\pm 0.0007}$
	Recall@100	$0.3522_{\pm 0.0005}$	$0.2205_{\pm 0.0003}$	$0.3547_{\pm 0.0009}$	$0.3366_{\pm 0.0006}$
	NCDG@20	$0.2213_{\pm 0.0004}$	$0.0817_{\pm 0.0003}$	$0.2208_{\pm 0.0005}$	$0.2053_{\pm 0.0009}$
	Recall@20	$0.2757_{\pm 0.0006}$	$0.1457_{\pm 0.0003}$	$0.2749_{\pm 0.0015}$	$0.2582_{\pm 0.0007}$

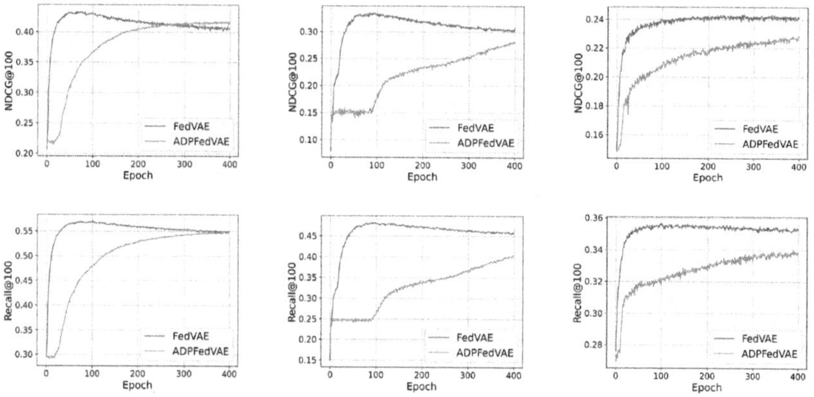

(a) ML-1M   (b) LastFm   (c) Steam

Fig. 3. Convergence speed of FedVAE and ADPFedVAE.

4.2 Results and Discussion

Performance Comparison. We first compare the recommendation performance of ADPFedVAE with centralized VAE [16], FCF [2], and FedVAE [26]. We perform these methods five times with different random seeds and report the mean and standard deviation of the performance metrics. The results on the three datasets are shown in Table 2. We further compare the training convergence speed of FedVAE and ADPFedVAE, the results of NDCG@100 and Recall@100 are shown in Fig. 3. From these results, we have the following observations:

- As a state-of-the-art deep recommender model, VAE performs superior to federated matrix factorization methods in both centralized and federated versions. This demonstrates that the strong fitting ability of the deep model can effectively improve the recommendation performance.

- Although the data is scattered across user devices, federated VAE can still perform similarly to the centralized recommendations. This shows that federated learning does not damage the performance of the recommendation model. Because the performance of a centrally trained VAE model will be the upper bound of federated VAE.
- After adding the AdpativeDP perturbations, the performance of ADPFedVAE drops slightly but is still competitive. This shows that ADPFedVAE provides fine-grained privacy protection through AdpativeDP at the cost of acceptable performance loss. There is always a utility-privacy tradeoff in federated recommender systems due to the *No Free Lunch Theorem.*
- During the training, ADPFedVAE achieves good convergence performance on both ML-1M and Steam but fluctuates at the early stage of training on LastFM. This is due to the small number of users of LastFM, which leads to weak model performance in the early training stage and is easily affected by perturbation noise.

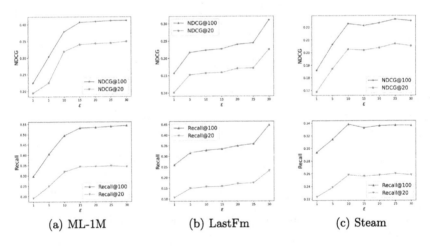

(a) ML-1M   (b) LastFm   (c) Steam

Fig. 4. Performance comparison with different budget ϵ.

Privacy-Utility Trade-Off. To evaluate the impact of privacy budget on the recommendation accuracy, we compare the performance of ADPFedVAE with under different privacy budget ϵ on each dataset. We conduct the experiment with $\epsilon \in \{5, 10, 15, 20, 25, 30\}$, and other settings remain the same. The results of NDCG and Recall are shown in Fig. 4. We can observe that as the privacy budget ϵ decreases, the model suffers more noise perturbations, and the performance drops too. Furthermore, ADPFedVAE is more robust on ML-1M and Steam due to their large scale of users. Whereas on LastFM, ADPFedVAE needs

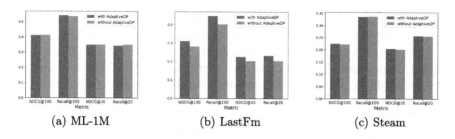

| (a) ML-1M | (b) LastFm | (c) Steam |

Fig. 5. Performance comparison between model with AdaptiveDP and without AdaptiveDP.

more privacy budget to maintain recommendation performance. It can be concluded that ADPFedVAE can achieve the trade-off between privacy and utility by setting an appropriate privacy budget ϵ.

Effect of AdaptiveDP. Finally, we compare the performance between models after adding noise with and without the AdaptiveDP mechanism under the same privacy budget. The results are shown in Fig. 5. In AdaptiveDP, the privacy budget is allocated by identifying the sensitivity of each layer of the model, and then adaptive noise is added. For models without AdaptiveDP, a fixed noise is added to each layer of the model. From the results, under the same privacy budget, ADP maintains recommendation performance while achieving fine-grained privacy protection through budget allocation.

5   Related Work

Federated Collaborative Filtering. To address the privacy risks arising from data collection in the centralized recommendation, Ammad-Ud-Din et al. [2] proposed the first federated collaborative filtering (FCF) approach based on implicit feedback for privacy-preserving collaborative filtering. In their work, the client locally performs a low-rank matrix factorization based on the alternating learning-to-square algorithm and shares the gradient of the item embedding matrix with the server. The server uses the Adam optimization algorithm to update the global model and achieves a similar performance to the centralized version. Following FCF, Flanagan et al. [9] proposed a multi-view joint matrix factorization to extend FCF to the scenario with multiple data sources. Chai et al. [4] proposed a secure explicit federated matrix factorization (FedMF) using homomorphic encryption. Subsequently, FedRec [18] and FedRec++ [17] are proposed for federated probabilistic matrix factorization with explicit feedback. Yang et al. [34] proposed federated collective matrix factorization to estimate user preferences using explicit and implicit feedback jointly. However, the above methods are all based on traditional matrix factorization to achieve personalized federated collaborative filtering. Combining deep learning with federated recommender systems is a promising direction.

Recently, several works [23,24] explored the combination of the neural collaborative filtering (NCF) and federated recommender systems and achieved excellent recommendation quality. Moreover, Muhammad et al. [23] proposed FedFast, which can cluster and sample clients to achieve fast convergence of federated training. Wu et al. [33] proposed FedGNN, the first federated recommendation system based on the graph neural network. Based on FedGNN, Liu et al. [20] integrated user social relations to train personalized models for each client. The work most similar to our method is FedVAE [26], which extends the variational autoencoder to the federated version. However, FedVAE ignores the privacy risks of the model and fails to provide any protection for training.

Private Federated Collaborative Filtering. Some works [17,18,20,33] randomly assign pseudo interactions or virtual ratings for items to obfuscate model updates. To provide privacy guarantee for aggregation, several works [4,24,34] employ homomorphic encryption to encrypt the transmitted model updates, but the computational cost is expensive. Compared to homomorphic encryption, differential privacy [7] is more suitable for federated recommender systems due to its flexibility. Wang et al. [32] utilize differential privacy protection based on Gaussian mechanism to perturb client model update. In [20,33,34], the Laplace noise is added to satisfy differential privacy. In addition, Minto et al. [22] provides stronger privacy guarantees by perturbing factors in uploaded item updates. However, the above works ignore the privacy risks contained in the model architecture. These limitations motivate us to propose a novel adaptive differential privacy mechanism for privacy-preserving federated collaborative filtering.

6   Conclusion

In this work, we propose a privacy-preserving federated collaborative filtering method ADPFedVAE, which combines the variational autoencoder model and adaptive differential privacy mechanism. ADPFedVAE first extends the VAE model to federated settings, which learns decentralized user interactions by collecting user model updates. Furthermore, we analyze the privacy risks existing in the federated VAE model and propose a novel adaptive differential privacy mechanism, AdaptiveDP. AdaptiveDP utilizes the model weight distribution to measure the privacy sensitivity of each layer during training, and allocates distinct privacy budgets to enhance privacy adaptively. Extensive experiments on three datasets show that our method can achieve competitive performance meanwhile providing fine-grained privacy protection. In future work, we will explore the privacy loss in multiple rounds of training and the privacy impact of different model architectures on federated recommendations.

References

1. Abadi, M., et al.: Deep learning with differential privacy. In: Proceedings of the 2016 ACM SIGSAC Conference on Computer and Communications Security, pp. 308–318 (2016)

2. Ammad-Ud-Din, M., et al.: Federated collaborative filtering for privacy-preserving personalized recommendation system. arXiv preprint arXiv:1901.09888 (2019)
3. Bobadilla, J., Ortega, F., Hernando, A., Gutiérrez, A.: Recommender systems survey. Knowl.-Based Syst. **46**, 109–132 (2013)
4. Chai, D., Wang, L., Chen, K., Yang, Q.: Secure federated matrix factorization. IEEE Intell. Syst. **36**(5), 11–20 (2020)
5. Chen, C., Campbell, N.: Understanding training-data leakage from gradients in neural networks for image classification. In: Workshop Privacy in Machine Learning, NeurIPS 2021 (2021)
6. Dwork, C., McSherry, F., Nissim, K., Smith, A.: Calibrating noise to sensitivity in private data analysis. J. Priv. Confidentiality **7**(3), 17–51 (2016)
7. Dwork, C., Roth, A., et al.: The algorithmic foundations of differential privacy. Found. Trends® Theor. Comput. Sci. **9**(3–4), 211–407 (2014)
8. Feng, T., Hashemi, H., Hebbar, R., Annavaram, M., Narayanan, S.S.: Attribute inference attack of speech emotion recognition in federated learning settings. arXiv preprint arXiv:2112.13416 (2021)
9. Flanagan, A., Oyomno, W., Grigorievskiy, A., Tan, K.E., Khan, S.A., Ammad-Ud-Din, M.: Federated multi-view matrix factorization for personalized recommendations. In: Hutter, F., Kersting, K., Lijffijt, J., Valera, I. (eds.) ECML PKDD 2020. LNCS (LNAI), vol. 12458, pp. 324–347. Springer, Cham (2021). https://doi.org/10.1007/978-3-030-67661-2_20
10. He, X., Liao, L., Zhang, H., Nie, L., Hu, X., Chua, T.S.: Neural collaborative filtering. In: Proceedings of the 26th International Conference on World Wide Web, pp. 173–182 (2017)
11. Herlocker, J.L., Konstan, J.A., Terveen, L.G., Riedl, J.T.: Evaluating collaborative filtering recommender systems. ACM Trans. Inf. Syst. (TOIS) **22**(1), 5–53 (2004)
12. Jalalirad, A., Scavuzzo, M., Capota, C., Sprague, M.: A simple and efficient federated recommender system. In: Proceedings of the 6th IEEE/ACM International Conference on Big Data Computing, Applications and Technologies, pp. 53–58 (2019)
13. Jordan, M.I., Ghahramani, Z., Jaakkola, T.S., Saul, L.K.: An introduction to variational methods for graphical models. Mach. Learn. **37**(2), 183–233 (1999)
14. Kairouz, P., Oh, S., Viswanath, P.: The composition theorem for differential privacy. In: International Conference on Machine Learning, pp. 1376–1385. PMLR (2015)
15. Kingma, D.P., Welling, M.: Auto-encoding variational bayes. arXiv preprint arXiv:1312.6114 (2013)
16. Liang, D., Krishnan, R.G., Hoffman, M.D., Jebara, T.: Variational autoencoders for collaborative filtering. In: Proceedings of the 2018 World Wide Web Conference, pp. 689–698 (2018)
17. Liang, F., Pan, W., Ming, Z.: FedRec++: lossless federated recommendation with explicit feedback. In: Proceedings of the AAAI Conference on Artificial Intelligence, vol. 35, pp. 4224–4231 (2021)
18. Lin, G., Liang, F., Pan, W., Ming, Z.: FedRec: federated recommendation with explicit feedback. IEEE Intell. Syst. **36**(5), 21–30 (2020)
19. Lin, M., et al.: Pruning networks with cross-layer ranking & k-reciprocal nearest filters. IEEE Trans. Neural Netw. Learn. Syst. (2022)
20. Liu, Z., Yang, L., Fan, Z., Peng, H., Yu, P.S.: Federated social recommendation with graph neural network. ACM Trans. Intell. Syst. Technol. (TIST) **13**(4), 1–24 (2022)

21. McMahan, B., Moore, E., Ramage, D., Hampson, S., Arcas, B.A.: Communication-efficient learning of deep networks from decentralized data. In: Artificial Intelligence and Statistics, pp. 1273–1282. PMLR (2017)

22. Minto, L., Haller, M., Livshits, B., Haddadi, H.: Stronger privacy for federated collaborative filtering with implicit feedback. In: Fifteenth ACM Conference on Recommender Systems, pp. 342–350 (2021)

23. Muhammad, K., et al.: FedFast: going beyond average for faster training of federated recommender systems. In: Proceedings of the 26th ACM SIGKDD International Conference on Knowledge Discovery & Data Mining, pp. 1234–1242 (2020)

24. Perifanis, V., Efraimidis, P.S.: Federated neural collaborative filtering. Knowl.-Based Syst. **242**, 108441 (2022)

25. Phan, N., Wu, X., Hu, H., Dou, D.: Adaptive laplace mechanism: differential privacy preservation in deep learning. In: 2017 IEEE International Conference on Data Mining (ICDM), pp. 385–394. IEEE (2017)

26. Polato, M.: Federated variational autoencoder for collaborative filtering. In: 2021 International Joint Conference on Neural Networks (IJCNN), pp. 1–8. IEEE (2021)

27. Qi, T., Wu, F., Wu, C., Huang, Y., Xie, X.: Privacy-preserving news recommendation model learning. arXiv preprint arXiv:2003.09592 (2020)

28. Reddi, S.J., et al.: Adaptive federated optimization. In: International Conference on Learning Representations (2020)

29. Schreyer, M., Sattarov, T., Borth, D.: Federated and privacy-preserving learning of accounting data in financial statement audits. arXiv preprint arXiv:2208.12708 (2022)

30. Sedhain, S., Menon, A.K., Sanner, S., Xie, L.: AutoRec: autoencoders meet collaborative filtering. In: Proceedings of the 24th International Conference on World Wide Web, pp. 111–112 (2015)

31. Srivastava, N., Hinton, G., Krizhevsky, A., Sutskever, I., Salakhutdinov, R.: Dropout: a simple way to prevent neural networks from overfitting. J. Mach. Learn. Res. **15**(1), 1929–1958 (2014)

32. Wang, Q., Yin, H., Chen, T., Yu, J., Zhou, A., Zhang, X.: Fast-adapting and privacy-preserving federated recommender system. VLDB J. **31**(5), 877–896 (2022)

33. Wu, C., Wu, F., Cao, Y., Huang, Y., Xie, X.: FedGNN: federated graph neural network for privacy-preserving recommendation. arXiv preprint arXiv:2102.04925 (2021)

34. Yang, E., Huang, Y., Liang, F., Pan, W., Ming, Z.: FCMF: federated collective matrix factorization for heterogeneous collaborative filtering. Knowl.-Based Syst. **220**, 106946 (2021)

35. Yang, Q., Liu, Y., Chen, T., Tong, Y.: Federated machine learning: concept and applications. ACM Trans. Intell. Syst. Technol. (TIST) **10**(2), 1–19 (2019)

36. Zhu, L., Liu, Z., Han, S.: Deep leakage from gradients. In: Advances in Neural Information Processing Systems, vol. 32 (2019)

Robust Clustered Federated Learning

Tiandi Ye[1], Senhui Wei[1], Jamie Cui[2], Cen Chen[1,3(✉)], Yingnan Fu[1], and Ming Gao[1,3]

[1] School of Data Science and Engineering, East China Normal University, Shanghai, China
{52205903002,51205903054,52175100004}@stu.ecnu.edu.cn
[2] Ant Group, Hangzhou, China
shanzhu.cjm@antgroup.com
[3] KLATASDS-MOE, School of Statistics, East China Normal University, Shanghai, China
{cenchen,mgao}@dase.ecnu.edu.cn

Abstract. Federated learning (FL) is a special distributed machine learning paradigm, where decentralized clients collaboratively train a model under the orchestration of a global server while protecting users' data privacy. Concept shift across clients as a specific type of data heterogeneity challenges the generic federated learning methods, which output the same model for all clients. Clustered federated learning is a natural choice for addressing concept shift. However, we empirically show that existing state-of-the-art clustered federated learning methods cannot match some personalized learning methods. We attribute it to the fact they group clients based on their entangled signals, which results in poor clustering. To tackle the problem, in this paper, we devise a lightweight disentanglement mechanism, which explicitly captures client-invariant and client-specific patterns. Incorporating the disentanglement mechanism into clients' local training, we propose a robust clustered federated learning framework (RCFL), which groups the clients based on their client-specific signals. We conduct extensive experiments on three popular benchmark datasets to show the superiority of RCFL over the competitive baselines, including personalized federated learning methods and clustered federated learning methods. Additional experiments demonstrate its robustness against several sensitive factors. The ablation study verifies the effectiveness of our introduced components in RCFL.

Keywords: Federated Learning · Personalization · Clustering

1 Introduction

Federated learning (FL) is a special distributed machine learning paradigm, which targets coordinating the users (clients) to collaboratively train a model without compromising their privacy and data security [16,25]. In general, the training of FL proceeds in communication rounds between the server and the clients until convergence or reaching a termination criterion. Specifically, it starts with initializing a model on the server. In each of the subsequent communication

X. Wang et al. (Eds.): DASFAA 2023, LNCS 13943, pp. 677–692, 2023.
https://doi.org/10.1007/978-3-031-30637-2_45

rounds, three steps sequentially execute. First, the server randomly selects a subset of clients and distributes the latest global model to the selected clients. Second, each selected client updates its local model on the private dataset and sends model updates (or weights) back to the server. Third, the server aggregates the model updates into the global model.

FL has demonstrated impressive success in many fields, such as federated question answering system [5] and federated recommendation system [22]. However, there are still many open problems to be explored, among which data heterogeneity issue is prominent. Some factors might lead to heterogeneous datasets across clients, such as label distribution skew, quantity imbalance, and concept shift [9,16]. *In this paper, we focus on addressing concept shift.* Concept shift means the conditional distributions $\mathcal{P}_u(x|y)$ vary across clients, which is quite common in federated learning scenarios [9,16]. For example, client preferences vary in the recommender system [23] and the styles of wedding dresses vary widely around the world. We further illustrate the concept shift in Fig. 1. The images in Fig. 1 are all giraffes, but their features are quite different in terms of visual style.

Fig. 1. Illustration of concept shift. Although the four images are all giraffes, their visual characteristics vary.

The gap between clients' conditional distributions induces optimization difficulty for generic federated learning methods, like FedAvg [25] and FedProx [21], which output the same global model for all clients. The capability of personalized federated learning (PFL) methods and clustered federated learning (CFL) methods in addressing data heterogeneity has attracted significant attention. PFL methods [1,6,7,15,20] leverage the data heterogeneity and then output customized model for each client. Different from PFL methods, CFL methods utilize the affinity between clients to cluster clients with similar data distribution, then perform sub-population federated learning. Until now, CFL is the first choice for addressing the concept shift [9]. However, few works have compared the performance of PFL and CFL methods in the presence of concept shift. And we empirically show that instead of exhibiting absolute advantages over PFL methods, even existing state-of-the-art CFL methods underperform some PFL methods (see Table 2). We attribute it to the fact that existing CFL methods suffer from inaccurate clustering.

Existing CFL methods perform clustering based on clients' model gradients [3,9,27], weights [12,31] or local optima [11,24]. These signals capture client-specific (such as lines and colors) as well as client-invariant patterns (such as the giraffe's long neck, see the concept shift illustration in Fig. 1), and the latter might dominate the clustering, resulting in blurred boundaries between clients with different data distributions. We believe that the main challenge in CFL is to extract the semantically meaningful and discriminative signals, on which clients with different data distributions are quite differentiated, in a privacy-preserving manner for more robust clustering.

To tackle the above-mentioned challenge, we devise a lightweight disentanglement mechanism to explicitly model client-invariant and client-specific patterns. Specifically, we design a dual-encoder architecture consisting of $G_i(\cdot|\theta_i)$ and $G_s(\cdot|\theta_s)$ capable of capturing client-invariant and client-specific patterns, respectively. Incorporating the disentanglement mechanism into clients' local training, we propose a novel and practical CFL framework, namely Robust Clustered Federated Learning (RCFL), where the server performs clustering over clients based on collected client's weights θ_s (*the weights indirectly reflect the client-specific patterns*).

The contributions of this paper are summarized as follows.

- CFL is a natural choice for addressing concept shift [9]. However, we conduct extensive experiments and empirically find that even state-of-the-art CFL methods cannot match some PFL methods. We believe that it's because existing CFL methods employ entangled signals for clustering, which capture client-specific and client-invariant patterns, that leads to inaccurate clustering and poor convergence accuracy.
- We design a lightweight disentanglement mechanism, through which the server can leverage client-specific signals for better (more precise and more robust) clustering and better CFL further.
- We conduct extensive experiments to evaluate the performance of RCFL. The results show its superiority over existing PFL and CFL methods. Compared to existing CFL methods, further experiments demonstrate the robustness of RCFL in clustering against several factors. The ablation study verifies the effectiveness of our introduced components in RCFL.

2   Related Work

2.1   Personalized Federated Learning

PFL methods regard data heterogeneity as a blessing and exploit it to customize local models for clients. Existing methods can be mainly divided into the following three types: mixing models [1,4,6], multi-task learning [20,29] and meta-learning [7,15]. For example, FedPer [1] and FedRep [6] learn shared representation for all clients and personalize local heads for each client. MOCHA [29] encourages related clients to learn similar models. Ditto [20] achieves personalization through regularized local finetuning. Inspired by MAML [8], Per-FedAvg [7] aims to learn a great initialization, which enables clients quickly adapt to their private data distribution.

2.2 Clustered Federated Learning

The core idea of CFL is clustering clients with similar data distribution and performing sub-population federated learning. ClusteredFL [27], FL+HC [3] and CIC-FL [9] perform clustering over clients according to their gradients (model updates). Specifically, ClusteredFL [27] divides the clients into several clusters through multi-round bipartite separation based on clients' gradients. Every bipartition on a cluster requires the corresponding cluster model to converge, which results in inefficient clustering and slow convergence. FL+HC [3] introduces a hierarchical clustering step to separate clusters of clients by the similarity of their local model updates. CIC-FL [9] focuses on coping with the class imbalance between clients in clustering, which, similar to ClusteredFL [27], partitions a cluster into two parts based on a special gradient-derived feature LEGLD. IFCA [11] hosts multiple cluster models on the server. Before local training, each client downloads all the cluster models and estimates its cluster identity in terms of its empirical risk, which brings heavy computation overhead and communication costs. FeSEM [31] formulates CFL as an optimization problem. The objective is to minimize the empirical risk and the distance in parameter space between the local model and its nearest center model.

These works performs clustering based on entangled signals (model weights, gradients and local optima), which poses great difficulty to clustering. Refined more discriminative signals is more conducive to better clustering. In this paper, we introduce a lightweight disentanglement mechanism to extract the client-specific information, making more precise clustering and better convergence performance. Besides, RCFL is a hot-plug framework, which is complementary to several existing CFL methods, such as FeSEM [31] and ClusteredFL [27].

3   Our Proposed RCFL Framework

3.1   Problem Statement

Assume there are N clients, denoted as $C_1, \ldots, C_N$. Each client C_u has a private dataset $D_u = \{x_j, y_j\}_{j=1}^{|D_u|}$ sampled from $\mathcal{P}_u(x, y) = \mathcal{P}_u(x|y)\mathcal{P}_u(y)$ and is assumed to participate at each round, which is quite common in cross-silo FL setting. Due to various factors such as geography or personal preference, concept shift exists between clients, i.e., $\mathcal{P}_u(x|y)$ varies across clients. We assume there are K inherent partitions in the population. The clients can be naturally grouped into K clusters, denoted by $\mathcal{C}_1, \ldots, \mathcal{C}_K$. The goal of CFL is to group the clients precisely into K clusters and maximize the overall performance of all clients. Specifically, the global server divides the clients into K clusters using information collected during the federated learning process, such as model weights or gradients. And then, in each cluster, clients execute a cluster-level federated learning algorithm, which can be FedAvg [25] or other methods orthogonal to CFL.

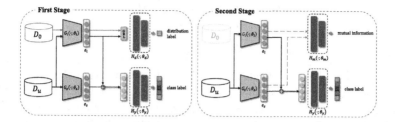

Fig. 2. The workflow of our proposed disentanglement mechanism. D_0 and D_u are the client C_u's local dataset and public dataset, respectively. $G_i(\cdot|\theta_i)$ and $G_s(\cdot|\theta_s)$ denote the client-invariant and client-specific pattern encoder, respectively. e_i and e_s are client-invariant and client-specific pattern embeddings. And $H_d(\cdot|\theta_d)$, $H_y(\cdot|\theta_y)$ and $H_m(\cdot|\theta_m)$ are the distribution classifier, label classifier, and mutual information neural estimator. The mechanism runs in two stages. In the first stage, e_i is fed into H_d after a gradient reversal layer (abbr. GRL). For samples from D_u, the summation of e_i and e_s is fed into H_y. In the second stage, H_m takes e_i and e_s as input and estimates the mutual information between them.

3.2 Proposed Framework

In this section, we present our framework RCFL. While optimizing its local empirical risk, each client runs a disentanglement mechanism to extract the client-specific signals. After receiving clients' client-specific signals, the server group the clients into K clusters and aggregate the latest update into the corresponding cluster model.

Disentanglement Mechanism. The workflow of our proposed disentanglement mechanism is shown in Fig. 2. It runs aided by a public auxiliary dataset, denoted as D_0, which is unlabelled, sampled from the global distributions, and can be accessed by all clients. The public dataset can be obtained from the public data or generated using some generative models [13,17,32]. So, each client C_u has access to both D_0 and D_u. The local model mainly consists of five components: domain-invariant pattern encoder $G_i(\cdot|\theta_i)$, domain-specific pattern encoder $G_s(\cdot|\theta_s)$, label classifier $H_y(\cdot|\theta_y)$, distribution classifier $H_d(\cdot|\theta_d)$, and mutual information neural estimator $H_m(\cdot|\theta_m)$. e_i and e_s are client-invariant and client-specific pattern embeddings captured by G_i and G_s, respectively. Particularly, $H_d(\cdot|\theta_d)$ is a binary classifier, of which the task is to determine whether a sample is from the local distribution or the global distribution. To promote G_i capture client-invariant pattern, we utilize a gradient reversal layer (abbr. GRL) proposed by [10], which reverses the gradient of distribution classification loss w.r.t. θ_i.

Each client adopts a two-stage alternating optimization within one round to better capture client-invariant patterns and client-specific patterns, respectively. In the first stage, e_i is fed to the distribution classifier H_d. And for samples from the private dataset, the element-wise summation of e_i and e_s is fed to the label classifier H_y. G_i, H_d and H_y are optimized for E_1 epochs with G_s fixed. The updates of θ_i, θ_d and θ_l follow:

$$\theta_i \longleftarrow \theta_i + \alpha \nabla_{\theta_i} (L_y - \beta L_d)$$
$$\theta_d \longleftarrow \theta_d - \alpha \nabla_{\theta_d} L_d \qquad , \qquad (1)$$
$$\theta_y \longleftarrow \theta_y - \alpha \nabla_{\theta_y} L_y$$

where L_y is the label classification loss, L_d is the distribution classification loss (binary cross entropy), α is the learning rate, and β is the balancing coefficient. The updates can be interpreted as updating θ_i and θ_y to minimize local empirical risk, updating θ_d for better classify whether a sample from the local or global distribution, and updating θ_i against H_d to promote client-invariant pattern.

In the second step, we extract the client-specific patterns. Specifically, while optimizing the label classification performance, we minimize the mutual information [2] between e_i and e_s with G_i fixed (fix the client-invariant patterns). We update θ_s, θ_y and θ_m for E_2 epochs as follows:

$$\theta_s \longleftarrow \theta_s - \alpha \nabla_{\theta_s} (L_y + \gamma L_m)$$
$$\theta_y \longleftarrow \theta_y - \alpha \nabla_{\theta_y} L_y \qquad , \qquad (2)$$
$$\theta_m \longleftarrow \theta_m - \alpha \nabla_{\theta_m} L_m$$

where L_m is the mutual information measuring the relationship between e_i and e_s, and γ balances L_y and L_m. Formally, L_m is defined in Eq. 3, where $\mathbb{P}_{XY}$ denotes the joint probability distribution, $\mathbb{P}_X$ and $\mathbb{P}_Y$ are the marginal distributions.

$$L_m(X; Y) = \int_{\mathcal{X} \times \mathcal{Y}} \log \frac{d\mathbb{P}_{XY}}{d\mathbb{P}_X \otimes \mathbb{P}_Y} d\mathbb{P}_{XY} \qquad (3)$$

Finally, the client performs minimization of local risk while decoupling client-invariant and client-specific patterns through two-stage alternating optimization.

Clustering and Intra-cluster Aggregation. At the beginning of each communication round, the server synchronizes clients with corresponding cluster models. As we have no information on the clients before the federated training, the server initializes K cluster models with the same weights and randomly broadcasts to the clients. After clients finish local training and send back their model updates/weights, the server groups the clients into K clusters $\{C_k\}_{k=1}^{K}$ using K-Means or other clustering algorithms based on their client-specific signals $\{\theta_{u,s}\}_{u=1}^{N}$ followed by intra-cluster model aggregation. We present the overview of RCFL in Fig. 3.

Cold Start Problem. Clustered federated learning framework should be flexible to handle new coming clients. Which cluster model should the server broadcast to a fully new client is the main challenge, which we call the cold start problem. To tackle this, we design a practical and easy-to-implement cold start mechanism. We record the client-specific signals collected after the first communication round and represent each cluster with the average of θ_s of the clients belonging to the same cluster, i.e., $\omega_k = \frac{1}{|C_k|} \sum_{u \in C_k} \theta_{u,s}$. Once a new client C_{N+1}

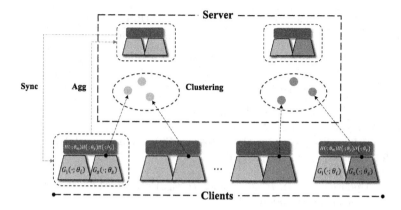

Fig. 3. The overview of RCFL. Sync and Agg denote synchronization and aggregation, respectively. RCFL sequentially runs the following three steps: (1) each client synchronizes its local model with the corresponding cluster model, (2) each client runs the disentanglement mechanism shown in Fig. 2, and (3) the server performs clustering over clients based on their client-specific weights $\{\theta_{u,s}\}_{u=1}^{N}$ and aggregates their latest model weights to corresponding cluster models.

participates in training, the client runs the disentanglement mechanism for a few epochs, and then sends back its weights $\theta_{N+1,s}$ to the server. Through calculating the euclidean distance between $\theta_{N+1,s}$ and $\{\omega_k\}_{k=1}^{K}$, the server broadcasts the nearest cluster model to the client to help it keep track of the latest cluster model. Thanks to the elegant performance of RCFL in efficient clustering, which is verified in Sect. 4.4, RCFL can handle the cold start problem easily.

Model Summary. The workflow of our proposed method RCFL is presented in Algorithm 1. First, the server synchronizes the clients with corresponding cluster models. Then, each client performs the disentanglement mechanism for $E_1 + E_2$ epochs to explicitly capture the client-invariant and client-specific patterns. When clients finish local training, the server collects their latest updated models and performs clustering over clients based on $\{\theta_{u,s}\}_{u=1}^{N}$. Finally, the server aggregates clients' model weights within a cluster and updates the corresponding cluster model.

4   Experiments

In this section, we conduct extensive experiments to verify the effectiveness of our proposed framework RCFL and compare it with existing competitive baselines. Further, we evaluate the robustness of RCFL against local epochs and the number of clusters K. Finally, we conduct an additional ablation study to validate the effectiveness of the introduced components in our framework.

Algorithm 1: Robust Clustered Federated Learning (RCFL)

Input : initialized K cluster models with same weights $\left\{\hat{\theta}_{k,i}, \hat{\theta}_{k,s}, \hat{\theta}_{k,y}, \hat{\theta}_{k,d}, \hat{\theta}_{k,m}\right\}_{k=1}^{K}$, total communication round T, epochs of the first stage E_1, epochs of the second stage E_2, learning rate α, and balance coefficient β and γ.

Output: $\left\{\hat{\theta}_{k,i}^{T}, \hat{\theta}_{k,s}^{T}, \hat{\theta}_{k,y}^{T}, \hat{\theta}_{k,d}^{T}, \hat{\theta}_{k,m}^{T}\right\}_{k=1}^{K}$

1 **for** *each round* $t = 0, 1, ..., T-1$ **do**
2 | [**Synchronize clients with corresponding cluster models**]
3 | **for** *client* $C_u \in \{C_1, ..., C_N\}$ **do**
4 | | [**First Stage**]
5 | | Update $G(\cdot|\theta_{u,i})$, $H(\cdot|\theta_{u,d})$ and $H(\cdot|\theta_{u,y})$ with fixed $G(\cdot|\theta_{u,s})$ for E_1 epochs using updates (1).
6 | | [**Second Stage**]
7 | | Update $G(\cdot|\theta_{u,s})$, $H(\cdot|\theta_{u,y})$ and $H(\cdot|\theta_{u,m})$ with fixed $G(\cdot|\theta_{u,i})$ for E_2 epochs using updates (2).
8 | | [**Return**]
9 | | Send back the updated parameters $\left\{\theta_{u,i}^{t+1}, \theta_{u,s}^{t+1}, \theta_{u,y}^{t+1}, \theta_{u,d}^{t+1}, \theta_{u,m}^{t+1}\right\}$
10 | **end**
11 | [**Clustering**]
12 | Cluster $\left\{\theta_{u,s}^{t+1}\right\}_{u=1}^{N}$ into $\mathcal{C}_1, ..., \mathcal{C}_K$
13 | [**Intra-cluster Aggregation**]
14 | **for** $k \in [K]$ **do**
15 | | $\left\{\hat{\theta}_{k,i}^{t+1}, \hat{\theta}_{k,s}^{t+1}, \hat{\theta}_{k,y}^{t+1}, \hat{\theta}_{k,d}^{t+1}, \hat{\theta}_{k,m}^{t+1}\right\} = \frac{1}{|\mathcal{C}_k|}\sum_{u \in \mathcal{C}_k} \theta_u^{t+1}$
16 | **end**
17 **end**

4.1 Datasets and Models

We simulate the concept shift setting, where $\mathcal{P}_u(x|y)$ varies across clients, with three widely used benchmark datasets, PACS [19], Office-Home [30] and Digit-Five. Datasets visualization is shown in Fig. 4.

- Digit-Five is a 10-class digit-recognition dataset collected from five datasets belonging to different domains, including MNIST [18], MNIST-M [10], SVHN [26], USPS [14] and SYN [10]. Each domain dataset is equally divided into five clients, and a total of 25 clients are simulated.
- PACS is a 7-class dataset with four domains: art painting, cartoon, photo, and sketch. We create 14 clients, of which the number of clients belonging to art painting, cartoon, photo and sketch are 3, 3, 2 and 6 respectively.
- Office-Home is a 65-class object recognition dataset, which consists of images from 4 different domains: artistic images, clip art, product images and real-world images. We simulate 14 clients, of which the number of clients belonging to different domains are 2, 4, 4, and 4 respectively.

(a) Dataset: Digit-Five   (b) Dataset: PACS   (c) Dataset: Office-Home

Fig. 4. We use three public datasets in our experiments: (1) Digit-Five dataset, which includes five domains: MNIST (mt), MNIST-M (mm), SVHN (sv), USPS (up) and Synthetic (sy). (2) PACS dataset, which includes four domains: art painting, cartoon, photo and sketch. (3) Office-Home dataset, which includes four domains: artistic images (AR), clip art (CA), product images (PD) and real-world images (RW).

We simulate a small public dataset by randomly sampling about 10% from all the domain datasets. And the size of the private dataset of all clients is the same under our split. Detailed statistics of our simulated dataset are presented in Table 1. For all experiments, we adopt a CNN-based model structure.

Table 1. Datasets statistics and the corresponding partitioning information.

Datasets	Domains	Classes	Clients	Smaples/C	Public Samples
Digit-Five	5	10	25	1166	4126
PACS	4	7	14	500	992
Office-Home	4	65	14	800	1588

4.2 Baselines and Evaluations

We compare our method against various baselines to show its effectiveness, including generic federated learning methods, PFL methods, and CFL methods.

- Standalone denotes local training, where clients optimize local models independently on their private data.
- Generic methods
  - FedAvg [25] is a vanilla federated learning method, which coordinates the clients to optimize a server model.
  - FedProx [21] is a re-parameterization of FedAvg, which adds a proximal term to the local empirical risk to prevent the local model from deviating far from the global model.
- Personalized Methods
  - FedPer [1] learns a unified representation (base model) for all clients and personalizes the local model with private heads.

- Per-FedAvg [7], a MAML-like PFL method, searches for a suitable initialization, which enables clients fastly personalize local models through fewer finetune steps.
- FedRep [6], similar to FedPer [1], adopts the decomposition of the base model and the personalized heads, which alternately updates the personalized heads and the base model.
- Ditto [20] learns a generic and personalized models simultaneously, and the former acts as an regularizer to client's personlaized model training.
- Clustered Methods
  - ClusteredFL [27] is the first solution to CFL, which iteratively divides a cluster into two parts based on the cosine similarity between clients' gradients.
  - FeSEM [31] is a full-parameter-based clustering method, which derives the optimal matching of clients and clusters through expectation maximization.
  - IFCA [11] is a recently proposed state-of-the-art CFL method. Before local optimization, each client synchronizes multiple cluster models from the server and optimizes the model which performs the best on its private dataset.

For evaluation, we use the *weighted average accuracy* over all clients on testsets. All experiment results are aggregated with three random runs.

4.3   Implementation and Hyperparameters

For all methods, we set the learning rate α to 0.01 with a learning rate decay of 0.99 per round, local training batch size 50, and communication rounds 100. We perform a grid search for the best β and γ from the candidate set {0.2, 0.4, 0.6, 0.8, 1}. All the methods are trained with Adam optimizer. Due to a lack of sufficient private data, to obtain a better H_d, we warm up RCFL in the first 20 communication rounds in FedAvg [25] manner. We set the local update epochs to 5 for the baselines. For fairness, E_1 is set to 3 for the first stage, and E_2 is set to 2 for the second stage in RCFL. The sizes of G_i and G_s are half that of the base encoder used for the baslines. In addition, we assume that the number of potential partitions of the population is known to the server. We fix the dataset splits and random seeds for reproducibility. All models are implemented in PyTorch and run on 1 T V100 GPU. For reproducibility, we make our source code publicly available at: https://github.com/tdye24/robust-cfl.

4.4   Overall Performance

The results of the eleven baselines and RCFL on three datasets are shown in Table 2. It can be observed that our proposed method *significantly* outperforms all the baselines across all datasets. We perform a Student's paired t-test for the results. The advantage of RCFL over the best baseline is statistically significant with $p < 0.01$ on all datasets. Due to the existence of severe concept shift,

Table 2. Overall performance comparison with all baselines. Best results are highlighted in bold, while the runner-ups' are underlined. † means the advantage of RCFL over the best baseline is statistically significant by the Student's paired t-test with $p < 0.01$.

Type	Method	Digit-Five	PACS	Office-Home
Centralized	Standalone	90.94 ± 0.21	60.69 ± 0.43	18.21 ± 0.56
Generic	FedAvg	87.86 ± 0.34	61.70 ± 0.47	28.48 ± 0.64
	FedProx	87.02 ± 0.56	55.15 ± 0.28	24.21 ± 0.39
Personalized	FedPer	85.77 ± 0.77	44.62 ± 0.18	14.04 ± 0.11
	Per-FedAvg	87.54 ± 0.82	65.25 ± 0.28	$\underline{30.36 \pm 0.14}$
	FedRep	53.24 ± 0.34	33.70 ± 0.18	3.93 ± 0.21
	Ditto	$\underline{93.62 \pm 0.24}$	$\underline{69.40 \pm 0.18}$	30.07 ± 0.32
Clustered	ClusteredFL	88.80 ± 0.15	63.39 ± 0.30	16.50 ± 0.55
	FeSEM	85.03 ± 0.42	48.69 ± 3.47	15.30 ± 0.63
	IFCA	92.12 ± 0.78	67.66 ± 1.03	28.50 ± 0.18
Ours	**RCFL**	$\mathbf{95.31 \pm 0.25^{\dagger}}$	$\mathbf{73.36 \pm 0.62^{\dagger}}$	$\mathbf{36.57 \pm 0.57^{\dagger}}$

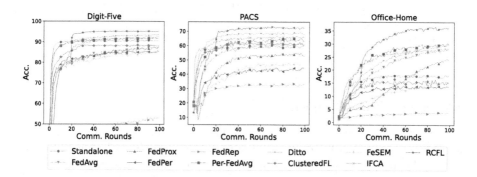

Fig. 5. Convergence comparison between RCFL and the baselines.

Standalone gives better performance on Digit-Five compared to FedAvg and Fed-Prox. Per-FedAvg and Ditto outperform Standalone and Generic methods by a large margin, demonstrating the strength of PFL methods in addressing concept shift. In the presence of concept shift, it is difficult to learn a universal representation for all clients, which leads to poor performance of FedPer and FedRep. More importantly, finetuning the private head without collaborative aggregation makes it limited for FedPer and FedRep to leverage the knowledge from related clients. We infer it is why PerPer and FedRep underperform FedAvg. Although existing CFL methods are generally effective, it is clear that the best existing CFL method IFCA is still inferior to Ditto by a large margin. And as a CFL method, RCFL beats all competitors. Furthermore, we plot the convergence curves of all methods on all datasets in Fig. 5. The point where the accuracy of RCFL increases sharply corresponds to that where we switch to normal RCFL from warm-up. Compared to the baselines, RCFL converges to a better point with faster speed.

To gain insight into the results of CFL methods, we evaluate the clustering performance using the commonly used clustering metric NMI [28], which has a value between 0–1, with larger values indicating better clustering. We plot the NMI curves of all CFL methods in Fig. 6. RCFL can precisely cluster the clients, while other methods suffer from poor clustering. Combined with Table 2, generally, the better the clustering results, the better the overall convergence performance.

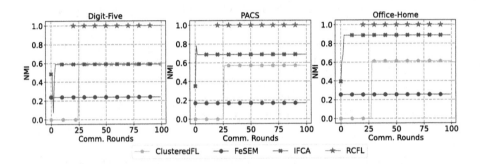

Fig. 6. Clustering results comparison between CFL methods on three datasets. The larger NMI, the better clustering.

4.5   Effect of Local Epochs

Federated learning expects more local updates to reduce communication costs. However, excessive local updates can easily lead to overfitting, which might result in very different weights or model updates for even similar clients.

In this section, we test the robustness of the CFL methods against local epochs. We evaluate all methods on Digit-Five with varying local epochs $E \in [2, 5, 10, 15, 20]$. In particular, for fair comparison, let $(E_1, E_2) \in \{(1, 1), (3, 2), (5, 5), (8, 7), (10, 10)\}$ in RCFL. Experiment results are shown in Table 3. We observe that RCFL consistently outperforms ClusteredFL, FeSEM, and IFCA across all local epochs candidates. The accuracy of RCFL is stable at about 95%, which shows its robustness against varying local epochs.

4.6   Effect of K

We have no information about the potential partitions for clients, so it is a critical task to set K, the number of clusters. To demonstrate the robustnessof

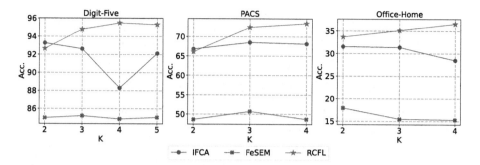

Fig. 7. Effects of the number of clusters K.

Table 3. Effects of local epochs. Best results are highlighted in bold, while the runner-ups' are underlined. † means the advantage of RCFL over the best baseline is statistically significant by the Student's paired t-test with $p < 0.01$.

Method	K				
	2	5	10	15	20
ClusteredFL	86.38 ± 0.25	88.80 ± 0.15	89.29 ± 0.28	89.51 ± 0.31	89.84 ± 0.27
FeSEM	85.57 ± 0.44	85.03 ± 0.42	85.46 ± 0.17	85.30 ± 0.23	85.88 ± 0.85
IFCA	93.56 ± 0.52	92.12 ± 0.78	93.44 ± 0.31	91.45 ± 0.32	93.13 ± 0.10
RCFL	**95.25 ± 0.11**†	**95.31 ± 0.25**†	**95.04 ± 0.12**†	**95.44 ± 0.03**†	**95.09 ± 0.21**†

RCFL against K, we run the CFL methods among our baselines with varying values of K on all datasets, where K is less than or equal to the number of potential partitions of each dataset. We don't compare ClusteredFL in this experiment as it's an adaptive clustering method, where K is adjusted by the algorithm itself. As shown in Fig. 7, RCFL exceeds IFCA and FeSEM in most cases. Interestingly, on Digit-Five, our proposed method achieves the best performance when K is 4, less than the number of potential partitions. Through analysis of the intermediate experiment results, we find that clients belonging to SVHN and SYN are clustered into the same cluster. The algorithm optimizes to a better solution because the difference between SVHN and SYN is very slight (see datasets visualization in Fig. 4), and the federation of clients from SVHN and SYN enhances each other.

4.7 Ablation Study

We conduct an additional ablation study to examine the effectiveness of the introduced components in RCFL.

Alternate Optimization. In the local training, we adopt an alternate optimization (AO) strategy between G_i and G_s to explicitly capture client-invariant and client-specific patterns, respectively. To demonstrate the benefit of this strategy, we run RCFL without alternate optimization on all datasets. From Table 4, we can see that RCFL w/o. AO experiences a severe performance drop on

Table 4. Ablation study on alternate optimization (AO) and mutual information minimization (MIM). † means the difference between the result and that of RCFL is statistically significant by the Student's paired t-test with $p < 0.01$.

Datasets	Digit-Five	PACS	Office-Home
RCFL	95.31 ± 0.25	73.36 ± 0.62	36.57 ± 0.57
RCFL w/o. AO	95.15 ± 0.11	72.47 ± 0.15†	33.86 ± 0.21†
RCFL w/o. MIM	94.38 ± 0.12†	69.25 ± 0.13†	33.36 ± 0.34†

PACS and Office-Home. We attribute it to that RCFL w/o. AO has trouble decomposing the original pattern into client-invariant and client-specific patterns. Although it is less affected on Digit-Five, there is still an accuracy discount.

Mutual Information Minimization. We further run RCFL without mutual information minimization (MIM) on all datasets. The performance on PACS and Office-Home degrades drastically, by about 4% and 3% respectively. The accuracy on Digit-Five dataset is also reduced by 1%.

5   Future Work

Some generative methods can help generate a synthetic but related public dataset for our framework. For example, the server can train a generative model with supervision information (such as a classifier) of clients [32]. In addition, some general knowledge can be shared across clusters, for example, the task-invariant patterns encoder G_i in our framework. We leave it for future work.

6   Conclusion

In this paper, we studied the concept shift problem in federated learning. We adequately compared the capabilities of PFL and CFL methods in the presence of concept shift. In particular, we highlighted the common problem of existing CFL methods, i.e., clustering clients based on clients' entangled signals, and proposed a novel and robust clustered federated learning framework called RCFL, which clusters the clients based on their extracted client-specific signals. We also provided a set of experiments on benchmark datasets to illustrate the superiority of RCFL over existing PFL and CFL methods. Finally, we conducted a series of experiments to show the robustness of RCFL and ran two variants of RCFL to validate the effectiveness of our introduced components.

Acknowledgments. This work was supported by the National Natural Science Foundation of China (grant numbers 62202170, U1911203, and 61977025), and the Ant Group.

References

1. Arivazhagan, M.G., Aggarwal, V., Singh, A.K., Choudhary, S.: Federated learning with personalization layers. arXiv preprint arXiv:1912.00818 (2019)
2. Belghazi, M.I., et al.: MINE: mutual information neural estimation. arXiv preprint arXiv:1801.04062 (2018)
3. Briggs, C., Fan, Z., Andras, P.: Federated learning with hierarchical clustering of local updates to improve training on non-IID data. In: 2020 International Joint Conference on Neural Networks (IJCNN), pp. 1–9. IEEE (2020)
4. Chen, C., Ye, T., Wang, L., Gao, M.: Learning to generalize in heterogeneous federated networks. In: Proceedings of the 31st ACM International Conference on Information & Knowledge Management, pp. 159–168 (2022)
5. Chen, J., Zhang, R., Guo, J., Fan, Y., Cheng, X.: FedMatch: federated learning over heterogeneous question answering data. In: Proceedings of the 30th ACM International Conference on Information & Knowledge Management, pp. 181–190 (2021)
6. Collins, L., Hassani, H., Mokhtari, A., Shakkottai, S.: Exploiting shared representations for personalized federated learning. In: International Conference on Machine Learning, pp. 2089–2099. PMLR (2021)
7. Fallah, A., Mokhtari, A., Ozdaglar, A.: Personalized federated learning with theoretical guarantees: a model-agnostic meta-learning approach. Adv. Neural. Inf. Process. Syst. **33**, 3557–3568 (2020)
8. Finn, C., Abbeel, P., Levine, S.: Model-agnostic meta-learning for fast adaptation of deep networks. In: International Conference on Machine Learning, pp. 1126–1135. PMLR (2017)
9. Fu, Y., Liu, X., Tang, S., Niu, J., Huang, Z.: CIC-FL: enabling class imbalance-aware clustered federated learning over shifted distributions. In: Jensen, C.S., et al. (eds.) DASFAA 2021. LNCS, vol. 12681, pp. 37–52. Springer, Cham (2021). https://doi.org/10.1007/978-3-030-73194-6_3
10. Ganin, Y., Lempitsky, V.: Unsupervised domain adaptation by backpropagation. In: International Conference on Machine Learning, pp. 1180–1189. PMLR (2015)
11. Ghosh, A., Chung, J., Yin, D., Ramchandran, K.: An efficient framework for clustered federated learning. In: NeurIPS, vol. 33 (2020)
12. Ghosh, A., Hong, J., Yin, D., Ramchandran, K.: Robust federated learning in a heterogeneous environment. arXiv preprint arXiv:1906.06629 (2019)
13. Goodfellow, I.J., et al.: Generative adversarial nets. In: NIPS (2014)
14. Hull, J.J.: A database for handwritten text recognition research. IEEE Trans. Pattern Anal. Mach. Intell. **16**(5), 550–554 (1994)
15. Jiang, Y., Konečný, J., Rush, K., Kannan, S.: Improving federated learning personalization via model agnostic meta learning. arXiv preprint arXiv:1909.12488 (2019)
16. Kairouz, P., et al.: Advances and open problems in federated learning. Found. Trends® Mach. Learn. **14**(1–2), 1–210 (2021)
17. Karras, T., Laine, S., Aila, T.: A style-based generator architecture for generative adversarial networks. In: 2019 IEEE/CVF Conference on Computer Vision and Pattern Recognition (CVPR), pp. 4396–4405 (2019)
18. LeCun, Y., Bottou, L., Bengio, Y., Haffner, P.: Gradient-based learning applied to document recognition. Proc. IEEE **86**(11), 2278–2324 (1998)
19. Li, D., Yang, Y., Song, Y.Z., Hospedales, T.M.: Deeper, broader and artier domain generalization. In: Proceedings of the IEEE International Conference on Computer Vision, pp. 5542–5550 (2017)

20. Li, T., Hu, S., Beirami, A., Smith, V.: Ditto: fair and robust federated learning through personalization. In: International Conference on Machine Learning, pp. 6357–6368. PMLR (2021)
21. Li, T., Sahu, A.K., Zaheer, M., Sanjabi, M., Talwalkar, A., Smith, V.: Federated optimization in heterogeneous networks. arXiv preprint arXiv:1812.06127 (2018)
22. Lin, Y., et al.: Meta matrix factorization for federated rating predictions. In: Proceedings of the 43rd International ACM SIGIR Conference on Research and Development in Information Retrieval, pp. 981–990 (2020)
23. Luo, S., Xiao, Y., Song, L.: Personalized federated recommendation via joint representation learning, user clustering, and model adaptation. In: Proceedings of the 31st ACM International Conference on Information & Knowledge Management, pp. 4289–4293 (2022)
24. Mansour, Y., Mohri, M., Ro, J., Suresh, A.T.: Three approaches for personalization with applications to federated learning. arXiv preprint arXiv:2002.10619 (2020)
25. McMahan, B., Moore, E., Ramage, D., Hampson, S., y Arcas, B.A.: Communication-efficient learning of deep networks from decentralized data. In: Artificial Intelligence and Statistics, pp. 1273–1282. PMLR (2017)
26. Netzer, Y., Wang, T., Coates, A., Bissacco, A., Wu, B., Ng, A.Y.: Reading digits in natural images with unsupervised feature learning (2011)
27. Sattler, F., Müller, K.R., Samek, W.: Clustered federated learning: model-agnostic distributed multitask optimization under privacy constraints. IEEE Trans. Neural Netw. Learn. Syst. **32**, 3710–3722 (2020)
28. Shannon, C.E.: A mathematical theory of communication. Bell Syst. Tech. J. **27**(3), 379–423 (1948)
29. Smith, V., Chiang, C.K., Sanjabi, M., Talwalkar, A.S.: Federated multi-task learning. Adv. Neural Inf. Process. Syst. **30** (2017)
30. Venkateswara, H., Eusebio, J., Chakraborty, S., Panchanathan, S.: Deep hashing network for unsupervised domain adaptation. In: Proceedings of the IEEE Conference on Computer Vision and Pattern Recognition, pp. 5018–5027 (2017)
31. Xie, M., et al.: Multi-center federated learning. arXiv preprint arXiv:2005.01026 (2020)
32. Zhu, Z., Hong, J., Zhou, J.: Data-free knowledge distillation for heterogeneous federated learning. In: International Conference on Machine Learning, pp. 12878–12889. PMLR (2021)

Privacy Preserving Federated Learning Framework Based on Multi-chain Aggregation

Yingchun Cui and Jinghua Zhu[✉]

School of Computer Science and Technology, Heilongjiang University,
Harbin 150080, China
zhujinghua@hlju.edu.cn

Abstract. Federated Learning is a promising machine learning paradigm for collaborative learning while preserving data privacy. However, attackers can derive the original sensitive data from the model parameters in Federated Learning with the central server because model parameters might leak once the server is attacked. To solve the above server attack challenge, in this paper, we propose a novel server-free Federated Learning framework named MChain-SFFL which performs multi-chain parallel communication in a fully distributed way to update the model to achieve more secure privacy protection. Specifically, MChain-SFFL first randomly selects multiple participants as the chain heads to initiate the model parameter aggregation process. Then MChain-SFFL leverages the single-masking and chained-communication mechanisms to transfer the masked information between participants within each serial chain. In this way, the masked local model parameters are gradually aggregated along the chain nodes. Finally, each chain head broadcasts the aggregated local model to the other nodes and this propagation process stops until convergence. The experimental results demonstrate that for Non-IID data, MChain-SFFL outperforms the compared methods in model accuracy and convergence speed. For IID data, the accuracy and convergence speed of MChain-SFFL are close to Chain-PPFL and FedAVG.

Keywords: Federated Learning · Privacy Preservation · Parallel Multi-chain

1 Introduction

Federated Learning (FL) is a promising collaborative learning paradigm proposed by Google in 2016, which only collects model parameters trained locally instead of raw data directly [7,12]. While FL allows participants to keep raw data locally, existing work has shown that it is susceptible to inference attacks and poisoning attacks [2,16]. For instance, adversaries who launch the membership inference attack can infer whether a user's data was used for training the model.

This work was supported by the Natural Science Foundation of Heilongjiang Province of China, LH2022F045.

Differential Privacy (DP) [1,15] achieves a trade-off between the level of privacy protection and model accuracy by adding noise to the original gradients. Secure Multi-party Computation (SMC) [4,6] uses cryptography techniques to protect privacy, which does not impair accuracy but increases the communication and computation overhead. Other methods combine these two approaches [18,21], which consume significant communication overhead and need a trusted third-party server.

In [10], the authors propose a chained FL framework named Chain-PPFL, which transfers masked information between participants with a serial chain to achieve privacy protection without impairing accuracy. However, it requires a third-party server. When the server is attacked or fails, the whole network stops working. Moreover, the convergence speed of this model is slow. BrainTorrent [14] allows clients to communicate directly between clients without a central server. However, it does not consider privacy protection in the process of client communication, and there is a risk of privacy leakage.

In this paper, we propose a novel multi-chain aggregation server-free federated learning framework (MChain-SFFL) to solve the above problems. In MChain-SFFL, no federated server is needed and each participant has an equal chance to be a temporary server. Firstly, MChain-SFFL randomly selects multiple participants as the chain head nodes to initiate the multi-chain model aggregation process. Secondly, multiple chains are formed in parallel between the participants to transfer encrypted parameter values. Each chain head node will obscure its own parameter values with a unique *Token* with each chain. The descendent node will be selected randomly by the participant who currently owns the *Token* from the other neighbors without getting their mask information. Then the current participant sends its output to the descendent node. Finally, each chain head updates the model parameters for that chain and broadcasts the parameters to all participants. Then each participant will average the received model parameters as the model update for this iteration. The main contributions of this paper are summarized as follows:

- We propose MChain-SFFL, a privacy-preserving federated learning framework based on multi-chain aggregation without any third-party server. This framework improves system security and multi-chain parallel computing improves the convergence speed of the model.
- We enhance the privacy protection level. Each chain head node generates a unique random number to mask the weights of the participants, which can protect the privacy of all parties without lowering the accuracy of the model.
- We experimentally verify the effectiveness of MChain-SFFL in terms of model accuracy and convergence speed.

In the next section, we introduce the related work of FL. In Sect. 3, we present MChain-SFFL in detail. Section 4 analyzes experimental results. Finally, we conclude this paper.

2   Related Work

2.1   Federated Learning

Federated Learning (FL) supports decentralized collaborative machine learning over a number of devices or companies [7,12]. FedAVG [13] is a baseline approach to FL, first applied to the Google Keyboard App. Using this method the server aggregates the received model parameters and then broadcasts the updated global model to all participants for the next iteration. FL can be divided into two categories: centralized and distributed. In centralized FL, the entire network will stop working when the central server suffers from a single point of failure and bottleneck problem. This problem can be solved by using the decentralized network [8,11].

In distributed FL, the participants exchange their model parameters directly in a peer-to-peer manner [9]. Gossip protocol is widely used in distributed systems [3], each participant sends the message to a group of other participants, and the message propagates participant by participant throughout the entire network. DeepChain [19] is a framework based on Block-chain which leverages incentive mechanism to ensure privacy and fairness in the collaborative training process. In the decentralized network setting, the SDTF proposed in [17] ensures the privacy of local data with a lower communication bandwidth cost.

2.2   Privacy Protection

Differential Privacy (DP) [1,15] is a commonly used privacy protection method in FL, which perturbs the original information by adding noise to the model parameters to protect privacy. Popular methods to implement DP include the Laplace mechanism and the Gaussian mechanism. An attacker cannot obtain accurate information about the model parameters by applying DP to FL. However, DP sacrifices the accuracy of the model. Moreover, the DLG method proposed by [24] experimentally concludes that adding noise to the DP algorithm causes gradient leakage when the variance of the noise is less than 10^{-4} and protects the gradient with loss of accuracy when the variance of the noise added is greater than 10^{-3}. Secure Multi-party Computation (SMC) [4,6] is another privacy-preserving method commonly used for FL, which mainly includes Homomorphic Encryption and Secret Sharing. Homomorphic Encryption (HE) [5,18] computes the ciphertext and the result of decryption is the same as the result of operating under plaintext. The Secret Sharing protocol proposed by Shamir [22] divides the secret value into multiple shares and distributes these shares to multiple participants, each of whom has only one share of the secret, and eventually all or a certain number of shared values can reconstruct the original secret value. Although SMC can ensure that model accuracy is not impaired, it requires expensive computation and communication resources. In this paper, we propose the MChain-SFFL that combines the advantages of DP and SMC to protect the model privacy without impairing accuracy and reduce computation load.

3   Method

In this Section, we introduce the details of the MChain-SFFL. Table 1 presents the symbols used in this paper and their corresponding meanings.

Table 1. List of symbols.

Symbols	Meaning
n	the number of users
t	training round
m	the number of chains running in parallel
l	the length of chain
T_c	the chain head of chain c
R_c^t	a random number generated by the chain c in round t
U_i^t	the information sent by user i in round t
n_p	a counter which counts the number of participants
$token_i^t$	the $token$ constructed by user i in round t
p_c^t	the update of the model parameters of chain c in round t

In this paper, we propose the MChain-SFFL with a novel multi-chain aggregation mechanism. There are n participants User 1, User 2,..., User n, each has their own dataset $D_1, D_2, ..., D_n$ and train their local model $w_1, w_2, ..., w_n$ on these datasets correspondingly. In the peer-to-peer network, we can use the method proposed by [23] to build the neighbor list. Each node maintains a neighbor list containing the name and location id of the node's neighbor nodes. Because this is not the main issue discussed in this paper, we assume that every participant has established their neighbor lists before training.

MChain-SFFL does not need the third-party server and builds m chains in parallel for each round to obtain a better global model. Specifically, as shown in Fig. 1, in each round, MChain-SFFL randomly selects m participants as chain heads (corresponding to yellow users), and each chain head generates a unique *Token*. The current participants owning the *Token* will randomly select a user from the rest of the neighbors as the descendent node without obtaining their mask information. When a specific chain length is reached, the last participant sends the model parameters accumulated on the chain to the corresponding chain head according to the id in the *Token*. Each chain head subtracts the random mask from accumulated parameters and does the average as the model update for that chain. Then each chain head broadcasts the parameter to all users. Each user aggregates the received parameters from m chain heads as the model update for this round. Because the participants on the chain transmit masked information, even if the same participant appears on multiple chains in the same round, the attacker cannot derive any sensitive information without collusion.

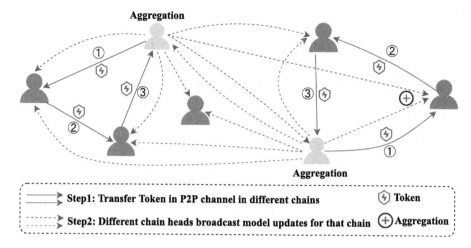

Fig. 1. Framework of MChain-SFFL. Note that, we only present the case where chain number $m=2$, chain length $l=3$ and the user number $n=7$ here, and the yellow users represent the heads of different chains. (Color figure online)

In this paper, we assume that all participants are honest-but-curious. There is no centralized trusted third party, and the communication channel is reliable and secure. The procedure of MChain-SFFL in round t is as follows:

Step 1: Suppose that $m(m < n)$ participants $T_1, T_2, ..., T_m$ are first selected randomly from n users and each of them serves as a chain head to initiate the training process. Each chain head node $T_c(c = 1, 2, ..., m)$ will use the pseudo-random generator to generate a random number $R_c^t(R_c^t \in R^d$, d is the dimension of w). Then R_c^t is added to local model parameter $w_{T_c}^t$ as a mask, and let $U_0^t = w_{T_c}^t + R_c^t$. Next, every chain head constructs $token_0^t\left(T_c^{id}, t, U_0^t, n_p\right)$, where T_c^{id} is the location id of the chain head T_c, and n_p is a counter to count the number of participants. Each chain performs the following operations in parallel.

Step 2: The chain head node sends $token_0^t$ to the first participant(named User 1) selected randomly from the neighbor list in this round. The parameter w_1^t is the local update for User 1, and the local training algorithm is the same as FedAVG. Then, User 1 uses U_0^t to mask its local update w_1^t.

$$U_1^t = w_1^t + U_0^t = w_1^t + w_{T_c}^t + R_c^t \tag{1}$$

$$n_p = n_p + 1 \tag{2}$$

User 1 updates $token_1^t(T_c^{id}, t, U_1^t, n_p)$ and selects a participant randomly from its neighbor-list to transmit the masked result.

......

User i uses the output of User $i-1$ as the mask to compute

$$U_i^t = w_i^t + U_{i-1}^t = \sum_{j=1}^{i} w_j^t + w_{T_c}^t + R_c^t, \tag{3}$$

$$n_p = n_p + 1. \tag{4}$$

Assuming that at this time $n_p = l - 1$, where l is the chain length. Then User i sends the updated $token_i^t$ to the chain head T_c corresponding to the chain.

Step 3: The chain head node T_c updates the model parameter for chain c to P_c^t according to the output of the last participant User i.

$$P_c^t = \frac{U_i^t - R_c^t}{n_p + 1} \tag{5}$$

T_c broadcasts P_c^t to all users. When the chain head node T_c receives P_c^t sent by the other chain head nodes, it updates the global model in the following way.

$$w^t = \frac{P_1^t + P_2^t + ... + P_m^t}{m} \tag{6}$$

Every user except the chain head will receive the model parameters broadcasted by m chain head nodes. Each user updates the model parameter w^t by (6).

Until now, one multi-chain aggregation calculation is completed, then return to step 1 for the next iteration.

4   Experiment

In our experiment, we compare the proposed MChain-SFFL with three methods: Chain-PPFL algorithm [10], FedAVG algorithm [13], and DP-based FL algorithm (Laplacian mechanism) [20]. The effectiveness of the proposed MChain-SFFL in accuracy and convergence speed is demonstrated. We use PyTorch as the experimental platform, and the simulation experiments are run on a computer with an Intel Core i5-6200U CPU @ 2.40 GHz.

4.1   Experimental Setup

Model and Parameter Settings: In this paper, we use two training models: (a) MLP, a simple multilayer-perceptron with 2-hidden layers with 200 units each using ReLU activations; (b) CNN, a CNN with two 5×5 convolution layers, a fully connected layer with a 512-unit and ReLU activation, and a softmax layer; The federated setting is set as follows: number of users $n = 100$, number of local epochs $E = 5$, local batch size $B = 16$, learning rate $\eta = 0.1$, DP-based FL respectively fixes $\epsilon = 8$ and $\epsilon = 1$.

Datasets: We use the MNIST dataset [12] to test the accuracy and convergence speed for two types of data, IID data and Non-IID data. IID data is shuffled MNIST, then partitioned into 100 users, each receiving 600 examples. Non-IID data is divided into 200 shards of size 300 by digit label. Each user has 2 shards.

Table 2. Training rounds required to reach 95% accuracy with different chain numbers and chain lengths.

m	l			
	10	15	25	40
2	281	252	188	156
3	248	196	181	161
4	242	187	153	180
5	185	169	147	178
6	175	173	164	194

Table 3. Accuracy of different chain numbers and chain lengths for 300 training rounds.

m	l			
	10	15	25	40
2	95.59	95.60	95.74	95.73
3	95.75	95.71	95.64	95.80
4	95.65	95.74	95.69	95.53
5	95.73	**95.90**	95.79	95.69
6	95.71	95.71	95.84	95.72

Hyperparameters m and l. Table 2 gives the number of rounds required for MChain-SFFL to train the MLP model with the MNIST(Non-IID) dataset to reach an accuracy of 95%. When chain number $m = 2$ and 3, the training rounds decrease with increasing chain length l; when $m = 4, 5, 6$, and $l \leq 25$, the training rounds also decrease with increasing chain length. In addition, when $l = 10$, 15, and 25, the training rounds decrease with the increase of m to reach 95% accuracy (except for $m = 6$, $l = 15$, and $m = 6$, $l = 25$). On the contrary, when $l = 40$, the larger m is, the more training rounds are almost needed. When $m = 1$, this is the case of Chain-PPFL. When $m = n$, $l = 1$, each user needs to broadcast n-1 times model parameters to others per round, i.e., MChain-SFFL needs to broadcast $n(n-1)$ times per round. There is no doubt that the communication cost is vast.

Table 3 gives the experimental results of MChain-SFFL using MNIST (Non-IID) dataset to train the MLP model for 300 rounds. The accuracy of each case is above 95%, which shows that MChain-SFFL has stability. Moreover, Table 3 shows that the accuracy is highest for $m = 5$, $l = 15$. Combining Table 2 and Table 3, we fix $m = 5$ and $l = 15$ in the following experiments, thus achieving a good balance between accuracy, convergence rate, and communication cost.

Table 4. Accuracy of different methods.

	CNN Non-IID	CNN IID	MLP Non-IID	MLP IID
FL-DP(ϵ=1)	98.52	99.18	94.76	95.48
FL-DP(ϵ=8)	98.77	99.23	95.07	96.20
FedAVG	98.88	99.31	95.27	96.31
Chain-PPFL	98.89	99.32	95.29	96.36
Ours	99.07	99.32	95.90	96.38

4.2 Experimental Results

Accuracy. Figure 2 shows the accuracy for Chain-PPFL, FedAVG, DP-based FL, and our MChain-SFFL of different rounds. Table 4 shows the best test accuracies for different configurations in the experiments. For the Non-IID data, the model accuracy of MChain-SFFL is better than other comparison methods and the trend is stable. Chain-PPFL and FedAVG show comparable accuracy. MChain-SFFL can train more participants per round compared with Chain-PPFL, so it can significantly improve the accuracy of Non-IID data under the same rounds. For the IID data, MChain-SFFL, Chain-PPFL, and FedAVG show comparable accuracy. Because MChain-SFFL removes mask information before aggregating model parameters, the privacy of model parameters can be protected without compromising model accuracy. DP-based FL ($\epsilon=1$ and $\epsilon=8$) adds noise to the model parameters, which results in lower accuracy than other methods. And the smaller ϵ is, the lower the accuracy.

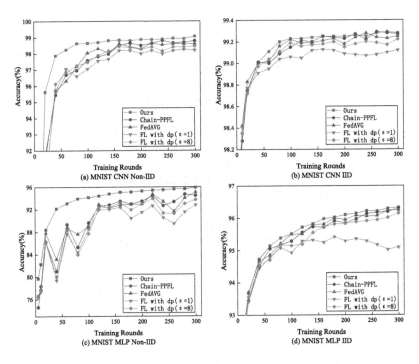

Fig. 2. Comparison of accuracy of different methods.

Convergence Speed. Figure 2 shows that in the Non-IID partition, the accuracy of MChain-SFFL can reach 98% in 40 rounds for the CNN model, while Chain-PPFL needs more than 80 rounds to achieve the same accuracy. For the MLP model, MChain-SFFL reaches an accuracy of 92% in only 50 rounds, while Chain-PPFL requires about 108 rounds. MChain-SFFL can train more users' data per round by using the multi-chain parallel mechanism compared with Chain-PPFL. Therefore, MChain-SFFL can effectively improve the convergence performance of Non-IID data. For the IID data, the convergence speed of

MChain-SFFL and Chain-PPFL is comparable for the CNN and MLP models. [10] shows that the convergence speed of FedAVG and Chain-PPFL is similar. And DP-based FL ($\epsilon=1$ and $\epsilon=8$) converges slower than these two methods due to adding noise during the process of model training, i.e. slower than MChain-SFFL.

5   Conclusion

In this paper, we propose a new framework named MChain-SFFL, which uses a multi-chain aggregation mechanism to improve the model accuracy and convergence speed. The experimental results demonstrate the effectiveness of MChain-SFFL. For Non-IID data, the accuracy of MChain-SFFL is better than other comparison methods, and MChain-SFFL can effectively improve the convergence speed of the model. For IID data, the accuracy and convergence speed of MChain-SFFL are close to Chain-PPFL and FedAVG.

References

1. Abadi, M., et al.: Deep learning with differential privacy. In: Proceedings of the 2016 ACM SIGSAC Conference on Computer and Communications Security, pp. 308–318 (2016)
2. Bagdasaryan, E., Veit, A., Hua, Y., Estrin, D., Shmatikov, V.: How to backdoor federated learning. In: International Conference on Artificial Intelligence and Statistics, pp. 2938–2948. PMLR (2020)
3. Baraglia, R., Dazzi, P., Mordacchini, M., Ricci, L.: A peer-to-peer recommender system for self-emerging user communities based on gossip overlays. J. Comput. Syst. Sci. **79**(2), 291–308 (2013)
4. Bonawitz, K., et al.: Practical secure aggregation for privacy-preserving machine learning. In: Proceedings of the 2017 ACM SIGSAC Conference on Computer and Communications Security, pp. 1175–1191 (2017)
5. Brakerski, Z.: Fully homomorphic encryption without modulus switching from classical GapSVP. In: Safavi-Naini, R., Canetti, R. (eds.) CRYPTO 2012. LNCS, vol. 7417, pp. 868–886. Springer, Heidelberg (2012). https://doi.org/10.1007/978-3-642-32009-5_50
6. Chen, V., Pastro, V., Raykova, M.: Secure computation for machine learning with SPDZ (2019). arXiv preprint arXiv:1901.00329
7. Konečný, J., McMahan, H.B., Yu, F.X., Richtárik, P., Suresh, A.T., Bacon, D.: Federated learning: strategies for improving communication efficiency (2016). arXiv preprint arXiv:1610.05492
8. Kuo, T.T., Ohno-Machado, L.: Modelchain: decentralized privacy-preserving healthcare predictive modeling framework on private blockchain networks (2018). arXiv preprint arXiv:1802.01746
9. Li, T., Sahu, A.K., Talwalkar, A., Smith, V.: Federated learning: challenges, methods, and future directions. IEEE Signal Process. Mag. **37**(3), 50–60 (2020)
10. Li, Y., Zhou, Y., Jolfaei, A., Yu, D., Xu, G., Zheng, X.: Privacy-preserving federated learning framework based on chained secure multi-party computing. IEEE Internet Things J. **8**(8), 6178–6186 (2020)

11. Lyu, L., Yu, J., Nandakumar, K., Li, Y., Ma, X., Jin, J.: Towards fair and decentralized privacy-preserving deep learning with blockchain, pp. 1–13 (2019). arXiv preprint arXiv:1906.01167

12. Mcmahan, H.B., Moore, E., Ramage, D., Hampson, S., Arcas, B.: Communication-efficient learning of deep networks from decentralized data. In: Proceeding of the 20th International Conference on Artificial Intelligence and Statistics (2016)

13. McMahan, H.B., Moore, E., Ramage, D., y Arcas, B.A.: Federated learning of deep networks using model averaging, vol. 2 (2016). arXiv preprint arXiv:1602.05629

14. Roy, A.G., Siddiqui, S., Pölsterl, S., Navab, N., Wachinger, C.: Braintorrent: A peer-to-peer environment for decentralized federated learning (2019). arXiv preprint arXiv:1905.06731

15. Shokri, R., Shmatikov, V.: Privacy-preserving deep learning. In: Proceedings of the 22nd ACM SIGSAC Conference on Computer and Communications Security, pp. 1310–1321 (2015)

16. Shokri, R., Stronati, M., Song, C., Shmatikov, V.: Membership inference attacks against machine learning models. In: 2017 IEEE Symposium on Security and Privacy (SP), pp. 3–18. IEEE (2017)

17. Tran, A.T., Luong, T.D., Karnjana, J., Huynh, V.N.: An efficient approach for privacy preserving decentralized deep learning models based on secure multi-party computation. Neurocomputing **422**, 245–262 (2021)

18. Truex, S., et al.: A hybrid approach to privacy-preserving federated learning. In: Proceedings of the 12th ACM Workshop on Artificial Intelligence and Security, pp. 1–11 (2019)

19. Weng, J., Weng, J., Zhang, J., Li, M., Zhang, Y., Luo, W.: Deepchain: auditable and privacy-preserving deep learning with blockchain-based incentive. IEEE Trans. Dependable Secure Comput. **18**(5), 2438–2455 (2019)

20. Wu, N., Farokhi, F., Smith, D., Kaafar, M.A.: The value of collaboration in convex machine learning with differential privacy. In: 2020 IEEE Symposium on Security and Privacy (SP), pp. 304–317. IEEE (2020)

21. Xu, R., Baracaldo, N., Zhou, Y., Anwar, A., Ludwig, H.: Hybridalpha: an efficient approach for privacy-preserving federated learning. In: Proceedings of the 12th ACM Workshop on Artificial Intelligence and Security, pp. 13–23 (2019)

22. Yao, A.C.C.: How to generate and exchange secrets. In: 27th Annual Symposium on Foundations of Computer Science (Sfcs 1986), pp. 162–167. IEEE (1986)

23. Zhao, H., Wang, C., Zhu, Y., Lin, W.: P2p network based on neighbor-neighbor lists. In: Journal of Physics: Conference Series, vol. 1168. IOP Publishing (2019)

24. Zhu, L., Liu, Z., Han, S.: Deep leakage from gradients. Adv. Neural. Inf. Process. Syst. **32**, 14747–14756 (2019)

FedGR: Federated Learning with Gravitation Regulation for Double Imbalance Distribution

Songyue Guo[1], Xu Yang[1], Jiyuan Feng[1], Ye Ding[2], Wei Wang[3], Yunqing Feng[4], and Qing Liao[1,5(✉)]

[1] Harbin Institute of Technology, Shenzhen, China
{guosongyue,xuyang97,fengjy}@stu.hit.edu.cn, liaoqing@hit.edu.cn
[2] Dongguan University of Technology, Dongguan, China
dingye@dgut.edu.cn
[3] Huazhong University of Science and Technology, Wuhan, China
weiwangw@hust.edu.cn
[4] Shanghai Pudong Development Bank, Shanghai, China
fengyq5@spdb.com.cn
[5] Pengcheng Laboratory, Shenzhen, China

Abstract. Federated Learning (FL) is a well-known framework for distributed machine learning that enables mobile phones and IoT devices to build a shared machine learning model via only transmitting model parameters to preserve sensitive data. However, existing Non-IID FL methods always assume data distribution of clients are under a single imbalance scenario, which is nearly impossible in the real world. In this work, we first investigate the performance of the existing FL methods under double imbalance distribution. Then, we present a novel FL framework, called **Fed**erated Learning with **G**ravitation **R**egulation (FedGR), that can efficiently deal with the double imbalance distribution scenario. Specifically, we design an unbalanced softmax to deal with the quantity imbalance in a client by adjusting the forces of positive and negative features adaptively. Furthermore, we propose a gravitation regularizer to effectively tackle the label imbalance among clients by facilitating collaborations between clients. At the last, extensive experimental results show that FedGR outperforms state-of-the-art methods on CIFAR-10, CIFAR-100, and Fashion-MNIST real-world datasets. Our code is available at https://github.com/Guosy-wxy/FedGR.

Keywords: Federated learning · Double imbalance distribution · Non-IID · Softmax · Regularizer term

1 Introduction

Despite the success of deep learning in numerous fields [12,22], a data center training model is typically required. In some real-world applications, individual participant data cannot be located on the same device due to data privacy [1].

X. Wang et al. (Eds.): DASFAA 2023, LNCS 13943, pp. 703–718, 2023.
https://doi.org/10.1007/978-3-031-30637-2_47

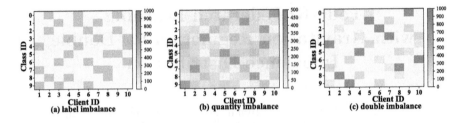

Fig. 1. Different imbalance distribution scenarios on CIFAR-10 dataset. (a) label imbalance, (b) quantity imbalance, (c) double imbalance.

Federated Learning (FL) [9,17,26] is designed for data privacy protection and efficient distributed training.

The advent of FL enables different clients to collectively build a robust global model without broadcasting local private data to the server. FL has demonstrated its ability to facilitate real-world applications in several domains, e.g., natural language processing [7], credit card fraud detection [28] and medical healthcare [6,25].

FL, however, also confronts the challenge of imbalance distribution [17,20]. The imbalance distribution of Non-IID data between clients brings serious performance degradation problems for FL [14,27]. This performance degradation is attributed to the phenomenon of client drift. Some recent works aim to deal with this problem, e.g., FedProx [14] included l_2 regularizer term to prevent local models from deviating too far from the global model, PerFedAvg [4] utilized contrastive learning, and MOON [13] used multi-task learning for fast client local adaptation to mitigate the impact of client drift. However, most existing studies focus on single imbalance distribution [21].

In this work, we focus on double imbalance distribution scenario, which is more common in the real world. We first define the imbalance distribution into two categories:

1) **Label imbalance.** According to Fig. 1(a), we simulate this scenario with 10 clients on CIFAR-10 dataset. The majority of clients have part labels of the whole, and client's labels are mostly different from others, while the quantity of each label in client is equal. For example, client 1 owns labels 4, 7, 9, but client 2 owns labels 0, 2, 8. Most recent works like FedProx [14], SCAFFOLD [10] and FedNova [23] only care about label imbalance.
2) **Quantity imbalance.** We still use 10 clients on CIFAR-10 to describe this imbalance distribution. Based on Fig. 1(b), each client owns an entire set of labels, but the quantity of each label in client varies, i.e., client 1 has 10 labels, and the number of label 9 is about 450, but the number of label 0 and 1 is approximately 0.

In this study, we focus on *double imbalance distribution* like Fig. 1(c), each client owns a partition of entire labels, and the quantity of each class in client varies, e.g., client 1 only possesses labels 4, 7, 9, and each class' sample number is imbalanced. It is clear that the double imbalance distribution scenario is more

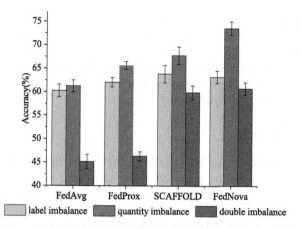

Fig. 2. Comparison with existing FL methods under different imbalance distribution.

in line with reality than any single imbalance scenario. However, existing works mostly omit the real scenario of double imbalance distribution scenario.

In order to investigate the performance of existing IID and Non-IID FL algorithms for double imbalance scenario, we use client distribution as Fig. 1 to design an observation experiment. We use TFCNN[1] as client's base model and implement all compared FL methods with the same model for a fair comparison. All performance results are expressed by accuracy of an average of five times. The experiment results are summarized in Fig. 2. From the performance results shown, we can find that no matter what FL algorithms gets a significant performance loss on the double imbalance, compared to the left two scenes. For example, the accuracy of FedAvg declines by about 16%. Apparently, the double imbalance scenario brings a new challenge for existing FL algorithms, which is the goal of our study tries to solve.

Motivated by the above observation experiment of double imbalance distribution, we propose a novel FL algorithm called **Fede**rated Learning with **G**ravitation **R**egulation (FedGR) to deal with this problem. We define a novel softmax function called unbalanced softmax to balance the importance of classes under quantity imbalance in clients. In addition, we propose an efficient gravitation regularizer to deal with label imbalance among clients by encouraging collaboration among clients. Combining these two components, we can correct the gradient of traditional loss function of typical FL methods. The contributions of this paper can be summarized as follows:

- We propose a novel federated learning method FedGR to effectively deal with the performance degradation problem caused by double imbalance distribution scenario.
- We design an unbalanced softmax function, which can solve the problem of unbalanced number of samples within the client by adjusting the forces of positive and negative samples on classes.

---

[1] https://www.tensorflow.org/tutorials/images/cnn.

- We propose a gravitation regularizer to alleviate the impact of label imbalance between clients by introducing cross-client forces to encourage the collaboration of different clients.
- Extensive experiments show that FedGR significantly outperforms the state-of-the-art federated learning methods on several benchmark datasets under double imbalance distribution scenario.

2   Related Work

Recently, federated learning on imbalance data distribution has drawn much interest in machine learning research. Zhao *et al.* [27] shared a limited public dataset across clients to relieve the degree of imbalance between various clients. FedProx [14] introduced a proximal term to limit the dissimilarity between the global model and local models. SCAFFOLD [10] used variance reduction to alleviate the effect of client drifting that causes weight divergence between the local and global models. FedNova [23] changed the aggregation phase by allocating different number of local steps per round to different client participants which have different computational capabilities to eliminate original objective inconsistency problem caused by imbalance data. PerFedAvg [4] used meta-learning to learn a new task quickly and effectively for quick local adaptation. Dinh *et al.* [20] proposed pFedme, introducing l_2-norm regularization to PerFedAvg which can control the balance between personalization and generalization performance. Li *et al.* [13] proposed MOON, which applied constrastive learning to make local representation closer to the global model's representation for better performance. Huang *et al.* [8] proposed FedAMP, an attention-based mechanism that enforces stronger pairwise collaboration amongst FL clients with similar data distributions. APFL [2] utilized model interpolation to adaptively control the mixture of global and local model. Astraea [3] created the mediator to reschedule the training of clients based on Kullback-Leibler divergence (KLD) of their data distribution for label imbalance. FedGC [18] introduced a softmax-based regularizer term to correct the loss function to be similar to the standard softmax in conventional central learning. Ghosh *et al.* proposed IFCA [5], a clustering FL framework that has many global models and assign each client to one of the K clusters the global model of which achieves the lowest loss value on the client's data. FedRS [15] proposed a restricted softmax to limit the update of missing classes' parameters during the local procedure.

However, existing methods ignore considering double imbalance distribution scenario, which are not applicable on real complicated FL scenarios.

3   FedGR

In this section, we will first explicitly introduce the whole structure of the proposed FedGR. Then, we first define a novel softmax function to deal with quantity imbalance in client. Third, we design a gravitation regularizer in server to deal with label imbalance between clients. At the last we present the algorithm of FedGR.

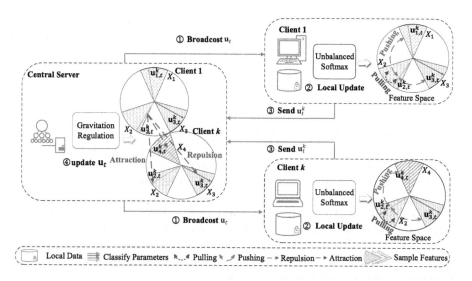

Fig. 3. Framework of FedGR. FedGR contains two components: unbalanced softmax in client and gravitation regularizer in server. Unbalanced Softmax aims to deal with the quantity imbalance in client. Graviation Regularizer aims to tackle the label imbalance among clients.

3.1 Framework of FedGR

On the basis of the aforementioned observation, we propose a method called **Fed**erated Learning with **G**ravitation **R**egulation (**FedGR**) to address the problem of double imbalance distribution. As shown in Fig. 3, FedGR have two components: clients and central server. In client, we design a novel *Unbalanced Softmax* to address quantity imbalance. Meanwhile, in server, we propose an efficient regularizer term called *Gravitation Regularizer* to solve label imbalance cross clients.

3.2 Unbalanced Softmax

In this subsection, we first promote the shortcoming of standard softmax faced with quantity imbalance. Then, we define a simple but efficient softmax function called unbalanced softmax to adjust the importance of classes under quantity imbalance situation. Finally, we analyze the benefits of unbalanced softmax under quantity imbalance scenario.

According to traditional FL algorithms [14,17], the cross-entropy function of client k can be formulated as:

$$\mathcal{L}^k = - \sum_{(\mathbf{x}_i, y_i) \in \mathcal{D}^k} \log p^k_{i,y_i}, \tag{1}$$

where $\mathcal{L}^k$ means the loss function of client k, $(\mathbf{x}_i, y_i)$ denotes i-th sample in training dataset, $\mathbf{x}_i$ is a vector of training data, and y_i is the label of i-th sample,

$\mathcal{D}^k$ denotes the local data, p_{i,y_i}^k is the probability of i-th samples belonging to y_i class. The probability is typically calculated by the standard softmax function by normalizing each class's score:

$$p_{i,y_i}^k = \frac{e^{\mathbf{u}_{y_i}^{k^T}\mathbf{v}_i^k}}{\sum_{y_j=1}^{C^k} e^{\mathbf{u}_{y_j}^{k}{}^T\mathbf{v}_i^k}}, \tag{2}$$

where $\mathbf{u}_{y_i}^k$ denotes the classification parameters of y_i in client k, $\mathbf{v}_i^k$ means the extracted feature of i-th sample, C^k is the number of labels in client k.

However, the standard softmax may not work well when faced with quantity imbalance because standard softmax gives each class the same weight. In fact, some classes possess insufficient data samples, like the tail class in long-tailed distribution. Hence, the classification parameters of head class will pull tail class to an error feature region, so the performance of client's model directly drops.

In order to solve quantity imbalance problem in client, we introduce a balance factor γ in unbalanced softmax to balance the importance of different classes. The unbalanced softmax can be formulated as:

$$\hat{p}_{i,y_i}^k = \frac{e^{\gamma_{y_i}^k \mathbf{u}_{y_i}^{k^T}\mathbf{v}_i^k}}{\sum_{y_j=1}^{C^k} e^{\gamma_{y_j}^k \mathbf{u}_{y_j}^{k}{}^T\mathbf{v}_i^k}}, \tag{3}$$

where $\gamma_{y_i}^k = \frac{N^k}{N_{y_i}^k}$ denotes as balance factor of each class, N^k means the number of total samples in client k, and $N_{y_i}^k$ means the number of label y_i samples. We utilize this balance factor to increase the importance of tail classes in client.

Theoretical Analysis. In order to show the efficient of the proposed unbalanced softmax, we analyze the benefits of it for quantity imbalance in clients as below:

First, the loss function of client k in FL can be rewritten by replacing the standard softmax as unbalanced softmax:

$$\mathcal{L}^k = -\sum_{(\mathbf{x}_i,y_i)\in\mathcal{D}^k} \log \hat{p}_{i,y_i}^k = -\sum_{(\mathbf{x}_i,y_i)\in\mathcal{D}^k} \log \frac{e^{\gamma_{y_i}^k \mathbf{u}_{y_i}^{k}{}^T\mathbf{v}_i^k}}{\sum_{y_j=1}^{C^k} e^{\gamma_{y_j}^k \mathbf{u}_{y_j}^{k}{}^T\mathbf{v}_i^k}}. \tag{4}$$

Second, the computation of the gradient of $\frac{\partial \mathcal{L}^k}{\partial \mathbf{u}_{y_i}^k}$ is formulated as below:

$$\frac{\partial \mathcal{L}^k}{\partial \mathbf{u}_{y_i}} = -\sum_{i=1,y\neq y_i}^{N} \gamma_{y_i}^k \hat{p}_{i,y_i}^k \mathbf{v}_i^k + \sum_{i=1,y=y_i}^{N} \gamma_{y_i}^k \left(1 - \hat{p}_{i,y_i}^k\right) \mathbf{v}_i^k. \tag{5}$$

Final, we use gradient descent with learning rate η to update the classification parameters of label y_i and decompose this update process into the pushing and pulling forces:

$$\mathbf{u}_{y_i}^k = \mathbf{u}_{y_i}^k \underbrace{-\eta \sum_{i=1,y\neq y_i}^{N} \gamma_y^k \hat{p}_{i,y_i}^k \mathbf{v}_i^k}_{\text{weighted pushing force}} + \underbrace{\eta \sum_{i=1,y=y_i}^{N} \gamma_{y_i}^k \left(1 - \hat{p}_{i,y_i}^k\right) \mathbf{v}_i^k}_{\text{weighted pulling force}}, \tag{6}$$

where the pulling force come from positive samples which have the same label y_i, while the pushing force is from negative samples whose labels are not y_i. The pulling force aims to pull the classification parameters close to the feature region of positive samples, while the pushing force aims to pushing the classification parameters far away from the feature region of negative samples.

From Eq. (6), it is obvious that pushing force and pulling force are weighted by our balance factor $\gamma_{y_i}^k$. If y_i is a tail class in client k, $\gamma_{y_i}^k$ will be larger. Consequently, pushing force and pulling force are more efficient, bringing the classification parameters of tail class closer to their feature regions.

3.3   Gravitation Regularizer

In this subsection, we first analyze the drawbacks of traditional global optimization objective in FL when faced with label imbalance. Then, we define a novel regularizer called gravitation regularizer to encourage the collaboration of clients under label imbalance situation. Final, we analyze the benefits of gravitation regularizer under label imbalance scenario.

In typical FL scenario, traditional FL considers to train a global model by the following optimization objective:

$$\min_{\mathbf{u}} F(\mathbf{u}) \triangleq \sum_{k=1}^{K} \alpha^k \mathcal{L}^k, \tag{7}$$

where K is the number of clients, α^k is the aggregation weight of client k. We define α^k as $\frac{N^k}{N}$, where N is the number of total data samples. Then, $\sum_{k=1}^{K} \alpha^k = 1$. We denote the distributed optimization objective as *global optimization*.

However, the typical optimization objective in FL might cause performance decrease when it only considers optimization within the client and ignores optimization cross clients. For instance, client k_1 has label 1, 2, 3 and client k_2 has label 2, 3, 4. In the client k_1, the label 1 should be pushed far away from the label 2 and label 3. On the other hand, the label 4 should be also pushed far away from the label 2 and label 3 in client k_2. Hence, the feature space of label 1 might incorrectly overlap to the feature space of label 4 if the objective function ignores the cross-client optimization.

In order to introduce cross-client optimization into the objective function for handling label imbalance, we design a new regularizer term of FedGR:

$$\min_{\mathbf{u}} F(\mathbf{u}) \triangleq \sum_{k=1}^{K} \alpha^k \mathcal{L}^k + \lambda \cdot \text{Gravitation-Reg}(\mathbf{u}), \tag{8}$$

where λ is a hyper-parameter to control the weight of the gravitation regularization term.

The gravitation regularizer term contains two components: Attraction regularizer and Repulsion regularizer, which are defined as follows:

$$\text{Gravitation-Reg}(\mathbf{u}) = -\sum_{k=1}^{K}\sum_{y_i=0}^{C^k}(\log\underbrace{\frac{\sum_{z\neq k}e^{\mathbf{u}_{y_i}^{z^T}\mathbf{u}_{y_i}^{k'}}}{\sum_{z\neq k}\sum_{y_j=1}^{C^z}e^{\mathbf{u}_{y_j}^{z^T}\mathbf{u}_{y_i}^{k'}}}}_{\text{Attraction}}$$

$$+\log\underbrace{\frac{e^{\mathbf{u}_{y_i}^{k^{T'}}\mathbf{u}_{y_i}^{k'}}}{e^{\mathbf{u}_{y_i}^{k^{T'}}\mathbf{u}_{y_i}^{k'}}+\sum_{z\neq k}\sum_{y_j=1,j\neq i}^{C^z}e^{\mathbf{u}_{y_j}^{z^T}\mathbf{u}_{y_i}^{k'}}}}_{\text{Repulsion}}),\tag{9}$$

where $\mathbf{u}_{y_i}^{k'},\mathbf{u}_{y_i}^{k^{T'}}$ suggests the gradient is set to be zero, which means the gradient is not required for these vectors. Attraction regularizer aims to increase similarity among the same labels' classification parameters across clients, while Repulsion regularizer aims to decrease similarity among different labels' classification parameters across clients. Hence, the Gravitation Regularizer can correct the gradient of the optimization loss function with Attraction regularizer and Repulsion regularizer.

Theoretical Analysis. In order to show how gravitation regularizer works, we will theoretically discuss the effort of it for label imbalance among clients as follows:

According to Eq. (8), optimization objective of FedGR equals the empirical risk respect to the loss function $\mathcal{L}_G^k$:

$$F(\mathbf{u}) = \sum_{k=1}^{K}\alpha^k\mathcal{L}^k + \lambda\cdot\text{Gravitation-Reg}(\mathbf{u})$$

$$= -\frac{1}{N}\sum_{k=1}^{K}\sum_{(\mathbf{x}_i,y_i)\in\mathcal{D}^k}(\hat{p}_{i,y_i}^k + \lambda\cdot\text{Attraction}(\mathbf{u}) + \lambda\cdot\text{Repulsion}(\mathbf{u}))\tag{10}$$

$$= -\frac{1}{N}\sum_{k=1}^{K}\mathcal{L}_G^k.$$

To show that the added gravitation regularizer complement ignored cross-client optimization, we correspondingly calculate the gradient of $\mathcal{L}_G^k$ to cross-client classification parameters $\frac{\partial\mathcal{L}_G^k}{\partial\mathbf{u}_{y_i}^z}$ and $\frac{\partial\mathcal{L}_G^k}{\partial\mathbf{u}_{y_j}^z}$ of FedGR. Then, the related gradient can be calculated as:

$$\frac{\partial\mathcal{L}_G^k}{\partial\mathbf{u}_{y_i}^z} = \left(\frac{\sum_{z\neq k}e^{\mathbf{u}_{y_j}^{z^T}\mathbf{u}_{y_i}^k}}{\sum_{z\neq k}\sum_{y_j=1}^{C^z}e^{\mathbf{u}_{y_j}^{z^T}\mathbf{u}_{y_i}^k}}-1\right)\mathbf{u}_{y_i}^k,\tag{11}$$

$$\frac{\partial \mathcal{L}_G^k}{\partial \mathbf{u}_{y_j}^z} = \frac{e^{\mathbf{u}_{y_j}^{z^T} \mathbf{u}_{y_i}^k}}{e^{\mathbf{u}_{y_i}^{k^T} \mathbf{u}_{y_i}^k} + \sum_{z \neq k} \sum_{y_j=1, j \neq i}^{C_z} e^{\mathbf{u}_{y_j}^{z^T} \mathbf{u}_{y_i}^k}} \mathbf{u}_{y_i}^k, (j \neq i) \tag{12}$$

Due to the ease of convergence on local data, the features of local client data are effectively trained on each client. Therefore, the distance between classification parameters $\mathbf{u}_{y_i}^k$ and features $\mathbf{v}_i^k$ tends to be zero: $\mathbf{u}_{y_i}^k \to \mathbf{v}_i^k$. Hence, we can get the following approximations:

$$\frac{\partial \mathcal{L}_G^k}{\partial \mathbf{u}_{y_i}^z} \approx \left(\frac{\sum_{z \neq k} e^{\mathbf{u}_j^{z^T} \mathbf{u}_{y_i}^k}}{\sum_{z \neq k} \sum_{j=1}^{C_z} e^{\mathbf{u}_j^{z^T} \mathbf{u}_{y_i}^k}} - 1 \right) \mathbf{v}_i^k, \tag{13}$$

$$\frac{\partial \mathcal{L}_G^k}{\partial \mathbf{u}_{y_j}^z} \approx \frac{e^{\mathbf{u}_{y_j}^{z^T} \mathbf{u}_{y_i}^k}}{e^{\mathbf{u}_{y_i}^{k^T} \mathbf{u}_{y_i}^k} + \sum_{z \neq k} \sum_{y_j=1, j \neq i}^{C_z} e^{\mathbf{u}_{y_j}^{z^T} \mathbf{u}_{y_i}^k}} \mathbf{v}_i^k, (j \neq i). \tag{14}$$

According to Eq. (6), Eq. (13) and Eq. (14), the gradient of $\mathcal{L}_G^k$ to $\mathbf{u}_{y_i}^z$ and $\mathbf{u}_{y_j}^z$ are similar to the pulling and pushing force of positive and negative samples respectively. Different with Eq. (6) which only considers the pulling and pushing force in client, the Eq. (13) and Eq. (14) introduce attraction and repulsion force cross client.

Finally, the updated classification parameters of y_i can be decomposed as in-client force and cross-client gravitation force based on Eq. (6), Eq. (13) and Eq. (14):

$$\mathbf{u}_{y_i,t+1}^z = \mathbf{u}_{y_i,t}^z \underbrace{-\eta \sum_{i=1, y \neq y_i}^{N_z} \gamma_{y_i}^z \hat{p}_{i,y_i}^z \mathbf{v}_{i,t}^z}_{\text{in-client pushing force}} + \underbrace{\eta \sum_{i=1, y=y_i}^{N_z} \gamma_{y_i}^z \left(1 - \hat{p}_{i,y_i}^z\right) \mathbf{v}_{i,t}^z}_{\text{in-client pulling force}}$$

$$\underbrace{- \eta \lambda \sum_{k=1, k \neq z}^{K} \sum_{(\mathbf{x}_i, y_i) \in \mathcal{D}^k} \left(\frac{\sum_{z \neq k} e^{\mathbf{u}_{j,t}^{z^T} \mathbf{u}_{y_i,t}^k}}{\sum_{z \neq k} \sum_{j=1}^{C_z} e^{\mathbf{u}_{j,t}^{z^T} \mathbf{u}_{y_i,t}^k}} - 1 \right) \mathbf{v}_{i,t}^k}_{\text{cross-client Attraction force}} \tag{15}$$

$$\underbrace{+ \eta \lambda \sum_{k=1, k \neq z}^{K} \sum_{(\mathbf{x}_i, y_i) \in \mathcal{D}^k} \frac{e^{\mathbf{u}_{j,t}^{z^T} \mathbf{u}_{y_i,t}^k}}{e^{\mathbf{u}_{y_i,t}^{k^T} \mathbf{u}_{y_i,t}^k} + \sum_{z \neq k} \sum_{j=1, j \neq i}^{C_z} e^{\mathbf{u}_{j,t}^{z^T} \mathbf{u}_{y_i,t}^k}} \mathbf{v}_{i,t}^k}_{\text{cross-client Repulsion force}}.$$

where η is the learning rate of gradient descent, λ is the weight of regularizer term. Related forces (in-client forces and cross-client forces) can be all formed like $\beta \mathbf{v}_i^k$.

Algorithm 1: FedGR

Input: Number of clients K; total communication rounds T; learning rate η;
   participate rate Q, regularizer weight λ; dataset of client k $\mathcal{D}^k$.
Output: model parameters $\mathbf{u}_T^k$.

1 Server initialized $\mathbf{u}_0$;
2 **for** *round $t = 0, \ldots, T-1$* **do**
3   Server select a subset of clients $\mathcal{S}(t) \leftarrow Q \cdot K$;
    // Clients Update:
4   **foreach** *participate client $k \in \mathcal{S}(t)$* **do**
5   download parameters from server $\mathbf{u}_t^k \leftarrow \mathbf{u}_t$;
6   update parameters $\mathbf{u}_{t+1}^k \leftarrow \mathbf{u}_t^k - \eta \frac{\partial \mathcal{L}^k}{\partial \mathbf{u}_{y_i}}$ (Eq. 6);
7   send $\mathbf{u}_{t+1}^k$ to central server;
8   **end**
    // Server Update:
9   aggregate the parameters $\tilde{\mathbf{u}}_{t+1} = \left[\mathbf{u}_{t+1}^k, \ldots, \mathbf{u}_{t+1}^K\right]^T$;
10   update parameters $\mathbf{u}_{t+1} \leftarrow \tilde{\mathbf{u}}_{t+1} - \lambda\eta\nabla_{\tilde{\mathbf{u}}_{t+1}} \text{Gravitation-Reg}\left(\tilde{\mathbf{u}}_{t+1}\right)$
    (Eq.15);
11 **end**

3.4 Training Process of FedGR

The overall process of the proposed FedGR algorithm is illustrated in Algorithm 1. There are two main processes: clients update (line 4–7) and server update (line 9–10). In the process of clients update, each client starts their local-update process on their datasets $\mathcal{D}^k$ in parallel. First, clients download the parameters $\mathbf{u}_t$ broadcast by central server in line 5. Then they update model parameters $\mathbf{u}_t^k$ by unbalanced softmax to obtain $\mathbf{u}_{t+1}^k$ from line 6. At last, client k transfers updated parameters $\mathbf{u}_{t+1}^k$ to the central server in line 7. In the process of server update, server updates the global model parameters $\mathbf{u}_{t+1}$ via gravitation regularizer in line 9–10.

4 Experiments

In this section, we first introduce basic experiment settings. Second, we show the ability of FedGR to deal with double imbalance distribution on several benchmark datasets, compared with start-of-the-art FL algorithms. Third, we make a ablation study of each component in FedGR. Fourth, we analyze the selection of hyper-parameter λ. Finally, we show the visualization comparisons conducted on feature level.

4.1 Experimental Setup

Datasets. We conduct experiments on three real-world image datasets: CIFAR-10, CIFAR-100 and Fashion-MNIST. The information of all datasets is listed in Table 1.

Table 1. Details of CIFAR-10, CIFAR-100 and Fashion-MNIST

Datasets	Class Number	Image Size	Training Samples	Test Samples
CIFAR-10 [11]	10	32×32	50,000	10,000
CIFAR100 [11]	100	32×32	50,000	10,000
Fashion-MNIST [24]	10	28×28	50,000	10,000

Table 2. Performance compared with state-of-the-art algorithms on CIFAR-10/100 dataset under double imbalance distribution. The best results are in **bold**, and the secondary optimal results are mark as underline. CIFAR-10 (2) means each client owns two labels, which is similar to CIFAR-10 (3), CIFAR-100 (20) and CIFAR-100 (30).

Algorithms	CIFAR-10 (2)		CIFAR-10 (3)		CIFAR-100 (20)		CIFAR-100 (30)	
	Acc(%)	F1(%)	Acc(%)	F1(%)	Acc(%)	F1(%)	Acc(%)	F1(%)
FedAvg [17]	50.36	48.27	53.79	49.42	36.15	34.10	42.19	40.42
FedProx [14]	48.84	46.96	54.94	53.85	36.24	34.42	42.21	41.09
FedNova [23]	56.33	54.59	68.63	66.09	38.63	37.72	45.35	45.59
SCAFFOLD [10]	57.37	54.43	67.32	62.44	38.43	37.76	46.82	45.44
PerFedAvg [4]	44.67	42.56	54.87	53.73	35.98	34.76	40.14	40.33
pFedMe [20]	45.81	44.35	50.18	50.24	35.36	33.59	40.18	40.54
FedOpt [19]	62.37	60.68	70.63	69.79	42.37	40.68	49.63	49.79
MOON [13]	61.45	<u>60.71</u>	72.91	70.45	40.53	41.46	47.91	48.76
FedRS [15]	<u>63.22</u>	60.13	<u>73.56</u>	70.13	<u>42.76</u>	<u>42.21</u>	<u>50.73</u>	50.31
FedGC [18]	62.91	60.35	72.11	<u>70.64</u>	42.11	40.35	50.21	<u>50.46</u>
FedGR (ours)	**67.84**	**65.62**	**77.86**	**75.32**	**45.44**	**44.85**	**53.16**	**53.32**
	(4.53↑)	(4.91↑)	(4.3↑)	(4.68↑)	(2.68↑)	(2.64↑)	(2.43↑)	(2.86↑)

Data Segmentation. According to related works, we first followed [13] to reshape the original balanced datasets to a quantity imbalance distribution, which means the number of samples of each label on the client side follows a power-law distribution. After that, we followed [14,15,17] to simulate the label imbalance distribution by giving each client a fixed number of labels.

Models and Hardware Settings. We use TFCNN for CIFAR-10/100 and Fashion-MNIST as the base model. All experiments are run by PyTorch on two NVIDIA GeForce V100 GPUs. By default, we run 1000 communication rounds. We set the number of total clients at 100 and a client participate ratio 10 % in each round. For local optimization, we set the batch size is 64. For server optimization, we set the weight of gravitation regularizer λ is 0.5. We use SGD with a learning rate 0.1 and a weight decay of 5e$-$4 as the optimizer for all optimization process.

4.2   Performance Comparison with State-of-the-Art Algorithms

We compare FedGR with several imbalance-oriented methods like FedNova [23], SCAFFOLD [10], PerFedAvg [4], pFedMe [20], FedOpt [19], MOON [13], FedRS [15] and FedGC [18] under different degrees of double imbalance.

Table 3. Performance compared with state-of-the-art algorithms on Fashion-MNIST under double imbalance distribution. The best results are in **bold**, and the secondary optimal results are mark as <u>underline</u>.

Algorithms	Fashion-MNIST (2)		Fashion-MNIST (3)	
	Accuracy(%)	Macro-F1(%)	Accuracy(%)	Macro-F1(%)
FedAvg (2018) [17]	51.36	50.34	54.34	50.42
FedProx (2020) [14]	53.94	52.75	55.14	54.97
FedNova (2020) [23]	58.23	56.57	63.35	62.59
SCAFFOLD (2020) [10]	56.32	59.44	65.32	63.94
PerFedAvg (2020) [4]	48.87	45.73	56.91	55.41
pFedMe (2020) [20]	47.61	46.47	52.68	52.13
FedOpt (2021) [19]	60.72	60.78	68.93	68.78
MOON (2021) [13]	60.37	<u>60.85</u>	72.67	70.52
FedRS (2021) [15]	<u>62.52</u>	60.54	<u>73.16</u>	<u>70.26</u>
FedGC (2022) [18]	62.11	60.74	73.01	70.13
FedGR (ours)	**65.62**(3.1↑)	**63.95**(3.24↑)	**76.11**(2.95↑)	**73.23**(2.97↑)

Table 4. Communication rounds compared with state-of-art algorithms on CIFAR-10 and Fashion-MNIST under double imbalance distribution.

Algorithms	CIFAR-10 (3)		CIFAR-100 (30)		Fahsion-MNIST (3)	
	#rounds	speedup	#rounds	speedup	#rounds	speedup
FedAvg [17]	1000	1×	1000	1×	1000	1×
FedProx [14]	800	1.25×	850	1.17×	860	1.16×
FedNova [23]	600	1.67×	650	1.53×	540	1.85×
SCAFFOLD [10]	860	1.16×	830	1.20×	810	1.24×
FedOpt [19]	390	2.56×	450	2.22×	360	2.77×
FedRS [15]	350	2.56×	430	2.22×	340	2.77×
FedGC [18]	590	1.69×	650	1.53×	610	1.63×
FedGR(ours)	**330**	**3.03×**	**390**	**2.56×**	**300**	**3.33×**

For each dataset, we simulate two different double imbalance scenarios for experimental generalizability. We use the average accuracy or F1 score of the last 50 rounds to represent the performance of one experiment. We also test all methods five times to reduce random errors.

Table 2, 3 show the Top-1 accuracy and macro-F1 of compared baselines on CIFAR-10, CIFAR-100 and Fashion-MINST datasets. It can be seen that our proposed *FedGR* achieves the highest performance on both accuracy and macro-F1 under different degrees of double imbalance. Compared with baselines, the highest performance gain of FedGR appears on CIFAR-10 datasets where each client only owns two labels (around 4.53%, 4.91% improvement on accuracy and F1 for the secondary best results). pFedMe achieves the lowest performance in most double imbalance scenarios, even lower than FedAvg. The possible reason is that meta-learning is not useful for dealing with serious quantity imbalance

Table 5. Ablation Study of FedGR.

Datasets	Unbalanced Softmax	Gravitation Regularizer	FedGR
CIFAR-10 (3)	✓		71.34 ± 0.24
		✓	70.52 ± 0.41
	✓	✓	**77.86 $\pm$ 0.35**
Fashion-MNIST (3)	✓		71.13 ± 0.36
		✓	70.31 ± 0.38
	✓	✓	**76.11 $\pm$ 0.21**

in client. FedRS achieves the secondary best on most of the scenes because it restricts the error update of missing classes, but the quantity imbalance is still a challenge for it.

We also compare the convergence speed with baselines. We choose the accuracy after a thousand rounds of FedAvg as the standard, then compare the number of communication rounds required by other methods to reach this accuracy. The results are shown in Table 4. We delete the PerFedAvg and pFedMe in Table 4 due to these two methods cannot achieve the accuracy of FedAvg. According to Table 4, FedGR can get the least communication rounds (from 300 to 390 rounds) to achieve FedAvg accuracy on all datasets, which is around two times faster than FedAvg. The reason why FedGR needs fewer rounds is that the gravitation regularizer encourages the effective collaboration of different clients by cross-client forces, which is not considered by existing FL methods.

4.3 Ablation Study

In this subsection, we design an experiment on CIFAR-10 and Fashion-MNIST to investigate the effect of each component of FedGR. According to Table 5, when FedGR with only unbalanced softmax, the performance is still higher than most existing FL methods because unbalanced softmax efficiently addresses the quantity imbalance problem in client. Similarly, we also find that gravitation regularizer of FedGR improves performance by enhancing the cross-client collaboration. In addition, the results show that the performance of FedGR with unbalanced softmax and gravitation regularizer obtains a significant improvement around 6% compared with FedGR with only unbalance softmax or gravitation regularizer. Therefore, our proposed FedGR actually have the capacity to handle the double imbalance distribution.

4.4 Parameter Analysis

One important hyper-parameter in FedGR is the weight of gravitation regularizer. We analyze the effect of λ. We evaluate its influence by experiments on CIFAR-10 and Fashion-MNIST under different degrees of double imbalance. It can be observed from Fig. 4 that FedGR gets the best performance when $\lambda =$

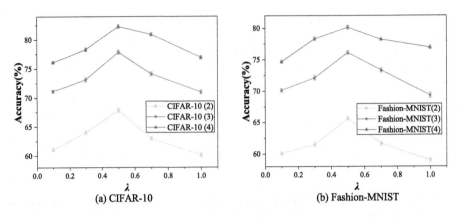

Fig. 4. Performance comparison with different selection of λ on CIFAR-10 and Fashion-MNIST under different degrees of double imbalance. (a) comparison on CIFAR-10, (b) comparision on Fashion-MNIST.

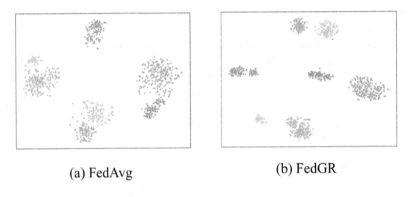

Fig. 5. Visualization results on feature level with t-SNE. (a) Results acquired by FedAvg, (b) Results acquired by FedGR.

0.5. When λ approaches 0, the influence of the gravitation regularizer is diminished, and only unbalanced softmax is effective. Consequently, FedGR tends to face the label imbalance issue. When λ is quite large (near 1), the gravitation regularizer harms the functioning of unbalanced softmax. Also, as the number of labels a client owns increases, the influence of different choices of λ decreases as a result of the label imbalance relief.

4.5   Visualization Results

We visualize the samples of CIFAR-10 dataset by t-SNE [16]. In Fig. 5, points in different colors refer to the features of samples in different classes. Samples are much closer within the same class cluster means better performance. It can be seen that FedGR clearly reduces the distance between the features of samples with the same label. This suggests that FedGR is more successful in accurate classification than FedAvg.

5   Conclusion

In this paper, we first show that existing FL algorithms face serious performance drop problem under double imbalance distribution. Based on this observation, we propose a novel FL algorithm called federated learning with gravitation regulation (FedGR) to deal with the problem of double imbalance distribution. We design a simple but effective unbalanced softmax by introducing a balance factor to balance the importance of classes for tackling quantity imbalance in client. Moreover, we propose a novel gravitation regularizer to call for the forces between clients for dealing with label imbalance among clients. Experiments have shown that FedGR outperforms the state-of-the-art FL methods under double imbalance distribution scenario.

Acknowledgement. This work was supported by National Key R & D Program of China (No. 2022YFF0606303) and National Natural Science Foundation of China (No. 62076079).

References

1. Chen, M., et al.: Federated learning of N-gram language models. arXiv preprint arXiv:1910.03432 (2019)
2. Deng, Y., Kamani, M.M., Mahdavi, M.: Adaptive personalized federated learning. arXiv preprint arXiv:2003.13461 (2020)
3. Duan, M., Liu, D., Chen, X., Liu, R., Tan, Y., Liang, L.: Self-balancing federated learning with global imbalanced data in mobile systems. IEEE Trans. Parallel Distrib. Syst. (2021)
4. Fallah, A., Mokhtari, A., Ozdaglar, A.: Personalized federated learning with theoretical guarantees: a model-agnostic meta-learning approach. In: Advances in Neural Information Processing Systems, vol. 33, pp. 3557–3568 (2020)
5. Ghosh, A., Chung, J., Yin, D., Ramchandran, K.: An efficient framework for clustered federated learning. In: Advances in Neural Information Processing Systems, vol. 33, pp. 19586–19597 (2020)
6. Gupta, O., Raskar, R.: Distributed learning of deep neural network over multiple agents. J. Netw. Comput. Appl. **116**, 1–8 (2018)
7. Huang, H., Shang, F., Liu, Y., Liu, H.: Behavior mimics distribution: combining individual and group behaviors for federated learning. arXiv preprint arXiv:2106.12300 (2021)
8. Huang, Y., et al.: Personalized cross-silo federated learning on non-IID data. In: AAAI, pp. 7865–7873 (2021)
9. Kairouz, P., et al.: Advances and open problems in federated learning. Found. Trends® Mach. Learn. **14**(1–2), 1–210 (2021)
10. Karimireddy, S.P., Kale, S., Mohri, M., Reddi, S., Stich, S., Suresh, A.T.: Scaffold: stochastic controlled averaging for federated learning. In: International Conference on Machine Learning, pp. 5132–5143. PMLR (2020)

11. Krizhevsky, A., Hinton, G., et al.: Learning multiple layers of features from tiny images (2009)
12. Krizhevsky, A., Sutskever, I., Hinton, G.E.: ImageNet classification with deep convolutional neural networks. Commun. ACM **60**(6), 84–90 (2017)
13. Li, Q., He, B., Song, D.: Model-contrastive federated learning. In: Proceedings of the IEEE/CVF Conference on Computer Vision and Pattern Recognition, pp. 10713–10722 (2021)
14. Li, T., Sahu, A.K., Zaheer, M., Sanjabi, M., Talwalkar, A., Smith, V.: Federated optimization in heterogeneous networks. In: Proceedings of Machine Learning and Systems, vol. 2, pp. 429–450 (2020)
15. Li, X.C., Zhan, D.C.: FedRS: federated learning with restricted softmax for label distribution non-IID data. In: Proceedings of the 27th ACM SIGKDD Conference on Knowledge Discovery & Data Mining, pp. 995–1005 (2021)
16. Van der Maaten, L., Hinton, G.: Visualizing data using t-SNE. J. Mach. Learn. Res. **9**(11) (2008)
17. McMahan, B., Moore, E., Ramage, D., Hampson, S., Arcas, B.A.: Communication-efficient learning of deep networks from decentralized data. In: Artificial Intelligence and Statistics, pp. 1273–1282. PMLR (2017)
18. Niu, Y., Deng, W.: Federated learning for face recognition with gradient correction. In: Proceedings of the AAAI Conference on Artificial Intelligence, vol. 36, pp. 1999–2007 (2022)
19. Reddi, S., et al.: Adaptive federated optimization. arXiv preprint arXiv:2003.00295 (2020)
20. T Dinh, C., Tran, N., Nguyen, J.: Personalized federated learning with Moreau envelopes. In: Advances in Neural Information Processing Systems, vol. 33, pp. 21394–21405 (2020)
21. Tan, A.Z., Yu, H., Cui, L., Yang, Q.: Towards personalized federated learning. IEEE Trans. Neural Netw. Learn. Syst. (2022)
22. Vaswani, A., et al.: Attention is all you need. In: Advances in Neural Information Processing Systems, vol. 30 (2017)
23. Wang, J., Liu, Q., Liang, H., Joshi, G., Poor, H.V.: Tackling the objective inconsistency problem in heterogeneous federated optimization. In: Larochelle, H., Ranzato, M., Hadsell, R., Balcan, M., Lin, H. (eds.) Advances in Neural Information Processing Systems, vol. 33, pp. 7611–7623. Curran Associates, Inc. (2020)
24. Xiao, H., Rasul, K., Vollgraf, R.: Fashion-MNIST: a novel image dataset for benchmarking machine learning algorithms. arXiv preprint arXiv:1708.07747 (2017)
25. Xu, J., Xu, Z., Walker, P., Wang, F.: Federated patient hashing. In: Proceedings of the AAAI Conference on Artificial Intelligence, vol. 34, pp. 6486–6493 (2020)
26. Yang, Q., Liu, Y., Chen, T., Tong, Y.: Federated machine learning: concept and applications. ACM Trans. Intell. Syst. Technol. (TIST) **10**(2), 1–19 (2019)
27. Zhao, Y., Li, M., Lai, L., Suda, N., Civin, D., Chandra, V.: Federated learning with non-IID data. arXiv preprint arXiv:1806.00582 (2018)
28. Zheng, W., Yan, L., Gou, C., Wang, F.Y.: Federated meta-learning for fraudulent credit card detection. In: Proceedings of the Twenty-Ninth International Conference on International Joint Conferences on Artificial Intelligence, pp. 4654–4660 (2021)

Federated Learning with Emerging New Class: A Solution Using Isolation-Based Specification

Xin Mu[(⊠)], Gongxian Zeng, and Zhengan Huang

Peng Cheng Laboratory, Shenzhen 518000, China
{mux,zenggx01}@pcl.ac.cn

Abstract. This paper investigates an important classification problem in federated learning (FL), i.e., federated learning with emerging new class or *FLENC*, where instances with new class that has not been used to train a classifier, may emerge in the testing process. As based on the assumption of the fixed class set, the common FL classification approaches will face the challenge that the predictive accuracy severely degrades if they are used to classify instances with a previously unseen class in *FLENC* problem. To address this issue, we explore a new FL framework under the one-shot setting, called the federated model with *specification*. This framework is to learn the local model with the specification which can explain the purpose or specialty of the model, and learn the global federated model in only a single round of communication. Further, we introduce an implementation FLIK by employing a data-dependent isolation-based kernel as the specification. FLIK is the first attempt to propose a unified method to address the following two important aspects of FL: (i) new class detection and (ii) known class classification. We report evaluations demonstrating the effectiveness of our proposed method in *FLENC* problem.

Keywords: Federated learning · Emerging new class · Isolation kernel

1 Introduction

Federated learning, which aims to train a high-quality machine learning model across multiple decentralized devices holding local data samples, without exchanging them, is a widely studied topic with well-recognized practical values [14,20,33]. Gboard[1] on Android, the Google Keyboard, is a typical example that enables mobile phones to collaboratively learn a shared prediction model while keeping all the training data on device, without storing the data in the central cloud.

In this paper, we investigate an important problem in FL, i.e., Federated Learning with Emerging New Class (*FLENC*), which is a branch of heterogeneous data problem in FL. This problem is illustrated in Fig. 1. Specifically, Party 1~3 federally learn a digital image classification model based on each local data, and no one can access others' data. After training, we can deploy it to classify a digital image. However, instances belonging to previously unseen class (six) may emerge as the environment changes (i.e., class 6 in Party 3), or the model has been deployed in a new party that has a new unseen

---

[1] https://blog.google/products/search/gboard-now-on-android/.

© The Author(s), under exclusive license to Springer Nature Switzerland AG 2023
X. Wang et al. (Eds.): DASFAA 2023, LNCS 13943, pp. 719–734, 2023.
https://doi.org/10.1007/978-3-031-30637-2_48

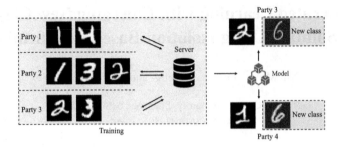

Fig. 1. Federated learning with emerging new class.

class (i.e., class 6 in Party 4). Ideally, we would like a new class to be detected by the FL model as soon as they emerge; and only instances that are likely to belong to the known class are passed to the FL classifier for prediction.

A problem closely related to the *FLENC* is Heterogeneous Federated Learning (HFL) [4,12], where the assumption is that there are multiple local parties which can characterize different model architectures based on local data with heterogeneous data distribution, i.e., either unique or overlapping classes. The previous works [4,12] often require an available public data set for all parties, where the public data set contains all classes, or assume the problem is under the fixed label set assumption. This is different from our problem, as the *FLENC* problem assumes data distribution is different in the training and testing process, and there is no public data set for all parties. The *FLENC* problem needs to face more challenges, i.e., a requirement is to accomplish two tasks: (1) detecting emerging new class and (2) classifying known class simultaneously.

To tap this challenge, we introduce a novel FL framework, *federated model with specification*. The main idea is that each party trains not only a local classifier, but also a specification function which more naturally explains the purpose or specialty of local model. This framework also is based on one-shot federated learning (OFL) setting [5], which emphasizes that a global model is learned from data in the network using only a single round of communication between the devices and the central server. Based on this framework, we propose a method named FLIK, Federated Learning with new class emerging by Isolation Kernel modelling. We employ an efficient data dependent point kernel technique *isolation kernel* as model specification, which is the sparse and finite-dimensional representation and enables an efficient dot product implementation [28, 30]. In FLIK, each party first learns its local machine learning model and a specification function, and sends them to the server. The server performs aggregation and sends back the aggregated results to parties. A party can use the global federated model combined specification function to complete new class detection and known class classification tasks on his own. We finally conduct empirical studies on the FL benchmark data sets to show the effectiveness and efficiency of our approach.

We summarize our key contributions as follows:

- This paper studies an important problem of heterogeneous data in federated learning, i.e., federated learning with emerging new class (*FLENC*), which needs to handle two tasks in the FL setting: detection for new class and classification for known class. To the best of our knowledge, this is the first time that detection for the new class and classification for the known class are considered simultaneously in the federated learning setting.

- To solve the *FLENC* problem, we propose a novel framework, federated model with *specification*, to combine multiple local models. Compared to methods based on iterative optimization techniques, this framework can learn a global model from data in the network using only a single round of communication. We further propose FLIK implementation under this framework, which employs a data-dependent point kernel implementations as model specification with no class information.
- We choose five representative datasets, varying from two common image evaluation datasets, two text datasets to a real-world news dataset, to comprehensively evaluate the performance. The experimental results show the proposed framework works effectively. Additionally, we also provide an analysis of the sensitivity of the main parameters in FLIK.

2   Related Work

The problem of **emerging new class** has attracted much attention in the machine learning and data mining community in recent years [18,21,22]. It strengthens a previously trained classifier to deal with emerging new class. Learning with Augmented Class (LAC) [2] uses the set of unlabelled instances to build new class detection model. In [18], Masud et. al. present time constraints for delayed classification for new class detection. The *SENCForest* [21] tackles the new class detection and classification problems by introducing a tree-based model. SAND [6] is a semi-supervised framework for the problem of new class emerging. Cai et. al. [1] introduce a geometric distance for this problem. In our paper, we aim to put forth a method that can solve the problem of new class emerging in the federated learning scenario. The existing works of new class emerging are exactly based on a single node. As far as we know, there is no algorithm focusing on this problem in the federated learning setting.

The field of **out-of-distribution (OOD) detection** is major research on allowing model to reject inputs that are semantically different from the training distribution and only make a prediction on in-distribution (ID) data. In fact, the methodologies and approaches for OOD detection entail two necessary aspects: (i) unseen class detection and (ii) seen class classification. An effective research direction for OOD detection is the output-based method, which is to detect OOD data by carefully designing a mechanism to look at the model output when introducing a mechanism on perturbation input data. For example, Hendrycks et al. [8] presented a simple OOD baseline that utilizes probabilities from softmax distributions. Liang and Li [16] used temperature scaling and input perturbation to amplify the ID/OOD separability. Techniques such as Mahalanobis distance [11] and Gram Matrix [26] are implemented for detecting OOD samples. However, these studies encountered the situation of not being easy to adopt a federated environment as they are commonly based on one single node assumption.

Federated learning is to collaboratively train a model across multiple participants on the premise of ensuring information security, and maintain a shared global model on a server [20,33]. Typically, federated learning is classified into three categories: horizontally federated learning, vertically federated learning and federated transfer learning. Horizontally federated learning or sample-based federated learning, is introduced in the scenarios where data sets share the same feature space but are different in samples [3,20]. Vertical federated learning or feature-based federated learning is applicable to the cases where two data sets share the same sample ID space but differ in feature space [7,23]. Federated transfer learning has been applied to the scenarios where the two data sets differ not only in samples but also in feature space [25]. In most of these studies, training a federated model by iterative optimization techniques requires numerous rounds of communication between the devices and the central server.

Recently, Guha et al. [5] introduce a new federated learning paradigm, *one-shot federated learning*, where a central server learns a global model over a network of federated devices in a single round of communication. This idea is effective and efficient in large networks with thousands of devices. Distilled one-shot federated learning distills each client's private dataset and sends the synthetic data (e.g. images or sentences) to the server [34]. Kasturiet et al. [10] present an alternative federated model, where the distribution parameters of the client's data along with its local model parameters are sent to the server. In [27], the server collects other devices' encoded data samples, and decodes them. Then, it uses the recovered data to train a federated model.

A common challenge of federated learning is the heterogeneity of the data distribution (i.e., non-IID) among distributed databases [13]. Because the distribution of each local dataset is highly different from the global distribution, and the local objective is inconsistent with the global optima, non-IID data can influence the accuracy of traditional FedAvg. Although there have been some FL algorithms to address the learning effectiveness under non-IID data settings [9,15,31], one potential shortcoming is that most existing solutions, built with a fixed label set, face serious challenges when dealing with data with the new class.

3   Preliminaries

Before introducing the detail of our proposed algorithm, we give some important concepts in this paper.

3.1   Federated Learning with Emerging New Class (FLENC)

We use $\{\mathcal{F}_1, \mathcal{F}_2, \cdots, \mathcal{F}_N\}$ to denote N data parties, all of whom wish to train a machine learning model by consolidating their respective data $\{\mathcal{D}_1, \mathcal{D}_2, \cdots, \mathcal{D}_N\}$. The federated learning system is a learning process in which the data owners collaboratively train a model $\mathcal{M}_{fed}$, and any data party $\mathcal{F}_i$ does not expose his data $\mathcal{D}_i$ to others. In each party $\mathcal{F}_i$, $\mathcal{D}_i = \{x_j, y_j\}^{|\mathcal{D}_i|}$, where $x_j \in \mathcal{X} = \mathbb{R}^{d_i}$, d_i is the number of features and $y_i \in \mathcal{Y}_i \in \mathcal{Y} = \{1, 2, \ldots, c\}$. In this paper, we consider the assumption of horizontal federated learning (HFL), $d_1 = \cdots = d_N$, and d is used for simplicity. Note that $\mathcal{Y}_i$ may be different from $\mathcal{Y}_j$, for $i \neq j$.

The goal of *FLENC* is to learn a federated model $\mathcal{M}_{fed}$ with N source parties, and any of source party $\mathcal{F}_i$ does not share his data $\mathcal{D}_i$ to others. The federated model $\mathcal{M}_{fed}$ is employed as follow,

$$y_{test} \leftarrow \mathcal{M}_{fed}(x_{test}),$$

where $x_{test} \in \mathbb{R}^d$, and label as $y_{test} \in \mathcal{Y}' = \{1, 2, \ldots, c, c+1, \ldots, M\}$ with $M > c$. Note that multiple new classes may exist in a test data set, and we deal with them as a single new meta-class.

Conceptually, the *FLENC* problem can be decomposed into two subproblems: (1) detecting emerging new class; (2) classifying known class. For every test instance, $\mathcal{M}_{fed}$ needs to act as a detector to determine whether it is likely to belong to a new class, and act as a classifier to produce a prediction for known class. Thus, the challenges in the *FLENC* problem are to detect emerging new class and classify instances of known class both with high accuracy. Note that multiple new classes may exist in the test data set. The proposed method deals with them as a single new meta-class.

3.2 One-Shot Federated Learning (OFL)

As shown in previous studies [19, 24], the training process of such a FL system often includes the following four steps:

- Each party locally computes the training model and sends it to the server;
- The server performs an aggregation method to complete parameter fusion;
- The server sends back the aggregated results to parties;
- Parties update their respective models by using new aggregated results.

Iterations through the above steps continue until convergence, thus completing the entire training process.

One-shot federated learning emphasizes that the central server learns a global model over a network of federated devices in a single round of communication [5]. It was undertaken primarily to the known bottleneck of communication in this traditional FL setting. The learning process also contains four steps, but it only needs one time. Thus, a grand challenge is how to find models that are helpful for the current task, without accessing the raw training data. Ensemble learning [35] is commonly used to combine multiple learners to improve predictive performance, which can be naturally applied in OFL. Guha et.al. [5] introduced three ensemble selection strategies: (1) Cross-Validation Selection: the server selects k best performing models from this subset of local models; (2) Data Selection: the server selects models from these local models trained on the top k largest data sets; (3) Random Selection: the server randomly selects k parties to create an ensemble model.

4 The Proposed Method

4.1 Federated Model with Specification

In this section, we introduce a new one-shot FL framework for *FLENC*, called Federated Model with *Specification* (**FMS**). The *specification* can be viewed as statistic descriptions or even a few simplified training samples, and it reveals the target to which the model aims [36]. Or it can be used to specify whether a machine learning model is compatible with the current task.

We define the *specification* in **FMS** as a function:

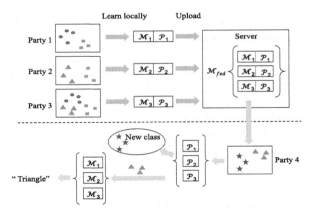

Fig. 2. An illustration of FMS.

Definition 1 *[Specification Function]: Given a classification model f learned from data set D, we define the specification as a function $\mathcal{P}()$, which is learned from data set D. The function $\mathcal{P}()$ produces a score for an instance x, which determines x as adapting on the current model f.*

The **FMS** can be described as follows: (1) Each party first learns its local machine learning mode $\mathcal{M}_i$ and a specification function $\mathcal{P}_i$ under each local data set. Here we assume that all providers are competent, and the local datasets are sufficient to solve their tasks. (2) Then, $\{\mathcal{M}_i, \mathcal{P}_i\}$ will be sent to the server. The server performs aggregation and sends back the aggregated result to parties. (3) Finally, one of the parties can use model specification function $\mathcal{P}_i$ and mode $\mathcal{M}_i$ to complete the classification task on his own party. The global federated model is defined as

$$\mathcal{M}_{fed} = \{\mathcal{M}_i, \mathcal{P}_i\}_{i=1}^N \tag{1}$$

where the model $\mathcal{M}_i$ and specification $\mathcal{P}_i$ are learned from a private local dataset $\mathcal{D}_i = \{(x_j, y_j)\}_{j=1}^{|\mathcal{D}|}$. Note that we learn a global model from data only in a single round of communication between each party and the server.

An illustrative example is provided in Fig. 2. Party 1 to 3 federally train a model, and Party 4 (who has a new class, i.e., "star") would like to use the final federated model. Note that "star" is unseen by Party 1 to 3. Party 1 to 3 first locally learn a model and a specification as $\{\mathcal{M}_i, \mathcal{P}_i\}$ based on local data, respectively. Then, they send them to the server. The server, as an aggregation center, assembles all models as $\mathcal{M}_{fed}$ and sends it to Party 4 who requires $\mathcal{M}_{fed}$ to work on a classification task. After that, in Party 4, instances of a new class can be detected by $\mathcal{M}_{fed}$ as soon as they emerge, and only instances belonging to the known class are passed to $\mathcal{M}_{fed}$ for prediction.

In this paper, we explore a data-dependent implementation, Isolation Kernel (IK) [30], as the specification function under this framework. This implementation is solely dependent on data distribution, requiring neither class information nor explicit learning to be a density estimator. It can be viewed as a data dependent kernel adapting to the structure of a data set. Recent works have also shown that it is able to improve the predictive accuracy and anomaly detection performance [29,32]. In fact, we further explore IK for new class detection task in the *FLENC* problem. In the following, we names this implementation as **FLIK** and the details are as follows.

4.2 Isolation Kernel as Specification Function

Let $D \in \mathbb{R}^d$ be a dataset sampled from the distribution $\mathcal{Q}$. $\mathbb{H}(D)$ denotes the set of all partitions over the data space. Each partition $\theta[z] \in \mathbb{H}$ isolates a point $z \in D$ from the rest of the points in D. For any two points $x, y \in D$, *Isolation Kernel* is defined to be the expectation taken over the probability distribution on all partitions $\mathbb{H}$ that both x and y fall into the same isolating partition $\theta \in H$. It can be formalized

$$K(x, y|D) = \mathbb{E}_{\mathbb{H}(D)}[\mathbb{1}(x, y \in \theta[z]|\theta[z] \in \mathbb{H})] \tag{2}$$

where $\mathbb{1}(\cdot)$ be an indicator function. If x and y belong to the same partition, the indicator will output 1; otherwise 0. In practice, we employ a finite number of partitions $H_i \in \mathbb{H}, i = 1, \cdots, t$ as follows,

$$K(x, y|D) = \frac{1}{t} \sum_{i=1}^{t} \mathbb{1}(x, y \in \theta | \theta \in H_i), \tag{3}$$

where each H_i is generated by using randomly chosen subset $D_i \in D$, $\psi = |D_i|$. θ is a shorthand for $\theta[z]$.

We employ a tree structure method to generate *partications* $\mathbb{H}(D)$, e.g., iForest [17]. It is done recursively by randomly selecting an axis-parallel split to subdivide the data into two non-empty subsets until every point is isolated. Each partition H_i produces ψ isolating partitions θ, and each partition contains a single data in D.

Hence, the model specification function $\mathcal{P}$ is as follows: given D and partitions H_i produced from sampled data ψ, we denote $\Phi_i(\cdot)$ as a ψ-dimensional binary column vector representing all partitions $\theta_j \in H_i, j = 1, \cdots, \psi$. If x falls into partition θ_j, the j-component of the vector equals 1, i.e., $\Phi_{ij}(x) = \mathbb{1}(x \in \theta_j | \theta_j \in H_i)$. Given t partitions, $[\Phi(x)]$ is the concatenation of $\Phi_1(x), \cdots, \Phi_t(x)$. Therefore, the model specification function $\mathcal{P}$ on whole data set D is formalized as

$$\mathcal{P}(D) = [\Phi(D)] \in \{0, 1\}^{N \times t\psi}. \tag{4}$$

$\Phi(x)$ can also be the reduced set by computing mean of $\Phi_1(x), \cdots, \Phi_t(x)$, and the model specification function becomes $\mathcal{P}(D) = [\Phi(D)] \in \{0, 1\}^{N \times \psi}$, where ψ can be controlled by a user.

Note that there are two reasons to choose IK as the model specification: (1) It is not practical to obtain much class information in the FL setting. In contrast, IK is a new data dependent method, which needs neither learning nor class information; (2) IK has one property: two points, as measured by IK derived in a sparse region, are more similar than the same two points, as measured by IK derived in a dense region [29]. This property works well for the heterogeneity of the data distribution problem, e.g., measuring the similarity between different classes with different density.

Specification for New Class Detection. As mentioned before, model specification based on IK is a new way to measure the data distribution. It can be regarded as a representation of distribution. Therefore, it is natural to consider that data with the known class should be more approximated well by the model specification function. On the contrary, the new class data which is unseen before may not be approximated well by the model specification.

We define a measurement S to calculate the similarity between a testing instance and model specification. The inner product is used as the measurement as follows:

$$S(x) = \frac{1}{N} \sum_{i=1}^{N} \langle \Phi(x), [\mathcal{P}(D)]_{i,:} \rangle, \quad i \in \{1, \ldots, N\} \tag{5}$$

where $\langle \cdot, \cdot \rangle$ is the inner product of two vectors, and $\Phi(\cdot)$ is isolation kernel mapping. We assume that D is from distribution $\mathcal{Q}$. If $x \in \mathcal{Q}$, $S(x, D)$ is large, which can be interpreted as x is likely to be part of $\mathcal{Q}$. If $x \notin \mathcal{Q}$, $S(x, D)$ is small, which can be interpreted as x is not likely to be part of $\mathcal{Q}$.

In this paper, we follow the assumption mentioned in [18,21] that there exists a difference between the distribution of data with the new class and data with the known class. And the data with the new class should have low similarity with respect to the model specification which is generated from a training dataset.

To sum up, the new class determination can be described as follows: FLIK first maps each x to $\Phi(x)$, and computes the similarity with regard to $\mathcal{P}$, i.e., $\langle \Phi(x), \mathcal{P} \rangle$. Then, new class points with smaller similarity value are less similar to $\mathcal{P}$.

Specification for known class classification. The proposed method provides a strategy to combine multiple parties' model results for known class classification. The main idea is to take advantage of the fact that, the more similar the testing data x and the model's training data are, the more confident the final prediction is. Basically, it means that one model should be selected to make the final prediction for a test instance x when x is more approximated well by $\mathcal{P}(D)$. Hence, the final prediction is output by considering r^* models which have the top r^* specification scores. Or we can treat the specification score as an ensemble weight to make the final prediction as follows

$$\mathcal{M}_{fed}(x) = \sum_{i=1}^{r^*} \mathcal{M}_i(x) \quad or \quad \sum_{i=1}^{N} S(x)\mathcal{M}_i(x) \tag{6}$$

where the local model $\mathcal{M}_i$ can be a common classification model like SVM, decision tree or deep neural network. The final output is the majority of class as

$$\max_{j \in \{1, \ldots, c\}} F[j]$$

where $F[j]$ is the class frequency for class j from Eq. (6).

4.3   Deployment

Training Phase. The process of training FLIK has two major steps: **1. Building local model**. The local model mainly contains two parts: the local classification model is

---

Algorithm 1: Training Phase

Input : N data parties $\{\mathcal{D}_1, \mathcal{D}_2, \cdots, \mathcal{D}_N\}$, $\mathcal{D}_i = \{x, y\}$.
Output: $\mathcal{M}_{fed}$
`// Local training`
1 **for** $i = 1, \ldots, N$ **do**
2 $\quad$ $\mathcal{M}_i \leftarrow f(D_i)$ `// training based model`;
3 $\quad$ $\mathcal{P}_i \leftarrow \Phi(D_i)$ `// using Eqn. (4)` ;
4 $\quad$ send $\{\mathcal{M}_i, \mathcal{P}_i\}$ to server.
5 **end**
`// Server aggregation`
6 Server aggregates each local model as $\mathcal{M}_{fed}$;
7 Server sends $\mathcal{M}_{fed}$ to each party.

Algorithm 2: Testing Phase

Input : $\mathcal{M}_{fed}$, testing data set D_{test}, buffer $\mathcal{B}$
Output: y - class label for instance x
1 **for** *each x in D_{test}* **do**
2 $\quad$ $s \leftarrow \frac{1}{N}\sum \mathcal{P}_i(x)$ `// #using Eqn.(5)`
3 $\quad$ $\mathcal{B} \leftarrow \mathcal{B} \cup \{x, s\}$
4 **end**
5 **for** *each $\{(x, s)\}$ in $\mathcal{B}$* **do**
6 $\quad$ **if** $s \leq Threshold$ **then** $y \leftarrow$ new class;
7 $\quad$ **else** $y \leftarrow \mathcal{M}(x), y \in \{1, \ldots, c\}$ `// using Eqn. (6)`;
8 **end**

---

first trained from a given training set D_i on each party, and the model specification is learned based on the data-dependent isolation kernel method. **2. Model aggregation**. Each party sends its local model to the server. The server performs an aggregation method to combine their local models as $\mathcal{M}_{fed}$, and sends it back to each party. Note that this procedure is only in a single round of communication.

The training procedure of **FLIK** is detailed in Algorithms 1. Lines 1~5 show the local training on each party: In lines 2 and 3, local data set D_i is implemented for building the specification model and classification model; In line 4, both of them will be sent to the server. On the side of the server, the server collects local models from all the parties as federated model $\mathcal{M}_{fed}$, and sends it back to each party. This completes the setup of **FLIK**.

Testing Phase. The test phase is deployed on one party as shown in Algorithm 2. Given an instance x, federated model $\mathcal{M}_{fed}(x)$ produces specification value s by using Eqn. (5), and $\{x, s\}$ is placed in buffer $\mathcal{B}$. Then sort all computed value s in $\mathcal{B}$, new class data is less similar to $\mathcal{P}(D)$ which will be smaller value s. Note that we employ a threshold for this process in line 6. Otherwise, it outputs the majority of class using Eqn. (6).

Note that we can employ a threshold determination method in [21] to choose this parameter. The intuition is to use the difference in standard deviations to separate these two types of distributions in buffer $\mathcal{B}$. We first produce a list Q which orders all value s's

---

Algorithm 3: Determining Threshold

Input : Q - the list of value s in $\mathcal{B}$, m - size of Q, τ^* - initialize to a larger value. $\sigma(\cdot)$ - standard deviations calculation

Output: t^* - threshold

1 **for** $i = 1, ..., m$ **do**

2      $Q_l \leftarrow Q[1:i]$;

3      $Q_r \leftarrow Q[i:m]$;

4      $\tau \leftarrow |\sigma(Q[1:i]) - \sigma(Q[i:m])|$;

5   **if** $\tau < \tau^*$ **then**

6          $t^* \leftarrow Q[i]$;

7          $\tau^* \leftarrow \tau$;

8   **end**

9 **end**

---

in $\mathcal{B}$. A threshold in this list yields two sub-lists Q_l and Q_r. To find the best threshold, we minimise the difference in standard deviations

$$\tau (= min|\sigma(Q^r) - \sigma(Q^l)|),$$

where $\sigma(x)$ is the standard deviation of x. Algorithm 3 shows the detail.

5 Experiment

In this section, we conduct experiments to evaluate **FLIK** on the effectiveness of addressing the *FLENC* problem and the analysis of the parameter sensitivity.

5.1 Experimental Setup

Data Sets. We use the following data sets to assess the performance of all methods in comparison. They are **MNIST**, **CIFAR10**, **20 Newsgroups**, **Sentiment140**[2] and **New York Times**[3]. We use the "word2vec[4]" to process each text item as a 300-dimension feature vector.

Table 1. Methods used in the empirical evaluation.

Method	Detection	Classification
LM + MaS	MaS	Local Model
DS + MaS		Data Share
ES + MaS		Ensemble without Selection
RS + MaS		Random Selection
FedAvg + MaS		FedAvg
FLIK	FLIK	

---

[2] http://help.sentiment140.com/home.

[3] http://developer.nytimes.com/.

[4] https://radimrehurek.com/gensim/index.html.

Competing Algorithms. To compare with FLIK, we choose five FL methods, and the details are as follows. The complete list of the methods used for detection and classification is shown in Table 1. Note that because some of these methods (e.g., LM, DS, etc.) have no ability of detection new class, a state-of-the-art detector (i.e., MaS [22]) is employed to detect instances with the new class, and we combine them for comparison.

1. **Local Model (LM)**: The server randomly selects one party's model as the classification model.
2. **Data Share (DS)**: Data Share is not limited to communication and privacy. The server can train a "global" model on data aggregated across all parties.
3. **Ensemble without Selection (ES)**: ES is an ensemble algorithm by assembling all parties' models as the classification model.
4. **Random Selection (RS)**: The server randomly selects k parties' models as an ensemble classification model for prediction.
5. **FedAvg [19]**: FedAvg is a most widely used FL method. FedAvg updates the model in each party by local gradient and sends the latest model to the central server. The central server conducts a weighted average over the model parameters received from active devices and broadcasts the latest averaged model to all devices. This process is done through several rounds of iterations.
6. **MaS [22]**: A detection method is based on the principal component of original data. MaS contains low-dimensional matrix sketches for approximating the original global, which is produced on the whole local data set for new class detection.

In this paper, we verify the different base classifiers, e.g., random forest (**RF**), **SVM** and a fully connected neural network with 2 hidden layers (**NN**). In the FLIK framework, we represent them as **FLIK-RF, FLIK-SVM, FLIK-NN**.

Evaluation Measures. Two measurements are used in this paper. One is *Accuracy*, *Accuracy (Acc)* $= \frac{A_n + A_o}{N}$, where N is the total number of testing instances, A_n is the number of emerging class instances identified correctly, and A_o is the number of known class instances classified correctly. The other is **AUC**, which is used for evaluating on new class detection.

Parameter Settings. All experiments are implemented in Python on Intel Core CPU machine with 16 GB memory and 3.2 GHz clock speed. The following implementations are used: In LM, DS, ES, and RS, the base classifier is SVM with RBF kernel and parameters are set by default values. FedAvg using NN structure is the same as FLIK-NN, and its codes are completed based on the authors' paper. MaS uses the codes as released by the corresponding authors. In RS, k is 80% of the number of parties. Random forest and NN both use default parameter values. In FLIK and Random forest, the number of trees is set to 1000 and $\psi = 512$.

5.2   FLIK on Synthetic Data

We simulate a federated learning environment using a two dimensional synthetic dataset as shown in Fig. 3, and the results of the synthetic dataset are shown in Table 2. The four classes are distributed into four parties. Party 1 has green and pink classes, Party 2 has

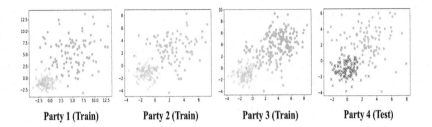

Party 1 (Train)   Party 2 (Train)   Party 3 (Train)   Party 4 (Test)

Fig. 3. An illustration on a synthetic dataset.

Table 2. The results of the synthetic dataset.

	Party 1	Party 2	Party 3	Party 4
Model architecture	SVM	SVM	SVM	FLIK-SVM
Class information	green, pink	green, yellow	green, yellow, pink	yellow, red
Number of instance	100,100	100,100	100,100,100	100,100
Training Performance	Acc : 0.975	Acc : 0.960	Acc : 0.960	＼
Testing Performance	Acc : 0.975	Acc : 0.965	Acc : 0.955	**Acc : 0.940; AUC : 0.912**

green and yellow classes, Party 3 has green, yellow and pink classes and Party 4 has red and yellow classes. Party 1∼3 federally learn a classification FL model. Then the model is deployed in a new Party 4 which has a new unseen class (red). Note that the classes in each local party can either be unique or overlapping across parties. In Party 4, the proposed method **FLIK** achieves high classification accuracy and performs well in the detection of emerging new class. This result shows that (i) the proposed method **FLIK** achieves high classification accuracy; and (ii) **FLIK** performs well in the detection of emerging new class.

5.3   FLIK on Real-World Datasets

We choose five representative datasets for evaluation, varying from two common image evaluation datasets (MNIST and CIFAR10), two text datasets (Sentiment140 and 20 Newsgroups) to a real-time news dataset (New York Times). The experimental results show the proposed framework works effectively.

Simulation. In the following experiments, each dataset is used to simulate a federated learning environment over ten trials, and the average result of ten trials is reported. For the sake of clarity, we first show an example of MNIST in Table 3. We simulate 5 parties, Party 1–4 federally train model $\mathcal{M}_{fed}$, and Party 5 uses $\mathcal{M}_{fed}$ for the test. Then one class and corresponding data are randomly selected from the whole dataset, which is set in Party 5 as the new class, e.g., class 9. After that, the new five classes from the rest of the classes are randomly selected into Party 1–4, e.g., class 1–5. The model architecture is based on SVM. We also show each local accuracy and testing phase performance in Table 3. AUC is an illustration of new class detection performance, so it does not show in lines 1–4. Note that in the testing phase, as the error of new class detection exists, accuracy could be affected.

Table 3. An example on the MNIST dataset.

	Party 1	Party 2	Party 3	Party 4	Party 5
Model architecture	SVM	SVM	SVM	SVM	FLIK-SVM
Class information	2,4,3	1,2,3	1,2,5	3,4,5	1,3,9
Number of instance	3000	3000	3000	3000	3000
Training performance	Acc : 0.735	Acc : 0.733	Acc : 0.727	Acc : 0.850	＼
Testing performance	Acc : 0.720	Acc : 0.751	Acc : 0.735	Acc : 0.750	**Acc : 0.753; AUC : 0.705**

Table 4. The results on different datasets.

Algorithm	20 Newsgroups	Sentiment140	NY Times	CIFAR10	MINST
LM + MaS	Acc : 0.501 ± 0.10 AUC : 0.511 ± 0.02	0.554 ± 0.02 0.599 ± 0.02	0.533 ± 0.05 0.540 ± 0.02	0.483 ± 0.09 0.525 ± 0.03	0.553 ± 0.04 0.523 ± 0.03
DS + MaS	Acc : 0.721 ± 0.05 AUC : 0.572 ± 0.02	**0.652 ± 0.06** 0.602 ± 0.04	0.683 ± 0.04 0.581 ± 0.01	**0.607 ± 0.02** 0.552 ± 0.01	**0.804 ± 0.03** 0.621 ± 0.03
ES + MaS	Acc : 0.687 ± 0.03 AUC : 0.562 ± 0.03	0.604 ± 0.02 0.603 ± 0.03	0.644 ± 0.03 0.583 ± 0.02	0.595 ± 0.04 0.551 ± 0.02	0.762 ± 0.06 0.619 ± 0.03
RS + MaS	Acc : 0.586 ± 0.12 AUC : 0.513 ± 0.04	0.601 ± 0.05 0.600 ± 0.02	0.542 ± 0.16 0.580 ± 0.07	0.582 ± 0.10 0.539 ± 0.04	0.722 ± 0.13 0.615 ± 0.02
FLIK-SVM	Acc : **0.729 ± 0.03** AUC : **0.651 ± 0.01**	0.629 ± 0.07 **0.652 ± 0.04**	**0.689 ± 0.05** **0.617 ± 0.05**	0.605 ± 0.04 **0.572 ± 0.03**	0.753 ± 0.05 **0.705 ± 0.01**
FedAvg + MaS	Acc : 0.702 ± 0.03 AUC : 0.577 ± 0.02	0.631± 0.03 0.602± 0.02	0.661 ± 0.03 0.582 ± 0.04	0.593 ± 0.03 0.523 ± 0.05	0.763± 0.03 0.623± 0.03
FLIK-SVM has #wins/#draws/#losses	7/1/2				

Note that in Sentiment140 experiments, one class is randomly selected as new class and the other two are set into parties 1–4. The setting of other datasets is the same as that of MNINT. We will run ten independent experiments with different simulations on each dataset. Table 4 shows the mean and the standard variance of the performance.

Results. In terms of new class detection, FLIK-SVM maintains a good performance in new class detection in all three datasets than any other methods. Although MaS, which is based on the principal component of the original data, can detect new class in *FLENC* setting, it may expose the original data. It has difficulty in finding the right parameters as well. MaS also performs worse than our method, which can in turn affect accuracy. In terms of classification results, LM and RS present worse performance in some cases. The more important reason is that randomly selected models may not handle the current classification task. DS is an algorithm which is provided all data to train the model, and it performs better than other methods. ES is similar to our framework, but it achieves worse results than FLIK-SVM. This explains the effectiveness of model specification. What has greatly contributed to the practical values of FLIK-SVM is that it achieves results comparable with state-of-the-art methods, and it has fewer communications in the model training.

In addition, Fig. 4 shows results on FLIK with different base classifiers on MNIST dataset, e.g., random forest (**RF**), **SVM** and a fully connected neural network with 2 hidden layers (**NN**). In the FLIK, we represent them as FLIK-RF, FLIK-SVM, FLIK-NN. FLIK-NN achieves good performance and close results with FedAvg. It demonstrates that the better the base classifier it has, the higher performance it achieves. Notice, however, that NN often comes at high computational costs in the training process.

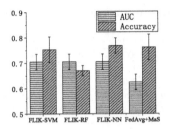

Fig. 4. The results on different classifiers.

Table 5. The results of FLIK on different model architecture.

	Party 1	Party 2	Party 3	Party 4	Party 4
Model architecture	NN	3-Layer CNN	SVM	FLIK	ES
Class information	2,4,3	1,2,3	1,2,5	1,4,7	1,4,7
Number of instance	3000	3000	3000	3000	3000
Training performance	Acc : 0.757	Acc : 0.762	Acc : 0.715	＼	＼
Testing performance	Acc : 0.721	Acc : 0.751	Acc : 0.705	**Acc : 0.733**	Acc : 0.591

5.4   FLIK on Different Model Architectures

In some real applications, different local parties may have different model architectures, so how to handle different parties with different model architectures is another challenge in the *FLENC* problem. Note that because FLIK imposes no limit on the architecture of base classifiers (i.e., not requiring all classifiers to have the same architecture), it can be naturally performed in this case.

We set three different model architectures for the three different parties as shown in Table 5. The three model architectures are set to a simple two-layer NN model, a 3-Layer CNN, and an SVM model, respectively. Table 5 shows that (1) a detector is necessary for the *FLENC* problem, e.g., ES performs worse than FLIK because of poor classification performance on new class; and (2) FLIK can perform well on the challenge of different parties with different model architectures.

5.5   Parameters Analysis

We study the influences of parameters in FLIK, i.e., the number of parties, the number of trees, and the number of ψ. We evaluate FLIK on the MNIST data set with different settings of one parameter while the other parameters are fixed.

Figure 5 (a) shows results of different size of parties, and the performance of FLIK is basically stable between 4 to 10 parties. Figure 5 (b) and Fig. 5 (c) show the number of trees and the number of ψ in FLIK. Similar to IK, FLIK has better performance when the tree is larger. In Fig. 5 (c), we also observe that the accuracy of FLIK converges at a small ratio. Figure 5 (d) shows the proposed method can be more efficient in the training phase. Note that similar results are also observed in the other data sets.

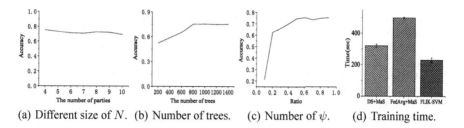

(a) Different size of N. (b) Number of trees. (c) Number of ψ. (d) Training time.

Fig. 5. The results on the sensitivity of parameters. In (c), the X-axis represents a ratio between the sample size and the original data size.

6 Conclusion

This paper addresses an important problem in federated learning, i.e., federated learning with emerging new class. We introduce a new one-shot federated learning framework, where each party learns a local model and a model specification, and the global federated model is learned only in a single round of communication. Based on this framework, we explore a data-dependent implementation, **FLIK**, to handle new class detection and known class classification in *FLENC*. The experiments on real-world data sets have demonstrated the superiority of **FLIK** over the existing methods.

Acknowledgements. This work was supported by the National Science Foundation of China (62106114).

References

1. Cai, X., Zhao, P., Ting, K., Mu, X., Jiang, Y.: Nearest neighbor ensembles: an effective method for difficult problems in streaming classification with emerging new classes. In: ICDM, pp. 970–975 (2019)
2. Da, Q., Yu, Y., Zhou, Z.H.: Learning with augmented class by exploiting unlabeled data. In: AAAI, pp. 1760–1766 (2014)
3. Gao, D., Ju, C., Wei, X., Liu, Y., Chen, T., Yang, Q.: HHHFL: hierarchical heterogeneous horizontal federated learning for electroencephalography. CoRR abs/1909.05784 (2019)
4. Gudur, G.K., Perepu, S.K.: Federated learning with heterogeneous labels and models for mobile activity monitoring. CoRR abs/2012.02539 (2020)
5. Guha, N., Talwalkar, A., Smith, V.: One-shot federated learning. CoRR abs/1902.11175 (2019)
6. Haque, A., Khan, L., Baron, M.: Sand: Semi-supervised adaptive novel class detection and classification over data stream. In: AAAI, pp. 1652–1658 (2016)
7. Hardy, S., et al.: Private federated learning on vertically partitioned data via entity resolution and additively homomorphic encryption. CoRR abs/1711.10677 (2017)
8. Hendrycks, D., Gimpel, K.: A baseline for detecting misclassified and out-of-distribution examples in neural networks. In: ICLR (2017)
9. Karimireddy, S.P., Kale, S., Mohri, M., Reddi, S.J., Stich, S.U., Suresh, A.T.: SCAFFOLD: stochastic controlled averaging for on-device federated learning. In: ICML (2020)
10. Kasturi, A., Ellore, A.R., Hota, C.: Fusion learning: a one shot federated learning. In: ICCS, pp. 424–436 (2020)

11. Lee, K., Lee, K., Lee, H., Shin, J.: A simple unified framework for detecting out-of-distribution samples and adversarial attacks. In: NeurIPS, pp. 7167–7177 (2018)
12. Li, D., Wang, J.: Fedmd: heterogenous federated learning via model distillation. CoRR abs/1910.03581 (2019)
13. Li, Q., Diao, Y., Chen, Q., He, B.: Federated learning on non-iid data silos: an experimental study. CoRR abs/2102.02079 (2021)
14. Li, T., Sahu, A.K., Talwalkar, A., Smith, V.: Federated learning: challenges, methods, and future directions. IEEE Signal Process. Mag. **37**(3), 50–60 (2020)
15. Li, T., Sahu, A.K., Zaheer, M., Sanjabi, M., Talwalkar, A., Smith, V.: Federated optimization in heterogeneous networks. In: Proceedings of Machine Learning and Systems (2020)
16. Liang, S., Li, Y., Srikant, R.: Enhancing the reliability of out-of-distribution image detection in neural networks. In: ICLR (2018)
17. Liu, F.T., Ting, K.M., Zhou, Z.H.: Isolation forest. In: ICDM, pp. 413–422 (2008)
18. Masud, M., Gao, J., Khan, L., Han, J., Thuraisingham, B.M.: Classification and novel class detection in concept-drifting data streams under time constraints. IEEE TKDE **23**(6), 859–874 (2011)
19. McMahan, B., Moore, E., Ramage, D., Hampson, S., y Arcas, B.A.: Communication-efficient learning of deep networks from decentralized data. In: AISTATS (2017)
20. McMahan, H.B., Moore, E., Ramage, D., y Arcas, B.A.: Federated learning of deep networks using model averaging. CoRR abs/1602.05629 (2016)
21. Mu, X., Ting, K.M., Zhou, Z.H.: Classification under streaming emerging new classes: a solution using completely-random trees. IEEE TKDE **29**(8), 1605–1618 (2017)
22. Mu, X., Zhu, F., Du, J., Lim, E., Zhou, Z.H.: Streaming classification with emerging new class by class matrix sketching. In: AAAI, pp. 2373–2379 (2017)
23. Nock, R., et al.: Entity resolution and federated learning get a federated resolution. CoRR abs/1803.04035 (2018)
24. Reddi, S.J., et al.: Adaptive federated optimization. In: ICLR (2021)
25. Saha, S., Ahmad, T.: Federated transfer learning: concept and applications. CoRR abs/2010.15561 (2020)
26. Sastry, C.S., Oore, S.: Detecting out-of-distribution examples with gram matrices. In: ICML (2020)
27. Shin, M., Hwang, C., Kim, J., Park, J., Bennis, M., Kim, S.: XOR mixup: privacy-preserving data augmentation for one-shot federated learning. CoRR abs/2006.05148 (2020)
28. Ting, K.M., Wells, J.R., Washio, T.: Isolation kernel: the X factor in efficient and effective large scale online kernel learning. CoRR abs/1907.01104 (2019)
29. Ting, K.M., Xu, B., Washio, T., Zhou, Z.H.: Isolation distributional kernel: a new tool for kernel based anomaly detection. In: KDD, pp. 198–206 (2020)
30. Ting, K.M., Zhu, Y., Zhou, Z.H.: Isolation kernel and its effect on SVM. In: KDD, pp. 2329–2337 (2018)
31. Wang, J., Liu, Q., Liang, H., Joshi, G., Poor, H.V.: Tackling the objective inconsistency problem in heterogeneous federated optimization. In: NeurIPS (2020)
32. Xu, B., Ting, K.M., Zhou, Z.H.: Isolation set-kernel and its application to multi-instance learning. In: KDD, pp. 941–949 (2019)
33. Yang, Q., Liu, Y., Chen, T., Tong, Y.: Federated machine learning: concept and applications. ACM TIST **10**(2), 12:1-12:19 (2019)
34. Zhou, Y., Pu, G., Ma, X., Li, X., Wu, D.: Distilled one-shot federated learning. CoRR abs/2009.07999 (2020)
35. Zhou, Z.H.: Ensemble Methods: Foundations and Algorithms. CRC Press, Boca Raton (2012)
36. Zhou, Z.H.: Learnware: on the future of machine learning. Front. Comp. Sci. **10**(4), 589–590 (2016). https://doi.org/10.1007/s11704-016-6906-3

A Static Bi-dimensional Sample Selection for Federated Learning with Label Noise

Qian Rong[1], Ling Yuan[1(✉)], Guohui Li[2], Jianjun Li[1], Lu Zhang[3], and Xuanang Ding[1]

[1] School of Computer Science and Technology,
Huazhong University of Science and Technology, Wuhan, China
{qianrong,cherryyuanling,jianjunli,dingxuanang}@hust.edu.cn
[2] School of Software Engineering,
Huazhong University of Science and Technology, Wuhan, China
guohuili@hust.edu.cn
[3] School of Cyber Science and Engineering,
Huazhong University of Science and Technology, Wuhan, China
luzhang_cs@hust.edu.cn

Abstract. In real-world Federated learning(FL), client training data may contain label noise, which can harm the generalization performance of the global model. Most existing noisy label learning methods rely on sample selection strategies that treat small-loss samples as correctly labeled ones, and large-loss samples as mislabeled ones. However, large-loss samples may also be valuable hard samples with correct labels. Also, these sample selection strategies have an issue in FL setting: Moreover sample selection strategies require training two models simultaneously and are executed in every mini-batch, which increases the communication cost and client computation. In this paper, we propose an efficient multi-stage federated learning framework to tackle label noise. Firstly, we use self-supervision pre-training to extract category-independent features without using any labels. Secondly, we propose a federated static two-dimensional sample selection(FedSTSS) method, which uses category probability entropy and loss as separation metrics to identify samples. Notably, to improve the accuracy of recognition samples, we add a weight entropy minimization term to the cross-entropy loss function. Finally, we use a semi-supervised method to finetune the global model on identified clean samples and mislabeled samples. Extensive experiments on multiple synthetic and real-world noisy datasets demonstrate that our method outperforms the state-of-the-art methods.

Keywords: Federated learning · Label noise · Sample selection · Self-supervised learning

Supported by the National Natural Science Foundation of China under Grant No. 62272180. The computation is completed in the HPC Platform of Huazhong University of Science and Technology.

X. Wang et al. (Eds.): DASFAA 2023, LNCS 13943, pp. 735–744, 2023.
https://doi.org/10.1007/978-3-031-30637-2_49

1   Introduction

Federated learning(FL) is a distributed machine learning paradigm that has attracted growing attention from academia and industry, protecting the privacy of the client's training data by collaborative training between the client and the server [6]. However, in real-world FL scenarios, client training data may contain label noise due to diverse annotators' skills, bias, and hardware reliability [12,13]. Also, the label noise ratios of clients are heterogeneous,e.g., they contain different noise ratios across clients. Noisy labels can harm the generalization performance of the global model due to deep neural networks easily overfit mislabeled samples. Although previous research has proposed many methods that focus on addressing label noise for centralized learning, these methods are not suitable for the FL setting [11,13]. Specifically, the core of existing competitive noisy label learning methods [5,8,14] is the sample selection strategy that treats small-loss samples as correctly labeled and large-loss samples as mislabeled samples. However, these sample selection strategies require training two models simultaneously and are executed in every mini-batch, which increases the communication cost and client computation in FL setting. Also, the loss is not sufficient as a separation metric, since hard samples with correct labels will also exhibit large losses during training.

In this paper, we propose an efficient multi-stage federated learning framework to tackle label noise, without any assumptions about noise; see Fig. 1 for an overview. In order to reduce the effect of noisy labels, we first use self-supervision pre-training to extract category-independent features without using any labels, which guarantees that feature extraction is not affected by noisy labels. Then, we use category probability entropy and loss as separation metrics to identify correctly labeled samples from noisy labels. Specifically, the loss and the class probability entropy can measure the closeness of the model's prediction of the sample to the observed label and the model's knowledge of the sample, respectively. If the model knows enough about the sample but makes predictions far from the observed label, then the observed label is likely to be incorrect. Notably, to improve the separability of the separation metrics, we add a weight entropy minimization term to the cross-entropy loss function, which can strengthen the model's predictions with lower entropy. Finally, we use semi-supervised to fine-tune the global model on identified clean samples and mislabeled samples.

Our main contributions are the following:

1) We propose a multi-stage federated learning framework to tackle label noise, which is robust to both high noise levels, asymmetric noise, and non-independent identically distributed data.

2) We introduce loss and category probability entropy as separation metrics to separate noisy label samples from clean samples. Furthermore, we propose a federated static two-dimensional sample selection(FedSTSS) method, which statically divides client data into label noise samples and clean samples.

3) To improve the separability of the separation metrics, we add a weighted entropy minimization term to the cross-entropy loss function, which can make

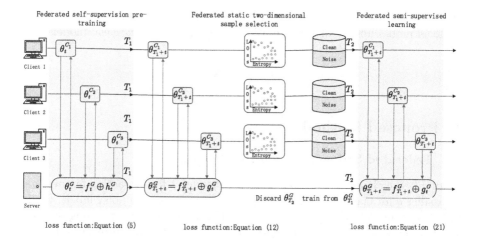

Fig. 1. An overview of our method.

the model output lower entropy predictions and thus improve the performance of FedSTSS.

4) We demonstrate that our multi-stage federated learning framework outperforms state-of-the-art federated noisy learning methods on multiple datasets.

2   Problem Definition

Consider a dataset $D = \{D_k\}_{k=1}^{K}$ that is independently stored in K clients with C categories, and each local client k has a local dataset $D_k = (x_k^i, y_k^i)_{i=1}^{n_k}$. y_k^i is the label given for the i/th training sample x_i^k, which can be the correct or noisy label. θ and θ_k denote the global model parameters and the model parameters of the local client k, respectively. The global model parameters θ aim at minimizing the empirical risk $\mathcal{L}(\theta)$,i.e.,

$$\min_{\theta} \mathcal{L}(\theta) = \sum_{k=1}^{N} \frac{n_k}{N} \mathcal{L}_k(\theta), N = \sum_{i=1}^{K} n_k \tag{1}$$

where $\mathcal{L}_k(\theta)$ is the empirical risk for the k/th client in D_k,i.e.,

$$\mathcal{L}_k(\theta) = \frac{1}{n_k} \sum_{i=1}^{n_k} Loss((x_k^i, y_k^i); \theta) \tag{2}$$

where $Loss$ is loss function and $Loss((x_k^i, y_k^i); \theta)$ denote the loss of sample (x_k^i, y_k^i). Most federated learning methods solve the optimization problem by iteratively performing local model updates and global model aggregation. The client updated the local model by:

$$\theta_k \leftarrow \theta_k - \eta \nabla_\theta \mathcal{L}_k(\theta_k) \tag{3}$$

where η is the local learning rate. The server selects the client set S_t to perform the local model update at t/th iteration, then aggregates the local models by:

$$\theta^{t+1} \leftarrow \sum_{S_t} \frac{n_k}{N} \theta_k^t, N = \sum_{i=1}^{K} n_k \qquad (4)$$

3   Proposed Method

3.1   Federated Self-supervision Pretraining

We divide the classification model into an encoder f for extracting features and a classifier g for classifying. To avoid the negative impact of noisy labels, we use Simsiam [2] model to pre-train the encoder, since contrastive learning does not require sample labels. Simsiam contains an encoder f and a projection head h, and the overall idea is to train the encoder by maximizing the similarity between the different perspectives of the samples. The objective function is as follows:

$$\min_{\theta} \mathcal{L}_{sp}(f \oplus h) = \sum_{k=1}^{N} \frac{n_k}{N} \mathcal{L}_k(f \oplus h), N = \sum_{i=1}^{K} n_k \qquad (5)$$

where $\oplus$ denotes neural network tandem operation and $f \oplus h$ indicates that the output of f is the input of h.

$$\mathcal{L}_k(f \oplus h) = \frac{1}{2} Sim(p_1, stopgrad(z_2)) + \frac{1}{2} Sim(p_2, stopgrad(z_1)) \qquad (6)$$

$$Sim(p_1, z_2) = -\frac{p_1}{||p_1||} \cdot \frac{z_2}{||z_2||} \qquad (7)$$

where $|| \cdot ||$ denotes the L_2 norm, $p_1 = h(f(x_1))$ and $z_2 = f(x_2)$, x_1 and x_2 are different perspectives of the sample x, respectively, which can generally be obtained by taking different random data augmentations of the sample x. The equivalent of z_2 in $Sim(p_1, stopgrad(z_2))$ is a constant, which is not involved in the gradient calculation in the above equation.

3.2   Federated Static Two-Dimensional Sample Selection

The learning process of neural network models is from random uniform prediction to prediction with some certainty, and the whole process is entropy-decreasing. When a neural network does not learn enough about a sample, the category prediction for that sample is close to a random guess. In other words, the degree of learning of a sample by a neural network is inversely proportional to the entropy of the class probability of the sample. Hard samples are likelier to exhibit large loss and large entropy of the predicted category probability. In contrast, noisy label samples are likelier to exhibit large cross-entropy loss and small information entropy of the predicted category probability. Therefore, we use loss and category probability entropy as separation metrics to jointly separate samples. For the classification problem, the output of the classification model is as follows.

$$o_i = softmax(b_i) \tag{8}$$

$$b_i = \mathcal{M}(x_i : \theta) \tag{9}$$

$$\mathcal{M} = (f \oplus g) \tag{10}$$

$\mathcal{M}$ denotes the classification model, containing the encoder f and classifier g, b_i is a C-dimensional vector, and the j/th term o_{ij} of o_i denotes the sample x_i belongs to the j/th probability of classification, and the entropy of x_i is calculated as follows.

$$entropy(x_i) = -\sum_{j=0}^{C} o_{ij} log(o_{ij}) \tag{11}$$

The neural network's predictions of partially correctly labeled samples are hesitant for a few classes. Therefore, to improve the separability of the separation metrics, we add a weight entropy minimization term to the cross-entropy loss function, which allows the neural network to make predictions with greater certainty. The specific loss function is as follows,

$$\min_{\theta} \mathcal{L}_{wce}(f \oplus g) = \sum_{k=1}^{N} \frac{n_k}{N} \mathcal{L}_k(f \oplus g), N = \sum_{i=1}^{K} n_k \tag{12}$$

$$\mathcal{L}_k(f \oplus g) = -\frac{1}{n_k} \sum_{i}^{n_k} \sum_{j=1}^{C} (y_k^{ij} log(o_{ij}) - w o_{ij} log(o_{ij})) \tag{13}$$

w is a control parameter, the value range is greater than or equal to 0.

The client calculates the cross-entropy loss for each local sample as follows,

$$CE(x_k^i, y_k^i) = Loss(y_k^i, g(f(x_k^i))) \tag{14}$$

Splice cross-entropy loss and category probability entropy into bi-dimensional sample separation metrics as follows,

$$e_k^i = (entropy(x_k^i), CE(x_k^i, y_k^i)) \tag{15}$$

The bi-dimensional separation metrics of all samples on the client side constitute the separation metrics dataset

$$E_k = \{e_k^0, e_k^1, e_k^2, \cdots, e_k^{n_k}\} \tag{16}$$

E_k denotes the dataset consisting of all sample separation metrics of the local client k. The client performs k-means clustering on E_k to classify all local samples into two classes

$$z_k^i = k/means(e_k^i, 2) \tag{17}$$

z_k^i denotes the id of the cluster predicted by the clustering algorithm $k/means$ for e_k^i, $z_k^i \in \{0,1\}$.

$$x_k^i \in \begin{cases} D_k^c \text{ if } CD_1 \wedge z_k^i = 0 \text{ or } CD_2 \wedge z_k^i = 1 \\ D_k^n \text{ if } CD_1 \wedge z_k^i = 1 \text{ or } CD_2 \wedge z_k^i = 0 \end{cases} \tag{18}$$

$$CD_1 \Leftrightarrow ||center(0)||_2 < ||center(1)||_2 \tag{19}$$

$$CD_2 \Leftrightarrow ||center(0)||_2 > ||center(1)||_2 \tag{20}$$

$center(0)$ and $center(1)$denotes the center of the 0-th and 1-th cluster, respectively. $|| \cdot ||$ denotes the L_2 norm, $\wedge$ and $\Leftrightarrow$ denote "and" and equivalently, D_k^c and D_k^n is the set of clean samples and the set of labeled noise samples on client k, respectively.

3.3 Federated Semi-supervised Finetune

We consider D_k^c and D_k^n as the labeled and unlabeled datasets on client k, respectively, and then perform semi-supervised learning Mixmatch [1] on client k. The objective function is as follows:

$$\min_{\theta} \mathcal{L}_{ss}(f \oplus g) = \sum_{k=1}^{N} \frac{n_k}{N} \mathcal{L}_k(f \oplus g), N = \sum_{i=1}^{K} n_k \tag{21}$$

where $f \oplus g$ indicates that a classifier g is connected to the back of the encoder f.

$$\mathcal{L}_k(f \oplus g) = \mathcal{L}_{X_i} + \frac{1}{\lambda}\mathcal{L}_{U_i} \tag{22}$$

$$\mathcal{L}_{X_i} = \frac{1}{|X'|} \sum_{x,p \in X'} H(p, f \oplus g(x)) \tag{23}$$

$$\mathcal{L}_{U_i} = \frac{1}{L|U'|} \sum_{u,q \in U'} ||q - f \oplus g(u)||_2^2 \tag{24}$$

$$X', U' = MixMatch(D_k^c, D_k^n, T, D, \alpha) \tag{25}$$

H denotes cross-entropy and λ_U, T, D, α are the hyperparameters.

4 Experiments

4.1 Experiment Setup

Dataset. Our experiments were conducted on the CIFAR10 [4], CIFAR100 [4], CIFAR10N [9], CIFAR100N [9], and Clothing1M [10] datasets.

Baselines. We compare our algorithm with the following baselines: (1) methods to tackle label noise in CL(DivideMix [5] and JoSRC [14]) applied to local clients;(2)classic FL methods(FedAvg [7] and FedProx [6]);(3) FL methods designed to be robust to label noise(RoFL [13], CsFL [12] and FedCorr [11]).

4.2   Noise Model

In our experiments, the label noise level at the client obeyed uniform, gaussian and constant distributions. Succinctly, the noise level of client k (for $k = 1, \cdots, N$) is

$$u_k = \begin{cases} p \sim U(0,1) & \text{if uniform distributions} \\ p = c & \text{if degenerate distributions} \\ p \sim N(\mu, \sigma) & \text{if gaussian distributions} \end{cases} \tag{26}$$

where $c \in [0,1]$ is a constant, μ and σ denote the mean and standard deviation, respectively. The label noise on each client is divided into two types: symmetric noise [8] and asymmetric noise [8]. We consider both IID [11] and non-IID [11] heterogeneous data partitions in this work.

4.3   Implementation Details

We set 100 clients for all datasets and sample 10% of clients per round of FL. As an encoder, we use the ResNet18 network [3] for CIFAR10 and CIFAR10N and the ResNet34 network [3] for CIFAR10 and CIFAR100N. For Clothing1M, we use ResNet-50 network [3] pre-trained on ImageNet as encoder. For the self-supervised pre-training stage, we adopt Simaiam [2] with the feature representations that have 2048 dimensions. For the federated static two-dimensional sample selection stage and semi-supervised learning stage, we train encoder f and classifier g with a learning rate of Cosine annealing from 30 to 0.01 for encoder f and a learning rate of 0.0001 for classifier g. We set ω to 1 for the CIFAR10 and CIFAR10N dataset and 0.1 for the CIFAR100, CIFAR100N and Clothing1M datasets, T is 0.5, α is 0.75, λ_u is 75.

4.4   Comparison with State-of-the-Art

Table 1. The best accuracies of various methods on CIFAR100 dataset with IID and symmetric label noise setting at different noise type and ratio.

Noise type	degenerate c			gaussian (μ, σ)					uniform
Noise ratio	0.2	0.4	0.6	(0.2,0.1)	(0.2,0.2)	(0.4,0.2)	(0.4,0.4)	(0.5,0.2)	0.5
FedAvg	0.5578	0.3918	0.239	0.606	0.6005	0.4903	0.4982	0.446	0.3892
FedProx	0.5421	0.3910	0.2354	0.598	0.6123	0.4910	0.5021	0.4452	0.3724
DivideMix	0.5231	0.3211	0.1956	0.5542	0.6012	0.4733	0.5120	0.4512	0.3632
JoSRC	0.5221	0.3852	0.2217	0.5596	0.6214	0.4932	0.5257	0.4532	0.3715
RoFL	0.5201	0.3815	0.2121	0.5432	0.6285	0.4936	0.5112	0.4332	0.3652
CsFL	0.5012	0.3621	0.1952	0.5242	0.5996	0.4963	0.5212	0.4396	0.3721
FedCorr	0.6524	0.5495	0.3198	0.6624	0.6784	0.5432	0.6051	0.5084	0.4732
ours	**0.7739**	**0.751**	**0.6756**	**0.7749**	**0.7661**	**0.749**	**0.7428**	**0.726**	**0.7328**

Table 2. The best accuracies of various methods on CIFAR100 dataset with non-IID and asymmetric label noise setting at different noise type and ratio.

Noise type	degenerate c			gaussian (μ, σ)					uniform
Noise ratio	0.2	0.4	0.6	(0.2,0.1)	(0.2,0.2)	(0.4,0.2)	(0.4,0.4)	(0.5,0.2)	0.5
FedAvg	0.5084	0.3159	0.1862	0.5132	0.5329	0.4436	0.4691	0.353	0.3261
FedProx	0.4916	0.3124	0.1963	0.5056	0.5321	0.4563	0.4752	0.365	0.3312
DivideMix	0.4825	0.3024	0.1852	0.5023	0.5326	0.4552	0.4637	0.3531	0.3212
JoSRC	0.4925	0.3237	0.1996	0.5233	0.5410	0.4595	0.4752	0.3621	0.3328
RoFL	0.4852	0.3096	0.1823	0.5022	0.5265	0.4425	0.4601	0.3472	0.3315
CsFL	0.4931	0.3132	0.1896	0.5054	0.5163	0.4596	0.4685	0.3751	0.3421
FedCorr	0.5793	0.4515	0.3207	0.5877	0.6043	0.4646	0.526	0.4132	0.4781
ours	**0.7743**	**0.7358**	**0.7173**	**0.7737**	**0.7679**	**0.7496**	**0.7515**	**0.7319**	**0.7414**

Table 3. The best accuracies of various methods on real world noisy labeled dataset with IID setting.

dataset	CIFAR10N				CIFAR100N			Clothing1M
Noise type	Aggregate	Random 1	Random 2	Random 3	Worst	Coarse	Fine	*
Noise ratio	9.03%	17.23%	18.12%	17.64%	40.21%	25.6%	40.20%	*
FedAvg	0.9118	0.8664	0.8689	0.8632	0.7703	0.5732	0.3841	0.6947
FedProx	0.9056	0.8594	0.8603	0.8621	0.7796	0.5764	0.3965	0.7025
DivideMix	0.9023	0.8594	0.8603	0.8621	0.7796	0.5825	0.4056	0.7011
JoSRC	0.9189	0.8755	0.8733	0.8775	0.7852	0.5924	0.4121	0.7152
RoFL	0.8903	0.8551	0.8502	0.8526	0.7804	0.5857	0.4012	0.7039
CsFL	0.8756	0.8231	0.8265	0.8233	0.7326	0.5512	0.3754	0.6852
FedCorr	0.9243	0.8924	0.8896	0.8943	0.8122	0.5652	0.5257	0.7255
ours	**0.9451**	**0.9485**	**0.9478**	**0.9463**	**0.9468**	**0.774**	**0.7528**	**0.7453**

We use the same configuration to compare our method with multiple baselines at different noise types and levels on CIFAR100 dataset and multiple real-world noisy labeled datasets with an iid setting; see Table 1, Table 2 and Table 3 for an overview. Experiments on multiple datasets show that our method has the best test accuracy, especially for high noise levels. At the same time, our method is still robust to non-independent homogeneous distribution, and asymmetric noise. We believe there are two reasons for this. Firstly, the high prediction accuracy of the local sample division for clean samples with fewer noisy labels is involved in training. And then the fact that labeled noisy samples are only involved in the fine-tuning phase of training has less impact on the neural network model.

4.5   Ablation Study

Table 4 shows the effects of the components in our method. $G(0.2, 0.1)$ denotes a normal distribution with a mean of 0.2 and a variance of 0.1, and $U(0\ 1)$ is uniformly distributed. SP+OS denotes fine-tuning using the original samples

Table 4. Ablation study results

Dataset	CIFAR10		CIFAR100		CIFAR10N		CIFAR100N	
noise type	$G(0.2,0.1)$	$U(0\ 1)$	$G(0.2,0.1)$	$U(0\ 1)$	Aggregate	Worst	Coarse	Fine
noise ratio	21.8%	49.2%	21.9%	49.5%	9.03%	40.21%	25.6%	40.2%
FedAvg	0.8711	0.7734	0.606	0.3892	0.9118	0.7703	0.5732	0.3841
SP+OS	0.9263	0.9116	0.7014	0.6483	0.9243	0.8842	0.7318	0.6148
SP+CS	0.9269	0.9122	0.7019	0.6462	0.9242	0.8878	0.7356	0.6142
ours	**0.9451**	**0.9434**	**0.7749**	**0.7328**	**0.9463**	**0.9468**	**0.774**	**0.7528**

containing noise labels after self-supervised pre-training. SP+CS refers to fine-tuning only using clean samples D_c^k after self-supervised pre-training. We can observe from Table 4 that self-supervised pre-training can substantially improve the robustness of the neural network model against noisy labels. We use a semi-supervised method to finetune the global model to make full use of all the data, so we can get better performance.

Table 5. Performance(%) of FedSTSS

Dataset		CIFAR10		CIFAR100		CIFAR10N		CIFAR100N	
noise type		$G(0.2,0.1)$	$U(0\ 1)$	$G(0.2,0.1)$	$U(0\ 1)$	Aggregate	Worst	Coarse	Fine
noise ratio		21.8%	49.2%	21.9%	49.5%	9.03%	40.21%	25.6%	40.2%
Accuracy	GL	99.98	95.09	81.21	49.72	99.78	96.58	96.67	66.38
	KLE	99.25	96.89	99.36	91.83	97.52	81.78	88.9	78.9
	FedSTSS	98.9	96.03	98.74	95.96	96.85	84.21	84.52	76.03
recall	GL	78.76	79.7	10	9.9	79.19	88.14	73.1	12.17
	KLE	97.22	94.44	87.4	75.11	98.04	98.61	93.5	87.07
	FedSTSS	98.76	97.53	95.89	90.34	99.17	99.12	98.03	96.14
F1-score	GL	88.11	86.71	17.82	16.63	88.3	**92.16**	83.24	20.56
	KLE	98.22	95.64	92.98	82.63	97.77	89.4	91.14	82.78
	FedSTSS	**98.83**	**96.77**	**97.29**	**93.07**	**98**	91.06	**91.29**	**84.91**

We utilize different methods to identify clean samples and noisy labeled samples, Table 5 shows the accuracy, recall and F1-score of identifying clean samples. GL represents the use of Gaussian Mixture Model to identify clean samples with loss as separation metric, KLE means using the K-means Model to identify clean samples with loss and entropy as separation measures, but does not add a weight entropy minimization term to the cross-entropy loss function, FedSTSS is our proposed Federated Static Two-Dimensional Sample Selection method.

5   Conclusion

In this paper, we propose an efficient multi-stage federated learning framework to tackle label noise. We firstly use self-supervised pre-training to extract category-independent features, which can avoid the negative effect of noisy labels and improve the robustness of our method. Then, we propose a federated static two-dimensional sample selection(FedSTSS) method, which introduces loss and class probability entropy as separation metrics to jointly identify clean samples and noisy labeled samples. In particular, we add a weight entropy minimization term to the cross-entropy loss function to make the model output predictions with less entropy. Finally, we use semi-supervised to finetune the global model on identified clean samples and mislabeled samples. Extensive experiments demonstrate that our method is robust to label noise and outperforms the state-of-the-art methods at multiple noise levels, especially in the case of high noise levels.

References

1. Berthelot, D., Carlini, N., Goodfellow, I., Papernot, N., Oliver, A., Raffel, C.A.: Mixmatch: a holistic approach to semi-supervised learning. In: Advances in Neural Information Processing Systems, vol. 32 (2019)
2. Chen, X., He, K.: Exploring simple siamese representation learning. In: Proceedings of the IEEE/CVF Conference on Computer Vision and Pattern Recognition, pp. 15750–15758 (2021)
3. He, K., Zhang, X., Ren, S., Sun, J.: Deep residual learning for image recognition. In: Proceedings of the IEEE Conference on Computer Vision and Pattern Recognition, pp. 770–778 (2016)
4. Krizhevsky, A., Hinton, G., et al.: Learning multiple layers of features from tiny images (2009)
5. Li, J., Socher, R., Hoi, S.C.: Dividemix: learning with noisy labels as semi-supervised learning (2020). arXiv preprint arXiv:2002.07394
6. Li, T., Sahu, A.K., Talwalkar, A., Smith, V.: Federated learning: challenges, methods, and future directions. IEEE Signal Process. Mag. **37**(3), 50–60 (2020)
7. McMahan, B., Moore, E., Ramage, D., Hampson, S., y Arcas, B.A.: Communication-efficient learning of deep networks from decentralized data. In: Artificial Intelligence and Statistics, pp. 1273–1282. PMLR (2017)
8. Tan, C., Xia, J., Wu, L., Li, S.Z.: Co-learning: learning from noisy labels with self-supervision. In: Proceedings of the 29th ACM International Conference on Multimedia, pp. 1405–1413 (2021)
9. Wei, J., Zhu, Z., Cheng, H., Liu, T., Niu, G., Liu, Y.: Learning with noisy labels revisited: a study using real-world human annotations. In: International Conference on Learning Representations. ICLR (2022)
10. Xiao, T., Xia, T., Yang, Y., Huang, C., Wang, X.: Learning from massive noisy labeled data for image classification. In: Proceedings of the IEEE Conference on Computer Vision and Pattern Recognition, pp. 2691–2699 (2015)
11. Xu, J., Chen, Z., Quek, T.Q., Chong, K.F.E.: Fedcorr: multi-stage federated learning for label noise correction (2022). arXiv preprint arXiv:2204.04677
12. Yang, M., Qian, H., Wang, X., Zhou, Y., Zhu, H.: Client selection for federated learning with label noise. IEEE Trans. Veh. Technol. **71**(2), 2193–2197 (2021)
13. Yang, S., Park, H., Byun, J., Kim, C.: Robust federated learning with noisy labels. IEEE Intell. Syst. **37**(2), 35–43 (2022)
14. Yao, Y., et al.: Jo-src: a contrastive approach for combating noisy labels. In: Proceedings of the IEEE/CVF Conference on Computer Vision and Pattern Recognition, pp. 5192–5201 (2021)

An Information Theoretic Perspective for Heterogeneous Subgraph Federated Learning

Jiayan Guo[1], Shangyang Li[2(✉)], and Yan Zhang[1]

[1] School of Intelligence Science and Technology, Peking University, Beijing, China
{guojiayan,zhyzhy001}@pku.edu.cn
[2] Peking-Tsinghua Center for Life Sciences, IDG/McGovern Institute for Brain Research, Academy for Advanced Interdisciplinary Studies, Peking University, Beijing, China
syli@pku.edu.cn

Abstract. Mining graph data has gained wide attention in modern applications. With the explosive growth of graph data, it is common to see many of them collected and stored in different distinction systems. These local graphs can not be directly shared due to privacy and bandwidth concerns. Thus, *Federated Learning* approach needs to be considered to collaboratively train a powerful generalizable model. However, these local subgraphs are usually heterogeneously distributed. Such heterogeneity brings challenges for subgraph federated learning. In this work, we analyze subgraph federated learning and find that sub-optimal objectives under the FedAVG training setting influence the performance of GNN. To this end, we propose InfoFedSage, a federated subgraph learning framework guided by Information bottleneck to alleviate the non-iid issue. Experiments on public datasets demonstrate the effectiveness of InfoFedSage against heterogeneous subgraph federated learning.

1 Introduction

Graphs are ubiquitous in the real world. Mining graph data has gained wide attention in multiple modern application domains like recommender systems [5,7,8,30], social network analysis [25], and natural language processing [26]. Recently, there has been explosive growth in graph data. In a resource-limited environment, a giant graph is split into subgraphs and stored in disjoint local storage systems [29]. Moreover, they can not be directly shared due to privacy and bandwidth concerns. In such a situation, graph machine learning models can only be directly trained on local graphs, and the overall generalization abilities of these models are limited. Thus we need to consider the *Federated Learning* (FL) approach to combine the knowledge from these local models to form an overall generalizable model. There has been preliminary research on subgraph federated learning [29]. However, due to the randomness and biases

---

J. Guo and S. Li—Equal contribution.

X. Wang et al. (Eds.): DASFAA 2023, LNCS 13943, pp. 745–760, 2023.
https://doi.org/10.1007/978-3-031-30637-2_50

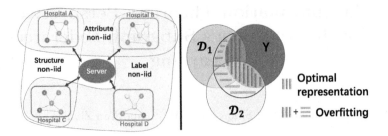

Fig. 1. (Left) Real world subgraphs are usually heterogeneously distributed. (Right) InfoFedSage is optimized to to capture the optimal representation, so as to prevent overfitting and improve generalization ability.

of the data collection process [1], these subgraphs are usually heterogeneously distributed in labels, attributes, and structures. For example, in Fig. 1 left, four subgraphs of patients are stored in four different hospitals. Hospital A and Hospital B have similar structural patterns, while the attributes (*e.g.* symptom and gender) of their patients may vary due to different specialists. Hospital A and Hospital C share similar patient attributes, but due to their different volumes, their subgraphs' structural distribution (*e.g.* average node degree) may be different. As for the labels, hospitals may have different distributions of diseases (*e.g.* Hospital B, C, and D in Fig. 1). These heterogeneously distributed subgraphs in different clients make it even harder for graph-based models like Graph Neural Networks (GNN) to generalize well for all patients. This problem is non-trivial and previous researches [14] also point out that non-iid distributed data in clients may make the classical federated learning algorithm, FedAVG, fails to increase the generalization ability of the global model.

In this work, we rethink how to get a "good" representation in such scenarios. Especially, the Information Bottleneck (IB) theory [24] has shown great power as an essential principle for representation learning from the perspective of information theory [2,6,27]. The representation is encouraged to involve as much information about the target to obtain high prediction accuracy and discard redundant information irrelevant to the target. In specific, we extend the IB principle to heterogeneous subgraph federated learning and make a theoretical analysis of the current graph federated learning approach under FedAVG training. We find out that the graph-based models trained by FedAVG optimization, like FedSage, could not have good generalization capability under non-iid distributed subgraphs. Further, we also point out the key point that affects prior graph federation learning algorithms is the suboptimal optimization objective. To increase the generalization ability of the global model, we propose InfoFedSage, a novel subgraph federate learning method guided by the IB, to get the optimal representation like Fig. 1 right. InfoFedSage first train a graph generator conditioned on global label distribution with the collaborative supervision of client models. Then the trained generator is distributed to local clients to generate graph samples that contain the global graph information. The local clients update the local parameters by distilling the knowledge from the global model and the generator. Moreover, to prevent overfitting of the local model, we use an information bottleneck regularizer to squeeze out irrelevant information. Finally, we conduct extensive experiments under graph benchmarks to demonstrate the effectiveness of our method against heterogeneous federated learning. In summary, our contributions are as follows:

- We analyze the bottleneck of subgraph federal learning from the perspective of information theory. In specific, the main limitation is the sub-optimal objective under the FedAVG training.
- Based on the analysis, we propose InfoFedSage, a novel framework for subgraph federated learning combining conditional generative learning and IB regularization.
- We design different non-iid splitting methods by using four public graph benchmarks. Experimental results show the effectiveness of InfoFedSage and its variants.

2 Related Work

2.1 Graph Federated Learning

There have been previous research studies on graph federated learning [11,20], which come from three main domains based on the learning objects: subgraph-level FL, node-level FL, and graph-level FL. Subgraph-level FL has the problem of missing links between nodes from different clients due to data transportation constraints. To address this, some previous work [29] has designed a graph mending mechanism to generate missing neighbors and links to expand local graph data, but this still requires sending local gradient information from clients, which can raise privacy concerns. Other research in this area has focused on heterogeneous graph data in clients. For node-level federated learning, data is stored through ego networks, while for graph-level FL, a cluster-based method [28] has been proposed to deal with non-IID graph data and aggregate client models with adaptive clustering.

2.2 Heterogeneous Federated Learning

Federated learning often faces the issue of data heterogeneity, where data distributions from different clients can vary greatly [21]. There is a significant amount of research on this topic, with previous work focusing on two main perspectives. One approach involves stabilizing local training by regularizing the local model to deviate from the global model [13], but this may not take full advantage of cross-user models and ignores the diversity of local data, indicating differences in the information structure of that data. This is an area worth further investigation. The other main line of research aims to improve the effectiveness of model aggregation through the use of knowledge distillation [17], either using proxy data or a data-free approach. However, the reliance on proxy data for

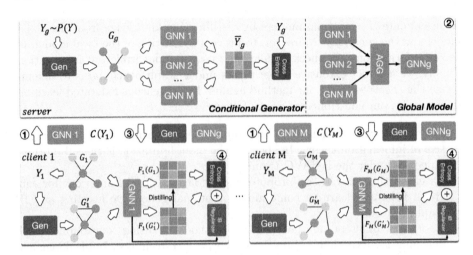

Fig. 2. Overview of the InfoFedSage framework. ① The client first sends its label count and local parameters to the server. ② The server trains a label-conditioned generator and composes the local parameters to form a global model. ③ The server distributes the generator and the global model to local clients. ④ Local clients update their local parameters based on the knowledge from the global model, conditional generator, and the information bottleneck regularizer.

this approach can make it infeasible for many applications where well-designed datasets may not be available on the server. To address this, data-free knowledge distillation [31] has been introduced by training a global generative model on the server and using it to produce samples for clients, but this does not consider the graph domain and generated samples may introduce extraneous noise that can interrupt the training of client models.

3 Problem Formulation

Heterogeneous subgraph federated learning and federated learning have the same basic settings, except that the input data types are different. We denote the global graph as $G = \{V, E, X\}$ with n nodes, where $V = [n] = \{1, 2, \cdots n\}$ represents the node set, $E \subseteq V \times V$ represents the edge set and $X \in \mathbb{R}^{n \times f}$ represents the node feature set. In the subgraph federated learning system, there is a central server S and m clients with distributed subgraphs $G_i = \{V_i, E_i, X_i\}$, in which i is the id of client and the total number of i is m. We assume $V = V_1 \cup \cdots \cup V_m$. For the central server S, there is no data stored and only a global model. In this work, we focus on the subgraph node classification task, where each node is associated with the label $Y \in [K]^n$. The system hopes to learn a global model F parameterized by θ that is collaboratively learned on isolated subgraphs in all data owners, and the model can perform well in tasks.

4 Information Perspective on Subgraph Federated Learning

We have introduced the definition of the distributed subgraph system federated learning. Based on this system, we formally define the objective function of IB in a subgraph federated learning scenario. Then, we analyze the shortcomings of the classical federated learning algorithm FedAVG.

4.1 Information Bottleneck

The IB principle tells us how to learn a good representation, it forces the representation to capture the relevant factors about the prediction and diminish the irrelevant parts. When it is expanded to graph data, the input data type is global graph data. Let $A \in \mathbb{R}^{n \times n}$ denote the adjacency matrix of G, the input data can be represented as $\mathcal{D} = (A, X)$. Formally, the information bottleneck has the following objective function:

$$\mathcal{L}_{\text{IB}} \triangleq \beta I(Z_X, \mathcal{D}) - I(Z_X, Y), \tag{1}$$

where Z_X is the node representation, Y is the node label, β is the Lagrangian multiplier to control the trade-off between sufficiency and minimality, and $I(\cdot, \cdot)$ denotes mutual information between variables. For subgraph federated learning, the central server S only maintains a global model with no input data stored. Applying IB to subgraph federated learning, Eq. 1 should be written as:

$$\mathcal{L}_{\text{IB-SFL}} \triangleq \beta(\frac{1}{m} \sum_{i=1}^{m} I(Z_{X_i}, \mathcal{D}_i)) - I(Z'_X, Y'), \tag{2}$$

where Z_{X_i} represents the node representation in data owner i, $Z'_X = Z_{X_1} \cup \cdots \cup Z_{X_m}$ and $Y' = Y_1 \cup \cdots \cup Y_m$. This is the form of IB objective in subgraph federated learning. We desire to learn an optimal global model under the constraint that the input data is stored locally, which requires the set of model outputs from each client to minimize the information from the local graph-structured data $\mathcal{D}_i$ (compression) and maximize the information to global label Y (prediction). The challenge is how to design a model to better approximate Eq. 2.

4.2 Information Perspective for FedAVG on Graph Data

[18] proposes FedAVG, which is the first basic federated learning algorithm. It is commonly used as the cornerstone of more advanced algorithms. [29] applies FedAVG on graph data and proposes the FedSage model. In their model, the weight of F is obtained by minimizing the aggregated risk

$$\min_{\theta} \frac{1}{m} \sum_{i=1}^{m} \mathbb{E}_{(\mathcal{D}_i, Y_i)}[\ell(F(\theta; \mathcal{D}_i), Y_i))], \tag{3}$$

where ℓ is the loss function. In practice, Eq. 3 could be viewed as

$$\mathcal{L}_{\text{FedAVG}} = -\frac{1}{m} \sum_{i=1}^{m} \frac{1}{|V_i|} \sum_{j=1}^{|V_i|} \left[\log p \left(y_i^{(j)} \mid F(v_i)^{(j)} \right) \right] \tag{4}$$

with data samples $\{v^{(j)}, y^{(j)}\}_{j=1}^{|V_i|}$ in different client. In order to understand FedAVG from the perspective of information theory, we introduce an lower bound of Eq. 4, as shown in Proposition 4.1.

Proposition 4.1 (The lower bound of $\mathcal{L}_{\text{FedAVG}}$). For $Z'_X = Z_{X_1} \cup \cdots \cup Z_{X_m}$ and $Y' = Y_1 \cup \cdots \cup Y_m$, we have

$$\mathcal{L}_{\text{FedAVG}} \geq -\frac{1}{m} \sum_{i=1}^{m} I(Z_{X_i}, Y^i) \geq -I(Z'_X, Y'), \tag{5}$$

The proof is given in Appendix A. Proposition 4.1 indicates that the objective function of FedAVG is the upper bound of the second term of Eq. 2. Obviously, it could not have good generalization capability under a non-iid setting. In other words, FedAVG is a suboptimal model.

5 InfoFedSage

In this section, we elaborate on InfoFedSage, a novel graph federated learning framework guided by the information bottleneck principle. We have understood that the objective function of the FedAVG model is the upper bound of the second term of IB objective in subgraph federated learning. Developing more efficient models requires approaching tighter bound. Based on the analysis, we introduce a tractable upper bound for our IB objective and design our model.

5.1 Get a Tighter Bound via Conditional Generative Graph Learning

Here, the first idea is to get information about the global view of the whole graph data distribution that is non-observable in the conventional FedAVG framework and then use it to guide local model learning. In short, the objective will be the following bound

$$\mathcal{L}_{\text{FedAVG}} \geq \mathcal{L}_{\text{GL}} \geq -\frac{1}{m} \sum_{i=1}^{m} I(Z'_{X_i}, Y'_i) \geq -I(Z'_X, Y'), \tag{6}$$

where $Z_{X_i} \subseteq Z'_{X_i} \subseteq Z'_X$ and $Y_i \subseteq Y'_i \subseteq Y'$. To achieve it, we learn a label conditional graph generative model $\Phi(G_g|Y')$ to sample graph attributes and structures that share the global graph information. The generated graph $G_g = (V_g, E_g, X_g)$ is used to expand local data. In specific, inspired by [31], the graph generative model is trained on the server and further distributed to clients. Given the global label distribution $P(Y')$, and the generative learning batch size n_g, we first sample a label matrix $Y_s \sim P(Y')$ where $Y_g \in \mathbb{R}^{n_g \times |Y'|}$ represents the one-hot label vectors of the nodes. Then the generative model takes the label matrix as input to output the distribution of the generative node attribute $X_g \sim P(X_g|Y')$, which is parameterized by a Gaussian. To enable the generative model to learn structural properties, we sample $X_g \in \mathbb{R}^{n_g \times d}$ via reparameterization trick [4], and use it to generate the adjacency matrix $A_g \in \mathbb{R}^{n_g \times n_g}$, which is parameterized by a Bernoulli. We then use the model from local clients to guide the learning, by:

$$\mathcal{L}_{\text{Gen}} = \mathbb{E}_{Y_s \sim P(Y')} \mathbb{E}_{X_g \sim P(X_g|Y_s)} \mathbb{E}_{A_g \sim B(A_g|X_g)} [$$
$$l(\sigma(\frac{1}{K} \sum_{i=1}^{K} f_i(X_g, A_g)), Y_s)] \tag{7}$$

where f_i and σ are the logit-output of client model i and the activation function. Given an arbitrary sample y, optimizing Eq. 7 requires access to the user models. Specifically, to enable diverse outputs from $\Phi(\cdot|y)$, we introduce a noise vector $\epsilon \sim \mathcal{N}(0, I)$ to the generator, which is resemblant to the reparameterization technique proposed by prior work [12], so that $G_g \sim \Phi(\cdot|Y') \equiv \Phi(G_g, \epsilon|\epsilon \sim \mathcal{N}(0, I))$. Since the graph generative model Φ is trained on the server and further distributed to local data owners to help with local model training, it provides some global information to clients. Therefore, the objective could be written as

$$\mathcal{L}_{\text{GL}} = \frac{1}{m} \sum_{i=1}^{m} \mathbb{E}_{(\mathcal{D}'_i, Y'_i)} [\ell(F(\mathcal{D}'_i), Y'_i))] \tag{8}$$

where $\mathcal{D}_i \subseteq \mathcal{D}'_i \subseteq \mathcal{D}$ and $Y_i \subseteq Y'_i \subseteq Y'$.

5.2 Get a Better Local Model via Information Bottleneck Regularizer

Our second idea is to get a better local model to have a tighter bound of Eq. 2. We now consider the first term of Eq. 2,

$$\frac{1}{m} \sum_{i=1}^{m} I(Z_{X_i}, \mathcal{D}_i) = \frac{1}{m} \sum_{i=1}^{m} (\sum \sum p(z_{X_i}, v_i) \log p(z_{X_i} \mid v_i)$$
$$- \sum p(z_{X_i}) \log p(v_i)). \tag{9}$$

However, computing the marginal distribution of $p(z_{X_i})$, might be difficult. Inspired by [2], let $r(z_{X_i})$ be a variational approximation to this marginal. Since $\text{KL}[p(z_{X_i}), r(z_{X_i})] \geq 0 \implies \sum p(z_{X_i}) \log p(z_{X_i}) \geq \sum p(z_{X_i}) \log r(z_{X_i})$, we have the following upper bound of Eq. 9:

$$\frac{1}{m} \sum_{i=1}^{m} I(Z_{X_i}, \mathcal{D}_i) \leq \frac{1}{m} \sum_{i=1}^{m} \text{KL}[p(Z_{X_i} \mid \mathcal{D}_i), r(Z_{X_i})] \tag{10}$$

In practice, we add an information bottleneck regularizer to force the distribution of the hidden state $p(z_{X_i} \mid v_i)$ of the last layer approaches to a prior distribution. We model $p(z_{X_i} \mid v_i)$ by an Gaussian and we use a standard Gaussian $\mathcal{N}(0, I)$ as $r(Z_{X_i})$. Then the IB regularizer can be termed as

$$\mathcal{L}_{\text{IB-regularizer}} = \frac{1}{m} \sum_{i=1}^{m} \text{KL}\left[\mathcal{N}(\mu_{z_i}, \sigma_{z_i}), \mathcal{N}(0, I)\right] \tag{11}$$

where μ_{z_i} and σ_{z_i} are the parameters of $p(z_{X_i} \mid v_i)$. The parameters are computed through one GraphSage model.

5.3   The Framework of InfoFedSage

The overall framework of InfoFedSage is illustrated in Fig. 2. As mentioned above, our model is unique in the generated modules and IB regularizer. On each communication round, each client i sends its local model GNN_i and the label count $C(Y_i)$ to the server. Then the server aggregates all the client models to form a global model. Besides, the server also collects the local label count to form the label distribution $P(Y)$. Then we sample labels to get $Y_g \in \mathbb{R}^{n_g \times |Y|}$. The generator in the server takes in Y_g and generates graph $G_g = (V_g, E_g, X_g)$ that contains global information. Then the generator is trained under the supervision of client models as in Eq. 7. After that, the server sends back the global model GNN_g and the generator Gen to the client. The local client i, $i = 1, ..., M$ uses the generator and its own labels to generate a synthetic subgraph G_i'. Then the local client i updates its parameter by the Knowledge Distillation loss, the supervision signal on the original graph G_i, and the synthetic graph G_i'. Moreover, we use an IB regularizer on the hidden state before the last layer of the local model to squeeze out irrelevant information brought by the generator.

By combining Eq. 6 and Eq. 8, we can minimize the upper bound of Eq. 2 by:

$$\mathcal{L}_{\text{InfoFedSage}} = \beta\left(\frac{1}{m} \sum_{i=1}^{m} \text{KL}[p(z_{X_i} \mid \mathcal{D}_i) | r(z_{X_i})]\right) \\ - \frac{1}{m} \sum_{i=1}^{m} \mathbb{E}_{(\mathcal{D}_i', Y_i')}[\ell(F(\mathcal{D}_i'), Y_i'))]. \tag{12}$$

where $\ell(F(\mathcal{D}_i'), Y_i'))$ can be written as

$$\ell(F(\mathcal{D}_i'), Y_i')) = \mathcal{L}_{\text{ce}}(\mathcal{D}_i, Y_i) + \mathcal{L}_{\text{ce}}(\mathcal{D}_{g_i}, Y_{g_i}) + \alpha \mathcal{L}_{\text{KD}}, \tag{13}$$

$\mathcal{L}_{\text{ce}}(\mathcal{D}_i, Y_i)$ is the cross-entropy between local input and local label, $\mathcal{L}_{\text{ce}}(\mathcal{D}_{g_i}, Y_{g_i})$ is the cross-entropy between generated input and generated label in client and $\mathcal{L}_{\text{KD}}$ is Knowledge Distillation loss. We can formulate the Knowledge Distillation loss as:

$$\mathcal{L}_{\text{KD}} = \text{KL}(F_i(G_i) \| F_i(G_i')), \tag{14}$$

where $F_i(G_i)$ is the output of the local model F_i on G_i and $F_i(G_i')$ is the output of the local model F_i on the generated graph G_i'.

Table 1. Statistics of the datasets and the synthesized distributed subgraph system with M = 3, 5 and 10. #C row shows the number of classes, $|V_i|$ and $|E_i|$ rows show the averaged number of nodes and edges in all subgraphs and ΔE shows the total number of missing cross-subgraph links. $dist_l$, $dist_s$ and $dist_a$ are metric values to evaluate non-iid of datasets.

	Data	Cora			Citeseer			PubMed			MSAcademic		
Origin	#C	7			6			3			15		
	—V—	2708			3312			19717			18333		
	—E—	5429			4550			44338			81894		
DD	M	3	5	10	3	5	10	3	5	10	3	5	10
iid	$|V_i|$	903	542	271	1104	662	331	6572	3943	1972	6111	3667	1833
	$|E_i|$	1675	968	450	1465	864	442	12932	7630	3789	23584	13949	5915
	ΔE	403	589	929	155	230	310	5543	6189	6445	11141	12151	22743
Label non-iid	$|V_i|$	903	542	271	1109	665	332	6572	3943	1972	6111	3667	1833
	$|E_i|$	1410	717	235	1076	461	147	8497	4023	1004	21807	12151	4965
	ΔE	1199	1844	3079	1322	2245	3080	18847	24223	34298	16473	21139	32244
Attribute non-iid	$|V_i|$	1222	765	391	1201	753	397	9007	6039	1972	6111	3667	1833
	$|E_i|$	1759	1055	527	1517	910	455	27298	16378	3789	27298	16378	8189
	ΔE	0	0	0	0	0	0	0	0	0	0	0	0

6   Experiments

In this section, we compare InfoFedSageGen, InfoFedSageIB, and InfoFedSage with baselines on subgraph federated learning. We will introduce the experimental setup in Sect. 6.1, the experimental result under different split settings in Sect. 6.2, the convergence analysis under different split settings in Sect. 6.3, and hyper-parameter studies in Sect. 6.4.

6.1   Experimental Setup

Datasets. We use four well known graph datasets, Cora [22], Citeseer [22], Pubmed [19] and MSAcademic [23], to conduct experiments. We apply various graph splitting methods to synthesize different non-iid subgraph data in distributed subgraph federated learning to set. For **iid** split, following [29] we use Louvian [3], a classical community detection algorithm to split the data into several non-overlapping clusters, which are then used as the client data. For **Label non-iid** split, following the typical non-iid data splitting method in federated learning, we sort the nodes according to their labels and split the graph into different subgraphs according to the order of the nodes. For **Attribute non-iid** setting, we compute the feature of an edge as the feature summation of its nodes. Then we use K-means [10], a classical clustering method to cluster and assign edges to different clients. The statistics of datasets under different synthetic settings are shown in Fig. 1. Following [29] the training, validation and testing ratios are set to 60%/20%/20%.

From Fig. 1, we find the splitting methods well reflect the non-iid effect from different perspectives. As the metric proposed for each non-iid is prominent for the corresponding splitting method.

Table 2. Experimental results under iid split setting with $M = 3, 5, 10$. Besides averaged accuracy, we also provide the corresponding std.

Model	Cora			Citeseer		
	$M = 3$	$M = 5$	$M = 10$	$M = 3$	$M = 5$	$M = 10$
FedSage	86.21 ± 0.71	85.69 ± 0.72	82.00 ± 1.08	75.01 ± 0.81	74.47 ± 1.97	73.82 ± 1.47
FedSage+	86.51 ± 0.54	85.73 ± 0.49	82.07 ± 0.37	76.14 ± 0.56	75.68 ± 1.32	75.03 ± 1.28
FedSageProx	86.00 ± 0.92	87.02 ± 0.94	81.47 ± 1.59	75.63 ± 0.82	74.75 ± 0.74	74.16 ± 1.18
InfoFedSageGEN	86.40 ± 1.10	85.66 ± 0.86	$\mathbf{84.33 \pm 1.73}$	74.22 ± 1.98	75.57 ± 1.21	73.79 ± 1.48
InfoFedSageIB	87.34 ± 1.22	86.17 ± 1.97	79.44 ± 0.51	75.27 ± 0.81	74.55 ± 1.50	74.83 ± 0.70
InfoFedSage	$\mathbf{87.39 \pm 2.03}$	$\mathbf{87.19 \pm 0.47}$	83.76 ± 0.92	$\mathbf{76.39 \pm 0.74}$	$\mathbf{76.16 \pm 1.43}$	$\mathbf{75.02 \pm 1.05}$
GlobSage	87.32 ± 0.23			75.78 ± 0.35		
Model	PubMed			MSAcademic		
	$M = 3$	$M = 5$	$M = 10$	$M = 3$	$M = 5$	$M = 10$
FedSage	88.10 ± 0.45	88.28 ± 0.52	87.87 ± 0.40	93.52 ± 0.31	93.52 ± 0.22	92.31 ± 0.34
FedSage+	88.34 ± 0.42	88.45 ± 0.49	88.23 ± 0.36	93.72 ± 0.15	93.65 ± 0.33	92.56 ± 0.23
FedSageProx	88.38 ± 0.23	88.32 ± 0.31	87.63 ± 0.36	93.61 ± 0.24	93.44 ± 0.42	92.35 ± 0.32
InfoFedSageGEN	88.51 ± 0.51	88.09 ± 0.32	88.02 ± 0.39	93.67 ± 0.35	93.61 ± 0.62	92.47 ± 0.31
InfoFedSageIB	$\mathbf{88.73 \pm 0.34}$	$\mathbf{88.58 \pm 0.52}$	88.24 ± 0.38	93.07 ± 0.50	92.91 ± 0.32	92.11 ± 0.65
InfoFedSage	88.39 ± 0.29	88.42 ± 0.61	$\mathbf{88.40 \pm 0.67}$	$\mathbf{93.76 \pm 0.14}$	$\mathbf{93.70 \pm 0.35}$	$\mathbf{92.77 \pm 0.43}$
GlobSage	88.68 ± 0.21			96.84 ± 0.13		

Comparison Methods. We conduct comprehensive evaluations on the following seven models, i.e., (1) GlobalSage: one GraphSage [9] model trained on the whole graph, which can be seen as an upper bound of FedSage and FedSage+, (2) FedSage: the federated subgraph learning algorithm use GraphSage [9] and FedAVG [16] proposed in [29], (3) FedSage+: the FedSage baseline with graph mending algorithm proposed in [29]. (4) FedSageProx: the FedSage baseline with Fedprox optimization [15]. (5) InfoFedSageGEN: Our proposed framework only considers the generated modules and does not include the IB regularizer. (6) InfoFedSageIB: Our proposed framework only considers IB regularizer modules and does not include the generated modules. (7) InfoFedSage: The whole model we proposed shown in Fig. 2. For baselines, we use the best hyper-parameters used in the original papers. For our methods, we tune α in $\{0.0001, 0.001, 0.01, 0.1\}$ and β in $\{0.0001, 0.001, 0.01, 0.1\}$. Our model is based on GraphSage. The number of GNN layers is fixed to 2 and the hidden size is fixed to 256. We use Adam optimizer with learning rate 1e−3 and weight decay 5e−4 for all methods. The number of communication rounds for all methods is set to 20.

Evaluation Methods. The metric used in our experiments is the node classification accuracy on the queries from the testing nodes on the whole graph. We report the average accuracy (%) over five random repetitions for each method.

6.2   Experimental Results

Result Under Iid. The experimental result under the iid setting is shown in Table 2. We find that InfoFedSage outperforms other methods in most situa-

Table 3. Experimental results under label non-iid split setting with $M = 3, 5, 10$. Besides averaged accuracy, we also provide the corresponding std.

Model	Cora			Citeseer		
	M = 3	M = 5	M = 10	M = 3	M = 5	M = 10
FedSage	84.49 ± 1.34	57.98 ± 1.98	33.89 ± 0.69	65.17 ± 1.92	51.94 ± 1.26	39.88 ± 1.05
FedSage+	85.65 ± 0.76	58.38 ± 1.67	34.23 ± 1.02	66.63 ± 1.32	54.23 ± 1.06	41.37 ± 1.12
FedSageProx	86.34 ± 1.51	73.52 ± 2.13	38.30 ± 1.65	71.25 ± 1.61	52.83 ± 2.12	32.02 ± 1.48
InfoFedSageGEN	84.04 ± 0.59	59.01 ± 1.80	34.68 ± 1.41	68.20 ± 0.96	55.10 ± 1.00	43.31 ± 1.67
InfoFedSageIB	80.99 ± 4.18	72.00 ± 5.54	44.96 ± 9.75	68.62 ± 2.54	56.34 ± 1.01	49.55 ± 1.44
InfoFedSage	**88.40 ± 1.75**	**81.83 ± 1.85**	**49.22 ± 1.46**	**71.20 ± 1.28**	**60.81 ± 1.26**	**49.76 ± 1.10**

Model	PubMed			MSAcademic		
	M = 3	M = 5	M = 10	M = 3	M = 5	M = 10
FedSage	70.17 ± 1.03	62.43 ± 0.94	48.11 ± 1.60	89.57 ± 1.38	80.36 ± 3.41	74.71 ± 1.16
FedSage+	81.12 ± 0.57	63.35 ± 0.69	51.35 ± 1.43	93.87 ± 1.23	87.87 ± 3.23	84.59 ± 1.12
FedSageProx	80.58 ± 1.23	59.41 ± 1.94	52.03 ± 1.58	94.01 ± 0.23	88.62 ± 0.21	84.68 ± 0.55
InfoFedSageGEN	**82.18 ± 0.61**	65.80 ± 1.47	50.34 ± 0.92	94.04 ± 0.43	89.47 ± 3.07	85.27 ± 0.97
InfoFedSageIB	80.64 ± 1.73	55.29 ± 0.51	50.78 ± 0.32	70.19 ± 2.44	65.59 ± 4.97	82.42 ± 6.38
InfoFedSage	81.88 ± 1.75	**72.41 ± 1.41**	**63.31 ± 1.76**	**94.60 ± 0.35**	**92.85 ± 0.90**	**89.81 ± 1.03**

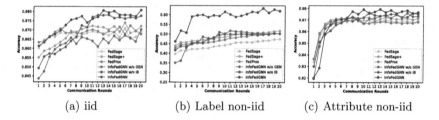

(a) iid   (b) Label non-iid   (c) Attribute non-iid

Fig. 3. Convergence Result of Different Methods on PubMed with 10 Clients.

tions, which verifies the effectiveness of the InfoFedSageGEN and the information bottleneck regularizer. Further, for $M = 10$ of Cora, the InfoFedSageGEN performs better than InfoFedSageIB and InfoFedSage, indicating that when the local graph is sparse the InfoFedSageIB will squeeze too much information thus decreasing the performance. We also observe that for $M = 3, 5$ of PubMed, the InfoFedSageIB outperforms InfoFedSageGEN and InfoFedSage. It is because the subgraphs of PubMed are relatively dense when there are not many clients, thus may bring extra noise and the model under IB regularizer is harder to overfit. Overall, although the relative performance of InfoFedSageGEN and InfoFedSageIB depends on the sparsity of the subgraphs, they all perform better than FedSage, FedSage+, and FedSageProx (Fig. 3).

Result Under Label Non-iid. The results under the label non-iid setting are shown in Table 3. Label non-iid has a significant impact on all model performance, but on the whole, our proposed models have achieved significantly better performance than the benchmark. It is worth noting that for experiments on Cora and MSAcademic, InfoFedSage outperforms all methods remarkably. In the Core experiment, InfoFedSage surpasses FedSage and FedSage+ by at

Table 4. Experimental results under attribute non-iid split setting with $M = 3, 5, 10$. Besides averaged accuracy, we also provide the corresponding std.

Model	Cora			Citeseer		
	M = 3	M = 5	M = 10	M = 3	M = 5	M = 10
FedSage	84.77 ± 1.25	82.19 ± 1.03	76.35 ± 1.20	72.62 ± 1.65	71.55 ± 1.88	71.86 ± 1.60
FedSage+	84.96 ± 1.12	82.43 ± 0.78	76.43 ± 0.97	72.73 ± 1.42	71.76 ± 1.57	72.03 ± 1.23
FedSageProx	85.32 ± 1.24	83.83 ± 0.56	80.34 ± 1.59	72.53 ± 1.47	71.55 ± 2.11	70.07 ± 1.51
InfoFedSageGEN	83.92 ± 0.97	83.22 ± 0.57	79.09 ± 1.56	**73.39 ± 1.61**	71.24 ± 1.60	72.51 ± 1.02
InfoFedSageIB	85.02 ± 0.42	83.06 ± 0.88	79.26 ± 1.28	72.14 ± 1.64	**72.19 ± 1.54**	70.83 ± 1.91
InfoFedSage	**85.50 ± 0.65**	**84.06 ± 0.88**	**80.59 ± 1.31**	73.06 ± 0.99	71.93 ± 1.37	**73.54 ± 0.24**

Model	PubMed			MSAcademic		
	M = 3	M = 5	M = 10	M = 3	M = 5	M = 10
FedSage	88.09 ± 0.37	86.93 ± 0.58	86.26 ± 0.59	91.76 ± 0.36	90.50 ± 0.29	88.33 ± 0.15
FedSage+	88.23 ± 0.45	87.22 ± 0.33	86.32 ± 0.67	91.89 ± 0.31	91.12 ± 0.15	88.46 ± 0.12
FedSageProx	88.35 ± 0.69	86.97 ± 0.62	86.20 ± 1.33	92.01 ± 0.33	90.98 ± 0.42	88.45 ± 0.27
InfoFedSageGEN	88.25 ± 0.44	87.50 ± 0.41	87.20 ± 0.39	92.44 ± 0.35	91.21 ± 0.24	89.00 ± 0.21
InfoFedSageIB	**88.63 ± 0.33**	87.58 ± 0.56	87.06 ± 0.74	91.98 ± 0.54	90.94 ± 0.21	88.71 ± 0.64
InfoFedSage	88.55 ± 0.47	**87.72 ± 0.47**	**87.41 ± 0.13**	**92.57 ± 0.39**	**91.53 ± 0.28**	**89.19 ± 0.23**

most 45.2% in 10 clients setting on average. For MSAcademic, InfoFedSage surpasses FedSage and FedSage+ by at most 20.21% in 10 clients setting on average. When the number of clients gets larger, the average results of our methods are still a bit higher. It is because the conditional generator models more precise label-conditioned distribution and thus produces well-qualified subgraphs that capture global distribution.

Result Under Attribute Non-iid. The results under the attribute non-iid setting are shown in Table 4. Our models achieve the best on attribute non-iid experiments. We see that compared with FedSage, InfoFedSage improves the classification accuracy by an average of 5.5% in 10 clients setting for the Core experiment. For other experiments, our models also get similar results.

Overall, InfoFedSage greatly improved compared with the benchmark in both iid and non-iid cases. In most cases, Infofedsage achieves the best performance, which is consistent with our theoretical analysis, as it combines generative learning and IB regularizer terms, and then has a tighter bound.

6.3 Convergence Analysis

There have been previous research studies on graph federated learning, which come from three main domains based on the learning objects: subgraph-level FL, node-level FL, and graph-level FL. Subgraph-level FL has the problem of missing links between nodes from different clients due to data transportation constraints. To address this, some previous work has designed a graph mending mechanism to generate missing neighbors and links to expand local graph data, but this still requires sending local gradient information from clients, which can raise privacy concerns. Other research in this area has focused on heterogeneous graph data in clients. For node-level federated learning, data is stored through ego networks,

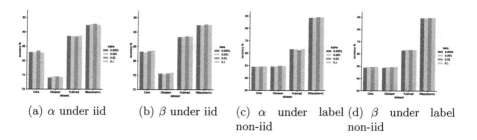

(a) α under iid   (b) β under iid   (c) α under label non-iid   (d) β under label non-iid

Fig. 4. Hyper-parameter Study of InfoFedSage

while for graph-level FL, a cluster-based method has been proposed to deal with non-IID graph data and aggregate client models with adaptive clustering.

6.4 Hyper-parameter Study

We study the impact of two different parameters: α, which determines the influence of the generated graph, and α, which is the coefficient that affects the influence of the information bottleneck regularizer. We will conduct these studies under both iid and non-iid synthetic settings, in order to compare and contrast the effects of these parameters under different data conditions. To study the impact of α, we will fix β to 0.001 and vary among the values $\{0.0001, 0.001, 0.01, 0.1\}$. To study the impact of, we will fix to 0.0001 and vary among the same set of values. The results of these studies, conducted under 10 clients, are shown in Fig. 4. Overall, we have observations that reflect the impact of these different parameters on our model. α controls the influence of the global knowledge distilled from the generated graph. We can observe from Fig. 4(a) that in the iid split setting, increasing α will boost the performance of InfoFedSage on Cora, PubMed, and MSacademic, indicating the effectiveness of the global knowledge. Further, in the label non-iid data as shown in Fig. 4(c), increasing α benefits the performance of Cora, Citeseer, and MSacademic. We can conclude that α influences the performance and such influence varies among different datasets. Besides, slightly increasing β will have some positive influence. As InfoFedSage distills the knowledge from the global generator to enhance the local clients, some noise may be brought by the generated graph. Thus increasing the influence of the IB regularizer will further avoid over-fitting of the local clients. It is more obvious in the non-iid synthetic data, as generators trained on non-iid data may generate more noise biased from the global distribution.

7 Conclusion and Discussion

In this work, we analyze graph federated learning from the perspective of information theory and find out the restrictions that affect current federated learning methods on graphs. Based on the analysis, we propose InfoFedSage to enhance the federated learning performance. Experimental results demonstrate the effectiveness of the proposed model and elucidate the rationality of the theoretical analysis. In the future, we will extend our work in dynamic graph settings.

A   Appendix

A.1   Proof for Proposition 4.1

We first state a lemma.

Lemma A.1. Given X have n states $(x_1, x_2, \cdots x_n)$ and x_1 can be divided into k sub states $(x_{11}, x_{12}, \cdots x_{1n})$, Y has m states $(y_1, y_2, \cdots y_n)$, we have

$$
\begin{aligned}
& I\left(x_{11}, x_{12}, \cdots, x_{1k}, x_2, \cdots, x_n; Y\right) \\
& = p\left(x_1\right) \cdot I\left(x_{11}, x_{12}, \cdots, x_{1k}; Y\right) + I\left(x_1, x_2, \cdots, x_n; Y\right)
\end{aligned}
\tag{15}
$$

Proof. The mutual information between $(x_{11}, x_{12}, \cdots x_{1n})$ and Y:

$$
\begin{aligned}
& I\left(x_{11}, x_{12}, \cdots, x_{1k}; Y\right) \\
& = H\left(x_{11}, x_{12}, \cdots, x_{1k}\right) - H\left(x_{11}, x_{12}, \cdots, x_{1k}/Y\right)
\end{aligned}
\tag{16}
$$

Then, we have

$$
\begin{aligned}
& I\left(x_{11}, x_{12}, \cdots, x_{1k}, x_2, \cdots, x_n; Y\right) \\
& = -\sum_{t=1}^{k} p\left(x_{1t}\right) \log \frac{p\left(x_{1t}\right)}{p\left(x_1\right)} + \sum_{t=1}^{k} \sum_{j=1}^{m} p\left(x_{1t} y_j\right) \log \frac{p\left(x_{1t} y_j\right)}{p\left(x_1 y_j\right)} \\
& + I\left(x_1, x_2, \cdots, x_n; Y\right) \\
& = p\left(x_1\right) \cdot I\left(x_{11}, x_{12}, \cdots, x_{1k}; Y\right) + I\left(x_1, x_2, \cdots, x_n; Y\right)
\end{aligned}
\tag{17}
$$

Then, we can get **Corollary A.1:**

$$
I\left(x_{11}, x_{12}, \cdots, x_{1k}, x_2, \cdots, x_n; Y\right) \geq I\left(x_1, x_2, \cdots, x_n; Y\right)
\tag{18}
$$

We restate **Proposition 4.1:** For $Z'_X = Z_{X1} \cup \cdots \cup Z_{Xm}$ and $Y' = Y_1 \cup \cdots \cup Y_m$, we have

$$
\mathcal{L}_{\text{FedAVG}} \geq -\frac{1}{m} \sum_{i=1}^{m} I(Z_{X_i}, Y_i) \geq -I(Z'_X, Y').
\tag{19}
$$

Proof. We first consider the first inequality, since the definition of mutual information has the form

$$
I(Z_{X_i}, Y_i) = \sum p(y_i, z_{X_i}) \log p(y_i \mid z_{X_i}) + H(Y_i)
\tag{20}
$$

Notice that the entropy of labels $H(Y_i)$ is independent of our optimization procedure and can be ignored. Therefore, we have

$$
\mathcal{L}_{\text{FedAVG}} = -\frac{1}{m} \sum_{i=1}^{m} I(Z_{X_i}; Y_i) + \frac{1}{m} H(Y) \geq -\frac{1}{m} \sum_{i=1}^{m} I(Z_{X_i}, Y_i).
\tag{21}
$$

Then, we consider the second inequality. Directly, based on our **Corollary A.1**, we can get

$$\mathcal{L}_{\text{FedAVG}} \geq -\frac{1}{m}\sum_{i=1}^{m} I(Z_{X_i}, Y_i)$$
$$\geq -max(I(Z_{X_i}, Y_i)) \geq -I(Z', Y_i) \geq -I(Z', Y') \tag{22}$$

References

1. Achille, A., Soatto, S.: Information dropout: learning optimal representations through noisy computation. IEEE Trans. Pattern Anal. Mach. Intell. **40**, 2897–2905 (2018)
2. Alemi, A.A., Fischer, I., Dillon, J.V., Murphy, K.: Deep variational information bottleneck. arXiv preprint arXiv:1612.00410 (2016)
3. Blondel, V.D., Guillaume, J.L., Lambiotte, R., Lefebvre, E.: Fast unfolding of communities in large networks. J. Stat. Mech: Theory Exp. **2008**, 10008 (2008)
4. Blum, A., Haghtalab, N., Procaccia, A.D.: Variational dropout and the local reparameterization trick. In: NIPS (2015)
5. Gao, C., et al.: Graph neural networks for recommender systems: challenges, methods, and directions. ArXiv abs/2109.12843 (2021)
6. Guo, J.N., Li, S., Zhao, Y., Zhang, Y.: Learning robust representation through graph adversarial contrastive learning. In: Bhattacharya, A., et al. (eds.) DASFAA 2022. LNCS, vol. 13245, pp. 682–697. Springer, Cham (2022). https://doi.org/10.1007/978-3-031-00123-9_54
7. Guo, J., et al.: Learning multi-granularity user intent unit for session-based recommendation. In: Proceedings of the 15'th ACM International Conference on Web Search and Data Mining, WSDM 2022 (2022)
8. Guo, J., Zhang, P., Li, C., Xie, X., Zhang, Y., Kim, S.: Evolutionary preference learning via graph nested GRU ode for session-based recommendation. In: Proceedings of the 31st ACM International Conference on Information & Knowledge Management (2022)
9. Hamilton, W.L., Ying, Z., Leskovec, J.: Inductive representation learning on large graphs. In: NIPS (2017)
10. Hartigan, J.A., Wong, M.A.: A k-means clustering algorithm (1979)
11. He, C., et al.: FedGraphNN: a federated learning system and benchmark for graph neural networks. ArXiv abs/2104.07145 (2021)
12. Kingma, D.P., Welling, M.: Auto-encoding variational bayes. CoRR abs/1312.6114 (2014)
13. Li, D., Wang, J.: FedMD: heterogenous federated learning via model distillation. ArXiv abs/1910.03581 (2019)
14. Li, T., Sahu, A.K., Talwalkar, A.S., Smith, V.: Federated learning: challenges, methods, and future directions. IEEE Signal Process. Mag. **37**, 50–60 (2020)
15. Li, T., Sahu, A.K., Zaheer, M., Sanjabi, M., Talwalkar, A., Smith, V.: Federated optimization in heterogeneous networks. Proc. Mach. Learn. Syst. **2**, 429–450 (2020)
16. Li, X., Huang, K., Yang, W., Wang, S., Zhang, Z.: On the convergence of FedAvg on non-IID data. ArXiv abs/1907.02189 (2020)
17. Lin, T., Kong, L., Stich, S.U., Jaggi, M.: Ensemble distillation for robust model fusion in federated learning. ArXiv abs/2006.07242 (2020)

18. McMahan, H.B., Moore, E., Ramage, D., Hampson, S., y Arcas, B.A.: Communication-efficient learning of deep networks from decentralized data. In: AISTATS (2017)
19. Namata, G., London, B., Getoor, L., Huang, B.: Query-driven active surveying for collective classification (2012)
20. Qiu, Y., Huang, C., Wang, J., Huang, Z., Xiao, J.: A privacy-preserving subgraph-level federated graph neural network via differential privacy. In: Memmi, G., Yang, B., Kong, L., Zhang, T., Qiu, M. (eds.) KSEM 2022. LNCS, vol. 13370, pp. 165–177. Springer, Cham (2022). https://doi.org/10.1007/978-3-031-10989-8_14
21. Sahu, A.K., Li, T., Sanjabi, M., Zaheer, M., Talwalkar, A.S., Smith, V.: Federated optimization in heterogeneous networks. arXiv Learning (2020)
22. Sen, P., Namata, G., Bilgic, M., Getoor, L., Gallagher, B., Eliassi-Rad, T.: Collective classification in network data (2008)
23. Shchur, O., Mumme, M., Bojchevski, A., Günnemann, S.: Pitfalls of graph neural network evaluation. ArXiv abs/1811.05868 (2018)
24. Tishby, N., Zaslavsky, N.: Deep learning and the information bottleneck principle. In: 2015 IEEE Information Theory Workshop (ITW), pp. 1–5 (2015)
25. Wasserman, S., Faust, K.: Social network analysis - methods and applications. In: Structural Analysis in the Social Sciences (2007)
26. Wu, L., et al.: Graph neural networks for natural language processing: a survey. ArXiv abs/2106.06090 (2021)
27. Wu, T., Ren, H., Li, P., Leskovec, J.: Graph information bottleneck. ArXiv abs/2010.12811 (2020)
28. Xie, H., Ma, J., Xiong, L., Yang, C.: Federated graph classification over non-IID graphs. Adv. Neural. Inf. Process. Syst. **34**, 18839–18852 (2021)
29. Zhang, K., Yang, C., Li, X., Sun, L., Yiu, S.M.: Subgraph federated learning with missing neighbor generation. ArXiv abs/2106.13430 (2021)
30. Zhang, P., et al.: Efficiently leveraging multi-level user intent for session-based recommendation via atten-mixer network. arXiv preprint arXiv:2206.12781 (2022)
31. Zhu, Z., Hong, J., Zhou, J.: Data-free knowledge distillation for heterogeneous federated learning. Proc. Mach. Learn. Res. **139**, 12878–12889 (2021)

Author Index

A
Ao, Xiang   3

B
Bai, Lei   151
Bi, Xin   237

C
Cai, Yuzheng   114
Chang, Jian   519
Chao, Pingfu   380
Chen, Cen   677
Chen, Chen   97
Chen, Derong   562
Chen, Jiazun   132
Chen, Jin   431
Chen, Jinchuan   509
Chen, Lei   300, 317, 351
Chen, Lulu   46
Chen, Rui   627
Chen, Wei   380
Cui, Jamie   677
Cui, Jie   644
Cui, Ningning   644
Cui, Yingchun   693

D
Deng, Liwei   431, 448
Deng, Zheli   644
Ding, Xiaoou   253
Ding, Xuanang   661, 735
Ding, Ye   703
Dong, Dong   535
Dong, Renzhi   604
Du, Xiaoyong   509
Du, Yali   604

F
Fang, Junhua   380
Feng, Dan   578
Feng, Jiyuan   703

F
Feng, Xinwei   3
Feng, Yunqing   703
Fu, Pengyu   578
Fu, Xiaoming   334
Fu, Yingnan   677

G
Gao, Heran   483
Gao, Jun   132
Gao, Ming   677
Gao, Xin   414
Geng, Jinkun   535
Guo, Jiayan   745
Guo, Songyue   703

H
Han, Qilong   627
Han, Xiaoyu   205
Han, Yinjun   132
He, Qing   3
He, Zhenying   205
Hou U, Leong   604
Hou, Huiwen   97
Hou, Yeqiao   464
Hu, Die   578
Hu, Gang   562
Hu, Haibo   627
Hu, Jia   414
Hu, Lei   237
Hu, Songlin   177
Hua, Wen   396
Huang, Jia   509
Huang, Shuai   80
Huang, Xuechun   237
Huang, Yibo   46
Huang, Zhengan   719

J
Jiang, Hongxu   535
Jiang, Wenbin   3
Jin, Hai   519

X. Wang et al. (Eds.): DASFAA 2023, LNCS 13943, pp. 761–763, 2023.
https://doi.org/10.1007/978-3-031-30637-2

L

Lai, Haoseng   617
Li, Binhong   519
Li, Bohan   334
Li, Fanxiao   617
Li, Guohui   661, 735
Li, Guoliang   80
Li, Hao   562
Li, Jiajia   396
Li, Jianjun   735
Li, Lei   396
Li, Man   644
Li, Maocheng   300, 317
Li, Pengfei   380
Li, Qing   300
Li, Ruixuan   177
Li, Shangyang   745
Li, Xiaobin   535
Li, Xin   535
Li, Yunchuan   62
Li, Zhixu   283
Li, Zongpeng   464
Liang, Shuang   562
Liao, Qing   703
Lin, Licheng   519
Lin, Yang   132
Liu, An   283, 380
Liu, Clement   205
Liu, Furui   604
Liu, Guanfeng   283
Liu, Renyang   617
Liu, Yida   253
Liu, Yuanjun   283
Luo, Diaohan   483
Luo, Xianqiang   562
Lv, Weifeng   351
Lyu, Yajuan   3

M

Ma, Lin-Tao   167
Ma, Nabo   509
Ma, Qiang   221
Ma, Shuai   3
Ma, Yuhang   535
Ma, Yuliang   644
Ma, Zhongyu   414
Miao, Hao   132
Mu, Xin   719

N

Niu, Yuean   546

P

Pan, Qingfeng   19

Q

Qi, Zhi   188
Qiao, Lianpeng   97
Qiu, Tao   35, 273

R

Rong, Qian   661, 735

S

Shang, Mingsheng   188, 588
Shao, Jie   562
She, Qiaoqiao   3
Shi, Jianfeng   483
Shi, Xiaoyu   588
Shi, Yexuan   351
Song, BingBing   617
Song, Yichen   253
Song, Yingze   205
Sun, Hao   448
Sun, Lin   464
Sun, Shangyi   46
Sun, Yinbo   167
Sun, Yongjiao   237

T

Tang, Xuehai   177
Tian, Fangzheng   535
Tian, Meng   3
Tian, Ran   414
Tong, Yongxin   351

W

Wang, Bin   35, 273
Wang, Bingchao   588
Wang, Binwu   151
Wang, Chu   414
Wang, Guoren   97, 499
Wang, Haojie   535
Wang, Hongzhi   253
Wang, Jiaqiang   546
Wang, Junkai   499
Wang, Leixia   627
Wang, Mingshen   396

Wang, Pengkun   151
Wang, Shiyu   167
Wang, Shuoxin   617
Wang, Wei   703
Wang, X. Sean   205
Wang, Xu   151
Wang, Yan   167
Wang, Yang   151
Wang, Yiming   604
Wang, Yong   80
Wang, Ziwei   62
Wei, Senhui   677
Wu, Heng   483
Wu, Jie   46
Wu, Sixing   617
Wu, Yao   46

X

Xia, Huahui   509
Xia, Xiufeng   35, 273
Xiao, Jiang   499, 519
Xiao, Meichun   273
Xie, Hong   188
Xie, Jiandong   431, 448
Xie, Yulai   578
Xin, Kexuan   396
Xu, Chen   19, 546
Xu, Han   562
Xu, Jianqiu   334
Xu, Ke   351
Xu, Mo   132
Xu, Shuai   334
Xu, Yi   351
Xu, Zhizhen   546

Y

Yan, Ming   46
Yan, Yukun   627
Yang, Changjie   448
Yang, Donghua   253
Yang, Ming   604
Yang, Ruixin   62
Yang, Xiaochun   35, 273
Yang, Xu   703
Yang, Zhenhua   19
Yang, Ziang   177
Ye, Chengyang   221

Ye, Fei   205
Ye, Qingqing   627
Ye, Tiandi   677
Yu, Jiong   617
Yuan, George Y.   237
Yuan, Ling   661, 735
Yuan, Mingxu   237
Yuan, Ye   97, 499

Z

Zeng, Gongxian   719
Zeng, Yuxiang   300, 317, 351
Zhang, James   167
Zhang, Kaining   351
Zhang, Lu   661, 735
Zhang, Rui   46
Zhang, Runhua   535
Zhang, Wei   237
Zhang, Wenbo   483
Zhang, Yan   745
Zhang, Yaying   369
Zhang, Yudong   151
Zhang, Yupu   431
Zhang, Zhiwei   499
Zhao, Lei   283
Zhao, Shixun   578
Zhao, Shuai   499
Zhao, Xiangguo   237
Zhao, Yan   62, 431, 448
Zheng, Kai   62, 431, 448
Zheng, Libin   300
Zheng, Qi   369
Zheng, Weiguo   114
Zheng, YangFei   167
Zheng, Zhiming   351
Zhong, Hong   644
Zhou, Biyu   177
Zhou, Fan   167
Zhou, Mingliang   604
Zhou, Rui   62
Zhou, Wei   617
Zhou, Xiaofang   396
Zhou, Zimu   351
Zhu, Chenhui   535
Zhu, Jinghua   693
Zhu, Rui   35, 273
Zong, Chuanyu   35, 273

Printed in the United States
by Baker & Taylor Publisher Services